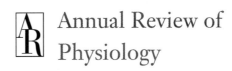

Annual Review of
Physiology

Editorial Committee (2012)

Peter Aronson, Yale University
David E. Clapham, Harvard Medical School
Holly A. Ingraham, University of California, San Francisco
David Julius, University of California, San Francisco
Roger Nicoll, University of California, San Francisco
Marlene Rabinovitch, Stanford University School of Medicine
Linda Samuelson, University of Michigan
Jo Rae Wright, Duke University

Responsible for the Organization of Volume 74 (Editorial Committee, 2010)

James M. Anderson
Richard C. Boucher, Jr.
David E. Clapham
Gerhard H. Giebisch
Holly A. Ingraham
David Julius
Roger Nicoll
Marlene Rabinovitch

Production Editor: Shirley S. Park
Bibliographic Quality Control: Mary A. Glass
Electronic Content Coordinator: Suzanne K. Moses
Illustration Editor: Glenda Lee Mahoney

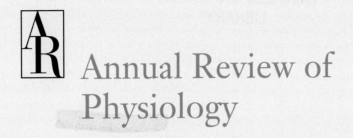

Annual Review of Physiology

Volume 74, 2012

David Julius, *Editor*
University of California, San Francisco

David E. Clapham, *Associate Editor*
Harvard Medical School

www.annualreviews.org • science@annualreviews.org • 650-493-4400

Annual Reviews
4139 El Camino Way • P.O. Box 10139 • Palo Alto, California 94303-0139

Annual Reviews
Palo Alto, California, USA

COPYRIGHT © 2012 BY ANNUAL REVIEWS, PALO ALTO, CALIFORNIA, USA. ALL RIGHTS RESERVED. The appearance of the code at the bottom of the first page of an article in this serial indicates the copyright owner's consent that copies of the article may be made for personal or internal use, or for the personal or internal use of specific clients. This consent is given on the condition that the copier pay the stated per-copy fee of $20.00 per article through the Copyright Clearance Center, Inc. (222 Rosewood Drive, Danvers, MA 01923) for copying beyond that permitted by Section 107 or 108 of the U.S. Copyright Law. The per-copy fee of $20.00 per article also applies to the copying, under the stated conditions, of articles published in any *Annual Review* serial before January 1, 1978. Individual readers, and nonprofit libraries acting for them, are permitted to make a single copy of an article without charge for use in research or teaching. This consent does not extend to other kinds of copying, such as copying for general distribution, for advertising or promotional purposes, for creating new collective works, or for resale. For such uses, written permission is required. Write to Permissions Dept., Annual Reviews, 4139 El Camino Way, P.O. Box 10139, Palo Alto, CA 94303-0139 USA.

International Standard Serial Number: 0066-4278
International Standard Book Number: 978-0-8243-0374-7
Library of Congress Catalog Card Number: 39-15404

All Annual Reviews and publication titles are registered trademarks of Annual Reviews.

∞ The paper used in this publication meets the minimum requirements of American National Standards for Information Sciences—Permanence of Paper for Printed Library Materials, ANSI Z39.48-1992. The paper contains 20% postconsumer recycled content.

Annual Reviews and the Editors of its publications assume no responsibility for the statements expressed by the contributors to this *Annual Review*.

TYPESET BY APTARA
PRINTED AND BOUND BY SHERIDAN BOOKS, INC., CHELSEA, MICHIGAN

Dear Readers,

It gives me great pleasure to introduce Volume 74 of the *Annual Review of Physiology*. We begin with a Perspective from Elwood Jensen, whose pioneering work on estrogen and its mechanism of action gave birth to the field of nuclear hormone receptors—a monumental contribution to modern physiology and medicine by any account. Dr. Jensen takes us through a journey of discovery that truly led from bench to bedside, establishing the physiological and pharmacological framework for treating hormone-responsive breast cancer. Following on a recent and successful *ARP* innovation, Dr. Jensen's life story has been recounted to us through his conversation with a noted scientist and expert in the field. For this we owe a great debt of thanks to David Moore (Professor of Molecular and Cellular Biology, Baylor College of Medicine), who sat with Dr. Jensen to produce this wonderful perspective. A video of this interview can be viewed at **http://www.annualreviews.org/r/Jensen_interview**.

Each year, we single out an area of physiology that merits attention as a Special Topic section based on recent exciting advances, novelty, broad interest, or all of the above. In this volume, we have chosen to highlight germ cells in a compendium of four chapters that bring us up to date on molecular events leading up to gamete interaction and fertilization, with perspectives on how advances in this field shape future approaches to contraception, or diagnosing or treating infertility and birth defects. Credit for organizing this fantastic Special Topics section goes to two members of the *ARP* Editorial Board, Holly Ingraham and David Clapham. Thank you Holly and David!

Accompanying these highlights is a superb collection of chapters within our standing sections devoted to specific organ systems. As is often the case, however, mechanistic themes cut across physiology. For example, this volume features a number of chapters that focus on ion recognition and transport in systems ranging from calcium sensing and homeostasis to lysosomal acidification, electrolyte transport in the kidney, vesicular packaging of neurotransmitters, and control of neuronal excitability. Others chapters focus on epigenetic mechanisms that influence normal development and disease within cardiovascular and endocrine systems. Still other contributions bring us up to date on recent advances and ideas pertaining to roles for innate and adaptive immunity in intestinal and pulmonary injury and gut-microbiome interactions.

As always, many thanks go out to our colleagues who have taken their precious time and energy to produce insightful and illuminating reviews that enable specialists and nonspecialists alike to be brought up to date on important and impactful advances that fit under the broad and fascinating tent of cellular and systems physiology. Many thanks, as well, to members of the *ARP* editorial team, who each year make the process both enjoyable and successful!

D. Julius
San Francisco

Annual Review of Physiology

Volume 74, 2012

Contents

PERSPECTIVES, David Julius, Editor

A Conversation with Elwood Jensen
David D. Moore 1

CARDIOVASCULAR PHYSIOLOGY, Marlene Rabinovitch, Section Editor

Epigenetic Control of Smooth Muscle Cell Differentiation
and Phenotypic Switching in Vascular Development and Disease
Matthew R. Alexander and Gary K. Owens 13

Epigenetics and Cardiovascular Development
Ching-Pin Chang and Benoit G. Bruneau 41

CELL PHYSIOLOGY, David E. Clapham, Section Editor

Lysosomal Acidification Mechanisms
Joseph A. Mindell 69

ENDOCRINOLOGY, Holly A. Ingraham, Section Editor

Biology Without Walls: The Novel Endocrinology of Bone
Gerard Karsenty and Franck Oury 87

Fetal Programming and Metabolic Syndrome
Paolo Rinaudo and Erica Wang 107

Nuclear Sphingolipid Metabolism
Natasha C. Lucki and Marion B. Sewer 131

GASTROINTESTINAL PHYSIOLOGY, James M. Anderson, Section Editor

Adenosine and Hypoxia-Inducible Factor Signaling in Intestinal Injury
and Recovery
Sean P. Colgan and Holger K. Eltzschig 153

Toll-Like Receptor–Gut Microbiota Interactions:
   Perturb at Your Own Risk!
   *Frederic A. Carvalho, Jesse D. Aitken, Matam Vijay-Kumar,
      and Andrew T. Gewirtz* 177

NEUROPHYSIOLOGY, *Roger Nicoll, Section Editor*

The Calyx of Held Synapse: From Model Synapse to Auditory Relay
   J. Gerard G. Borst and John Soria van Hoeve 199

Neurotransmitter Corelease: Mechanism and Physiological Role
   Thomas S. Hnasko and Robert H. Edwards 225

Small-Conductance Ca^{2+}-Activated K^+ Channels: Form and Function
   John P. Adelman, James Maylie, and Pankaj Sah 245

RENAL AND ELECTROLYTE PHYSIOLOGY, *Gerhard H. Giebisch, Section Editor*

The Calcium-Sensing Receptor Beyond Extracellular Calcium
   Homeostasis: Conception, Development, Adult Physiology,
   and Disease
   Daniela Riccardi and Paul J. Kemp 271

Cell Biology and Pathology of Podocytes
   Anna Greka and Peter Mundel 299

A New Look at Electrolyte Transport in the Distal Tubule
   Dominique Eladari, Régine Chambrey, and Janos Peti-Peterdi 325

Renal Function in Diabetic Disease Models: The Tubular System
   in the Pathophysiology of the Diabetic Kidney
   Volker Vallon and Scott C. Thomson 351

RESPIRATORY PHYSIOLOGY, *Richard C. Boucher, Jr., Section Editor*

Autophagy in Pulmonary Diseases
   *Stefan W. Ryter, Kiichi Nakahira, Jeffrey A. Haspel,
      and Augustine M.K. Choi* 377

Stop the Flow: A Paradigm for Cell Signaling Mediated by Reactive
   Oxygen Species in the Pulmonary Endothelium
   Elizabeth A. Browning, Shampa Chatterjee, and Aron B. Fisher 403

SPECIAL TOPIC, GERM CELLS IN REPRODUCTION, *David E. Clapham and Holly A. Ingraham, Special Topic Editors*

The Molecular Control of Meiotic Chromosomal Behavior:
   Events in Early Meiotic Prophase in *Drosophila* Oocytes
   Cathleen M. Lake and R. Scott Hawley 425

The Control of Male Fertility by Spermatozoan Ion Channels
Polina V. Lishko, Yuriy Kirichok, Dejian Ren, Betsy Navarro, Jean-Ju Chung, and David E. Clapham 453

Sperm-Egg Interaction
Janice P. Evans 477

Genetics of Mammalian Reproduction: Modeling the End of the Germline
Martin M. Matzuk and Kathleen H. Burns 503

Indexes

Cumulative Index of Contributing Authors, Volumes 70–74 529

Cumulative Index of Chapter Titles, Volumes 70–74 532

Errata

An online log of corrections to *Annual Review of Physiology* articles may be found at http://physiol.annualreviews.org/errata.shtml

Other Reviews of Interest to Physiologists

From the *Annual Review of Biochemistry*, Volume 80 (2011)

  From Serendipity to Therapy
   Elizabeth F. Neufeld

  Transmembrane Communication: General Principles and Lessons from the Structure and Function of the M2 Proton Channel, K^+ Channels, and Integrin Receptors
   Gevorg Grigoryan, David T. Moore, and William F. DeGrado

  Regulation of Phospholipid Synthesis in the Yeast *Saccharomyces cerevisiae*
   George M. Carman and Gil-Soo Han

  Sterol Regulation of Metabolism, Homeostasis, and Development
   Joshua Wollam and Adam Antebi

  Structural Biology of the Toll-Like Receptor Family
   Jin Young Kang and Jie-Oh Lee

  Amino Acid Signaling in TOR Activation
   Joungmok Kim and Kun-Liang Guan

  Caspase Substrates and Cellular Remodeling
   Emily D. Crawford and James A. Wells

From the *Annual Review of Biophysics*, Volume 40 (2011)

  P-Type ATPases
   Michael G. Palmgren and Poul Nissen

From the *Annual Review of Biomedical Engineering*, Volume 13 (2011)

  Bioengineering Heart Muscle: A Paradigm for Regenerative Medicine
   Gordana Vunjak-Novakovic, Kathy O. Lui, Nina Tandon, and Kenneth R. Chien

  Nuclear Mechanics in Disease
   Monika Zwerger, Chin Yee Ho, and Jan Lammerding

From the *Annual Review of Cell and Developmental Biology*, Volume 27 (2011)

The Role of Atg Proteins in Autophagosome Formation
Noboru Mizushima, Tamotsu Yoshimori, and Yoshinori Ohsumi

From the *Annual Review of Genomics and Human Genetics*, Volume 12 (2011)

Putting Medical Genetics into Practice
Malcolm A. Ferguson-Smith

Transitions Between Sex-Determining Systems in Reptiles and Amphibians
Stephen D. Sarre, Tariq Ezaz, and Arthur Georges

From the *Annual Review of Immunology*, Volume 29 (2011)

Immunoglobulin Responses at the Mucosal Interface
Andrea Cerutti, Kang Chen, and Alejo Chorny

Inflammatory Mechanisms in Obesity
Margaret F. Gregor and Gökhan S. Hotamisligil

Human TLRs and IL-1Rs in Host Defense: Natural Insights from Evolutionary, Epidemiological, and Clinical Genetics
Jean-Laurent Casanova, Laurent Abel, and Lluis Quintana-Murci

Integration of Genetic and Immunological Insights into a Model of Celiac Disease Pathogenesis
Valérie Abadie, Ludvig M. Sollid, Luis B. Barreiro, and Bana Jabri

The Inflammasome NLRs in Immunity, Inflammation, and Associated Diseases
Jenny P.-Y. Ting, Beckley K. Davis, and Haitao Wen

From the *Annual Review of Medicine*, Volume 62 (2011)

Antiestrogens and Their Therapeutic Applications in Breast Cancer and Other Diseases
Simak Ali, Laki Buluwela, and R. Charles Coombes

Mechanisms of Endocrine Resistance in Breast Cancer
C. Kent Osborne and Rachel Schiff

Pharmacogenetics of Endocrine Therapy for Breast Cancer
Michaela J. Higgins and Vered Stearns

Therapeutic Approaches for Women Predisposed to Breast Cancer
Katherine L. Nathanson and Susan M. Domchek

New Approaches to the Treatment of Osteoporosis
Barbara C. Silva and John P. Bilezikian

Regulation of Bone Mass by Serotonin: Molecular Biology and Therapeutic Implications
Gerard Karsenty and Vijay K. Yadav

Alpha-1-Antitrypsin Deficiency: Importance of Proteasomal
and Autophagic Degradative Pathways in Disposal of Liver
Disease–Associated Protein Aggregates
David H. Perlmutter

Interactions Between Gut Microbiota and Host Metabolism Predisposing
to Obesity and Diabetes
Giovanni Musso, Roberto Gambino, and Maurizio Cassader

From the ***Annual Review of Microbiology***, Volume 65 (2011)

The Human Gut Microbiome: Ecology and Recent Evolutionary Changes
Jens Walter and Ruth Ley

From the ***Annual Review of Nutrition***, Volume 31 (2011)

Interaction Between Obesity and the Gut Microbiota: Relevance in Nutrition
Nathalie M. Delzenne and Patrice D. Cani

From the ***Annual Review of Pharmacology and Toxicology***, Volume 52 (2012)

AMPK and mTOR in Cellular Energy Homeostasis and Drug Targets
Ken Inoki, Joungmok Kim, and Kun-Liang Guan

From the ***Annual Review of Pathology: Mechanisms of Disease***, Volume 6 (2011)

Disorders of Bone Remodeling
Xu Feng and Jay M. McDonald

The Acute Respiratory Distress Syndrome: Pathogenesis and Treatment
Michael A. Matthay and Rachel L. Zemans

The HIF Pathway and Erythrocytosis
Frank S. Lee and Melanie J. Percy

Retinoids, Retinoic Acid Receptors, and Cancer
Xiao-Han Tang and Lorraine J. Gudas

From the ***Annual Review of Plant Biology***, Volume 62 (2011)

Anion Channels/Transporters in Plants: From Molecular Bases
to Regulatory Networks
*Hélène Barbier-Brygoo, Alexis De Angeli, Sophie Filleur, Jean-Marie Frachisse,
Franco Gambale, Sébastien Thomine, and Stefanie Wege*

In Vivo Imaging of Ca^{2+}, pH, and Reactive Oxygen Species Using Fluorescent
Probes in Plants
Sarah J. Swanson, Won-Gyu Choi, Alexandra Chanoca, and Simon Gilroy

From the ***Annual Review of Phytopathology***, Volume 49 (2011)

What Can Plant Autophagy Do for an Innate Immune Response?
Andrew P. Hayward and S.P. Dinesh-Kumar

Annual Reviews is a nonprofit scientific publisher established to promote the advancement of the sciences. Beginning in 1932 with the *Annual Review of Biochemistry*, the Company has pursued as its principal function the publication of high-quality, reasonably priced *Annual Review* volumes. The volumes are organized by Editors and Editorial Committees who invite qualified authors to contribute critical articles reviewing significant developments within each major discipline. The Editor-in-Chief invites those interested in serving as future Editorial Committee members to communicate directly with him. Annual Reviews is administered by a Board of Directors, whose members serve without compensation.

2012 Board of Directors, Annual Reviews

Richard N. Zare, *Chairperson of Annual Reviews, Marguerite Blake Wilbur Professor of Natural Science, Department of Chemistry, Stanford University*
Karen S. Cook, *Vice-Chairperson of Annual Reviews, Director of the Institute for Research in the Social Sciences, Stanford University*
Sandra M. Faber, *Vice-Chairperson of Annual Reviews, Professor of Astronomy and Astronomer at Lick Observatory, University of California at Santa Cruz*
John I. Brauman, *J.G. Jackson-C.J. Wood Professor of Chemistry, Stanford University*
Peter F. Carpenter, *Founder, Mission and Values Institute, Atherton, California*
Susan T. Fiske, *Eugene Higgins Professor of Psychology, Princeton University*
Eugene Garfield, *Emeritus Publisher, The Scientist*
Samuel Gubins, *President and Editor-in-Chief, Annual Reviews*
Steven E. Hyman, *Professor of Neurobiology, Harvard Medical School, and Distinguished Service Professor, Harvard University*
Roger D. Kornberg, *Professor of Structural Biology, Stanford University School of Medicine*
Sharon R. Long, *Wm. Steere-Pfizer Professor of Biological Sciences, Stanford University*
J. Boyce Nute, *Palo Alto, California*
Michael E. Peskin, *Professor of Particle Physics and Astrophysics, SLAC, Stanford University*
Claude M. Steele, *Dean of the School of Education, Stanford University*
Harriet A. Zuckerman, *Senior Fellow, The Andrew W. Mellon Foundation*

Management of Annual Reviews

Samuel Gubins, President and Editor-in-Chief
Paul J. Calvi Jr., Director of Technology
Steven J. Castro, Chief Financial Officer and Director of Marketing & Sales
Jennifer L. Jongsma, Director of Production
Laurie A. Mandel, Corporate Secretary
Jada Pimentel, Director of Human Resources

Annual Reviews of

Analytical Chemistry
Anthropology
Astronomy and Astrophysics
Biochemistry
Biomedical Engineering
Biophysics
Cell and Developmental Biology
Chemical and Biomolecular Engineering
Clinical Psychology
Condensed Matter Physics
Earth and Planetary Sciences
Ecology, Evolution, and Systematics
Economics
Entomology
Environment and Resources
Financial Economics
Fluid Mechanics
Food Science and Technology
Genetics
Genomics and Human Genetics
Immunology
Law and Social Science
Marine Science
Materials Research
Medicine
Microbiology
Neuroscience
Nuclear and Particle Science
Nutrition
Pathology: Mechanisms of Disease
Pharmacology and Toxicology
Physical Chemistry
Physiology
Phytopathology
Plant Biology
Political Science
Psychology
Public Health
Resource Economics
Sociology

SPECIAL PUBLICATIONS
Excitement and Fascination of Science, Vols. 1, 2, 3, and 4

Elwood V. Jensen

A Conversation with Elwood Jensen

David D. Moore

Department of Molecular and Cellular Biology, Baylor College of Medicine, Houston, Texas 77030; email: moore@bcm.edu

Keywords

estrogen receptor, estradiol, tamoxifen, breast cancer, Matterhorn, alternative approach

VIDEO

Please visit **http://www.annualreviews.org/r/Jensen_interview** for a video of this interview.

INTRODUCTION

Elwood Jensen first described the estrogen receptor in 1958, opening up a new field of hormone action (1). This field expanded along with the nuclear receptor superfamily to incorporate the actions of a wide and still growing range of biological regulators that includes diverse steroids, thyroid hormone, retinoids, bile acids, and even heme. The 48 members of this nuclear receptor family encoded in the human genome impact nearly every facet of our biology. Elwood is a member of the National Academy of Sciences and was a recipient of the 2004 Albert Lasker Basic Medical Research Award "for the discovery of the superfamily of nuclear hormone receptors and elucidation of a unifying mechanism that regulates embryonic development and diverse metabolic pathways." At the Lasker Award ceremony, Nobel laureate Dr. Joseph L. Goldstein described Elwood as "the patriarch of the field" on the basis of "his pioneering work on the estrogen receptor, the matriarch of the superfamily" (2).

In the summer of 2010, the *Annual Review of Physiology* gave me the opportunity to discuss with Elwood this seminal discovery, along with other aspects of his long and illustrious career. As he describes in the interview, his work completely overturned the conventional view of hormone action at the time. It now seems quite odd that the prevailing "wisdom" was that estradiol acted as an enzyme cofactor in redox reactions that resulted in the transfer of hydrogen from NADH to NADPH (3). Elwood and his postdoctoral fellow Herbert Jacobson used tritium gas to reduce a double bond in an appropriate precursor, generating estradiol labeled to very high specific activity. When they administered physiological doses of this tracer to immature female rats, they found that it was not chemically altered, as predicted by the then current models, but was instead specifically retained in known estrogen target tissues such as the uterus. Elwood correctly deduced that the tracer was held there by a specific protein, which he termed estrophilin and which we now know as the estrogen receptor. Subsequent work by Elwood and colleagues (4) and also the late Jack Gorski and colleagues (5, 6) indicated that estrogen binds the estrogen receptor in the cytoplasm, and the complex then moves to the nucleus. Bert O'Malley was the first to clearly demonstrate that estrogen and progesterone act in the nucleus to induce specific messenger RNAs (7).

Elwood went on to purify the estrogen receptor and obtained polyclonal as well as monoclonal antibodies (8, 9). In collaboration with Pierre Chambon, Elwood used these antibodies to isolate the initial estrogen receptor cDNA clones (10), a major step in the elucidation of the nuclear receptor superfamily. Elwood and colleagues also contributed significantly to the development of diagnostic measurements of estrogen receptor in breast cancer specimens and to the development of estrogen receptor antagonist therapies (11). Thus, Elwood's discoveries not only unraveled fundamentals of molecular endocrinology but also led directly to major advances in breast cancer diagnosis and therapy.

As Elwood recounts in the interview, as well as in his Lasker Award essay (12) and other reminiscences, his pioneering studies benefited from what he terms an "alternative approach." He illustrates this with his remarkable story of ascending the Matterhorn, despite his complete lack of mountain climbing experience, and of only later learning that it had been the last major European peak to be scaled (**Figure 1**). In retrospect, as is often the case in science, what had appeared to be the most straightforward route was actually not simple at all, but insightful analysis revealed the feasibility of a seemingly intractable alternative. Edward Whymper was the English climber who deduced that the seemingly sheer northeast face could be climbed. Elwood's insightful alternative approaches to studying estrogen action allowed him to accurately follow the fate of the very small amounts of the hormone required for physiological responses and to identify, and later characterize, the estrogen receptor.

Figure 1

The genesis of "alternative approach": the Matterhorn ascent. In the top left of the photo, taken from Zermatt, the starting point of the final phase of Elwood Jensen's Matterhorn climb is the steep Swiss face, which meets the more moderate Italian side at the peak. Until the British climber Edward Whymper climbed the Matterhorn via the steep Swiss side, all previous ascents to the peak had been attempted from the Italian side. Jensen followed Edward Whymper's alternative route to the summit. This also inspired Jensen to try alternative approaches in his research that led to the discovery of the estrogen receptor. Shown in the picture are Mary, Jensen's late wife, who accompanied her husband up to this last section of the ascent, and Kyle Packer, the student from Colorado who had arranged for the guide, whom they were to join in Zermatt.

The George and Elizabeth Wile Chair for Cancer Research, Elwood was 90 years old when I interviewed him at the Vontz Center for Molecular Studies at the University of Cincinnati College of Medicine on July 19, 2010. He was recruited there in 2002, remaining scientifically active, as demonstrated by his description, with collaborators Sohaib Khan and Thomas Burris, that the estrogen antagonist tamoxifen can bind to two distinct sites on the receptor (13, 14). Elwood had suffered some health problems, including being hit by a truck, but was getting around on an electric scooter and was full of energy and very sharp during the interview. In fact, after my somewhat garbled first question, I barely got a word in for the rest of the 38-min interview. It is appropriate to clarify some of the events that he referred to in the period around his discovery of the estrogen receptor, however. The meeting in Vienna where Elwood first presented his major discovery on the estrogen receptor to the three other speakers plus two additional attendees was in 1958. The progesterone receptor work by O'Malley and colleagues (15) that he mentions was published in 1970. John Baxter and colleagues' 1975 description (16) of chromatin binding properties of the thyroid hormone receptor was based on earlier studies by Oppenheimer and colleagues (17) demonstrating the existence of such a receptor. Ron Evans, who shared the Lasker Award with Elwood and Chambon, began his postdoctoral fellowship with James Darnell in 1975. But in 1986 Evans and colleagues (18), along with Bjorn Vennstrom and colleagues (19), showed that the cellular proto-oncogene c-*erbA* encodes the thyroid hormone receptor.

It was a thrill and a privilege to be able to talk with Elwood about his work and his life and to be able to share it with you. I hope that you enjoy the interview, too.

INTERVIEW AT THE VONTZ CENTER FOR MOLECULAR STUDIES AT THE UNIVERSITY OF CINCINNATI COLLEGE OF MEDICINE, JULY 19, 2010

Moore: So, Elwood, it's a great pleasure for me to have a chance to have a chat with you today. I thought that maybe we would get started if you could tell us how things were in terms of hormone action, what people were thinking about estrogen and other hormones as to how they worked, and the problems that were being addressed and your point of view towards them.

Jensen: Well, in the 1950s, that was the big era of enzymes. Biochemistry was largely endocrinology, and transcription, RNA synthesis, and that didn't come in for another ten years. So the current thinking of steroid hormones, estradiol, the principal one, was that it had to be reacting with enzymes. And the conventional model that they had is that the 17-hydroxy group of estradiol was oxidized by enzymatic oxidation using one coenzyme, and then with that it was oxidized to a ketone group to estrone, and then it was reduced by the oxidized form of another coenzyme, therefore—thereby transmitting or translocating a reduced hydrogen from one form to a reduced form of what was then called NADP. And that was known to be an important factor in many biosynthetic reactions.

And coming into the field and having joined the Department of Surgery at the University of Chicago, and Dr. Huggins, who had formed that in the middle '60s and brought in a couple of, two or three, young people to join it. He was an urologist who won the Nobel Prize for work on prostate cancer, but he was interested in breast cancer and estrogens too. And he showed me how very tiny amounts of estradiol—say, in a test animal, such as the rat, you could make the female reproductive tract grow to six times or so its size in the immature animal or the castrated animal. Then everybody thought it had to be enzymes.

So that fascinated me. How could such a tiny amount of an organic compound cause such a remarkable growth response? And as I said, the prevailing theory then was that it was oxidized by one coenzyme reduced by another, but then on the other hand, you had a synthetic estrogen that was diethylstilbestrol, which could not be oxidized. It had no aliphatic hydroxyl groups. That's a perfectly good hormone.

So we said, well, maybe we should attack this from a different angle. A little background for all of this was my experience there in 1947 in Zurich learning about steroid chemistry with Professor Ruzicka, who won the Nobel Prize and was head of biochemistry there in the so-called ETH, Eidgenössische Technische Hochschule, in Zurich. While I was there, I promised my late wife of 41 years that I would not try to climb any mountains because I had no experience. So when I came to Switzerland, I saw the Matterhorn standing all alone. And most peaks are just a little bit of elevation along a ridge of mountains, but the Matterhorn stands all by itself regally there, with the northeast face facing Switzerland and looking like a sheer wall of rock. It was the last major mountain in Europe to be climbed.

So I saw it, and I was approached by a student from Colorado, a young man who was getting his PhD there in Zurich, and he was an expert climber. He had lined up a guide to accompany him. You're allowed to have two guides to climb there now. Used to be you didn't have [to]—and in the first climb finally when someone climbed the Matterhorn, four out of seven people were killed.

So the law requires no more than two climbers to a guide. They wanted someone else to share the cost of the guide. I was in good physical shape from tennis and boxing and judo. I was fascinated by the mountain, so my wife released me from my promise, and we started out at 6:00 in the morning to get to the top before the sun melted the snow and made landslides.

It was the hardest physical thing I ever did in my life because I had no experience—although I was in good physical shape, but the atmosphere was rather oxygen free up there. So we got to the top about 8:00 in the morning, starting out early when it was still dark. I was going to stay—I had to stay up there for a while, and no one was there. And after a half-hour, the guide said: No, we have to start back. I'm glad we did because on the way down, where we'd been about an hour before, there was a landslide, and two days later two people were killed in such a landslide. So a long-winded background here. This idea of—I got ahead of myself a little bit. So it was the—when we got back I started reading a little bit about the history of the Matterhorn, and I learned that it was the last mountain to be climbed.

I wondered how it could be the last mountain to be climbed if even a novice could do it with the help of a guide. Everybody tried to go up on the Italian side, which looks—the Italian border goes right over the head of the top of the Matterhorn, the Swiss on one side, and Italy in one, and the French border coming in almost there. So it's right at the corner of three countries. And everybody tried to come up from the Italian side because that looks more gradual. It turns out there's one place in there where it was not—only a very few experts have ever made it up there, and people hadn't—as I said, it was the last one to be climbed. And an Englishman engraver named Edward Whymper who had tried the Italian side a couple of times—his hobby was mountain climbing—failed.

He came with binoculars and studied that northeast face of the Matterhorn, the different faces, and said maybe it isn't as difficult as one thinks. So they started out to try the northeast face, which looks like a sheer wall of rock, but there are crevasses and things in there. And they made the first climb, 1865, this actually was, and that was the first climb of the Matterhorn on the descent. He went up with six colleagues all roped together. On the descent the rope broke between number three and four, and four, five, six, and seven fell to their deaths. So now, you can have two people to a guide.

Anyway, this idea of alternative approach, try to do it differently than people are—the conventional way, that didn't seem to be working. So when we looked at the estradiol and making the uterus grow this way, everybody thought that it was being oxidized and reduced, but no one could really prove it well. It didn't explain diethylstilbestrol. So we decided to take an alternative approach. My graduate student Herbert Jacobson, now a professor at Albany Medical School, and I, we decided to try to determine not what the hormone does to the tissue, which everybody was looking at—what results from giving the hormone, what about the enzyme—but what happens to the hormone itself. Well, the reason probably people hadn't tried this so much was that its very low dose, as I mentioned, that to form—to find the actually physiological dose and to give a hyperphysiological dose wouldn't really be significant. Luckily, we were at the University of

Chicago then, and the Fermilab, where the plutonium project was developed and everything, had the facilities to handle carrier-free tritium.

If you tried to bring that on campus, they wouldn't allow it. But we were able to go out to Fermilab and take 6-dehydroestradiol and developed an apparatus that we could measure the uptake of tritium by a catalytic reduction of a double bond. We wanted to be sure that it was complete because we thought maybe there'd be an adverse catalytic effect of the higher isotope—tritium, three times as heavy as hydrogen. Actually, it went a little bit faster with tritium than with hydrogen itself. So with this apparatus, we were able to reduce—get 6-dehydroestradiol, or we could actually detect a billionth of a gram—a trillionth of a gram.

Really, you need about a billionth of a gram to be able to study it well. We could find a trillionth of a gram, so we came back, administered this to immature rats or castrated rats, and find out to our surprise nothing happens to it. It's not changed chemically.

Moore: So I think that you started the synthesis with 30 curies of carrier-free tritium; is that right?

Jensen: Well, about 60, I think, is what it took in our apparatus. We used up all of that and then pumped the rest back again for the next time.

Moore: Not millicuries, curies?

Jensen: Curies, yeah. So they had the facilities in the Fermilab. A side effect I should say [is] that later in Chicago I had the pleasure of living next door to the Fermis for a couple of years and going to their daughter's wedding. Wonderful gentleman and deserving of all the plutonium work he did.

So in any case, what we could do is to find what happened to the hormone in the tissue—the answer being nothing—but it did bind to a hitherto unrecognized protein there, an extranuclear protein there in its target tissues, and caused it to move to the nucleus and act as a transcription factor for synthesis of specific RNAs. But people—biochemists didn't like that. They said, where were the enzymes? How dare you call it a receptor when it isn't in a membrane? People thought receptors were in membranes.

But about six years later, Bert O'Malley down in Houston actually tried—got interested in this and tried similar experiments with progesterone and found out that its target tissues had a receptor protein which the hormone caused to move to the nucleus and synthesized specific RNAs. About the same time, Ron Evans—and John Baxter a little bit later—showed with thyroid hormone a similar receptor protein being caused by using the hormone to move to the nucleus and act as a transcription factor. So then about ten years after our first report at a meeting in Vienna, where only five people were present because everybody was there to learn how estradiol worked in a different—our session conflicted with a plenary session where 1,000 people were [there] to hear how estradiol now we know doesn't work—and five, three of whom were other speakers were there to learn about estrogen receptor. So that's the background of estrogen receptor.

And it stuck in my mind this idea of alternative approach: If it isn't working by conventional methods, try something differently.

Moore: An alternative approach—

Jensen: An alternative approach, yeah. That founded the field of nuclear receptors, as it's known, estrogen being the first, progesterone the second. Now there are about 49 was the last figure I've seen of biochemical regulators that all follow the same pattern, so it did pay off.

Moore: Forty-nine in mice, but humans only have 48. Mice have one extra.

Jensen: [Laughter] Anyway, then we had other challenges. One had the receptor protein, one tried to make antibodies to it, so one could use immunochemical methods to study the receptor and measure it a lot better than the binding technique—and without any success. People also have actually proposed—these are immunochemists—it was proposed maybe estradiol was so ubiquitous of a hormone that binds to its receptor and—but the receptor is not recognized as a foreign protein because it's present to some extent in so many tissues.

But to us then coming in with a little background in alternative approach and what the tissue did to the hormone, we decided we would use an unconventional method. We thought if we had the receptor protein and added what we thought might be the antibody preparation from the immune response and then we had the radioactive hormone bind to the receptor as a marker, if you had an antibody, it would make it move faster then because the antibody [was] about the same size as a receptor protein.

Moore: So this was in a sucrose gradient?

Jensen: Yes, sucrose gradient in an ultracentrifuge, and it worked just like a charm. We could show that you could detect the antibodies there. You could see that different antibodies could react, different preparations of antibody could react, with different places in the receptor protein because if you could saturate it with one, you could add the second antibody, and then you would get further movement on the gradient. So we were able to get the first antibodies to the receptor and use those then to measure the receptor, study it, and determine its amount in biological methods, including those from humans.

And the third alternative approach we had then was for studying the response of breast cancer to hormonal-type therapy. So it was known that there was some connection as early as the 1890s. Sir George Beatson took the ovaries out of—he was struck by the fact that breast cancer patients with large tumors because then they didn't find it as early enough, and you had very large tumors, that during menopause, the tumor would grow and then retract again. So there must be some connection, something between the ovary and the breast cancer. They didn't know about steroid hormones, of course, until about 1920.

So back in the late 1890s, the idea of something, some connection between the ovary and breast cancer was established by Beatson, who was knighted for this, Sir George Beatson, then. There are places that have Beatson Institutes. So anyway, people then started taking out the ovaries for breast cancer, in the younger woman taking them out, and in older woman taking out their adrenal glands because that's a source of estrogens after menopause, or taking out then the pituitary gland, which controls all of these hormone factories. But the only problem is only about 1/3 or a little bit less—1/3 of the patients will respond to this kind of therapy. And you're taking out these tissues, organs out of [ten] patients, and about three would respond. And all that time when they didn't respond, you'd have to take about eight months to see if they did.

The cancer would be growing and growing all the time. So one had some idea to identify which are the ones that are going to respond because then the ones that aren't going to respond, if you predict this, then you can use chemotherapy right away when the [tumor] is still much smaller. So we decided there's an alternative approach. Let's measure the receptor content of the breast tumors of patients who are going to undergo adrenalelectomy or a hypophysectomy or ovariectomy in a younger patient and see if there's any correlation there.

And it was again quite clear by this approach. You could see that those that didn't, that had escaped from the hormone dependency of the original breast tissue, they didn't need the hormone anymore. So they didn't make the receptor, and they did not respond to taking away the receptor because they didn't need it anymore. They had gotten an escape. But those that responded, they still had the hormone because the hormone was making them grow, and taking it away, they went away. So this gave you a good way to predict which [patients] should have adrenalectomy or hypophysectomy and which should be put on chemotherapy right away before the tumor had a chance to grow anymore after it was first detected.

Moore: So this led to tamoxifen, right?

Jensen: This was taking out the tissues.

Moore: That was before, right?

Jensen: Yeah. And there were some antiestrogens that were known, tamoxifen being one of them, but not much study was done. Then a man named Craig Jordan and I—and he was at Northwestern University at that time before he went to New York—we thought maybe we could try and see if tamoxifen wouldn't do the same thing as taking out the tissue, taking out the hormone-producing tissue. And sure enough, there was a pretty good correlation between the amount of receptor that was there and their response. So just as it was by taking out the hormone, you didn't have to do that. You just give them the antiestrogen tamoxifen. Craig and I used to delight the audiences when we would give talks by a little jingle that [went]:

A lady with growth neoplastic
thought castration was just a bit drastic.
She preferred that her ill could be cured with a pill.
Today it's no longer fantastic.

Anyway, that's the background of tamoxifen, and that does put the—it may prevent the estrogen from acting. Then along came other people studying this and found that getting something to prevent the synthesis of the estrogen in the first place rather than its action on the tissue might have some effect. And along came the so-called aromatase inhibitors preventing the synthesis of the hormone, putting the factory out of business rather than the action of the product at the target level. So today, some patients do better on tamoxifen. Some do better on [the aromatase inhibitor] Arimidex, but most do a little better on Arimidex, and people are taking it more and more there.

My wife, who you just met, is a hormone responder. She had her breast cancer removed, and it was receptor positive. Every breast cancer now has [levels of] hormone receptor, estrogen receptor, determined on it, and she had it. It was hormone receptor positive, and they then gave her—they took it off [and checked] with a mammogram, but to make sure if there were any metastases that were already there when they took out the primary tumor, they put her on Arimidex. So she's finishing her fourth year of Arimidex, and the receptor is gone.

So this is in a way an alternative approach to do with a pill instead of put the surgeon out of business and use a chemical to inhibit the action of estrogen by keeping estrogen from being there to act. So this then was what I like to say was made one especially for my wife's benefit from it. It makes one feel good.

And it did bring joy to my heart in Shanghai when the 50th anniversary of our presentation in Vienna where five people were present. The 50th anniversary of that was 2008, and Shanghai had a celebration of this as a surprise for me. They did not tell me. They asked me please to come and give a lecture and bring your slides along and bring some photos along if you can too and a CV. So when I got there, there was this beautiful brochure dedicated to the 50th anniversary here. And on this one they had to figure, they said 100,000 ladies every year are spared, prevented from dying due to our work. So those made me feel especially good.

Moore: I would think so. That's spectacular, really.

Jensen: Especially when my wife had the same. This was 52 years now, and I think a lot has been learned since then, but we do still have a problem with cancer. Early detection is important, but in many of the patients, we can do well with hormonal therapy. Unfortunately, those that have escaped the chemotherapy, we're getting better agents too in chemotherapy, but those remissions are not quite as good, and the chemotherapeutic agents are pretty toxic.

Moore: So I think that you and Sohaib [Khan] are doing some more on tamoxifen now, right?

Jensen: Right. That's one of the things that we're studying right now since I came here. He's the one that came when I had retired from Chicago because they had retirement age back when I reached 70—that went up when I was 73. When I was 70, you had to retire at age 70. Of course, that isn't as old as it was 30 years ago.

Moore: Right.

Jensen: Anyway, I had to retire, but we spent a year, my present wife, in New York at Memorial Sloan Kettering, which is interesting. I couldn't have lived 35 years like I did in Chicago in New York, I think, but it was interesting for a year being right there with Memorial Sloan Kettering and Rockefeller University right on this corner there in Manhattan. Then they got a job in Hamburg ready for me, and [the] Albert [von] Humboldt Foundation set up a professorship for me to go to Hamburg for a while. I spent about seven and a half years in Hamburg. And then I was going to retire, but they wanted me to come to the Karolinska for a year because Jan-Åke Gustafsson [now] from Houston had discovered—he had discovered a second estrogen receptor, namely, now, the ER-β is what it's called. And what we had was ER-α. So ER-β and ER-α they were studying quite a lot there. And I went there for a year and ended up spending three years and had about four publications with Jan-Åke on ER-β. And knowing that both of the two different ER receptors, you can get a better picture than you could with one receptor alone, so it all adds up.

Then when I was in—that's in Stockholm, Sohaib Khan came up to—was in Stockholm for some other reason. He came over and said, when you get back, won't you come to Cincinnati for a while? That was very tempting because I grew up in Springfield, Ohio. I was born in North Dakota but moved to Springfield when I was four, and up there Cincinnati was the big city. I knew it pretty well coming down to watch the Reds play baseball or come to the opera, symphony. So I took tennis lessons at Camargo Country Club in college and in high school, actually, at the Hyde Park Country Club. So I said, I'll come for seven months. And I'm here eight years now and not smart enough to know when to leave.

But in any case, that's a little bit of how I got here and how I got to interact with other people like Jan-Åke Gustafsson, who's another Texan—who had to retire at age 65 in Stockholm. I tried to get him to come here to the University of Cincinnati, and he was going to think about coming here. But somehow, our administration got him mixed up and thought that he had to come right away or they couldn't have him. We weren't ready to have him right away. No, he had until another three-quarters of a year, and so they told him no. Then when they found out that he could come, they already had someone else. So Jan-Åke went to—Jan-Åke had already decided to come to Texas. He's happy there.

Moore: He seems to be happy there.

Jensen: That's a little background of the whole history of estrogens, estrogen receptor, and breast cancer. And still a lot of women do—are not found early enough or do not get the proper therapy. But more and more they are doing this, and many are alive today. It would not have been [the case] even ten years ago.

Moore: That's really spectacular. So I think that the readers and the viewers of the Web site of the *Annual Review of Physiology* will be thrilled to have had this opportunity to hear your stories. Do you have any other stories for them? Any other things that you'd like to pass along?

Jensen: Let me think.

Moore: I'm thinking about boxing.

Jensen: About what?

Moore: About boxing. When you were much younger, of course, you, I understand, were in the golden gloves.

Jensen: Oh! Yeah. I was. My mother was a grade school teacher. I was in Minnesota until I was four, born in North Dakota. She taught me to read and write when I was four. I came to start school in Springfield—that first year, that's what you learned, how to read and write.

So they put me ahead right away and to the end of the second year, so a year and a half ahead of time, and that made me much younger than everybody else, and I was kind of tall, skinny, and socially immature. So this was a handicap. I'm thinking how I got started on boxing. Anyway, I guess it was—

Moore: At Wittenberg?

Jensen: It must've been at when I was in Wittenberg, and they had an intramural boxing tournament there—that was how it was. And being tall and skinny gave me an advantage. I fought in the 140-pound class of welterweight. That gave me an advantage of reach [over] some of the more stocky individuals.

So in the Memorial Hall in Springfield, they would have each spring—about three or four weeks in a row, they would have what they called the golden gloves for three counties, not only Springfield, Ohio, but they had a big what's called then a CCC camp. They had a lot of pretty rough people working—during the Depression, this was. So they had people coming up in Memorial Hall every Friday night for a while. So I do remember that first Friday night. Here I came out, this skinny, young Norwegian Dane from North Dakota originally. And the fellow, he dropped his guard a little bit, and I caught him right on the chin, and he went down for the count.

I raised my hand there with all this—I guess 3,000 people out there Friday night. That sort of changed things. The next Friday night, they came up against another pretty tough-looking guy, but he raised his guard a bit, and I caught him in his solar plexus. He went down to the count. So two knockouts in a row.

Actually my late wife, Mary, whose father was a boxing fan, she was in the audience there in Memorial Hall. I didn't know her, of course, at that time. She was somewhat older, five years older than I, but she was there to see me win.

In the finals of the three-county tournament, I somewhat fortunately, I think, came down with the flu epidemic of 1939. They did have a flu epidemic then, and I recovered enough to go down and fight, but I did not recover enough. I lost the decision, and my opponent went on and won. He was really quite good. He went on and won the state and the national [tournaments and] turned professional.

So anyway, it was this boxing that you reminded me of that helped me a lot with self-confidence that I had because of being younger than other people. So I did have a good career at Wittenberg. I had no problem going into intramural boxing there, and I played on the tennis team there. I did enjoy Wittenberg and was president at my fraternity there. It all kind of—a lot of it went back to the boxing and all of that.

Moore: It all worked out very well.

Jensen: It all worked out very well, right.

Moore: Well, thank you, Elwood.

Jensen: You're welcome.

Moore: I really enjoyed talking with you, and I hope that you have many more years of success as things continue to move forward.

DISCLOSURE STATEMENT

David D. Moore is not aware of any affiliations, memberships, funding, or financial holdings that might be perceived as affecting the objectivity of this review.

LITERATURE CITED

1. Jensen EV, Jacobson HI. 1960. Fate of steroid estrogens in target tissues. In *Biological Activities of Steroids in Relation to Cancer*, ed. G Pincus, E Vollmer, pp. 161–74. New York: Academic

2. Goldstein JL. 2004. Towering science: An ounce of creativity is worth a ton of impact. *Nat. Med.* 10:1015–17
3. Talalay P, Williams-Ashman HG. 1958. Activation of hydrogen transfer between pyridine nucleotides by steroid hormones. *Proc. Natl. Acad. Sci. USA* 44:15–26
4. Jensen EV, Suzuki T, Kawashima T, Stumpf WE, Jungblut PW, DeSombre ER. 1968. A two-step mechanism for the interaction of estradiol with rat uterus. *Proc. Natl. Acad. Sci. USA* 59:632–38
5. Toft D, Gorski J. 1966. A receptor molecule for estrogens: isolation from the rat uterus and preliminary characterization. *Proc. Natl. Acad. Sci. USA* 55:1574–81
6. Gorski J, Toft D, Shyamala G, Smith D, Notides A. 1968. Hormone receptors: studies on the interaction of estrogen with the uterus. *Recent Prog. Horm. Res.* 24:45–80
7. O'Malley BW, Means AR. 1974. Female steroid hormones and target cell nuclei. *Science* 183:610–20
8. Greene GL, Closs LE, Fleming H, DeSombre ER, Jensen EV. 1977. Antibodies to estrogen receptor: immunochemical similarity of estrophilin from various mammalian species. *Proc. Natl. Acad. Sci. USA* 74:3681–85
9. Greene GL, Fitch FW, Jensen EV. 1980. Monoclonal antibodies to estrophilin: probes for the study of estrogen receptors. *Proc. Natl. Acad. Sci. USA* 77:157–61
10. Green S, Walter P, Greene G, Krust A, Goffin C, et al. 1986. Cloning of the human oestrogen receptor cDNA. *J. Steroid Biochem.* 24:77–83
11. Jensen EV, Jordan VC. 2003. The estrogen receptor: a model for molecular medicine. *Clin. Cancer Res.* 9:1980–89
12. Jensen EV. 2004. From chemical warfare to breast cancer management. *Nat. Med.* 10:1018–21
13. Jensen EV, Khan SA. 2004. A two-site model for antiestrogen action. *Mech. Ageing Dev.* 125:679–82
14. Wang Y, Chirgadze NY, Briggs SL, Khan S, Jensen EV, Burris TP. 2006. A second binding site for hydroxytamoxifen within the coactivator-binding groove of estrogen receptor β. *Proc. Natl. Acad. Sci. USA* 103:9908–11
15. O'Malley BW, Sherman MR, Toft DO. 1970. Progesterone "receptors" in the cytoplasm and nucleus of chick oviduct target tissue. *Proc. Natl. Acad. Sci. USA* 67:501–8
16. Charles MA, Ryffel GU, Obinata M, McCarthy BJ, Baxter JD. 1975. Nuclear receptors for thyroid hormone: evidence for nonrandom distribution within chromatin. *Proc. Natl. Acad. Sci. USA* 72:1787–91
17. Oppenheimer JH, Koerner D, Schwartz HL, Surks MI. 1972. Specific nuclear triiodothyronine binding sites in rat liver and kidney. *J. Clin. Endocrinol. Metab.* 35:330–33
18. Weinberger C, Thompson CC, Ong ES, Lebo R, Gruol DJ, Evans RM. 1986. The *c-erbA* gene encodes a thyroid hormone receptor. *Nature* 324:641–46
19. Sap J, Munoz A, Damm K, Goldberg Y, Ghysdael J, et al. 1986. The c-erb-A protein is a high-affinity receptor for thyroid hormone. *Nature* 324:635–40

Epigenetic Control of Smooth Muscle Cell Differentiation and Phenotypic Switching in Vascular Development and Disease

Matthew R. Alexander and Gary K. Owens*

Robert M. Berne Cardiovascular Research Center, University of Virginia School of Medicine, Charlottesville, Virginia 22908; email: mra2e@virginia.edu, gko@virginia.edu

Keywords

arteriogenesis, vascular injury and repair, atherosclerosis, mesenchymal stem cells, pluripotency genes

Abstract

The vascular smooth muscle cell (SMC) in adult animals is a highly specialized cell whose principal function is contraction. However, this cell displays remarkable plasticity and can undergo profound changes in phenotype during repair of vascular injury, during remodeling in response to altered blood flow, or in various disease states. There has been extensive progress in recent years in our understanding of the complex mechanisms that control SMC differentiation and phenotypic plasticity, including the demonstration that epigenetic mechanisms play a critical role. In addition, recent evidence indicates that SMC phenotypic switching in adult animals involves the reactivation of embryonic stem cell pluripotency genes and that mesenchymal stem cells may be derived from SMC and/or pericytes. This review summarizes the current state of our knowledge in this field and identifies some of the key unresolved challenges and questions that we feel require further study.

VASCULAR SMOOTH MUSCLE CELLS EXHIBIT EXTENSIVE PHENOTYPIC PLASTICITY

The vascular smooth muscle cell (SMC) in mature animals is a highly specialized cell whose principal function is contraction and regulation of blood vessel diameter, blood pressure, and blood flow distribution [reviewed in Owens et al. (1)]. Investment of nascent blood vessels with SMCs or SMC-like pericytes, and the subsequent differentiation of these cells, is required for appropriate vascular maturation and formation of functioning vascular networks during embryogenesis, as well as in vascular remodeling in adult organisms (2–4). During vascular development, vascular SMCs and pericytes also play a critical role in secretion of extracellular matrix components, including collagen and elastin, which determine the mechanical properties of mature blood vessels (5, 6). Differentiated SMCs in adult blood vessels proliferate at an extremely low rate, exhibit low synthetic activity, and express a unique repertoire of contractile proteins, ion channels, and signaling molecules required for the cell's contractile function (7, 8). However, results of lineage-tracing studies [9–11; reviewed in Hoofnagle et al. (12, 13)] have provided definitive evidence that SMCs within adult animals retain remarkable plasticity and can undergo profound and reversible changes in phenotype, a process referred to as phenotypic switching, in response to vascular injury or disease [reviewed in Owens (7)].

Although there are likely many alternative phenotypic states of SMCs (**Figure 1**), in general SMC phenotypic switching is characterized by markedly reduced expression of SMC-selective differentiation marker genes and increased SMC proliferation, migration, and synthesis of extracellular matrix components required for vascular repair [reviewed in Owens et al. (1)]. Indeed, the extensive plasticity exhibited by the fully mature SMC is an inherent property of the cell that likely evolved in higher organisms because it conferred a survival advantage. That is, mutations that compromised the ability of the SMC to participate in vascular repair were likely detrimental to the organism and did not persist. However, an unfortunate consequence of this plasticity is that it predisposes the SMC to environmental cues/signals that can induce adverse phenotypic switching and contribute to development and/or progression of vascular disease. Indeed, there is

Figure 1

Smooth muscle cells (SMC) are highly plastic cells that may undergo phenotypic modulation to distinct phenotypes within atherosclerotic lesions. Platelet-derived growth factor (PDGF) stimulation increases migration and proliferation of cultured SMC (157–160), and there is evidence that in vivo PDGF signaling promotes SMC investment of developing atherosclerotic plaque fibrous caps (161, 162), suggesting that PDGF promotes SMC phenotypic modulation to a migroproliferative state that enhances fibrous cap formation. Transforming growth factor beta (TGFβ) dramatically promotes the production of extracellular matrix, particularly collagen, by SMC, leading to a matrigenic phenotypic state (163), whereas oxidized phospholipids (oxPLs) appear to promote an intermediate phenotype by promoting SMC migration and proliferation and enhancing expression of collagen VIII (102, 134, 164). Inflammatory cytokines such as tumor necrosis factor alpha (TNFα) and interleukin (IL)-1 promote expression of adhesion molecules such as intercellular adhesion molecule-1 (ICAM-1) (165), inflammatory cytokines such as IL-6 (166), and multiple matrix metalloproteinases (MMPs) (167–169), leading to the formation of inflammatory-state SMC. Expression of ICAM-1 by these cells promotes adhesion to monocyte/macrophages and T cells through binding to the integrin lymphocyte function–associated antigen 1 (LFA-1) (170). Additionally, increased levels of inorganic phosphate (P_i) promote the formation of osteochondrogenic SMCs that promote calcification in vitro and are found within calcified vessels of matrix Gla protein–deficient mice in vivo (11). Although the studies cited in this caption suggest that these stimuli can give rise to distinct SMC phenotypes, further studies directly comparing the effects of these factors on genome-wide gene expression and functional characteristics of SMC will be needed to determine whether these phenotypes are truly distinct. In addition, a major deficiency in the field is that there is a lack of definitive lineage-tracing studies showing that SMC give rise to these distinct phenotypes in vivo in the context of atherosclerosis or repair of a damaged blood vessel. Finally, because SMC-like pericytes possess properties of mesenchymal stem cells (22), mesenchymal stem cell–like cells derived from SMC may exist within atherosclerotic lesions and may represent an intermediate, undetermined state that can contribute to the extensive plasticity of SMC within atherosclerosis and other disease states.

strong evidence that phenotypic switching of the SMC plays a critical role in a large number of major diseases in humans, including atherosclerosis, asthma, hypertension, and cancer (1).

Let us start by considering the potential role of SMC phenotypic switching in the pathogenesis of cancer, as this is a poorly understood and understudied area. A long-standing dogma among pathologists is that tumors having a predominance of immature dilated leaky blood vessels with poor perivascular cell (i.e., SMC and pericyte) investment have a high metastatic potential (14). Consistent with these observations, Ramaswamy et al. (15) showed that four out of nine genes whose downregulation is part of a 17-gene molecular signature of highly metastatic human tumors are markers of differentiated SMCs and pericytes. Although not discussed by these authors, these results provide compelling evidence that failed SMC-pericyte investment and differentiation within tumor blood vessels are highly linked with tumor metastatic potential. There may be a correlation between poor perivascular cell investment and metastasis because immature tumor vessels are leaky, with a resulting increase in blood cells passing into the tumor interstitial tissue (thus explaining why most highly metastatic tumors are also hemorrhagic), and because tumor

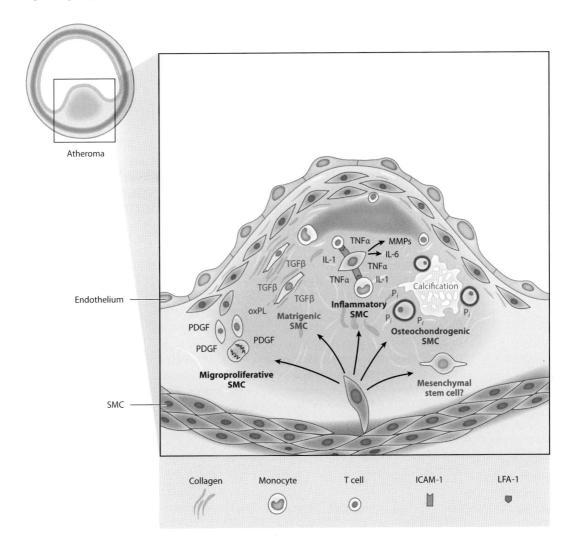

cells are shed at an increased rate into the bloodstream (16). However, there is no direct evidence for these events, and the mechanisms responsible for disrupting the maturation of tumor blood vessels are poorly understood.

The best-known example of a disease in which SMC phenotypic switching plays a key role is atherosclerosis, a disease currently responsible for >40% of all deaths in Western civilization. Atherosclerosis is a complex disease involving many cell types, including macrophages, lymphocytes, endothelial cells (ECs), and vascular SMCs (17). Of interest, the role of the SMC in the pathogenesis of atherosclerosis appears to vary, depending on the stage of the disease. The SMC has a maladaptive role in lesion development and progression; up to 70% of the mass of lesions is estimated to be SMCs or SMC derived (e.g., extracellular matrix) (7, 17). Paradoxically, at the same time that intimal SMCs within eccentric atherosclerotic lesions are contributing to luminal encroachment, medial SMCs play a key role in adaptive outward remodeling, a process beneficial in maintaining blood flow (18). Within the fibrous cap of advanced atherosclerotic plaques, SMCs may play either a critical beneficial role or a detrimental role in determining plaque stability, depending on the cell's phenotypic state (19, 20). That is, synthetic-state SMCs are the primary cells responsible for stabilizing fibrous caps by virtue of their proliferation and production of extracellular matrix proteins. However, in response to environmental cues that are not well understood, these cells may undergo apoptosis and/or activate expression of various matrix metalloproteinases and/or inflammatory mediators that can act in concert with those produced by macrophages and promote end-stage disease events such as plaque rupture and thrombosis (19, 20). Major challenges for the field are to define the extent of the plasticity of phenotypically modulated SMCs and to identify molecular mechanisms that regulate critical transitions in SMC phenotype following vascular injury or in cardiovascular disease.

EXPRESSION OF SMOOTH MUSCLE CELL–SELECTIVE DIFFERENTIATION MARKER GENES IS DEPENDENT ON INTEGRATION OF LOCAL ENVIRONMENTAL CUES

The model that has evolved to explain SMC differentiation is that it is highly dependent on the cell integrating complex local environmental cues/signals, including mechanical forces, neuronal influences, extracellular matrix components, and various soluble cytokines and growth factors that influence the expression of the repertoire of genes that determine SMC phenotype (reviewed in References 1, 7, and 21). That is, SMCs constantly integrate the complex signals present in their local environment, and these signals in aggregate determine the appropriate patterns of gene expression and cell function. This model is not unlike that which is believed to describe the control of the differentiation of virtually all cell types. However, the SMC is somewhat unique in the extent of plasticity that it can and must exhibit even in adult organisms compared with many other cell types. Indeed, as discussed below, there is recent evidence that phenotypically modulated SMCs and pericytes may be the primary source of multipotential mesenchymal stem cells (MSCs) (22). Key challenges in understanding the control of SMC differentiation are (*a*) to identify the critical environmental cues/signals that influence differentiation and maturation of the SMC, (*b*) to determine the molecular mechanisms and signaling pathways whereby these environmental cues influence the expression of genes characteristic of the SMC lineage, and (*c*) to determine how these processes are altered during vascular injury-repair or disease.

An initial step in elucidating molecular controls of SMC differentiation was to identify appropriate marker genes that encode for proteins that are selective or specific for SMCs and that are required for the cell's differentiated functions. Indeed, there has been considerable progress in this area over the past several decades, and we now have a relatively large repertoire of SMC-selective marker genes with which to study SMC differentiation and maturation [reviewed in Owens et al. (1)]. Appropriate markers include a variety of contractile proteins important for the differentiated function of the SMC, including smooth muscle (SM) α-actin (23, 24), SM myosin heavy chains (SM MHCs) (25–27), SM myosin light chains (28), h_1-calponin, and SM α-tropomyosin (29). In addition, differentiated SMCs express a number of proteins that are part of the cytoskeleton and/or are believed to be involved in the regulation of contraction such as h-calponin (30), SM22α (30), h-caldesmon (31), β-vinculin (32), metavinculin (32), telokin (33), smoothelin (34), LPP (35), and desmin (36). All these proteins show at least some degree of SMC specificity and selectivity (reviewed in References 1, 7, and 21). Virtually all, if not all, of these SMC differentiation marker genes can also be expressed in non-SMCs under some conditions, with the possible exception of the SM MHC isoforms, which appear to be the most specific markers of differentiated SMCs. As such, rigorous assessment of the differentiated state of the SMC depends on examining multiple marker genes, including those that are both increased and decreased with differentiation, maturation, and phenotypic switching. For a more complete discussion of these markers, please see our previous detailed review of this topic (1).

A major advance for this field was the identification of promoter-enhancer regions of a number of the SMC differentiation marker genes that confer SMC-specific/selective expression in vivo in transgenic mice (36–40). Studies by Li et al. (37) and Kim et al. (39) were the first to identify sufficient regions of the SM22α promoter that conferred expression in transgenic mice. Of interest, they showed that a promoter construct containing just 441 bp of the SM22α 5′ promoter region conferred expression in cardiac, skeletal, and arterial SMCs in transgenic mouse embryos, but not in gastrointestinal or other SMC tissues that normally express the endogenous SM22α gene. Our laboratory identified regions of the SM MHC (40) and SM α-actin (38) promoters that could drive expression in vivo in transgenic mice in a manner that recapitulated expression of the endogenous genes. Moreover, we subsequently characterized multiple regulatory modules and *cis* elements required for SMC-specific/selective expression of these genes in transgenic mice (38, 40–44), including several that mediate responsiveness of these genes to environmental signals thought to be important in the control of SMC differentiation and maturation (41, 45–47; reviewed in Reference 48). This includes the activation of SMC marker genes by TGFβ1 and contractile agonists such as angiotensin II as well as repression by PDGF BB, PDGF DD, and proatherogenic oxidized phospholipids (the primary active components of oxidized low-density lipoproteins) (see below).

SMOOTH MUSCLE CELL–SPECIFIC TRANSCRIPTIONAL REGULATION DEPENDS ON UNIQUE COMPLEX COMBINATORIAL INTERACTIONS OF MULTIPLE *CIS* ELEMENTS AND THEIR *TRANS* BINDING FACTORS

Tremendous progress has been made in the past decade in identifying mechanisms that contribute to transcriptional regulation of SMC marker genes [see reviews by Owens and coworkers (1, 48, 49), Firulli & Olson (50), Majesky (51), and Miano (52)]. Due to space constraints and the large amount of work in this area, we cannot discuss these regulatory pathways in detail. However, suffice to say that the model that has emerged is that regulation of SMC-selective gene expression is not dependent on any single factor completely specific for SMCs but rather depends on unique combinatorial interactions of multiple factors that are either ubiquitously expressed or selective for SMCs (**Figure 2**). We consider one example, CArG-SRF-dependent regulation (where SRF denotes serum response factor), to illustrate this general model because this is the best-characterized regulatory paradigm in SMC differentiation.

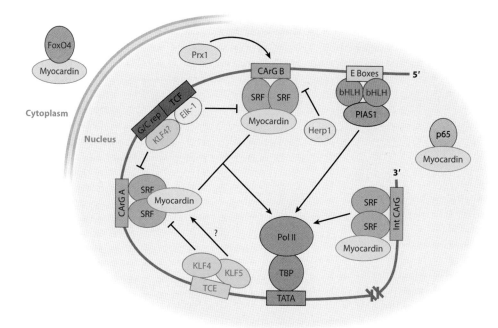

Figure 2

Coordinate expression of SMC-selective genes is mediated through complex combinatorial interactions of multiple conserved *cis*-regulatory elements and transcription factors. Myocardin plays an important role in promoting interactions of serum response factor (SRF) homodimers with conserved CArG boxes within the promoters of SMC-selective contractile and cytoskeletal genes to promote the recruitment of RNA polymerase II (Pol II) and to drive expression of these genes (1, 73, 74, 78). Additional factors cooperate to promote myocardin-SRF-CArG interactions and to drive SMC-selective gene expression, including paired-related homeobox gene 1 (Prx1) (62) and complexes of the protein inhibitor of activated Stat1 (PIAS1) with basic helix-loop-helix (bHLH) factors such as E2-2 and E12 at E-box *cis*-regulatory elements (42, 153). In contrast, transcription factors such as Krupple-like factor 4 (KLF4), E26 ETS-like transcription factor 1 (Elk-1), and HES-related repressor protein 1 (Herp1) repress SMC gene expression, at least in part by disrupting myocardin-SRF-CArG interactions (68, 73, 91, 154). Elk-1 is recruited to SMC marker gene promoters through binding to ternary complex factor (TCF) sites (68), whereas KLF4 is recruited to SMC promoters through binding to TGFβ control elements (TCE) (41, 44) and G/C-rich repressor elements (G/C rep) (M.S. Angulo & G.K. Owens, unpublished data) found within multiple SMC marker gene promoters. Like KLF4, Krupple-like factor 5 (KLF5) has been demonstrated to bind at the TCE and to promote SMC marker gene expression (41, 44), although other studies have provided evidence that KLF5 represses SMC marker gene expression (155). Finally, transcription factors such as forkhead box O4 (FoxO4) and the p65 subunit of NFκB repress SMC gene expression through inhibiting myocardin-SRF interactions by blocking myocardin entry into the nucleus and by blocking SRF binding sites, respectively (92, 156).

Site-directed mutagenesis studies in transgenic mice have shown that expression of most SMC marker genes identified to date depends on one or more CArG elements [i.e., a $CC(AT)_6GG$ motif] found within their promoter and/or intronic sequences (36–39, 43, 53). For example, we demonstrated that the region of the SM α-actin promoter from −2,560 to +2,784 completely recapitulated expression patterns of the endogenous SM α-actin gene in vivo in transgenic mice. However, expression of this >5,300-bp promoter enhancer was completely abolished by a 2- or 4-bp mutation of any one of three highly conserved CArG elements contained within it (38).

CArG elements bind the transcription factor SRF, a MADS (MCM1, Agamous, Deficiens, SRF)-box transcription factor, that was first named because of its ability to confer serum inducibility to the growth-responsive gene, c-*fos*, through binding to a sequence known as the serum response element (SRE) (or CArG box). SRF binds CArG boxes as a dimer; dimerization and DNA binding occur through the MADS-box domain (54). In addition to regulating growth-responsive genes such as c-*fos* and multiple SMC marker genes (55), SRF binding to CArG boxes also regulates numerous skeletal muscle–specific and cardiac muscle–specific genes (56, 57). A long-standing question has been to determine how SRF, a ubiquitously expressed transcription factor, can regulate both growth-responsive and cell-specific genes in SMCs and muscle and non-muscle cell types. Investigators have proposed several possible mechanisms, including a number that appear unique to SMCs. Importantly, these mechanisms are not mutually exclusive, and SMC selectivity is likely the result of some combination of the following mechanisms and/or others yet to be discovered (48, 52). These mechanisms, reviewed in detail in several of our recent reviews (1, 58, 59) as well as in a review by Miano (52), include (*a*) alterations in SRF expression levels (60, 61); (*b*) regulation of SRF binding affinity to the CArG elements found in many SMC promoters by homeodomain factors such as Prx1/Mhox (47, 62); (*c*) cooperative interactions between the multiple CArG elements found in most SMC marker genes but not in ubiquitously expressed genes such as c-*fos* that contain a single CArG (36, 38, 39, 43, 53, 63, 64) [although an exception is the telokin promoter, which contains a single functional CArG element (65)]; (*d*) posttranscriptional modifications of SRF or SRF cofactors (66–69); (*e*) cooperative interaction with other *cis*-regulatory elements and their binding factors, including CRP1/2, GATA4, and Nkx3.2 (70, 71); (*f*) regulation of SRF's ability to bind to CArG elements though modulation of chromatin structure (72, 73), a topic considered in more detail below; and (*g*) interaction with SMC-specific/selective SRF coactivators, including myocardin and MRTF (MKL-1/2) A/B (67, 74–81).

Of major interest, the SM α-actin CArG elements are 100% conserved across all species in which the gene has been cloned, including chickens and humans, which diverge by more than 500 million years of evolution. Remarkably, this includes conservation of a single G or C substitution within the central 10-bp A/T-rich core of the CArG motif, changes that we have shown dramatically reduce SRF binding affinity but that also make the SMC CArGs subject to regulation of SRF binding through interaction with the homeodomain protein Prx1 (47, 82). Of interest, activation of CArG-dependent SMC marker genes by contractile agonists is mediated, at least in part, by Prx1-dependent enhanced SRF binding to these conserved degenerate CArG elements. This paradigm of conserved degenerate CArG elements is not unique to SM α-actin but is a component of many of the SMC marker genes. So why did Mother Nature equip SMC genes with "imperfect" or degenerate CArG elements and completely conserve them through evolution? To address this question, we tested the effects of gain-of-function mutations of the CArG boxes (i.e., mutations of the G or C base pairs to A or T, referred to as SRE substitution SM α-actin mutations) that increase SRF binding. Whereas the wild-type SM α-actin promoter exhibited a high degree of SMC specificity when tested in cultured cells, the gain-of-function SRE mutant SM α-actin promoter showed that complete loss of SMC specificity was expressed in virtually any cell type tested, including ECs that did not express their endogenous SM α-actin gene (83). On this basis, we predicted that the SRE mutant SM α-actin promoter LacZ transgenic mouse would also show complete lack of SMC specificity in vivo, i.e., result in ubiquitous expression in all or most mouse tissues. However, what we found was complete retention of cell selectivity. For example, the gene was expressed at exactly the same time as the endogenous SM α-actin gene and was transiently expressed in cardiomyocytes and skeletal myoblasts during early development but was then silenced during late embryogenesis in these tissues and became SMC restricted (another example of the complete dissociation of results in cultured cells versus in vivo; more on this below) (82). However, the SRE substitution mutant SM α-actin transgene failed to be suppressed following vascular injury. Thus, it appears that the SMC evolved degenerate CArG elements, and completely conserved them through evolution, not to confer SMC-specific gene regulation but to modulate expression of these genes during reversible phenotypic switching. This possibility is highly interesting and has key implications in terms of mechanisms that control SMC plasticity, a recurring theme throughout this review.

One of the most significant and exciting advances for the field of SMC differentiation in the past decade was the discovery of myocardin by Olson and coworkers (77). Myocardin is an extremely potent SRF coactivator that is exclusively expressed in cardiac and differentiated SMCs (57, 77, 78). Moreover, mouse embryos homozygous for a myocardin loss-of-function mutation die by embryonic day 10.5 and show impaired vascular SMC differentiation on the basis of detailed in situ analysis that revealed the lack of perivascular cells expressing SMC-selective genes within the dorsal aorta, where the first SMC forms during embryogenesis (84). However, a confounding aspect of this study was that the authors also found marked retardation of growth of myocardin-null embryos and gross abnormalities in yolk sac vasculogenesis, suggesting that effects on SMC development could have been indirect. Indeed, in collaboration with the Olson lab, we showed that cell-autonomous myocardin is not required for SMC development in that lineage-tagged myocardin-null embryonic stem cells (ESCs) can form aortic SMCs in the context of chimeric knockout (KO) mice (85). However, we recently extended these studies and showed that myocardin-null cells show markedly reduced contributions to SMCs within the bladder and gastrointestinal SMCs of chimeric KO mice, indicating that myocardin dependence varies between different SMC subtypes (86). These studies also showed that myocardin-null ESCs were defective in their ability to form ventricular but not atrial cardiomyocytes, thus adding further credence to the possibility that defects in SMC development in conventional myocardin KO mice are not primary but secondary to other defects, including defective yolk sac vasculogenesis and/or defective heart function.

Interpretation of results of studies in conventional myocardin KO mice is also likely confounded by the possible activation of the myocardin-related transcription factors MRTF A and B. There has been much recent progress in the elucidation of the relative functions of myocardin versus those of the MRTFs, including the demonstration of their importance in mediating rho kinase and cytoskeletal signaling pathways (76, 87) and differential roles in mediating the development of SMCs from different embryonic lineages (80, 81). However, the study that we believe provides the most clarity regarding the normal requirement of SMCs for myocardin in vivo was published by Parmacek and coworkers (88). In brief, they showed that selective KO of myocardin in neural crest–derived SMCs (by crossing floxed myocardin mice with either Pax3-cre or Wnt1-cre mice) resulted in 100% perinatal lethality due to failed closure of the ductus arteriosis secondary to impaired SMC differentiation. This is an interesting observation in that compensatory activation of the myocardin-related transcription factors MRTF A and B was unlikely because there was presumably no selection pressure for activation of these genes within the ductus until birth, at which time there was insufficient time for compensation to be effective. As such, these studies provide compelling evidence that myocardin is normally required for SMC differentiation, at least in the ductus arteriosus.

There is also much known regarding the mechanisms by which myocardin and MRTFs activate CArG-dependent SMC genes. Myocardin potently and selectively induces expression of all CArG-dependent SMC marker genes tested to date, including those encoding SM α-actin, SM MHC, SM22α, and calponin in cultured SMCs and cultured embryonic fibroblasts (57, 74, 77, 78). Of interest, myocardin appears to be most efficacious in activating those genes that contain multiple CArG elements, and Olson and colleagues (79) have presented evidence for a model whereby the leucine zipper motif of myocardin may bridge adjacent CArG elements and unmask myocardin's activation domain, observations consistent with our findings that the phasing and spacing of SM α-actin 5′ CArG elements are critical for transcriptional activation (55). The spacing of the SM α-actin 5′ CArG elements is 40 bp and is 100% conserved across all species cloned to date. This places these elements on the same DNA face, consistent with the formation of a myocardin dimer–containing SRF higher-order complex. We found that either the deletion or the addition of a 5-bp region between the CArGs resulted in complete loss of activity, whereas the insertion of 10 bp (which maintains phasing) resulted in only modest decreases in activity (55). Expression of the SM MHC and SM22α promoters also involves cooperative interaction of multiple CArG elements and the formation of myocardin or MRTF containing SRF higher-order complexes such that this appears to represent a general model for SMC-selective and coordinate activation of multiple SMC marker genes (**Figure 1** presents a model for cooperative interaction of SMC CArG elements through a myocardin dimer).

We (74) and others (79) have shown that adenovirus-mediated overexpression of myocardin activates multiple CArG-dependent SMC differentiation markers in ESCs, or embryonic fibroblast systems, but significantly does not activate CArG-independent SMC genes such as those encoding smoothelin, α1-integrin, and FRNK, indicating that myocardin alone is not sufficient to activate the entire SMC differentiation program [see review by Parmecek (89)]. Nevertheless, a number of groups, including ours, have shown that the administration of a myocardin siRNA or adenovirus-mediated overexpression of a myocardin dominant negative construct significantly reduced expression of multiple SMC marker genes, including those encoding SM MHC, SM22α, and SM α-actin, by up to 80% in cultured SMCs. These findings thus provide direct evidence that endogenous myocardin plays a key role in the regulation of expression of multiple CArG-dependent SMC marker genes (57, 74, 77–79). Interestingly, although both the telokin and c-*fos* gene promoters contain a single CArG element rather than multiple CArG elements, myocardin activates only telokin (90), indicating that multiplicity of CArG elements alone does not appear to be sufficient to explain SMC-specific transcriptional activation and that other factors and mechanisms are required.

There has also been considerable progress in elucidating the mechanisms by which myocardin mediates responses of SMCs to environmental cues that either activate or repress SMC gene expression in vivo. For example, we showed that myocardin plays a key role in both angiotensin II–induced increases (62), as well as in PDGF BB–induced decreases (44, 73, 91) in the expression of SMC marker genes. Wang et al. (68) demonstrated that PDGF BB–induced suppression of SMC genes involved phosphorylation of Elk-1, which competed with myocardin for CArG-SRF binding, and more recent evidence has demonstrated that the proinflammatory transcription factor NFκB can also bind myocardin and inhibit interactions with CArG-SRF (92). Olson and coworkers (93) also showed that myocardin-induced activation of SMC and cardiomyocyte-selective genes involved myocardin-dependent recruitment of histone acetyl transferases, histone hyperacetylation, and the formation of euchromatin permissive for transcriptional activation. For a more complete review of the mechanisms whereby myocardin activates CArG-dependent SMC genes, and the relationship between myocardin and MRTFs A and B, please see previous reviews by our group (1) and others (89, 94, 95).

EVIDENCE THAT CHROMATIN STRUCTURE AND EPIGENETIC MECHANISMS PLAY A KEY ROLE IN REGULATING SMOOTH MUSCLE CELL DIFFERENTIATION AND PHENOTYPIC SWITCHING

Genomic DNA is packaged into a compact structure known as chromatin. Chromatin is a dynamic polymer composed primarily of DNA and protein, the structure of which is regulated both epigenetically (e.g., by DNA methylation, histone modifications, or histone-binding proteins) and by *trans*-acting DNA-binding proteins (e.g., transcription factors/repressors or polymerase machinery) (96). The fundamental unit of chromatin is the nucleosome, composed of 146 bp of DNA wrapped around an octamer of histone proteins (two copies each of histones H2A, H2B, H3, and H4). Nucleosomes in turn are connected by linker chromatin, composed primarily of DNA and histone H1. Epigenetics and chromatin structure strongly impact virtually all DNA-templated processes, especially transcription/gene expression. Recent studies in non-SMC systems have shown that diverse chromatin structures are stably heritable through mitosis and, in some cases, through meiosis after such structures are established in response to transient environmental/developmental cues (97). These results raise important questions regarding the interplay between signals from the extracellular environment and the regulation of chromatin structure/gene expression in normal cell differentiation versus processes that perturb cell differentiation in disease. Different combinations of histone N-terminal tail modifications are an integral epigenetic component of chromatin architecture. These tail regions are not bound to the nucleosome core particle (98). Rather, they are freely exposed to the nuclear environment and undergo a plethora of modifications, including acetylation, phosphorylation, methylation, ubiquitination, and ADP-ribosylation (96). There is clear evidence that these modifications, especially acetylation, phosphorylation, and methylation of histones H3 and H4, regulate the binding of sequence-specific transcription factors to DNA within the context of intact cells and chromatin.

Although the mechanisms underlying these processes are uncertain, the current opinion held by most researchers in the chromatin field is that histone modifications either directly (through changes in histone tail–DNA interactions) or indirectly (through histone-binding proteins) alter chromatin conformation and thereby regulate how accessible genomic DNA is to sequence-specific transcriptional activators/repressors. Our laboratory was one of the first to explore the role of epigenetic regulation of chromatin structure on transcription of SMC genes: We provided evidence that a key step in the control of induction of CArG-dependent genes in a retinoic acid (RA)-A404 model of the early stages of SMC differentiation involved alterations in the ability of SRF to bind to CArG-containing regions of SMC genes (72). Of major interest, we found that SRF and myocardin were expressed in a unique SMC precursor line (A404) that had previously been described by our laboratory (72), in the absence of detectable expression of any other known SMC marker, including the earliest known markers, SM α-actin and SM22α (74). In contrast, myocardin expression was absent from P19 embryonal carcinoma stem cells, the parental line from which A404 cells were derived. Treatment of A404 cells with all-*trans*-RA, which induced expression of all known SMC marker genes (72), was associated with marked increases in myocardin expression. In addition, we showed that, although SRF was highly expressed in A404 cells, it was unable to bind to the CArG-containing regions of SMC genes within intact chromatin and did bind to the constitutively expressed c-*fos* CArG promoter region or to SMC CArG regions in gel shift assays (72). Treatment of A404 cells with RA resulted in the association of SRF with CArG-containing regions of SMC promoters within intact chromatin, as well as in the hyperacetylation of histones associated with these regions (**Figure 3**). Taken together, these results support a model in which myocardin and SRF are expressed in A404 progenitor cells but are unable to bind to CArG-containing regions of SMC genes because of spatial restrictions associated with chromatin structure that are selective for SMC promoter regions. However, upon treatment with RA, the chromatin organization within SMC promoter regions is relaxed at least in part by histone acetylation. SRF then binds and recruits myocardin and other possible coactivators and activates the expression of multiple CArG-dependent SMC genes. Consistent with these results, Qiu & Li (99) presented evidence that CREB-CArG-dependent expression of the SM22α gene in cultured SMCs depends on histone acetyltransferase (HAT) activity. In brief, they found that treatment of cells with trichostatin A—a histone deacetylase (HDAC) inhibitor—increased,

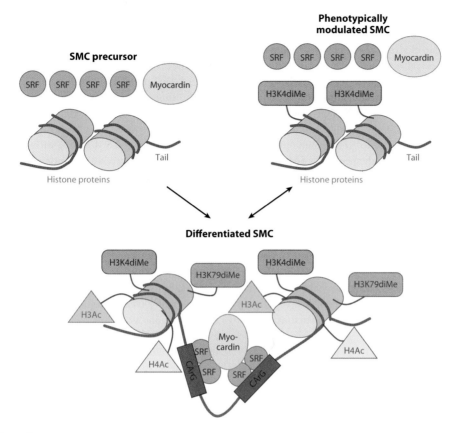

Figure 3
Epigenetic mechanisms play a key role in SMC differentiation, as well as in phenotypic switching in response to vascular injury or atherosclerotic disease. During SMC differentiation, the tails of histone proteins associated with the promoters of SMC-selective genes such as those encoding SM α-actin and SM MHC are posttranslationally modified through alterations such as acetylation of histones 3 and 4 (H3Ac and H4Ac) and dimethylation of lysines 4 and 79 on histone 3 (H3K4diMe and H3K79diMe) (72, 73). These modifications are thought to open up the chromatin within these promoters to permit binding of SRF-myocardin complexes to CArG-box elements and to drive expression of SMC-selective genes (59). After vascular injury in vivo as well as in response to stimuli promoting SMC phenotypic modulation in vitro such as PDGF and oxidized phospholipids, the repression of SMC marker genes is associated with the loss of activating histone modifications such as H3/H4 acetylation and H3K79 dimethylation and reduced accessibility of SRF-myocardin complexes to CArG boxes (73, 100, 117). Interestingly, however, such phenotypically modulated SMC do not exhibit loss of H3K4diMe within their promoter regions (73), suggesting that this histone modification may represent a type of lineage memory that permits SMC to reactivate expression of their contractile and cytoskeletal proteins upon resolution of vascular injury or disease.

whereas overexpression of HDACs decreased, SM22 promoter activity. In addition, Cao et al. (93) provided evidence that myocardin can bind both HATs and HDACs and can induce acetylation of histones surrounding CArG-containing regions of SMC marker genes. Finally, we showed that RA treatment of A404 cells is associated with enriched acetylation of histones H3 and H4 at SM α-actin and SM MHC CArG-containing promoter regions (72). Results are consistent with a model wherein epigenetic regulation of chromatin structure plays a key role in controlling transcriptional activation of SMC genes, including the regulation of CArG-SRF transcription by control of access of SRF to CArG-containing regions.

Since completion of our initial studies in A404 cells (72), we have carried out extensive further characterization of the role of histone modifications in the control of SMC gene expression in multiple in vivo and in vitro model systems that provide evidence that SMC-selective and gene locus–selective changes in chromatic structure play a key role in the control of SMC differentiation and SMC phenotypic switching either in vitro in response to PDGF BB (100), PDGF DD (101), or proatherogenic oxidized phospholipids (102) or in vivo following vascular injury (73). Using a novel quantitative real-time RT-PCR-based chromatin immunoprecipitation (ChIP) assay adapted from Litt et al. (103), we showed that enrichment of SRF binding to CArG-containing regions of multiple SMC marker genes, including those encoding SM α-actin and SM MHC, within intact chromatin was highly specific for SMCs; i.e., enrichment was observed in SMCs but not in multiple non-SMCs tested, including ECs, skeletal myoblasts, ESCs, and 10T1/2 embryonic fibroblasts. Consistent with these results in cultured cells, we also showed marked enrichment of SRF binding to the CArG regions of SM α-actin and SM MHC in SMCs versus non-SMCs in vivo [see McDonald et al. (73)].

Of major significance, we observed a unique pattern of histone modifications in SMCs compared with non-SMCs. For example, we observed marked enrichment of a number of histone modifications associated with chromatin relaxation and gene activation (103), including H3K4 dimethylation, H4 acetylation, H3 K79 dimethylation, and H3K9 acetylation, at the SM α-actin and SM MHC gene loci in SMCs versus non-SMCs, including ECs, fibroblasts, and ESCs. In contrast, non-SMCs showed enrichment of histone modifications associated with transcriptional silencing, including H4K20 dimethylation. Meanwhile, ECs showed enrichment of H4 acetylation and H3K4 dimethylation at the EC-selective vascular endothelial cadherin gene locus compared with ESCs and SMCs. SMCs showed increased H4K20 dimethylation of this locus, consistent with transcriptional silencing. Furthermore, micrococcal nuclease digestion experiments demonstrated that the CArG-box chromatin of SMC gene promoters exists in a form that is much more accessible to digestion in SMCs than in non-SMCs, suggesting that SMC genes are euchromatin like in SMCs and are heterochromatin like in non-SMCs (73). Taken together, these results provide evidence in support of a general model wherein cell type–selective and locus-selective epigenetic modifications play a key role in the regulation of cell determination and differentiation [reviewed in Gan et al. (104)]. However, there are a number of critical unresolved questions.

First, is SMC-selective epigenetic patterning a cause or a consequence of cellular differentiation? Thus far, our studies and by others in the field have shown simply a correlation between specific histone modifications and gene activation or repression but have not shown if these modifications play a causal role in mediating these changes. Indeed, this is a general major weakness of virtually all epigenetic studies of gene expression in mammalian systems studied thus far. That is, although a variety of studies have shown that inhibitors of histone-modifying enzymes can alter gene expression, such approaches result in global genome-wide changes in histones such that it is impossible to ascertain if effects on a given gene are direct or indirect. Thus, new approaches are needed to test if conditional gene locus–selective changes in histones (or for that matter other epigenetic controls such as DNA methylation) directly alter expression of the gene at issue.

Second, what are the mechanisms that regulate SMC-specific and SMC gene–selective histone modifications? Although we and others have shown that myocardin binds to p300 HAT and mediates acetylation of CArG-dependent SMC genes, nothing is known regarding how the SMC marker gene loci are initially modified to make them permissive for transcriptional activation during the early stages of SMC differentiation from pluripotential embryonic cells in the developing embryo. Moreover, even more intriguing, we demonstrated that H3K4 dimethylation and H4 hyperacetylation of the SM α-actin gene locus in SMC do not depend on the CArG boxes and thus are also independent of either SRF or myocardin/MRTF binding (73). In brief, we showed that a stably integrated SM α-actin promoter-enhancer transgene containing inactivating mutations of all three CArG elements still demonstrated marked enrichment of H3K4 dimethylation and H4 hyperacetylation, despite complete loss of SRF binding and being transcriptionally silent. These changes were selective in that the CArG mutant transgene did show complete loss of H3K79 and H3K9 acetylation, whereas the control wild-type SM α-actin promoter-enhancer transgene and the endogenous SM α-actin promoter locus in the same cells showed enrichment of all these marks as well as of SRF. Taken together, these findings provide compelling evidence that acquisition of the activating histone modifications of H3K4 dimethylation and H4 acetylation lies upstream of SRF and myocardin binding to the CArG elements. As such, key questions include the following:

1. When during the process of SMC specification from multipotential embryonic cells does epigenetic programming of SMC lineage occur?
2. What are the mechanisms that mediate cell type–specific and gene locus–selective epigenetic modification of SMC marker gene loci?
3. How stable are these changes when SMCs undergo reversible phenotypic switching such as that which occurs during repair of vascular injury?

MECHANISMS THAT CONTRIBUTE TO PHENOTYPIC SWITCHING OF SMOOTH MUSCLE CELLS

Phenotypic modulation is a critical process in vascular injury-repair and contributes to a multitude of human diseases. Yet investigators have identified relatively few specific factors or regulatory pathways that selectively and directly promote SMC phenotypic modulation in vivo, although many potential regulatory mechanisms have been identified on the basis of studies in cultured SMCs [reviewed in Owens et al. (1)]. The reasons for the paucity of studies in this area are likely the following: (*a*) the incorrect belief that SMC phenotypic modulation is simply secondary to growth stimulation, i.e., the old and incorrect adage that differentiation and proliferation are mutually exclusive processes, and (*b*) the assumption that phenotypic modulation of SMCs is a passive rather than an active process and due simply to loss of positive SMC differentiation factors. It is now well established that differentiation and proliferation are not mutually exclusive and that many factors other than SMC proliferation status influence the SMC differentiation state. We extensively reviewed this topic in 1995 (7), and we only briefly summarize several relevant observations here. First, during late embryogenesis and postnatal development, SMCs have an extremely high rate of proliferation (105), yet at this time they undergo the most rapid rate of induction of expression of multiple SMC differentiation marker genes (24). Second, SMCs within advanced atherosclerotic lesions show a very low rate of proliferation that approaches that of fully differentiated SMCs, yet the former SMCs are highly phenotypically modulated, as evidenced by marked reductions in the expression of SMC marker genes (106, 107). These results show that cessation of proliferation alone is not sufficient to promote SMC differentiation and suggest that other SMC differentiation cues are absent and/or that active repressors of SMC differentiation are present.

Consistent with the hypothesis that SMC phenotypic modulation may be controlled actively and is not due simply to loss of positive differentiation signals, we (108–110) and others (111–115) have shown that treatment of cultured SMCs with PDGF BB, PDGF DD, lysophosphatidic acid, inorganic phosphate, proatherogenic oxidized phospholipids, and inflammatory cytokines such as IL-1β can induce rapid downregulation of expression of multiple SMC differentiation marker genes, including those encoding SM α-actin, SM MHC, SM22α, and h_1-calponin. Moreover, studies have implicated a wide range of signaling pathways in the responses to these agents, including pathways involving ERK, p38 MAPKs, and Akt (68, 101, 116–118). These studies have been reviewed in detail (1, 119, 120) and provide valuable insight into potential regulatory mechanisms that may contribute to SMC phenotypic switching. However, a major limitation of such studies is that they were done on cultured SMCs that had already undergone profound phenotypic switching as a result of cell isolation and culture and the loss of many of the critical environmental cues that mediate differentiation and maturation of SMCs in vivo. Although the cells retain expression of many SMC marker genes, their morphology and functions are far different from those of their in vivo counterparts. As such, results of studies of phenotypic switching in these cells must be viewed at best with caution and at worse with healthy skepticism. Let us cite just a couple examples to illustrate the reason for our concerns. First, we showed that expression of the SM α-actin gene in vivo in transgenic mice depends completely on an intronic CArG enhancer region. We observed no activity of promoter constructs that lacked the first intron or where the intronic CArG element was mutated (38). In contrast, the entire intronic region, including the intronic CArG enhancer, was completely dispensable for high-level expression in cultured SMCs (121). Second, there are many examples in the SMC literature in which completely opposite results can be obtained, depending on the laboratory doing the studies; the reasons for these contradictory results are unclear. For example, studies from multiple laboratories have shown that PDGF, angiotensin II, and TGFβ1 can either promote (45, 47, 122–124) or inhibit (68, 101, 108, 125–128) differentiation of cultured vascular SMCs, depending on the SMC culture line or methods employed. Indeed, our own laboratory found that TGFβ1 markedly inhibited growth and differentiation of aortic SMCs derived from Sprague-Daley rats (45) but had the opposite effects on aortic SMCs derived by the same methods from spontaneously hypertensive (SHR) rats (126). Similarly, we published a series of experiments showing that aortic SMCs in SHR rats undergo a high rate of endoreduplication and formation of polyploid SMCs compared with the normotensive Wistar Kyoto (WKY) rats from which the SHR strain was derived (129). However, in cultured aortic SMCs derived from these strains, we found the complete opposite: WKY cells showed a nearly 100% conversion to polyploidy in vitro (reviewed in References 130 and 131). In view of these marked discrepancies, it is critically important that results of studies in cultured SMCs be validated through studies in vivo.

As an alternative to attempting to individually dissect which of the wide plethora of altered environmental cues associated with vascular injury (or experimental atherogenesis) directly contribute to phenotypic modulation of the SMC phenotype in vivo, we first carried out a series of vascular injury experiments in our SM α-actin, SM MHC, and SM22α promoter-enhancer LacZ transgenic mice (132). Following vascular injury, we observed nearly complete loss of expression of all three transgenes, thus showing for the first time that SMC phenotypic modulation in vivo is mediated at least in part by transcriptional repression. Of interest, we also observed that after 14 days of injury, subpopulations of cells showed reinduction of SM α-actin and SM MHC, further documenting the dynamic nature of SMC phenotypic modulation in vivo. We subsequently demonstrated that mutation of a conserved G/C repressor element located 5′ to the proximal CArG element in the SM22α promoter (**Figure 2**), and also found in the promoters of many other SMC marker genes, nearly completely abolished downregulation of SM22α in vivo in response to vascular injury (132) or within intimal lesions of ApoE −/− mice (133). Indeed, these studies were the first studies and to date are the only studies to our knowledge that have identified a specific molecular mechanism that is clearly required for SMC phenotypic switching in vivo. The challenge was then to identify the factor and mechanisms that regulate the activity of the G/C repressor.

Of major significance, using siRNA suppression studies and KO SMC culture lines, we found that transcriptional repression of SMC marker genes in cultured SMCs in response to treatment with PDGF BB, PDGF DD, or proatherogenic oxidized phospholipids such as POVPC was dependent on KLF4 (101, 102, 117, 134), a transcription factor we originally identified using a modified yeast one-hybrid screen of factors expressed in phenotypically modulated SMCs that bind to G/C-rich *cis* elements found in SMC marker gene promoters (41). In addition, we showed the following. (*a*) KLF4 was not expressed in differentiated SMCs in normal blood vessels but was rapidly induced following vascular injury (91). (*b*) KLF4 overexpression was associated with profound inhibition of expression of the potent SMC-selective SRF coactivator myocardin and both CArG-dependent and CArG-independent SMC marker genes (91). (*c*) KLF4 expression was increased in lesions of ApoE −/− mice on a Western diet (134), as well as in atherosclerotic lesions within human coronary arteries (M. Salmon & G.K. Owens, unpublished observations). (*d*) KLF4 bound to several conserved G/C-rich *cis* elements within SMC promoters, including the G/C repressor element (91), as well as to a TGFβ control element in vitro in DNA-binding assays and gel shift assays (41, 44; M. Salmon & G.K. Owens, unpublished observations). (*e*) Conditional KO of *KLF4* was associated with a transient delay in repression of SM α-actin and SM22α following carotid artery ligation–induced vascular injury in vivo, but there was subsequent hyperproliferation of SMCs and a >300% increase in neointima formation. (*f*) There was increased binding of KLF4 to the SM22α and SM α-actin promoters in vivo following vascular injury on the basis of ChIP analysis of the mouse carotid (135). (*g*) KLF4 induces SMC growth arrest at least in part through p53-dependent activation of the cell cycle–inhibitory gene *p21* (135). (*h*) Overexpression of *KLF4* in cultured SMCs resulted in profound activation of expression of multiple other induced pluripotential stem (iPS) cell pluripotency factors, including Oct4 and Sox2, but not Nanog, suggesting that SMC phenotypic switching involves the activation of multiple pluripotency genes in addition to *KLF4* (O. Cherepanova, O. Sarmento & G.K. Owens, unpublished results). (*i*) Effects of KLF4 in suppressing SMC marker gene expression were mediated at least in part through epigenetic changes associated with transcriptional silencing, including reduced H4 histone acetylation mediated through KLF4-dependent recruitment of HDACs 2 and 5 and nearly complete loss of SRF binding to the SMC promoter CArG elements within intact chromatin (73, 100, 117).

Taken together, the preceding results provide compelling evidence that KLF4 plays a critical role in the regulation of SMC phenotypic switching in cultured SMCs and in vivo following ligation-induced vascular injury and that it mediates its effects at least in part through inducing epigenetic changes of SMC marker gene loci associated with the formation of heterochromatin and transcriptional silencing [**Figure 4**; see also McDonald & Owens (59) for a more complete review of this area]. However, no studies have defined the role or mechanisms by which KLF4 contributes to the development of atherosclerosis, nor is it clear if the effects of conditional KO of *KLF4* observed in our previous studies of ligation-induced vascular injury (135) were the direct result of loss of KLF4 in SMCs versus other cell types.

Whereas there is clear evidence that KLF4 expression is induced in cultured SMCs in response to treatment with PDGF BB, PDGF DD, proatherogenic oxidized phospholipids, and a number of other factors that induce SMC phenotypic switching, relatively little is known regarding specific mechanisms that activate the transcription of *KLF4*. Previous studies by our laboratory have shown that PDGF BB–induced activation of KLF4 in cultured SMCs depends on Sp1 and on multiple Sp1 binding sites in the *KLF4* promoter (125). Moreover, we showed increased Sp1 binding to the *KLF4* promoter within intact chromatin in vivo on the basis of ChIP assays following carotid ligation injury in mice. However, there is no direct evidence that the induction of KLF4 in SMCs in vivo is Sp1 dependent, and given the widespread expression of *Sp1* in many cell types and under a wide variety of conditions, the specificity of KLF4 induction in SMCs during phenotypic switching must involve additional regulatory mechanisms.

Of interest, studies by several groups (136–138) have shown that the microRNAs (miRNAs) *miR-145* and *miR-143* play a key role in regulating SMC phenotypic switching in response to

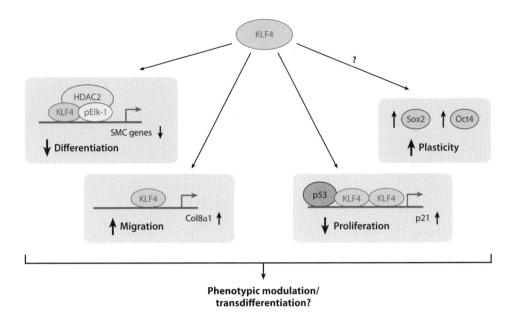

Figure 4

KLF4 regulates multiple aspects of SMC phenotypic modulation. KLF4 levels are increased in SMC in response to vascular injury in vivo, as well as in response to stimuli promoting SMC phenotypic modulation in vitro such as PDGF and oxidized phospholipids (91, 102). Increased KLF4 levels result in multiple changes to SMC gene expression and function. For example, KLF4 cooperates with phosphorylated Elk-1 (pElk-1) and histone deacetylase 2 (HDAC2) to repress the expression of SMC-selective contractile and cytoskeletal proteins (117, 135). KLF4 also promotes SMC migration through binding to the collagen VIII alpha 1 (Col8α1) promoter and enhancing the production of this extracellular matrix molecule, and KLF4 cooperates with p53 on the p21 promoter to increase p21 levels and to reduce SMC proliferation (134, 135). Finally, because KLF4 promotes expression of the pluripotency-associated transcription factors Sox2 (suppressor of cytokine signaling 2) and Oct4 (octamer-binding transcription factor 4) in embryonic stem cells (171, 172) and is a member of the class of proteins that can induce pluripotency in somatic cells (148), KLF4 may modulate SMC phenotype through enhancing cell pluripotency as a means of aiding in repair during vascular and other disease states.

vascular injury, and there is evidence that these miRNAs act in part by targeting KLF4 degradation, thereby promoting SMC differentiation. Cordes et al. (136) showed that *miR-145* and *miR-143* were selectively expressed in multipotential murine cardiac progenitor cells before becoming localized to SMCs in late embryogenesis and that they target the degradation of a network of transcription factors known to suppress SMC differentiation, including KLF4 and Elk-1. Cordes et al. (136) and Xin et al. (137) also showed that *miR-143* and *miR-145* were activated by myocardin, Nkx-2.5, and SRF in differentiated SMCs, thus providing a potential positive-feedback mechanism to stabilize the SMC differentiated state. In contrast, these miRNAs were downregulated in injured or atherosclerotic vessels. Of major importance, Xin et al. (137) deleted *miR-143* and *miR-145* either singly or in combination in mice. Mice lacking both miRNAs were viable but did not show any overt abnormalities in SMC differentiation, suggesting that these miRNAs are not required for normal SMC differentiation. However, neointima formation in response to vascular injury was profoundly inhibited due at least in part to the disarray of actin stress fibers and diminished SMC migratory activity. Consistent with these findings, Cheng et al. (138) found that *miR-145* was abundantly expressed in normal differentiated blood vessels, was downregulated following balloon injury of the rat carotid artery and that adenovirus-mediated overexpression of *miR-145* could partially restore downregulation of SMC marker genes and neointima formation following balloon injury of the rat carotid artery. They also presented evidence in cultured SMCs that *miR-145* targets KLF5 degradation and that this loss of KLF5 was required for PDGF BB–induced phenotypic switching of these cells. However, they did not present direct evidence for KLF5 involvement in the vascular injury response in vivo, although previous studies by Shindo et al. (139) presented evidence that KLF5 heterozygous KO mice showed diminished neointima formation following vascular injury compared with wild-type mice. Taken together, these studies indicate that *miR-143/145* play a key role in SMC phenotypic switching in response to vascular injury and that the effects of these miRNAs are likely mediated, at least in part, through targeting degradation of multiple transcription factors, including KLF4, KLF5, and Elk-1, that are linked to the repression of SMC differentiation. Despite the lack of evidence for a developmental phenotype in *miR-143/145* KO mice, additional studies are needed to clarify these miRNAs' roles and mechanism of action in SMC development from multipotential embryonic cells, particularly given clear evidence that upregulation of expression of these miRNAs is required for ESC differentiation at least in part by their targeting degradation of Oct4, Sox2, and KLF4, thus resulting in loss of pluripotency (140). Indeed, we have been unable to detect KLF4 expression within SM α-actin+ perivascular cells (i.e., newly formed SMCs) in the dorsal aortae of developing embryos (T. Yoshida, G. Owens & S. Sinha, unpublished data), consistent with a model wherein KLF4 is already depleted prior to embryonic cells differentiating into SMCs. Moreover, the failure of Xin et al. (137) to see defects in SMC development in *miR-143/145* KO mice suggests redundant pathways that can downregulate KLF4 when ESCs undergo initial differentiation and well before the specification of SMC lineages. Additional studies are needed in this area to determine if *miR-143/145*-induced loss of KLF4 is a late event in SMC specification and/or if this requirement varies between SMCs of different embryological origins. In addition, further studies are needed to better define the SMC selectivity of *miR-143/145*, given extensive evidence that they are widely expressed in a large number of different tumor and myofibroblast cell lines (141–143) that also show KLF4-dependent growth and differentiation.

Several additional miRNAs have recently been implicated in the control of SMC differentiation and/or phenotypic switching. Davis et al. (144) presented evidence showing that *miR-21* mediates the induction of a contractile phenotype in human vascular SMCs by TGFβ and bone morphogenetic proteins (BMPs). Of interest, they found that posttranscriptional controls mediated TGFβ- and BMP4-induced activation of *miR-21* and that *miR-21* suppressed expression of the tumor suppressor gene *PDCD4* (programmed cell death 4), which acted as a negative regulator of SMC gene expression. In contrast, this same group showed that PDGF BB, which suppressed Tribbles-like protein 3, induced *miR-24*. This suppression coincided with reduced expression of Smad proteins, thereby promoting a synthetic SMC phenotype (145). These observations suggest a mechanism that contributes to the antagonism between PDGF and TGFβ in the control of SMC phenotype. Leeper et al. (146) recently identified *miR-26a* as an antagonist of TGFβ-dependent SMC differentiation: Inhibition of this miRNA resulted in accelerated SMC differentiation while inhibiting SMC migration and proliferation through a Smad-1-dependent mechanism. However, there is no direct evidence validating that *miR-21*, *miR-24*, or *miR-26a* regulates SMC differentiation or phenotypic switching in vivo, and further studies in this interesting area are needed.

If the above seems very confusing and complex, it is, so let us seek some possible unifying hypotheses and mechanisms. At the time of our laboratory's initial report in 2000 that KLF4 was a key factor regulating SMC phenotypic switching (41), there were perhaps no more than a half-dozen labs worldwide studying this transcription factor, including Yang's lab (147), which initially discovered that KLF4 mediated growth arrest of gut epithelial cells. However, interest in KLF4 has increased dramatically since 2006 on the basis of the discovery by Takahashi & Yamanaka

(148) that *KLF4* is one of four genes, along with *Oct4*, *Sox2*, and c-*myc*, that when overexpressed using retroviruses can reprogram a variety of somatic cells, including dermal fibroblasts, into ESC-like cells or iPS cells (148, 149). Moreover, there is now evidence that *KLF4* is part of a network of pluripotency genes, including *Oct4*, *Sox2*, *Nanog*, and several other genes, that play a key role in the maintenance of pluripotency and self-renewal in normal ESCs (150). These latter observations are of major interest, given observations by Peault's group from UCLA (22) suggesting that multipotential MSCs derived from human tissues, including adipose tissue and skeletal muscle, appear to be derived nearly exclusively from perivascular SMCs-pericytes. Taken together, these observations have several major and potentially profound implications.

First, these findings raise the possibility that MSCs represent phenotypically modulated SMCs and/or pericytes, i.e., SMCs-pericytes that have reactivated pluripotency gene networks in response to injury or stress as a means of increasing their plasticity (**Figures 1** and **4**). Although this hypothesis is intriguing, little is known regarding factors and mechanisms that regulate phenotypic switching of pericytes and whether they are similar to those in SMCs derived from large vessels, the focus of virtually all studies thus far in this area. In addition, there is no evidence that phenotypically modulated SMCs or pericytes give rise to MSCs in vivo and subsequently differentiate into different cell types, although some intriguing lineage-tracing studies show that SMCs can transdifferentiate into osteochondrocytes, albeit in the setting of calcifying arteries of matrix Gla protein–deficient mice (11). A related interesting question is the relationship between fibroblasts, activated myofibroblasts, and tumor stromal cells, which express many SMC-pericyte markers and play a critical role in injury-repair as well as in disease processes (151).

Second, the above results potentially explain why and how iPS cell induction works because such induction is very artificial in that it involves retrovirus-mediated gross overexpression of multiple pluripotency genes in cultured fibroblasts. That is, our results indicate that SMCs, and presumably other cell types, have evolved mechanisms to reactivate at least some pluripotency gene networks as a means of increasing cellular plasticity and enhancing injury-repair processes critical for survival of the organism in question. Intriguingly, because all tissues are vascularized, this places SMCs-pericytes in close proximity to any site in which multipotential cells might be needed. However, much additional work is needed to test the preceding ideas. First, there is a need for definitive in vivo lineage-tracing studies to determine the potency and plasticity of phenotypically modulated SMCs and pericytes to form other cell lineages in vivo. Such studies should include models of vascular injury-repair because this process has been subject to selection, optimization, and conservation through evolution. However, studies of many other tissue and cell types are also needed. Second, further studies are needed to determine the epigenetic mechanisms that control phenotype transitions and/or possible transdifferentiation to alternative cell types. A key question is whether pericytes and/or SMCs can give rise to alternative cell types and in so doing undergo reprogramming of their chromatin to the new cell type.

Studies in our laboratory have provided compelling evidence that epigenetic changes play a critical role not only in the initial specification of SMC lineage during development (72, 73) but also in SMC phenotypic switching in response to vascular injury in vivo (73, 135) or in cultured SMCs in response to treatment with PDGF BB (73, 100), PDGF DD (101), and oxidized phospholipids (102) or by adenovirus-mediated overexpression of KLF4 (73) [reviewed in McDonald & Owens (59)]. Of major interest, a common feature in all these models is the persistence of H3K4 dimethylation of SMC marker gene promoter loci despite these genes being transcriptionally silenced due to the loss of SRF-myocardin binding, the formation of heterchromatin, and the loss of H3/H4 hyperacetylation (**Figure 3**). This raises the intriguing possibility that H3K4 dimethylation serves as a mechanism of epigenetic cell lineage memory, i.e., a mechanism for phenotypically modulated SMCs to remain permissive for redifferentiation into SMCs during reversible phenotypic switching. Consistent with this hypothesis, we determined changes in histone patterning when SMCs are induced to undergo reversible phenotypic switching in response to transient or repeated treatment with PDGF BB in vitro or vascular injury in vivo (73, 152). Results of quantitative ChIP assays showed that treatment of cultured SMCs for either 24 or 72 h resulted in marked decreases in the association of SRF with CArG-containing regions of the SM α-actin and SM MHC SMC genes as well as resulting in decreased H4 acetylation, but no change in H3K4 dimethylation occurred. SMCs treated with PDGF BB for 24 h followed by 48 h in the absence of PDGF BB showed nearly complete restoration of expression of both SM α-actin and SM MHC as well as SRF binding and H4 acetylation, but such treatment had no effect on H3K4 dimethylation of these SMC gene loci. Similarly, vascular injury in vivo was also associated with decreased expression of SMC marker genes, decreased SRF binding to CArG regions, and reduced H4 histone acetylation (73). In addition, we showed that adenovirus-mediated overexpression of KLF4 profoundly suppressed SMC marker gene expression in cultured SMCs, reduced SRF binding to CArG elements within intact chromatin, decreased nuclease sensitivity (a measure of chromosome condensation), and reduced H4 acetylation.

However, as was the case with PDGF BB treatment, there was no effect on H3K4 dimethylation of SMC marker gene CArG-containing promoter regions [see figure 4 of McDonald et al. (73)]. To determine mechanisms whereby H3K4 dimethylation might promote SMC gene expression, we tested if this histone modification might serve as a docking site for stabilizing myocardin binding to CArG regions. Of major interest, results of peptide-binding assays showed that myocardin selectively bound to H3K4-modified peptides but not to peptides containing a variety of alternative histone modifications. Finally, we showed that PDGF BB–induced repression of SMC marker genes was associated with the selective recruitment of HDACs 2, 4, and 5 to SMC marker gene loci and was markedly inhibited by siRNA-induced suppression of these HDACs (117, 152). Proatherogenic oxidized phospholipids induced similar changes and also increased the association of phosphorylated Elk-1 to SMC promoters, a change that Olson and coworkers (68) have shown inhibits SRF-myocardin binding to SMC genes and contributes to selective transcriptional repression of CArG-dependent SMC genes in response to PDGF BB. Moreover, this group has shown that class II HDACs interact with a domain of myocardin distinct from the p300-binding domain and suppress SMC gene activation by myocardin (93). Taken together, the preceding studies provide strong evidence that epigenetic controls likely play a critical role in the control of SMC phenotypic switching and that KLF4 plays an important role in this process. However, as is the case with nearly all epigenetic studies to date, the evidence is correlative rather than causal in nature. New experimental approaches are needed to directly test if and how specific histone modifications such as H3K4 dimethylation of SMC genes modulate SMC phenotype.

SUMMARY, CONCLUSIONS, AND FUTURE DIRECTIONS

The SMC is a fascinating cell type that can exhibit a wide range of different phenotypes in development and disease. Although there has been much progress over the past decade in elucidating mechanisms that control SMC differentiation and phenotypic switching, including the role of iPS cell pluripotency genes like *KLF4* and epigenetic mechanisms that regulate chromatin structure, there are still many unanswered questions and challenges, including the following. (*a*) What is the relationship between SMCs, pericytes, myofibroblasts, and MSCs? (*b*) What are the mechanisms that contribute to epigenetic programming of SMCs during the transition of multipotential embryonic cells to SMC lineages, and which changes are stable during SMC phenotypic switching during vascular injury-repair or in diseases such as atherosclerosis? (*c*) Can phenotypically modulated SMCs transdifferentiate into alternative cell lineages, and if so, into what cell types can they transdifferentiate, and what mechanisms control these transitions? (*d*) Can we exploit SMC plasticity and identify novel therapeutic targets to promote the endogenous stem cell–like properties of SMCs for the purposes of promoting the stabilization of atherosclerotic plaques, enhancing wound repair, reducing tumor cell metastasis, or promoting the development of collateral vessels following a heart attack or for treatment of peripheral vascular disease? Such challenges, just a few of the many we face, highlight the critical importance of continued studies in this field.

DISCLOSURE STATEMENT

The authors are not aware of any affiliations, memberships, funding, or financial holdings that might be perceived as affecting the objectivity of this review.

LITERATURE CITED

1. Owens GK, Kumar MS, Wamhoff BR. 2004. Molecular regulation of vascular smooth muscle cell differentiation in development and disease. *Physiol. Rev.* 84(3):767–801
2. Carmeliet P. 2000. Mechanisms of angiogenesis and arteriogenesis. *Nat. Med.* 6(4):389–95
3. Hanahan D. 1997. Signaling vascular morphogenesis and maintenance. *Science* 277(5322):48–50
4. Hungerford JE, Little CD. 1999. Developmental biology of the vascular smooth muscle cell. *J. Vasc. Res.* 36:2–27
5. Wagenseil JE, Mecham RP. 2009. Vascular extracellular matrix and arterial mechanics. *Physiol. Rev.* 89(3):957–89
6. Li DY, Brooke B, Davis EC, Mecham RP, Sorensen LK, et al. 1998. Elastin is an essential determinant of arterial morphogenesis. *Nature* 393(6682):276–80
7. Owens GK. 1995. Regulation of differentiation of vascular smooth muscle cells. *Physiol. Rev.* 75(3):487–517
8. Somlyo AP, Somlyo AV. 2003. Calcium sensitivity of smooth muscle and non-muscle myosin II: modulation by G proteins, kinases, and myosin phosphatase. *Physiol. Rev.* 88:1325–68
9. Bentzon JF, Weile C, Sondergaard CS, Hindkjaer J, Kassem M, Falk E. 2006. Smooth muscle cells in atherosclerosis originate from the local vessel wall and not circulating progenitor cells in ApoE knockout mice. *Arterioscler. Thromb. Vasc. Biol.* 26(12):2696–702
10. Bentzon JF, Sondergaard CS, Kassem M, Falk E. 2007. Smooth muscle cells healing atherosclerotic plaque disruptions are of local, not blood, origin in apolipoprotein E knockout mice. *Circulation* 116(18):2053–61
11. Speer MY, Yang HY, Brabb T, Leaf E, Look A, et al. 2009. Smooth muscle cells give rise to osteochondrogenic precursors and chondrocytes in calcifying arteries. *Circ. Res.* 104(6):733–41
12. Hoofnagle MH, Thomas JA, Wamhoff BR, Owens GK. 2006. Origin of neointimal smooth muscle: We've come full circle. *Arterioscler. Thromb. Vasc. Biol.* 26(12):2579–81
13. Hoofnagle MH, Wamhoff BR, Owens GK. 2004. Lost in transdifferentiation. *J. Clin. Investig.* 113(9):1249–51
14. Morikawa S, Baluk P, Kaidoh T, Haskell A, Jain RK, McDonald DM. 2002. Abnormalities in pericytes on blood vessels and endothelial sprouts in tumors. *Am. J. Pathol.* 160(3):985–1000
15. Ramaswamy S, Ross KN, Lander ES, Golub TR. 2003. A molecular signature of metastasis in primary solid tumors. *Nat. Genet.* 33(1):49–54
16. Carmeliet P, Jain RK. 2000. Angiogenesis in cancer and other diseases. *Nature* 407(6801):249–57
17. Ross R. 1993. The pathogenesis of atherosclerosis: a perspective for the 1990s. *Nature* 362:801–9
18. Glagov S, Weisenberg E, Zarins CK, Stankunavicius R, Kolettis GJ. 1987. Compensatory enlargement of human atherosclerotic coronary arteries. *N. Engl. J. Med.* 316:1371–75
19. Galis ZS, Khatri JJ. 2002. Matrix metalloproteinases in vascular remodeling and atherogenesis: the good, the bad, and the ugly. *Circ. Res.* 90(3):251–62

20. Galis ZS, Sukhova GK, Lark MW, Libby P. 1994. Increased expression of matrix metalloproteinases and matrix degrading activity in vulnerable regions in human atherosclerotic plaques. *J. Clin. Investig.* 94:2493–503
21. Hungerford JE, Little CD. 1999. Developmental biology of the vascular smooth muscle cell: building a multilayered vessel wall. *J. Vasc. Res.* 36:2–27
22. Crisan M, Yap S, Casteilla L, Chen CW, Corselli M, et al. 2008. A perivascular origin for mesenchymal stem cells in multiple human organs. *Cell Stem Cell* 3(3):301–13
23. Gabbiani G, Schmid E, Winter S, Chaponnier C, De Chastonay C, et al. 1981. Vascular smooth muscle cells differ from other smooth muscle cells: predominance of vimentin filaments and a specific-type actin. *Proc. Natl. Acad. Sci. USA* 78:298–300
24. Owens GK, Thompson MM. 1986. Developmental changes in isoactin expression in rat aortic smooth muscle cells in vivo. Relationship between growth and cytodifferentiation. *J. Biol. Chem.* 261:13373–80
25. Rovner AS, Murphy RA, Owens GK. 1986. Expression of smooth muscle and nonmuscle myosin heavy chains in cultured vascular smooth muscle cells. *J. Biol. Chem.* 261:14740–45
26. Rovner AS, Thompson MM, Murphy RA. 1986. Two different heavy chains are found in smooth muscle myosin. *Am. J. Physiol. Cell Physiol.* 250:861–70
27. Miano JM, Cserjesi P, Ligon K, Perisamy M, Olson EN. 1994. Smooth muscle myosin heavy chain marks exclusively the smooth muscle lineage during mouse embryogenesis. *Circ. Res.* 75:803–12
28. Hasegawa Y, Ueda Y, Watanabe M, Morita F. 1992. Studies on amino acid sequences of two isoforms of 17-kDa essential light chain of smooth muscle myosin from porcine aorta media. *J. Biochem.* 111:798–803
29. Hansson GK, Jonasson L, Holm J, Claesson-Welsh L. 1986. Class II MHC antigen expression in the atherosclerotic plaque: Smooth muscle cells express HLA-DR, HLA-DQ and the invariant gamma chain. *Clin. Exp. Immunol.* 64:261–68
30. Winder SJ, Sutherland C, Walsh MP. 1991. Biochemical and functional characterization of smooth muscle calponin. *Adv. Exp. Med. Biol.* 304:37–51
31. Sobue K, Sellers JR. 1991. Caldesmon, a novel regulatory protein in smooth muscle and nonmuscle actomyosin systems. *J. Biol. Chem.* 266:12115–18
32. Geiger B, Tokuyasu KT, Dutton AH, Singer SJ. 1980. Vinculin, an intracellular protein localized at specialized sites where microfilament bundles terminate at cell membranes. *Proc. Natl. Acad. Sci. USA* 77:4127–31
33. Herring BP, Smith AF. 1996. Telokin expression is mediated by a smooth muscle cell-specific promoter. *Am. J. Physiol. Cell Physiol.* 270:1656–65
34. van der Loop FTL, Gabbiani G, Kohnen G, Ramaekers FCS, van Eys GJJM. 1997. Differentiation of smooth muscle cells in human blood vessels as defined by smoothelin, a novel marker for the contractile phenotype. *Arterioscler. Thromb. Vasc. Biol.* 17:665–71
35. Gorenne I, Nakamoto RK, Phelps CP, Beckerle MC, Somlyo AV, Somlyo AP. 2003. LPP, a LIM protein highly expressed in smooth muscle. *Am. J. Physiol. Cell Physiol.* 285(3):674–85
36. Mericskay M, Parlakian A, Porteu A, Dandre F, Bonnet J, et al. 2000. An overlapping CArG/octamer element is required for regulation of *desmin* gene transcription in arterial smooth muscle cells. *Dev. Biol.* 226:192–208
37. Li L, Miano JM, Mercer B, Olson EN. 1996. Expression of the SM22α promoter in transgenic mice provides evidence for distinct transcriptional regulatory programs in vascular and visceral smooth muscle cells. *J. Cell Biol.* 132(5):849–59
38. Mack CP, Owens GK. 1999. Regulation of SM α-actin expression in vivo is dependent upon CArG elements within the 5′ and first intron promoter regions. *Circ. Res.* 84:852–61
39. Kim S, Ip HS, Lu MM, Clendenin C, Parmacek MS. 1997. A serum response factor-dependent transcriptional regulatory program identifies distinct smooth muscle cell sublineages. *Mol. Cell. Biol.* 17(4):2266–78
40. Madsen CS, Regan CP, Hungerford JE, White SL, Manabe I, Owens GK. 1998. Smooth muscle–specific expression of the smooth muscle myosin heavy chain gene in transgenic mice requires 5′-flanking and first intronic DNA sequence. *Circ. Res.* 82:908–17
41. Adam PJ, Regan CR, Hautmann MB, Owens GK. 2000. Positive and negative acting krupple-like transcription factors bind a transforming growth factor beta control element required for expression of the smooth muscle differentiation marker SM22α in vivo. *J. Biol. Chem.* 275(4):37798–806

42. Kumar MS, Hendrix J, Johnson AD, Owens GK. 2003. The smooth muscle α-actin gene requires two E-boxes for proper expression in vivo and is a target of class I basic helix-loop-helix proteins. *Circ. Res.* 92:840–47
43. Manabe I, Owens GK. 2001. CArG elements control smooth muscle subtype-specific expression of smooth muscle myosin in vivo. *J. Clin. Investig.* 107:823–34
44. Liu Y, Sinha S, Owens GK. 2003. A transforming growth factor-β control element required for SM α-actin expression in vivo also partially mediates GKLF-dependent transcriptional repression. *J. Biol. Chem.* 278(48):48004–11
45. Hautmann M, Madsen CS, Owens GK. 1997. A transforming growth factor β(TGF) control element drives TGF-induced stimulation of SM α-actin gene expression in concert with two CArG elements. *J. Biol. Chem.* 272(16):10948–56
46. Hautmann M, Adam PJ, Owens GK. 1999. Similarities and differences in smooth muscle α-actin induction by transforming growth factor β in smooth muscle versus non-muscle cells. *Arterioscler. Thromb. Vasc. Biol.* 19:2049–58
47. Hautmann M, Thompson MM, Swartz EA, Olson EN, Owens GK. 1997. Angiotensin II-induced stimulation of smooth muscle α-actin expression by serum response factor and the homeodomain transcription factor MHox. *Circ. Res.* 81:600–10
48. Kumar MS, Owens GK. 2003. Combinatorial control of smooth muscle–specific gene expression. *Arterioscler. Thromb. Vasc. Biol.* 23:737–47
49. Owens GK, Vernon SM, Madsen CS. 1996. Molecular regulation of smooth muscle cell differentiation. *J. Hypertens.* 14(Suppl. 5):55–64
50. Firulli AB, Olson EN. 1997. Modular regulation of muscle gene transcription: a mechanism for muscle cell diversity. *Trends Genet.* 13(9):364–69
51. Majesky MW. 2003. Decisions, decisions. SRF coactivators and smooth muscle myogenesis. *Circ. Res.* 92(8):824–26
52. Miano JM. 2003. Serum response factor: toggling between disparate programs of gene expression. *Mol. Cell. Cardiol.* 35(6):577–93
53. Yano H, Hayashi K, Momiyama T, Saga H, Haruna M, Sobue K. 1995. Transcriptional regulation of the chicken caldesmon gene. Activation of gizzard-type caldesmon promoter requires a CArG box-like motif. *J. Biol. Chem.* 270(40):23661–66
54. Shore P, Sharrocks AD. 1994. The transcription factors Elk-1 and serum response factor interact by direct protein-protein contacts mediated by a short region of Elk-1. *Mol. Cell. Biol.* 14:3283–91
55. Mack CP, Thompson MM, Lawrenz-Smith S, Owens GK. 2000. Smooth muscle α-actin CArG elements coordinate formation of a smooth muscle cell-selective, serum response factor-containing activation complex. *Circ. Res.* 86:221–32
56. Sartorelli V, Kurabayashi M, Kedes L. 1993. Muscle-specific gene expression. A comparison of cardiac and skeletal muscle transcription strategies. *Circ. Res.* 72:925–31
57. Du KL, Ip HS, Li J, Chen MM, Dandre F, et al. 2003. Myocardin is a critical serum response factor cofactor in the transcriptional program regulating smooth muscle cell differentiation. *Mol. Cell. Biol.* 23(7):2425–37
58. Yoshida T, Owens GK. 2005. Molecular determinants of vascular smooth muscle cell diversity. *Circ. Res.* 96(3):280–91
59. McDonald OG, Owens GK. 2007. Programming smooth muscle plasticity with chromatin dynamics. *Circ. Res.* 100(10):1428–41
60. Croissant JD, Kim JH, Eichele G, Goering L, Lough J, et al. 1996. Avian serum response factor expression restricted primarily to muscle cell lineages is required for α-actin gene transcription. *Dev. Biol.* 177(1):250–64
61. Belaguli NS, Schildmeyer LA, Schwartz RJ. 1997. Organization and myogenic restricted expression of the murine serum response factor gene. A role for autoregulation. *J. Biol. Chem.* 272(29):18222–31
62. Yoshida T, Hoofnagle MH, Owens GK. 2004. Myocardin and Prx1 contribute to angiotensin II-induced expression of smooth muscle α-actin. *Circ. Res.* 94(8):1075–82
63. Li L, Miano JM, Cserjesi P, Olson EN. 1996. SM22α, a marker of adult smooth muscle, is expressed in multiple myogenic lineages during embryogenesis. *Circ. Res.* 78(2):188–95

64. Miano JM, Carlson MJ, Spencer JA, Misr RP. 2000. Serum response factor-dependent regulation of the smooth muscle calponin gene. *J. Biol. Chem.* 275(13):9814–22
65. Hoggatt AM, Simon GM, Herring BP. 2002. Cell-specific regulatory modules control expression of genes in vascular and visceral smooth muscle tissues. *Circ. Res.* 91(12):1151–59
66. Manak JR, Prywes R. 1993. Phosphorylation of serum response factor by casein kinase II: evidence against a role in growth factor regulation of *fos* expression. *Oncogene* 8(3):703–11
67. Han Z, Li X, Wu J, Olson EN. 2004. A myocardin-related transcription factor regulates activity of serum response factor in *Drosophila*. *Proc. Natl. Acad. Sci. USA* 101(34):12567–72
68. Wang Z, Wang DZ, Hockemeyer D, McAnally J, Nordheim A, Olson EN. 2004. Myocardin and ternary complex factors compete for SRF to control smooth muscle gene expression. *Nature* 428(6979):185–89
69. Rivera VM, Miranti CK, Misra RP, Ginty DD, Chen RH, et al. 1993. A growth factor-induced kinase phosphorylates the serum response factor at a site that regulates its DNA-binding activity. *Mol. Cell. Biol.* 13:6260–73
70. Chang DF, Belaguli NS, Iyer D, Roberts WB, Wu SP, et al. 2003. Cysteine-rich LIM-only proteins CRP1 and CRP2 are potent smooth muscle differentiation cofactors. *Dev. Cell* 4(1):107–18
71. Nishida W, Nakamura M, Mori S, Takahashi M, Ohkawa Y, et al. 2002. A triad of serum response factor and the GATA and NK families governs the transcription of smooth and cardiac muscle genes. *J. Biol. Chem.* 277(9):7308–17
72. Manabe I, Owens GK. 2001. Recruitment of serum response factor and hyperacetylation of histones at smooth muscle-specific regulatory regions during differentiation of a novel P19-derived in vitro smooth muscle differentiation system. *Circ. Res.* 88(11):1127–34
73. McDonald OG, Wamhoff BR, Hoofnagle MH, Owens GK. 2006. Control of SRF binding to CArG box chromatin regulates smooth muscle gene expression in vivo. *J. Clin. Investig.* 116:36–48
74. Yoshida T, Sinha S, Dandre F, Wamhoff BR, Hoofnagle MH, et al. 2003. Myocardin is a key regulator of CArG-dependent transcription of multiple smooth muscle marker genes. *Circ. Res.* 92:856–64
75. Cen B, Selvaraj A, Burgess RC, Hitzler JK, Ma Z, et al. 2003. Megakaryoblastic leukemia 1, a potent transcriptional coactivator for serum response factor (SRF), is required for serum induction of SRF target genes. *Mol. Cell. Biol.* 23(18):6597–608
76. Du KL, Chen MM, Li J, Lepore JJ, Mericko P, Parmacek MS. 2004. Megakaryoblastic leukemia factor-1 transduces cytoskeletal signals and induces smooth muscle cell differentiation from undifferentiated embryonic stem cells. *J. Biol. Chem.* 279(17):17578–86
77. Wang D, Chang PS, Wang Z, Sutherland L, Richardson JA, et al. 2001. Activation of cardiac gene expression by myocardin, a transcriptional cofactor for serum response factor. *Cell* 105(7):851–62
78. Chen J, Kitchen CM, Streb JW, Miano JM. 2002. Myocardin: a component of a molecular switch for smooth muscle differentiation. *J. Mol. Cell. Cardiol.* 34(10):1345–56
79. Wang Z, Wang DZ, Pipes GC, Olson EN. 2003. Myocardin is a master regulator of smooth muscle gene expression. *Proc. Natl. Acad. Sci. USA* 100(12):7129–34
80. Li J, Zhu X, Chen M, Cheng L, Zhou D, et al. 2005. Myocardin-related transcription factor B is required in cardiac neural crest for smooth muscle differentiation and cardiovascular development. *Proc. Natl. Acad. Sci. USA* 102(25):8916–21
81. Oh J, Richardson JA, Olson EN. 2005. Requirement of myocardin-related transcription factor-B for remodeling of branchial arch arteries and smooth muscle differentiation. *Proc. Natl. Acad. Sci. USA* 102(42):15122–27
82. Hendrix J, Wamhoff BR, McDonald T, Sinha S, Yoshida T, Owens GK. 2005. 5′ CArG degeneracy in smooth muscle α-actin is required for injury-induced gene suppression in vivo. *J. Clin. Investig.* 115(2):418–27
83. Hautmann MB, Madsen CS, Owens GK. 1998. Substitution of the degenerate SM (smooth muscle) α-actin CArG elements with c-*fos* SREs results in increased basal expression but relaxed specificity and reduced angiotensin II inducibility. *J. Biol. Chem.* 273(14):8398–406
84. Li S, Wang DZ, Wang Z, Richardson JA, Olson EN. 2003. The serum response factor coactivator myocardin is required for vascular smooth muscle development. *Proc. Natl. Acad. Sci. USA* 100(16):9366–70

85. Pipes GC, Sinha S, Qi X, Zhu C, Gallardo TD, et al. 2005. Stem cells and their derivatives can bypass the requirement of myocardin for smooth muscle gene expression. *Dev. Biol.* 288:502–13
86. Hoofnagle MH, Neppl RL, Berzin EL, Pipes GC, Olson EN, et al. 2011. Myocardin is differentially required for the development of smooth muscle cells and cardiomyocytes. *Am. J. Physiol. Heart Circ. Physiol.* 300:H1707–21
87. Lockman K, Hinson JS, Medlin MD, Morris D, Taylor JM, Mack CP. 2004. Sphingosine 1-phosphate stimulates smooth muscle cell differentiation and proliferation by activating separate serum response factor co-factors. *J. Biol. Chem.* 279(41):42422–30
88. Huang J, Cheng L, Li J, Chen M, Zhou D, et al. 2008. Myocardin regulates expression of contractile genes in smooth muscle cells and is required for closure of the ductus arteriosus in mice. *J. Clin. Investig.* 118(2):515–25
89. Parmacek MS. 2004. Myocardin—not quite MyoD. *Arterioscler. Thromb. Vasc. Biol.* 24(9):1535–37
90. Zhou J, Herring BP. 2005. Mechanisms responsible for the promoter-specific effects of myocardin. *J. Biol. Chem.* 280(11):10861–69
91. Liu Y, Sinha S, McDonald OG, Shang Y, Hoofnagle MH, Owens GK. 2005. Kruppel-like factor 4 abrogates myocardin-induced activation of smooth muscle gene expression. *J. Biol. Chem.* 280(10):9719–27
92. Tang Rh, Zheng XL, Callis TE, Stansfield WE, He J, et al. 2008. Myocardin inhibits cellular proliferation by inhibiting NF-κB(p65)-dependent cell cycle progression. *Proc. Natl. Acad. Sci. USA* 105(9):3362–67
93. Cao D, Wang Z, Zhang CL, Oh J, Xing W, et al. 2005. Modulation of smooth muscle gene expression by association of histone acetyltransferases and deacetylases with myocardin. *Mol. Cell. Biol.* 25(1):364–76
94. Parmacek MS. 2008. Myocardin: dominant driver of the smooth muscle cell contractile phenotype. *Arterioscler. Thromb. Vasc. Biol.* 28(8):1416–17
95. Wang DZ, Olson EN. 2004. Control of smooth muscle development by the myocardin family of transcriptional coactivators. *Curr. Opin. Genet. Dev.* 14(5):558–66
96. Cheung P, Allis CD, Sassone-Corsi P. 2000. Signaling to chromatin through histone modifications. *Cell* 103(2):263–71
97. Hall IM, Shankaranarayana GD, Noma K, Ayoub N, Cohen A, Grewal SI. 2002. Establishment and maintenance of a heterochromatin domain. *Science* 297(5590):2232–37
98. Suto RK, Clarkson MJ, Tremethick DJ, Luger K. 2000. Crystal structure of a nucleosome core particle containing the variant histone H2A.Z. *Nat. Struct. Biol.* 7(12):1121–24
99. Qiu P, Li L. 2002. Histone acetylation and recruitment of serum responsive factor and CREB-binding protein onto SM22 promoter during SM22 gene expression. *Circ. Res.* 90(8):858–65
100. Yoshida T, Gan Q, Shang Y, Owens GK. 2007. Platelet-derived growth factor-BB represses smooth muscle cell marker genes via changes in binding of MKL factors and histone deacetylases to their promoters. *Am. J. Physiol. Cell Physiol.* 292(2):886–95
101. Thomas JA, Deaton RA, Hastings NE, Shang Y, Moehle CW, et al. 2009. PDGF-DD, a novel mediator of smooth muscle cell phenotypic modulation, is upregulated in endothelial cells exposed to atherosclerosis-prone flow patterns. *Am. J. Physiol. Heart Circ. Physiol.* 296(2):442–52
102. Pidkovka NA, Cherepanova OA, Yoshida T, Alexander MR, Deaton RA, et al. 2007. Oxidized phospholipids induce phenotypic switching of vascular smooth muscle cells in vivo and in vitro. *Circ. Res.* 101:792–801
103. Litt MD, Simpson M, Gaszner M, Allis CD, Felsenfeld G. 2001. Correlation between histone lysine methylation and developmental changes at the chicken β-globin locus. *Science* 293(5539):2453–55
104. Gan Q, Yoshida T, McDonald OG, Owens GK. 2007. Epigenetic mechanisms contribute to pluripotency and cell lineage determination of embryonic stem cells. *Stem Cells* 25:2–9
105. Cook CL, Weiser MC, Schwartz PE, Jones CL, Majack RA. 1994. Developmentally timed expression of an embryonic growth phenotype in vascular smooth muscle cells. *Circ. Res.* 74:189–96
106. O'Brien ER, Alpers CE, Stewart DK, Ferguson M, Tran N, et al. 1993. Proliferation in primary and restenotic coronary atherectomy tissue. Implications for antiproliferative therapy. *Circ. Res.* 73:223–31
107. Wilcox JN. 1992. Analysis of local gene expression in human atherosclerotic plaques. *J. Vasc. Surg.* 15:913–16

108. Blank RS, Owens GK. 1990. Platelet-derived growth factor regulates actin isoform expression and growth state in cultured rat aortic smooth muscle cells. *J. Cell Physiol.* 142:635–42
109. Corjay MH, Thompson MM, Lynch KR, Owens GK. 1989. Differential effect of platelet-derived growth factor- versus serum-induced growth on smooth muscle α-actin and nonmuscle β-actin mRNA expression in cultured rat aortic smooth muscle cells. *J. Biol. Chem.* 264:10501–6
110. Holycross BJ, Blank RS, Thompson MM, Peach MJ, Owens GK. 1992. Platelet-derived growth factor-BB-induced suppression of smooth muscle cell differentiation. *Circ. Res.* 71:1525–32
111. Thyberg J, Palmberg L, Nilsson J, Ksiazek T, Sjolund M. 1983. Phenotypic modulation in primary cultures of arterial smooth muscle cells: on the role of platelet-derived growth factor. *Differentiation* 25:156–67
112. Somasundaram C, Kallmeier RC, Babij P. 1995. Regulation of smooth muscle myosin heavy chain gene expression in cultured vascular smooth muscle cells by growth factors and contractile agonists. *Basic Appl. Mycol.* 6:31–36
113. Li X, Van Putten V, Zarinetchi F, Nicks M, Thaler S, et al. 1997. Suppression of smooth muscle α-actin expression by platelet-derived growth factor in vascular smooth-muscle cells involves Ras and cytosolic phospholipase A2. *Biochem. J.* 327:709–16
114. Speer MY, Li X, Hiremath PG, Giachelli CM. 2010. Runx2/Cbfa1, but not loss of myocardin, is required for smooth muscle cell lineage reprogramming toward osteochondrogenesis. *J. Cell Biochem.* 110(4):935–47
115. Clement N, Gueguen M, Glorian M, Blaise R, Andreani M, et al. 2007. Notch3 and IL-1β exert opposing effects on a vascular smooth muscle cell inflammatory pathway in which NF-κB drives crosstalk. *J. Cell Sci.* 120(Pt. 19):3352–61
116. Chen CN, Li YS, Yeh YT, Lee PL, Usami S, et al. 2006. Synergistic roles of platelet-derived growth factor-BB and interleukin-1β in phenotypic modulation of human aortic smooth muscle cells. *Proc. Natl. Acad. Sci. USA* 103(8):2665–70
117. Yoshida T, Gan Q, Owens GK. 2008. Krüppel-like factor 4, Elk-1, and histone deacetylases cooperatively suppress smooth muscle cell differentiation markers in response to oxidized phospholipids. *Am. J. Physiol. Cell Physiol.* 295(5):1175–82
118. Hayashi K, Takahashi M, Nishida W, Yoshida K, Ohkawa Y, et al. 2001. Phenotypic modulation of vascular smooth muscle cells induced by unsaturated lysophosphatidic acids. *Circ. Res.* 89(3):251–58
119. Majesky MW, Dong XR, Regan JN, Hoglund VJ. 2011. Vascular smooth muscle progenitor cells: building and repairing blood vessels. *Circ. Res.* 108(3):365–77
120. Kawai-Kowase K, Owens GK. 2007. Multiple repressor pathways contribute to phenotypic switching of vascular smooth muscle cells. *Am. J. Physiol. Cell Physiol.* 292(1):59–69
121. Shimizu RT, Blank RS, Jervis R, Lawrenz-Smith SC, Owens GK. 1995. The smooth muscle α-actin gene promoter is differentially regulated in smooth muscle versus non-smooth muscle cells. *J. Biol. Chem.* 270:7631–43
122. Landerholm TE, Dong X-R, Belaguli N, Schwartz RJ, Majesky MW. 1999. A role for serum response factor in coronary smooth muscle differentiation from proepicardial cells. *Development* 126(10):2053–62
123. Geisterfer AA, Peach MJ, Owens GK. 1988. Angiotensin II induces hypertrophy, not hyperplasia, of cultured rat aortic smooth muscle cells. *Circ. Res.* 62:749–56
124. Simper D, Stalboerger PG, Panetta CJ, Wang S, Caplice NM. 2002. Smooth muscle progenitor cells in human blood. *Circulation* 106(10):1199–204
125. Deaton RA, Gan Q, Owens GK. 2009. Sp1-dependent activation of KLF4 is required for PDGF-BB-induced phenotypic modulation of smooth muscle. *Am. J. Physiol. Heart Circ. Physiol.* 296(4):1027–37
126. Stouffer GA, Owens GK. 1994. TGF-β promotes proliferation of cultured SMC via both PDGF-AA-dependent and PDGF-AA-independent mechanisms. *J. Clin. Investig.* 93:2048–55
127. Boerth NJ, Dey NB, Cornwell TL, Lincoln TM. 1997. Cyclic GMP-dependent protein kinase regulates vascular smooth muscle cell phenotype. *J. Vasc. Res.* 34(4):245–59
128. Berk BC, Vekshtein V, Gordon H, Tsuda T. 1989. Angiotensin II stimulated protein synthesis in cultured vascular smooth muscle cells. *Hypertension* 13:305–14
129. Owens GK, Schwartz SM. 1982. Alterations in vascular smooth muscle mass in the spontaneously hypertensive rat. Role of cellular hypertrophy, hyperploidy and hyperplasia. *Circ. Res.* 51:280–89

130. Owens GK. 1993. Determinants of angiotensin II-induced hypertrophy versus hyperplasia in vascular smooth muscle. *Drug Dev. Res.* 29(2):83–87
131. Owens GK, Rabinovitch PS, Schwartz SM. 1981. Smooth muscle cell hypertrophy versus hyperplasia in hypertension. *Proc. Natl. Acad. Sci. USA* 78:7759–63
132. Regan CP, Adam PJ, Madsen CS, Owens GK. 2000. Molecular mechanisms of decreased smooth muscle differentiation marker expression after vascular injury. *J. Clin. Investig.* 106(9):1139–47
133. Wamhoff BR, Hoofnagle MH, Burns A, Sinha S, McDonald OG, Owens GK. 2004. A G/C element mediates repression of the SM22α promoter within phenotypically modulated smooth muscle cells in experimental atherosclerosis. *Circ. Res.* 95(10):981–88
134. Cherepanova OA, Pidkovka NA, Yoshida T, Gan Q, Adiguzel E, et al. 2009. Oxidized phospholipids induce type VIII collagen expression and vascular smooth muscle cell migration. *Circ. Res.* 104:609–18
135. Yoshida T, Kaestner KH, Owens GK. 2008. Conditional deletion of Krüppel-like factor 4 delays downregulation of smooth muscle cell differentiation markers but accelerates neointimal formation following vascular injury. *Circ. Res.* 102(12):1548–57
136. Cordes KR, Sheehy NT, White MP, Berry EC, Morton SU, et al. 2009. miR-145 and miR-143 regulate smooth muscle cell fate and plasticity. *Nature* 460:705–10
137. Xin M, Small EM, Sutherland LB, Qi X, McAnally J, et al. 2009. MicroRNAs miR-143 and miR-145 modulate cytoskeletal dynamics and responsiveness of smooth muscle cells to injury. *Genes Dev.* 23(18):2166–78
138. Cheng Y, Liu X, Yang J, Lin Y, Xu DZ, et al. 2009. MicroRNA-145, a novel smooth muscle cell phenotypic marker and modulator, controls vascular neointimal lesion formation. *Circ. Res.* 105(2):158–66
139. Shindo T, Manabe I, Fukushima Y, Tobe K, Aizawa K, et al. 2002. Krüppel-like zinc-finger transcription factor KLF5/BTEB2 is a target for angiotensin II signaling and an essential regulator of cardiovascular remodeling. *Nat. Med.* 8(8):856–63
140. Xu N, Papagiannakopoulos T, Pan G, Thomson JA, Kosik KS. 2009. MicroRNA-145 regulates OCT4, SOX2, and KLF4 and represses pluripotency in human embryonic stem cells. *Cell* 137(4):647–58
141. Cho WC, Chow AS, Au JS. 2011. MiR-145 inhibits cell proliferation of human lung adenocarcinoma by targeting EGFR and NUDT1. *RNA Biol.* 8(1):125–31
142. Chen Z, Zeng H, Guo Y, Liu P, Pan H, et al. 2010. miRNA-145 inhibits non-small cell lung cancer cell proliferation by targeting c-Myc. *J. Exp. Clin. Cancer Res.* 29:151
143. Zaman MS, Chen Y, Deng G, Shahryari V, Suh SO, et al. 2010. The functional significance of microRNA-145 in prostate cancer. *Br. J. Cancer* 103(2):256–64
144. Davis BN, Hilyard AC, Nguyen PH, Lagna G, Hata A. 2010. Smad proteins bind a conserved RNA sequence to promote microRNA maturation by Drosha. *Mol. Cell* 39(3):373–84
145. Chan MC, Hilyard AC, Wu C, Davis BN, Hill NS, et al. 2010. Molecular basis for antagonism between PDGF and the TGFβ family of signalling pathways by control of miR-24 expression. *EMBO J.* 29(3):559–73
146. Leeper NJ, Raiesdana A, Kojima Y, Chun HJ, Azuma J, et al. 2011. MicroRNA-26a is a novel regulator of vascular smooth muscle cell function. *J. Cell Physiol.* 226(4):1035–43
147. Shields JM, Christy RJ, Yang VW. 1996. Identification and characterization of a gene encoding a gut enriched Krüppel-like factor expressed during growth arrest. *J. Biol. Chem.* 271(33):20009–17
148. Takahashi K, Yamanaka S. 2006. Induction of pluripotent stem cells from mouse embryonic and adult fibroblast cultures by defined factors. *Cell* 126(4):663–76
149. Wernig M, Meissner A, Foreman R, Brambrink T, Ku M, et al. 2007. In vitro reprogramming of fibroblasts into a pluripotent ES-cell-like state. *Nature* 448(7151):318–24
150. Jiang J, Chan YS, Loh YH, Cai J, Tong GQ, et al. 2008. A core Klf circuitry regulates self-renewal of embryonic stem cells. *Nat. Cell Biol.* 10(3):353–60
151. Gabbiani G. 2003. The myofibroblast in wound healing and fibrocontractive diseases. *J. Pathol.* 200(4):500–3

152. Yoshida T, Gan Q, Shang Y, Owens GK. 2007. Platelet-derived growth factor-BB represses smooth muscle cell marker genes via changes in binding of MKL factors and histone deacetylases to their promoters. *Am. J. Physiol. Cell Physiol.* 292:886–95
153. Kawai-Kowase K, Kumar MS, Hoofnagle MH, Yoshida T, Owens GK. 2005. PIAS1 activates the expression of smooth muscle cell differentiation marker genes by interacting with serum response factor and class I basic helix-loop-helix proteins. *Mol. Cell. Biol.* 25(18):8009–23
154. Doi H, Iso T, Yamazaki M, Akiyama H, Kanai H, et al. 2005. HERP1 inhibits myocardin-induced vascular smooth muscle cell differentiation by interfering with SRF binding to CArG box. *Arterioscler. Thromb. Vasc. Biol.* 25(11):2328–34
155. Fujiu K, Manabe I, Ishihara A, Oishi Y, Iwata H, et al. 2005. Synthetic retinoid Am80 suppresses smooth muscle phenotypic modulation and in-stent neointima formation by inhibiting KLF5. *Circ. Res.* 97(11):1132–41
156. Liu ZP, Wang Z, Yanagisawa H, Olson EN. 2005. Phenotypic modulation of smooth muscle cells through interaction of Foxo4 and myocardin. *Dev. Cell* 9(2):261–70
157. Banai S, Wolf Y, Golomb G, Pearle A, Waltenberger J, et al. 1998. PDGF-receptor tyrosine kinase blocker AG1295 selectively attenuates smooth muscle cell growth in vitro and reduces neointimal formation after balloon angioplasty in swine. *Circulation* 97(19):1960–69
158. Jawien A, Bowen Pope DF, Lindner V, Schwartz SM, Clowes AW. 1992. Platelet-derived growth factor promotes smooth muscle migration and intimal thickening in a rat model of balloon angioplasty. *J. Clin. Investig.* 89:507–11
159. Kenagy RD, Hart CE, Stetler-Stevenson WG, Clowes AW. 1997. Primate smooth muscle cell migration from aortic explants is mediated by endogenous platelet-derived growth factor and basic fibroblast growth factor acting through matrix metalloproteinases 2 and 9. *Circulation* 96(10):3555–60
160. Martinet Y, Bitterman PB, Mornex JF, Grotendorst GR, Martin GR, Crystal RG. 1986. Activated human monocytes express the c-sis proto-oncogene and release a mediator showing PDGF-like activity. *Nature* 319(6049):158–60
161. Kozaki K, Kaminski WE, Tang J, Hollenbach S, Lindahl P, et al. 2002. Blockade of platelet-derived growth factor or its receptors transiently delays but does not prevent fibrous cap formation in ApoE null mice. *Am. J. Pathol.* 161(4):1395–407
162. Sano H, Sudo T, Yokode M, Murayama T, Kataoka H, et al. 2001. Functional blockade of platelet-derived growth factor receptor-β but not of receptor-α prevents vascular smooth muscle cell accumulation in fibrous cap lesions in apolipoprotein E-deficient mice. *Circulation* 103(24):2955–60
163. Amento EP, Ehsani N, Palmer H, Libby P. 1991. Cytokines and growth factors positively and negatively regulate interstitial collagen gene expression in human vascular smooth muscle cells. *Arterioscler. Thromb.* 11:1223–30
164. Johnstone SR, Ross J, Rizzo MJ, Straub AC, Lampe PD, et al. 2009. Oxidized phospholipid species promote in vivo differential Cx43 phosphorylation and vascular smooth muscle cell proliferation. *Am. J. Pathol.* 175(2):916–24
165. Couffinhal T, Duplaa C, Moreau C, Lamaziere JM, Bonnet J. 1994. Regulation of vascular cell adhesion molecule-1 and intercellular adhesion molecule-1 in human vascular smooth muscle cells. *Circ. Res.* 74:225–34
166. Loppnow H, Libby P. 1990. Proliferating or interleukin 1-activated human vascular smooth muscle cells secrete copious interleukin 6. *J. Clin. Investig.* 85(3):731–38
167. Fabunmi RP, Baker AH, Murray EJ, Booth RF, Newby AC. 1996. Divergent regulation by growth factors and cytokines of 95 kDa and 72 kDa gelatinases and tissue inhibitors or metalloproteinases-1, -2, and -3 in rabbit aortic smooth muscle cells. *Biochem. J.* 315(Pt. 1):335–42
168. Keen RR, Nolan KD, Cipollone M, Scott E, Shively VP, et al. 1994. Interleukin-1β induces differential gene expression in aortic smooth muscle cells. *J. Vasc. Surg.* 20(5):774–84
169. Galis ZS, Muszynski M, Sukhova GK, Simon-Morrissey E, Unemori EN, et al. 1994. Cytokine-stimulated human vascular smooth muscle cells synthesize a complement of enzymes required for extracellular matrix digestion. *Circ. Res.* 75(1):181–89

170. Marx N, Neumann FJ, Zohlnhöfer D, Dickfeld T, Fischer A, et al. 1998. Enhancement of monocyte procoagulant activity by adhesion on vascular smooth muscle cells and intercellular adhesion molecule-1-transfected Chinese hamster ovary cells. *Circulation* 98(9):906–11
171. Niwa H, Ogawa K, Shimosato D, Adachi K. 2009. A parallel circuit of LIF signalling pathways maintains pluripotency of mouse ES cells. *Nature* 460(7251):118–22
172. Boyer LA, Lee TI, Cole MF, Johnstone SE, Levine SS, et al. 2005. Core transcriptional regulatory circuitry in human embryonic stem cells. *Cell* 122(6):947–56

Epigenetics and Cardiovascular Development

Ching-Pin Chang[1] and Benoit G. Bruneau[2,3]

[1]Division of Cardiovascular Medicine, Department of Medicine, Stanford Cardiovascular Institute, Institute for Stem Cell Biology and Regenerative Medicine, Stanford University School of Medicine, Stanford, California 94305; email: chingpin@stanford.edu

[2]Gladstone Institute of Cardiovascular Disease, San Francisco, California 94158

[3]Department of Pediatrics and Cardiovascular Research Institute, University of California, San Francisco, California 94158

Keywords

chromatin remodeling, histone modification, gene expression, congenital heart disease, vascular disease

Abstract

The cardiovascular system is broadly composed of the heart, which pumps blood, and the blood vessels, which carry blood to and from tissues of the body. Heart malformations are the most serious common birth defect, affecting at least 2% of newborns and leading to significant morbidity and mortality. Severe heart malformations cause heart failure in fetuses, infants, and children, whereas milder heart defects may not trigger significant heart dysfunction until early or midadulthood. Severe vasculogenesis or angiogenesis defects in embryos are incompatible with life, and anomalous arterial patterning may cause vascular aberrancies that often require surgical treatment. It is therefore important to understand the underlying mechanisms that control cardiovascular development. Understanding developmental mechanisms will also help us design better strategies to regenerate cardiovascular tissues for therapeutic purposes. An important mechanism regulating genes involves the modification of chromatin, the higher-order structure in which DNA is packaged. Recent studies have greatly expanded our understanding of the regulation of cardiovascular development at the chromatin level, including the remodeling of chromatin and the modification of histones. Chromatin-level regulation integrates multiple inputs and coordinates broad gene expression programs. Thus, understanding chromatin-level regulation will allow for a better appreciation of gene regulation as a whole and may set a fundamental basis for cardiovascular disease. This review focuses on how chromatin-remodeling and histone-modifying factors regulate gene expression to control cardiovascular development.

INTRODUCTION

The developing heart arises from lateral plate mesoderm early in embryogenesis and begins to emerge shortly after gastrulation. Following rapid specification and initiation of differentiation into the specific cell types that will form the heart, a series of complex morphogenetic events ensues (**Figure 1**). The formation of a simple heart tube is quickly followed by the origami-like folding of this structure as it grows. The result is a fully functional organ composed of four chambers separated by septa and valves. The proper formation and function of the heart are essential for embryonic survival, and defects in heart formation cause significant morbidity and mortality in prenatal or postnatal life (1–3). The vasculature begins its development shortly after the heart starts to form. The earliest vessels are the arterial precursors, which are followed closely by primitive veins. Once the major vascular beds have been established, intricate branching patterns are formed as the vasculature expands to supply a growing embryo. The end result is a complex network of specialized vessels that allow for the efficient delivery of oxygenated blood and the expedient return of unoxygenated blood to the heart. Aberrant vascular development may cause lethality or a vascular condition that requires surgical treatment (4).

The regulation of cardiovascular development has been well studied. In particular, signaling pathways that modulate different events during cardiovascular development are well defined. There is also a rich knowledge of the transcription factors that dictate cardiovascular gene expression and control important morphogenetic steps. Two areas of application have exemplified the importance of understanding cardiovascular transcriptional regulation. The first is the genetics of congenital heart disease: Most of the inherited forms of congenital heart disease are a result of mutations in cardiovascular transcription factor genes. The second is the powerful harnessing of these transcription factors to reprogram somatic cells toward a cardiomyocyte-like state, paving the way for potential regenerative therapies.

A deeper understanding of the regulatory mechanisms controlling cardiovascular gene expression will be critical for deciphering the mechanisms underlying congenital heart defects, as

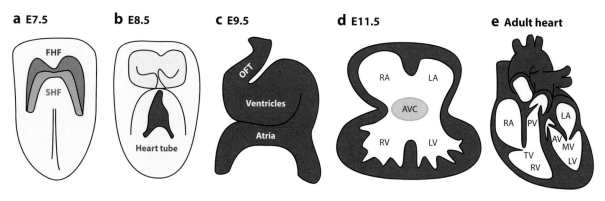

Figure 1

Heart morphogenesis in mice. (*a*) Two mesodermal heart fields, termed the first and secondary heart fields (FHF and SHF), give rise to cardiac progenitor cells. (*b*) After initial differentiation into cardiac cells, these two fields form a heart tube in the ventral midline. (*c*) The heart tube undergoes rightward looping and contains atria, ventricles, and the cardiac tract (OFT). FHF contributes primarily to the left ventricle, and SHF to the atria, right ventricle, and OFT. (*d*) Later, the atrial and ventricular chambers are septated to form a four-chamber heart, and the cardiac OFT divides into left and right ventricular outflow tracts, as well as into aorta and the main pulmonary artery. (*e*) Heart valves develop at the junction of atrial and ventricular chambers (atrioventricular valves) and at the junction of ventricular outflow tracts and great arteries (semilunar valves). Embryonic dates of development in mice are denoted (E denotes embryonic day). Other abbreviations: AV, aortic valve; AVC, atrioventricular cushion; LA, left atrium; LV, left ventricle; MV, mitral valve; PV, pulmonic valve; RA, right atrium; RV, right ventricle; TV, tricuspid valve.

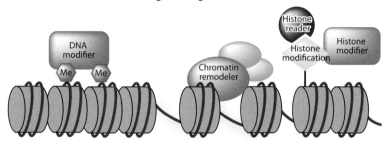

Figure 2

Chromatin and its regulation. (*a*) The information of genes is recorded and stored in the DNA sequence. (*b*) Chromatin in different cells indexes and organizes this information in distinct ways to control gene expression. Chromatin structure can be dynamically regulated by three classes of chromatin regulators: DNA methylation (Me) by DNA modifiers, nucleosome positioning by chromatin remodelers, and histone modification by histone modifiers.

well as for helping to design more powerful and specific regenerative therapies for heart disease. Controlling gene expression in specific tissues and developmental windows is essential for cardiovascular development in embryos. Genes are tightly packed in chromatin, which not only serves as a packaging mechanism for DNA but, importantly, acts as a major controller of gene expression (**Figure 2**). Chromatin is a dynamic structure that provides a DNA scaffold responsive to external cues to regulate gene expression. The fundamental unit of the chromatin is the nucleosome, consisting of 147 base pairs of DNA that envelop an octameric core of histone proteins. This histone core is composed of two molecules of each of the H2A, H2B, H3, and H4 histones. These repeating units of nucleosomes are tightly organized into higher-order scaffolds to further condense the DNA. The position of each nucleosome is stabilized by weak histone-DNA interactions between the positively charged residues in the histones and the negatively charged phosphate backbone of the DNA (5). Chromatin-level gene regulation is accomplished using three processes: DNA methylation, ATP-dependent chromatin remodeling, and covalent histone modifications. This review focuses on the roles of chromatin-remodeling and histone modification factors in cardiovascular development. The ATP-dependent chromatin-remodeling complexes contain an ATPase subunit and use the energy from ATP hydrolysis to restructure nucleosomes: They can move, destabilize, eject, or alter the composition of nucleosomes. Certain subunits of chromatin-remodeling complexes can recognize specific residues of DNA or histones that are covalently modified by a group of enzymes. Histone-modifying enzymes, such as histone methyltransferases and histone deacetylases, catalyze the covalent modification of histone proteins to alter the histone-DNA contacts and thereby loosen or tighten the chromatin to control the availability of genomic loci to transcriptional or other regulation. Histones can be modified by many different chemical processes, including methylation, acetylation, phosphorylation, ubiquitination, sumoylation, and ADP-ribosylation. These different chromatin modifications add to the diversity of chromatin structure that can respond to various pathophysiological signals. Furthermore, the combination of specific histone modification and nucleosome structure at specific gene loci may provide a mark, or histone code, on the genome (see Reference 6 for a review). These marks can act as beacons or docking sites recognized by transcription factors or other chromatin regulators, thereby introducing specificity and complexity of gene expression control.

By controlling cardiovascular gene expression, chromatin regulation plays critical roles in cardiovascular development. Proteins that regulate chromatin structure function in a time- and tissue-specific manner to direct cardiovascular development, and some have been directly implicated in human cardiovascular diseases.

DIFFERENT CLASSES OF CHROMATIN-REGULATING ENZYMES

Although there are myriad types of chromatin regulators, the ones currently implicated in regulating cardiovascular development fall into two broad categories: chromatin remodelers and histone modifiers. The general activities of these two categories of chromatin regulators are described below.

The ATP-Dependent Chromatin-Remodeling Complexes

There are four different families of SWI-like, ATP-dependent chromatin-remodeling complexes: the SWI/SNF (switching defective/sucrose nonfermenting), ISWI (imitation switch), CHD (chromodomain, helicase, DNA binding), and INO80 (inositol requiring 80) complexes (see References 5 and 7 for reviews). These families share an evolutionarily conserved SWI-like ATPase catalytic domain, but each one has its own unique functional domains. The ATPase domain serves as a molecular motor to disengage histone-DNA contacts for nucleosome restructuring. Other functional domains may recognize covalently modified histones, modulate ATPase activity, or interact with regulatory factors. The targeting specificities of each family of chromatin remodelers may thus be determined by interactions among ATPase and other functional domains of the remodeler.

Histone Modifiers

Several classes of enzymes effect chemical modification of histone tails. Histone acetyltransferases (HATs) acetylate histones, and histone deacetylases (HDACs) reverse this acetylation process. The Sirtuin deacetylases have no sequence homology to the classical HDACs and require NAD^+ (nicotinamide adenine dinucleotide) for their deacetylase activity. Histone methyltransferases (HMTs), which compose a broader family of enzymes, add methyl groups to histone tails at a broad but specific set of amino acid residues. More recent findings have identified the enzymes that reverse histone methylation, and these enzymes include a family of proteins that share a conserved domain termed the Jumonji domain. Jumonji proteins are named after the first-discovered member of the family, which was identified in a genetrap mutation that resulted in cardiovascular defects. Moreover, the polycomb repressive complexes silence gene expression partly through histone modification, including histone methylation and ubiquitylation. The poly(ADP-ribose) polymerases (PARPs) catalyze the transfer of ADP-ribose units from NAD^+ to histone proteins to regulate chromatin structure and gene expression.

Table 1 Vascular development and angiogenesis

Class	Chromatin factor	Gene modification	Phenotype	Comments	Reference(s)
SWI/SNF	Brg1	Deletion in the endothelium	Yolk sac vascular defects, hypotrabeculation, lethality at E10.5–11.5	Brg1 activates Fzd receptors and Wnt target genes in yolk sac endothelial cells.	8, 11
	Baf180	Germline deletion	Coronary and myocardial defects, lethality at E12.5–15.5	Baf180 is required for the expression of genes essential for coronary vessel formation.	16
Class II histone deacetylase (HDAC)	Hdac7	Germline deletion	Bleeding from leaky and dilated blood vessels, fewer smooth muscle cells around vessels	Hdac7 suppresses *Mmp10* to maintain vessel wall integrity. Hdac7 interacts with Vegf and β-catenin to regulate angiogenesis.	9, 17, 18
Sirtuin	Sirt1	Morpholino knockdown in zebra fish	Abnormal trunk vasculature, defective endothelial sprouting, bleeding	Sirt1 deacetylates Foxo1 to inhibit its transcriptional activity and its antiangiogenic activity.	19
		Deletion in the endothelium	No apparent vascular defects in embryos, delayed postnatal retinal angiogenesis	Sirt1 deacetylates NICD to promote NICD degradation and to facilitate endothelial branching and proliferation.	20
Histone methyltransferase	Hypb/Setd2	Germline deletion	Dilated capillaries, failure to form a mature vascular network, lethality at E10.5–11.5	Hypb methylates H3K36. Hypb silencing impairs endothelial cell migration and tubular formation.	21
	Dot1-like protein (Dot1L)	Germline deletion	Yolk sac angiogenesis defects, cardiac dilatation, lethality at E9.5–10.5	Dot1L methylates H3K79 and regulates telomere elongation and cell proliferation.	136

VASCULAR DEVELOPMENT

Whether specific epigenetic events are required for vascular development remains unknown. However, recent studies show that chromatin-regulating molecules such as the Brg1/Brm-associated-factor (BAF) complex and the Hdac7 HDAC are crucial for vascular development in mice (**Table 1**) (8, 9).

Brg1 and Yolk Sac Vascular Remodeling

The BAF complex, consisting of 12 protein subunits, is a major SWI/SNF type of ATP-dependent chromatin-remodeling complex in vertebrates (7, 10). The ATPase subunit of the BAF complex is encoded by one of two highly homologous genes, *Brg1* and *Brm*. Tissue-specific deletion of *Brg1* in the endothelium causes lethality at approximately embryonic day 10.5 (E10.5) of mouse development (8). Mutant embryos display severe anemia and vascular defects in the yolk sac, but not in the embryo proper (8, 11). The yolk sac of mutant embryos fails to form interconnected vessels to establish a mature vascular tree (8, 11), indicating abnormal vascular sprouting or pruning. In contrast to *Brg1*, *Brm* appears dispensable for yolk sac vascular development. *Brm*-null mice are viable and do not have apparent developmental defects (12). *Brm*- and *Brg1*-null double mutations in the endothelium do not exacerbate yolk sac vascular defects caused by *Brg1*-null mutation alone (11), suggesting that these two ATPases of the BAF complex are not functionally redundant in vascular development.

Brg1 regulates endothelial Wnt signaling in the yolk sac through transcriptional activation of a number of Wnt Fzd receptor and target genes (13). In the absence of Brg1, yolk sac endothelial cells show reduced Wnt signaling with β-catenin degradation. It is unclear why this Brg1-based Wnt/β-catenin regulation affects endothelial development only in the yolk sac, given that endothelial β-catenin is essential for vascular development in both the yolk sac and the embryo (14).

Baf180 and Coronary Vascular Development

Baf180 (alias polybromo) is a subunit of the BAF complex. It contains six bromodomains that recognize acetylated histone tails. Baf180 is widely expressed in developing embryos (15) and is essential for coronary vessel and cardiac chamber development (16). The epicardium of *Baf180*-null mice fails to properly undergo epithelial-to-mesenchymal transformation (EMT), and the heart of *Baf180*-null mice displays downregulation of many factors that promote EMT, vasculogenesis, or angiogenesis (16). The role of Baf180 in cardiac chamber development is discussed in later sections.

Histone Deacetylases and Vascular Development

Hdac7 is required for the development of embryonic vasculature (9). Mouse embryos lacking *Hdac7* die at E11.5 from vascular defects (9). The dorsal aortae and cardinal veins of *Hdac7*-null mice are dilated, leaky, and surrounded by fewer smooth muscle cells. Hdac7 normally interacts with Mef2 in the endothelium to repress *Mmp10* expression (9). In the absence of *Hdac7*, *Mmp10* is derepressed in the endothelium, releasing the Mmp10 matrix metalloproteinase to weaken the blood vessel wall. Concurrent with *Mmp10* upregulation, tissue inhibitor of metalloproteinase 1 (*TIMP1*) is downregulated in *Hdac7*-null endothelial cells, further enhancing Mmp10 activity and exacerbating vascular destruction.

Studies of HDACs in endothelial cell cultures suggest that HDACs may play a role during embryonic endothelial differentiation and angiogenesis. Silencing of *Hdac7* in endothelial cells alters endothelial cell morphology, migration, and capillary tube formation (17). In endothelial cells, Hdac7 complexes with β-catenin and prevents the translocation of β-catenin into the nuclei (18). Interestingly, vascular endothelial growth factor (Vegf) triggers the degradation of endothelial Hdac7 and the disruption of the Hdac7/β-catenin complex, thereby enhancing the nuclear localization of β-catenin (18). Hdac7 therefore plays a role in Vegf- and β-catenin-mediated endothelial cell growth and differentiation.

Sirtuins and Angiogenesis

Sirt1, highly expressed in the growing vasculature, is essential for the angiogenic activity of endothelial cells (19). Knockdown of *sirt1* in zebra fish causes abnormal endothelial cell migration, leading to defective angiogenesis, abnormal patterning of trunk vasculature, and spontaneous hemorrhages (19). In mice, however, disruption of *Sirt1* in endothelial cells does not cause apparent vascular phenotypes in embryos or in early postnatal life. Instead, the mutant mice display blunted neovascularization in response to hindlimb ischemia (19). Many genes involved in vascular remodeling or endothelial differentiation are misregulated when *Sirt1* is silenced in endothelial cells. Sirt1 deacetylates Foxo1 to inhibit the transcriptional activity and to thereby restrain the antiangiogenic activity of Foxo1 (19). Sirt1 also deacetylates Notch1 intracellular domain (NICD) to promote NICD degradation and to facilitate endothelial branching and proliferation during postnatal retinal angiogenesis in mice or embryonic angiogenesis in zebra fish (20). Therefore, Sirt1 regulates endothelial angiogenesis at least partly through its deacetylation of Foxo1 and NICD. Further studies are necessary to determine the role of histone deacetylation by Sirt1 during vascular development.

Hypb/Setd2 and Vascular Remodeling

Hypb/Setd2 is a histone H3K36 methyltransferase essential for vascular development (21). Hypb2 knockout mice die at E10.5–11.5 with defects in vascular remodeling. Capillaries of the mutant embryo and yolk sac are dilated and fail to remodel into larger blood vessels or to form intricate vascular networks. Because silencing of endothelial Hypb impairs endothelial cell migration and tubular formation, Hypb may function cell autonomously in endothelial cells to regulate vascular remodeling.

Vascular Shear Stress and Histone Modification

Shear stress induces histone modifications in cultured endothelial cells (22). Stress-induced histone changes include the acetylation of H3K14 within the promoters of c-*Fos* and c-*Jun*. Preventing H3K14 acetylation inhibits shear-dependent endothelial gene expression (22). Therefore, during embryonic vasculogenesis or angiogenesis, vascular shear stress may induce similar histone modifications on gene promoters to regulate endothelial gene expression.

GREAT ARTERY AND CARDIAC OUTFLOW TRACT

Septation of the arterial trunk and outflow tract of the embryonic heart to form the aorta and main pulmonary artery, as well as the left and right ventricular outlets, is essential to separate the systemic circulation from the pulmonary circulation. This septation process requires interactions among multiple cell types, including endothelial, smooth muscle, endocardial, myocardial, and neural crest cells. Abnormalities in the septation cause cardiac shunting, leading to heart failure and/or cyanosis. Several chromatin-regulating molecules are important in the septation of the great arteries and the cardiac outflow tract (**Table 2**).

Brg1 and Ductus Arteriosus

The ductus arteriosus is the arterial conduit connecting the main pulmonary artery and the descending aorta. The patency of the ductus arteriosus is essential for fetal circulation, and its closure in the neonatal period converts fetal circulation to adult circulation. Contraction of smooth muscle cells of the ductus arteriosus mediates this neonatal ductal closure. Persistence of patent ductus

Table 2 Great artery and cardiac outflow tract development

Class	Chromatin factor	Gene modification	Phenotype	Comments	Reference(s)
SWI/SNF	Brg1	Deletion in smooth muscle cells	One-third of mutant mice have patent ductus arteriosus and die of heart failure.	Brg1 interacts with myocardin to activate the expression of genes encoding smooth muscle contractile proteins.	23–26
		Deletion in secondary heart field myocardial progenitors	Hypoplastic outflow tract and right ventricle, lethality at E10.5	Brg1 activates *Bmp10* to promote cell proliferation of the outflow tract and the right ventricle.	40
CHD	Chd7	Heterozygous germline mutation in mice	Aortic arch interruptions in $Chd7^{+/-}$ mice	Chd7 interacts with *Tbx1* in pharyngeal endoderm to regulate aortic arch formation.	30
		Chd7 knockdown in frog embryos	Abnormal position of truncus arteriosus	CHD7 associates with BRG1/PBAF to control *Sox9*, *Twist*, *Slug*, and neural crest cell activation.	31
Histone demethylase	Jumonji/Jarid2	Germline deletion	Double-outlet right ventricle (DORV), myocardial defects	Defects may result from misregulation of cardiac gene expression because of repressive functions of Jarid2 and Jmjd6.	32, 33
	Jmjd6/Ptdsr	Germline deletion	Pulmonary artery hypoplasia, DORV, ventricular septal defects		34
Polycomb repressive complex 1	Rae28	Germline deletion	Tetralogy of Fallot, DORV, aortic valve stenosis	Rae28 maintains *Nkx2.5* expression. Rae28 overexpression in cardiomyocytes causes dilated cardiomyopathy.	36, 37
Histone methyltransferase	Mixed lineage leukemia 2 (MLL2)	Mutated in patients with Kabuki syndrome	Aortic coarctation, atrial and ventricular septal defects	MLL2 haploinsufficiency may impair histone methylation and reduce target gene expression.	132, 133

arteriosus (PDA) after birth causes left-to-right cardiac shunting, pulmonary hypertension, and congestive heart failure.

The BAF complex plays an important role in smooth muscle development. Knockdown of Brg1 or Brm reduces the expression of smooth muscle–specific contractile proteins in cultured smooth muscle cells (23, 24). Approximately one-third of mice lacking smooth muscle *Brg1* display cyanosis and dilated cardiac chambers within 3 days after birth (25) due to the presence of PDA. The intestines of *Brg1*-mutant mice have disorganized smooth muscle layers with reduced expression of smooth muscle contractile proteins. Although not directly tested, the mutant ductus arteriosus may have similar smooth muscle abnormalities, preventing its perinatal closure. Brg1 facilitates the binding of the myocardin family of SRF coactivators to gene promoters to induce the expression of

Patent ductus arteriosus (PDA): a heart defect in which the ductus arteriosus fails to close after birth

smooth muscle contractile genes (23, 24, 26), a crucial step in the maturation of ductus arteriosus muscle. Deletion of *Brm* alleles in mice, however, does not cause neonatal lethality or PDA, nor does it exacerbate the PDA phenotype of *Brg1*-mutant mice. Brg1 is therefore the dominant ATPase of the BAF complex for smooth muscle maturation.

Chd7 and Neural Crest Cell Development

Haploinsufficiency of CHD7, one of nine members of the CHD chromatin remodeler family (7), is the major cause of CHARGE syndrome (27). CHARGE syndrome, an autosomal dominant disorder, is characterized by coloboma, heart defects, atresia choanae, retarded growth and development, genital hypoplasia, and ear abnormalities/deafness (reviewed in References 28 and 29). Clinical features of CHARGE syndrome are variable and incompletely penetrant. Most cardiovascular anomalies are related to pharyngeal arch artery defects or cardiac outflow tract malformations such as tetralogy of Fallot, PDA, atrial septal defects, and double-outlet right ventricle (29). Some defects are recapitulated in $Chd7^{+/-}$ mice, which display aortic arch interruption because of hypoplastic pharyngeal arch arteries (30). CHD7 may function in neural crest cells to regulate cardiac outflow tract development (31). In frog embryos, knockdown of *Chd7* or overexpression of the dominant-negative form of Chd7 causes multiple defects characteristic of CHARGE syndrome, including outflow tract malformation. Studies using frog embryos or cultured cells suggest that CHD7 associates with BRG1/PBAF to control *Sox9*, *Twist*, and *Slug* expression and neural crest cell activation (31). Tissue-specific gene deletions in mice will be essential to determine whether Chd7 functions cell autonomously in neural crest cells to regulate great artery and cardiac outflow tract development.

Jumonji and Outflow Tract Septation

Jarid2/Jumonji and Jmjd6/Ptdsr are JmjC domain–containing histone demethylases that can silence gene expression. Both Jarid2 and Jmjd6 contribute to cardiac outflow tract septation. *Jarid2*-null mice have cardiac malformations that include myocardial noncompaction, ventricular septal defects, and double-outlet right ventricle (32, 33). *Jmjd6*-deficient mice show ventricular septal defects, pulmonary artery hypoplasia, and double-outlet right ventricle (34). The cardiac defects of *Jarid2*- or *Jmjd6*-null mice may result from misregulation of cardiac gene expression because of repressive functions of Jarid2 and Jmjd6.

Rae28 and Outflow Tract Septation

Rae28 is a member of the polycomb repressive complex 1 (PRC1), which catalyzes monoubiquitinylation of histone H2A and regulates gene silencing (35). Deletion of *Rae28* in mice causes tetralogy of Fallot, double-outlet right ventricle, and aortic valve stenosis (36, 37), indicating a requirement of Rae28 for cardiac outflow tract septation and valve development. Although Rae28 is essential for maintaining *Nkx2.5* expression (36), Rae28 does not seem to function in cardiomyocytes to regulate outflow tract development. Rae28 overexpression in cardiomyocytes causes dilated cardiomyopathy but does not rescue the outflow tract defects of *Rae28*-null mice (38). Besides outflow tract malformations, *Rae28* deletion causes many neural crest–related tissue defects in the palate, thyroid, and parathyroid (37), suggesting that Rae28 may regulate neural crest development to control cardiac outflow tract septation.

Pharyngeal arch artery defects: abnormalities in the remodeling of pharyngeal arch arteries, whose development leads to the formation of the aortic arch, common carotid, brachiocephic, ductus arteriosus, and subclavian arteries. Defects of pharyngeal arch artery remodeling lead to a variety of vascular malformations

Tetralogy of Fallot: a heart defect characterized by right ventricular outflow tract obstruction, right ventricular hypertrophy, ventricular septal defects, and overriding aorta

Atrial septal defect: a heart defect resulting from incomplete atrial septation

Double-outlet right ventricle: a heart defect in which both the aorta and the main pulmonary artery arise from the right ventricle

Aortic arch interruption: a heart defect in which a segment of the aortic arch artery is occluded or absent

Ventricular septal defect: a heart defect resulting from incomplete ventricular septation

HEART MUSCLE AND CHAMBER DEVELOPMENT

The formation of heart muscle and the cardiac chamber is essential for the pumping function of the heart. Abnormalities in heart muscle or chamber development result in cardiomyopathy and heart failure. Many different classes of chromatin regulators play crucial roles in the regulation of cardiac muscle differentiation, maturation, and morphogenesis (**Table 3**). The functions of these cardiac chromatin regulators are described below.

The BAF Complex and Heart Development

Many protein subunits of the BAF complex regulate heart muscle and chamber development. These BAF subunits interact with transcription factors and other chromatin regulators to control cardiac gene expression during heart muscle development.

Endocardial Brg1 and myocardial trabeculation. Brg1 acts both cell autonomously and non–cell autonomously to regulate heart muscle development. In the endocardium, Brg1 controls the expression of the secreted metalloproteinase *Adamts1* (a disintegrin and metalloproteinase with thrombospondin motif gene 1) to regulate the composition of cardiac jelly, which is extracellular matrix essential for myocardial trabeculation (8). *Adamts1* is normally repressed in the endocardium from E9.5 to E11.5, a developmental window with abundant cardiac jelly and active trabecular growth of the myocardium (8). Later, from E12.5 to E14.5, *Adamts1* is activated in the endocardium, releasing the protease into the extracellular space to degrade versican and other matrix proteins. This results in the dissipation of cardiac jelly with consequent termination of trabecular growth at E13.5–14.5 (8). However, in the absence of endocardial Brg1, *Adamts1* is prematurely activated at E9.5 in the endocardium, leading to the early degradation of versican and hyaluronan, disappearance of cardiac jelly, and thus hypotrabeculation of the myocardium (8). The transcriptional repression of *Adamts1* by Brg1 at E9.5 allows for the establishment of cardiac jelly to support the initiation of myocardial trabeculation during early heart development. In contrast, the derepression of *Adamts1* from E12.5 to E14.5 triggers cardiac jelly degradation to terminate trabeculation. Such Brg1-based dynamic regulation of *Adamts1* is important because either inadequate or excessive trabeculation can cause cardiomyopathy and heart failure (39). This is an important example of how one tissue layer (the endocardium) can regulate the morphogenesis of an adjacent tissue layer (the myocardium) through a secreted metalloproteinase to modulate the extracellular matrix microenvironment. To further understand the molecular mechanism of such chromatin-based developmental regulation, it will be essential to determine which transcription factors Brg1 interacts with to repress *Adamts1* and how *Adamts1* is derepressed later in heart development.

Myocardial Brg1 in cardiac growth, differentiation, and hypertrophy. Brg1 also functions in the myocardium to control myocardial gene expression, tissue growth, and differentiation (40, 41). Early deletion of *Brg1* in developing myocardium leads to severe defects in heart development that are associated with disruption of the expression of important regulators of heart formation (41). Slightly later deletion leads to thin myocardium and absent interventricular septum (40). These defects of the myocardium are caused by a failure of myocardial cell proliferation due to a deficiency in Bmp10 and ectopic expression of $p57^{kip2}$, a cyclin-dependent kinase inhibitor that prevents cell cycle progression (40). Bmp10 normally suppresses $p57^{kip2}$ expression in the developing myocardium to promote cell proliferation. Because Bmp10 can rescue the proliferation defects of myocardial *Brg1*-null embryos, Bmp10 functions downstream of Brg1 to control myocardial proliferation. Beyond controlling cell proliferation through Bmp10, Brg1 interacts with other factors to maintain embryonic cardiomyocytes in a fetal state of differentiation. Brg1 forms an epigenetic complex with two other classes of chromatin-modifying factors—HDACs and PARPs—to transcriptionally repress α-myosin heavy chain (α-*MHC*) and activate β-*MHC* expression in embryonic hearts (40). In mice, α-*MHC* is the primary *MHC* isoform expressed in adult heart ventricles, whereas β-*MHC* is the major isoform in embryonic heart ventricles. By repressing α-*MHC* and activating β-*MHC*, the Brg1/HDAC/PARP protein complex maintains a fetal state of *MHC* expression (40). In the absence of Brg1, embryonic cardiomyocytes switch *MHC* expression from β-*MHC* to α-*MHC*, indicating a premature differentiation of embryonic cardiomyocytes. Inhibition of HDAC or PARP activity in cultured embryos triggers a premature MHC switch from β-*MHC* to α-*MHC* expression without causing myocardial proliferation defects. In contrast, Bmp10 rescues myocardial proliferation defects without affecting the premature MHC switch of myocardial *Brg1*–null embryos (40). These observations indicate that Brg1 commands two separate molecular pathways to independently regulate myocardial cell proliferation and differentiation in embryos.

The control of *MHC* isoform expression by Brg1, first observed in embryonic heart development, is mechanistically relevant to the pathogenesis of adult hypertrophic cardiomyopathy (40). Brg1, although highly expressed in embryos, is downregulated in adult cardiomyocytes in mice, coinciding with the physiological switch to α-MHC in adult hearts. However, when the adult heart is stressed by pressure overload, Brg1 is reactivated in the cardiomyocytes to assemble a fetal epigenetic complex—Brg1/HDAC/PARP—that represses α-*MHC* and activates β-*MHC*, causing a pathological switch from α-MHC to β-MHC in stressed adult hearts. Deletion of *Brg1* in adult hearts to prevent stress-induced Brg1 reactivation reduces cardiac hypertrophy, abolishes cardiac fibrosis, and reverses the pathological MHC switch. Interestingly, upregulation of *BRG1* is also observed in heart tissues obtained from patients with hypertrophic cardiomyopathy, and the level of BRG1 upregulation correlates strongly with the severity of hypertrophy and MHC switch (40). These findings thus suggest a causal role of Brg1-mediated chromatin regulation in the pathogenesis of adult cardiomyopathy and an epigenetic link between fetal heart development and adult heart disease.

It remains unclear how Brg1 differentially regulates the expression of α-*MHC* and β-*MHC*. Chromatin immunoprecipitation studies show that Brg1 binds to the promoters of both α- and β-*MHC* in fetal hearts as well as in adult myopathic hearts. The differential activities of Brg1 on *MHC* promoters may result from its association with different transcription factors, histone modifiers, or other chromatin remodelers.

Brg1 and right heart development. Deletion of *Brg1* in precursors of the right ventricle and cardiac outflow tract results in right ventricular chamber hypoplasia and outflow tract shortening at E10.5, and the mutant embryos die at E11.5 (40). In contrast, deletion of *Brg1* in myocardial cells after initial cardiomyocyte differentiation and expression of Sm22α causes thin myocardium in the ventricles without apparent reduction of right ventricular chamber size (40). The difference in the right ventricular phenotypes may indicate an early time window when Brg1 acts in the secondary heart field to regulate right ventricular and cardiac outflow tract morphogenesis. It will be interesting to determine what downstream genes, other than those encoding Bmp10 and p57^{kip2}, in the myocardial precursors are controlled by Brg1 to regulate right heart development.

Brg1 and zebra fish heart development. The role of Brg1 in heart muscle development is conserved in zebra fish (41). The characterization of the *young* mutant, in which *brg1* is mutated, reveals important and specific roles for brg1 in zebra fish heart development. Mutation of *brg1* in

Table 3 Heart muscle and chamber development

Class	Chromatin factor	Gene modification	Phenotype	Comments	Reference(s)
SWI/SNF	Brg1	Deletion in the endocardium	Hypotrabeculation, absent cardiac jelly, lethality at E10.5–11.5	Brg1 represses *Adamts1* to prevent premature degradation of cardiac jelly and to promote myocardial trabeculation.	8
		Deletion in the myocardium	Thin myocardium, absent interventricular septum, lethality at E11.5	Brg1 activates *Bmp10* to promote myocardial proliferation. The Brg1/HDAC (histone deacetylase)/PARP [poly(ADP ribose) polymerase] complex controls *MHC* isoform expression.	40
		Deletion in secondary heart field myocardial progenitors	Hypoplastic right ventricle and outflow tract, lethality at E10.5	Brg1 activates *Bmp10* to promote myocardial proliferation of the right ventricle and the outflow tract.	40
		Point mutation, morpholino knockdown in zebra fish	Abnormal looping, hypoplastic myocardium, reduced contractility	Conserved Brg1 function in zebra fish heart development	41
	Baf45c	Knockdown in zebra fish	Abnormal looping, reduced contractility, disarrayed muscle fibers	Baf45c knockout mice have no significant developmental defects.	47, 50
	Baf60c	Knockdown in mouse hearts	Abnormal looping, hypoplastic ventricles, shortened outflow tract, lethality at E10–11	Baf60c regulates many cardiac genes to control heart development. Baf60c activates Notch and *Nodal* for looping.	42, 52
		Overexpression in noncardiac mesoderm	Differentiation of noncardiac mesoderm into cardiomyocytes	Baf60c interacts with Tbx5, Nkx2.5, and Gata4 to activate early cardiogenesis.	51
		Overexpression in zebra fish	Enlarged heart	Baf60c interacts with Gata to activate cardiac gene expression.	53
	Baf180	Germline deletion	Hypoplastic ventricles, ventricular septal defects, coronary defects, lethality at E12.5–15.5	Baf180 regulates the retinoid acid pathway to control cardiac chamber development.	15
INO80	Pontin (Rvb1, Tip49, Tip49a) Reptin (Rvb2, Tip48, Tip49b)	Knockdown of pontin or activation of reptin in zebra fish	Cardiac muscle hyperplasia	Pontin and Reptin antagonistically regulate the β-catenin pathway to control myocardial proliferation.	61

Category	Gene	Manipulation	Phenotype	Description	References
Class I HDAC	Hdac1	Deletion in the myocardium	No apparent cardiac defects	Hdac1 is redundant with Hdac2.	78
	Hdac2	Germline deletion	Thickened myocardium, hypotrabeculation, perinatal lethality	Hdac2/Hopx suppress SRF- and Gata4-dependent gene expression to inhibit cardiac proliferation.	78, 80, 84
		Deletion in the myocardium	No apparent cardiac defects	Hdac2 is redundant with Hdac1.	78
		Deletion of Hdac1 and Hdac2 in the myocardium	Arrhythmias, dilated cardiomyopathy, lethality within 2 weeks after birth	Hdac1 and Hdac2 repress fetal genes involved in calcium handling and contractility.	78
	Hdac3	Deletion in the myocardium	Cardiac hypertrophy, abnormal fatty acid metabolism, lethality at 3–4 months of age	Hdac3 suppresses peroxisome proliferator–activated receptor activity on the promoters of metabolic genes.	87
		Overexpression in the myocardium	Cardiac hyperplasia	Hdac3 suppresses the expression of several cell cycle inhibitors.	90
Class II HDAC	Hdac5, Hdac9	Germline deletion of both Hdac5 and Hdac9	Frequent ventricular septal defects, occasional thin myocardium, variable lethality starting at E15.5	Hdac5 and Hdac9 are functionally redundant. There is variable penetrance of lethality and cardiac phenotypes.	91
Sirtuin	Sirt1	Germline deletion	Atrial and ventricular septal defects, valve defects, lethality at birth	Sirt1 deacetylates p53, preventing p53 from triggering cell apoptosis.	93, 94
Histone acetyltransferase	p300	Germline deletion or point mutation that ablates p300 acetyltransferase activity	Thin myocardium, hypotrabeculation, atrial and ventricular septal defects	p300 is essential for cell proliferation and is a transcriptional coactivator of Gata4.	97
Histone demethylase	Jarid2/Jumonji	Germline deletion	Hypertrabeculation, noncompaction, double-outlet right ventricle	Variable lethality and embryonic defects depend on the strain.	32, 33, 108
		Deletion in the endocardium	Hypertrabeculation	Jarid2 suppresses endocardial Notch1 and Nrg1 to regulate trabeculation.	109
Histone methyltransferase	Smyd1	Germline deletion	Malformed right ventricle	Smyd1 binds skNAC and regulates Hand2 and Irx4.	120, 124
		Knockdown in zebra fish	Disrupted myofibril formation	Smyd1 is required for myofibril organization.	122
	Wolf-Hirschhorn syndrome candidate 1 (Whsc1)	Germline deletion	Atrial and ventricular septal defects, lethality within 10 days after birth	Whsc1 associates with Nkx2.5 to repress gene transcription.	131
	Mixed lineage leukemia 2 (MLL2)	Mutated in patients with Kabuki syndrome	Atrial and ventricular septal defects, aortic coarctation	MLL2 haploinsufficiency may impair histone methylation and reduce target gene expression.	132, 133
	Dot1-like protein (Dot1L)	Germline deletion	Cardiac dilatation, yolk sac angiogenesis defects, lethality at E9.5–10.5	Dot1L methylates H3K79 and regulates telomere elongation and cell proliferation.	136
PARP	Parp-1, Parp-2	Germline deletion	Parp-1 and Parp-2 double-knockout mice die at E7–8.	Parp binds with Brg1 to control MHC isoform expression in embryonic hearts and in stressed adult hearts.	40, 148

fish causes abnormal cardiac looping, narrowing of the cardiac chamber, and reduction of cardiac contractility. At a cellular level, mutation of *brg1* in zebra fish results in malaligned and abnormally shaped cardiomyocytes. These myocardial phenotypes are consistent with those observed in mice when *Brg1* is deleted in the heart, indicating a highly conserved role for BAF complexes in heart development.

Overriding aorta: a heart defect in which the aorta straddles and receives blood from both the left and right ventricles

Interaction of Brg1 and cardiac transcription factors. Dosage of Brg1 is critical for normal heart development. $Brg1^{+/-}$ mice survive at a sub-Mendelian ratio with variable heart defects, including ventricular septal defects, patent foramen ovale, conduction abnormalities, and cardiac dilatation (42). Congenital heart defects in humans can result from haploinsufficiency of transcription factors such as TBX5, NKX2-5, and GATA4 (43). Brg1 interacts with these factors to effect cardiac gene expression (42). Genetically, *Brg1* interacts with cardiac transcription factors in a dose-dependent manner to regulate heart development in mice (41). When $Brg1^{+/-}$ is introduced along with an *Nkx2.5* or *Tbx20* heterozygous mutation that alone causes mild or no cardiac phenotypes (44–46), severe cardiac defects with atrial and ventricular septal defects occur, and none of the compound mutants survive beyond E14.5. Similarly, compound *Brg1* and *Tbx5* heterozygous mice exhibit more severe cardiac defects than do mice with either mutation alone. Transcriptional profiling of E11.5 hearts with *Brg1*, *Tbx5*, or *Nkx2.5* heterozygous mutations suggests that Brg1 and these transcription factors govern a complex genetic program to regulate gene expression, depending on the dosage of Brg1 and transcription factors on the promoters of genes essential for heart development (41). These findings strongly suggest that the molecular mechanism underlying transcription factor haploinsufficiency in congenital heart disease is impaired recruitment of chromatin-remodeling complexes.

Other BAF Protein Subunits and Heart Muscle Development

The composition of the BAF complex may be very diverse, varying in different cell types and under different pathophysiological conditions. The actual protein subunits within the BAF complex may influence how BAF interacts with different transcription or chromatin regulators and how BAF interacts with chromatins on genomic loci. The temporal and spatial activities of a BAF subunit may determine the functional specificities of the BAF complex during cardiovascular development. Therefore, it is important to understand the roles of various BAF subunits in heart development.

Baf45c. Baf45c (or Dpf3) is a subunit of the BAF complex that contains two PHD (plant homeodomain) domains, which recognize acetylated and methylated lysine residues of histones H3 and H4 (47, 48). BAF45c/DPF3 expression is increased in the hypertrophic right ventricular myocardium of patients with tetralogy of Fallot (49), which is characterized by overriding aorta, right ventricular outflow tract obstruction, ventricular septal defects, and right ventricular hypertrophy. The outflow tract obstruction causes pressure overload and consequent hypertrophy of the right ventricle in these patients. Because pressure overload activates Brg1/BAF to promote cardiac hypertrophy (40), the elevation of BAF45c is likely essential for the development of right ventricular hypertrophy in patients. Further studies are necessary to define if BAF45c is a crucial component of the BAF complex that controls cardiac hypertrophy in congenital and acquired heart diseases.

Studies in zebra fish suggest a role for Baf45c in heart muscle development. Morpholino knockdown of *dpf3* in zebra fish results in abnormal cardiac looping, a poorly defined atrioventricular boundary, reduced cardiac contractility, and disarrayed muscle fibers (47). These phenotypes are similar to those of zebra fish with Brg1 knockdown or mutation (41), suggesting an essential role of the Brg1 and Baf45c subunits of the BAF complex in zebra fish heart development. However, loss of Baf45c in mice does not result in any significant developmental defects (50).

BAF45c/DPF3 may recognize modified histones and help recruit BRG1 to muscle-relevant gene loci that contain acetylated or methylated H3 and H4 (47). The binding of DPF3 to histone H3 is regulated by two different histone modifications: H3K14ac (lysine-14-acetylated histone H3) and H3K4me3 (lysine-4-methylated histone H3) (48). H3K14ac recognized by PHD1 promotes such binding, whereas H3K4me3 recognized by PHD2 inhibits such binding. This regulation produces two opposing effects on the transcriptional activation of DPF3 target genes (48). Studies support a model in which H3K14 acetylation marks a gene locus for recruiting DPF3/BAF to preinitiate gene transcription, whereas H3K4 methylation dissociates DPF3/BAF from the locus, allowing the entry of transcriptional machinery to initiate and activate gene transcription (48). These findings suggest that H3K14 and H3K4 modifications are essential for the expression of BAF target genes essential for heart development.

Baf60c. Baf60c is a facultative member of BAF complexes that has no known biochemical function. Baf60c is required for early cardiogenesis (51), cardiac looping (52), and subsequent heart morphogenesis (42). Baf60c directly interacts with transcription factors Tbx5, Nkx2.5, and Gata4 to activate cardiac-specific genes for cardiogenesis. The interaction of Baf60c with key cardiac transcription factors presumably allows for the efficient recruitment of BAF complexes to target loci. Ectopic expression of *Baf60c*, *Tbx5*, and *Gata4* in noncardiogenic mesodermal cells induces these cells to differentiate into beating cardiomyocytes, indicating an important instructive role for BAF complexes in cardiac differentiation (51). These findings have been extended in zebra fish, in which Baf60c not only interacts with GATA factors to induce cardiac gene expression but also allows cardiac precursors to migrate to their normal location in the developing heart (53).

Baf180. Baf180 (polybromo), a subunit of the BAF complex, contains six bromodomains that recognize acetylated histone tails. In the absence of Baf180, mouse embryos die between E12.5 and E15.5 with placental defects, ventricular septal defects, and hypoplastic ventricles (15). These phenotypes are similar to those observed in embryos lacking RXRα, a component of the retinoic acid (RA) pathway (54, 55). Baf180 is required for the expression of a subset of RA target genes such as those encoding RARβ2 and CRABPII (15), suggesting that Baf180 interacts with the RA pathway to regulate cardiac chamber formation. A tissue-specific gene deletion of *Baf180* in different tissues of the heart such as endocardium and myocardium will further elucidate the spatial roles of Baf180 in heart muscle development.

INO80 and Heart Muscle Development

The INO80 chromatin-remodeling complexes include INO80, SRCAP, and p400/Tip60 (reviewed in References 5, 7, and 56). These multiprotein complexes contain two DNA helicases—Pontin (alias Rvb1, Tip49, Tip49a) and Reptin (alias Rvb2, Tip48, Tip49b)—whose activities are required for INO80 complex–medicated chromatin remodeling (57–60). Pontin and Reptin regulate heart muscle development in zebra fish (61), suggesting a role of the INO80 complex in heart development. In zebra fish, both morpholino knockdown of pontin and a gain-of-function mutation of the gene encoding reptin cause hyperplasia of cardiomyocytes (61). This antagonistic function of pontin and reptin during heart muscle development appears to be mediated partly through their opposing effects on the β-catenin pathway (61). It remains to be determined whether Pontin, Reptin, or the INO80 complex in general plays any role in mammalian heart development.

ISWI Complex in Heart Muscle Development

Five ISWI chromatin-remodeling complexes—NURF, ACF, CHRAC, NoRC, and WICH—are present in mammals (see References 7, 62, and 63 for reviews). Two different types of ATPases, SNF2H and SNF2L, are present in the different ISWI complexes (62). *Snf2h*-null mouse embryos die between E5.5 and E7.5 (64), precluding the study of Snf2h in cardiovascular development. However, indirect evidence suggests that Snf2h is involved in heart development. The Snf2h-containing WICH complex includes Williams syndrome transcription factor (Wstf), which is essential for heart development. *Wstf* mutations in mice cause aortic coarctation, ventricular hypotrabeculation, and ventricular and atrial septal defects (65). Tissue-specific deletion of *Snf2h* in mice will be necessary to define the role of Snf2h in cardiovascular development.

Histone Deacetylases and Heart Muscle Development

HDACs are a class of chromatin modification factors that remove acetyl groups from conserved lysine residues of histone H3 and H4 N-terminal tails, and they usually repress gene expression (66, 67). HDACs are classified into four subfamilies (classes I, IIa, IIb, and IV) on the basis of their protein structure, enzymatic activity, subcellular localization, and expression pattern (67, 68). Class I and IIa proteins are regarded as classic HDACs and are the most extensively studied. Class I HDACs are widely expressed, reside predominantly in the cell nucleus, and display high enzymatic activity toward histone substrates. Class IIa HDACs have a more restricted pattern of expression, function as signal transducers that shuttle between the cytoplasm and the nucleus, and possess only weak enzymatic activity (69–72). Most class I HDAC proteins are subunits of multiprotein nuclear complexes crucial for transcriptional repression. In contrast, gene repression by class IIa HDACs is mediated by their interactions with class I HDACs and other transcriptional repressors such as nuclear receptor corepressor (N-CoR) and silencing mediator of RA and thyroid hormone receptors (SMRT), heterochromatin protein 1, and C-terminal-binding protein (69, 73–76).

Class I HDACs consist of HDAC1, -2, -3, and -8 proteins. HDAC1 and HDAC2 often coexist in a variety of repressive complexes, including the Swi-independent 3, nucleosome remodeling and deacetylase, corepressor of RE1-silencing transcription factor (REST, alias neuron-restrictive silencer factor), and polycomb repressive complex 2 (PRC2) complexes (67). HDAC3 proteins are associated with two homologous transcriptional corepressors, N-CoR and SMRT (77, 78), as well as with the retinoblastoma (Rb) complex (79). No complex has been described for HDAC8 (40, 80). Class IIa HDACs contain HDAC4, -5, -7, and -9 proteins (9). Class IIa proteins have large N-terminal domains with binding sites for the MEF2 transcription factor and the 14-3-3 chaperone proteins, and these HDACs shuttle between the cytoplasm and the nucleus. Class III HDACs, also termed sirtuins, have no sequence similarity to class I and II HDACs. Sirtuins require NAD^+ for deacetylation, and a key function of sirtuins is the regulation of transcriptional repression. Several HDAC genes have been knocked out in mice, revealing the important role of HDAC proteins in heart muscle development.

HDAC1 and HDAC2. Hdac1 and Hdac2 have redundant functions in the myocardium for the regulation of cardiac gene expression and cardiomyocyte differentiation (78). *Hdac1*-null mice die in utero before E10.5 with reduced cell proliferation and general growth retardation, possibly due to increased expression of cyclin-dependent kinase inhibitors p21 and p27 (78). Myocardium-specific deletion of either *Hdac1* or *Hdac2* in mice results in no apparent cardiac phenotype, and mutant mice survive to adulthood (78). However, mice lacking both *Hdac1* and *Hdac2* in the myocardium die within 2 weeks after birth due to cardiac arrhythmias and dilated cardiomyopathy (78). These abnormalities are likely caused by the upregulation of genes that encode fetal calcium channels and skeletal muscle–specific contractile proteins (78). These fetal genes, including hyperpolarization-activated nonselective cation current (I_f), the T-type Ca^{2+} current ($I_{Ca,T}$), and α-skeletal actin (α-SA), are transcriptionally repressed in the normal adult myocardium by REST through the recruitment of class I and class IIa HDACs (81–83). Mice overexpressing dominant-negative REST phenocopy *Hdac1/2* doubly null mice and develop dilated cardiomyopathy, ventricular arrhythmias, and sudden cardiac death (81). The combined loss of Hdac1 and Hdac2 may therefore result in the failure of REST to repress fetal genes involved in calcium handling and contractility, thereby causing cardiac arrhythmias and cardiomyopathy.

Global deletion of *Hdac2* in mice results in neonatal lethality within 24 h after birth (78). Similarly, *Hdac2*-deficient mice produced from a genetrap embryonic stem cell line exhibit partial perinatal lethality (80). Myocardium is thickened in these mice, suggesting that Hdac2 inhibits myocardial growth. This inhibition of myocardial growth requires the homeodomain-only protein (Hop, also known as Hopx) (84). Hdac2 and Hop form a repressive complex to inhibit SRF-dependent gene expression that contributes to myocardial proliferation and differentiation (85, 86). Also, the Hdac2/Hop complex deacetylates Gata4 to reduce its transcriptional activity on target genes essential for myocardial proliferation (84). Although these findings indicate that the Hdac2/Hop complex is a cell autonomous regulator of myocardial growth in embryos, the absence of cardiac phenotype in myocardium-specific *Hdac2* knockout mice (78) suggests an additional, non–cell autonomous role of Hdac2 for heart muscle development. Tissue-specific deletion of *Hdac2* will provide the answers.

HDAC3. Mice with myocardium-specific deletion of *Hdac3* develop severe cardiac hypertrophy and fibrosis and die at approximately 3–4 months of age (87). Mutant hearts show lipid accumulation due to changes in metabolic gene expression. In mutant mice, many genes involved in fatty acid uptake, β-oxidation, and oxidative phosphorylation are upregulated, whereas genes that govern glucose utilization are downregulated. Loss of Hdac3 on the promoters of metabolic genes may disinhibit peroxisome proliferator–activated receptor α (PPARα) activity on those promoters (87), thereby leading to abnormal gene expression with consequent metabolic derangements. Mice with deletion of *Hdac1*, *Hdac2*, or other *Hdac* genes (78, 87) do not display such metabolic abnormalities, indicating a unique role of Hdac3 in myocardial energy metabolism. PPARs are central regulators of cardiac fatty acid metabolism (88, 89), and thus these studies suggest that PPARs recruit Hdac3 to the promoters of metabolic genes to facilitate transcriptional repression. Because Hdac3 is a component of the N-CoR/SMRT and Rb repressive complexes (77–79), it will be interesting to investigate the roles of Hdac3 and these complexes in cardiac metabolism during embryonic heart development.

Hdac3 overexpression in cardiomyocytes stimulates myocardial proliferation and produces myocardial thickening in newborn mice (90). In neonatal hearts, Hdac3 suppresses the expression of several cyclin-dependent kinase inhibitors, including Cdkn1a, Cdkn1b, Cdkn1c, Cdkn2b, and Cdkn2c (90). These studies support a role of Hdac3 as a regulator of myocardial proliferation.

HDAC5 and HDAC9. Hdac5 and Hdac9 are functionally redundant for heart muscle development. *Hdac5* or *Hdac9* knockout mice survive to adulthood without apparent cardiac defects (91), whereas compound *Hdac5*- and *Hdac9*-null mutations cause lethality starting at E15.5, with few mutant mice surviving to postnatal day 7 (91). Most compound mutants exhibit ventricular septal defects, and some have thin myocardium. These myocardial defects may result from enhanced transcriptional activities of Mef2, SRF, myocardin, or CAMTA2 (92) due to the loss of Hdac suppression.

SIRT1. Deletion of *Sirt1* in mice causes perinatal lethality and cardiac malformations (93, 94), including atrial septal, ventricular septal, and heart valve defects. Sirt1 deacetylates p53, preventing p53 from triggering cellular senescence and apoptosis due to DNA damage and stress (93, 95). It is unclear whether this pathway also contributes to heart development. Also, recent studies show that Sirt1 deacetylates Foxo1 and NICD to regulate angiogenesis (19, 20). It will be interesting to see whether such Sirt1 functions are essential for heart development.

Histone Acetyltransferases and Heart Muscle Development

HATs add acetyl groups to histone tails, resulting in a decrease in histone-histone and histone-DNA interactions and thereby loosening the chromatin and increasing transcriptional activation. HATs counteract HDACs to determine the acetylation state of chromatin and the level of gene expression (96).

Mouse embryos lacking *p300*, a well-defined HAT member, die between E9.5 and E11.5 (97). These embryos have a poorly vascularized yolk sac, severe pericardial effusion, and myocardial defects that include thin myocardium, diminished trabeculation, and low expression of MHC and α-actinin (97). *p300*-null embryos have a generalized reduction in cell proliferation (97), which probably underlies the myocardial growth defects. Furthermore, a single amino acid mutation of p300, which ablates its acetyltransferase activity, results in embryonic lethality between E12.5 and E15.5 and in multiple heart defects, including thin compact myocardium, ventricular and atrial septal defects, and reduced coronary vessels (98). Misregulation of the transcriptional activity of Gata4, a transcription factor essential for heart development, may contribute to heart defects in these mice. p300 acts as a transcriptional coactivator of Gata4 by acetylating Gata4 to enhance its DNA binding and transcriptional activity (99, 100). Therefore, p300 may control heart muscle development partly through its acetylation of Gata4. The role of histone acetylation by p300 during heart development is not yet clear, although p300 occupancy correlates with the acetylation of several cardiac gene promoters that are differentially expressed between the left and right ventricles (101).

Histone Methylation and Heart Development

Histone methylation at arginine and lysine residues offers an important avenue of epigenetic control. The effect of histone methylation on gene expression is context dependent. Histone H3 can be methylated on K4, K9, K27, K36, and K79. The methylations of K4, K36, and K79 are associated with transcriptional activation, whereas K9 and K27 methylations correlate with gene repression. Histone methylation has not been clearly implicated in cardiovascular development, but mouse genetics findings indicate that histone methylation is a major regulator of heart development, as it is in other organ systems.

Jumonji. Jarid2/Jumonji is the founding member of the Jumonji family of histone demethylases that contain the conserved Jumonji C (JmjC) domain (see Reference 102 for a review). Although the JmjC proteins generally function as histone demethylases, Jarid2 contains mutations at key amino acids necessary for enzymatic function and may not have demethylase activity (102–107). Mice lacking *Jarid2* exhibit various degrees of lethality and embryonic defects, depending on the genetic background. *Jarid2* mutations may cause neurulation defects, double-outlet right ventricle, and myocardial hypertrabeculation (32, 33, 108). Jarid2 functions in the endocardium to control myocardial trabeculation (109). Jarid2 transcriptionally represses the expression of *Notch1* and its downstream gene *Nrg1* (Neuregulin1) in the endocardium (109). Because both Notch1-RBPJk

(110, 111) and Nrg1-ErbB2/ErbB4 (112–114) signals are essential for myocardial trabeculation, enhanced *Notch1* and *Nrg1* expression in *Jarid2*-mutant mice may trigger excessive trabeculation. Such trabecular hyperplasia may be caused by enhanced Bmp10 activated by endocardial Notch signaling (40, 110, 115) and by increased cyclin D1 activated by Nrg1-ErbB2/B4 signaling (116). Jarid2 may therefore suppress endocardial Notch1 and Nrg1 to prevent excessive trabeculation of the myocardium.

Regulation of cardiac jelly composition may also contribute to hypertrabeculation in embryos lacking endocardial *Jarid2*. Endocardial cells are essential for the establishment of cardiac jelly matrix in the space between the endocardium and the myocardium to promote myocardial trabeculation (8). In the absence of endocardial Brg1, *Adamts1* is derepressed in the endocardium, causing premature jelly degradation and thus hypotrabeculation (8). Strikingly, in contrast to the endocardial *Brg1* mutation, the absence of endocardial *Jarid2* results in excessive cardiac jelly formation associated with abundant trabecular growth (109). These contrasting phenotypes suggest that the extent of cardiac jelly and myocardial trabeculation is determined by a balance between Jarid2 and Brg1 actions in the endocardium. Because both hypotrabeculation and hypertrabeculation cause human cardiomyopathy, it will be essential to determine how Jarid2 counteracts Brg1 in the endocardium to control cardiac jelly production and trabecular growth.

Jmjd6/Ptdsr is another JmjC domain–containing protein essential for heart development. *Jmjd6*-deficient mice show a delay in terminal differentiation of multiple organs (117) and perinatal lethality due to cardiac malformations that include ventricular septal defects, pulmonary artery hypoplasia, and double-outlet right ventricle (34). The cardiac defects of *Jmjd6*-null mice may result from the misregulation of cardiac gene expression due to the loss of Jmjd6-mediated gene repression.

SMYD. Smyd proteins are HMTs that contain conserved SET domains required for methyltransferase activity (118) and an MYND domain that mediates interaction with HDACs to repress gene expression (119). Two members of the Smyd family have been studied in heart development. Deletion of *Smyd1/Bop* in mice causes cardiac enlargement and embryonic lethality (120). *Smyd1*-null mice have a hypoplastic right ventricle with reduced expression of *Hand2* and *Irx4*, two genes required for myocardial growth and differentiation (120, 121). In zebra fish, *smyd1* knockdown produces defects in the function and structure of cardiac and skeletal muscle (122). Smyd1 is a transcriptional target of Mef2c (123), and Smyd1 interacts with the muscle-specific transcription factor skNAC to control gene expression (124, 125). Like *Smyd1*-null mice, *skNAC* knockout mice display ventricular hypoplasia with decreased cardiomyocyte proliferation (124), indicating that Smyd1/skNAC interactions are essential for heart muscle development. *Smyd2*, although highly expressed in cardiomyocytes, is dispensable for heart muscle development and function (126). Myocardial deletion of *Symd2* causes no observable cardiac abnormalities, suggesting redundancy among Smyd family members.

WHSC1. The Wolf-Hirschhorn syndrome candidate (*WHSC1*) gene encodes an HMT that trimethylates lysine 36 of histone H3 (H3K36me3) among other histone targets (127). *WHSC1* resides in a region of human chromosome 4 (4p16.3) termed the Wolf-Hirschhorn critical region (128), deletion of which is associated with Wolf-Hirschhorn syndrome. This syndrome is characterized by mental retardation, epilepsy, craniofacial abnormalities, and cardiac septal defects (129, 130). Most *Whsc1* knockout mice die neonatally with growth retardation, bone defects, and heart abnormalities (atrial and ventricular septal defects) (131). Heterozygous *Whsc1* mice display at a low frequency certain features of Wolf-Hirschhorn syndrome. Such *Whsc1* haploinsufficiency is enhanced by concomitant *Nkx2.5* mutations. One-third of $Whsc1^{+/-};Nkx2.5^{+/-}$ mice exhibit atrial and ventricular septal defects, whereas mice with a single $Whsc1^{+/-}$ or $Nkx2.5^{+/-}$ mutation have minimal or no cardiac lesions (131). Whsc1 associates with Nkx2.5 and occupies genetic loci to repress gene transcription, possibly through H3K36me3 modifications (131). Whsc1 therefore functions with cardiac transcription factors to regulate heart development.

MLL2. Mixed lineage leukemia (MLL) complexes methylate lysine 4 of histone H3. A role for MLL complexes in heart development has been suggested from the finding that *MLL2* is mutated in patients with Kabuki syndrome, which is characterized by a spectrum of birth defects that include cardiac septation defects (132, 133). Several functions have been identified, and it is thought that the mechanism of disease is haploinsufficiency of *MLL2*, which would be predicted to result in impaired histone methylation and therefore reduced expression of MLL2 target genes. It is not known how MLL2 is recruited to specific genes, but one of its partner proteins, Ptip, is an important mediator of this gene-specific recruitment. Loss of Ptip in the adult heart leads to arrhythmias because of abnormal regulation of the ion channel subunit–encoding gene *Kcnip2* (134).

DOT1L. The DOT1-like (DOT1L) protein is a SET domain–negative methyltransferase that catalyzes the methylation of H3K79 (135). H3K79 methylation is globally lost in Dot1L-deficient embryonic stem cells that display aberrant telomere elongation and proliferation defects (136). Dot1L-null mice are stunted in growth and die at E9.5–10.5 with cardiac dilatation and angiogenesis defects in the yolk sac (136). Further studies are needed to determine whether such cardiac dilatation is a primary defect or is secondary to yolk sac angiogenesis defects.

Polycomb Repressive Complex and Heart Development

Gene silencing by the polycomb repressive complex relies mostly on the regulation of chromatin structure, partly through histone modification (see Reference 138 for a review). PRC1 is responsible for monoubiquitylation of H2AK119 through Ring1A and Ring1B ubiquitin ligases, whereas PRC2 methylates H3K27 through its enzymatic subunits Ezh1 and Ezh2. The PRC1 component Pc (known as CBX in mammals) binds specifically to H3K27me3, the product of PRC2 catalysis, suggesting an interaction between PRC1 and PRC2 to regulate gene expression. The Rae28 component of PRC1 is essential for cardiac outflow tract septation as described above (36, 37). The SUMO-specific protease 2 (Senp2), which desumoylates Pc2/CBX4 and reduces its recruitment to H3K27me3 sites of gene promoters, is essential for heart development (137). Mice lacking Senp2 die at E10 with hypoplastic endocardial cushions and thin myocardium (137). Senp2 disruption enhances PRC1 occupancy on *Gata* promoters, leading to the repression of *Gata4* and *Gata6* (137), two genes required for heart development (139–142). Misregulation of *Gata* genes may thus contribute to the endocardial cushion and myocardial defects in *Senp2*-null mice. These studies suggest an important role of PRC1 in heart development. Future studies are essential to further determine the respective roles of PRC1 and PRC2 and their interactions during cardiovascular development.

Poly(ADP-Ribose) Polymerases and Heart Muscle Development

PARPs catalyze the transfer of ADP-ribose units from NAD^+ to the carboxyl group of glutamic acid, aspartic acid, or lysine residues of acceptor proteins, which include PARP itself, histone H1, and histone H2B (see Reference 143 for a review). Once activated, PARP transfers ADP-ribose to itself and to histone proteins. Such poly(ADP-ribosylation) of PARP and histone proteins releases

PARP from DNA, strips histones from chromatin, and relaxes the chromatin superstructure, thereby triggering chromatin opening to allow transcriptional regulation (144–147). PARP activity is essential for heart muscle development (40). Parp-1 and Parp-2 are functionally redundant in early embryogenesis. *Parp-1*-null mice are viable, whereas *Parp-1* and *Parp-2* double-knockout mice die before E7–8 (148). Such early embryonic lethality is likely caused by cardiovascular abnormalities (149), suggesting a possible role of PARP in cardiovascular development. Studies using whole-embryo cultures and chemical inhibitors to knockdown PARP enzymatic activity suggest that PARP is essential for maintaining embryonic β-*MHC* expression in the heart (40). Cultured embryos treated with PARP inhibitors show increased expression of α-*MHC* (the primary isoform in the adult heart) and reduced expression of β-*MHC* (the primary isoform in the embryonic heart). Despite the *MHC* isoform switch, PARP inhibition does not affect myocardial proliferation or myocardial thickness of cultured embryos. In embryonic hearts Parp-1 associates with Brg1 and/or Hdac proteins on cardiac *MHC* promoters to repress α-*MHC* and to activate β-*MHC* (40), providing a mechanism for maintaining β-*MHC* expression in embryonic ventricles. Interestingly, this mechanism of *MHC* control by BRG/HDAC/PARP is silenced in adult hearts (40), causing a physiological MHC switch to predominantly α-*MHC* expression in adult ventricles. However, this BRG/HDAC/PARP embryonic complex is reactivated in adult hearts under stress, triggering a pathological MHC switch from adult α-*MHC* to fetal β-*MHC* (40), a crucial step in the myopathic process (40, 150). Therefore, the BRG/HDAC/PARP chromatin-remodeling complex, essential for heart development, also plays a crucial role in the pathogenesis of adult cardiomyopathy, indicating an epigenetic link between developmental and pathological gene expression.

CONCLUSIONS

Epigenetic regulation of cardiovascular development is an important aspect of normal organ formation as well as of defective function in disease situations. Such regulation is particularly apparent in congenital defects of cardiovascular development, strongly pointing to the importance of chromatin-level regulation in embryonic development. Future studies will be critical to understanding the breadth of epigenetic regulation in cardiovascular development. Such studies of epigenetic mechanisms will be important not only for our understanding of the etiology of congenital cardiovascular defects but also for potential applications to regenerative medicine strategies.

DISCLOSURE STATEMENT

The authors are not aware of any affiliations, memberships, funding, or financial holdings that might be perceived as affecting the objectivity of this review.

ACKNOWLEDGMENTS

We thank Drs. Chen-Hao Chen, Pei Han, Calvin T. Hang, Wei Li, Chieh-Yu Lin, and Chien-Jung Lin at Stanford University for assisting with manuscript preparation. C.-P.C. is supported by funds from the Oak Foundation, March of Dimes Foundation, National Institutes of Health (NIH), Lucile Packard Foundation, California Institute of Regenerative Medicine (CIRM), and American Heart Association Established Investigator Award (AHA EIA). B.G.B. is supported by funds from the NIH, CIRM, Lawrence J. and Florence A. DeGeorge Charitable Trust, and AHA EIA.

LITERATURE CITED

1. Pierpont ME, Basson CT, Benson DW Jr, Gelb BD, Giglia TM, et al. 2007. Genetic basis for congenital heart defects: current knowledge: a scientific statement from the American Heart Association Congenital Cardiac Defects Committee, Council on Cardiovascular Disease in the Young: endorsed by the American Academy of Pediatrics. *Circulation* 115:3015–38
2. Brickner ME, Hillis LD, Lange RA. 2000. Congenital heart disease in adults. First of two parts. *N. Engl. J. Med.* 342:256–63
3. Brickner ME, Hillis LD, Lange RA. 2000. Congenital heart disease in adults. Second of two parts. *N. Engl. J. Med.* 342:334–42
4. Sandler T. 2004. *Langman's Medical Embryology*. Philadelphia: Williams & Wilkins
5. Clapier CR, Cairns BR. 2009. The biology of chromatin remodeling complexes. *Annu. Rev. Biochem.* 78:273–304
6. Jenuwein T, Allis CD. 2001. Translating the histone code. *Science* 293:1074–80
7. Ho L, Crabtree GR. 2010. Chromatin remodelling during development. *Nature* 463:474–84
8. Stankunas K, Hang CT, Tsun ZY, Chen H, Lee NV, et al. 2008. Endocardial Brg1 represses ADAMTS1 to maintain the microenvironment for myocardial morphogenesis. *Dev. Cell* 14:298–311
9. Chang S, Young BD, Li S, Qi X, Richardson JA, Olson EN. 2006. Histone deacetylase 7 maintains vascular integrity by repressing matrix metalloproteinase 10. *Cell* 126:321–34
10. Han P, Hang CT, Yang J, Chang CP. 2011. Chromatin remodeling in cardiovascular development and physiology. *Circ. Res.* 108:378–96
11. Griffin CT, Brennan J, Magnuson T. 2008. The chromatin-remodeling enzyme BRG1 plays an essential role in primitive erythropoiesis and vascular development. *Development* 135:493–500
12. Reyes JC, Barra J, Muchardt C, Camus A, Babinet C, Yaniv M. 1998. Altered control of cellular proliferation in the absence of mammalian brahma (SNF2α). *EMBO J.* 17:6979–91
13. Griffin CT, Curtis CD, Davis RB, Muthukumar V, Magnuson T. 2011. The chromatin-remodeling enzyme BRG1 modulates vascular Wnt signaling at two levels. *Proc. Natl. Acad. Sci. USA* 108:2282–87
14. Cattelino A, Liebner S, Gallini R, Zanetti A, Balconi G, et al. 2003. The conditional inactivation of the β-catenin gene in endothelial cells causes a defective vascular pattern and increased vascular fragility. *J. Cell Biol.* 162:1111–22
15. Wang Z, Zhai W, Richardson JA, Olson EN, Meneses JJ, et al. 2004. Polybromo protein BAF180 functions in mammalian cardiac chamber maturation. *Genes Dev.* 18:3106–16
16. Huang X, Gao X, Diaz-Trelles R, Ruiz-Lozano P, Wang Z. 2008. Coronary development is regulated by ATP-dependent SWI/SNF chromatin remodeling component BAF180. *Dev. Biol.* 319:258–66
17. Mottet D, Bellahcene A, Pirotte S, Waltregny D, Deroanne C, et al. 2007. Histone deacetylase 7 silencing alters endothelial cell migration, a key step in angiogenesis. *Circ. Res.* 101:1237–46
18. Margariti A, Zampetaki A, Xiao Q, Zhou B, Karamariti E, et al. 2010. Histone deacetylase 7 controls endothelial cell growth through modulation of β-catenin. *Circ. Res.* 106:1202–11
19. Potente M, Ghaeni L, Baldessari D, Mostoslavsky R, Rossig L, et al. 2007. SIRT1 controls endothelial angiogenic functions during vascular growth. *Genes Dev.* 21:2644–58
20. Guarani V, Deflorian G, Franco CA, Kruger M, Phng LK, et al. 2011. Acetylation-dependent regulation of endothelial Notch signalling by the SIRT1 deacetylase. *Nature* 473:234–38
21. Hu M, Sun XJ, Zhang YL, Kuang Y, Hu CQ, et al. 2010. Histone H3 lysine 36 methyltransferase Hypb/Setd2 is required for embryonic vascular remodeling. *Proc. Natl. Acad. Sci. USA* 107:2956–61
22. Illi B, Nanni S, Scopece A, Farsetti A, Biglioli P, et al. 2003. Shear stress-mediated chromatin remodeling provides molecular basis for flow-dependent regulation of gene expression. *Circ. Res.* 93:155–61
23. Zhou J, Zhang M, Fang H, El-Mounayri O, Rodenberg JM, et al. 2009. The SWI/SNF chromatin remodeling complex regulates myocardin-induced smooth muscle-specific gene expression. *Arterioscler. Thromb. Vasc. Biol.* 29:921–28
24. Zhang M, Fang H, Zhou J, Herring BP. 2007. A novel role of Brg1 in the regulation of SRF/MRTFA-dependent smooth muscle-specific gene expression. *J. Biol. Chem.* 282:25708–16
25. Zhang M, Chen M, Kim JR, Zhou J, Jones RE, et al. 2011. SWI/SNF complexes containing Brahma or Brahma-related gene 1 play distinct roles in smooth muscle development. *Mol. Cell. Biol.* 31:2618–31

26. Huang J, Cheng L, Li J, Chen M, Zhou D, et al. 2008. Myocardin regulates expression of contractile genes in smooth muscle cells and is required for closure of the ductus arteriosus in mice. *J. Clin. Investig.* 118:515–25
27. Vissers LE, van Ravenswaaij CM, Admiraal R, Hurst JA, de Vries BB, et al. 2004. Mutations in a new member of the chromodomain gene family cause CHARGE syndrome. *Nat. Genet.* 36:955–57
28. Layman WS, Hurd EA, Martin DM. 2010. Chromodomain proteins in development: lessons from CHARGE syndrome. *Clin. Genet.* 78:11–20
29. Zentner GE, Layman WS, Martin DM, Scacheri PC. 2010. Molecular and phenotypic aspects of *CHD7* mutation in CHARGE syndrome. *Am. J. Med. Genet. A* 152:674–86
30. Randall V, McCue K, Roberts C, Kyriakopoulou V, Beddow S, et al. 2009. Great vessel development requires biallelic expression of Chd7 and Tbx1 in pharyngeal ectoderm in mice. *J. Clin. Investig.* 119:3301–10
31. Bajpai R, Chen DA, Rada-Iglesias A, Zhang J, Xiong Y, et al. 2010. CHD7 cooperates with PBAF to control multipotent neural crest formation. *Nature* 463:958–62
32. Lee Y, Song AJ, Baker R, Micales B, Conway SJ, Lyons GE. 2000. Jumonji, a nuclear protein that is necessary for normal heart development. *Circ. Res.* 86:932–38
33. Jung J, Mysliwiec MR, Lee Y. 2005. Roles of JUMONJI in mouse embryonic development. *Dev. Dyn.* 232:21–32
34. Schneider JE, Böse J, Bamforth SD, Gruber AD, Broadbent C, et al. 2004. Identification of cardiac malformations in mice lacking *Ptdsr* using a novel high-throughput magnetic resonance imaging technique. *BMC Dev. Biol.* 4:16
35. Nomura M, Takihara Y, Shimada K. 1994. Isolation and characterization of retinoic acid-inducible cDNA clones in F9 cells: One of the early inducible clones encodes a novel protein sharing several highly homologous regions with a *Drosophila Polyhomeotic* protein. *Differentiation* 57:39–50
36. Shirai M, Osugi T, Koga H, Kaji Y, Takimoto E, et al. 2002. The *Polycomb*-group gene *Rae28* sustains *Nkx2.5/Csx* expression and is essential for cardiac morphogenesis. *J. Clin. Investig.* 110:177–84
37. Takihara Y, Tomotsune D, Shirai M, Katoh-Fukui Y, Nishii K, et al. 1997. Targeted disruption of the mouse homologue of the *Drosophila polyhomeotic* gene leads to altered anteroposterior patterning and neural crest defects. *Development* 124:3673–82
38. Koga H, Kaji Y, Nishii K, Shirai M, Tomotsune D, et al. 2002. Overexpression of *Polycomb*-group gene *rae28* in cardiomyocytes does not complement abnormal cardiac morphogenesis in mice lacking *rae28* but causes dilated cardiomyopathy. *Lab. Investig.* 82:375–85
39. Jenni R, Rojas J, Oechslin E. 1999. Isolated noncompaction of the myocardium. *N. Engl. J. Med.* 340:966–67
40. Hang CT, Yang J, Han P, Cheng HL, Shang C, et al. 2010. Chromatin regulation by Brg1 underlies heart muscle development and disease. *Nature* 466:62–67
41. Takeuchi JK, Lou X, Alexander JM, Sugizaki H, Delgado-Olguin P, et al. 2011. Chromatin remodelling complex dosage modulates transcription factor function in heart development. *Nat. Commun.* 2:187
42. Lickert H, Takeuchi JK, Von Both I, Walls JR, McAuliffe F, et al. 2004. Baf60c is essential for function of BAF chromatin remodelling complexes in heart development. *Nature* 432:107–12
43. Bruneau BG. 2008. The developmental genetics of congenital heart disease. *Nature* 451:943–48
44. Biben C, Weber R, Kesteven S, Stanley E, McDonald L, et al. 2000. Cardiac septal and valvular dysmorphogenesis in mice heterozygous for mutations in the homeobox gene *Nkx2-5*. *Circ. Res.* 87:888–95
45. Jay PY, Harris BS, Maguire CT, Buerger A, Wakimoto H, et al. 2004. *Nkx2-5* mutation causes anatomic hypoplasia of the cardiac conduction system. *J. Clin. Investig.* 113:1130–37
46. Stennard FA, Costa MW, Lai D, Biben C, Furtado MB, et al. 2005. Murine T-box transcription factor Tbx20 acts as a repressor during heart development, and is essential for adult heart integrity, function and adaptation. *Development* 132:2451–62
47. Lange M, Kaynak B, Forster UB, Tonjes M, Fischer JJ, et al. 2008. Regulation of muscle development by DPF3, a novel histone acetylation and methylation reader of the BAF chromatin remodeling complex. *Genes Dev.* 22:2370–84
48. Zeng L, Zhang Q, Li S, Plotnikov AN, Walsh MJ, Zhou MM. 2010. Mechanism and regulation of acetylated histone binding by the tandem PHD finger of DPF3b. *Nature* 466:258–62

49. Kaynak B, von Heydebreck A, Mebus S, Seelow D, Hennig S, et al. 2003. Genome-wide array analysis of normal and malformed human hearts. *Circulation* 107:2467–74
50. Mertsalov IB, Ninkina NN, Wanless JS, Buchman VL, Korochkin LI, Kulikova DA. 2008. Generation of mutant mice with targeted disruption of two members of the *d* 4 gene family: *neuro-d*4 and *cer-d*4. *Dokl. Biochem. Biophys.* 419:65–68
51. Takeuchi JK, Bruneau BG. 2009. Directed transdifferentiation of mouse mesoderm to heart tissue by defined factors. *Nature* 459:708–11
52. Takeuchi JK, Lickert H, Bisgrove BW, Sun X, Yamamoto M, et al. 2007. Baf60c is a nuclear Notch signaling component required for the establishment of left-right asymmetry. *Proc. Natl. Acad. Sci. USA* 104:846–51
53. Lou X, Deshwar AR, Crump JG, Scott IC. 2011. Smarcd3b and Gata5 promote a cardiac progenitor fate in the zebrafish embryo. *Development* 138:3113–23
54. Kastner P, Grondona JM, Mark M, Gansmuller A, LeMeur M, et al. 1994. Genetic analysis of RXRα developmental function: convergence of RXR and RAR signaling pathways in heart and eye morphogenesis. *Cell* 78:987–1003
55. Sucov HM, Dyson E, Gumeringer CL, Price J, Chien KR, Evans RM. 1994. RXRα mutant mice establish a genetic basis for vitamin A signaling in heart morphogenesis. *Genes Dev.* 8:1007–18
56. Bao Y, Shen X. 2007. INO80 subfamily of chromatin remodeling complexes. *Mutat. Res.* 618:18–29
57. Gallant P. 2007. Control of transcription by Pontin and Reptin. *Trends Cell Biol.* 17:187–92
58. Jonsson ZO, Jha S, Wohlschlegel JA, Dutta A. 2004. Rvb1p/Rvb2p recruit Arp5p and assemble a functional Ino80 chromatin remodeling complex. *Mol. Cell* 16:465–77
59. Jonsson ZO, Dhar SK, Narlikar GJ, Auty R, Wagle N, et al. 2001. Rvb1p and Rvb2p are essential components of a chromatin remodeling complex that regulates transcription of over 5% of yeast genes. *J. Biol. Chem.* 276:16279–88
60. Shen X, Mizuguchi G, Hamiche A, Wu C. 2000. A chromatin remodelling complex involved in transcription and DNA processing. *Nature* 406:541–44
61. Rottbauer W, Saurin AJ, Lickert H, Shen X, Burns CG, et al. 2002. Reptin and pontin antagonistically regulate heart growth in zebrafish embryos. *Cell* 111:661–72
62. Dirscherl SS, Krebs JE. 2004. Functional diversity of ISWI complexes. *Biochem. Cell Biol.* 82:482–89
63. Corona DF, Tamkun JW. 2004. Multiple roles for ISWI in transcription, chromosome organization and DNA replication. *Biochim. Biophys. Acta* 1677:113–19
64. Stopka T, Skoultchi AI. 2003. The ISWI ATPase Snf2h is required for early mouse development. *Proc. Natl. Acad. Sci. USA* 100:14097–102
65. Yoshimura K, Kitagawa H, Fujiki R, Tanabe M, Takezawa S, et al. 2009. Distinct function of 2 chromatin remodeling complexes that share a common subunit, Williams syndrome transcription factor (WSTF). *Proc. Natl. Acad. Sci. USA* 106:9280–85
66. Backs J, Olson EN. 2006. Control of cardiac growth by histone acetylation/deacetylation. *Circ. Res.* 98:15–24
67. Haberland M, Montgomery RL, Olson EN. 2009. The many roles of histone deacetylases in development and physiology: implications for disease and therapy. *Nat. Rev. Genet.* 10:32–42
68. Ekwall K. 2005. Genome-wide analysis of HDAC function. *Trends Genet.* 21:608–15
69. Fischle W, Dequiedt F, Hendzel MJ, Guenther MG, Lazar MA, Voelter W, Verdin E. 2002. Enzymatic activity associated with class II HDACs is dependent on a multiprotein complex containing HDAC3 and SMRT/N-CoR. *Mol. Cell* 9:45–57
70. Lahm A, Paolini C, Pallaoro M, Nardi MC, Jones P, et al. 2007. Unraveling the hidden catalytic activity of vertebrate class IIa histone deacetylases. *Proc. Natl. Acad. Sci. USA* 104:17335–40
71. Jones P, Altamura S, De Francesco R, Gallinari P, Lahm A, et al. 2008. Probing the elusive catalytic activity of vertebrate class IIa histone deacetylases. *Bioorg. Med. Chem. Lett.* 18:1814–19
72. Schuetz A, Min J, Allali-Hassani A, Schapira M, Shuen M, et al. 2008. Human HDAC7 harbors a class IIa histone deacetylase-specific zinc binding motif and cryptic deacetylase activity. *J. Biol. Chem.* 283:11355–63
73. Guenther MG, Barak O, Lazar MA. 2001. The SMRT and N-CoR corepressors are activating cofactors for histone deacetylase 3. *Mol. Cell. Biol.* 21:6091–101

74. Zhang CL, McKinsey TA, Lu JR, Olson EN. 2001. Association of COOH-terminal-binding protein (CtBP) and MEF2-interacting transcription repressor (MITR) contributes to transcriptional repression of the MEF2 transcription factor. *J. Biol. Chem.* 276:35–39
75. Dressel U, Bailey PJ, Wang SC, Downes M, Evans RM, Muscat GE. 2001. A dynamic role for HDAC7 in MEF2-mediated muscle differentiation. *J. Biol. Chem.* 276:17007–13
76. Zhang CL, McKinsey TA, Olson EN. 2002. Association of class II histone deacetylases with heterochromatin protein 1: potential role for histone methylation in control of muscle differentiation. *Mol. Cell. Biol.* 22:7302–12
77. Lagger G, O'Carroll D, Rembold M, Khier H, Tischler J, et al. 2002. Essential function of histone deacetylase 1 in proliferation control and CDK inhibitor repression. *EMBO J.* 21:2672–81
78. Montgomery RL, Davis CA, Potthoff MJ, Haberland M, Fielitz J, et al. 2007. Histone deacetylases 1 and 2 redundantly regulate cardiac morphogenesis, growth, and contractility. *Genes Dev.* 21:1790–802
79. Fajas L, Egler V, Reiter R, Hansen J, Kristiansen K, et al. 2002. The retinoblastoma-histone deacetylase 3 complex inhibits PPARγ and adipocyte differentiation. *Dev. Cell* 3:903–10
80. Trivedi CM, Luo Y, Yin Z, Zhang M, Zhu W, et al. 2007. Hdac2 regulates the cardiac hypertrophic response by modulating Gsk3β activity. *Nat. Med.* 13:324–31
81. Kuwahara K, Saito Y, Takano M, Arai Y, Yasuno S, et al. 2003. NRSF regulates the fetal cardiac gene program and maintains normal cardiac structure and function. *EMBO J.* 22:6310–21
82. Kuwahara K, Saito Y, Ogawa E, Takahashi N, Nakagawa Y, et al. 2001. The neuron-restrictive silencer element-neuron-restrictive silencer factor system regulates basal and endothelin 1-inducible atrial natriuretic peptide gene expression in ventricular myocytes. *Mol. Cell. Biol.* 21:2085–97
83. Nakagawa Y, Kuwahara K, Harada M, Takahashi N, Yasuno S, et al. 2006. Class II HDACs mediate CaMK-dependent signaling to NRSF in ventricular myocytes. *J. Mol. Cell. Cardiol.* 41:1010–22
84. Trivedi CM, Zhu W, Wang Q, Jia C, Kee HJ, et al. 2010. Hopx and Hdac2 interact to modulate Gata4 acetylation and embryonic cardiac myocyte proliferation. *Dev. Cell* 19:450–59
85. Chen F, Kook H, Milewski R, Gitler AD, Lu MM, et al. 2002. Hop is an unusual homeobox gene that modulates cardiac development. *Cell* 110:713–23
86. Shin CH, Liu ZP, Passier R, Zhang CL, Wang DZ, et al. 2002. Modulation of cardiac growth and development by HOP, an unusual homeodomain protein. *Cell* 110:725–35
87. Montgomery RL, Potthoff MJ, Haberland M, Qi X, Matsuzaki S, et al. 2008. Maintenance of cardiac energy metabolism by histone deacetylase 3 in mice. *J. Clin. Investig.* 118:3588–97
88. Watanabe K, Fujii H, Takahashi T, Kodama M, Aizawa Y, et al. 2000. Constitutive regulation of cardiac fatty acid metabolism through peroxisome proliferator-activated receptor α associated with age-dependent cardiac toxicity. *J. Biol. Chem.* 275:22293–99
89. Cheng L, Ding G, Qin Q, Huang Y, Lewis W, et al. 2004. Cardiomyocyte-restricted peroxisome proliferator-activated receptor-δ deletion perturbs myocardial fatty acid oxidation and leads to cardiomyopathy. *Nat. Med.* 10:1245–50
90. Trivedi CM, Lu MM, Wang Q, Epstein JA. 2008. Transgenic overexpression of Hdac3 in the heart produces increased postnatal cardiac myocyte proliferation but does not induce hypertrophy. *J. Biol. Chem.* 283:26484–89
91. Chang S, McKinsey TA, Zhang CL, Richardson JA, Hill JA, Olson EN. 2004. Histone deacetylases 5 and 9 govern responsiveness of the heart to a subset of stress signals and play redundant roles in heart development. *Mol. Cell. Biol.* 24:8467–76
92. Song K, Backs J, McAnally J, Qi X, Gerard RD, et al. 2006. The transcriptional coactivator CAMTA2 stimulates cardiac growth by opposing class II histone deacetylases. *Cell* 125:453–66
93. Cheng HL, Mostoslavsky R, Saito S, Manis JP, Gu Y, et al. 2003. Developmental defects and p53 hyperacetylation in Sir2 homolog (SIRT1)-deficient mice. *Proc. Natl. Acad. Sci. USA* 100:10794–99
94. Bordone L, Cohen D, Robinson A, Motta MC, van Veen E, et al. 2007. SIRT1 transgenic mice show phenotypes resembling calorie restriction. *Aging Cell* 6:759–67
95. Smith J. 2002. Human Sir2 and the "silencing" of p53 activity. *Trends Cell Biol.* 12:404–6
96. Wang Z, Zang C, Cui K, Schones DE, Barski A, et al. 2009. Genome-wide mapping of HATs and HDACs reveals distinct functions in active and inactive genes. *Cell* 138:1019–31

97. Yao TP, Oh SP, Fuchs M, Zhou ND, Ch'ng LE, et al. 1998. Gene dosage-dependent embryonic development and proliferation defects in mice lacking the transcriptional integrator p300. *Cell* 93:361–72
98. Shikama N, Lutz W, Kretzschmar R, Sauter N, Roth JF, et al. 2003. Essential function of p300 acetyltransferase activity in heart, lung and small intestine formation. *EMBO J.* 22:5175–85
99. Yanazume T, Hasegawa K, Morimoto T, Kawamura T, Wada H, et al. 2003. Cardiac p300 is involved in myocyte growth with decompensated heart failure. *Mol. Cell. Biol.* 23:3593–606
100. Gusterson RJ, Jazrawi E, Adcock IM, Latchman DS. 2003. The transcriptional co-activators CREB-binding protein (CBP) and p300 play a critical role in cardiac hypertrophy that is dependent on their histone acetyltransferase activity. *J. Biol. Chem.* 278:6838–47
101. Mathiyalagan P, Chang L, Du XJ, El-Osta A. 2010. Cardiac ventricular chambers are epigenetically distinguishable. *Cell Cycle* 9:612–17
102. Klose RJ, Kallin EM, Zhang Y. 2006. JmjC-domain-containing proteins and histone demethylation. *Nat. Rev. Genet.* 7:715–27
103. Whetstine JR, Nottke A, Lan F, Huarte M, Smolikov S, et al. 2006. Reversal of histone lysine trimethylation by the JMJD2 family of histone demethylases. *Cell* 125:467–81
104. Shirato H, Ogawa S, Nakajima K, Inagawa M, Kojima M, et al. 2009. A jumonji (Jarid2) protein complex represses cyclin D1 expression by methylation of histone H3-K9. *J. Biol. Chem.* 284:733–39
105. Shen X, Kim W, Fujiwara Y, Simon MD, Liu Y, et al. 2009. Jumonji modulates *polycomb* activity and self-renewal versus differentiation of stem cells. *Cell* 139:1303–14
106. Li G, Margueron R, Ku M, Chambon P, Bernstein BE, Reinberg D. 2010. Jarid2 and PRC2, partners in regulating gene expression. *Genes Dev.* 24:368–80
107. Landeira D, Sauer S, Poot R, Dvorkina M, Mazzarella L, et al. 2010. Jarid2 is a PRC2 component in embryonic stem cells required for multi-lineage differentiation and recruitment of PRC1 and RNA polymerase II to developmental regulators. *Nat. Cell Biol.* 12:618–24
108. Takeuchi T, Kojima M, Nakajima K, Kondo S. 1999. *jumonji* gene is essential for the neurulation and cardiac development of mouse embryos with a C3H/He background. *Mech. Dev.* 86:29–38
109. Mysliwiec MR, Bresnick EH, Lee Y. 2011. Endothelial Jarid2/Jumonji is required for normal cardiac development and proper Notch1 expression. *J. Biol. Chem.* 286:17193–204
110. Grego-Bessa J, Luna-Zurita L, del Monte G, Bolos V, Melgar P, et al. 2007. Notch signaling is essential for ventricular chamber development. *Dev. Cell* 12:415–29
111. Oka C, Nakano T, Wakeham A, de la Pompa JL, Mori C, et al. 1995. Disruption of the mouse RBP-Jκ gene results in early embryonic death. *Development* 121:3291–301
112. Meyer D, Birchmeier C. 1995. Multiple essential functions of neuregulin in development. *Nature* 378:386–90
113. Lee KF, Simon H, Chen H, Bates B, Hung MC, Hauser C. 1995. Requirement for neuregulin receptor erbB2 in neural and cardiac development. *Nature* 378:394–98
114. Gassmann M, Casagranda F, Orioli D, Simon H, Lai C, et al. 1995. Aberrant neural and cardiac development in mice lacking the ErbB4 neuregulin receptor. *Nature* 378:390–94
115. Chen H, Shi S, Acosta L, Li W, Lu J, et al. 2004. BMP10 is essential for maintaining cardiac growth during murine cardiogenesis. *Development* 131:2219–31
116. Timms JF, White SL, O'Hare MJ, Waterfield MD. 2002. Effects of ErbB-2 overexpression on mitogenic signalling and cell cycle progression in human breast luminal epithelial cells. *Oncogene* 21:6573–86
117. Böse J, Gruber AD, Helming L, Schiebe S, Wegener I, et al. 2004. The phosphatidylserine receptor has essential functions during embryogenesis but not in apoptotic cell removal. *J. Biol.* 3:15
118. Rea S, Eisenhaber F, O'Carroll D, Strahl BD, Sun ZW, et al. 2000. Regulation of chromatin structure by site-specific histone H3 methyltransferases. *Nature* 406:593–99
119. Gelmetti V, Zhang J, Fanelli M, Minucci S, Pelicci PG, Lazar MA. 1998. Aberrant recruitment of the nuclear receptor corepressor-histone deacetylase complex by the acute myeloid leukemia fusion partner ETO. *Mol. Cell. Biol.* 18:7185–91
120. Gottlieb PD, Pierce SA, Sims RJ, Yamagishi H, Weihe EK, et al. 2002. *Bop* encodes a muscle-restricted protein containing MYND and SET domains and is essential for cardiac differentiation and morphogenesis. *Nat. Genet.* 31:25–32

121. Srivastava D, Cserjesi P, Olson EN. 1995. A subclass of bHLH proteins required for cardiac morphogenesis. *Science* 270:1995–99
122. Tan X, Rotllant J, Li H, De Deyne P, Du SJ. 2006. SmyD1, a histone methyltransferase, is required for myofibril organization and muscle contraction in zebrafish embryos. *Proc. Natl. Acad. Sci. USA* 103:2713–18
123. Phan D, Rasmussen TL, Nakagawa O, McAnally J, Gottlieb PD, et al. 2005. BOP, a regulator of right ventricular heart development, is a direct transcriptional target of MEF2C in the developing heart. *Development* 132:2669–78
124. Park CY, Pierce SA, von Drehle M, Ivey KN, Morgan JA, et al. 2010. skNAC, a Smyd1-interacting transcription factor, is involved in cardiac development and skeletal muscle growth and regeneration. *Proc. Natl. Acad. Sci. USA* 107:20750–55
125. Sims RJ 3rd, Weihe EK, Zhu L, O'Malley S, Harriss JV, Gottlieb PD. 2002. m-Bop, a repressor protein essential for cardiogenesis, interacts with skNAC, a heart- and muscle-specific transcription factor. *J. Biol. Chem.* 277:26524–29
126. Diehl F, Brown MA, van Amerongen MJ, Novoyatleva T, Wietelmann A, et al. 2010. Cardiac deletion of *Smyd2* is dispensable for mouse heart development. *PLoS ONE* 5:e9748
127. Marango J, Shimoyama M, Nishio H, Meyer JA, Min DJ, et al. 2008. The MMSET protein is a histone methyltransferase with characteristics of a transcriptional corepressor. *Blood* 111:3145–54
128. Wright TJ, Ricke DO, Denison K, Abmayr S, Cotter PD, et al. 1997. A transcript map of the newly defined 165 kb Wolf-Hirschhorn syndrome critical region. *Hum. Mol. Genet.* 6:317–24
129. Bergemann AD, Cole F, Hirschhorn K. 2005. The etiology of Wolf-Hirschhorn syndrome. *Trends Genet.* 21:188–95
130. Hirschhorn K, Cooper HL, Firschein IL. 1965. Deletion of short arms of chromosome 4–5 in a child with defects of midline fusion. *Humangenetik* 1:479–82
131. Nimura K, Ura K, Shiratori H, Ikawa M, Okabe M, et al. 2009. A histone H3 lysine 36 trimethyltransferase links Nkx2-5 to Wolf-Hirschhorn syndrome. *Nature* 460:287–91
132. Ng SB, Bigham AW, Buckingham KJ, Hannibal MC, McMillin MJ, et al. 2010. Exome sequencing identifies *MLL2* mutations as a cause of Kabuki syndrome. *Nat. Genet.* 42:790–93
133. Digilio MC, Marino B, Toscano A, Giannotti A, Dallapiccola B. 2001. Congenital heart defects in Kabuki syndrome. *Am. J. Med. Genet.* 100(4):269–74
134. Stein AB, Jones TA, Herron TJ, Patel SR, Day SM, et al. 2011. Loss of H3K4 methylation destabilizes gene expression patterns and physiological functions in adult murine cardiomyocytes. *J. Clin. Investig.* 121:2641–50
135. Feng Q, Wang H, Ng HH, Erdjument-Bromage H, Tempst P, et al. 2002. Methylation of H3-lysine 79 is mediated by a new family of HMTases without a SET domain. *Curr. Biol.* 12:1052–58
136. Jones B, Su H, Bhat A, Lei H, Bajko J, et al. 2008. The histone H3K79 methyltransferase Dot1L is essential for mammalian development and heterochromatin structure. *PLoS Genet.* 4:e1000190
137. Kang X, Qi Y, Zuo Y, Wang Q, Zou Y, et al. 2010. SUMO-specific protease 2 is essential for suppression of *polycomb* group protein-mediated gene silencing during embryonic development. *Mol. Cell* 38:191–201
138. Margueron R, Reinberg D. 2011. The polycomb complex PRC2 and its mark in life. *Nature* 469:343–49
139. Rivera-Feliciano J, Lee KH, Kong SW, Rajagopal S, Ma Q, et al. 2006. Development of heart valves requires *Gata4* expression in endothelial-derived cells. *Development* 133:3607–18
140. Moskowitz IP, Wang J, Peterson MA, Pu WT, Mackinnon AC, et al. 2011. Transcription factor genes *Smad4* and *Gata4* cooperatively regulate cardiac valve development. *Proc. Natl. Acad. Sci. USA* 108:4006–11
141. Tian Y, Yuan L, Goss AM, Wang T, Yang J, et al. 2010. Characterization and in vivo pharmacological rescue of a Wnt2-Gata6 pathway required for cardiac inflow tract development. *Dev. Cell* 18:275–87
142. Pikkarainen S, Tokola H, Kerkela R, Ruskoaho H. 2004. GATA transcription factors in the developing and adult heart. *Cardiovasc. Res.* 63:196–207
143. Schreiber V, Dantzer F, Ame JC, de Murcia G. 2006. Poly(ADP-ribose): novel functions for an old molecule. *Nat. Rev. Mol. Cell Biol.* 7:517–28
144. Poirier GG, de Murcia G, Jongstra-Bilen J, Niedergang C, Mandel P. 1982. Poly(ADP-ribosyl)ation of polynucleosomes causes relaxation of chromatin structure. *Proc. Natl. Acad. Sci. USA* 79:3423–27

145. Realini CA, Althaus FR. 1992. Histone shuttling by poly(ADP-ribosylation). *J. Biol. Chem.* 267:18858–65
146. Kim MY, Mauro S, Gevry N, Lis JT, Kraus WL. 2004. NAD^+-dependent modulation of chromatin structure and transcription by nucleosome binding properties of PARP-1. *Cell* 119:803–14
147. Tulin A, Spradling A. 2003. Chromatin loosening by poly(ADP)-ribose polymerase (PARP) at *Drosophila* puff loci. *Science* 299:560–62
148. Menissier de Murcia J, Ricoul M, Tartier L, Niedergang C, Huber A, et al. 2003. Functional interaction between PARP-1 and PARP-2 in chromosome stability and embryonic development in mouse. *EMBO J.* 22:2255–63
149. Conway SJ, Kruzynska-Frejtag A, Kneer PL, Machnicki M, Koushik SV. 2003. What cardiovascular defect does my prenatal mouse mutant have, and why? *Genesis* 35:1–21
150. Lompre AM, Schwartz K, d'Albis A, Lacombe G, Van Thiem N, Swynghedauw B. 1979. Myosin isoenzyme redistribution in chronic heart overload. *Nature* 282:105–7

Lysosomal Acidification Mechanisms*

Joseph A. Mindell

Membrane Transport Biophysics Section, National Institute of Neurological Disorders and Stroke, National Institutes of Health, Bethesda, Maryland 20892; email: mindellj@ninds.nih.gov

*This is a work of the U.S. Government and is not subject to copyright protection in the United States.

Keywords

V-ATPase, transporter, ClC-7, channel, counterion

Abstract

Lysosomes, the terminal organelles on the endocytic pathway, digest macromolecules and make their components available to the cell as nutrients. Hydrolytic enzymes specific to a wide range of targets reside within the lysosome; these enzymes are activated by the highly acidic pH (between 4.5 and 5.0) in the organelles' interior. Lysosomes generate and maintain their pH gradients by using the activity of a proton-pumping V-type ATPase, which uses metabolic energy in the form of ATP to pump protons into the lysosome lumen. Because this activity separates electric charge and generates a transmembrane voltage, another ion must move to dissipate this voltage for net pumping to occur. This so-called counterion may be either a cation (moving out of the lysosome) or an anion (moving into the lysosome). Recent data support the involvement of ClC-7, a Cl^-/H^+ antiporter, in this process, although many open questions remain as to this transporter's involvement. Although functional results also point to a cation transporter, its molecular identity remains uncertain. Both the V-ATPase and the counterion transporter are likely to be important players in the mechanisms determining the steady-state pH of the lysosome interior. Exciting new results suggest that lysosomal pH may be dynamically regulated in some cell types.

LYSOSOMAL ACIDIFICATION MECHANISMS

Mammalian cells use H^+ for a broad range of physiological functions, which is not surprising given the ubiquity and reactivity of the bare proton (or hydronium ion). Protons lie at the center of bioenergetics because mitochondria use a gradient of these ions across their inner membranes as a key intermediate in oxidative phosphorylation. Perhaps taking a cue from these cellular batteries, a wide range of other membrane-bound intracellular organelles, including the Golgi complex, secretory vesicles, endosomes, and lysosomes, also generate transmembrane proton gradients (**Figure 1**). Key cellular processes depend on the luminal pH inside given organelles; such processes include posttranslational modification in the secretory pathway, ligand targeting in the endosomal pathway, and macromolecule degradation in the lysosome. Each organelle maintains a characteristic internal pH, which is essential for facilitating organelle function. Indeed, disorders that affect organellar acidification can lead to a range of diseases, many of which are severe or life threatening. Perhaps the most extreme example of organellar acidification in mammalian cells is the lysosome. This digestive organelle depends on maintaining a highly acidic pH (less than pH 5.0) in its lumen to successfully perform its digestive function and to drive efflux of digested materials.

Evidence for an acidic lysosomal lumen initially came from the convergence of multiple approaches. Actually, the earliest evidence for acidification of intracellular compartments predates the discovery of the lysosome. In 1893, Metchnikoff (1) found that paramecia could ingest particles of pH-sensitive litmus paper and that upon internalization the paper changed color, consistent with entry into an acidic compartment. In the modern era, early indications of acidic lysosomes came from observations by Coffey & De Duve (2), who noted that the hydrolytic enzymes contained in the lysosome share acidic pH optima. Studies on the distribution of radioactively or fluorescently labeled weak bases (like acridine orange) on isolated lysosomes provided direct evidence that the organelles could maintain an internal pH more acidic than the pH of the bathing medium.

It is now well established that this pH gradient is generated by the action of a V-type ATPase, a proton-pumping membrane protein that uses the free energy of ATP hydrolysis to drive protons against their electrochemical gradient into the lysosome lumen (3). However, with

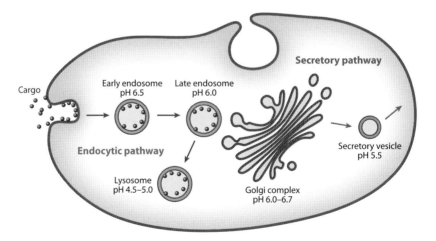

Figure 1

Interior pHs of intracellular organelles. pH gradually drops along the endocytic pathway (*blue organelles*) from early endosomes to late endosomes to lysosomes; lysosomes are the most acidic. In contrast, pH gradually increases along the secretory pathway (*yellow organelles*): The pH of the Golgi complex is between pH 6.7 and 6.0 (*cis* to *trans*), and that of secretory vesicles is more acidic (pH 5.5).

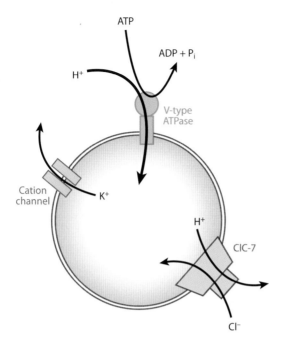

Figure 2

Lysosomal transporters involved in pH homeostasis. The V-type ATPase (*green*) uses the metabolic energy of ATP hydrolysis to drive protons into the lumen. This process builds a net positive charge inside the lumen of the lysosome. This charge can be dissipated by K^+ efflux through a cation channel or transporter (*blue*) or by Cl^- influx through ClC-7, a Cl^-/H^+ antiporter (*yellow*).

each proton pumped the ATPase also generates a voltage difference across the lysosome membrane; this voltage difference inhibits further pumping. Thus, for the ATPase to effectively acidify the lysosome interior, proton movement must be accompanied by the movement of a counterion to dissipate the transmembrane voltage generated by the ATPase. In theory, this counterion movement could be generated by entry of a cytoplasmic anion into the lysosome interior, by exit of a cation from the lumen to the cytoplasm, or by both (**Figure 2**). The identity of this counterion remains controversial. Yet the process of counterion movement may be an important element of the acidification process. Although many organelles use the same V-type ATPase to acidify their interiors, each is able to maintain a stable, characteristic internal pH; varying counterion mechanisms may account for this diversity. The mechanisms of this pH regulation are poorly understood, especially at a quantitative level. Because the movement of counterions is essential for acidification, the molecules involved present potential elements of such organellar pH-regulatory mechanisms. The lysosome is an ideal organelle to probe these mechanisms because it is relatively easily studied both in living cells and in isolation. Here, I review the known and hypothesized elements of the lysosomal acidification mechanism and consider the experimental results that provide insight into the counterion movement. I also highlight exciting new results that could pertain to ultimately understanding the regulation and maintenance of lysosomal pH.

THE V-TYPE PROTON ATPASE

The primary driver of acidification throughout the endocytic pathway, including in the lysosome, is the V-type proton ATPase. This ATPase harvests free energy from ATP hydrolysis to drive protons uphill into the lysosome. The V-type ATPase is structurally similar to the F_0F_1 ATPases involved in mitochondrial oxidative phosphorylation. However, whereas the F_0F_1 ATPases can either synthesize or hydrolyze ATP, the V-type ATPases appear to be optimized for proton pumping because they apparently work only in the hydrolytic direction in vivo.

Structure and Mechanism

The V-ATPases are multisubunit complexes composed of a soluble V_1 subcomplex (analogous to the F_1 portion of the F-ATPase) and a membrane-embedded V_0 subcomplex (analogous to F_0) (4). Each subcomplex is composed of multiple protein subunits (**Figure 3**). The soluble V_1 domain includes at least eight subunits (A–H) and includes the loci catalyzing ATP hydrolysis. The membrane-embedded V_0 subcomplex includes subunits a, c, c′, c″, d, and e and is required for proton translocation across the membrane. In the intact complex, the V_1 and V_0 portions are connected by a central stalk similar to stalks seen in the F_0F_1 ATPase (including the D and F subunits) as well as by several peripheral stalks (three in most eukaryotic V-ATPases; these stalks are formed by the E and G subunits).

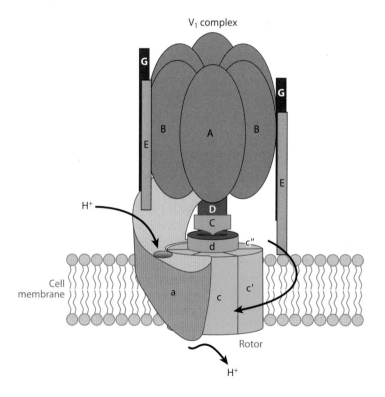

Figure 3

Structure of the V-type proton ATPase. The soluble V_1 complex (*red*) is made up of ATP-hydrolyzing A and B subunits. The rotor, made up of the c, c′, and c″ subunits (*green*), is embedded in the membrane and carries protons from uptake sites on the cytoplasmic side to release sites on the luminal side of the membrane. The a subunit (*brown*) contains access pathways to both sides of the membrane as well as a critical, conserved arginine residue required for transport. The D subunit of the V_1 domain and the d subunit of the V_0 domain compose the central stalk (*blue*), which connects the rotor to the ATPase domains. The E and G subunits (*light purple* and *dark purple*, respectively) compose the peripheral stalks.

Like the F_0F_1 ATPases, the V-ATPase is a rotary proton transport motor. Hydrolysis of ATP at the interfaces between the A and B subunits in the catalytic domain drives the rotation of the central stalk, a driveshaft (5, 6). The stalk is coupled to the proton-translocating ring of the c subunit, which is located within the membrane. This structure in turn mediates proton translocation. Protons are thought to ride around the c subunit–based rotor on the acidic side chain of a strictly conserved glutamate residue located midway along a membrane-spanning helix on each c subunit. The protons access this group through a pair of hemichannels; each hemichannel provides an aqueous pathway halfway through the membrane through different regions of the a subunit (reviewed in Reference 4). A conserved positive arginine resides between the hemichannels, forcing the proton to take the long way around the rotor and presumably helping it dissociate through the luminal hemichannel. This mechanism can explain proton pumping by the ATPase: The rotor is driven uniquely in one direction by the irreversible hydrolysis of ATP, and protons access the rotor only from the cytoplasm-facing hemichannel and leave it only from the lumen-facing hemichannel, resulting in unidirectional proton flux.

The ability of a V-type ATPase to generate a pH gradient depends critically on the ratio of ATP hydrolysis to proton transport. In this respect, the V-ATPases seem to be optimized for proton pumping compared with the F_0F_1 ATPases, which can be driven in either direction (ATP synthesis or hydrolysis) in vivo. With a hexameric A/B complex with three ATP-binding sites, and the likely six-membered ring formed by the three varieties of proton-translocating c subunits (including c, c′, and c″) (7), a stoichiometry of two protons translocated per ATP hydrolyzed is expected. Such stoichiometries have indeed been measured using both kinetic and thermodynamic methods (8, 9). On the basis of the free energy of ATP hydrolysis, an ideal proton ATPase that pumps two protons per ATP could generate the proton gradient of more than four pH units (10). Indeed, very large pH gradients can be generated by the ATPase in the vacuoles of citrus fruits, near the ideal gradient (11), although other transporters may also be involved in this system.

The V-ATPase Is Electrogenic

On theoretical grounds, a transporter that moves a single ion unidirectionally across the membrane should build up a gradient across that membrane, i.e., a voltage difference. The sign of this voltage should be such that it inhibits further transport: Ultimately, the voltage should build to the point that net transport is fully inhibited and no further flux occurs (absent dissipation of that charge by some counterion movement; see below). As a unidirectional proton transporter, the V-ATPase is thus predicted to be electrogenic. This prediction was first tested using voltage-sensitive probes, like $diS-C_3(5)$, a member of the carbocyanine family. These hydrophobic cationic fluorescent dyes accumulate in compartments with negative transmembrane voltages, quenching their fluorescence. Using these dyes, Ohkuma and colleagues (12) and Harikumar & Reeves (13) demonstrated that, whereas isolated lysosomes initially showed negative membrane potentials, these voltages became more positive upon the addition of ATP to activate the V-ATPase. An active electrogenic transporter also generates electrical currents across its membrane for both plant (14) and yeast (15) transporters, as patch clamp techniques have verified. These methods clearly establish the electrogenicity of the V-ATPase. As noted above, for an electrogenic transporter to effectively move its substrate, the voltage it generates must be dissipated by another mechanism. Possible mechanisms for this dissipation in lysosomes are discussed below.

Role in pH Regulation

The V-type ATPase generates different pH gradients in different organelles and on the plasma membrane. Although the mechanisms that set these pHs remain unknown, a reasonable hypothesis is that regulation of the ATPase may account for at least some of these differences. Several forms of regulation have been observed for the V-ATPase protein complex, although none of them have been shown to directly influence organellar acidification.

The most dramatic form of regulation observed for a V-type ATPase is the reversible dissociation of the enzyme complex observed in both yeast and the tobacco hornworm, *Manduca sexta*. In both cases, nutrient restriction (glucose deprivation in yeast and molt or starvation in *Manduca*) led to the dissociation of the V-ATPase into membrane-embedded V_0 and soluble V_1 (16, 17). Notably, this process, which turns off both ATPase activity and proton translocation, is completely reversible with the restoration of nutrients in yeast (17). Furthermore, the process is not dependent on translation of new protein (17). The ATPase dissociation is modulated by interactions with other proteins and protein complexes (18–20), including the glycolytic enzyme aldolase (21), as well as with microtubules (22).

In addition to these metabolic regulatory mechanisms, targeting of the ATPase can be regulated by varying its subunit composition. This process has been characterized in detail in yeast, in which all the ATPase subunits are encoded by single genes except for the a subunit, which has two coding genes. These two genes, *VPH1* and *STV1*, are 54% identical but have very different effects on the complex; whereas Vph1p targets the ATPase to vacuoles, Stv1p targets the complex instead to the Golgi complex (23, 24). When expressed in a strain with both the *STV1* and *VPH1* genes disrupted, overexpression of the Golgi complex–targeted Stv1p results in some targeting of the V-ATPase complex to the vacuole membrane, allowing direct comparison of the functional properties of the two ATPase isoforms (24). Forgac's group (25) took advantage of this to compare the functional properties of the two isoforms in an otherwise very similar environment. This group found that the two isoforms have similar kinetic properties, both showing K_M for ATP of ~250 μM. An approximately eight-fold reduction in V_{max} was attributed to fewer assembled ATPase complexes. The property most pertinent here is the coupling of ATP to proton transport, with Stv1p-containing complexes showing a four- to five-fold-lower coupling of ATP hydrolysis to proton transport compared with Vph1p-containing complexes. Thus, the ATPase can slip; it can hydrolyze ATP without pumping protons. Such a mechanism seems wasteful of ATP, but the observation is consistent with the lower pH observed in the vacuole compared with that in the Golgi complex and implies that pH may be differentially regulated in the different compartments targeted by these subunits.

Like the yeast protein, mammalian V-ATPase subunits are encoded by multiple genes, and the complexes are found with a range of isoforms (reviewed in Reference 26). Generally, a single isoform of each subunit is widely expressed, with alternate subunits expressed in limited ranges of tissues. For example, the d1 subunit is ubiquitously expressed, whereas the d2 is expressed at high levels only in osteoclasts, the kidney, and the lung (27). Indeed, the phenotype of mice with knockouts of the d2 subunit was limited to bone; the animals had osteopetrosis (increased bone density) (28), suggesting a tissue-specific function. Other subunit isoforms are highly expressed in kidney tubules and in the inner ear (29), both tissues with highly specialized transport needs. Notably, though, for each of these examples the alternate subunit is associated with an unusual targeting of the V-ATPase to the plasma membrane, rather than to an alternate organelle (as is observed in yeast). To this author's knowledge, there are no known cases of mammalian V-ATPase isoforms associated with targeting to alternate organelles (reviewed in Reference 26). Thus, the question remains as to whether the V-ATPase, the essential driver of organellar acidification, is directly involved in the determination and regulation of organellar pH in general and of lysosomal pH in particular. To understand the mechanism of organellar pH regulation, we must consider other possible mechanisms.

COUNTERION MOVEMENT AND LYSOSOMAL ACIDIFICATION

As noted above, for a V-ATPase to effectively acidify an organelle, its action must be supplemented with a mechanism to dissipate the luminal-positive transmembrane voltage it generates. This process is described by the term counterion pathway. Two general mechanisms can provide this dissipation. First, cation permeability may carry cations out of the organelle, with one cation removed for each proton translocated. Alternatively, anion permeability may move an anion into the lumen for each proton. A combination of these mechanisms would also be effective. Experimental analysis of lysosomal counterion pathways has a long and varied history. Many of the pioneering studies of lysosomal pH examined ion effects on acidification in isolated lysosomes. These results often pertain to understanding the counterion pathway, although they yield limited insight into the molecular basis of this path. More recently, a number of channels and transporters have been proposed to play roles in the counterion mechanism. However, the identity of the counterion remains controversial, as I discuss below. Nevertheless, if counterion movement is a rate-limiting step in the acidification process, then the counterion pathway may be extremely important in the regulation of lysosomal pH.

The Counterion Pathway in Isolated Lysosomes

Many of the pioneering papers that explored the lysosomal acidification process presented data relevant to understanding the role of counterions in this mechanism. In a 1979 study of acidification in isolated lysosomes using the fluorescent weak base acridine orange as an indicator, Dell'Antone (30) found that substituting external Cl^- with SO_4^{2-} completely inhibited dye accumulation (and hence acidification), an effect reversed by adding Cl^- back to the bathing medium. Later, in the study that firmly established that lysosomal acidification is driven by an ATPase, Ohkuma et al. (3) prepared lysosomes containing a fluorescent dextran derivative by injecting the labeled dextran into rats and subsequently isolating lysosomes from those animals. After a short time, the dextran accumulated exclusively in lysosomes, and the fluorescein moiety provided a pH-dependent fluorescent signal. In these experiments, the removal of external Cl^- dramatically slowed acidification, whereas external cation replacement had little effect, supporting the involvement of anions as counterions. Of course, a cation-conducting counterion path would move cations out of the lysosomal lumen, but even in this case, changing the K^+ gradient might be expected to affect the rate of cation exit and therefore the acidification rate.

Further information on the role of cations came from experiments adding the K^+ ionophore valinomycin and the H^+ ionophore FCCP in different orders. Each of these agents makes the lysosome membrane highly and specifically permeable to its ion of choice. In experiments adding these ionophores, Ohkuma et al. (3) found that adding either one alone had minimal effects on lysosomal pH. In contrast, adding both agents rapidly dissipated the pH gradient. If there was a substantial K^+ permeability in the lysosome membrane before valinomycin addition, then FCCP alone should have dissipated the pH gradient. Together, these experiments suggest that anion movements are the primary path for counterions in this preparation.

Similarly, experiments measuring the lysosomal voltage point to significant Cl^- permeability and limited K^+ permeability. Using the voltage-sensitive dye diS-C_3(5), Harikumar & Reeves (13) found small but measurable effects of K^+ addition on dissipating the lysosomal membrane potential, whereas Cl^- and other anions were more effective. Although similar experiments by Ohkuma et al. (12) revealed a higher K^+ permeability, such experiments were performed with tritosomes, lysosomes with altered buoyant density resulting from the injection of triton

WR-1339 into rats. The authors point out that this manipulation changes the K^+ permeability; the ion is much less permeant in untreated lysosomes (12). Finally, using the anionic voltage–sensitive dye merocyanine 540, Cuppoletti and coworkers (31) found that ATP addition induced a large positive shift in membrane potential that was reversed in a concentration-dependent manner by the addition of Cl^-, but not by the addition of SO_4^{2-}. A thorough study of endosome and lysosome acidification by Van Dyke (32) supports the previous general conclusions regarding the counterion pathway, with acidification possible but slowed in media with either K^+ or Cl^- replaced with an impermeant ion and a generally low permeability of the lysosomal membrane to other physiological ions. Together, these experiments support the conclusion that both Cl^- and K^+ facilitate lysosomal acidification, with Cl^- perhaps the primary ion contributing in some circumstances and both ions contributing in others.

Remarkably, despite many demonstrations of important roles for Cl^- in lysosomal acidification, none of these studies directly demonstrated Cl^- flux in the organelle. Such flux was recently established by work from my lab (33) in isolated rat liver lysosomes and was shown, surprisingly, to result not from the action of a Cl^-, as had been assumed, but by a transporter exchanging two Cl^- ions moving in one direction for a single proton moving in the opposite direction. These experiments also revealed that this Cl^-/H^+ antiporter is the primary pathway for Cl^- movement across the lysosomal membrane and facilitated identification of the molecular basis for this activity, as I discuss below.

In summary, multiple studies using isolated lysosomes to examine functional properties of the lysosomal acidification process agree that the primary permeability of native lysosomal membranes is to monovalent anions and cations, with both types of ions potentially serving as counterions for the V-ATPase. The relative contributions of these ions remain to be clearly established, but promising new methods may provide the means to determine these contributions.

The Counterion Pathway in Lysosomes in Intact Cells

Most efforts to probe lysosomal counterion conductances have used isolated lysosomes. These preparations have great advantages insofar as the primary membrane in the sample is lysosomal and the ionic compositions, at least outside and sometimes on both sides, of the membrane can be well controlled. However, the ideal situation to study acidification is in lysosomes within living cells. Studies of lysosomal ion dynamics in vivo have been limited by the difficulty of adjusting cytoplasmic conditions across an intact plasma membrane. Recent work from Grinstein's lab (34) sought to improve the study of lysosomes in intact cells (RAW macrophages) by using a creative approach to breach the plasma membrane. These cells natively express the $P2X_7$ receptor, an ATP-activated ion channel (35). Under normal activating conditions, $P2X_7$ receptors are ligand-activated, cation-selective ion channels. However, upon prolonged stimulation with ATP, these channels open a larger pore, possibly formed by a pannexin protein, that is permeable to molecules of up to ~900 Da (36). By activating $P2X_7$ in RAW cells and by bathing the cells in varying solutions, the investigators could dialyze the cytoplasm and significantly alter the cells' ion compositions.

Using this method, Grinstein's group (34) attempted to test the importance of Cl^- for lysosomal acidification by dialyzing into their cells a Cl^--free medium. Under these conditions, lysosomes whose pH gradients had been dissipated with FCCP could reacidify equally well with or without Cl^- in the bathing medium. On the basis of this observation, the investigators concluded that cytoplasmic Cl^- is not required for lysosomal acidification. However, quantitative chemical analysis revealed nearly 10 mM residual Cl^- remaining in the dialyzed cytoplasm of the Cl^--free cells. On the basis of the Cl⁻ dependence of acidification (32), this is likely to be a sufficient concentration for Cl⁻ to serve as a counterion.

Further experiments using the $P2X_7$ system explored the role of cations in lysosomal acidification. However, because the relevant ions in this case must reside in the lysosome lumen, further manipulations were required to manipulate their concentrations. To change luminal cation concentrations, the investigators permeabilized the plasma membrane as described above, bathed the cells in solutions containing altered cation concentrations, resealed the PM by deactivating the $P2X_7$ receptors, and then added a dipeptide, Gly-Phe-β-naphthylamide, to the cells. This membrane-permeant peptide diffuses into the cells and into lysosomes, where it is cleaved by the peptidase cathepsin C (found only in lysosomes) (37). The products are thought to accumulate in lysosomes and, because they are osmotically active, to cause the organelles to swell and partially rupture, thereby allowing the cytoplasmic solution to equilibrate with the lysosomal interior (37, 38). Remarkably, removal of the peptide seems to restore lysosomal integrity and to allow the organelles to reacidify (34). Using these methods, Grinstein and colleagues (34) determined that removing permeant cations (primarily K^+) from the lysosome lumen raised its luminal pH significantly, supporting earlier indications of a cation conductance contributing to lysosomal acidification. Although the methods used in this paper require further validation and may be limited in applicability to cell types expressing the appropriate proteins, this general approach is very promising, offering the possibility of studying detailed lysosomal transport mechanisms in conditions approaching the cells' native state.

THE COUNTERION PATHWAY IN THE MOLECULAR ERA

The results discussed above reveal a great deal about the functional aspects of the counterion conductance, but they do not directly address the molecular identity of the transporters and channels that might be involved. Over the past two decades, a variety of molecules, including both Cl⁻ channels and transporters and cation channels, have been proposed to play this role, with varying degrees of experimental support.

The Cystic Fibrosis Transmembrane Regulator

The cystic fibrosis transmembrane regulator (CFTR) is the protein mutated in patients with the disease cystic fibrosis. This protein is a member of the very broad ATP-binding cassette (ABC) transporter family, whose members use ATP hydrolysis to drive a huge array of substrates uphill against their concentration gradients (39). CFTR, however, is unique among ABC transporters as an ATP-activated and phosphorylation-activated Cl⁻ channel, facilitating downhill anion flux (40). As the first Cl⁻ channel to be identified at the molecular level, CFTR proved to be a tempting candidate for an organellar anion shunt. However, it has had a checkered history in that role (nicely reviewed in Reference 41). Over the years several groups have reported observed changes in organellar pH in cells from cystic fibrosis patients (42) and in phagosomes from $Cftr^{-/-}$ mice (43, 44). However, further work has questioned these results; several careful studies revealed no pH changes in endosomes or the Golgi complex due to CFTR mutations (45–47). The methodology used in the study of phagosomal CFTR was also questioned, and further examination revealed no pH change in phagosomes in several cell lines with pharmacological inhibition of CFTR or knockout of the gene (48, 49). Furthermore, the limited tissue distribution of CFTR expression could at best account for acidification in a very limited subset of lysosomes. Given the overall picture, CFTR is unlikely to be an important player in organellar acidification, particularly in the lysosome.

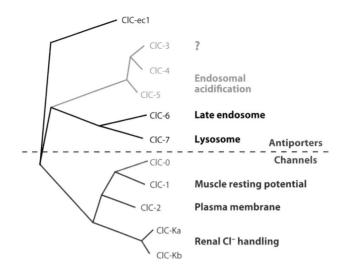

Figure 4
Phylogenetic tree of the mammalian CLC proteins. The antiporters are shown in blue, with different shades for the two antiporter subfamilies, and the channels in red; the antiporters and channels are separated by a dashed line. For reference, ClC-ec1, an antiporter from *Escherichia coli*, and ClC-0, the first ClC discovered (from *Torpedo*), are also indicated. Physiological functions are noted where known.

The CLC Family of Chloride Channels and Transporters

Another candidate for the anion conductance in intracellular organelles, including the lysosome, is the CLC family of Cl^- channels and transporters. The first known member of this family was originally identified as a voltage-dependent Cl^- channel in the electric organ of the *Torpedo* ray by White & Miller (50) and was cloned by Jentsch et al. (51). Since that time, the CLCs have grown to be the largest known family of Cl^- transporters (**Figure 4**), with nine known mammalian isoforms as well as homologs in yeast, plants, invertebrates, and a wide range of bacteria and archaea (52). CLC proteins have been implicated in a range of human diseases from myotonia to disorders of renal transport to osteopetrosis (53). X-ray structures of bacterial (54) and eukaryotic (55) CLCs reveal the proteins to be homodimers, consistent with functional (56) and biochemical (57) data. Recently, Accardi, Miller, and colleagues (58, 59) found that a bacterial CLC homolog is not a channel, but rather a Cl^-/H^+ antiporter that couples the downhill movement of two Cl^- ions to the uphill movement of a single proton (or vice versa). Indeed, some of the mammalian CLC isoforms are also antiporters (**Figure 4**) (60, 61). Notably, these proteins, ClC-4 and ClC-5, are localized to the endosome and may serve as counterion pathways in that organelle (62–64).

Could ClC-7 Be a Lysosomal Counterion Pathway?

Members of the CLC family of Cl^- channels and transporters were first proposed to have a role in lysosomal acidification on the basis of the results of a mouse knockout study of ClC-7 from the Jentsch lab (65). ClC-7 (and the closely related ClC-6) was cloned in 1995 and was shown to be broadly expressed on the basis of Northern blots (66). However, neither of these proteins could be functionally expressed, and so their function remained uncertain. This situation changed when the Jentsch lab (65) created a ClC-7 knockout mouse. These mice are gravely ill and die within 30 days of birth. Careful analysis of their pathology revealed severe osteopetrosis, or hypercalcification of bone, leading to growth retardation and deformation. This pathology results from a loss of function of bone-resorbing osteoclasts. In wild-type animals, these cells form a large acidic compartment in contact with the bone matrix (the so-called ruffled border). In knockout animals, osteoclasts develop but do not acidify the ruffled border (on the basis of acridine orange fluorescence); therefore, such osteoclasts do not resorb bone. The authors noted that the widely expressed ClC-7 is localized to lysosomes in many cell types. Because the ruffled border is formed from the fusion of acidic lysosome-related organelles with the plasma membrane, these investigators suggested that ClC-7 contributes to the counterion pathway in the ruffled border and, by analogy, in lysosomes. Both severe and benign human osteopetroses can also result from mutations in ClC-7 (65, 67–71).

Using isolated HeLa cell lysosomes and intact HeLa cells, my lab (33) recently tested the hypothesis that ClC-7 is part of the lysosomal counterion pathway. Because ClC-7 is ubiquitously expressed (as would be expected for a counterion conductance), we used siRNA to transiently knock down baseline ClC-7 expression. Lysosomes in HeLa cells have a Cl^-/H^+ antiporter activity similar to the one described above in rat liver lysosomes. Knockdown of ClC-7 expression with siRNA abolished this antiport activity. If this antiporter participates in the counterion pathway, then its knockdown (or inhibition) should inhibit lysosomal acidification. Indeed, staining living HeLa cells with LysoTracker Green (a weak base that is similar to acridine orange and that concentrates in and stains acidic organelles) showed reduced staining in ClC-7 knockout cells compared with both wild-type and control siRNA-transfected cells (33). This reduced staining provides a crude assessment of lysosomal pH, suggesting a decrease in lysosomal acidity in the knockout cells, as would be predicted if ClC-7 were a major part of the counterion pathway in these cells.

Further results from ClC-7 knockout mice complicate the picture, however. Several sets of quantitative measurements (using Oregon Green 488 dextran) on cells from these animals (34, 72, 73) reveal no significant change in lysosomal pH for the ClC-7 knockout compared with wild-type cells. Also, cells defective in Ostm1, a β subunit needed for proper targeting of ClC-7 to lysosomes, also maintain acidic pH in their lysosomes (73). What could account for these differences? The dye used in the HeLa lysosome experiments is qualitative (74). However, like acridine orange, such dyes can be useful for simple conclusions. Furthermore, similar experiments with ratiometric dextran dyes yield similar results (S.B. Lioi & J.A. Mindell, unpublished observations). Several other explanations for these differences must be considered. First, knockout animals compensate for defects due to their lost protein expression (75, 76). Knockout mice may have pH-regulatory mechanisms in their lysosomes that allow them to adapt other transport proteins to maintain pH. Indeed, the Jentsch lab (77) found increased levels of retargeting of ClC-3 and ClC-6 in lysosomal fractions from the brain in the ClC-7 knockouts. Moreover, these experiments were performed in different cell types. Lysosomes in different tissues may use different combinations of channels and transporters to maintain pH in the context of the very different digestive demands placed on them. Little is known about how acidification might be tuned in a tissue-specific manner.

Given the lack of pH change in ClC-7 knockout cells, it is surprising that other functional changes are apparent. Wartosch et al. (78) injected a fluorescently labeled protein (β-lactoglobulin) into mice with a kidney-specific ClC-7 knockout. In these animals the kidneys are chimeric for the ClC-7 knockout. These researchers sacrificed the animals at varying times and monitored the degradation and release of the labeled protein using both Western blots and imaging of tissue slices. They found that, though the labeled protein was delivered efficiently to the lysosomes, the ClC-7 knockout cells in the kidney were substantially slower at degrading the β-lactoglobulin than were the wild-type cells. Given the lack of a pH change in their experiments, the authors suggested that ClC-7 may utilize the pH gradient to maintain Cl^- in the lysosome at concentrations higher than the equilibrium concentration. These authors point out that cathepsin C, a lysosomal protease, shows [Cl⁻]-dependent activity (79), raising the possibility that the regulation of intralysosomal Cl⁻ may be an important role for ClC-7.

In summary, there is broad agreement that ClC-7 is a $2Cl^-/1H^+$ antiporter localized in the lysosomal membrane and that it accounts for the observed Cl⁻ permeability of the lysosomal membrane. But how much it contributes to the essential counterion pathway or to regulating lysosomal [Cl⁻] remains an open question.

Role of Proton-Chloride Coupling in ClC-7 Function

Whatever the physiological task of ClC-7, one of the remarkable observations regarding the intracellular CLCs concerns the role of coupled H^+/Cl^- transport in their function. The CLC family includes both ion channels and antiporters (divided approximately half and half in mammals). Consistently, all the channels function in the plasma membrane, whereas all the antiporters function in intracellular organelle membranes. Thus, evolutionary pressures must have maintained this sorting integrity. How is proton coupling related to ClC-7 function? This question was recently addressed by the Jentsch lab, which created mice in which the wild-type ClC-7 antiporter was replaced with a mutant form of the protein lacking the essential gating glutamate (referred to as UNC, denoting uncoupled). This glutamate residue is essential for coupled H^+/Cl^- transport (58, 61) in the CLC antiporters and for voltage-dependent gating (80) in the CLC channels. Mutations at this site eliminate proton transport and essentially yield a passive Cl⁻ uniporter (58)—a transporter that moves Cl⁻ only down its electrochemical gradient. Remarkably, mice carrying this mutant transporter recapitulate much of the phenotype of the total ClC-7 knockout (81), including osteopetrosis, growth retardation, and the accumulation of lysosomal storage material, although the phenotype is milder than that of the full knockout. Analysis of acidification in lysosomes from the mutant mice revealed a complex picture. The organelles could not support measurable Cl^-/H^+ antiport but could acidify, as measured using Oregon Green dextran fluorescence. Indeed, in an in vitro assay on isolated lysosomes, the uncoupled mutants seemed to acidify to a slightly lower pH in the uncoupled lysosomes than did the wild type, whereas the full knockout acidified to a slightly higher pH. In living cells, however, lysosomal pH appeared identical in cells from wild-type, knockout, and UNC mice. Experiments using a novel fluorescent Cl⁻ indicator dye hint at higher luminal [Cl⁻] in the UNC and knockout lysosomes, but because the dye is not calibrated to known Cl⁻ concentrations, it is difficult to interpret these results.

This work presents a complex and difficult picture to interpret. The UNC ClC-7 causes almost as severe a phenotype as does the complete knockout. Taken as a whole, the data support models of ClC-7 function as a counterion pathway, a Cl⁻-concentrating mechanism, or both.

New Possibilities: Active Regulation of Lysosomal pH

Recent work from the Maxfield lab (82–84) may help clarify the role of ClC-7 as well as expanding our understanding of the dynamic features of lysosomal acidification. Majumdar and coworkers noted that conflicting results had been reported regarding the degradation of amyloid Aβ, a key molecule in the pathology of Alzheimer's disease. Specifically, whereas primary cultures of microglia (central nervous system macrophages) can internalize but cannot degrade Aβ peptide, either microglia activated by passive immunization or macrophages can both internalize and degrade the Aβ peptide (82). Investigating this difference, Majumdar et al. (83) found that microglia contain higher levels of lysosomal proteases than do similar macrophages but that the microglia lysosomes were substantially more basic than were lysosomes of macrophages (pH ∼ 6 versus pH ∼ 5, respectively). However, when the microglia were activated by treatment with macrophage colony–stimulating factor (MCSF) or interleukin-6, their lysosomal pH dropped to ∼5, and they became more effective at digesting Aβ peptide (83). Probing the mechanism of this effect, the authors further found that ClC-7 in the quiescent microglia was not targeted primarily to lysosomes; instead, it was apparently destined for degradation by the proteasome (84). However, upon activation ClC-7 was recruited to lysosomes. Furthermore, knockdown of ClC-7 expression with siRNA prevented the effects of MCSF, including the retargeting of ClC-7, the reduction of Aβ degradation, and increased acidification (84).

These results have profound implications for our appreciation of the subtleties of lysosomal acidification. First, they provide further evidence that ClC-7 is an important part of the acidification mechanism, presumably due to its role as a counterion pathway. In addition, they suggest that lysosomal acidification is more dynamic and more regulated than previously considered.

Candidates for the Cation Pathway

Whereas there are well-established candidate transporters for the anion-moving portion of the lysosomal counterion pathway, candidates for the cation-transporting component are more tentative. One candidate is the cation channel TRPML1. This channel is a member of the TRP (which denotes transient receptor potential, reflecting the effect on the *Drosophila* retina of mutating the founding member) channel family. This family also includes the channels that sense heat and cold, among many others (85). The TRP channels are tetrameric cation channels with a six-transmembrane domain architecture similar to that of the voltage-gated K^+, Na^+, and Ca^{2+} channels and are activated by a wide range of stimuli, including G protein–coupled receptor interactions, ligand activation, and temperature (85). TRPML1 is encoded by the *MCOLN1* gene, which is mutated in the lysosomal storage disorder mucolipidosis type IV (MLIV). This autosomal recessive disease is characterized by a slowly progressing neurodegenerative phenotype (86). The TRPML1 protein is localized to lysosomes, and its disruption (in MLIV patient fibroblasts) increases (87), maintains (88), or even decreases (89) lysosomal pH, creating confusion as to its role in the acidification process. Furthermore, TRPML1 mediates iron release from endolysosomes (90) and plays a role in lysosomal Ca^{2+} release (91), and its dysfunction leads to a range of other lysosomal phenotypes. The actual role of TRPML1 remains to be clearly defined, although it may serve many of its proposed roles. Finally, a two-pore channel, TPC2, localizes to lysosomes and plays a role in Ca^{2+} release from the organelles (92). Whether this channel has a role in pH regulation remains to be seen.

CONCLUSIONS

In conclusion, although many of the molecular players in lysosomal acidification (and organellar acidification in general) are now known, the mechanisms by which these transporters work together (along with luminal buffers, membrane voltage, and other factors) remain unclear. New possibilities, like the dynamic regulation of lysosomal pH, are emerging and must be incorporated into any comprehensive framework.

SUMMARY POINTS

1. Multiple factors influence lysosomal acidification.
2. V-ATPases, the primary drivers of acidification, convert metabolic energy into proton gradients. Maximally efficient ATPases may acidify organelles to lower pHs than are observed in most compartments.

3. Because the V-ATPase is electrogenic, counterions must move to dissipate voltage and to facilitate bulk proton transport.
4. Extensive functional analysis on isolated lysosomes clearly demonstrates both anion and cation permeabilities, although the anion pathway is more consistently observed.
5. CFTR is unlikely to provide the anion pathway in most lysosomes.
6. ClC-7 is a good candidate for the anion pathway, but open questions remain as to its relevance, particularly the lack of pH change in knockout mouse lysosomes.
7. No cation channel has emerged as a consensus candidate for a cation-selective counterion pathway.

FUTURE ISSUES

1. The roles of V-ATPase isoforms need clarification.
2. The relative contributions of anion and cation transport to the counterion pathway should be quantitatively determined.
3. Tissue-specific differences in acidification mechanisms require exploration.
4. The molecular basis of the lysosomal cation permeability needs to be determined.
5. How does the proton-dependent transport of substrates out of lysosomes affect the acidification process?

DISCLOSURE STATEMENT

The author is not aware of any affiliations, memberships, funding, or financial holdings that might be perceived as affecting the objectivity of this review.

LITERATURE CITED

1. Metchnikoff E. 1968 (1893). *Lectures on the Comparative Pathology of Inflammation*. New York: Dover
2. Coffey JW, De Duve C. 1968. Digestive activity of lysosomes. I. The digestion of proteins by extracts of rat liver lysosomes. *J. Biol. Chem.* 243:3255–63
3. Ohkuma S, Moriyama Y, Takano T. 1982. Identification and characterization of a proton pump on lysosomes by fluorescein-isothiocyanate-dextran fluorescence. *Proc. Natl. Acad. Sci. USA* 79:2758–62
4. Forgac M. 2007. Vacuolar ATPases: rotary proton pumps in physiology and pathophysiology. *Nat. Rev. Mol. Cell Biol.* 8:917–29
5. Hirata T, Iwamoto-Kihara A, Sun-Wada GH, Okajima T, Wada Y, Futai M. 2003. Subunit rotation of vacuolar-type proton pumping ATPase: relative rotation of the G and C subunits. *J. Biol. Chem.* 278:23714–19
6. Yokoyama K, Nakano M, Imamura H, Yoshida M, Tamakoshi M. 2003. Rotation of the proteolipid ring in the V-ATPase. *J. Biol. Chem.* 278:24255–58
7. Arai H, Terres G, Pink S, Forgac M. 1988. Topography and subunit stoichiometry of the coated vesicle proton pump. *J. Biol. Chem.* 263:8796–802
8. Johnson RG, Beers MF, Scarpa A. 1982. H^+ ATPase of chromaffin granules. Kinetics, regulation, and stoichiometry. *J. Biol. Chem.* 257:10701–7

9. Schmidt AL, Briskin DP. 1993. Reversal of the red beet tonoplast H$^+$-ATPase by a pyrophosphate-generated proton electrochemical gradient. *Arch. Biochem. Biophys.* 306:407–14
10. Grabe M, Wang H, Oster G. 2000. The mechanochemistry of V-ATPase proton pumps. *Biophys. J.* 78:2798–813
11. Muller ML, Jensen M, Taiz L. 1999. The vacuolar H$^+$-ATPase of lemon fruits is regulated by variable H$^+$/ATP coupling and slip. *J. Biol. Chem.* 274:10706–16
12. Ohkuma S, Moriyama Y, Takano T. 1983. Electrogenic nature of lysosomal proton pump as revealed with a cyanine dye. *J. Biochem.* 94:1935–43
13. Harikumar P, Reeves JP. 1983. The lysosomal proton pump is electrogenic. *J. Biol. Chem.* 258:10403–10
14. Hedrich R, Kurkdjian A, Guern J, Flugge UI. 1989. Comparative studies on the electrical properties of the H$^+$ translocating ATPase and pyrophosphatase of the vacuolar-lysosomal compartment. *EMBO J.* 8:2835–41
15. Kettner C, Bertl A, Obermeyer G, Slayman C, Bihler H. 2003. Electrophysiological analysis of the yeast V-type proton pump: variable coupling ratio and proton shunt. *Biophys. J.* 85:3730–38
16. Wieczorek H, Grber G, Harvey WR, Huss M, Merzendorfer H, Zeiske W. 2000. Structure and regulation of insect plasma membrane H$^+$ V-ATPase. *J. Exp. Biol.* 203:127–35
17. Kane PM. 1995. Disassembly and reassembly of the yeast vacuolar H$^+$-ATPase in vivo. *J. Biol. Chem.* 270:17025–32
18. Seol JH, Shevchenko A, Deshaies RJ. 2001. Skp1 forms multiple protein complexes, including RAVE, a regulator of V-ATPase assembly. *Nat. Cell Biol.* 3:384–91
19. Smardon AM, Tarsio M, Kane PM. 2002. The RAVE complex is essential for stable assembly of the yeast V-ATPase. *J. Biol. Chem.* 277:13831–39
20. Smardon AM, Kane PM. 2007. RAVE is essential for the efficient assembly of the C subunit with the vacuolar H$^+$-ATPase. *J. Biol. Chem.* 282:26185–94
21. Lu M, Ammar D, Ives H, Albrecht F, Gluck SL. 2007. Physical interaction between aldolase and vacuolar H$^+$-ATPase is essential for the assembly and activity of the proton pump. *J. Biol. Chem.* 282:24495–503
22. Xu T, Forgac M. 2001. Microtubules are involved in glucose-dependent dissociation of the yeast vacuolar [H$^+$]-ATPase in vivo. *J. Biol. Chem.* 276:24855–61
23. Manolson MF, Proteau D, Preston RA, Stenbit A, Roberts BT, et al. 1992. The *VPH1* gene encodes a 95-kDa integral membrane polypeptide required for in vivo assembly and activity of the yeast vacuolar H$^+$-ATPase. *J. Biol. Chem.* 267:14294–303
24. Manolson MF, Wu B, Proteau D, Taillon BE, Roberts BT, et al. 1994. STV1 gene encodes functional homologue of 95-kDa yeast vacuolar H$^+$-ATPase subunit Vph1p. *J. Biol. Chem.* 269:14064–74
25. Kawasaki-Nishi S, Nishi T, Forgac M. 2001. Yeast V-ATPase complexes containing different isoforms of the 100-kDa a-subunit differ in coupling efficiency and in vivo dissociation. *J. Biol. Chem.* 276:17941–48
26. Toei M, Saum R, Forgac M. 2010. Regulation and isoform function of the V-ATPases. *Biochemistry* 49:4715–23
27. Smith AN, Jouret F, Bord S, Borthwick KJ, Al-Lamki RS, et al. 2005. Vacuolar H$^+$-ATPase d2 subunit: molecular characterization, developmental regulation, and localization to specialized proton pumps in kidney and bone. *J. Am. Soc. Nephrol.* 16:1245–56
28. Lee SH, Rho J, Jeong D, Sul JY, Kim T, et al. 2006. v-ATPase V$_0$ subunit d2-deficient mice exhibit impaired osteoclast fusion and increased bone formation. *Nat. Med.* 12:1403–9
29. Karet FE, Finberg KE, Nelson RD, Nayir A, Mocan H, et al. 1999. Mutations in the gene encoding B1 subunit of H$^+$-ATPase cause renal tubular acidosis with sensorineural deafness. *Nat. Genet.* 21:84–90
30. Dell'Antone P. 1979. Evidence for an ATP-driven "proton pump" in rat liver lysosomes by basic dyes uptake. *Biochem. Biophys. Res. Commun.* 86:180–89
31. Cuppoletti J, Aures-Fischer D, Sachs G. 1987. The lysosomal H$^+$ pump: 8-azido-ATP inhibition and the role of chloride in H$^+$ transport. *Biochim. Biophys. Acta* 899:276–84
32. Van Dyke RW. 1993. Acidification of rat liver lysosomes: quantitation and comparison with endosomes. *Am. J. Physiol. Cell Physiol.* 265:901–17
33. Graves AR, Curran PK, Smith CL, Mindell JA. 2008. The Cl$^-$/H$^+$ antiporter ClC-7 is the primary chloride permeation pathway in lysosomes. *Nature* 453:788–92

34. Steinberg BE, Huynh KK, Brodovitch A, Jabs S, Stauber T, et al. 2010. A cation counterflux supports lysosomal acidification. *J. Cell Biol.* 189:1171–86
35. Pelegrin P, Barroso-Gutierrez C, Surprenant A. 2008. P2X7 receptor differentially couples to distinct release pathways for IL-1β in mouse macrophage. *J. Immunol.* 180:7147–57
36. Pelegrin P, Surprenant A. 2006. Pannexin-1 mediates large pore formation and interleukin-1β release by the ATP-gated P2X7 receptor. *EMBO J.* 25:5071–82
37. Berg TO, Stromhaug E, Lovdal T, Seglen O, Berg T. 1994. Use of glycyl-L-phenylalanine 2-naphthylamide, a lysosome-disrupting cathepsin C substrate, to distinguish between lysosomes and prelysosomal endocytic vacuoles. *Biochem. J.* 300(Pt. 1):229–36
38. McGuinness L, Bardo SJ, Emptage NJ. 2007. The lysosome or lysosome-related organelle may serve as a Ca^{2+} store in the boutons of hippocampal pyramidal cells. *Neuropharmacology* 52:126–35
39. Jones PM, O'Mara ML, George AM. 2009. ABC transporters: a riddle wrapped in a mystery inside an enigma. *Trends Biochem. Sci.* 34:520–31
40. Gadsby DC, Vergani P, Csanady L. 2006. The ABC protein turned chloride channel whose failure causes cystic fibrosis. *Nature* 440:477–83
41. Haggie PM, Verkman AS. 2009. Defective organellar acidification as a cause of cystic fibrosis lung disease: reexamination of a recurring hypothesis. *Am. J. Physiol. Lung Cell Mol. Physiol.* 296:859–67
42. Barasch J, Kiss B, Prince A, Saiman L, Gruenert D, al-Awqati Q. 1991. Defective acidification of intracellular organelles in cystic fibrosis. *Nature* 352:70–73
43. Di A, Brown ME, Deriy LV, Li C, Szeto FL, et al. 2006. CFTR regulates phagosome acidification in macrophages and alters bactericidal activity. *Nat. Cell Biol.* 8:933–44
44. Deriy LV, Gomez EA, Zhang G, Beacham DW, Hopson JA, et al. 2009. Disease-causing mutations in the cystic fibrosis transmembrane conductance regulator determine the functional responses of alveolar macrophages. *J. Biol. Chem.* 284:35926–38
45. Seksek O, Biwersi J, Verkman AS. 1996. Evidence against defective trans-Golgi acidification in cystic fibrosis. *J. Biol. Chem.* 271:15542–48
46. Biwersi J, Verkman AS. 1994. Functional CFTR in endosomal compartment of CFTR-expressing fibroblasts and T84 cells. *Am. J. Physiol. Cell Physiol.* 266:149–56
47. Lukacs GL, Chang XB, Kartner N, Rotstein OD, Riordan JR, Grinstein S. 1992. The cystic fibrosis transmembrane regulator is present and functional in endosomes. Role as a determinant of endosomal pH. *J. Biol. Chem.* 267:14568–72
48. Haggie PM, Verkman AS. 2007. Cystic fibrosis transmembrane conductance regulator-independent phagosomal acidification in macrophages. *J. Biol. Chem.* 282:31422–28
49. Haggie PM, Verkman AS. 2009. Unimpaired lysosomal acidification in respiratory epithelial cells in cystic fibrosis. *J. Biol. Chem.* 284:7681–86
50. White MM, Miller C. 1979. A voltage-gated anion channel from the electric organ of *Torpedo californica*. *J. Biol. Chem.* 254:10161–66
51. Jentsch TJ, Steinmeyer K, Schwarz G. 1990. Primary structure of *Torpedo marmorata* chloride channel isolated by expression cloning in *Xenopus* oocytes. *Nature* 348:510–14
52. Jentsch TJ. 2008. CLC chloride channels and transporters: from genes to protein structure, pathology and physiology. *Crit. Rev. Biochem. Mol. Biol.* 43:3–36
53. Planells-Cases R, Jentsch TJ. 2009. Chloride channelopathies. *Biochim. Biophys. Acta* 1792:173–89
54. Dutzler R, Campbell EB, Cadene M, Chait BT, MacKinnon R. 2002. X-ray structure of a ClC chloride channel at 3.0 Å reveals the molecular basis of anion selectivity. *Nature* 415:287–94
55. Feng L, Campbell EB, Hsiung Y, Mackinnon R. 2010. Structure of a eukaryotic CLC transporter defines an intermediate state in the transport cycle. *Science* 330:635–41
56. Miller C. 1982. Open-state substructure of single chloride channels from *Torpedo electroplax*. *Philos. Trans. R. Soc. Lond. Ser. B* 299:401–11
57. Middleton RE, Pheasant DJ, Miller C. 1996. Homodimeric architecture of a ClC-type chloride ion channel. *Nature* 383:337–40
58. Accardi A, Miller C. 2004. Secondary active transport mediated by a prokaryotic homologue of ClC Cl^- channels. *Nature* 427:803–7

59. Accardi A, Kolmakova-Partensky L, Williams C, Miller C. 2004. Ionic currents mediated by a prokaryotic homologue of CLC Cl$^-$ channels. *J. Gen. Physiol.* 123:109–19
60. Scheel O, Zdebik AA, Lourdel S, Jentsch TJ. 2005. Voltage-dependent electrogenic chloride/proton exchange by endosomal CLC proteins. *Nature* 436:424–27
61. Picollo A, Pusch M. 2005. Chloride/proton antiporter activity of mammalian CLC proteins ClC-4 and ClC-5. *Nature* 436:420–23
62. Gunther W, Luchow A, Cluzeaud F, Vandewalle A, Jentsch TJ. 1998. ClC-5, the chloride channel mutated in Dent's disease, colocalizes with the proton pump in endocytotically active kidney cells. *Proc. Natl. Acad. Sci. USA* 95:8075–80
63. Mohammad-Panah R, Harrison R, Dhani S, Ackerley C, Huan LJ, et al. 2003. The chloride channel ClC-4 contributes to endosomal acidification and trafficking. *J. Biol. Chem.* 278:29267–77
64. Piwon N, Gunther W, Schwake M, Bosl MR, Jentsch TJ. 2000. ClC-5 Cl$^-$-channel disruption impairs endocytosis in a mouse model for Dent's disease. *Nature* 408:369–73
65. Kornak U, Kasper D, Bosl MR, Kaiser E, Schweizer M, et al. 2001. Loss of the ClC-7 chloride channel leads to osteopetrosis in mice and man. *Cell* 104:205–15
66. Brandt S, Jentsch TJ. 1995. ClC-6 and ClC-7 are two novel broadly expressed members of the CLC chloride channel family. *FEBS Lett.* 377:15–20
67. Cleiren E, Benichou O, Van Hul E, Gram J, Bollerslev J, et al. 2001. Albers-Schönberg disease (autosomal dominant osteopetrosis, type II) results from mutations in the *ClCN7* chloride channel gene. *Hum. Mol. Genet.* 10:2861–67
68. Campos-Xavier AB, Saraiva JM, Ribeiro LM, Munnich A, Cormier-Daire V. 2003. Chloride channel 7 (*CLCN7*) gene mutations in intermediate autosomal recessive osteopetrosis. *Hum. Genet.* 112:186–89
69. Henriksen K, Gram J, Schaller S, Dahl BH, Dziegiel MH, et al. 2004. Characterization of osteoclasts from patients harboring a G215R mutation in ClC-7 causing autosomal dominant osteopetrosis type II. *Am. J. Pathol.* 164:1537–45
70. Frattini A, Pangrazio A, Susani L, Sobacchi C, Mirolo M, et al. 2003. Chloride channel *ClCN7* mutations are responsible for severe recessive, dominant, and intermediate osteopetrosis. *J. Bone Miner. Res.* 18:1740–47
71. Kornak U, Ostertag A, Branger S, Benichou O, de Vernejoul MC. 2006. Polymorphisms in the *CLCN7* gene modulate bone density in postmenopausal women and in patients with autosomal dominant osteopetrosis type II. *J. Clin. Endocrinol. Metab.* 91:995–1000
72. Kasper D, Planells-Cases R, Fuhrmann JC, Scheel O, Zeitz O, et al. 2005. Loss of the chloride channel ClC-7 leads to lysosomal storage disease and neurodegeneration. *EMBO J.* 24:1079–91
73. Lange PF, Wartosch L, Jentsch TJ, Fuhrmann JC. 2006. ClC-7 requires Ostm1 as a β-subunit to support bone resorption and lysosomal function. *Nature* 440:220–23
74. DiCiccio JE, Steinberg BE. 2011. Lysosomal pH and analysis of the counter ion pathways that support acidification. *J. Gen. Physiol.* 137:385–90
75. Satyanarayana A, Kaldis P. 2009. Mammalian cell-cycle regulation: several Cdks, numerous cyclins and diverse compensatory mechanisms. *Oncogene* 28:2925–39
76. Rudmann DG, Durham SK. 1999. Utilization of genetically altered animals in the pharmaceutical industry. *Toxicol. Pathol.* 27:111–14
77. Poet M, Kornak U, Schweizer M, Zdebik AA, Scheel O, et al. 2006. Lysosomal storage disease upon disruption of the neuronal chloride transport protein ClC-6. *Proc. Natl. Acad. Sci. USA* 103:13854–59
78. Wartosch L, Fuhrmann JC, Schweizer M, Stauber T, Jentsch TJ. 2009. Lysosomal degradation of endocytosed proteins depends on the chloride transport protein ClC-7. *FASEB J.* 23:4056–68
79. Cigic B, Pain RH. 1999. Location of the binding site for chloride ion activation of cathepsin C. *Eur. J. Biochem.* 264:944–51
80. Dutzler R, Campbell EB, MacKinnon R. 2003. Gating the selectivity filter in ClC chloride channels. *Science* 300:108–12
81. Weinert S, Jabs S, Supanchart C, Schweizer M, Gimber N, et al. 2010. Lysosomal pathology and osteopetrosis upon loss of H$^+$-driven lysosomal Cl$^-$ accumulation. *Science* 328:1401–3
82. Majumdar A, Chung H, Dolios G, Wang R, Asamoah N, et al. 2008. Degradation of fibrillar forms of Alzheimer's amyloid β-peptide by macrophages. *Neurobiol. Aging* 29:707–15

83. Majumdar A, Cruz D, Asamoah N, Buxbaum A, Sohar I, et al. 2007. Activation of microglia acidifies lysosomes and leads to degradation of Alzheimer amyloid fibrils. *Mol. Biol. Cell* 18:1490–96
84. Majumdar A, Capetillo-Zarate E, Cruz D, Gouras GK, Maxfield FR. 2011. Degradation of Alzheimer's amyloid fibrils by microglia requires delivery of ClC-7 to lysosomes. *Mol. Biol. Cell* 22:1664–76
85. Ramsey IS, Delling M, Clapham DE. 2006. An introduction to TRP channels. *Annu. Rev. Physiol.* 68:619–47
86. Altarescu G, Sun M, Moore DF, Smith JA, Wiggs EA, et al. 2002. The neurogenetics of mucolipidosis type IV. *Neurology* 59:306–13
87. Bach G, Chen CS, Pagano RE. 1999. Elevated lysosomal pH in Mucolipidosis type IV cells. *Clin. Chim. Acta* 280:173–79
88. Pryor PR, Reimann F, Gribble FM, Luzio JP. 2006. Mucolipin-1 is a lysosomal membrane protein required for intracellular lactosylceramide traffic. *Traffic* 7:1388–98
89. Soyombo AA, Tjon-Kon-Sang S, Rbaibi Y, Bashllari E, Bisceglia J, et al. 2006. TRP-ML1 regulates lysosomal pH and acidic lysosomal lipid hydrolytic activity. *J. Biol. Chem.* 281:7294–301
90. Dong XP, Cheng X, Mills E, Delling M, Wang F, et al. 2008. The type IV mucolipidosis-associated protein TRPML1 is an endolysosomal iron release channel. *Nature* 455:992–96
91. Zhang F, Jin S, Yi F, Li P-L. 2009. TRP-ML1 functions as a lysosomal NAADP-sensitive Ca^{2+} release channel in coronary arterial myocytes. *J. Cell. Mol. Med.* 13:3174–85
92. Calcraft PJ, Ruas M, Pan Z, Cheng X, Arredouani A, et al. 2009. NAADP mobilizes calcium from acidic organelles through two-pore channels. *Nature* 459:596–600

Biology Without Walls: The Novel Endocrinology of Bone

Gerard Karsenty* and Franck Oury

Department of Genetics and Development, Columbia University, New York, NY 10032; email: gk2172@columbia.edu, fo2133@columbia.edu

Keywords

osteocalcin, serotonin, bone mass accrual, energy metabolism, reproduction

Abstract

Classical studies of vertebrate physiology have usually been confined to a given organ or cell type. The use of mouse genetics has changed this approach and has rejuvenated the concept of a whole-body study of physiology. One physiological system that has been profoundly influenced by mouse genetics is skeletal physiology. Indeed, genetic approaches have identified several unexpected organs that affect bone physiology. These new links have begun to provide a plausible explanation for the evolutionary involvement of hormones such as leptin with bone physiology. These genetic approaches have also revealed bone as a true endocrine organ capable of regulating energy metabolism and reproduction. Collectively, the body of work discussed below illustrates a new and unconventional role for bone in mammalian physiology.

*Corresponding author.

INTRODUCTION: THE OBVIOUS AND BEYOND

Physiology has been approached by two different methodologies. Molecular physiology is the most recent approach and deals with the function of one particular cell type or one protein (or group of proteins), usually with a focus on transcription factors, membrane surface receptors, or ion channels. This focused approach has provided a better understanding of molecular or cellular events and has contributed to novel rationales for effective treatments of human disease. However, the physiological approach that predates molecular methods is whole-organism physiology. Claude Bernard initially defined this aspect of physiology when he described the milieu interiéur. In the next century, W.C. Cannon forged the fundamental concept of homeostasis, and L.J. Henderson proposed that there is functional dependency between organs (1–4). Now, many decades later, the ability to spatially and temporally inactivate one gene in a particular cell type has provided critical experimental tools for studying and understanding whole-organism physiology. Below, we discuss how a whole-organism approach to physiology has influenced and transformed our understanding of bone physiology. This transformation was made possible by leveraging fully the concept of functional dependency and in all cases by providing a molecular basis for the novel functions that are described in detail in this review.

SPECIFIC FEATURES OF BONE AND THEIR IMPACT ON WHOLE-ORGANISM PHYSIOLOGY

To surmise which organs the skeleton and more precisely bones interact with, one needs to look at two characteristics of bone tissue. Bone is the only tissue in the body that contains a cell type whose function is to destroy (resorb) the host tissue, the osteoclast. Destruction of bone can be viewed as an autoimmune reaction and is required during bone modeling, which is responsible for linear growth in childhood, and during bone remodeling, which is responsible for maintenance of bone mass in adulthood (5–8).

Bone is also one of the largest tissues in the human body. This second important feature of bone implies that bone remodeling consumes a large amount of energy. For this reason, we hypothesized that bone remodeling must be coordinately regulated with energy metabolism (5, 9). Clinical evidence supports this hypothesis. For instance, anorexia nervosa in children leads to a complete arrest of skeletal growth. Likewise, adult anorectic patients develop osteoporosis, whereas adult obese patients display a higher bone mass that protects them from osteoporosis (10–14). Although these studies are subject to interpretation, they suggest a correlation between bone mass accrual and food intake.

Clinical experiences tell us one more thing. One of the most-established features of bone pathology is that osteoporosis, a low-bone-mass disease, appears after menopause (11, 15, 16). In other words, sex steroid hormones regulate bone mass. It is also possible that bone or bone-derived hormones reciprocate to regulate fertility. At a more global level, one wonders if bone mass accrual, energy metabolism, and reproduction are all coordinated by endocrine regulation (9). The possible cross talk between these three distinct physiological systems has several implications. First, such a coordinated regulation would begin with bony vertebrates because the energetic needs of bone modeling and remodeling justify its existence. Second, given the role of the brain in energy homeostasis, bone (re)modeling may be subject to central regulation. Finally, there are likely to be feedback loops that originate from bone and that affect energy metabolism. Current data supporting the hypothesis that bone mass, metabolism, and reproduction are linked are presented below.

BONE AS IT IS KNOWN TO BE: A TAKER

Coordinated Control of Bone Mass and Energy Metabolism: The Viewpoint of the Adipocyte (Part I)

Of all hormones known to regulate energy metabolism, leptin is the best candidate to test the hypothesis that bone mass, metabolism, and reproduction are linked. Leptin, an adipocyte-derived hormone, regulates appetite, energy expenditure, and fertility by signaling in the brain (9, 17–22). Thus, it fulfills many requirements of our hypothesis. A less obvious but equally important reason is that leptin appears during evolution with bone cells, not with appetite, reproduction, or adipocytes (23, 24). It is reasonable to assume that the appearance of a given gene during evolution coincides with the functional needs of a given organism. If we apply this general assumption to leptin, its appearance in bony vertebrates might suggest an endocrine link between the control of bone (re)modeling and the control of energy metabolism. This hypothesis has been addressed and was validated largely by using cell-specific loss-of-function leptin mouse models.

In addition to its role in energy metabolism, leptin is a powerful inhibitor of bone mass accrual. Hence, and in full agreement with the notion that there may be coregulation of bone (re)modeling and energy metabolism, leptin decreases both food intake and bone mass (5, 9, 22, 25, 26). Leptin is exceptionally powerful in inhibiting bone mass accrual. Indeed, mice or humans lacking leptin or its receptor develop a high-bone-mass phenotype even though they are hypogonadic, a condition that tends to greatly increase bone resorption. Only leptin signaling deficiency can achieve such a true biological tour de force. If this feature is taken at face value, it suggests that the inhibition of bone mass accrual may be a major function of leptin. This has been verified genetically through the use of a partial gain-of-function leptin signaling model that showed that leptin's regulation of bone mass requires a lower threshold of leptin signaling than that needed for regulation of energy metabolism and reproduction (9). This latter observation resonates with the above-mentioned fact that leptin and bone appear simultaneously during evolution.

Leptin regulation of bone mass accrual revealed for the first time the existence of central control of bone mass. There are two known mediators linking leptin signaling in the brain to the osteoblasts, the ultimate target cell of leptin. The first one is the sympathetic nervous system, signaling through the β_2-adrenergic receptor (Adrβ_2) present in osteoblasts (**Figure 1**) (20, 22). In the osteoblast, sympathetic tone recruits several transcriptional components of the molecular clock, cMyc and cAMP response element binding (CREB) protein, to inhibit cell proliferation (**Figure 1**) (27–34). Sympathetic tone also increases expression in osteoblasts of *RankL*, the most powerful osteoclast differentiation factor (**Figure 1**) (20, 28, 33). Thus, the sympathetic tone inhibits bone formation and favors bone resorption, which in turn reduces bone mass accrual (**Figure 1**) (5, 35–38). As a result, β-blockers antagonizing Adrβ_2 can cure osteoporosis in mice, rats, and humans (22, 39). The second mediator of leptin regulation of bone mass accrual is the cocaine amphetamine regulated transcript (CART), a peptide that is found in the brain and the general circulation and whose expression is regulated by leptin (20, 40–42). CART also acts on osteoblasts but inhibits *RankL* expression and bone resorption (**Figure 1**). This function of CART, the only one identified in CART-less or *cartpt*$^{-/-}$ mice maintained on a normal diet, is important because the absence of CART increases bone resorption, as seen in leptin signaling–deficient mice (**Figure 1**) (20, 43). That the sympathetic tone through Adrβ_2 and CART signaling is not involved in the control of appetite or energy expenditure in mice fed a normal diet or in fertility implies that if bone metabolism and energy metabolism are coregulated by the same molecules, such molecules must reside in the brain. The broader implication of these collective results is that the brain controls bone mass accrual (44, 45).

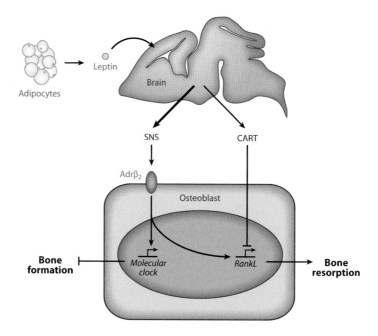

Figure 1

The sympathetic nervous system (SNS) and CART (cocaine amphetamine regulated transcript) mediate leptin signaling in the brain to the osteoblasts. The SNS inhibits bone formation and favors bone resorption. Following β_2-adrenergic receptor (Adrβ_2) activation in osteoblasts, the sympathetic tone favors expression of *RankL*, the most powerful osteoclast differentiation factor, and recruits several transcriptional components of the molecular clock, inhibiting bone formation. CART, the second mediator of the leptin regulation of bone mass accrual, also acts on osteoblasts, but by inhibiting *RankL* expression and bone resorption.

Coordinated Regulation of Bone Mass and Energy Metabolism: The Viewpoint of the Brain

The mechanism whereby leptin signaling in the brain affects bone physiology provides a rare but frightening example of how a genetics-only approach to a biological problem could have misled scientists. Indeed, as explained below, a genetics-only approach would have misled scientists about the role of hypothalamic neurons by suggesting that such neurons are not involved in leptin signaling.

Chemical lesion experiments of rat hypothalamic neurons performed in the 1940s resulted in hyperphagia and obesity similar to those observed in leptin signaling–deficient mice (22, 46–51). The signaling form of the leptin receptor is highly expressed in neurons of the ventromedial hypothalamus (VMH) nuclei and arcuate hypothalamus nuclei, a part of the brain involved in the regulation of many homeostatic functions (19, 52–56). On the basis of these observations, we and other investigators in the field hypothesized that leptin signals directly in the hypothalamus to regulate bone mass accrual. This working hypothesis was initially supported by many experiments seeking to verify this model. For instance, chemical lesioning of VMH neurons resulted in a high-bone-mass phenotype similar to the one seen in *ob/ob* mice, which lack leptin, whereas lesioning in *ob/ob* mice, followed by leptin intracerebroventricular (ICV) infusion, failed to rescue the bone phenotype observed in these mice (22). Such evidence indicated that VMH neurons regulate bone mass accrual in a leptin-dependent manner, but direct proof that leptin actually binds these neurons was lacking.

A couple of years later, another group of investigators carried out a genetic experiment that involved inactivation of the leptin receptor selectively in VMH nuclei, in arcuate nuclei, or in both nuclei (57, 58). Surprisingly, all these mutant mice had normal bone mass. Even more remarkable, appetite, energy metabolism, and body weight were normal when these mutant animals were fed a normal diet, the diet on which leptin signaling–deficient mice display hyperphagia. How could one reconcile these contradictory sets of data, and were they really contradictory?

One possible way to explain this paradox is to consider differences between chemical lesions and genetic lesions. The genetic approach appears to have the advantage because it allows for more precise deletion. A second and possibly more constructive way to approach these data is to consider that both experimental approaches are valid and that each has a valid set of data. The chemical lesioning experiments indicate that leptin requires the integrity of the VMH and arcuate neurons to regulate bone mass and energy metabolism, respectively. The genetic data, in contrast, showed that leptin does not need to bind to VMH or to arcuate neurons to fulfill its functions. Thus, the two sets of data are complementary and suggest a novel hypothesis: Leptin may not signal in the hypothalamus but may signal elsewhere in the brain to regulate the synthesis and/or secretion of a neuromediator(s) that then acts in hypothalamic neurons.

As shown for other biomedical mysteries, clinical observations greatly helped in identifying this hypothetical mediator. Serotonin reuptake inhibitors (SSRIs), a class of drugs preventing serotonin reuptake in neurons, are widely used to treat depression and other mood disorders. Like most drugs, SSRIs have side effects that include bone loss, hyperphagia, and body weight gain (59–63). These clinical observations indicated that brain serotonin affects, in ways yet to be defined, bone mass accrual, appetite, and perhaps other aspects of energy metabolism.

These clinical observations provided true insight in the search for this hypothetical mediator. Serotonin is a neuromediator made by brainstem neurons and is also a hormone synthesized by the enterochromaffin cells of the duodenum (58, 64–68). However, serotonin does not cross the blood-brain barrier, and thus each pool of serotonin behaves as a totally independent entity with conceivably different functions (58, 68). Embryonic or postnatal inactivation of tryptophan hydroxylase 2 (Tph2), the initial enzyme necessary for the synthesis of serotonin, showed that brain serotonin is a powerful activator of bone mass accrual (**Figure 2**). Because serotonin does not cross the blood-brain barrier, this experiment identified it as the first neuromediator to truly affect bone mass (58–69). Brain serotonin is also an activator of appetite and a regulator of energy expenditure. Axon tracing and cell-specific and time-specific gene inactivation showed that serotonin signals in VMH and arcuate neurons through distinct receptors to postnatally regulate bone mass and appetite, respectively (**Figure 2**). Serotonin favors bone mass accrual by decreasing sympathetic tone in VMH neurons, and it also enhances appetite by favoring expression in arcuate neurons of pro-opiomelanocortin-α (*Pomc*), melanocortin receptor 4 (*MC4R*), and other genes regulating appetite (**Figure 2**) (58, 68). In-depth molecular studies showed that in both hypothalamic nuclei serotonin fulfills its function through the transcription factor CREB (**Figure 2**) (68).

That serotonin influences bone mass and energy metabolism in a manner opposite that of leptin suggested a model whereby leptin coordinates the inhibition of bone mass accrual and appetite by decreasing serotonin synthesis and/or release (**Figure 2**) (58). This model has now been verified in vivo. Classical neurophysiology, expression analyses, genetic epistasis studies, cell-specific gene inactivation experiments, and pharmacological interventions all demonstrated that, indeed, leptin binds to serotonergic neurons and inhibits serotonin synthesis and release from these neurons. Data gathered so far indicate that inhibition of serotonin synthesis and release by brainstem neurons are the main mechanisms whereby leptin postnatally coordinates the regulation of bone mass accrual and appetite (68). In addition, an inhibitor of serotonin signaling efficiently decreased appetite and body weight in leptin-deficient mice, further verifying that serotonin is a target of leptin signaling in the brain (68). This work has important therapeutic implications.

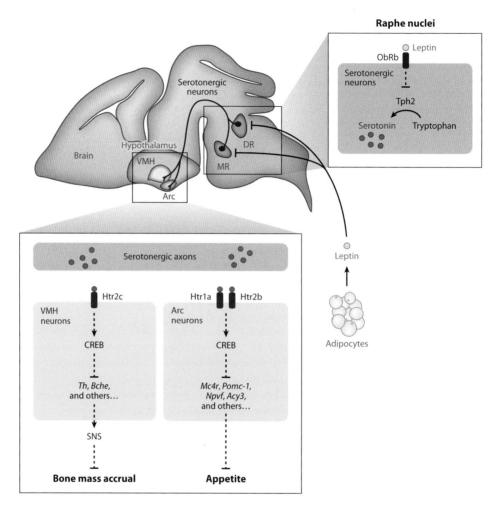

Figure 2

Brain-derived serotonin regulation of bone mass accrual and appetite. Brain-derived serotonin is synthesized by the hydroxylation of tryptophan, a rate-limiting reaction performed by the enzyme tryptophan hydroxylase 2 (Tph2) in the neurons of the dorsal raphe nuclei (DR) and median raphe nuclei (MR) in the brain stem. The axonal projections of serotonergic neurons reach ventromedial hypothalamus (VMH) and arcuate hypothalamus (Arc) neurons of the hypothalamus. Following its binding to the Htr2c receptor in neurons of the VMH nuclei, serotonin favors bone mass accrual, whereas following its binding to the Htr1a and Htr2b receptors in neurons of the Arc nuclei, serotonin favors appetite. The cAMP response element binding (CREB) protein is a crucial transcriptional effector of brain-derived serotonin in Arc neurons. CREB inhibits the expression of the genes encoding tyrosine hydroxylase (*Th*) and butyrylcholinesterase (*Bche*) in the VMH and the expression of several genes affecting appetite [melanocortin receptor 4 (*Mc4r*), pro-opiomelanocortin-α (*Pomc-1*), neuropeptide VF precursor (*Npvf*), aspartoacylase 3 (*Acy3*)] in Arc neurons. Leptin, an adipocyte-derived hormone, directly inhibits serotonin production and its release by the raphe nuclei neurons of the brain stem. The action of leptin is mediated by its receptor, ObRb, which is expressed on these neurons. SNS denotes sympathetic nervous system. Dashed lines indicate that regulation is not a primary signal (but direct); there may be other molecules in between.

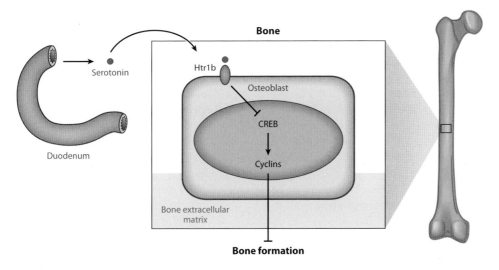

Figure 3
Gut-derived serotonin regulation of bone mass accrual. Gut-derived serotonin is synthesized in enterochromaffin cells of the duodenum and acts on osteoblasts through its receptor, Htr1b, and CREB to inhibit osteoblast proliferation.

These and other data gathered in several laboratories paint a richer and perhaps more lucid picture of leptin biology. Leptin coevolved with bone tens of million years after fertility and appetite and at a time when food sources were limiting. This suggests that leptin arose to coordinate the regulation of appetite and bone mass accrual so that bone growth does not occur in the absence of food or with low energy intake. Thus, leptin is the first of a small group of hormones tightly linking bone physiology and energy metabolism.

Coordinated Regulation of Bone Mass and Energy Metabolism, a Detour Through the Gut: Surprising and Yet Predictable

The importance of brain-derived serotonin during bone (re)modeling, along with the fact that serotonin does not cross the blood-brain barrier, begged the question as to whether gut-derived serotonin exerts any influence on bone (re)modeling. This is the first question we began to address. Only later did we realize that two humans with skeletal dysplasia might help answer this research question. Initially, as a control for specificity to understand how brain serotonin functions in bone physiology, we inactivated tryptophan hydroxylase 1 (Tph1), the counterpart of Tph2 in the gut. Normally, Tph1 is expressed in enterochromaffin cells of the gut (**Figure 3**) (58, 64, 66, 70). This experiment revealed that gut serotonin influences bone formation in a manner opposite that of brain serotonin: Gut serotonin inhibits rather than enhances bone formation by osteoblasts (68). In essence, gut serotonin acts as a hormone, binding to receptors on osteoblasts that are distinct from those on VMH neurons. Moreover, in contrast to brain serotonin's activity, gut serotonin inhibits the activity of the transcription factor CREB. As a result, it hampers osteoblast proliferation (**Figure 3**) (68). Thus, serotonin is a rare example of a single molecule that exerts totally opposite effects on the same physiological function, depending on its site of synthesis. Brain serotonin favors bone mass accrual, whereas gut serotonin inhibits bone formation and therefore bone mass accrual (68). That each pool of serotonin affects bone mass through CREB also identified serotonin as a major transcriptional regulator of bone (re)modeling by affecting transcriptional programs in both osteoblasts and hypothalamic neurons (**Figure 3**) (68, 69). One more surprise was the fact that removing serotonin on either side of the blood-brain barrier results in a bone phenotype identical to that observed after simply depleting the small amount of brain-derived serotonin (brain serotonin accounts for only 5% of total serotonin).

At the time this review was submitted, published studies showed that circulating serotonin levels are high in patients suffering from osteoporosis pseudolioma disease and low in patients affected with high-bone-mass syndrome (70, 71). The medical relevance and therapeutic implication of these observations go well beyond these two tragic but rare diseases. Indeed, these findings imply that inhibiting Tph1 activity in enterochromaffin cells of the duodenum could become an anabolic treatment for osteoporosis. This therapeutic application has been validated in rodents, and its broader implications in humans are currently being tested. That an inhibitor of gut serotonin synthesis increases bone formation may be the most emphatic verification of the importance of serotonin in bone mass regulation (72).

BONE'S CHANGE OF IDENTITY: FROM A TAKER TO A GIVER

That both energy metabolism and the gastrointestinal tract influence bone (re)modeling is a novel notion; the same is true for the central control of bone mass. Yet these two novel modes of regulation of bone mass fit well with the well-established notion that bones are recipients of hormonal inputs, despite the broadly accepted view that bones are calcified tubes with only structural properties. To dispel this latter notion and to truly change this narrow concept of bones, one needs to show that bone is not only a recipient of external influences but an endocrine organ affecting functions that have nothing to do with its own integrity. The second part of this review article addresses this aspect of bone physiology.

Coordinated Regulation of Bone (Re)modeling and Energy Metabolism: The Viewpoint of the Osteoblasts

Although we had long suspected that bone must be an endocrine organ regulating energy metabolism, it took almost 10 years and a stroke of luck to demonstrate that this is the case. We began by elucidating the function of genes encoding secreted or signaling molecules that are expressed exclusively in osteoblasts. *Esp* (*embryonic stem cell phosphatase*), one such gene, eventually revealed the endocrine nature of bone.

Esp encodes a large protein containing a long extracellular domain, a transmembrane domain, and an intracellular tyrosine phosphatase moiety (73–75). Remarkably, this gene is expressed in only two cell types, the osteoblast and the Sertoli cell of the testis. Its pattern of expression justified an in vivo functional analysis of this gene. This was done through two complementary strategies. The Smith laboratory knocked in a *Lac Z* allele in the *Esp* locus (73), whereas we removed the phosphatase domain of OST-PTP (osteoblast-testicular tyrosine phosphatase) in an osteoblast-specific manner (76). Both mutant mouse strains exhibited an identical metabolic phenotype, described below. This result shows that *Esp* influences insulin sensitivity through its expression in osteoblasts, as explained below.

The first phenotype noticed in both $Esp^{-/-}$ mice and $Esp_{osb}^{-/-}$ mice is that, although they were born at the expected Mendelian ratio, they had a strong tendency to die in the first 2 weeks of life. This pattern was so pronounced that at weaning we failed to obtain 25% of homozygous mutant mice when heterozygous mutant mice were intercrossed, despite a normal Mendelian ratio at birth. No obvious developmental defect of any kind could explain these postnatal deaths (74). In contrast, an extensive biochemical analysis showed that $Esp^{-/-}$ mice and $Esp_{osb}^{-/-}$ mice were hypoglycemic and hyperinsulinemic (74, 77). A more in-depth analysis showed that insulin secretion was increased, as was insulin sensitivity, in mice lacking *Esp* in osteoblasts, whereas mice overexpressing *Esp* in osteoblasts were glucose intolerant because of a decrease in insulin secretion and sensitivity. Thus, the analysis of *Esp* function showed unambiguously that the osteoblast influences insulin secretion from pancreatic β cells and alters insulin sensitivity in liver, muscle, and white adipose tissue (74). A simple yet powerful experiment confirmed this finding. In a coculture assay in which cells were separated by a filter, osteoblasts, but not a closely related cell type like fibroblasts, enhanced insulin secretion by islets or β cells (74). Therefore, the osteoblast is an endocrine cell favoring insulin secretion.

The protein encoded by *Esp* is not secreted and therefore cannot be a hormone. The search for the only known hormone that is made by osteoblasts and that regulates glucose metabolism was facilitated by what we thought a hormone should be and what we also knew about osteoblast biology. The requirement that hormones be cell-specific molecules narrowed the search dramatically because there is only one known secreted protein that is made only by osteoblasts: osteocalcin. Osteocalcin was an even more credible candidate to be a hormone regulating energy metabolism because *Osteocalcin*$^{-/-}$ mice exhibit an obvious increase in abdominal fat mass. Given the osteoblast-specific nature of osteocalcin and the fact that it is secreted, bone may affect energy metabolism (74).

Osteocalcin is extremely abundant in the bone extracellular matrix (ECM) and is a small protein (46 and 49 amino acids long in mice and in humans, respectively) that can be carboxylated on three glutamic acid residues (78). Carboxylation of glutamic acid residues is a posttranslational modification that increases a protein's affinity for mineral ions. This feature of osteocalcin and the fact that it is so abundant in a mineralized ECM suggested that this protein is involved in bone ECM mineralization (79). Yet loss- and gain-of-function mutations in *Osteocalcin* have unambiguously established that this is not the case (79).

Besides being present in the bone ECM, osteocalcin is also found in the general circulation. So osteocalcin may be an osteoblast-derived hormone that regulates glucose metabolism and other aspects of energy metabolism. This hypothesis was verified by showing that, unlike wild-type osteoblasts, *Osteocalcin*$^{-/-}$ osteoblasts cannot induce insulin secretion by pancreatic β cells. Accordingly, *Osteocalcin*$^{-/-}$ mice have a metabolic phenotype that is the mirror image of the one observed in *Esp*$^{-/-}$ mice; they are hyperglycemic, hypoinsulinemic, and insulin resistant in liver, muscle, and white adipose tissue. That the glucose intolerance phenotype of *Osteocalcin*$^{-/-}$ mice was corrected by removing one allele of *Esp* from these mice established that *Esp* acts upstream of *Osteocalcin*. In other words, *Esp*$^{-/-}$ mice are a gain-of-function model for osteocalcin. Remarkably, *Esp*$^{-/-}$ mice or wild-type mice receiving exogenous osteocalcin do not develop an obesity phenotype or a glucose intolerance phenotype when fed a high-fat diet (79). These results raise the prospect that osteocalcin may become a treatment for type 2 diabetes, a hypothesis being tested currently. Using multiple methods, we established that the form of osteocalcin responsible for its metabolic function is not the carboxylated form but the undercarboxylated form, which is the least abundant form of circulating osteocalcin (**Figure 4**).

In summary, this work demonstrated that bone is an endocrine organ regulating energy metabolism, a function that is critical for bone (re)modeling. Further work also showed that *Esp* and *Osteocalcin* expression is regulated by activating transcription factor 4, an osteoblast-enriched transcription factor. These collective studies established the importance of the osteoblast as a cell type and more generally the importance of the skeleton as a determinant of whole-body glucose metabolism (80–83). Since the initial description of osteocalcin metabolic function was reported in the mouse, numerous studies have indicated that in humans, as in mice, serum total and/or undercarboxylated osteocalcin is a marker of glucose tolerance (84–93).

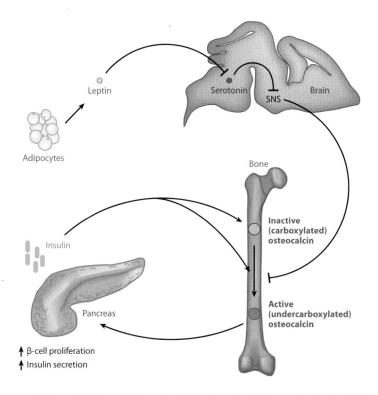

Figure 4

Endocrine regulation of energy metabolism by bone. Bone mediates such regulation by an osteoblast-specific secreted molecule, osteocalcin, that when undercarboxylated acts as a hormone favoring β-cell proliferation and insulin secretion in the pancreas. The mechanism by which osteocalcin may be activated is regulated in osteoblasts by insulin signaling, which favors osteocalcin bioavailability by promoting its undercarboxylation. In contrast, the sympathetic tone, which is regulated centrally by leptin, decreases osteocalcin bioactivation. SNS denotes sympathetic nervous system.

Coordinated Regulation of Bone Mass and Energy Metabolism: The Viewpoint of the Pancreas

It was quite unexpected to find two genes expressed in osteoblasts regulating glucose metabolism. However, that one of them encodes an intracellular phosphatase whereas the other one encodes a hormone that is not even the substrate of this phosphatase was puzzling. In addition, the regulation of insulin secretion by osteocalcin raised another question: Does insulin signaling in osteoblasts regulate the expression, secretion, or activation of osteocalcin? Such insulin signaling was found to regulate all three processes.

An efficient way to regulate the activity of tyrosine kinase receptors is through the use of intracellular tyrosine protein phosphatases. The insulin receptor, a tyrosine kinase receptor, operates in this manner, and its activity is negatively regulated in many insulin target cells by a tyrosine phosphatase, PTP-1B. This observation suggests that, if expressed in osteoblasts, the insulin receptor may be a substrate of OST-PTP (94, 95). Therefore, insulin signaling in osteoblasts may be necessary for glucose homeostasis. This is a worthwhile question to address because inactivating the insulin receptor in classical target tissues such as muscle and white adipose tissue did not result in glucose intolerance when mice were fed a normal diet (96–100). Such experiments raise the prospect that insulin signals in additional cell types to fulfill its metabolic functions.

As hypothesized, the insulin receptor is expressed in osteoblasts and is a substrate of ESP. Moreover, selective inactivation of the insulin receptor in osteoblasts results in glucose intolerance and in a decrease in insulin secretion (77). Various biochemical and genetic evidence showed that insulin signaling in osteoblasts favors osteocalcin activation by decreasing its carboxylation through an increase in bone (re)modeling. Indeed, osteoblasts are multifunctional cells that are responsible for bone formation and that, through at least two genes, determine osteoclast differentiation. Those two genes are *Rankl*, a positive regulator, and *Osteoprotegerin* (*Opg*), a soluble receptor sequestering RANKL and thus a negative regulator of this process (33, 101). An analysis of mice lacking the insulin receptor in osteoblasts showed that insulin signaling in this cell type favors bone resorption by inhibiting the expression of *Opg*. One gene expressed in osteoclasts and regulated by OPG, *Tcirg1*, contributes to the acidification of the extracellular space around osteoclasts (102–105). Thus, insulin signaling in osteoblasts favors acidification of bone ECM, a necessary component of bone resorption.

Because the only known mechanism to decarboxylate a protein outside a cell is an acid pH, it was hypothesized that insulin signaling in osteoblasts increases osteocalcin decarboxylation by stimulating bone resorption by osteoclasts (104, 105). Biochemical and genetics approaches indeed suggested that insulin signaling in osteoblasts promotes decarboxylation, i.e., activation of osteocalcin, through the activation of osteoclastic function. This ultimately favors insulin secretion. Thus, in a feed-forward loop, insulin signals in osteoblasts to enhance bone resorption, which then activates osteocalcin and upregulates *Insulin* expression and secretion.

The elucidation of the role of insulin signaling in osteoblasts raised another question: Does the endocrine function of bone also exist in humans? Because bone is one of the youngest organs to appear during evolution, it seemed likely that critical functions and regulatory mechanisms of bone might differ greatly between mice and humans. However, the pathways used by leptin, serotonin, and insulin/phosphatase/osteocalcin have similar function in mice and humans. Indeed, an analysis of osteopetrotic patients showed that a decrease in osteoclast function results in a decrease in the active form of osteocalcin and hypoinsulinemia (104, 105). The only difference between mice and humans is that *Esp*, which is a pseudogene in humans, is replaced in human osteoblasts by *PTP1B*, which encodes a tyrosine phosphatase that dephosphorylates the insulin receptor (94, 95).

Coordinated Regulation of Bone Mass and Energy Metabolism: The Viewpoint of the Adipocyte (Part II)

That osteocalcin bioactivity is enhanced by insulin signaling in osteoblasts also implies that there must be a hormone(s) that, unlike insulin, will inhibit osteocalcin expression, secretion, or bioactivity to maintain blood glucose levels within a normal range. Leptin is the only known negative regulator of the osteocalcin endocrine function; it does so by favoring *Esp* expression.

Among the many metabolic functions of leptin is the inhibition of insulin secretion through a neuronal relay. Study of cell-specific mutant mouse strains lacking either the leptin receptor, Adrβ_2, or *Esp* showed that leptin signaling in the brain relies on sympathetic signaling in osteoblasts to enhance *Esp* expression and to inhibit insulin secretion (106). This results in a decrease in osteocalcin bioactivity. That leptin regulates an aspect of energy metabolism through bone adds further credence to the notion that this hormone's main function is to coordinate the regulation of energy metabolism and bone physiology (**Figure 4**).

That insulin and leptin act directly and indirectly, respectively, in osteoblasts to regulate energy metabolism underscores the importance of bone as an important determinant of energy metabolism. This notion is validated by the fact that the broadly expressed transcription factor Foxo1, which regulates glucose metabolism, does so in part through its osteoblast expression

(107, 108). This body of work does not imply that the osteoblast is the most important cell involved in the regulation of energy metabolism. Instead, we suggest that it would be a mistake to ignore the importance of the osteoblast in this physiological process. Although we currently know of only one hormone that fulfills the metabolic functions of the osteoblast, other hormones that are made by osteoblasts and that regulate energy metabolism may exist.

Coordinated Regulation of Bone Mass and Fertility: The Viewpoint of the Osteoblasts

Although many important questions remain to be addressed regarding the regulation of energy metabolism by bone, there is a need to solidify our understanding of the endocrine role of bone, especially the link between bone remodeling and energy metabolism and the link between bone and reproduction.

Menopause favors bone loss (15, 16, 109, 110). This medical observation implies that gonads, mostly through sex steroid hormones, affect bone cell function (this aspect of bone physiology is not discussed here). The regulation of bone mass accrual by gonads also suggests that in turn bone, in its endocrine capacity, may affect reproductive function in one or both genders. Verifying this hypothesis would further enhance the emerging importance of bone as an endocrine organ.

Testing this hypothesis in vivo was greatly helped by a striking feature of mutant mice lacking osteocalcin: Whereas female *Osteocalcin*-deficient mice were normally fertile, the male mutant mice were rather poor breeders, whether their partners were wild type or *Osteocalcin* deficient. As was the case for energy metabolism, the demonstration that this phenotype betrays a true biological function of osteocalcin was made more complete and convincing because of the availability of gain-of-function ($Esp^{-/-}$) and loss-of-function ($Osteocalcin^{-/-}$) mutations for *Osteocalcin* (74). *Osteocalcin*-deficient mice showed decreased weights of the testes, epididymides, and seminal vesicles, whereas these organs' weights were increased in *Esp*-deficient mice. *Osteocalcin*-deficient males showed a 50% decrease in sperm count with a corresponding impairment of Leydig cell maturation, whereas *Esp*-deficient male mice showed a 30% increase in sperm count (111). These phenotypes suggested that osteocalcin may enhance testosterone synthesis. Coculture assays and subsequent in vivo experiments confirmed this suggestion (111).

The supernatant of cultured wild-type osteoblasts was able to increased testosterone production by Leydig cells to far greater levels than those observed for other mesenchymal cells. In contrast, this same osteoblast culture supernatant did not affect estrogen production by ovarian explants. As predicted, the supernatant of *Osteocalcin*-deficient osteoblast cultures was ineffective in promoting testosterone production in Leydig cells. Again, further cell-based and in vivo assays showed that osteocalcin increases expression of all genes necessary for testosterone biosynthesis in Leydig cells (111). Accordingly, circulating testosterone levels are low in $Osteocalcin^{-/-}$ mice and are high in $Esp^{-/-}$ mice. In contrast, circulating estrogen levels, as well as the expression of Cyp19A1 (an aromatase enzyme needed to convert testosterone to estrogen), are not affected in $Esp^{-/-}$ and $Osteocalcin^{-/-}$ mice (111). That $Osteocalcin^{-/-}$ mice develop a peripheral testicular insufficiency in the face of high levels of pituitary hormones, including luteinizing hormone and follicle-stimulating hormone, underscores the regulatory role of osteocalcin in male reproduction and suggests that some male patients with gonadal insufficiency may be defective in components of this bone-testis axis.

To formally establish that osteocalcin regulates testosterone production as a bone-derived hormone and not as a testis-secreted growth factor, investigators generated mice that lacked *Osteocalcin* only in osteoblasts. Male $Osteocalcin_{osb}^{-/-}$ mice exhibited the same defects in testosterone production as did $Osteocalcin^{-/-}$ mice; deletion of *Osteocalcin* in Leydig cells did not affect male

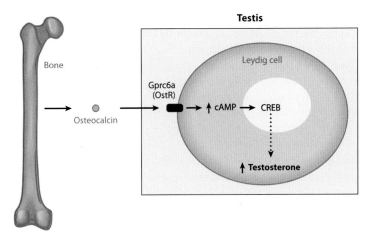

Figure 5

Endocrine regulation of male fertility by bone. Osteocalcin favors male fertility, increasing testosterone production by Leydig cells of the testes. By binding to a G protein–coupled receptor expressed in the Leydig cells of the testes, osteocalcin, an osteoblast-derived hormone, promotes testosterone production by the testes in a cAMP response element binding (CREB) protein–dependent manner. The dashed arrow indicates that regulation is not a primary signal (but direct); there may be other molecules in between.

fertility (111). Taken together, these experiments established that osteocalcin is a bone-derived hormone favoring fertility in male mice by promoting Leydig cell maturation and testosterone production (**Figure 5**). In other words, these experiments verified that for at least one gender the skeleton participates in endocrine regulation of reproduction. Such experiments also reveal that this novel aspect of reproductive endocrinology appears to be sexually dimorphic.

Osteocalcin's Molecular Mode of Action: Characterization of Its Receptor

In the molecular era, the identification of a novel hormone immediately begs the question of its mechanism of action. A prerequisite to answering this question is to characterize its cognate receptor on relevant target cells. In the case of osteocalcin, this was achieved through a two-step strategy that took advantage of the fact that osteocalcin regulates fertility in males but not in females (111).

The first step elucidated the signal transduction pathway affected by osteocalcin in two target cells, the β cell of the pancreas and the Leydig cell of the testis (74, 104, 105, 111). This approach identified cAMP production as the only intracellular signaling event triggered reproducibly by osteocalcin in these two cell types. We interpreted this result as suggesting that the osteocalcin receptor is probably a G protein–coupled receptor (GPCR) linked to adenylate cyclase. In the second step of this experimental strategy, we took advantage of the sexually dimorphic aspects of osteocalcin function by asking whether there were testis-specific orphan GPCRs. Out of more than 100 orphan GPCRs submitted to this test, 22 of them were expressed more highly in testes than in ovaries, and 4 were expressed predominantly or exclusively in Leydig cells (111). One of these 4 orphan GPCRs, Gprc6a, was a particularly good candidate to be an osteocalcin receptor because its inactivation in mice results in metabolic and reproduction phenotypes similar to those seen in $Osteocalcin^{-/-}$ mice (112). Furthermore, and although this was never tested through any binding assays, Gprc6a may be a calcium-serving receptor that functions better in the presence of osteocalcin (111).

Although the aforementioned result could not be reproduced, several criteria formally identified Gprc6a as an osteocalcin receptor present in Leydig cells (**Figure 5**) (111, 112). First, osteocalcin binds directly to wild-type cells, but not to *Gprc6a*-deficient Leydig cells. Second, osteocalcin increases cAMP production in wild-type cells, but not in *Gprc6a*-deficient Leydig cells. Third, and more convincingly, Leydig cell–specific deletion of *Gprc6a* revealed a reproduction phenotype caused by low testosterone production that was similar if not identical to the phenotype seen in the case of osteocalcin inactivation. Fourth, compound heterozygous mice lacking one copy of *Osteocalcin* and one copy of *Gprc6a* had a reproduction phenotype identical in all aspects to the one seen in *Osteocalcin*$^{-/-}$ or *Gprc6a*$^{-/-}$ mice. The identification of Gprc6a as an osteocalcin receptor led subsequently to the realization that CREB is a transcriptional effector of osteocalcin regulation of testosterone biosynthesis (**Figure 5**) (111). The identification of Gprc6a now allows us to specifically identify the functions of osteocalcin. It also enables us to perform a more sophisticated dissection of osteocalcin's molecular mode of action in known and yet-to-be-identified target cells.

WHAT DID WE LEARN, AND WHERE DO WE STAND?

Although studies on the endocrine function of bone tissue are still ongoing, there are several lessons to be learned from this body of work. The first and most stimulating lesson is that because so much was discovered in a short amount of time about a single organ, namely the skeleton, there is probably much more to be learned about other organs. Second, this work demonstrates that genetics can uncover intimate connections between organs and can provide an ideal approach to link multiple physiologies and medical disciplines. For the foreseeable future, mouse genetics is the most powerful tool to map out and to study unidentified interorgan connections that exist in vertebrates.

The whole-organism molecular genetic approach to skeleton physiology may explain why a hormone like leptin co-appeared with bony vertebrates during evolution. Moreover, because the skeleton affects glucose homeostasis, energy expenditure, and fertility, bone may affect many more organs and physiological functions outside of the skeleton. As such, we suggest that vertebrate physiology is best studied in animal models containing a bony skeleton. Last, the demonstration that the osteoblast is an important endocrine cell suggests that further studies probing the molecular aspects of bone mass loss over time are needed.

If we look to a particular aspect of the skeleton, osteocalcin, we are struck by the fact that this hormone affects functions that go awry during aging. This observation is important because traditionally the skeleton is considered a victim of the aging process, as evidenced by age-dependent osteoporosis. Interestingly, two functions already ascribed to osteocalcin identify it as a fitness hormone affecting processes that deteriorate or disappear with aging. These findings therefore lead us to hypothesize that bone may both determine the aging process and be a victim of the aging process. Further knowledge about whole-organism physiology will be revealed from the study of this particular organ.

DISCLOSURE STATEMENT

The authors are not aware of any affiliations, memberships, funding, or financial holdings that might be perceived as affecting the objectivity of this review.

LITERATURE CITED

1. Bernard C. 1865. *An Introduction to the Study of Experimental Medicine*. Paris: Flamarion
2. Cannon WB. 1929. Organization for physiological homeostasis. *Physiol. Rev.* 9:399–431

3. Cannon WB. 1932. *The Wisdom of the Body*. New York: Norton
4. Henderson LJ. 1913. *The Fitness of the Environment: An Inquiry into the Biological Significance of the Properties of Matter*. New York: Macmillan
5. Karsenty G. 2006. Convergence between bone and energy homeostases: leptin regulation of bone mass. *Cell Metab.* 4:341–48
6. Rodan GA, Martin TJ. 2000. Therapeutic approaches to bone diseases. *Science* 289:1508–14
7. Teitelbaum SL. 2000. Osteoclasts, integrins, and osteoporosis. *J. Bone Miner. Metab.* 18:344–49
8. Harada S, Rodan GA. 2003. Control of osteoblast function and regulation of bone mass. *Nature* 423:349–55
9. Ducy P, Amling M, Takeda S, Priemel M, Schilling AF, et al. 2000. Leptin inhibits bone formation through a hypothalamic relay: a central control of bone mass. *Cell* 100:197–207
10. Reid IR. 2002. Relationships among body mass, its components, and bone. *Bone* 31:547–55
11. Rigotti NA, Nussbaum SR, Herzog DB, Neer RM. 1984. Osteoporosis in women with anorexia nervosa. *N. Engl. J. Med.* 311:1601–6
12. Wolfert A, Mehler PS. 2002. Osteoporosis: prevention and treatment in anorexia nervosa. *Eat. Weight Disord.* 7:72–81
13. Zhao LJ, Liu YJ, Liu PY, Hamilton J, Recker RR, Deng HW. 2007. Relationship of obesity with osteoporosis. *J. Clin. Endocrinol. Metab.* 92:1640–46
14. Zipfel S, Seibel MJ, Lowe B, Beumont PJ, Kasperk C, Herzog W. 2001. Osteoporosis in eating disorders: a follow-up study of patients with anorexia and bulimia nervosa. *J. Clin. Endocrinol. Metab.* 86:5227–33
15. Khosla S. 2010. Update on estrogens and the skeleton. *J. Clin. Endocrinol. Metab.* 95:3569–77
16. Riggs BL, O'Fallon WM, Muhs J, O'Connor MK, Kumar R, Melton LJ 3rd. 1998. Long-term effects of calcium supplementation on serum parathyroid hormone level, bone turnover, and bone loss in elderly women. *J. Bone Miner. Res.* 13:168–74
17. Ahima RS. 2004. Body fat, leptin, and hypothalamic amenorrhea. *N. Engl. J. Med.* 351:959–62
18. Ahima RS, Saper CB, Flier JS, Elmquist JK. 2000. Leptin regulation of neuroendocrine systems. *Front. Neuroendocrinol.* 21:263–307
19. Clement K, Vaisse C, Lahlou N, Cabrol S, Pelloux V, et al. 1998. A mutation in the human leptin receptor gene causes obesity and pituitary dysfunction. *Nature* 392:398–401
20. Elefteriou F, Ahn JD, Takeda S, Starbuck M, Yang X, et al. 2005. Leptin regulation of bone resorption by the sympathetic nervous system and CART. *Nature* 434:514–20
21. Spiegelman BM, Flier JS. 2001. Obesity and the regulation of energy balance. *Cell* 104:531–43
22. Takeda S, Elefteriou F, Levasseur R, Liu X, Zhao L, et al. 2002. Leptin regulates bone formation via the sympathetic nervous system. *Cell* 111:305–17
23. Doyon C, Drouin G, Trudeau VL, Moon TW. 2001. Molecular evolution of leptin. *Gen. Comp. Endocrinol.* 124:188–98
24. Huising MO, Geven EJ, Kruiswijk CP, Nabuurs SB, Stolte EH, et al. 2006. Increased leptin expression in common carp (*Cyprinus carpio*) after food intake but not after fasting or feeding to satiation. *Endocrinology* 147:5786–97
25. Elefteriou F, Takeda S, Ebihara K, Magre J, Patano N, et al. 2004. Serum leptin level is a regulator of bone mass. *Proc. Natl. Acad. Sci. USA* 101:3258–63
26. Karsenty G, Oury F. 2010. The central regulation of bone mass, the first link between bone remodeling and energy metabolism. *J. Clin. Endocrinol. Metab.* 95:4795–801
27. Fu L, Patel MS, Bradley A, Wagner EF, Karsenty G. 2005. The molecular clock mediates leptin-regulated bone formation. *Cell* 122:803–15
28. Karsenty G, Kronenberg HM, Settembre C. 2009. Genetic control of bone formation. *Annu. Rev. Cell Dev. Biol.* 25:629–48
29. Lowrey PL, Takahashi JS. 2004. Mammalian circadian biology: elucidating genome-wide levels of temporal organization. *Annu. Rev. Genomics Hum. Genet.* 5:407–41
30. Morse D, Sassone-Corsi P. 2002. Time after time: inputs to and outputs from the mammalian circadian oscillators. *Trends Neurosci.* 25:632–37
31. Okamura H, Miyake S, Sumi Y, Yamaguchi S, Yasui A, et al. 1999. Photic induction of *mPer1* and *mPer2* in *Cry*-deficient mice lacking a biological clock. *Science* 286:2531–34

32. Perreau-Lenz S, Pevet P, Buijs RM, Kalsbeek A. 2004. The biological clock: the bodyguard of temporal homeostasis. *Chronobiol. Int.* 21:1–25
33. Teitelbaum SL, Ross FP. 2003. Genetic regulation of osteoclast development and function. *Nat. Rev. Genet.* 4:638–49
34. Zheng B, Albrecht U, Kaasik K, Sage M, Lu W, et al. 2001. Nonredundant roles of the *mPer1* and *mPer2* genes in the mammalian circadian clock. *Cell* 105:683–94
35. Pasco JA, Henry MJ, Sanders KM, Kotowicz MA, Seeman E, Nicholson GC. 2004. β-Adrenergic blockers reduce the risk of fracture partly by increasing bone mineral density: Geelong Osteoporosis Study. *J. Bone Miner. Res.* 19:19–24
36. Rejnmark L, Vestergaard P, Mosekilde L. 2006. Treatment with β-blockers, ACE inhibitors, and calcium-channel blockers is associated with a reduced fracture risk: a nationwide case-control study. *J. Hypertens.* 24:581–89
37. Schlienger RG, Kraenzlin ME, Jick SS, Meier CR. 2004. Use of β-blockers and risk of fractures. *JAMA* 292:1326–32
38. Turker S, Karatosun V, Gunal I. 2006. β-Blockers increase bone mineral density. *Clin. Orthop. Relat. Res.* 443:73–74
39. Bonnet N, Benhamou CL, Malaval L, Goncalves C, Vico L, et al. 2008. Low dose β-blocker prevents ovariectomy-induced bone loss in rats without affecting heart functions. *J. Cell Physiol.* 217:819–27
40. Asnicar MA, Smith DP, Yang DD, Heiman ML, Fox N, et al. 2001. Absence of cocaine- and amphetamine-regulated transcript results in obesity in mice fed a high caloric diet. *Endocrinology* 142:4394–400
41. Elias CF, Lee C, Kelly J, Aschkenasi C, Ahima RS, et al. 1998. Leptin activates hypothalamic CART neurons projecting to the spinal cord. *Neuron* 21:1375–85
42. Kristensen P, Judge ME, Thim L, Ribel U, Christjansen KN, et al. 1998. Hypothalamic CART is a new anorectic peptide regulated by leptin. *Nature* 393:72–76
43. Ahn JD, Dubern B, Lubrano-Berthelier C, Clement K, Karsenty G. 2006. *Cart* overexpression is the only identifiable cause of high bone mass in melanocortin 4 receptor deficiency. *Endocrinology* 147:3196–202
44. Abizaid A, Gao Q, Horvath TL. 2006. Thoughts for food: brain mechanisms and peripheral energy balance. *Neuron* 51(6):691–702
45. Sato S, Hanada R, Kimura A, Abe T, Matsumoto T, et al. 2007. Central control of bone remodeling by neuromedin U. *Nat. Med.* 13(10):1234–40
46. Hetherington AW, Ranson SW. 1940. Hypothalamic lesions and adiposity in the rat. *Anat. Rec.* 78:149–72
47. Hetherington AW, Ranson SW. 1942. The relation of various hypothalamic lesions to adiposity in the rat. *J. Comp. Neurol.* 76:475–99
48. Brobeck JR. 1946. Mechanisms of the development of obesity in animals with hypothalamic lesions. *Physiol. Rev.* 26:541–59
49. Anand BK, Brobeck JR. 1951. Localization of a feeding center in the hypothalamus of the rat. *Proc. Soc. Exp. Biol. Med.* 77:323–24
50. Debons AF, Silver L, Cronkite EP, Johnson HA, Brecher G, et al. 1962. Localization of gold in mouse brain in relation to gold thioglucose obesity. *Am. J. Physiol.* 202:743–50
51. Olney JW. 1969. Brain lesions, obesity, and other disturbances in mice treated with monosodium glutamate. *Science* 164:719–21
52. Chen H, Charlat O, Tartaglia LA, Woolf EA, Weng X, et al. 1996. Evidence that the diabetes gene encodes the leptin receptor: identification of a mutation in the leptin receptor gene in *db/db* mice. *Cell* 84:491–95
53. Fei H, Okano HJ, Li C, Lee GH, Zhao C, et al. 1997. Anatomic localization of alternatively spliced leptin receptors (Ob-R) in mouse brain and other tissues. *Proc. Natl. Acad. Sci. USA* 94:7001–5
54. Lee GH, Proenca R, Montez JM, Carroll KM, Darvishzadeh JG, et al. 1996. Abnormal splicing of the leptin receptor in diabetic mice. *Nature* 379:632–35
55. Tartaglia LA, Dembski M, Weng X, Deng N, Culpepper J, et al. 1995. Identification and expression cloning of a leptin receptor, OB-R. *Cell* 83:1263–71

56. Zhang Y, Proenca R, Maffei M, Barone M, Leopold L, et al. 1994. Positional cloning of the mouse obese gene and its human homologue. *Nature* 372:425–32
57. Balthasar N, Coppari R, McMinn J, Liu SM, Lee CE, et al. 2004. Leptin receptor signaling in POMC neurons is required for normal body weight homeostasis. *Neuron* 42:983–91
58. Yadav VK, Oury F, Suda N, Liu ZW, Gao XB, et al. 2009. A serotonin-dependent mechanism explains the leptin regulation of bone mass, appetite, and energy expenditure. *Cell* 138:976–89
59. Haney EM, Chan BK, Diem SJ, Ensrud KE, Cauley JA, et al. 2007. Association of low bone mineral density with selective serotonin reuptake inhibitor use by older men. *Arch. Intern. Med.* 167:1246–51
60. Kaye W, Gendall K, Strober M. 1998. Serotonin neuronal function and selective serotonin reuptake inhibitor treatment in anorexia and bulimia nervosa. *Biol. Psychiatry* 44:825–38
61. Laekeman G, Zwaenepoel L, Reyntens J, de Vos M, Casteels M. 2008. Osteoporosis after combined use of a neuroleptic and antidepressants. *Pharm. World Sci.* 30:613–16
62. Richards JB, Papaioannou A, Adachi JD, Joseph L, Whitson HE, et al. 2007. Effect of selective serotonin reuptake inhibitors on the risk of fracture. *Arch. Intern. Med.* 167:188–94
63. Ziere G, Dieleman JP, van der Cammen TJ, Hofman A, Pols HA, Stricker BH. 2008. Selective serotonin reuptake inhibiting antidepressants are associated with an increased risk of nonvertebral fractures. *J. Clin. Psychopharmacol.* 28:411–17
64. Gershon MD, Tack J. 2007. The serotonin signaling system: from basic understanding to drug development for functional GI disorders. *Gastroenterology* 132:397–414
65. Mann JJ, McBride PA, Brown RP, Linnoila M, Leon AC, et al. 1992. Relationship between central and peripheral serotonin indexes in depressed and suicidal psychiatric inpatients. *Arch. Gen. Psychiatry* 49:442–46
66. Walther DJ, Peter J-U, Bashammakh S, Hörtnagl H, Voits M, et al. 2003. Synthesis of serotonin by a second tryptophan hydroxylase isoform. *Science* 299:76
67. Yadav VK, Oury F, Tanaka K, Thomas T, Wang Y, et al. 2011. Leptin-dependent serotonin control of appetite: temporal specificity, transcriptional regulation, and therapeutic implications. *J. Exp. Med.* 208:41–52
68. Yadav VK, Ryu JH, Suda N, Tanaka KF, Gingrich JA, et al. 2008. Lrp5 controls bone formation by inhibiting serotonin synthesis in the duodenum. *Cell* 135:825–37
69. Oury F, Yadav VK, Wang Y, Zhou B, Liu XS, et al. 2010. CREB mediates brain serotonin regulation of bone mass through its expression in ventromedial hypothalamic neurons. *Genes Dev.* 24:2330–42
70. Boyden LM, Mao J, Belsky J, Mitzner L, Farhi A, et al. 2002. High bone density due to a mutation in LDL-receptor-related protein 5. *N. Engl. J. Med.* 346:1513–21
71. Gong Y, Slee RB, Fukai N, Rawadi G, Roman-Roman S, et al. 2001. LDL receptor-related protein 5 (LRP5) affects bone accrual and eye development. *Cell* 107:513–23
72. Yadav VK, Balaji S, Suresh PS, Liu XS, Lu X, et al. 2010. Pharmacological inhibition of gut-derived serotonin synthesis is a potential bone anabolic treatment for osteoporosis. *Nat. Med.* 16:308–12
73. Lee K, Nichols J, Smith A. 1996. Identification of a developmentally regulated protein tyrosine phosphatase in embryonic stem cells that is a marker of pluripotential epiblast and early mesoderm. *Mech. Dev.* 59:153–64
74. Lee NK, Sowa H, Hinoi E, Ferron M, Ahn JD, et al. 2007. Endocrine regulation of energy metabolism by the skeleton. *Cell* 130:456–69
75. Mauro LJ, Olmsted EA, Skrobacz BM, Mourey RJ, Davis AR, Dixon JE. 1994. Identification of a hormonally regulated protein tyrosine phosphatase associated with bone and testicular differentiation. *J. Biol. Chem.* 269:30659–67
76. Dacquin R, Mee PJ, Kawaguchi J, Olmsted-Davis EA, Gallagher JA, et al. 2004. Knock-in of nuclear localised β-galactosidase reveals that the tyrosine phosphatase Ptprv is specifically expressed in cells of the bone collar. *Dev. Dyn.* 229:826–34
77. Ferron M, Hinoi E, Karsenty G, Ducy P. 2008. Osteocalcin differentially regulates β cell and adipocyte gene expression and affects the development of metabolic diseases in wild-type mice. *Proc. Natl. Acad. Sci. USA* 105:5266–70
78. Hauschka PV, Lian JB, Cole DE, Gundberg CM. 1989. Osteocalcin and matrix Gla protein: vitamin K-dependent proteins in bone. *Physiol. Rev.* 69:990–1047

79. Ducy P, Desbois C, Boyce B, Pinero G, Story B, et al. 1996. Increased bone formation in osteocalcin-deficient mice. *Nature* 382:448–52
80. Kajimura D, Hinoi E, Ferron M, Kode A, Riley KJ, et al. 2011. Genetic determination of the cellular basis of the sympathetic regulation of bone mass accrual. *J. Exp. Med.* 388:34–42
81. Yang X, Karsenty G. 2004. ATF4, the osteoblast accumulation of which is determined post-translationally, can induce osteoblast-specific gene expression in non-osteoblastic cells. *J. Biol. Chem.* 279:47109–14
82. Yang X, Matsuda K, Bialek P, Jacquot S, Masuoka HC, et al. 2004. ATF4 is a substrate of RSK2 and an essential regulator of osteoblast biology; implication for Coffin-Lowry Syndrome. *Cell* 117:387–98
83. Yoshizawa T, Hinoi E, Jung DY, Kajimura D, Ferron M, et al. 2009. The transcription factor ATF4 regulates glucose metabolism in mice through its expression in osteoblasts. *J. Clin. Investig.* 119:2807–17
84. Aonuma H, Miyakoshi N, Hongo M, Kasukawa Y, Shimada Y. 2009. Low serum levels of undercarboxylated osteocalcin in postmenopausal osteoporotic women receiving an inhibitor of bone resorption. *Tohoku J. Exp. Med.* 218:201–5
85. Fernandez-Real JM, Izquierdo M, Ortega F, Gorostiaga E, Gomez-Ambrosi J, et al. 2009. The relationship of serum osteocalcin concentration to insulin secretion, sensitivity, and disposal with hypocaloric diet and resistance training. *J. Clin. Endocrinol. Metab.* 94:237–45
86. Hwang YC, Jeong IK, Ahn KJ, Chung HY. 2009. The uncarboxylated form of osteocalcin is associated with improved glucose tolerance and enhanced β-cell function in middle-aged male subjects. *Diabetes Metab. Res. Rev.* 25:768–72
87. Im JA, Yu BP, Jeon JY, Kim SH. 2008. Relationship between osteocalcin and glucose metabolism in postmenopausal women. *Clin. Chim. Acta* 396:66–69
88. Kanazawa I, Yamaguchi T, Yamamoto M, Yamauchi M, Kurioka S, et al. 2009. Serum osteocalcin level is associated with glucose metabolism and atherosclerosis parameters in type 2 diabetes mellitus. *J. Clin. Endocrinol. Metab.* 94:45–49
89. Kindblom JM, Ohlsson C, Ljunggren O, Karlsson MK, Tivesten A, et al. 2009. Plasma osteocalcin is inversely related to fat mass and plasma glucose in elderly Swedish men. *J. Bone Miner. Res.* 24:785–91
90. Levinger I, Zebaze R, Jerums G, Hare DL, Selig S, Seeman E. 2011. The effect of acute exercise on undercarboxylated osteocalcin in obese men. *Osteoporos. Int.* 2(5):1621–66
91. Pittas AG, Harris SS, Eliades M, Stark P, Dawson-Hughes B. 2009. Association between serum osteocalcin and markers of metabolic phenotype. *J. Clin. Endocrinol. Metab.* 94:827–32
92. Winhofer Y, Handisurya A, Tura A, Bittighofer C, Klein K, et al. 2010. Osteocalcin is related to enhanced insulin secretion in gestational diabetes mellitus. *Diabetes Care* 33:139–43
93. Yeap BB, Chubb SA, Flicker L, McCaul KA, Ebeling PR, et al. 2010. Reduced serum total osteocalcin is associated with metabolic syndrome in older men via waist circumference, hyperglycemia, and triglyceride levels. *Eur. J. Endocrinol.* 163:265–72
94. Delibegovic M, Bence KK, Mody N, Hong EG, Ko HJ, et al. 2007. Improved glucose homeostasis in mice with muscle-specific deletion of protein-tyrosine phosphatase 1B. *Mol. Cell. Biol.* 27:7727–34
95. Delibegovic M, Zimmer D, Kauffman C, Rak K, Hong EG, et al. 2009. Liver-specific deletion of protein-tyrosine phosphatase 1B (PTP1B) improves metabolic syndrome and attenuates diet-induced endoplasmic reticulum stress. *Diabetes* 58:590–99
96. Bluher M, Michael MD, Peroni OD, Ueki K, Carter N, et al. 2002. Adipose tissue selective insulin receptor knockout protects against obesity and obesity-related glucose intolerance. *Dev. Cell* 3:25–38
97. Bruning JC, Michael MD, Winnay JN, Hayashi T, Horsch D, et al. 1998. A muscle-specific insulin receptor knockout exhibits features of the metabolic syndrome of NIDDM without altering glucose tolerance. *Mol. Cell* 2:559–69
98. Konner AC, Janoschek R, Plum L, Jordan SD, Rother E, et al. 2007. Insulin action in AgRP-expressing neurons is required for suppression of hepatic glucose production. *Cell Metab.* 5:438–49
99. Kulkarni RN, Bruning JC, Winnay JN, Postic C, Magnuson MA, Kahn CR. 1999. Tissue-specific knockout of the insulin receptor in pancreatic β cells creates an insulin secretory defect similar to that in type 2 diabetes. *Cell* 96:329–39
100. Michael MD, Kulkarni RN, Postic C, Previs SF, Shulman GI, et al. 2000. Loss of insulin signaling in hepatocytes leads to severe insulin resistance and progressive hepatic dysfunction. *Mol. Cell* 6:87–97

101. Kong YY, Boyle WJ, Penninger JM. 1999. Osteoprotegerin ligand: a common link between osteoclastogenesis, lymph node formation and lymphocyte development. *Immunol. Cell Biol.* 77:188–93
102. Ferron M, Wei J, Yoshizawa T, Del Fattore A, DePinho RA, et al. 2010. Insulin signaling in osteoblasts integrates bone remodeling and energy metabolism. *Cell* 142:296–308
103. Fulzele K, Riddle RC, DiGirolamo DJ, Cao X, Wan C, et al. 2010. Insulin receptor signaling in osteoblasts regulates postnatal bone acquisition and body composition. *Cell* 142:309–19
104. Saftig P, Hunziker E, Wehmeyer O, Jones S, Boyde A, et al. 1998. Impaired osteoclastic bone resorption leads to osteopetrosis in cathepsin-K-deficient mice. *Proc. Natl. Acad. Sci. USA* 95:13453–58
105. Scimeca JC, Franchi A, Trojani C, Parrinello H, Grosgeorge J, et al. 2000. The gene encoding the mouse homologue of the human osteoclast-specific 116-kDa V-ATPase subunit bears a deletion in osteosclerotic (*oc/oc*) mutants. *Bone* 26:207–13
106. Hinoi E, Gao N, Jung DY, Yadav V, Yoshizawa T, et al. 2008. The sympathetic tone mediates leptin's inhibition of insulin secretion by modulating osteocalcin bioactivity. *J. Cell Biol.* 183:1235–42
107. Rached MT, Kode A, Silva BC, Jung DY, Gray S, et al. 2010. FoxO1 expression in osteoblasts regulates glucose homeostasis through regulation of osteocalcin in mice. *J. Clin. Investig.* 120:357–68
108. Rached MT, Kode A, Xu L, Yoshikawa Y, Paik JH, et al. 2010. FoxO1 is a positive regulator of bone formation by favoring protein synthesis and resistance to oxidative stress in osteoblasts. *Cell Metab.* 11:147–60
109. Nakamura T, Imai Y, Matsumoto T, Sato S, Takeuchi K, et al. 2007. Estrogen prevents bone loss via estrogen receptor α and induction of Fas ligand in osteoclasts. *Cell* 130:811–23
110. Manolagas SC, Kousteni S, Jilka RL. 2002. Sex steroids and bone. *Recent Prog. Horm. Res.* 57:385–409
111. Oury F, Sumara G, Sumara O, Ferron M, Chang H, et al. 2011. Endocrine regulation of male fertility by the skeleton. *Cell* 144:796–809
112. Pi M, Chen L, Huang MZ, Zhu W, Ringhofer B, et al. 2008. GPRC6A null mice exhibit osteopenia, feminization and metabolic syndrome. *PLoS ONE* 3:e3858

Fetal Programming and Metabolic Syndrome

Paolo Rinaudo[1,2] and Erica Wang[1]

[1]Division of Reproductive Endocrinology and Infertility, Department of Obstetrics, Gynecology and Reproductive Sciences, University of California, San Francisco, California 94115; email: rinaudop@obgyn.ucsf.edu

[2]Eli and Edythe Broad Center for Regeneration Medicine and Stem Cell Research, University of California, San Francisco, California 94143

Keywords

nutrition, developmental-origin-of-health-and-disease hypothesis, in utero stress

Abstract

Metabolic syndrome is reaching epidemic proportions, particularly in developing countries. In this review, we explore the concept—based on the developmental-origin-of-health-and-disease hypothesis—that reprogramming during critical times of fetal life can lead to metabolic syndrome in adulthood. Specifically, we summarize the epidemiological evidence linking prenatal stress, manifested by low birth weight, to metabolic syndrome and its individual components. We also review animal studies that suggest potential mechanisms for the long-term effects of fetal reprogramming, including the cellular response to stress and both organ- and hormone-specific alterations induced by stress. Although metabolic syndrome in adulthood is undoubtedly caused by multiple factors, including modifiable behavior, fetal life may provide a critical window in which individuals are predisposed to metabolic syndrome later in life.

MS: metabolic
syndrome

1. INTRODUCTION

Compelling evidence indicates that stressful environmental conditions during sensitive periods of early development cause a predisposition to chronic disease later in life (1). The premise of this concept, known as the developmental-origin-of-health-and-disease hypothesis, is that the fetus adapts to its environment. Specifically, the developing fetus senses the environment during specific windows of sensitivity and optimizes future metabolic responses by reprogramming its genome. This reprogramming favors early survival and reproductive success but potentially causes a predisposition to disease in later stages of life (2).

Here we summarize the epidemiological evidence linking prenatal stress to the development of metabolic syndrome (MS) and review animal studies that reveal potential mechanisms for the long-term effects of fetal stress. Excellent reviews on the topic exist (3–5). Our focus is on the primary mechanisms and potential molecular pathways by which organ dysfunction occurs and leads to components of MS.

2. FETAL REPROGRAMMING

A quarter of a century ago, Barker & Osmond (1) at the University of Southampton, England, crystallized the concept of fetal programming and early origin of adult disease by suggesting that stress in utero, as manifested by low birth weight (LBW), increases the risk of cardiovascular disease and stroke in specific areas of England and Wales. Prior researchers had formulated similar hypotheses on the basis of findings in humans (6, 7) and animals (8). Hales & Barker (9), however, provided a mechanistic explanation by proposing the thrifty phenotype hypothesis to complement the already existing thrifty genotype hypothesis (10). According to Hales & Barker's hypothesis, fetuses exposed to suboptimal conditions during uterine life program developmental processes in anticipation of similar suboptimal conditions in postnatal life. If postnatal conditions are instead optimal and resources are abundant, the organism is ill-prepared to cope with the different environment and hence is more susceptible to developing diseases (11).

Integral to the concept of reprogramming is the existence of critical windows of sensitivity, during which the organism is particularly sensitive to the environment. The Barker hypothesis identified the uterine period as a key period of developmental plasticity; however, it is now clear that there are additional windows of sensitivity, including the preconception period and early postnatal life (**Figure 1**) (12).

Indeed, some critics of the Barker hypothesis emphasize that early postnatal stress is at least as important as in utero stress in determining long-term effects and introduced an alternative hypothesis—that exposures or insults gradually accumulating through episodes of illness, adverse environmental conditions, and behaviors increase the risk of chronic disease and mortality (13, 150). In this so-called life course hypothesis, lifestyle factors in adulthood make a much greater contribution than in utero stress. According to this view, in utero stress causes a predisposition to disease that is manifested only if an additional stress occurs later in postnatal life (e.g., obesity or high-fat diet). This is exemplified by the increasing incidence of diabetes in low- and middle-income countries, where the combination of poor nutrition in utero and overnutrition in later life is common (14).

Another important concept is that the effects of in utero stress differ according to the timing of its occurrence. In rats and sheep, maternal undernourishment during the period of conception and implantation, for example, leads to hypertension and cardiovascular dysfunction in offspring (15–19). Embryos cultured in vitro during the preimplantation period have long-term health problems (19).

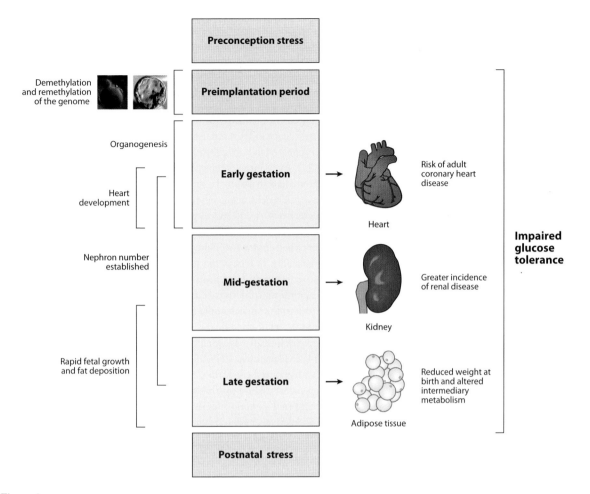

Figure 1

Critical windows of sensitivity during human pregnancy for the development of components of metabolic syndrome later in life. The different effects of stress during pregnancy can be explained by the cellular events that occur during particular periods of pregnancy. Data were obtained from the Dutch Famine Cohort (26). Glucose intolerance is a common consequence of in utero stress and is independent of the time of occurrence. Stress in early gestation leads to a greater risk of an abnormal atherogenic lipid profile, obesity in women, and coronary artery disease. Stress in the second trimester of pregnancy, during which the number of nephrons increases rapidly, is associated with a 3.2-fold increase in risk for microalbuminuria. Finally, stress during the third trimester of pregnancy, when fat deposition occurs, has a relatively higher effect in reducing birth weight. Additional windows of sensitivity include the preconception and the early postnatal periods. Surprisingly, animal studies show that even stress limited to the preimplantation period (5 days) can lead to impaired glucose intolerance (96, 97).

Other than evidence from the Dutch famine, when caloric intake was severely reduced for a short, well-defined short period, it is often difficult to identify a specific time frame in which in utero stress can be analyzed. The only additional evidence of a temporally limited stress derives from follow-up of children conceived by assisted reproductive technology (ART). More than 3 million children have been born worldwide as a result of ART. Data exist to support the hypothesis that stress in the form of embryo culture during the preimplantation period may be associated with early differences in metabolic profile. Epidemiological data are limited by the relatively short history of ART. Because the first child conceived by in vitro fertilization (IVF) was born in 1978, IVF children are at most in their early thirties and are rather young to manifest components of

ART: assisted reproductive technology

MS. However, there is compelling evidence that IVF offspring have a different metabolic profile starting at birth. Among term singleton infants, those conceived through ART have a 2.6-fold-greater risk of LBW than do those conceived spontaneously (20).

Recently, a cohort study investigated metabolic and pubertal measures in 233 IVF children ages 8–18 and a similar number of age- and gender-matched spontaneously conceived controls born to subfertile parents (21). IVF children had significant increases in peripheral body mass and percentage of peripheral fat compared with controls, as well as a trend toward a higher percentage of total body fat (21). Waist circumference did not differ between the two groups. Among pubertal girls, those conceived by IVF also appeared to have greater bone age than did controls, as evidenced by radiographs of the left hand (22). Even more interesting, systolic and diastolic blood pressure levels were higher in IVF children than in controls, and fasting glucose levels were higher in pubertal IVF children (23). These differences persisted despite adjustments for current body size, birth weight, and parental characteristics. A difference of 4 mm Hg in systolic blood pressure is clinically significant, as evidenced by the increase in hypertension with age (24).

On the contrary, one study described an improved metabolic profile at 6 years of age in 69 IVF children compared with 71 controls (25). IVF children were taller than controls after correction for parental height and had a more favorable lipid profile, with higher high-density lipoprotein and lower triglyceride levels. The different findings of these studies are likely secondary to the different age of the test subjects and to the small sample size of the two studies and underline the need for more research in the field.

Studies of the Dutch famine are particularly illuminating for evaluating in utero stress. During the 4-month duration of the famine (December 1944 to April 1945), the official daily ration was 400–800 calories (26). Fetuses undernourished during early gestation had an atherogenic lipid profile, increased risk for coronary heart disease in adulthood, and a decline in cognitive function (26, 27). Undernourishment during mid-gestation increased the incidence of microalbuminuria and obstructive airway disease. Finally, glucose tolerance was altered in all fetuses exposed to famine but was particularly evident after undernourishment in late gestation (**Figure 1**) (26, 28).

Another important point is the concept of sexually dimorphic effects after in utero stress (29). Women with less than average birth weight and higher weight at 1 year showed the highest incidence of cardiovascular death rates; this pattern was not present in men (30). Only in women with a birth weight greater than 4 kg (normal birth weight at term is between 2.5 and 4 kg) did systolic blood pressure increase in parallel with birth weight (31). In animal models, only female offspring of pregnant dams exposed to a high-fat diet exhibit hypertension in later life (32); in contrast, only adult male rats show alterations in triglycerides and expression of hepatic fatty acid enzymes after uterine artery ligation (33).

These sexually dimorphic effects are caused by multiple mechanisms that are not clearly understood. Apart from differences in sex steroids, these effects may reflect variations in activation of the hypothalamic-pituitary-adrenal (HPA) axis (32), different responses to oxidative stress (34), and the faster postnatal growth rate of male mice compared with female mice (29).

3. DEFINITION OF METABOLIC SYNDROME

MS is a constellation of metabolic risk factors that increase an individual's predisposition to atherosclerotic cardiovascular disease, hypertension, and type 2 diabetes. Estimates suggest that the population-attributable fraction for MS is approximately 6–7% for all-cause mortality, 12–17% for cardiovascular disease, and 30–52% for diabetes (35). Among adults in the United States, the age-adjusted prevalence of MS is 23.7% or 47 million people (36). Importantly, the prevalence increases from 6.7% among individuals aged 20–29 years to more than 40% in those 60 years or older (36).

In both adults and children, the definition of MS is controversial (37–40). The most widely used definitions in adults are from the National Cholesterol Education Program/Adult Treatment Panel III (NCEP/ATP III) (37) and the International Diabetes Federation (IDF) (38). Diagnostic signs include elevated blood pressure (>130/85 mm Hg), central obesity (>102 cm in men and >88 cm in women), dyslipidemia (serum triglycerides ≥150 mg dl^{-1} and high-density lipoprotein cholesterol <40 mg dl^{-1} in men and <50 mg dl^{-1} in women), and glucose intolerance (serum glucose ≥100 mg dl^{-1}). The main difference between the two definitions is that abdominal obesity, defined by ethnicity-specific waist circumference measurements, is a required component of the IDF definition. As a result, MS is more prevalent when diagnosed with the IDF definition than with the NCEP/ATP III definition (39% versus 34.5%) (36).

NCEP/ATP III: National Cholesterol Education Program/Adult Treatment Panel III

IDF: International Diabetes Federation

IUGR: intrauterine growth restriction

The definition of MS in children and adolescent is more controversial, and to date, no unified criteria exist (41). Attempts have been made to characterize MS in the pediatric population using modified criteria from NCEP/ATP III, the World Health Organization, and the European Group for the Study of Insulin Resistance (42). In a study that applied eight different criteria for MS to 1,289 children aged 4–16 years, the prevalence of the syndrome varied significantly, with figures between 6% and 39% (42). Application of age-modified NCEP/ATP III criteria to participants 12–19 years from the National Health and Nutrition Examination Survey suggested a 9.2% prevalence among adolescents (43). In overweight children and adolescents, the prevalence was much higher: 38.7% in moderately obese individuals and 49.7% in severely obese individuals (44).

The utility of diagnosing MS as a whole versus individual risk factor components for the prediction of cardiovascular disease remains to be shown. The literature provides conflicting data on this issue (40, 45–47). MS did not predict cardiovascular mortality independently of its individual components in a large cohort of U.S. men aged 50 years at baseline (47), but it did in another study that included men and women ≥65 years followed for 4 years (46).

Regardless of the criteria used to define MS, the prevalence of the diagnostic signs that constitute MS is increasing worldwide, affecting individuals at a progressively younger age. This phenomenon is particularly evident in developing countries, where an accelerated economic and cultural transition and the metabolically obese phenotype (i.e., normal body weight with increased abdominal adiposity) is common. In these countries, the combination of poor nutrition in utero and overnutrition in later life is likely responsible for the observed epidemic of MS and diabetes (14). The incidence of the syndrome is increasing; MS could affect 40% of the population by 2025.

4. EVIDENCE OF FETAL REPROGRAMMING OF METABOLIC SYNDROME FROM HUMAN STUDIES

An important methodological challenge in analyzing human data is to identify in utero stress. LBW has been used to classify fetuses below the expected standard of growth. Two definitions of LBW are often used: (*a*) <2,500 g regardless of gestational age and (*b*) small for gestational age, defined as below the tenth percentile of the population-specific growth curve. These definitions are prone to error. In particular, false negatives may occur. For example, a fetus below the tenth percentile, and therefore defined as small for gestation age, may be constitutionally small for genetic reasons (both parents were of small size) but may indeed have had a perfectly healthy gestation (a false positive). This fetus, although small, will not be predisposed to long-term consequences. Conversely, newborns above the tenth percentile and thus considered to be of "normal" weight may have been stressed in utero but may not have reached their ideal weight because of intrauterine growth restriction (IUGR) (a false negative). Here we use IUGR to indicate that fetuses were stressed in utero, regardless of their weight. Currently there are no clinical measures to differentiate between babies of the same birth weight who were or were not exposed to in utero stress.

Overall, the link between LBW and adult disease is broadly continuous. Heavier babies at birth (excluding overweight newborns of diabetic mothers) have a lower incidence of MS in adulthood than do those of normal birth weight. Epidemiological data support the use of birth weight as a continuous measure rather than as a dichotomous measure associated with long-term disease risk.

Epidemiological studies have convincingly linked a suboptimal gestational environment to an increased risk for components of adult-onset MS (e.g., hypertension, glucose intolerance, dyslipidemia, obesity). Fewer studies have specifically studied the association between in utero stress and MS as a whole (**Table 1**).

The evidence linking LBW and hypertension later in life is well documented, although few studies have questioned the strength of the association (48). In a few studies, for example, low birth weight was not associated with increased blood pressure (49) but was associated with other markers of IUGR, such as the ratio between placental weight and either newborn weight (49) or birth length (50). Yet an extensive, systematic review of 444,000 participants concluded that birth weight is inversely related to systolic blood pressure; the size of the effect is approximately 2 mm Hg kg^{-1} (51). Furthermore, the Bogalusa Heart Study suggests that higher blood pressure of black adolescents compared with white adolescents reflects a lower birth weight (52). A meta-analysis of 197,954 adults from 20 Nordic cohorts (birth years 1910–1987) confirmed the inverse association between birth weight and systolic blood pressure, even after adjusting for current body mass index (BMI) (31).

The literature also provides convincing data on the link between LBW and future risk for impaired glucose tolerance. The first evidence of this link came from a cohort of 468 men born in Hertfordshire, England, between 1920 and 1930 (53). Men with impaired glucose tolerance or undiagnosed diabetes on the basis of a 75-g glucose challenge had a lower mean birth weight and a lower weight at 1 year of age (53). Likewise, individuals who were in utero during the Dutch famine of 1944–1945 had both a lower birth weight and lower glucose tolerance at the age of 50 years than did individuals born the year before or after the famine. Those who were exposed during mid- to late gestation and who had an elevated BMI had the highest 2-h glucose levels (28). Smaller case-control studies have also concluded that thinness at birth as measured by ponderal index [mass (kg)/height (m^3)] and IUGR is associated with insulin resistance later in life (54, 55).

In terms of hypercholesterolemia, the evidence for the association of in utero stress and an abnormal lipid profile is mixed. In 219 middle-aged men and women, Barker et al. (56) showed that abdominal circumference at birth, but not birth weight, correlated inversely with serum levels of total cholesterol and apolipoprotein B. Studies from the Dutch Famine Birth Cohort showed that, although lipid profiles did not differ between exposed and unexposed individuals at age 58, those exposed in utero were more likely to be on lipid-lowering medications (21% versus 15%) (57). In a similar analysis of the Helsinki Birth Cohort, excluding those who used lipid-lowering medications, researchers found that a decrease of 1 kg m^{-2} in BMI at birth was associated with a slight increase in non-high-density-lipoprotein cholesterol and apolipoprotein B concentrations (58). A review of the literature involving 38 studies and 28,578 individuals did not find strong evidence for a link between birth weight and lipid profiles later in life (59).

Data on the link between birth weight and waist circumference are limited; however, several studies have examined obesity and body composition. The first study to observe the association between intrauterine malnutrition and later obesity was based on the Dutch Famine Birth Cohort of 300,000 19-year-old men (60). Participants who were exposed to in utero malnutrition in the third trimester of pregnancy had lower obesity rates, defined as weight for height ≥120% of standard, whereas those who were exposed in the first and/or second trimesters had significantly higher obesity rates. This study accounted for socioeconomic status, but not for other known confounders such as maternal BMI. In a much smaller case-control study of 32 white men

Table 1 Selected human studies linking either a well-identified pregnancy stress or a marker of stress identified at birth (like LBW) to MS or its individual components[a]

Type of stress	Stress marker	MS component	Details	Reference
IVF			RR 2.6 for low birth weight	20
IVF		HTN	Higher SBP: 109 ± 11 versus 105 ± 10 mm Hg	23
		IGT	Higher fasting glucose: 5.0 ± 0.4 versus 4.8 ± 0.4 mmol liter^{-1}	23
IVF		Obesity	Higher peripheral fat mass: 7.6 ± 4.2 versus 6.7 ± 3.2 kg	21
IVF		Obesity	Prevalence of BMI >25: 44.9%	147
		HTN, dyslipedemia, insulin resistance	Prevalence of 6.9% (12/173)	147
IVF		Dyslipidemia	Higher HDL: 1.67 ± 0.04 versus 1.53 ± 0.04 mmol liter^{-1}	25
		Dyslipidemia	Lower triglycerides: 0.65 ± 0.04 versus 0.78 ± 0.04 mmol liter^{-1}	25
	Birth weight	HTN	Inverse correlation with SBP by 2 mm Hg kg^{-1}	49
	Birth weight	HTN	Inverse correlation with SBP by 1.52 mm Hg kg^{-1} (men), 2.80 mm Hg kg^{-1} (women)	31
Biafran famine		HTN	OR 2.87 for systolic HTN in fetal-infant exposure	66
PIH		HTN	Higher childhood SBP by 2 mm Hg in exposed group	148
PIH		HTN	OR 1.88 for HTN in exposed group	79
	Weight at 1 year	IGT	Higher prevalence of IGT: 23% (<18 lb) versus 13% (>27 lb)	53
		IGT	Higher prevalence of DM: 17% (<18 lb) versus 0% (>27 lb)	53
Dutch famine		IGT	Higher mean 2-h glucose by 0.05 mmol liter^{-1} in exposed group	28
	Ponderal index	IGT	Inverse correlation with insulin resistance	54
	IUGR	IGT	Lower glucose-stimulated insulin uptake: 6.7 ± 2.9 versus 8.0 ± 1.9 mg kg^{-1} fat-free mass × min	55
Biafran famine		IGT	OR 1.65 for IGT in fetal-infant exposure	66
	Abdominal circumference at birth	Dyslipidemia	Inverse correlation with total cholesterol by 0.25 mmol liter^{-1} per 1 inch	56
Dutch famine		Dyslipidemia	OR 2.1 for high-fat diet in exposed group	57
	BMI at birth	Dyslipidemia	Inverse correlation with non-HDL cholesterol by 0.051 mmol liter^{-1} per 1 kg m^{-2}	58
	Birth weight	Dyslipidemia	No consistent association between birth weight and lipids	59
Dutch famine		Obesity	Exposure during third trimester decreased obesity rates (0.82% versus 1.32%)	60

(Continued)

Table 1 *(Continued)*

Type of stress	Stress marker	MS component	Details	Reference
Dutch famine		Obesity	Exposure during first and second trimester increased obesity rates (2.77% versus 1.45%)	60
	Birth weight	Obesity	Higher percentage body fat in low birth weight (29.3% versus 25.3%)	61
	Birth weight	Obesity	No association between birth weight and BMI after accounting for maternal weight	62
Biafran famine		Obesity	OR 1.41 for overweight in fetal-infant exposure	66
	Birth weight	Obesity	Significant association between birth weight and childhood obesity	149
Gestational DM		Obesity	OR 1.4 for adolescent overweight in exposed group	75
Diet-controlled gestational DM		Obesity	No difference in prevalence of childhood obesity	76
Dutch famine		MS	Exposure not associated with MS	63
	Birth weight	MS	RR 2.41 for MS associated with lowest birth weight tertile	64
	Birth weight	MS	OR 1.8 for MS associated with lowest birth weight tertile	65
Maternal diabetes		MS	Higher prevalence of childhood MS in exposed LGA (50% versus 21%)	73
Maternal obesity		MS	OR 1.81 for childhood MS in exposed group	73
	Birth weight	CAD	Decrease in standardized mortality ratios with increase in birth weight	67
	Birth weight	CAD	Decrease in standardized mortality ratios with increase in birth weight	30
	Birth weight	CAD	RR 1.49 for nonfatal CAD with birth weight <5 lb	68
	Birth weight	CAD	Higher prevalence of CAD in low birth weight (11% versus 3%)	69
	Birth weight	CAD	Age-adjusted HR 0.62 for CAD for every 1-kg increase	70
	Birth weight	CAD	Age-adjusted HR 0.67 for CAD for every 1-kg increase	71

[a]Abbreviations: BMI, body mass index; CAD, coronary artery disease; DM, diabetes mellitus; HDL, high-density lipoprotein; HR, hazard ratio; HTN, hypertension; IGT, impaired glucose tolerance; IUGR, intrauterine growth restriction; IVF, in vitro fertilization; LGA, large for gestational age; MS, metabolic syndrome; OR, odds ratio; PIH, pregnancy-induced hypertension; RR, relative risk; SBP, systolic blood pressure.

64–72 years of age, those with LBW (mean 2.76 kg) had a higher percentage of body fat and fat mass by dual-energy X-ray absorptiometry than did those with a higher birth weight (mean 4.23 kg) (61). Other studies have concluded that maternal or parental BMI as a confounder largely explains the association between birth weight and later obesity (62). The important role of maternal BMI also supports the importance of the in utero environment on future health.

The association between in utero stress and MS has also been explored. In an analysis of a subset of 783 men and women aged 57–59 years from the Dutch Famine Birth Cohort, the prevalence of MS, defined with NCEP/ATP III criteria, was not significantly greater among those exposed to malnutrition (63). In British cohorts, by contrast, in utero stress was associated with a higher prevalence of MS in both men and women in the seventh decade of life. This finding was confirmed in multiple study populations, including postmenopausal Caucasian women living in the United States (64), young adults aged 26–31 years in the Netherlands (65), and 40-year-old Nigerians who survived the famine in Biafra (66).

More important than MS or its individual components is the risk for actual cardiovascular disease. The landmark paper that fueled the study of birth weight as a marker for future disease risk came from a cohort of 5,654 men born between 1911 and 1930 in Hertfordshire, England. This study showed that men with the lowest weights at birth and at 1 year of age had the highest death rates from ischemic heart disease (67). The association between LBW and cardiovascular disease has been confirmed in different racial and ethnic study populations (68–70). Not surprisingly, the highest risk for coronary heart disease associated with LBW appears to be restricted to individuals who have a high BMI in adulthood (71, 72).

Other studies have examined the association of a suboptimal in utero environment, represented by gestational diabetes or pregnancy-related hypertension disorders, and risk of MS. Boney et al. (73) demonstrated that large-for-gestational-age offspring born to diabetic mothers were at significant risk of developing childhood MS, as were offspring of obese mothers. Even in a low-risk population of women with gestational diabetes, 6.9% of offspring had abnormal glucose metabolism (74). In contrast, studies have shown that the association between having gestational diabetes and being overweight in adolescence disappears after adjustment for birth weight and maternal BMI (75) and that prenatal exposure to diet-controlled gestational diabetes does not increase the prevalence of childhood obesity (76). Interestingly, a meta-analysis found that women with polycystic ovary syndrome (PCOS), a common endocrinopathy characterized by anovulation, insulin resistance, and androgen excess, have a higher incidence of gestational diabetes, pregnancy-induced hypertension, and preeclampsia (77). However, infant birth weight was not different between women with PCOS and controls (77). There are currently no studies evaluating long-term health of the PCOS offspring.

Last, studies have demonstrated that pregnancy-associated hypertension disorders are associated with increased systolic and diastolic blood pressure among 9-year-old offspring (78) and with an increased risk of antihypertensive drug use in offspring independent of birth weight or preterm delivery (79).

5. PHYSIOLOGICAL AND MOLECULAR MECHANISMS LEADING TO ABNORMAL REPROGRAMMING

Embryos, fetuses, and newborns show remarkable plasticity in their response to the environment. During the prenatal period, the human fetus rapidly evolves from a single cell to 10 trillion cells differentiated into more than 250 subtypes. Therefore, that various forms of stress during this time can lead to long-term problems is biologically plausible. Evidence from epidemiological studies is compelling but is limited by potential confounding variables throughout an individual's lifetime. Animal models allow evaluations of the effects of a specific controlled stress applied over a well-defined time. However, studies in animals are not immune to bias and errors (80–82).

Multiple animal models exist, including rodent and sheep models. Sheep physiology is closer to human physiology, but sheep have a long gestation (145 days) and usually carry one or two fetuses. The rodent model is the most frequently used due to low cost and short gestation. Thus, acknowledging the different physiological characteristics of rodents and humans is important (see sidebar on differences between rodents and humans in analyzing animal models of reprogramming) (83).

The use of animal models has allowed for important observations. Importantly, there is evidence of a transgenerational effect. Female rats that are food restricted during their perinatal life but

DIFFERENCES BETWEEN RODENTS AND HUMANS TO BE CONSIDERED WHEN ANALYZING ANIMAL MODELS OF REPROGRAMMING

First, the large litter size ($n = 8$–12) and relatively short gestation (21 days) in rodents imply very different energetic needs than do humans during pregnancy. For example, pregnant rodents require up to 30-fold-higher protein content in their diet than humans and large mammals do; in addition, postnatal growth in rodents is faster than in humans and sheep (144).

Second, rodents have unique physiology. They are nocturnal animals (normally more active during the dark cycle); exhibit unique dietary habits such as coprophagia, which can alter nutrient flux; and have functional brown adipose tissue throughout life, which can result in very different energy balance compared with humans (145).

Third, the HPA axis and the appetite regulatory networks develop after birth in rodents and during the third trimester of pregnancy in humans. In rodents, the kidneys and brown adipose start to develop during gestation but continue to develop in the neonatal period, whereas in humans the process is already completed at birth (83, 130). The rat pancreas develops in late gestation and undergoes an important remodeling phase at the time of weaning (2–3 weeks); however, in humans, pancreatic islet remodeling continues until age 4 (146).

exposed to a normal diet during pregnancy deliver offspring with reduced pancreatic β-cell mass and fewer cells expressing only insulin (84). A similar phenotype occurs in F2 offspring of diabetic pregnant rats (85). A low-protein diet leads to hypertension by decreasing the number of nephrons in F1 offspring (86); this phenotype is also present in the F2 generation (87). Epigenetic changes are likely responsible for these effects.

Certain in utero or postnatal interventions can reverse the effects of prenatal stress. For example, undernutrition in utero leads to obesity and hyperleptinemia in adult rats (88). These effects can be reversed by injecting leptin from day 3 to day 13 of postnatal life (89). In another study, the hypertensive effects of a low-protein diet were reversed by adding glycine to the diet (90).

Different experimental interventions have been used to study stress during pregnancy (**Figure 2**) (3, 5, 91). These include (*a*) nutritional stress, with changes in macronutrients (global caloric restriction up to 50–70% of control, low-protein isocaloric diets, high-fat diets) or micronutrients (iron, zinc, sodium, calcium); (*b*) surgical stress (bilateral uterine ligation in rats or reduction of placental areas); (*c*) pharmacological treatments, either to alter the HPA axis with dexamethasone or steroid synthesis blockers or to induce diabetes with streptozotocin; and (*d*) exposure to toxins (smoke, endocrine disruptors, medications during pregnancy).

The physiological and molecular mechanisms leading to abnormal reprogramming are diverse and only partially known. Different stresses can promote reprogramming via unique and stress-specific pathways. For example, adult offspring of semistarved dams show insulin resistance in the liver but not in the peripheral tissues (92), whereas offspring of diabetic mothers exhibit both hepatic and peripheral insulin resistance (92). Importantly, multiple mechanisms can occur at the same time.

Overall, the mechanisms of stress can be broken down into several categories: (*a*) the cellular response to stress [e.g., epigenetic changes, mitochondrial dysfunction, the unfolded protein response (UPR), oxidative stress, the differential expression of transcription factors], (*b*) alterations in adult organ morphology or cell number [e.g., by adapting to a suboptimal environment, the organism trades off the development of less essential organs, such as the kidney (nephron mass) and the pancreas (β-cell mass), for the development of more essential organs such as the brain], (*c*) tissue or systemic responses (e.g., alterations in the placenta or in endocrine pathways, in particular the HPA axis), and (*d*) a combination of the above. We discuss these mechanisms below.

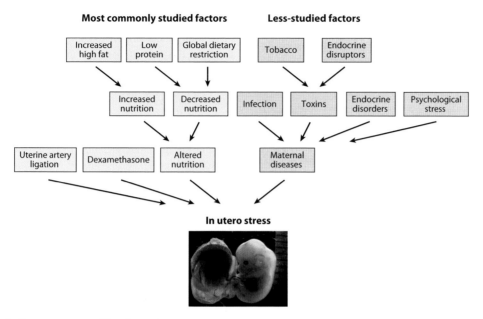

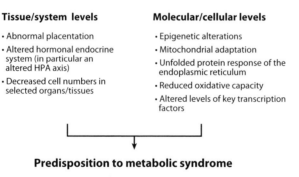

Figure 2

(*a*) Mechanisms of in utero stress. Different types of in utero stress can affect a fetus's health outcome. The most common experimental procedures used in animal models to study reprogramming are highlighted in yellow. The less-studied or more-difficult-to-study factors for reprogramming are in gray. (*b*) As a consequence of in utero stress, the organism reprograms its development at the cellular, tissue, and systemic levels. Overall, an organism stressed in utero will manifest a metabolism that favors lipid storage and increased cardiovascular reactivity, both of which are key components of metabolic syndrome. HPA axis denotes hypothalamic-pituitary-adrenal axis.

5.1. Cellular Response to Stress

Changes in energy and nutrient availability modify stress signaling pathways at the cellular level. Several of these stress pathways are also altered in MS and diabetes. It is therefore tempting to postulate that these pathways can be permanently altered following suboptimal conditions in utero.

5.1.1. Epigenetic changes. Epigenetic changes are changes in gene function that occur without changes in gene sequence. Epigenetic regulation can occur at the DNA level through the modification of cytosine bases (methylation of cytosine in CpG dinucleotides) or through chromatin conformational changes. Histone modifications can result in an open (active) or a closed (inactive) chromatin structure; DNA methylation of CpG results in the inability of transcription factors to bind to genes. The net result is either silencing or activation of transcription.

The preimplantation stage is particularly sensitive to epigenetic regulation. In fact, embryonic DNA methylation patterns proceed through defined phases during this stage of development. In the early murine embryo, a wave of DNA methylation is followed by remethylation at the morula or blastocyst stage. IVF and embryo culture can generate imprinting errors in mice by inducing abnormal DNA and histone methylation marks (93). The development rate and global patterns of gene expression are abnormal in mouse embryos cultured in vitro (94). Mouse blastocysts conceived in vitro have fewer trophectodermal cells than do in vivo controls. Mouse fetuses generated by IVF display delayed fetal development in comparison to controls. In particular, culture in suboptimal conditions resulted in a more severe phenotype than did culture in optimized conditions (95). Importantly, a recent paper found altered glucose parameters in adult mice conceived in vitro (96). Our group has found similar results (97). These animal studies would lend support to the metabolic findings of Ceelen et al. (23) in ART children.

Epigenetic changes follow different types of stress and different time exposure in different tissues. In the rat uterine artery ligation model, promoter methylation is increased, and expression of β-cell Pdx1 genes is decreased; these changes were associated with adult-onset diabetes (98). Uterine artery ligation is also associated with decreased methylation of the promoter of p53 genes in the kidney. The increase in p53 gene expression may lead to renal apoptosis, resulting in fewer nephrons (99). In an elegant experiment, the Meaney group (100) showed changes in the promoter of the glucocorticoid receptor in offspring exposed to different postnatal care. Epigenetic changes can also explain the transgenerational effects described above (85, 87).

5.1.2. Mitochondrial changes. Mitochondria interface the environmental calorie supply and the energy requirements of each organ; they generate ATP and regulate cellular processes such as signal transduction and apoptosis. Alterations in mitochondria function may therefore be a key cellular mechanism of reprogramming. Although studies on IUGR offspring have not been reported, caloric restriction evidently increases skeletal muscle mitochondrial biogenesis (by upregulating sirtuin 1, a protein upregulated in cases of increased insulin sensitivity) and decreases mitochondrial free-radical generation and oxidative damage to mitochondrial DNA in the rat heart (101). Mitochondrial density is reduced in insulin-resistant offspring of parents with type 2 diabetes (102). However, a definitive connection between mitochondrial dysfunction and insulin resistance is missing, and recent evidence argues against such a nexus (101).

5.1.3. Endoplasmic reticulum stress. The UPR is a cellular stress response that occurs in the lumen of the endoplasmic reticulum when there is excessive accumulation of proteins with altered tridimensional structure. Initially, the cell tries to repair the system by synthesizing molecular chaperones involved in protein folding and by halting additional protein translation. If repair does not occur, the cell initiates apoptosis. The UPR is thought to mediate inflammatory and stress signaling pathways and to be particularly important in regulating chronic metabolic diseases such as obesity, insulin resistance, and type 2 diabetes (103).

In sheep, in utero nutrient restriction and a postnatal high-fat diet cause organ-specific alteration of the UPR in different tissues. In particular, perirenal adipose tissue showed activation of the UPR and altered insulin signaling, whereas renal tissue showed reduced activation of the UPR and less histological damage (104).

5.1.4. Oxidative stress. Oxidative stress describes a condition of imbalance in the production of reactive oxygen species and the ability of antioxidant defenses to scavenge them. Oxidative stress may derive from the increased production of reactive oxygen species or from a decrease in antioxidant capacity. Pregnancy is a state of oxidative stress because of increased metabolic activity in placental mitochondria and decreased antioxidant-scavenging ability (105). In pregnancies complicated by hypertension and diabetes, the placenta shows signs of increased hypoxia and oxidative stress and activation of the endoplasmic reticulum stress response (106). β-Cells of adult offspring whose mothers were exposed to a low-protein diet in utero show evidence of increased oxidative stress, increased lipid peroxidation, and impairment of oxidative defense (107). This study was the first to link maternal protein restriction to age-associated increased oxidative stress, impairment of oxidative defense, and fibrosis in offspring.

5.1.5. Changes in transcription factor levels. An increase in gluconeogenesis in IUGR rats can be explained by an increase in the level of peroxisome proliferator–activated receptor-γ coactivator-1 (PGC-1) or PGC-1α (108). This transcription factor regulates hepatic glucose production by increasing the mRNA levels of three key metabolic enzymes: glucose-6-phosphatase, phosphoenolpyruvate carboxykinase, and fructose-1,6-bisphosphatase (108). Hepatic levels of glucokinase, a key glycolytic enzyme, are also decreased in IUGR rats (109).

5.2. Organ-Specific Alterations in Response to Prenatal Stress

Prenatal stress affects different organs in different ways. However, a decrease in cell number is often present, as described in depth elsewhere (3). Interestingly, prenatal stress and the resulting reprogramming decrease insulin sensitivity (pancreas, liver, muscle, and fat tissue), increase cardiovascular reactivity, and increase HPA responses to stress in adult life. These effects lead to hyperlipidemia, obesity, glucose intolerance and hypertension—all components of MS.

5.2.1. Pancreas. Abnormal programming of the fetal pancreas may be a prime mechanism for adult-onset MS and diabetes. Prenatal stress during pregnancy can decrease both β-cell mass and β-cell function; such decreases may result in a decreased insulin response to glucose (110). Different types of stress have recognizably different effects. For example, global caloric restrictions (50% of control diet) in rats at embryonic day 15 reduce the number and size of islets; however, the existing β cells have a normal proliferative capacity (111). Conversely, an isocaloric low-protein diet results in smaller islets and a relatively smaller proportion of β cells, owing to decreased β-cell proliferation and increased apoptosis (112). Different levels of corticosterone in the two models of fetal stress may explain these differences: A low-protein diet is associated with normal corticosterone levels, whereas global food restriction is associated with reduced corticosterone levels (113, 114). Additional mechanisms responsible for the reduced cell number include a decrease in the number of pancreatic stem cells (115) and the inappropriate expression of transcription or growth factors (e.g., insulin-like growth factor 2) (116).

5.2.2. Liver. Prenatal stress promotes increased hepatic gluconeogenesis and hepatic insulin resistance (117), which contribute to hyperglycemia. Evidence of hepatic reprogramming is important, as unsuppressed endogenous hepatic glucose production is a common component of the insulin resistance associated with type 2 diabetes. In addition, in male IUGR mice, triglyceride levels are elevated because of increased hepatic fatty acid synthesis and decreased β oxidation (33). Prenatal stress does not reduce the ratio of liver weight to total fetal weight (118) unless the prenatal stress is severe (119). However, livers of protein-restricted rats have fewer but larger lobules (118). From a mechanistic point of view, intrauterine stress may acutely decrease liver mitochondrial ATP production and NADH availability. Because ATP and NADH are two key regulators of intermediary metabolism, such changes can reduce fetal growth (120).

5.2.3. Skeletal muscle. Insulin-sensitive skeletal muscles are the major site of glucose uptake. Under euglycemic conditions, approximately 80% of total body glucose uptake occurs in skeletal muscles (121). Different experimental models in different species show that prenatal stress can alter muscle size and lead to insulin resistance (3).

Protein-restricted rat offspring show decreased muscle mass of both fast-twitch-type fibers (anterior tibialis) and slow-twitch-type fibers (soleus muscle) at postnatal day 21 (122). In contrast, sheep fetuses at embryonic day 125 had 40% higher weight of the fast-twitch plantaris muscle if they were cultured in vitro with granulosa cells from the zygote to the blastocyst. In particular, muscle of fetuses that were cultured in vitro as embryos showed both hyperplastic and hypertrophic changes, suggesting that myogenesis can be altered by preimplantation stress (123).

Muscles of IUGR rats are insulin resistant and show significantly decreased glycogen content and insulin-stimulated 2-deoxyglucose uptake (124). This insulin resistance may reflect significantly decreased pyruvate oxidation and ATP production in muscle mitochondria (124), which may decrease recruitment of the insulin-regulated glucose transporter GLUT4 to the cell surface. Muscles of IUGR rats also show increased expression of PGC-1 and its downstream gluconeogenesis pathway (125).

5.2.4. Adipose tissues. Worldwide, the increase in MS parallels the increase in obesity (126). Although adipose tissue contributes only approximately 3% to glucose disposal after an oral glucose tolerance test (121), it plays an important role in intermediate metabolism by releasing free fatty acids, leptin, and inflammatory cytokines (126).

Free fatty acids cause both insulin resistance and the release of proinflammatory cytokines in insulin-sensitive tissues such as skeletal muscle, liver, and endothelium. In particular, an increase in free fatty acids inhibits insulin-stimulated glucose uptake into muscle (126, 127).

Maternal undernutrition can increase retroperitoneal fat (88) and increase the proportion of large fat cells in visceral fat (128). In protein-restricted rats, insulin-stimulated glucose uptake and insulin-dependent lipolysis are reduced in adipocytes, indicating insulin resistance. These changes appear to be caused by a post–insulin receptor effect, as these rats have lower levels of phosphatidylinositol 3-kinase and protein kinase B but similar levels of insulin receptor and insulin receptor tyrosine phosphorylation compared with controls (129).

An important observation links maternal undernutrition to obesity and hyperleptinemia in adult offspring (88). Leptin is a trophic factor that is central in organizing the formation of hypothalamic circuitry. Perturbations in perinatal nutrition that alter leptin levels may therefore have lasting consequences for the formation and function of circuits regulating food intake and body weight (130).

5.2.5. Kidney. Different intrauterine stresses can lead to hypertension in later life by different mechanisms. In a classic experiment, Langley-Evans et al. (18) showed that, although a low-protein diet before conception was not associated with hypertension, blood pressure was increased in rats exposed to low protein during gestation or even for limited periods of pregnancy (days 0–7, 8–14, and 15–22). The increase elicited by these discrete periods of undernutrition was lower than that induced by feeding a low-protein diet throughout pregnancy. The effect in early gestation was significant only in males (18).

A low-protein diet or uterine artery ligation leads to hypertension by decreasing the abundance of nephrons (86). Interestingly, global caloric reduction in pregnancy has a less consistent effect on blood pressure. Brenner & Chertow (131) were the first to propose that hypertension could trail a congenital deficit of nephrons. These researchers proposed that a reduction in renal mass, and therefore in glomerular filtration surface area, leads to hypertension (131). Sustained exposure of nephrons to higher glomerular perfusion pressure gradually results in focal and segmental glomerular sclerosis, which leads to further glomerular loss, further reduced ability to excrete sodium, and a self-perpetuating cycle of increasing blood pressure and progressive kidney disease. Protein restriction also increases the expression of two ascending-limb sodium cotransporters, thereby facilitating sodium retention and hypertension (132) and suppressing the renin-angiotensin system (133).

5.2.6. Endothelium. Evidence from studies in humans and animals suggests that stress in utero can alter endothelium-dependent vasodilatation. Such a change may occur because of decreased levels of nitric oxide production (134, 135).

5.2.7. Heart. Stress in utero that leads to fetal hypoxia can alter myocardial structure (by inhibiting cardiomyocyte proliferation and increasing apoptosis) and reduce cardiac performance (136). In addition, hypoxia predisposes the developing heart to increased vulnerability to ischemia and reperfusion injury later in life (136), likely because of decreased expression of cardioprotective genes like those encoding protein kinase Cε, heat shock protein-70, endothelial nitric oxide synthase (136), and insulin-like growth factor 1 (137).

5.3. Tissue and Systemic Response to Stress

The placenta, HPA axis, and appetite regulatory networks share the characteristics of regulating the development and physiological response of multiple individual organs. It is therefore particularly important to study how these systems are altered by prenatal stress.

5.3.1. Abnormal placentation. By virtue of its transport, immune, and hormonal roles, the placenta is in a key position to play a direct role in fetal programming by changing the pattern or amount of substrate transported to the fetus. Aberration in placental function is probably the most frequent cause of IUGR because optimal placental development and the ability of the placenta to compensate for stimulus-induced injury are central to the promotion of normal fetal growth.

Abnormal placentation often results from reduced trophoblast invasion, which leads to hypoxia and increased oxidative stress. In humans, IUGR placentas are not simply smaller versions of a term placenta; they display alterations in vascularization, trophoblast expression of transporters, trophoblast enzyme activity, and hormone production. Furthermore, alteration of placenta-imprinted genes may play an important role. Imprinted genes are involved in growth regulation, controlling both the supply (placental side) and the demand (fetal side) of nutrients. Paternally derived imprinted genes enhance fetal growth, whereas maternally imprinted genes suppress fetal growth. Several imprinted genes encode specific transporters in trophoblasts. The placenta may function as a nutrient sensor, matching fetal growth rate to available nutrient resources by altering transport function. Such a function would explain fetal growth restriction and possibly growth enhancement in pregnancies with gestational diabetes (138).

The placenta is vital in moderating fetal exposure to maternal factors. Glucocorticoid levels are approximately 20% lower in the fetus than in the mother because the placenta expresses high levels of 11β hydroxysteroid dehydrogenase type 2 (11β-HSD-2), which converts the active cortisol/corticosterone to the inactive 11 keto steroids. A deficiency in 11β-HSD-2 leads to overexposure of the fetus to glucocorticoids. Indeed, low levels of placental 11β-HSD-2 activity have been found in LBW in humans and rodent models, though this has not always been reported (139). Placentas of pregnancies associated with preeclampsia or unexplained IUGR have an increase of the UPR (106).

5.3.2. Abnormal activation of the HPA axis. The HPA axis is fundamental in regulating the response of the individual to stress. Increased exposure to cortisol in utero (due to stress, pharmacological treatment, or impaired function of 11β-HSD-2) (140) has long-term effects. In rats, for example, prenatal glucocorticoid exposure leads to LBW, alters cardiac development, reduces nephron number, and alters glucose-insulin homeostasis by increasing the expression of hepatic gluconeogenesis enzymes and impairing β-cell function (139). The physiological effects of a low-protein diet in utero may be mediated in part by an increase in glucocorticoid levels (113).

Epigenetic changes are involved in altering the activation of the HPA axis. Rat pups exposed to increased maternal care during the first postnatal week exhibit a persistent lack of DNA methylation of a specific CpG base in the first exon of the glucocorticoid receptor (100). This epigenetic change decreases hypothalamic corticotrophin-releasing factor expression and results in adult animals manifesting a modest HPA response to stress.

5.3.3. Appetite regulatory network. One important potential locus of reprogramming is the hypothalamic appetite regulatory network (3), in which leptin plays an important role (130). Early postnatal life plays an important role in determining future food intake. Rats from small litters tend to be heavier than those from larger litters because of increased food intake by the former (141). The experimental evidence targeted to the pregnancy period is scarce because the majority of studies limit postnatal intervention or combine a postnatal dietary manipulation with a prenatal one. Undernutrition during gestation alone does not affect the weight, length, or fat content of offspring in adulthood (142). However, adult male, but not female, offspring of rats injected with insulin during the last week of gestation develop significant obesity as adults, accompanied by elevated extracellular norepinephrine levels in the paraventricular nucleus (143).

6. CONCLUSIONS

Evidence linking severe stress in utero to fetal reprogramming is convincing, and uterine stress predisposes individuals to components of MS later in life. Undeniably, individual behaviors such as diet and exercise also affect one's predisposition to adult MS; however, this review provides compelling evidence that the short duration of in utero exposure to stress plays a crucial role of relatively great magnitude.

Although animal data provide multiple biologically plausible mechanisms for fetal reprogramming, many questions remain unanswered. For example, the mechanisms leading to adult disease need to be further studied so that novel strategies can be developed to counteract the effects of in utero stress. In addition, the use of LBW as a marker for in utero stress is of inadequate value, and new strategies to discover biomarkers of in utero stress would be of great importance.

These findings have particular relevance for the populations of developing countries, where the combination of poor nutrition in utero and overnutrition in later life may drive the observed epidemics of obesity and diabetes. In addition, given that preimplantation embryo stress can activate many reprogramming mechanisms, children conceived by ART should be increasingly monitored over time to evaluate for potential occurrence of MS. Physicians should routinely incorporate information of pregnancy course and health into the individual medical history.

SUMMARY POINTS

1. There is a broadly continuous positive relationship between in utero stress and predisposition to MS or components of MS.
2. The use of a single marker of stress in utero (e.g., LBW) is inadequate and is potentially prone to error.
3. Different critical periods of sensitivity vary in different species and in males and females; overall, the prenatal and early postnatal periods represent critical periods during which reprogramming can occur. Stress occurring during these phases has a disproportionally greater effect in term of later life predisposition to diseases compared with the same stress occurring outside of a window of sensitivity.
4. Different types of stresses applied during the same period have different long-term effects. For example, adult rat offspring of semistarved dams show insulin resistance in the liver but not in the peripheral tissues (92). In contrast, offspring of diabetic mothers exhibit insulin resistance at both the hepatic and peripheral tissue levels (92).
5. Different cellular pathways are activated following in utero stress. Such activation leads to alterations in specific organs or systems.
6. Epigenetic changes occur following in utero stress and can explain transgenerational effects.
7. Health problems are amplified if there is a mismatch between the in utero and the postnatal environments; the most common scenario is the fetus being exposed to reduced energy in utero followed by a postnatal life rich in food resources. This concept (14) has particular relevance in developing countries.
8. Patient medical history should specify if ART was used to conceive offspring and should include detailed information on the health of maternal pregnancy.

FUTURE ISSUES

1. The preimplantation period is particularly sensitive to stress. ART constitutes a potential new form of preimplantation stress. Given that more than 3.5 million children have been conceived by ART, additional studies are needed to follow up children conceived by ART.
2. There is a need to discover novel biomarkers of in utero stress and to develop high-throughput technology to discern the cellular mechanisms that are altered after in utero stress.
3. Strategies to counteract the effect of in utero stress are needed. For example, the potential beneficial effect of specific nutrients (90) or hormones (89) on health should be studied. Such research could be particularly important for populations in developing countries.

DISCLOSURE STATEMENT

The authors are not aware of any affiliations, memberships, funding, or financial holdings that might be perceived as affecting the objectivity of this review.

ACKNOWLEDGMENTS

This work was made possible by funding from the National Institute of Child Health and Human Development (R01 HD062803-02) and the American Diabetes Association to P.R. The authors thank Drs. Annemarie Donjacour, Marcelle Cedars, and Robert Lustig for their valuable suggestions and comments. The authors apologize to colleagues whose work could not be cited because of space constraints.

LITERATURE CITED

1. Barker DJ, Osmond C. 1986. Infant mortality, childhood nutrition, and ischaemic heart disease in England and Wales. *Lancet* 1:1077–81
2. Gluckman PD, Lillycrop KA, Vickers MH, Pleasants AB, Phillips ES, et al. 2007. Metabolic plasticity during mammalian development is directionally dependent on early nutritional status. *Proc. Natl. Acad. Sci. USA* 104:12796–800
3. McMillen IC, Robinson JS. 2005. Developmental origins of the metabolic syndrome: prediction, plasticity, and programming. *Physiol. Rev.* 85:571–633
4. Symonds ME, Sebert SP, Hyatt MA, Budge H. 2009. Nutritional programming of the metabolic syndrome. *Nat. Rev. Endocrinol.* 5:604–10
5. Bertram CE, Hanson MA. 2001. Animal models and programming of the metabolic syndrome. *Br. Med. Bull.* 60:103–21
6. Kermack WO, McKendrick AG, McKinlay PL. 1934. Death-rates in Great Britain and Sweden: expression of specific mortality rates as products of two factors, and some consequences thereof. *J. Hyg.* 34:433–57
7. Wadsworth ME, Cripps HA, Midwinter RE, Colley JR. 1985. Blood pressure in a national birth cohort at the age of 36 related to social and familial factors, smoking, and body mass. *Br. Med. J.* 291:1534–38
8. McCance RA, Widdowson EM. 1974. The determinants of growth and form. *Proc. R. Soc. Lond. Ser. B* 185:1–17
9. Hales CN, Barker DJ. 1992. Type 2 (non-insulin-dependent) diabetes mellitus: the thrifty phenotype hypothesis. *Diabetologia* 35:595–601
10. Neel JV. 1962. Diabetes mellitus: a "thrifty" genotype rendered detrimental by "progress"? *Am. J. Hum. Genet.* 14:353–62
11. Bateson P, Barker D, Clutton-Brock T, Deb D, D'Udine B, et al. 2004. Developmental plasticity and human health. *Nature* 430:419–21
12. Srinivasan M, Patel MS. 2008. Metabolic programming in the immediate postnatal period. *Trends Endocrinol. Metab.* 19:146–52
13. Rasmussen KM. 2001. The "fetal origins" hypothesis: challenges and opportunities for maternal and child nutrition. *Annu. Rev. Nutr.* 21:73–95
14. Chan JC, Malik V, Jia W, Kadowaki T, Yajnik CS, et al. 2009. Diabetes in Asia: epidemiology, risk factors, and pathophysiology. *JAMA* 301:2129–40
15. Kwong W, Wild A, Roberts P, Willis A, Fleming T. 2000. Maternal undernutrition during the preimplantation period of rat development causes blastocyst abnormalities and programming of postnatal hypertension. *Development* 127:4195–202
16. Edwards L, McMillen I. 2002. Periconceptual nutrition programs development of the cardiovascular system in the fetal sheep. *Am. J. Physiol. Regul. Integr. Comp. Physiol.* 283:669–79
17. Gardner D, Pearce S, Dandrea J, Walker R, Ramsay M, et al. 2004. Peri-implantation undernutrition programs blunted angiotensin II evoked baroreflex responses in young adult sheep. *Hypertension* 43:1290–96

18. Langley-Evans SC, Welham SJ, Sherman RC, Jackson AA. 1996. Weanling rats exposed to maternal low-protein diets during discrete periods of gestation exhibit differing severity of hypertension. *Clin. Sci.* 91:607–15
19. Fernandez-Gonzalez R, Moreira P, Bilbao A, Jimenez A, Perez-Crespo M, et al. 2004. Long-term effect of in vitro culture of mouse embryos with serum on mRNA expression of imprinting genes, development, and behavior. *Proc. Natl. Acad. Sci. USA* 101:5880–85
20. Schieve LA, Meikle SF, Ferre C, Peterson HB, Jeng G, Wilcox LS. 2002. Low and very low birth weight in infants conceived with use of assisted reproductive technology. *N. Engl. J. Med.* 346:731–37
21. Ceelen M, van Weissenbruch MM, Roos JC, Vermeiden JP, van Leeuwen FE, Delemarre–van de Waal HA. 2007. Body composition in children and adolescents born after in vitro fertilization or spontaneous conception. *J. Clin. Endocrinol. Metab.* 92:3417–23
22. Ceelen M, van Weissenbruch MM, Vermeiden JP, van Leeuwen FE, Delemarre–van de Waal HA. 2008. Pubertal development in children and adolescents born after IVF and spontaneous conception. *Hum. Reprod.* 23:2791–98
23. Ceelen M, van Weissenbruch MM, Vermeiden JP, van Leeuwen FE, Delemarre–van de Waal HA. 2008. Cardiometabolic differences in children born after in vitro fertilization: follow-up study. *J. Clin. Endocrinol. Metab.* 93:1682–88
24. Law CM, de Swiet M, Osmond C, Fayers PM, Barker DJ, et al. 1993. Initiation of hypertension in utero and its amplification throughout life. *Br. Med. J.* 306:24–27
25. Miles HL, Hofman PL, Peek J, Harris M, Wilson D, et al. 2007. In vitro fertilization improves childhood growth and metabolism. *J. Clin. Endocrinol. Metab.* 92:3441–45
26. Roseboom T, de Rooij S, Painter R. 2006. The Dutch famine and its long-term consequences for adult health. *Early Hum. Dev.* 82:485–91
27. de Rooij SR, Wouters H, Yonker JE, Painter RC, Roseboom TJ. 2010. Prenatal undernutrition and cognitive function in late adulthood. *Proc. Natl. Acad. Sci. USA* 107:16881–86
28. Ravelli AC, van der Meulen JH, Michels RP, Osmond C, Barker DJ, et al. 1998. Glucose tolerance in adults after prenatal exposure to famine. *Lancet* 351:173–77
29. Symonds ME. 2007. Integration of physiological and molecular mechanisms of the developmental origins of adult disease: new concepts and insights. *Proc. Nutr. Soc.* 66:442–50
30. Osmond C, Barker DJ, Winter PD, Fall CH, Simmonds SJ. 1993. Early growth and death from cardiovascular disease in women. *Br. Med. J.* 307:1519–24
31. Gamborg M, Byberg L, Rasmussen F, Andersen PK, Baker JL, et al. 2007. Birth weight and systolic blood pressure in adolescence and adulthood: meta-regression analysis of sex- and age-specific results from 20 Nordic studies. *Am. J. Epidemiol.* 166:634–45
32. Khan IY, Taylor PD, Dekou V, Seed PT, Lakasing L, et al. 2003. Gender-linked hypertension in offspring of lard-fed pregnant rats. *Hypertension* 41:168–75
33. Lane RH, Kelley DE, Gruetzmacher EM, Devaskar SU. 2001. Uteroplacental insufficiency alters hepatic fatty acid-metabolizing enzymes in juvenile and adult rats. *Am. J. Physiol. Regul. Integr. Comp. Physiol.* 280:183–90
34. Tarry-Adkins JL, Joles JA, Chen JH, Martin-Gronert MS, van der Giezen DM, et al. 2007. Protein restriction in lactation confers nephroprotective effects in the male rat and is associated with increased antioxidant expression. *Am. J. Physiol. Regul. Integr. Comp. Physiol.* 293:1259–66
35. Ford ES. 2005. Risks for all-cause mortality, cardiovascular disease, and diabetes associated with the metabolic syndrome. *Diabetes Care* 28:1769–78
36. Ford ES. 2005. Prevalence of the metabolic syndrome defined by the International Diabetes Federation among adults in the US. *Diabetes Care* 28:2745–49
37. Grundy SM, Cleeman JI, Daniels SR, Donato KA, Eckel RH, et al. 2005. Diagnosis and management of the metabolic syndrome: an American Heart Association/National Heart, Lung, and Blood Institute Scientific Statement. *Circulation* 112:2735–52
38. Alberti KG, Zimmet P, Shaw J. 2005. The metabolic syndrome—a new worldwide definition. *Lancet* 366:1059–62

39. Alberti KG, Zimmet PZ. 1998. Definition, diagnosis and classification of diabetes mellitus and its complications. Part 1: diagnosis and classification of diabetes mellitus provisional report of a WHO consultation. *Diabet. Med.* 15:539–53
40. Reaven GM. 2006. The metabolic syndrome: Is this diagnosis necessary? *Am. J. Clin. Nutr.* 83:1237–47
41. Steinberger J, Daniels SR, Eckel RH, Hayman L, Lustig RH, et al. 2009. Progress and challenges in metabolic syndrome in children and adolescents: a scientific statement from the American Heart Association Atherosclerosis, Hypertension, and Obesity in the Young Committee of the Council on Cardiovascular Disease in the Young; Council on Cardiovascular Nursing; and Council on Nutrition, Physical Activity, and Metabolism. *Circulation* 119:628–47
42. Reinehr T, de Sousa G, Toschke AM, Andler W. 2007. Comparison of metabolic syndrome prevalence using eight different definitions: a critical approach. *Arch. Dis. Child* 92:1067–72
43. de Ferranti SD, Gauvreau K, Ludwig DS, Neufeld EJ, Newburger JW, Rifai N. 2004. Prevalence of the metabolic syndrome in American adolescents: findings from the Third National Health and Nutrition Examination Survey. *Circulation* 110:2494–97
44. Weiss R, Dziura J, Burgert TS, Tamborlane WV, Taksali SE, et al. 2004. Obesity and the metabolic syndrome in children and adolescents. *N. Engl. J. Med.* 350:2362–74
45. Bayturan O, Tuzcu EM, Lavoie A, Hu T, Wolski K, et al. 2010. The metabolic syndrome, its component risk factors, and progression of coronary atherosclerosis. *Arch. Intern. Med.* 170:478–84
46. Scuteri A, Najjar SS, Morrell CH, Lakatta EG. 2005. The metabolic syndrome in older individuals: prevalence and prediction of cardiovascular events: the Cardiovascular Health Study. *Diabetes Care* 28:882–87
47. Sundstrom J, Vallhagen E, Riserus U, Byberg L, Zethelius B, et al. 2006. Risk associated with the metabolic syndrome versus the sum of its individual components. *Diabetes Care* 29:1673–74
48. Huxley R, Neil A, Collins R. 2002. Unravelling the fetal origins hypothesis: Is there really an inverse association between birthweight and subsequent blood pressure? *Lancet* 360:659–65
49. Hemachandra AH, Klebanoff MA, Duggan AK, Hardy JB, Furth SL. 2006. The association between intrauterine growth restriction in the full-term infant and high blood pressure at age 7 years: results from the Collaborative Perinatal Project. *Int. J. Epidemiol.* 35:871–77
50. Menezes AM, Hallal PC, Horta BL, Araújo CL, de Fátima Vieira F, et al. 2007. Size at birth and blood pressure in early adolescence: a prospective birth cohort study. *Am. J. Epidemiol.* 165:611–16
51. Huxley RR, Shiell AW, Law CM. 2000. The role of size at birth and postnatal catch-up growth in determining systolic blood pressure: a systematic review of the literature. *J. Hypertens.* 18:815–31
52. Cruickshank JK, Mzayek F, Liu L, Kieltyka L, Sherwin R, et al. 2005. Origins of the "black/white" difference in blood pressure: roles of birth weight, postnatal growth, early blood pressure, and adolescent body size: the Bogalusa heart study. *Circulation* 111:1932–37
53. Hales CN, Barker DJ, Clark PM, Cox LJ, Fall C, et al. 1991. Fetal and infant growth and impaired glucose tolerance at age 64. *Br. Med. J.* 303:1019–22
54. Phillips DI, Barker DJ, Hales CN, Hirst S, Osmond C. 1994. Thinness at birth and insulin resistance in adult life. *Diabetologia* 37:150–54
55. Jaquet D, Gaboriau A, Czernichow P, Levy-Marchal C. 2000. Insulin resistance early in adulthood in subjects born with intrauterine growth retardation. *J. Clin. Endocrinol. Metab.* 85:1401–6
56. Barker DJ, Martyn CN, Osmond C, Hales CN, Fall CH. 1993. Growth in utero and serum cholesterol concentrations in adult life. *Br. Med. J.* 307:1524–27
57. Lussana F, Painter RC, Ocke MC, Buller HR, Bossuyt PM, Roseboom TJ. 2008. Prenatal exposure to the Dutch famine is associated with a preference for fatty foods and a more atherogenic lipid profile. *Am. J. Clin. Nutr.* 88:1648–52
58. Kajantie E, Barker DJ, Osmond C, Forsen T, Eriksson JG. 2008. Growth before 2 years of age and serum lipids 60 years later: the Helsinki Birth Cohort study. *Int. J. Epidemiol.* 37:280–89
59. Lauren L, Jarvelin MR, Elliott P, Sovio U, Spellman A, et al. 2003. Relationship between birthweight and blood lipid concentrations in later life: evidence from the existing literature. *Int. J. Epidemiol.* 32:862–76
60. Ravelli GP, Stein ZA, Susser MW. 1976. Obesity in young men after famine exposure in utero and early infancy. *N. Engl. J. Med.* 295:349–53

61. Kensara OA, Wootton SA, Phillips DI, Patel M, Jackson AA, Elia M. 2005. Fetal programming of body composition: relation between birth weight and body composition measured with dual-energy X-ray absorptiometry and anthropometric methods in older Englishmen. *Am. J. Clin. Nutr.* 82:980–87
62. Parsons TJ, Power C, Manor O. 2001. Fetal and early life growth and body mass index from birth to early adulthood in 1958 British cohort: longitudinal study. *Br. Med. J.* 323:1331–35
63. de Rooij SR, Painter RC, Holleman F, Bossuyt PM, Roseboom TJ. 2007. The metabolic syndrome in adults prenatally exposed to the Dutch famine. *Am. J. Clin. Nutr.* 86:1219–24
64. Yarbrough DE, Barrett-Connor E, Kritz-Silverstein D, Wingard DL. 1998. Birth weight, adult weight, and girth as predictors of the metabolic syndrome in postmenopausal women: the Rancho Bernardo Study. *Diabetes Care* 21:1652–58
65. Ramadhani MK, Grobbee DE, Bots ML, Castro Cabezas M, Vos LE, et al. 2006. Lower birth weight predicts metabolic syndrome in young adults: the Atherosclerosis Risk in Young Adults (ARYA)-study. *Atherosclerosis* 184:21–27
66. Hult M, Tornhammar P, Ueda P, Chima C, Edstedt Bonamy A-K, et al. 2010. Hypertension, diabetes and overweight: looming legacies of the Biafram famine. *PLoS ONE* 5:e13582
67. Barker DJ, Winter PD, Osmond C, Margetts B, Simmonds SJ. 1989. Weight in infancy and death from ischaemic heart disease. *Lancet* 2:577–80
68. Rich-Edwards JW, Stampfer MJ, Manson JE, Rosner B, Hankinson SE, et al. 1997. Birth weight and risk of cardiovascular disease in a cohort of women followed up since 1976. *Br. Med. J.* 315:396–400
69. Stein CE, Fall CH, Kumaran K, Osmond C, Cox V, Barker DJ. 1996. Fetal growth and coronary heart disease in south India. *Lancet* 348:1269–73
70. Lawlor DA, Ronalds G, Clark H, Smith GD, Leon DA. 2005. Birth weight is inversely associated with incident coronary heart disease and stroke among individuals born in the 1950s: findings from the Aberdeen Children of the 1950s prospective cohort study. *Circulation* 112:1414–18
71. Rich-Edwards JW, Kleinman K, Michels KB, Stampfer MJ, Manson JE, et al. 2005. Longitudinal study of birth weight and adult body mass index in predicting risk of coronary heart disease and stroke in women. *Br. Med. J.* 330:1115
72. Frankel S, Elwood P, Sweetnam P, Yarnell J, Smith GD. 1996. Birthweight, body-mass index in middle age, and incident coronary heart disease. *Lancet* 348:1478–80
73. Boney CM, Verma A, Tucker R, Vohr BR. 2005. Metabolic syndrome in childhood: association with birth weight, maternal obesity, and gestational diabetes mellitus. *Pediatrics* 115:e290–96
74. Malcolm JC, Lawson ML, Gaboury I, Lough G, Keely E. 2006. Glucose tolerance of offspring of mother with gestational diabetes mellitus in a low-risk population. *Diabet. Med.* 23:565–70
75. Gillman MW, Rifas-Shiman S, Berkey CS, Field AE, Colditz GA. 2003. Maternal gestational diabetes, birth weight, and adolescent obesity. *Pediatrics* 111:e221–26
76. Whitaker RC, Pepe MS, Seidel KD, Wright JA, Knopp RH. 1998. Gestational diabetes and the risk of offspring obesity. *Pediatrics* 101:E9
77. Boomsma CM, Eijkemans MJ, Hughes EG, Visser GH, Fauser BC, Macklon NS. 2006. A meta-analysis of pregnancy outcomes in women with polycystic ovary syndrome. *Hum. Reprod. Update* 12:673–83
78. Geelhoed JJ, Fraser A, Tilling K, Benfield L, Davey Smith G, et al. 2010. Preeclampsia and gestational hypertension are associated with childhood blood pressure independently of family adiposity measures: the Avon Longitudinal Study of Parents and Children. *Circulation* 122:1192–99
79. Palmsten K, Buka SL, Michels KB. 2010. Maternal pregnancy-related hypertension and risk for hypertension in offspring later in life. *Obstet. Gynecol.* 116:858–64
80. Neitzke U, Harder T, Schellong K, Melchior K, Ziska T, et al. 2008. Intrauterine growth restriction in a rodent model and developmental programming of the metabolic syndrome: a critical appraisal of the experimental evidence. *Placenta* 29:246–54
81. Walters E, Edwards RG. 2003. On a fallacious invocation of the Barker hypothesis of anomalies in newborn rats due to mothers' food restriction in preimplantation phases. *Reprod. Biomed. Online* 7:580–82
82. Jansson T, Lambert GW. 1999. Effect of intrauterine growth restriction on blood pressure, glucose tolerance and sympathetic nervous system activity in the rat at 3–4 months of age. *J. Hypertens.* 17:1239–48

83. Symonds ME, Sebert SP, Budge H. 2009. The impact of diet during early life and its contribution to later disease: critical checkpoints in development and their long-term consequences for metabolic health. *Proc. Nutr. Soc.* 68:416–21
84. Blondeau B, Avril I, Duchene B, Breant B. 2002. Endocrine pancreas development is altered in foetuses from rats previously showing intra-uterine growth retardation in response to malnutrition. *Diabetologia* 45:394–401
85. Holemans K, Aerts L, Van Assche FA. 1991. Evidence for an insulin resistance in the adult offspring of pregnant streptozotocin-diabetic rats. *Diabetologia* 34:81–85
86. Langley-Evans SC, Welham SJ, Jackson AA. 1999. Fetal exposure to a maternal low protein diet impairs nephrogenesis and promotes hypertension in the rat. *Life Sci.* 64:965–74
87. Harrison M, Langley-Evans SC. 2009. Intergenerational programming of impaired nephrogenesis and hypertension in rats following maternal protein restriction during pregnancy. *Br. J. Nutr.* 101:1020–30
88. Vickers MH, Breier BH, Cutfield WS, Hofman PL, Gluckman PD. 2000. Fetal origins of hyperphagia, obesity, and hypertension and postnatal amplification by hypercaloric nutrition. *Am. J. Physiol. Endocrinol. Metab.* 279:83–87
89. Vickers MH, Gluckman PD, Coveny AH, Hofman PL, Cutfield WS, et al. 2005. Neonatal leptin treatment reverses developmental programming. *Endocrinology* 146:4211–16
90. Jackson AA, Dunn RL, Marchand MC, Langley-Evans SC. 2002. Increased systolic blood pressure in rats induced by a maternal low-protein diet is reversed by dietary supplementation with glycine. *Clin. Sci.* 103:633–39
91. Langley-Evans SC. 2009. Nutritional programming of disease: unravelling the mechanism. *J. Anat.* 215:36–51
92. Holemans K, Van Bree R, Verhaeghe J, Meurrens K, Van Assche FA. 1997. Maternal semistarvation and streptozotocin-diabetes in rats have different effects on the in vivo glucose uptake by peripheral tissues in their female adult offspring. *J. Nutr.* 127:1371–76
93. Li T, Vu TH, Ulaner GA, Littman E, Ling JQ, et al. 2005. IVF results in de novo DNA methylation and histone methylation at an Igf2-H19 imprinting epigenetic switch. *Mol. Hum. Reprod.* 11:631–40
94. Giritharan G, Talbi S, Donjacour A, Di Sebastiano F, Dobson A, Rinaudo P. 2007. Effect of in vitro fertilization on gene expression and development of mouse preimplantation embryos. *Reproduction* 134:63–72
95. Delle Piane L, Lin W, Liu X, Donjacour A, Minasi P, et al. 2010. Effect of the method of conception and embryo transfer procedure on mid-gestation placenta and fetal development in an IVF mouse model. *Hum. Reprod.* 25:2039–46
96. Scott KA, Yamazaki Y, Yamamoto M, Lin Y, Melhorn SJ, et al. 2010. Glucose parameters are altered in mouse offspring produced by assisted reproductive technologies and somatic cell nuclear transfer. *Biol. Reprod.* 83:220–27
97. Rinaudo PF, Giritharan G, Delle Piane L, Donjacour A. 2009. *Mice conceived by in vitro fertilization (IVF) display reduced growth curve and glucose intolerance.* Presented at Soc. Gynecol. Investig., Glasgow, Scotland
98. Park JH, Stoffers DA, Nicholls RD, Simmons RA. 2008. Development of type 2 diabetes following intrauterine growth retardation in rats is associated with progressive epigenetic silencing of Pdx1. *J. Clin. Investig.* 118:2316–24
99. Pham TD, MacLennan NK, Chiu CT, Laksana GS, Hsu JL, Lane RH. 2003. Uteroplacental insufficiency increases apoptosis and alters p53 gene methylation in the full-term IUGR rat kidney. *Am. J. Physiol. Regul. Integr. Comp. Physiol.* 285:962–70
100. Weaver IC, Cervoni N, Champagne FA, D'Alessio AC, Sharma S, et al. 2004. Epigenetic programming by maternal behavior. *Nat. Neurosci.* 7:847–54
101. Schiff M, Benit P, Coulibaly A, Loublier S, El-Khoury R, Rustin P. 2011. Mitochondrial response to controlled nutrition in health and disease. *Nutr. Rev.* 69:65–75
102. Morino K, Petersen KF, Dufour S, Befroy D, Frattini J, et al. 2005. Reduced mitochondrial density and increased IRS-1 serine phosphorylation in muscle of insulin-resistant offspring of type 2 diabetic parents. *J. Clin. Investig.* 115:3587–93
103. Hotamisligil GS. 2010. Endoplasmic reticulum stress and the inflammatory basis of metabolic disease. *Cell* 140:900–17

104. Sharkey D, Gardner DS, Fainberg HP, Sebert S, Bos P, et al. 2009. Maternal nutrient restriction during pregnancy differentially alters the unfolded protein response in adipose and renal tissue of obese juvenile offspring. *FASEB J.* 23:1314–24
105. Myatt L. 2006. Placental adaptive responses and fetal programming. *J. Physiol.* 572:25–30
106. Burton GJ, Yung HW, Cindrova-Davies T, Charnock-Jones DS. 2009. Placental endoplasmic reticulum stress and oxidative stress in the pathophysiology of unexplained intrauterine growth restriction and early onset preeclampsia. *Placenta* 30(Suppl. A):43–48
107. Tarry-Adkins JL, Chen JH, Smith NS, Jones RH, Cherif H, Ozanne SE. 2009. Poor maternal nutrition followed by accelerated postnatal growth leads to telomere shortening and increased markers of cell senescence in rat islets. *FASEB J.* 23:1521–28
108. Lane RH, MacLennan NK, Hsu JL, Janke SM, Pham TD. 2002. Increased hepatic peroxisome proliferator-activated receptor-γ coactivator-1 gene expression in a rat model of intrauterine growth retardation and subsequent insulin resistance. *Endocrinology* 143:2486–90
109. Desai M, Byrne CD, Zhang J, Petry CJ, Lucas A, Hales CN. 1997. Programming of hepatic insulin-sensitive enzymes in offspring of rat dams fed a protein-restricted diet. *Am. J. Physiol. Gastrointest. Liver Physiol.* 272:1083–90
110. Simmons RA, Templeton LJ, Gertz SJ. 2001. Intrauterine growth retardation leads to the development of type 2 diabetes in the rat. *Diabetes* 50:2279–86
111. Garofano A, Czernichow P, Breant B. 1997. In utero undernutrition impairs rat β-cell development. *Diabetologia* 40:1231–34
112. Petrik J, Reusens B, Arany E, Remacle C, Coelho C, et al. 1999. A low protein diet alters the balance of islet cell replication and apoptosis in the fetal and neonatal rat and is associated with a reduced pancreatic expression of insulin-like growth factor-II. *Endocrinology* 140:4861–73
113. Langley-Evans SC, Gardner DS, Jackson AA. 1996. Maternal protein restriction influences the programming of the rat hypothalamic-pituitary-adrenal axis. *J. Nutr.* 126:1578–85
114. Lesage J, Blondeau B, Grino M, Breant B, Dupouy JP. 2001. Maternal undernutrition during late gestation induces fetal overexposure to glucocorticoids and intrauterine growth retardation, and disturbs the hypothalamo-pituitary adrenal axis in the newborn rat. *Endocrinology* 142:1692–702
115. Joanette EA, Reusens B, Arany E, Thyssen S, Remacle RC, Hill DJ. 2004. Low-protein diet during early life causes a reduction in the frequency of cells immunopositive for nestin and CD34 in both pancreatic ducts and islets in the rat. *Endocrinology* 145:3004–13
116. Hill DJ, Duvillie B. 2000. Pancreatic development and adult diabetes. *Pediatr. Res.* 48:269–74
117. Holemans K, Verhaeghe J, Dequeker J, Van Assche FA. 1996. Insulin sensitivity in adult female rats subjected to malnutrition during the perinatal period. *J. Soc. Gynecol. Investig.* 3:71–77
118. Burns SP, Desai M, Cohen RD, Hales CN, Iles RA, et al. 1997. Gluconeogenesis, glucose handling, and structural changes in livers of the adult offspring of rats partially deprived of protein during pregnancy and lactation. *J. Clin. Investig.* 100:1768–74
119. McMillen IC, Adams MB, Ross JT, Coulter CL, Simonetta G, et al. 2001. Fetal growth restriction: adaptations and consequences. *Reproduction* 122:195–204
120. Ogata ES, Swanson SL, Collins JW Jr, Finley SL. 1990. Intrauterine growth retardation: altered hepatic energy and redox states in the fetal rat. *Pediatr. Res.* 27:56–63
121. DeFronzo RA. 2004. Pathogenesis of type 2 diabetes mellitus. *Med. Clin. N. Am.* 88:787–835
122. Desai M, Crowther NJ, Lucas A, Hales CN. 1996. Organ-selective growth in the offspring of protein-restricted mothers. *Br. J. Nutr.* 76:591–603
123. Maxfield EK, Sinclair KD, Broadbent PJ, McEvoy TG, Robinson JJ, Maltin CA. 1998. Short-term culture of ovine embryos modifies fetal myogenesis. *Am. J. Physiol. Endocrinol. Metab.* 274:1121–23
124. Selak MA, Storey BT, Peterside I, Simmons RA. 2003. Impaired oxidative phosphorylation in skeletal muscle of intrauterine growth-retarded rats. *Am. J. Physiol. Endocrinol. Metab.* 285:130–37
125. Lane RH, Maclennan NK, Daood MJ, Hsu JL, Janke SM, et al. 2003. IUGR alters postnatal rat skeletal muscle peroxisome proliferator-activated receptor-γ coactivator-1 gene expression in a fiber specific manner. *Pediatr. Res.* 53:994–1000
126. Boden G. 2008. Obesity and free fatty acids. *Endocrinol. Metab. Clin. N. Am.* 37:635–46

127. Kieffer TJ, Habener JF. 2000. The adipoinsular axis: effects of leptin on pancreatic β-cells. *Am. J. Physiol. Endocrinol. Metab.* 278:1–14
128. Nguyen LT, Muhlhausler BS, Botting KJ, Morrison JL. 2010. Maternal undernutrition alters fat cell size distribution, but not lipogenic gene expression, in the visceral fat of the late gestation guinea pig fetus. *Placenta* 31:902–9
129. Ozanne SE, Dorling MW, Wang CL, Nave BT. 2001. Impaired PI 3-kinase activation in adipocytes from early growth-restricted male rats. *Am. J. Physiol. Endocrinol. Metab.* 280:534–39
130. Bouret SG, Simerly RB. 2004. Minireview: leptin and development of hypothalamic feeding circuits. *Endocrinology* 145:2621–26
131. Brenner BM, Chertow GM. 1993. Congenital oligonephropathy: an inborn cause of adult hypertension and progressive renal injury? *Curr. Opin. Nephrol. Hypertens.* 2:691–95
132. Manning J, Beutler K, Knepper MA, Vehaskari VM. 2002. Upregulation of renal BSC1 and TSC in prenatally programmed hypertension. *Am. J. Physiol. Ren. Physiol.* 283:202–6
133. Woods LL, Ingelfinger JR, Nyengaard JR, Rasch R. 2001. Maternal protein restriction suppresses the newborn renin-angiotensin system and programs adult hypertension in rats. *Pediatr. Res.* 49:460–67
134. Leeson CP, Kattenhorn M, Morley R, Lucas A, Deanfield JE. 2001. Impact of low birth weight and cardiovascular risk factors on endothelial function in early adult life. *Circulation* 103:1264–68
135. Nuyt AM, Alexander BT. 2009. Developmental programming and hypertension. *Curr. Opin. Nephrol. Hypertens.* 18:144–52
136. Patterson AJ, Zhang L. 2010. Hypoxia and fetal heart development. *Curr. Mol. Med.* 10:653–66
137. Sundgren NC, Giraud GD, Schultz JM, Lasarev MR, Stork PJ, Thornburg KL. 2003. Extracellular signal-regulated kinase and phosphoinositol-3 kinase mediate IGF-1 induced proliferation of fetal sheep cardiomyocytes. *Am. J. Physiol. Regul. Integr. Comp. Physiol.* 285:1481–89
138. Glazier JD, Cetin I, Perugino G, Ronzoni S, Grey AM, et al. 1997. Association between the activity of the system A amino acid transporter in the microvillous plasma membrane of the human placenta and severity of fetal compromise in intrauterine growth restriction. *Pediatr. Res.* 42:514–19
139. Meaney MJ, Szyf M, Seckl JR. 2007. Epigenetic mechanisms of perinatal programming of hypothalamic-pituitary-adrenal function and health. *Trends Mol. Med.* 13:269–77
140. Edwards CR, Benediktsson R, Lindsay RS, Seckl JR. 1993. Dysfunction of placental glucocorticoid barrier: link between fetal environment and adult hypertension? *Lancet* 341:355–57
141. Oscai LB, McGarr JA. 1978. Evidence that the amount of food consumed in early life fixes appetite in the rat. *Am. J. Physiol. Regul. Integr. Comp. Physiol.* 235:141–44
142. Stephens DN. 1980. Growth and the development of dietary obesity in adulthood of rats which have been undernourished during development. *Br. J. Nutr.* 44:215–27
143. Jones AP, Olster DH, States B. 1996. Maternal insulin manipulations in rats organize body weight and noradrenergic innervation of the hypothalamus in gonadally intact male offspring. *Brain Res. Dev. Brain Res.* 97:16–21
144. Widdowson EM. 1950. Chemical composition of newly born mammals. *Nature* 166:626–28
145. Lowell BB, S-Susulic V, Hamann A, Lawitts JA, Himms-Hagen J, et al. 1993. Development of obesity in transgenic mice after genetic ablation of brown adipose tissue. *Nature* 366:740–42
146. Fowden AL, Hill DJ. 2001. Intra-uterine programming of the endocrine pancreas. *Br. Med. Bull.* 60:123–42
147. Beydoun HA, Sicignano N, Beydoun MA, Matson DO, Bocca S, et al. 2010. A cross-sectional evaluation of the first cohort of young adults conceived by in vitro fertilization in the United States. *Fertil. Steril.* 94:2043–49
148. Geelhoed JJ, Fraser A, Tilling K, Benfield L, Davey Smith G, et al. Preeclampsia and gestational hypertension are associated with childhood blood pressure independently of family adiposity measures: the Avon Longitudinal Study of Parents and Children. *Circulation* 122:1192–99
149. Maffeis C, Micciolo R, Must A, Zaffanello M, Pinelli L. 1994. Parental and perinatal factors associated with childhood obesity in north-east Italy. *Int. J. Obes. Relat. Metab. Disord.* 18:301–5
150. Neitzke UTA, Harder T, Plagemann A. 2011. Intrauterine growth restriction and developmental programming of the metabolic syndrome: a critical appraisal. *Microcirculation* 18:304–11

Nuclear Sphingolipid Metabolism

Natasha C. Lucki[1] and Marion B. Sewer[2]

[1]School of Biology, Georgia Institute of Technology, Atlanta, Georgia 30332
[2]Skaggs School of Pharmacy & Pharmaceutical Sciences, University of California, San Diego, La Jolla, California 92093; email: msewer@ucsd.edu

Keywords

sphingomyelin, ceramide, sphingosine-1-phosphate, ganglioside, nucleus

Abstract

Nuclear lipid metabolism is implicated in various processes, including transcription, splicing, and DNA repair. Sphingolipids play roles in numerous cellular functions, and an emerging body of literature has identified roles for these lipid mediators in distinct nuclear processes. Different sphingolipid species are localized in various subnuclear domains, including chromatin, the nuclear matrix, and the nuclear envelope, where sphingolipids exert specific regulatory and structural functions. Sphingomyelin, the most abundant nuclear sphingolipid, plays both structural and regulatory roles in chromatin assembly and dynamics in addition to being an integral component of the nuclear matrix. Sphingosine-1-phosphate modulates histone acetylation, sphingosine is a ligand for steroidogenic factor 1, and nuclear accumulation of ceramide has been implicated in apoptosis. Finally, nuclear membrane–associated ganglioside GM1 plays a pivotal role in Ca^{2+} homeostasis. This review highlights research on the factors that control nuclear sphingolipid metabolism and summarizes the roles of these lipids in various nuclear processes.

INTRODUCTION

> **Nuclear matrix:** a filamentous protein network in the nucleus
>
> **Cyclic phosphatidic acid (PA):** a naturally occurring analog of lysophosphatidic acid, cyclic PA differs from phosphatidic acid in having a cyclic phosphate at the *sn*-2 and *sn*-3 positions of the glycerol carbons. This structure is critical for its biological activity (163)
>
> **Steroidogenic factor 1 (SF-1):** a nuclear receptor that regulates the transcription of genes involved in steroid hormone biosynthesis and endocrine development
>
> **Sphingolipids:** a family of glycolipids and phospholipids that are characterized by the presence of a common sphingoid base backbone

The nucleus is an organelle with a high capacity for lipid metabolism (1). In recent years, many studies have established that nuclear lipids play distinct roles in many cellular processes, including DNA replication, RNA processing, chromatin structure, and Ca^{2+} homeostasis (reviewed in References 2–5). Phosphatidylinositol phosphates (PIPs), the most extensively characterized nuclear lipids, have pivotal roles in chromatin remodeling, gene transcription, and mRNA export (6–8). Most lipids are localized to the nuclear envelope (NE), where in addition to providing structural support they participate in multiple signaling cascades. However, bioactive lipids are localized in other nuclear compartments, including chromatin (9–11) and the nuclear matrix (12). Significantly, the concentration of nuclear lipids can be dynamically altered by metabolic flux in response to signaling cascades that are often uncoupled from cytosolic processes. Similarly, extracellular stimuli can elicit lipid metabolism and signaling exclusively in the nucleus. For example, insulin growth factor 1 (IGF-1) induces the phosphorylation and activation of nuclear phospholipase C β1 (PLCβ1) (13), which consequently results in nuclear diacylglycerol (DAG) accumulation with a corresponding decrease in phosphatidylinositol biphosphate (14, 15). Furthermore, accumulation of nuclear DAG in response to IGF-1 stimulation promotes protein kinase C (PKC) nuclear translocation (15–17). Activation of this signaling pathway modulates various nuclear processes, including gene expression and cell proliferation (4).

In addition to PIPs, other classes of phospholipids, including phosphatidylcholine, phosphatidylethanolamine, and phosphatidylserine, also have varied nuclear functions (5, 18–24). Roles for diacylglycerol kinases (DGKs), a family of enzymes that convert DAG to phosphatidic acid (PA), in varied nuclear processes are well documented (25–28). For example, nuclear DGK-ζ expression regulates A172 cell growth by decreasing DAG concentrations (29). DGK-θ is localized in nuclear speckles (30) and is also activated by α-thrombin in IIC9 fibroblast nuclei (31) and by nerve growth factor in PC12 cells (32). We have shown that PA regulates steroidogenic gene transcription by serving as an agonist for the nuclear receptor steroidogenic factor 1 (SF-1) (33). SF-1 regulates the transcription of multiple genes in the endocrine system, including most genes that are required for steroid hormone biosynthesis and endocrine development (34, 35). DGK-θ directly interacts with SF-1, and activation of the cAMP pathway stimulates DGK activity in the nucleus of H295R adrenocortical cells (33). Consistent with these findings establishing roles for DAG/PA in nuclear processes, lipins (proteins that have phosphatidate phosphatase activity and catalyze the formation of DAG in the glycerol-3-phosphate pathway) are also emerging as regulators of gene expression (36). Lipin-1 binds to peroxisome proliferator–activated receptor (PPAR)α and serves as a coactivator in the expression of genes involved in fatty acid uptake, mitochondrial function, and lipid metabolism (37). Finally, recent studies have identified cyclic PA as a PPARγ antagonist that binds to the receptor with nanomolar activity and inhibits the expression of PPARγ target genes and adipogenesis (38).

In addition to phospholipids, the nucleus is emerging as a hub for sphingolipid metabolism. Sphingolipids comprise a large family of phospholipids and glycolipids (**Figure 1**) that share a common sphingoid base backbone (**Figure 2**). These molecules participate in many signal transduction pathways (39–41). To date, various sphingolipid species have been identified in multiple nuclear compartments, including chromatin, NE, and nuclear matrix (1, 42–52). In this review, we summarize studies that have identified a role for this class of lipids in regulating nuclear processes.

NUCLEUS ORGANIZATION AND ENDONUCLEAR DOMAINS

The nucleus is a well-organized substructure with a dynamic framework (53). It is composed of a well-defined NE that encapsulates several endonuclear domains, including the nuclear matrix,

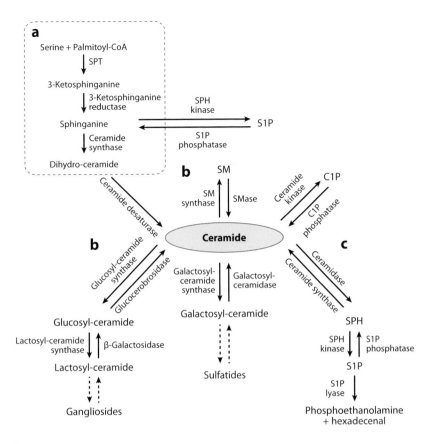

Figure 1

The sphingolipid metabolic pathway. De novo biosynthesis begins with the condensation of serine and palmitoyl-CoA and various fatty acyl-CoAs. Ceramide can be generated through (*a*) de novo biosynthesis, (*b*) degradation of sphingomyelin (SM) or glucosyl-ceramides and galactosyl-ceramides, or (*c*) acylation of sphingosine (SPH). Ceramide can be phosphorylated into ceramide-1-phosphate (C1P) or hydrolyzed to form SPH, which is phosphorylated into sphingosine-1-phosphate (S1P). S1P can be either dephosphorylated to form SPH or irreversibly cleaved into phosphoethanolamine and hexadecenal. SPT denotes serine palmitoyltransferase.

chromatin, and nucleolus. The NE is a bilayer whose outer and inner leaflets display unique lipid compositions. Although a detailed comparison of the relative distribution of lipid species between the two leaflets of the nuclear membrane has not been reported, cholesterol has been shown to reside in the outer membrane but not in the inner membrane (54), whereas the gangliosides GM1 and GD1a were detected in both (55). The outer membrane is continuous with the endoplasmic reticulum (ER) and thus shares certain lipidomic properties. Conversely, the inner membrane is closely associated with the nuclear lamina and has distinct lipid characteristics (**Figure 3**) (53).

Like plasma membranes, nuclear membranes have been suggested to contain many types of receptors, including inositol 1,4,5-triphosphate, PPAR, and retinoic acid (RA) receptors (56, 57). Some agonists activate signaling exclusively through nuclear membrane–localized receptors. RA, for example, activates phospholipase A2 (PLA2), PLC, and phospholipase D (PLD) only in the nucleus (58–60). Additionally, compelling new evidence suggests that nuclear

Ganglioside: a glycosphingolipid with one or more sialic acids (*n*-acetylneuraminic acid) linked to the carbohydrate chain

GM1: the major ganglioside subspecies in the nuclear envelope that binds to NCX. Regulates nuclear Ca^{2+} homeostasis

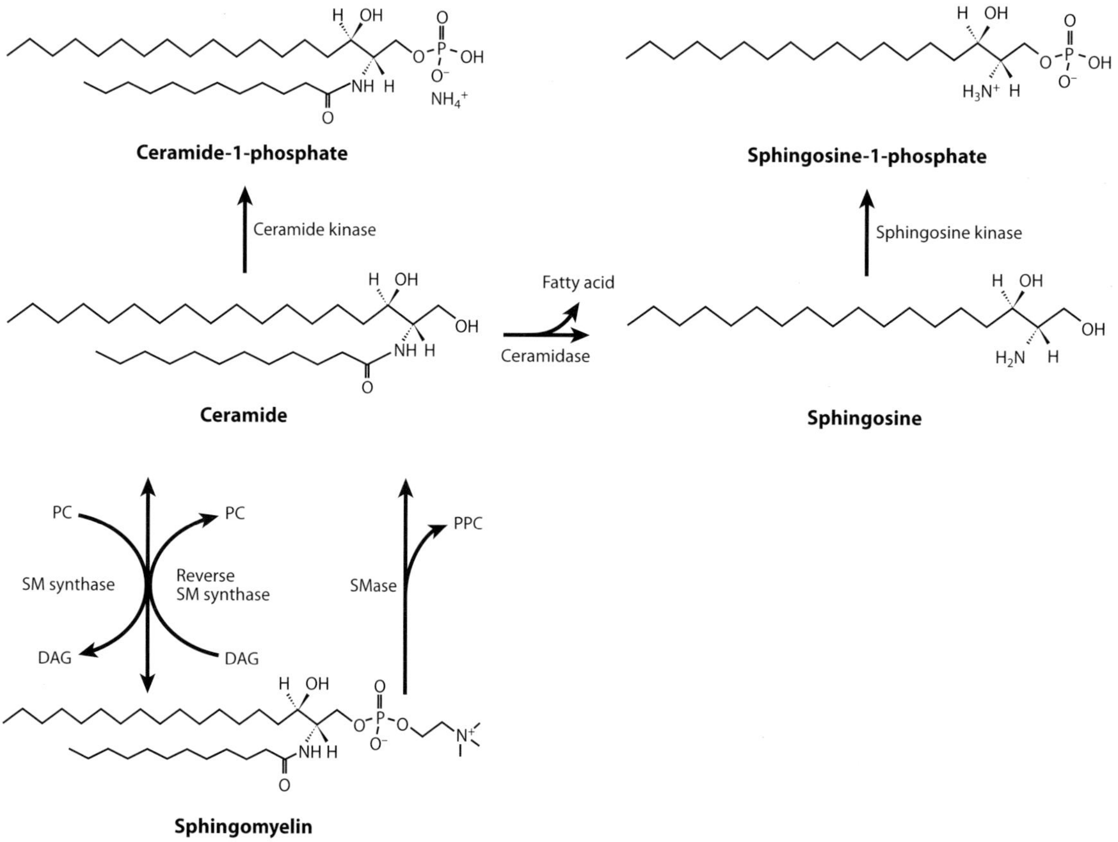

Figure 2

Nuclear sphingomyelin (SM) metabolism. SM synthase catalyzes the formation of SM from ceramide and phosphatidylcholine (PC). SM can be degraded by sphingomyelinase (SMase), which generates phosphorylcholine (PPC) as a by-product, or by reverse SM synthase, which catalyzes the transfer of the PPC group in SM to diacylglycerol (DAG). Ceramide formed by SM hydrolysis can be phosphorylated into ceramide-1-phosphate or hydrolyzed into sphingosine, which can then be converted into its phosphorylated form, sphingosine-1-phosphate.

membrane–associated enzymes have physicochemical properties that are different from those of their plasma membrane and/or cytosolic counterparts. For example, the kinetic parameters of nuclear PLC differ from those of cytoplasmic or plasma membrane–associated PLC (61).

The nuclear matrix is often viewed as the basic organizing structure of the nucleus that is responsible for maintaining nuclear shape. However, the nuclear matrix is also the site at which many processes, including DNA replication, gene transcription, and protein phosphorylation, occur (62–64). Many enzymes linked to PIP metabolism associate with the nuclear matrix (65), suggesting that the matrix is actively involved in nuclear lipid signaling cascades. Chromatin is closely associated with the nuclear matrix and exhibits a dynamic structure that is actively regulated by multiple interconnected mechanisms, including DNA methylation and histone modification (66, 67). Heterochromatin regions, which are transcriptionally inactive but contain many specific nuclear proteins that regulate gene transcription (68), are similarly organized by the nuclear matrix (**Figure 3**).

Heterochromatin: a tightly packed form of DNA that is usually associated with silenced gene regions

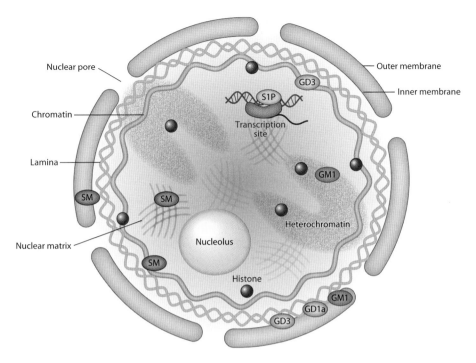

Figure 3

The localization of different sphingolipid species in subnuclear domains. The outer membrane is continuous with the endoplasmic reticulum, whereas the inner membrane is associated with the nuclear lamina. The nuclear pore allows passive flow of small molecules between the cytosol and the nucleoplasm. Abbreviations: S1P, sphingosine-1-phosphate; SM, sphingomyelin.

SPHINGOLIPID BIOSYNTHESIS AND METABOLISM

Sphingolipids are synthesized de novo from serine and palmitoyl-CoA to form a sphingoid base, which is further N-acylated with various fatty acyl-CoAs to make N-acylsphinganine (dihydroceramide) and is sequentially desaturated to form ceramide (**Figure 1**). Ceramide can be metabolized into more complex sphingolipids by the incorporation of O-linked head groups such as phosphorylcholine or carbohydrate moieties to form sphingomyelin (SM) (**Figure 2**) or glycosphingolipids (**Figure 4**), respectively. Alternatively, ceramide can be phosphorylated into ceramide-1-phosphate (C1P) by ceramide kinase (CERK) or hydrolyzed into sphingosine (SPH) by ceramidases (**Figure 2**). Sphingosine kinases (SKs) then phosphorylate SPH to form sphingosine-1-phosphate (S1P). Irreversible sphingolipid degradation occurs by the action of S1P lyase, which cleaves S1P into phosphoethanolamine and hexadecenal. Aside from being de novo synthesized, ceramide can also be formed by the hydrolysis of SM and glycosphingolipids through the salvage pathway or by the N-acylation of SPH through the action of ceramide synthases (**Figure 1**).

Cellular sphingolipid concentrations are tightly regulated by the actions of multiple enzymes that act to maintain sphingolipid homeostasis. Because several sphingolipid species have unique physiological functions, these enzymes are important not only to ensure optimal sphingolipid concentrations but also to regulate the capacity of these bioactive lipids to activate cell signaling. For example, because ceramidases control the ratio between ceramide and S1P (**Figure 2**), the activity of this family of hydrolases dictates whether a cell undergoes apoptosis or proliferates (69–71). Similarly, the activity of SK1 is strongly correlated with cell growth (72–75). Furthermore,

Sphingomyelin (SM): a sphingolipid with phosphocholine attached to the terminal OH group of ceramide

Glycosphingolipid: a sphingolipid with carbohydrate groups attached to the terminal OH group of ceramide

Salvage pathway: the regeneration of ceramide from the breakdown of complex sphingolipids

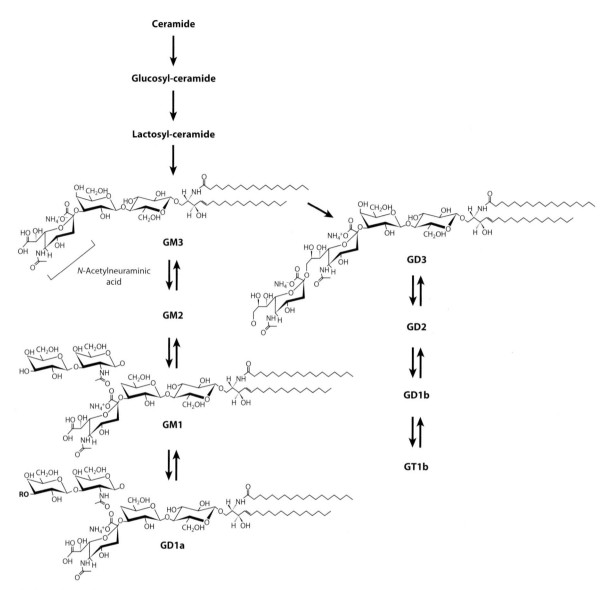

Figure 4
Ganglioside structure and biosynthetic pathway. Lactosyl-ceramide, formed by the addition of two carbohydrate moieties (glucose and galactose) to the terminal OH group of ceramide, is the precursor of gangliosides. GM1 has one terminal N-acetylneuraminic acid (sialic acid) group. GD1a, which has two terminal sialic acid groups, is converted into GM1 by neuraminidase.

sphingomyelinase (SMase), the enzyme that breaks the phosphodiester bond of SM to form ceramide (**Figure 2**), is activated by multiple apoptotic factors, including tumor necrosis factor α (TNFα) (76), ionizing radiation (77), and Fas ligand (78). Finally, because the length of the N-linked fatty acyl chain on ceramide is a key determinant of the cellular function of these molecules (53, 79–83), ceramide synthases play a central role in regulating the composition of sphingolipid pools (84).

Sphingolipid-metabolizing enzymes display distinct subcellular localizations and membrane topologies (85–89). Enzymes linked to de novo sphingolipid synthesis [e.g., serine palmitoyltransferase, ceramide synthase, dihydroceramide desaturase] are localized primarily in the ER lumen, whereas different isoforms of ceramidase and SMase are expressed at the plasma membrane, lysosomes, or mitochondria (90). Glycosphingolipids are synthesized by a series of enzymes residing in the Golgi apparatus. Due to the hydrophobic nature of most sphingolipid species, compartmentalization of sphingolipid metabolism and subsequently signaling are important themes in sphingolipid biology (90). In this manner, ceramide localized at the plasma membrane, for example, participates in signaling pathways distinct from those of mitochondrial ceramide (91–96). Adding to this complexity, different isoforms of the same enzyme may have distinct substrate specificities (85, 97–102).

NUCLEAR SPHINGOLIPIDS

Biochemical, analytical, and microscopic techniques have been utilized to identify sphingolipid-metabolizing enzymes in nuclei and to quantify the concentrations of sphingolipid species (97). SM is the most prominent sphingolipid in nuclei; its concentration in nuclear matrix is three times higher than in chromatin (103). SM is a major component of chromatin (10), where it plays a role in DNA replication and chromatin architecture (43, 50, 104). The catabolism of SM gives rise to ceramide and subsequently to SPH, S1P, and C1P (**Figure 2**), all of which have specific nuclear functions. S1P, which has recognized roles in cell proliferation, migration, and differentiation (39), regulates gene transcription by specifically binding to histone deacetylases (HDACs) 1 and 2 and inhibiting their enzymatic activity (52). Gangliosides are also prominent in the outer and inner membranes of the NE. The ganglioside GM1 (**Figure 4**), in particular, has been extensively studied for its key role in nuclear Ca^{2+} homeostasis (105). Collectively, studies of nuclear sphingolipids not only illustrate the multifaceted regulatory capabilities of these lipid mediators, which in most cases differ from their cytoplasmic functions, but also highlight the importance of location (i.e., chromatin versus NE versus nuclear matrix) in determining their nuclear functions.

The localization of sphingolipid enzymes in various subnuclear compartments facilitates the dynamic nuclear metabolism of sphingolipids. To date, enzymes involved in SM, ceramide, SPH, and glycosphingolipid metabolism have been identified in nuclear extracts isolated from different cell types (42–44, 46, 48, 106–108). SM was the first sphingolipid identified as a component of the nuclear matrix (109). However, the likelihood of nuclear sphingolipid metabolism became apparent only after enzymes that metabolize sphingolipids were found in the nucleus (12, 45, 110). In fact, the nuclear levels of distinct sphingolipid subspecies change under different cellular physiological states (50, 78, 111, 112), illustrating the intrinsic capacity for dynamic nuclear sphingolipid metabolism. For example, chromatin-associated SM synthase and SMase control dynamic oscillations in SM concentrations during the cell cycle (112–114). Furthermore, ganglioside-metabolizing enzymes, including neuraminidase (sialidase) (115) and GM2/GD2 synthase (116), play pivotal roles in regulating GM1 levels in the NE.

Sphingomyelin

SM was first identified in the NE of hepatocytes (117, 118). It was subsequently shown to be present in chromatin (10) and the nuclear matrix (12) and to be associated with double-stranded RNA (dsRNA) (**Figure 3**) (49). As discussed above, SM is the most abundant sphingolipid in the nucleus and the major phospholipid associated with chromatin (10), although it is enriched three

Table 1 Summary of reported nuclear functions for different sphingolipid species and their endonuclear localization[a]

Sphingolipid	Nuclear localization	Function
Sphingomyelin	Nuclear envelope Nuclear matrix Chromatin	DNA synthesis Chromatin assembly Membrane structure RNA stability
Ceramide	?	Apoptosis
Ceramide-1-phosphate	Perinuclear region	cPLA2α translocation
Sphingosine	?	SF-1 antagonist ligand
Sphingosine-1-phosphate	Chromatin	Histone acetylation
GM1	Inner nuclear membrane	Ca^{2+} homeostasis
GD3	?	Histone H1 phosphorylation Apoptosis
GD1a	Inner nuclear membrane	Reservoir for GM1

[a]Question marks indicate that the specific endonuclear domain has not been determined. Abbreviations: cPLA2α, α-type cytosolic phospholipase A2; SF-1, steroidogenic factor 1.

times higher in the nuclear matrix (103). Studies have demonstrated that nuclear SM levels are dynamic and oscillate in response to different cellular states (43, 45, 50, 106, 112, 119). Additionally, distinct cellular cues differentially affect SM concentration in distinct subnuclear compartments (e.g., nuclear matrix versus chromatin) (103). For example, SM levels in the nuclear matrix increase at the beginning of S phase of the cell cycle during hepatic regeneration (103), whereas chromatin-associated SM decreases during the same period (119). These studies not only highlight the specificity of SM metabolism in different nuclear domains but also suggest that the amount of SM associated with DNA is proportional to the state of chromatin condensation (**Table 1**).

SMase, the enzyme that catalyzes the degradation of SM to form ceramide and phosphorylcholine (**Figure 2**), has been detected in the nuclear matrix (42), the NE (43), and chromatin (112). SM synthase has also been identified in the NE and chromatin (113). These enzymes appear to have different physicochemical characteristics, depending on their intranuclear localization, and their activities change in response to different cellular cues, such as cell proliferation and apoptosis (50, 104, 107, 112). Chromatin-associated SM synthase differs in pH optimum and K_m from the same isoform localized in the NE (113). Moreover, plasma membrane–localized SMase may be involved in axonal growth (120), whereas nuclear SMase activation is associated with apoptosis (46, 78). Different isoforms of the same enzyme can also reside in different intranuclear compartments, as is believed to be the case for neutral SMase. This ceramide-generating enzyme (**Figure 2**) was proposed to reside in the NE of rat liver cells and to translocate to the nuclear matrix during DNA synthesis (43). However, a subsequent study used biochemical and immunohistochemical approaches to demonstrate that SMase possesses a nuclear export signal and is enriched in the nuclear matrix (121). It is thus believed that distinct isoforms of SMase participate in each of these processes.

SM plays a role in stabilizing DNA during the cell cycle (112), and the dynamic changes in the amount of chromatin-associated SM are due to the opposing activities of SMase and SM synthase (112, 119). Concomitant activation of SMase and inhibition of SM synthase at the beginning of S phase may lead to a decrease in SM levels, which facilitates DNA unwinding (5). Increased SM synthase activity at the end of S phase then facilitates double-helix restoration after DNA synthesis ends (5). Notably, these changes in SM concentrations occur selectively in chromatin (104).

Interestingly, reverse SM synthase, an enzyme that catalyzes the reverse reaction as that of SM synthase (**Figure 2**), has also been identified in chromatin (114). Similar to SMase, reverse SM synthase catalyzes the degradation of SM but with the key difference that it catalyzes the transfer of the phosphorylcholine group from SM to DAG, forming ceramide and phosphatidylcholine (**Figure 2**). Therefore, this enzyme not only promotes the accumulation of ceramide but also decreases DAG levels while increasing phosphatidylcholine. This is significant because the ceramide/DAG ratio is linked to cell proliferation and apoptosis. Although ceramide's role in cell fate has been described predominantly in whole-cell studies (70), ceramide is a well-known mediator of apoptosis (122). Nonetheless, the induction of apoptosis in rat liver occurs through the selective accumulation of nuclear ceramide due to the activation of neutral SMase at the NE (46). In contrast, numerous studies have shown that an increase in nuclear DAG mediates the recruitment of various PKC isoforms into the nucleus (123, 124). PKC and DAG may be necessary for the transitions from G1 to S phase (125) and from G2 to M phase (126, 127) of the cell cycle. Although the identification of direct nuclear and cytoplasmic targets for ceramide is still an area of active investigation, by analogy to cytosolic signaling, it has been proposed that nuclear ceramide concentrations and PKC activity are also directly related (128).

Sphingomyelin (SM) cycle: the activation of sphingomyelinase by extracellular stimuli that leads to sphingomyelin turnover to form ceramide

Early studies pointed to an interaction between SM and RNA by observing a significant reduction in SM levels after nuclear digestion with RNase (129). This association was strengthened by studies identifying a complex containing RNA, proteins, SM, phosphatidylcholine, SM synthase, and neutral SMase from hepatocyte nuclei that were sequentially treated with Triton X-100 and a DNAse/RNAse cocktail (45). RNAse-resistant RNA became sensitive to enzymatic hydrolysis when it was pretreated with SMase, suggesting that SM protects RNA from degradation (15). Furthermore, because RNA digestion was temperature sensitive (i.e., higher temperatures yielded more undigested RNA that was hydrolyzed by RNAse), this RNA was assumed to be dsRNA (49). Although the precise role of dsRNA-bound SM is unclear, SM may play a role in RNA maturation by associating with newly synthesized RNA and protecting it from enzymatic digestion prior to its export from the nucleus (5). SM may also stabilize dsRNA by forming a bridge between the two strands (**Table 1**) (49).

Ceramide and Ceramide-1-Phosphate

As depicted in **Figure 1**, ceramide sits at the hub of the sphingolipid metabolic pathway because it not only serves as the building block for more complex sphingolipids (e.g., SM and glycosphingolipids) but is also an intermediate metabolite for either sphingolipid degradation or the generation of phosphorylated species such as S1P and C1P. The presence of SMase and SM synthase in the nucleus suggests that ceramide is actively produced and consumed. Ceramidase activity in liver nuclear membranes was also reported (110), suggesting that the ceramide generated can be further metabolized. The nuclear localization of these enzymes suggests the existence of a nuclear SM cycle, although the regulation of ceramide concentrations in the nucleus is poorly understood. Akin to nuclear SM, spatially localized ceramide metabolism in different subnuclear domains may participate in distinct nuclear processes.

Nuclear ceramide accumulation has been associated with apoptosis in rat hepatocytes after hepatic vein ligation (46). [Portal vein ligation is a procedure that involves the ligation of hepatic lobes, which promotes atrophy of ligated lobes while inducing hypertrophy of nonligated lobes. This procedure can be used as a model of apoptosis of the ligated lobes, whereas hypertrophic hepatocytes can be studied for cell proliferation mechanisms (46, 47).] Increased SMase and ceramidase activity occurs after portal vein branch ligation, which correlates with DNA fragmentation and cell death (**Table 1**) (46). Additionally, SM degradation in chromatin at the beginning of S phase during hepatic regeneration suggests an accumulation of ceramide in this subnuclear domain (104). In Jurkat T cells, Fas ligand simultaneously stimulates neutral SMase and inhibits SM synthase activities in a caspase-3-dependent manner, which results in the time- and dose-dependent accumulation of nuclear ceramide (78). Although the precise molecular mechanisms involved have yet to be defined, these studies suggest a role for nuclear ceramide in cell proliferation. Recently, ceramide accumulation in rat hepatic nuclei as a result of neutral SMase activation was reported to occur in response to a high-fat diet (130). The basis for the accumulation of nuclear ceramide and the site at which this occurs are unknown. However, given that whole-cell studies have established a role for ceramide in insulin resistance, nuclear ceramide metabolism may have implications in insulin signaling (131, 132). Ceramide concentrations also increase in RAW 264.7 macrophages that have been activated by the Toll-like receptor 4–specific ligand Kdo_2-lipid A (133). Finally, ceramide regulates nuclear protein import in smooth muscle cells by inducing p38 mitogen-activated protein kinase activation and the subsequent relocalization of two nuclear transport proteins, importin A and cellular apoptosis susceptibility gene (123). The inhibition of nuclear import by exogenously supplemented ceramide resulted in diminished expression of proliferation protein markers, including cyclin A and proliferating cell nuclear antigen, and in reduced proliferative capacity (123).

Rovina et al. (62) recently identified nuclear export and import signals in the primary sequence of CERK. This finding suggests that nuclear ceramide can be further metabolized into C1P, which may harbor unique nuclear functions. Although these functions are yet to be reported, there is compelling evidence for a regulatory role for C1P in α-type cytosolic phospholipase A2 (cPLA2α) activity and arachidonic acid release in many cell types (64). C1P binds to the Ca^{2+}-binding regions in the C2 domain of cPLA2α and promotes its translocation to the perinuclear region of cells (134). Moreover, CERK activity is required for IL-1β-induced prostaglandin production (135), as does the interaction between C1P and cPLA2α (136). Nuclear C1P may represent a yet-to-be-established, key mediator of the inflammatory response.

Sphingosine and Sphingosine-1-Phosphate

The nuclear localization of ceramidase (46) enables the local hydrolysis of ceramide to SPH. We demonstrated that SPH plays an important role in steroid hormone production in the human adrenal cortex by serving as a ligand for SF-1 (137). SPH is bound to SF-1 under basal conditions and antagonizes receptor function (137). Activation of the cAMP signaling pathway, the major regulatory cascade in adrenal steroidogenesis (138, 139), reduces the amount of receptor-bound SPH and enables the transcription of genes involved in the conversion of cholesterol to steroid hormones (**Figure 5**) (137). Significantly, we have found that acid ceramidase directly binds to SF-1 in the nucleus of H295R adrenocortical cells (N.C. Lucki & M.B. Sewer, unpublished observations). This suggests that ligand formation and delivery are facilitated by a direct interaction between enzyme and receptor and provides support for a novel coregulatory role of this enzyme in controlling gene expression. We also found that cAMP rapidly decreases nuclear concentrations of ceramide and stimulates the nuclear localization of SK1 while concomitantly increasing SPH and S1P levels (D. Li & M.B. Sewer, unpublished observations). Collectively, these studies implicate signal-induced nuclear sphingolipid metabolism as a critical regulator of gene transcription.

The amount of intracellular SPH is regulated not only by the action of ceramidases and ceramide synthases (**Figure 1**) but also by its phosphorylation to S1P by SK (**Figure 2**). The two isoforms of SK have distinct subcellular localizations and physiological functions (72, 140). SK1 is cytoplasmic and is associated primarily with cell proliferation and growth, whereas SK2 is

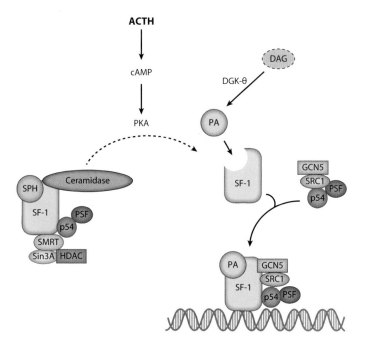

Figure 5
Model for the role of sphingosine (SPH) in controlling the transactivation potential of steroidogenic factor 1 (SF-1). Under basal conditions, SPH is bound to SF-1 and corepressory proteins. Adrenocorticotropin hormone (ACTH) signaling activates protein kinase A (PKA), which promotes the release of SPH from the receptor's ligand-binding pocket. Concomitantly, activation of the ACTH/cAMP pathway increases nuclear diacylglycerol kinase (DGK) activity, leading to increased phosphatidic acid (PA) biosynthesis. PA binding to SF-1 activates the receptor, thereby facilitating the recruitment of coactivator proteins and enabling interaction with the promoters of target genes. Abbreviations: GCN5, general control of amino acid synthesis protein 5; HDAC, histone deacetylase; PSF, polypyrimidine tract–binding protein–associated splicing factor; Sin3A, Sin3 homolog A; SMRT, silencing mediator for retinoid of thyroid hormone receptors; SRC1, steroid receptor coactivator 1.

mainly nuclear and is linked to apoptosis. Nuclear SK activity was first described in the NE and nucleoplasm of Swiss 3T3 cells (108). Platelet-derived growth factor was shown to upregulate nucleoplasmic SK activity that correlated with progression through S phase (108). This study provided an early indication that nuclear S1P production might be involved in cell cycle regulation. Of note, we found that cAMP promotes the phosphorylation and nuclear translocation of SK1 in H295R adrenocortical cells (D. Li & M.B. Sewer, unpublished observations). More recently, Hait et al. (52) reported that SK2 interacts with histone H3 in chromatin of MCF-7 breast cancer cells, establishing a role for endonuclear S1P in the epigenetic regulation of gene transcription (52). Expression of SK2 induced histone acetylation, which correlated with the formation of S1P and dihydro-S1P in the nuclei of these cells (**Figure 3**). Furthermore, the authors demonstrated that S1P and dihydro-S1P inhibited the activity of HDAC1 and -2 by binding to their active site (52). SK2 interacted with HDAC1 and HDAC2 and thus facilitated S1P transfer to the enzymes. Finally, SK2 associated with HDACs at the promoters of the cyclin-dependent kinase inhibitor p21 and c-*fos* genes, where it induced histone acetylation and gene transcription (**Table 1**) (52). These findings not only identified HDACs as nuclear targets of S1P but also uncovered a novel role for this multifaceted sphingolipid molecule as a regulator of histone posttranslational modification and global gene expression programs.

Gangliosides

NCX: a Na^+/Ca^{2+} exchanger localized at the plasma and nuclear membranes

Gangliosides are formed by a ceramide molecule linked to an oligosaccharide chain containing hexose and N-acetylneuraminic acid (sialic acid) groups (**Figure 4**). Many studies have established that gangliosides are intrinsic components of the nucleus and occur in both membranes of the NE (**Figure 3**). The first evidence for nuclear ganglioside localization came from subcellular fractionation studies of bovine mammary gland and rat liver cells (141, 142), in which nuclear ganglioside concentrations were found to be 10% of those of the plasma membrane (143). The major species identified were GM1 and GM3 in rat liver (142) and GM3, GD3, and GTb1 in bovine mammary gland cells (143). Subsequent studies in neuroblastoma and primary neuronal cells demonstrated by cytochemical analysis with cholera toxin B subunit linked to horseradish peroxidase that GM1 occurs at high concentrations in the NE of differentiating, but not quiescent, cells (44, 144). Similarly, GM1, GM3, and c-series gangliosides were observed in mature rat brain (145). However, developing rat brain comprised relatively more GM3 and GD3 (145), indicating that the synthesis of distinct ganglioside subspecies in the nucleus is differentially regulated during development.

Nuclear quantification of gangliosides from cultured neuro2A cells by two successive high-density sucrose gradient purifications found that GM1 and GD1a are the predominant gangliosides found in nuclei and are localized primarily in the NE (44). A more detailed characterization of the nuclear localization of gangliosides became possible by subjecting the isolated nuclei to mild treatment with a sodium citrate solution, which allows the separation of the inner and outer nuclear membranes (146, 147). The use of this technique revealed that GM1 and GD1a are present in both membranes of the NE in primary neurons (48). Interestingly, GD1a is converted to GM1 by neuraminidase (**Figure 4**), a membrane-associated enzyme that is present in both membranes of the NE (148, 149). Nuclear GD1a may serve as a storage reserve precursor for GM1.

The characterization of gangliosides in other endonuclear domains is relatively unexplored. However, some studies have determined the presence of ganglioside subspecies in heterochromatin and chromatin. GM1 may associate with heterochromatin from mouse epithelial cells (150), whereas immunocytochemical studies showed that GD3 colocalizes with chromatin in rat cortical neurons subjected to β-amyloid peptide (151). GD3 accumulation is concomitant with reduced levels of SM and increased activity of 2,8-sialyltransferase (GD3 synthase) (151), the enzyme that forms GD3 from GM3 (**Figure 4**). GD3 synthase knockdown by RNA interference prevented β-amyloid peptide–induced entry into S phase and apoptosis (151), supporting a role for GD3 in cell cycle activation and cell death. In addition, translocation of GD3 from the cytosol into the nucleus was observed in HUT-78 T-lymphoma cells (51). This translocation strongly correlated with histone H1 phosphorylation after activation of apoptosis (51), which suggests that GD3 may have an epigenetic role in the transcriptional regulation of specific genes (**Table 1**).

GM1 and Ca^{2+} homeostasis. GM1 associated with the inner NE plays a prominent role in nuclear Ca^{2+} homeostasis. This nuclear function of GM1 emerged after researchers discovered that this ganglioside is tightly associated with a Na^+/Ca^{2+} exchanger (NCX) (48). NCX mediates countertransport of three Na^+ ions for one Ca^{2+} ion against a Ca^{2+} gradient. On the basis of the topology of plasma membrane–associated NCX (152), it has been proposed that in the NE, negatively charged sialic acid groups on the GM1 oligosaccharide chain interact with positively charged amino acid residues on the polypeptide loop between transmembrane segments 5 and 6 of NCX at the nucleoplasm (153, 154). This interaction is thought to facilitate Ca^{2+} transport from the nucleoplasm (low $[Ca^{2+}]$) to the NE lumen (high $[Ca^{2+}]$).

Xie et al. (48) demonstrated the association between GM1 and NCX by immunoprecipitation of NE extracts with an antibody against NCX followed by sodium dodecyl sulfate polyacrylamide gel electrophoresis and Western blotting analysis of GM1. These investigators further determined by immunoblot analysis of the separate NE membranes that this association occurs specifically at the inner membrane of the NE (48). Interestingly, plasma membrane NCX showed no association with GM1, suggesting a difference between plasma membrane–localized NCX and NE-localized NCX (48). Given that splice variants of this exchanger have been described (155, 156), differential association between nuclear and cytoplasmic NCX with GM1 may be due to alternatively spliced NCX isoforms.

GM1 association with NCX potentiates Na^+ and Ca^{2+} exchange between the nucleoplasm and the NE lumen. This theory was first demonstrated by Ca^{2+} uptake experiments with isolated nuclei (48) and more recently using genetically encoded chameleon Ca^{2+} sensors (157). Subsequent studies employed a comparison of NE/ER Ca^{2+} elevation in GM1-expressing NG108-15 and GM1-deficient NG-CR72 cells (154, 157), which showed significantly higher NE/ER Ca^{2+} elevation in cells containing GM1. Similar results were observed in C6 cells (158), which contain NCX/GM1 in the NE but not in the plasma membrane, whereas no NE/ER Ca^{2+} elevation was observed in NCX-deficient Jurkat T cells (153). Because the NE lumen is continuous with the ER, these data support a function for the NCX/GM1 complex as an alternate mechanism for transferring cytosolic Ca^{2+} to the ER. Studies using knockout (KO) mice engineered to lack GM2/GD2 synthase, which results in deficient synthesis of GM2, GD2, and GM1 (**Figure 4**), have demonstrated key regulatory roles for nuclear membrane–associated GM1 in Ca^{2+} homeostasis (159, 160). These KO mice develop late-onset neurological disease (161) and display deficient Ca^{2+} regulatory capabilities in their cerebellar granule neurons (162). Supplementation of neuronal cultures from GM2/GD2 synthase KO mice with exogenous GM1 restored a normal phenotype (116), supporting a role for this ganglioside in regulating Ca^{2+} homeostasis. Moreover, most cell types studied to date express nuclear NCX/GM1, which suggests that this complex is a ubiquitous mechanism for Ca^{2+} homeostasis employed by cells.

CONCLUSIONS AND FUTURE PERSPECTIVES

Nuclear sphingolipid metabolism is an area of research undergoing significant progress. Emerging new data are paving the way toward a more comprehensive understanding of the unique roles for these bioactive lipids in nuclear processes. Analogous to their cytosolic functions, distinct sphingolipid species have unique nuclear functions and act via temporally and spatially specific mechanisms. Future studies aimed at elucidating the contributions that sphingolipid-metabolizing enzymes play in nuclear processes and at quantifying the nuclear concentrations of different sphingolipid species under different cellular conditions will allow for a more thorough understanding of how nuclear lipid metabolism coordinates global changes in cell function.

SUMMARY POINTS

1. Distinct subsets of sphingolipid-metabolizing enzymes catalyze sphingolipid turnover in varied subnuclear domains.

2. Sphingolipid concentrations at different intranuclear compartments fluctuate in response to diverse physiological cues. Some extracellular stimuli may affect nuclear sphingolipid turnover independently of cytosolic signaling.

3. Sphingomyelin plays structural and regulatory roles in chromatin architecture, DNA synthesis, and RNA stability.
4. Sphingosine regulates gene transcription by serving as a ligand for the nuclear receptor steroidogenic factor 1.
5. Sphingosine-1-phosphate plays an epigenetic role in gene expression by controlling histone acetylation.
6. The GM1 ganglioside modulates nuclear Ca^{2+} homeostasis by forming a complex with a Na^+/Ca^{2+} exchanger in the inner membrane of the nuclear envelope.

DISCLOSURE STATEMENT

The authors are not aware of any affiliations, memberships, funding, or financial holdings that might be perceived as affecting the objectivity of this review.

LITERATURE CITED

1. Ledeen RW, Wu G. 2008. Nuclear sphingolipids: metabolism and signaling. *J. Lipid Res.* 49:1176–86
2. Ledeen RW, Wu G. 2006. Sphingolipids of the nucleus and their role in nuclear signaling. *Biochim. Biophys. Acta* 1761:588–98
3. Tamiya-Koizumi K. 2002. Nuclear lipid metabolism and signaling. *J. Biochem.* 132:13–22
4. Barlow CA, Laishram RS, Anderson RA. 2011. Nuclear phosphoinositides: a signaling enigma wrapped in a compartmental conundrum. *Trends Cell Biol.* 20:25–35
5. Albi E, Viola Magni M. 2004. The role of intranuclear lipids. *Biol. Cell* 96:657–67
6. Krylova IN, Sablin EP, Moore J, Xu RX, Waitt GM, et al. 2005. Structural analyses reveal phosphatidyl inositols as ligands for the NR5 orphan receptors SF-1 and LRH-1. *Cell* 120:343–55
7. Sablin EP, Blind RD, Krylova IN, Ingraham JG, Cai F, et al. 2009. Structure of SF-1 bound by different phospholipids: evidence for regulatory ligands. *Mol. Endocrinol.* 23:25–34
8. Irvine RF. 2003. Nuclear lipid signalling. *Nat. Rev. Mol. Cell Biol.* 4:349–60
9. Pliss A, Kuzmin AN, Kachynski AV, Prasad PN. 2010. Nonlinear optical imaging and Raman microspectrometry of the cell nucleus throughout the cell cycle. *Biophys. J.* 99:3483–91
10. Albi E, Mersel M, Leray C, Tomassoni ML, Viola-Magni MP. 1994. Rat liver chromatin phospholipids. *Lipids* 29:715–19
11. Boronenkov IV, Loijens JC, Umeda M, Anderson RA. 1998. Phosphoinositide signaling pathways in nuclei are associated with nuclear speckles containing pre-mRNA processing factors. *Mol. Biol. Cell* 9:3547–60
12. Neitcheva T, Peeva D. 1995. Phospholipid composition, phospholipase A2 and sphingomyelinase activities in rat liver nuclear membrane and matrix. *Int. J. Biochem. Cell Biol.* 27:995–1001
13. Xu A, Suh PG, Marmy-Conus N, Pearson RB, Seok OY, et al. 2001. Phosphorylation of nuclear phospholipase C β1 by extracellular signal-regulated kinase mediates the mitogenic action of insulin-like growth factor I. *Mol. Cell. Biol.* 21:2981–90
14. Cocco L, Martelli AM, Gilmour RS, Ognibene A, Manzoli FA, Irvine RF. 1988. Rapid changes in phospholipid metabolism in the nuclei of Swiss 3T3 cells induced by treatment of the cells with insulin-like growth factor I. *Biochem. Biophys. Res. Commun.* 154:1266–72
15. Divecha N, Banfic H, Irvine RF. 1991. The polyphosphoinositide cycle exists in the nuclei of Swiss 3T3 cells under the control of a receptor (for IGF-I) in the plasma membrane, and stimulation of the cycle increases nuclear diacylglycerol and apparently induces translocation of protein kinase C to the nucleus. *EMBO J.* 10:3207–14

16. Martelli AM, Gilmour RS, Neri LM, Manzoli L, Corps AN, Cocco L. 1991. Mitogen-stimulated events in nuclei of Swiss 3T3 cells. Evidence for a direct link between changes of inositol lipids, protein kinase C requirement and the onset of DNA synthesis. *FEBS Lett.* 283:243–46
17. Banfic H, Zizak M, Divecha N, Irvine RF. 1993. Nuclear diacylglycerol is increased during cell proliferation in vivo. *Biochem. J.* 290(Pt. 3):633–36
18. Raben DM, Baldassare JJ. 2000. Nuclear envelope signaling: role of phospholipid metabolism. *Eur. J. Histochem.* 44:67–80
19. Banno Y, Tamiya-Koizumi K, Oshima H, Morikawa A, Yoshida S, Nozawa Y. 1997. Nuclear ADP-ribosylation factor (ARF)- and oleate-dependent phospholipase D (PLD) in rat liver cells. Increases of ARF-dependent PLD activity in regenerating liver cells. *J. Biol. Chem.* 272:5208–13
20. Manzoli FA, Capitani S, Mazzotti G, Barnabei O, Maraldi NM. 1982. Role of chromatin phospholipids on template availability and ultrastructure of isolated nuclei. *Adv. Enzyme Regul.* 20:247–62
21. Capitani S, Caramelli E, Felaco M, Miscia S, Manzoli FA. 1981. Effect of phospholipid vesicles on endogenous RNA polymerase activity of isolated rat liver nuclei. *Physiol. Chem. Phys.* 13:153–58
22. Li Y, Choi M, Cavey G, Daugherty J, Suino K, et al. 2005. Crystallographic identification and functional characterization of phospholipids as ligands for the orphan nuclear receptor steroidogenic factor-1. *Mol. Cell* 17:491–502
23. Ortlund EA, Lee Y, Solomon IH, Hager JM, Safi R, et al. 2005. Modulation of human nuclear receptor LRH-1 activity by phospholipids and SHP. *Nat. Struct. Biol.* 12:357–63
24. Wang W, Zhang C, Marimuthu A, Krupka HI, Tabrizizad M, et al. 2005. The crystal structures of human steroidogenic factor-1 and liver receptor homologue-1. *Proc. Natl. Acad. Sci. USA* 102:7505–10
25. Martelli AM, Bortul R, Bareggi R, Manzoli L, Narducci, Cocco L. 2002. Diacylglycerol kinases in nuclear lipid-dependent signal transduction pathways. *Cell Mol. Life Sci.* 59:1129–37
26. Martelli AM, Fala F, Faenza I, Billi AM, Cappellini A, et al. 2004. Metabolism and signaling activities of nuclear lipids. *Cell Mol. Life Sci.* 61:1143–56
27. Topham MK, Epand RM. 2009. Mammalian diacylglycerol kinases: molecular interactions and biological functions of selected isoforms. *Biochim. Biophys. Acta* 1790:416–24
28. Wattenberg BW, Pitson SM, Raben DM. 2006. The sphingosine and diacylglycerol kinase superfamily of signaling kinases: localization as a key to signaling function. *J. Lipid Res.* 47:1128–39
29. Topham MK, Bunting M, Zimmerman GA, McIntyre TM, Blackshear PJ, Prescott SM. 1998. Protein kinase C regulates the nuclear localization of diacylglycerol kinase-ζ. *Nature* 394:697–700
30. Tabellini G, Bortul R, Santi S, Riccio M, Baldini G, et al. 2003. Diacylglycerol kinase-θ is localized in the speckle domains of the nucleus. *Exp. Cell Res.* 287:143–54
31. Bregoli L, Baldassare JJ, Raben DM. 2001. Nuclear diacylglycerol kinase-θ is activated in response to α-thrombin. *J. Biol. Chem.* 276:23288–95
32. Tabellini G, Billi AM, Fala F, Cappellini A, Evagelisti C, et al. 2004. Nuclear diacylglycerol kinase-θ is activated in response to nerve growth factor stimulation of PC12 cells. *Cell Signal.* 16:1263–71
33. Li D, Urs AN, Allegood J, Leon A, Merrill AH Jr, Sewer MB. 2007. Cyclic AMP-stimulated interaction between steroidogenic factor 1 and diacylglycerol kinase θ facilitates induction of CYP17. *Mol. Cell. Biol.* 27:6669–85
34. Ingraham HA, Lala DS, Ikeda Y, Luo X, Shen WH, et al. 1994. The nuclear receptor steroidogenic factor 1 acts at multiple levels of the reproductive axis. *Genes Dev.* 8:2302–12
35. Parker KL, Rice DA, Lala DS, Ikeda Y, Luo X, et al. 2002. Steroidogenic factor 1: an essential mediator of endocrine development. *Recent Prog. Horm. Res.* 57:19–36
36. Csaki LS, Reue K. 2010. Lipins: multifunctional lipid metabolism proteins. *Annu. Rev. Nutr.* 30:257–72
37. Finck BN, Gropler MC, Chen Z, Leone TC, Croce MA, et al. 2006. Lipin 1 is an inducible amplifier of the hepatic PGC-1α/PPARα regulatory pathway. *Cell Metab.* 4:199–210
38. Tsukahara T, Tsukahara R, Fujiwara Y, Yue J, Cheng Y, et al. 2010. Phospholipase D2-dependent inhibition of the nuclear hormone receptor PPARγ by cyclic phosphatidic acid. *Mol. Cell* 39:421–32
39. Strub GM, Maceyka M, Hait NC, Milstien S, Spiegel S. 2010. Extracellular and intracellular actions of sphingosine-1-phosphate. *Adv. Exp. Med. Biol.* 688:141–55
40. Hannun YA, Obeid LM. 2009. Principles of bioactive lipid signaling: lessons from sphingolipids. *Nat. Rev. Mol. Cell Biol.* 9:139–50

41. Zeidan YH, Hannun YA. 2007. Translational aspects of sphingolipid metabolism. *Trends Mol. Med.* 13:327–36
42. Tamiya-Koizumi K, Umekawa H, Yoshida S, Kojima K. 1989. Existence of Mg^{2+}-dependent, neutral sphingomyelinase in nuclei of rat ascites hepatoma cells. *J. Biochem.* 106:593–98
43. Alessenko A, Chatterjee S. 1995. Neutral sphingomyelinase: localization in rat liver nuclei and involvement in regeneration/proliferation. *Mol. Cell. Biochem.* 143:169–74
44. Wu G, Lu ZH, Ledeen RW. 1995. GM1 ganglioside in the nuclear membrane modulates nuclear calcium homeostasis during neurite outgrowth. *J. Neurochem.* 65:1419–22
45. Micheli M, Albi E, Leray C, Magni MV. 1998. Nuclear sphingomyelin protects RNA from RNase action. *FEBS Lett.* 431:443–47
46. Tsugane K, Tamiya-Koizumi K, Nagino M, Nimura Y, Yoshida S. 1999. A possible role of nuclear ceramide and sphingosine in hepatocyte apoptosis in rat liver. *J. Hepatol.* 31:8–17
47. Mizuno S, Nimura Y, Suzuki H, Yoshida S. 1996. Portal vein branch occlusion induces cell proliferation of cholestatic rat liver. *J. Surg. Res.* 60:249–57
48. Xie X, Wu G, Lu ZH, Ledeen RW. 2002. Potentiation of a sodium-calcium exchanger in the nuclear envelope by nuclear GM1 ganglioside. *J. Neurochem.* 81:1185–95
49. Rossi G, Magni MV, Albi E. 2007. Sphingomyelin-cholesterol and double stranded RNA relationship in the intranuclear complex. *Arch. Biochem. Biophys.* 459:27–32
50. Albi E, Cataldi S, Rossi G, Viola Magni M, Toller M, et al. 2008. The nuclear ceramide/diacylglycerol balance depends on the physiological state of thyroid cells and changes during UV-C radiation-induced apoptosis. *Arch. Biochem. Biophys.* 478:52–58
51. Tempera I, Buchetti B, Lococo E, Gradini R, Mastronardi A, et al. 2008. GD3 nuclear localization after apoptosis induction in HUT-78 cells. *Biochem. Biophys. Res. Commun.* 368:495–500
52. Hait NC, Allegood J, Maceyka M, Strub GM, Harikumar KB, et al. 2009. Regulation of histone acetylation in the nucleus by sphingosine-1-phosphate. *Science* 325:1254–57
53. Dupuy F, Fanani ML, Maggio B. 2011. Ceramide N-acyl chain length: a determinant of bidimensional transitions, condensed domain morphology, and interfacial thickness. *Langmuir* 27:3783–91
54. Stiban J, Fistere D, Colombini M. 2006. Dihydroceramide hinders ceramide channel formation: implications on apoptosis. *Apoptosis* 11:773–80
55. Kotzerke J, Stibane C, Dralle H, Wiese H, Burchert W. 1989. Screening for pheochromocytoma in the MEN 2 syndrome. *Henry Ford Hosp. Med. J.* 37:129–31
56. Farooqui AA, Horrocks LA. 2005. Signaling and interplay mediated by phospholipases A2, C, and D in LA-N-1 cell nuclei. *Reprod. Nutr. Dev.* 45:613–31
57. Ondrias K, Lencesova L, Sirova M, Labudova M, Pastorekova S, et al. 2011. Apoptosis induced clustering of IP_3R1 in nuclei of nondifferentiated PC12 cells. *J. Cell. Physiol.* 226:3147–55
58. Martelli AM, Manzoli L, Cocco L. 2004. Nuclear inositides: facts and persepectives. *Pharmacol. Ther.* 101:47–64
59. Cocco L, Martelli AM, Gilmour RS, Rhee SG, Manzolli FA. 2001. Nuclear phospholipase C and signaling. *Biochim. Biophys. Acta Mol. Cell Biol. Lipids* 1530:1–14
60. Antony P, Freysz L, Horrocks LA, Farooqui AA. 2001. Effects of retinoic acid on the Ca^{2+}-independent phospholipase A2 in nuclei of LA-N-1 neuroblastoma cells. *Neurochem. Res.* 26:83–88
61. Stiban J, Caputo L, Colombini M. 2008. Ceramide synthesis in the endoplasmic reticulum can permeabilize mitochondria to proapoptotic proteins. *J. Lipid Res.* 49:625–34
62. Rovina P, Schanzer A, Graf C, Mechtcheriakova D, Jaritz M, Bornancin F. 2009. Subcellular localization of ceramide kinase and ceramide kinase-like protein requires interplay of their Pleckstrin Homology domain-containing N-terminal regions together with C-terminal domains. *Biochem. Biophys. Acta* 1791:1023–30
63. Schimizu M, Tada E, Makiyama T, Yasufuku K, Moriyama Y, et al. 2009. Effects of ceramide, ceramidase inhibition and expression of ceramide kinase on cytosolic phospholipase A2α; additional role for ceramide-1-phosphate in phosphorylation and Ca^{2+} signaling. *Cell Signal.* 21:440–47
64. Lamour N, Chalfant CE. 2008. Ceramide kinase and the ceramide-1-phosphate/cPLA2α interaction as a therapeutic target. *Curr. Drug Targets* 9:674–82

65. Samanta S, Stiban J, Maugel TK, Colombini M. 2011. Visualization of ceramide channels by transmission electron microscopy. *Biochim. Biophys. Acta* 1808:1196–201
66. Cheng X, Blumenthal RM. 2010. Coordinated chromatin control: structural and functional linkage of DNA and histone methylation. *Biochemistry* 49:2999–3008
67. Lelievre SA. 2009. Contributions of extracellular matrix signaling and tissue architecture to nuclear mechanisms and spatial organization of gene expression control. *Biochim. Biophys. Acta* 1790:925–35
68. Lamond AI, Earnshaw WC. 1998. Structure and function in the nucleus. *Science* 280:547–53
69. Mao C, Obeid LM. 2008. Ceramidases: regulators of cellular responses mediated by ceramide, sphingosine, and sphingosine-1-phosphate. *Biochim. Biophys. Acta* 1781:424–34
70. Pettus BJ, Chalfant CE, Hannun YA. 2002. Ceramide in apoptosis: an overview and current perspectives. *Biochim. Biophys. Acta* 1585:114–25
71. Strelow A, Bernardo K, Adam-Klages S, Linke T, Sandhoff K, et al. 2000. Overexpression of acid ceramidase protects from tumor necrosis factor-induced cell death. *J. Exp. Med.* 192:601–12
72. Spiegel S, Milstien S. 2007. Functions of the multifaceted family of sphingosine kinases and some close relatives. *J. Biol. Chem.* 282:2125–29
73. Takabe K, Paugh SW, Milstien S, Spiegel S. 2008. "Inside-out" signaling of sphingosine-1-phosphate: therapeutic targets. *Pharmacol. Rev.* 60:181–95
74. Sukocheva O, Wang L, Verrier E, Vadas MA, Xia P. 2009. Restoring endocrine response in breast cancer cells by inhibition of the sphingosine kinase-1 signaling pathway. *Endocrinology* 150:4484–92
75. Takabe K, Kim RH, Allegood JC, Mitra P, Ramachandran S, et al. 2010. Estradiol induces export of sphingosine 1-phosphate from breast cancer cells via ABCC1 and ABCG2. *J. Biol. Chem.* 285:10477–86
76. Henkes LE, Sullivan BT, Lynch MP, Kolesnick R, Arsenault D, et al. 2008. Acid sphingomyelinase involvement in tumor necrosis factor α-regulated vascular and steroid disruption during luteolysis in vivo. *Proc. Natl. Acad. Sci. USA* 105:7670–75
77. Kolesnick RN, Haimovitz-Friedman A, Fuks Z. 1994. The sphingomyelin signal transduction pathway mediates apoptosis for tumor necrosis factor, Fas, and ionizing radiation. *Biochem. Cell Biol.* 72:471–74
78. Watanabe M, Kitano T, Kondo T, Yabu T, Taguchi Y, et al. 2004. Increase of nuclear ceramide through caspase-3-dependent regulation of the "sphingomyelin cycle" in Fas-induced apoptosis. *Cancer Res.* 64:1000–7
79. Nybond S, Bjorkqvist YJ, Ramstedt B, Slotte JP. 2005. Acyl chain length affects ceramide action on sterol/sphingomyelin-rich domains. *Biochim. Biophys. Acta* 1718:61–66
80. Megha, Sawatzki P, Kolter T, Bittman R, London E. 2007. Effect of ceramide N-acyl chain and polar headgroup structure on the properties of ordered lipid domains (lipid rafts). *Biochim. Biophys. Acta* 1768:2205–12
81. Senkal CE, Ponnusamy S, Bielawski J, Hannun YA, Ogretmen B. 2010. Antiapoptotic roles of ceramide-synthase-6-generated C16-ceramide via selective regulation of the ATF6/CHOP arm of ER-stress-response pathways. *FASEB J.* 24:296–308
82. Senkal CE, Ponnusamy S, Rossi MJ, Bialewski J, Sinha D, et al. 2007. Role of human longevity assurance gene 1 and C18-ceramide in chemotherapy-induced cell death in human head and neck squamous cell carcinomas. *Mol. Cancer Ther.* 6:712–22
83. Mesicek J, Lee H, Feldman T, Jiang X, Skobeleva A, et al. 2010. Ceramide synthases 2, 5, and 6 confer distinct roles in radiation-induced apoptosis in HeLa cells. *Cell Signal.* 22:1300–7
84. Stiban J, Tidhar R, Futerman AH. 2010. Ceramide synthases: roles in cell physiology and signaling. *Adv. Exp. Med. Biol.* 688:60–71
85. Sun W, Jin J, Xu R, Hu W, Szulc ZM, et al. 2010. Substrate specificity, membrane topology, and activity regulation of human alkaline ceramidase 2 (ACER2). *J. Biol. Chem.* 285:8995–9007
86. Levy M, Futerman AH. 2010. Mammalian ceramide synthases. *IUBMB Life* 62:347–56
87. Gault CR, Obeid LM, Hannun YA. 2010. An overview of sphingolipid metabolism: from synthesis to breakdown. *Adv. Exp. Med. Biol.* 688:1–23
88. El Bawab S, Roddy P, Qian T, Bielawska A, Lemasters JJ, Hannun YA. 2000. Molecular cloning and characterization of a human mitochondrial ceramidase. *J. Biol. Chem.* 275:21508–13
89. Hwang YH, Tani M, Nakagawa T, Okino N, Ito M. 2005. Subcellular localization of human neutral ceramidase expressed in HEK293 cells. *Biochem. Biophys. Res. Commun.* 331:37–42

90. Breslow DK, Weissman JS. 2010. Membranes in balance: mechanisms of sphingolipid homeostasis. *Mol. Cell* 40:267–79
91. Mullen TD, Jenkins RW, Clarke CJ, Bielawski J, Hannun YA, Obeid LM. 2011. Ceramide synthase-dependent ceramide generation and programmed cell death: involvement of salvage pathway in regulating post-mitochondrial events. *J. Biol. Chem.* 286:15929–42
92. Novgorodov SA, Chudakova DA, Wheeler BW, Bielawski J, Kindy MS, et al. 2011. Developmentally regulated ceramide synthase 6 increases mitochondrial Ca^{2+} loading capacity and promotes apoptosis. *J. Biol. Chem.* 286:4644–58
93. Mimeault M. 2002. New advances on structural and biological functions of ceramide in apoptotic/necrotic cell death and cancer. *FEBS Lett.* 530:9–16
94. Siskind LJ, Kolesnick RN, Colombini M. 2006. Ceramide forms channels in mitochondrial outer membranes at physiologically relevant concentrations. *Mitochondrion* 6:118–25
95. Grassme H, Jekle A, Riehle A, Schwarz H, Berger J, et al. 2001. CD95 signaling via ceramide-rich membrane rafts. *J. Biol. Chem.* 276:20589–96
96. Bollinger CR, Teichgraber V, Gulbins E. 2005. Ceramide-enriched membrane domains. *Biochim. Biophys. Acta* 1746:284–94
97. Gupta S, Maurya MR, Merrill AH Jr, Glass CK, Subramaniam S. 2011. Integration of lipidomics and transcriptomics data towards a systems biology model of sphingolipid metabolism. *BMC Syst. Biol.* 5:26
98. Hornemann T, Penno A, Rutti MF, Ernst D, Kivrak-Pfiffner F, et al. 2009. The SPTLC3 subunit of serine palmitoyltransferase generates short chain sphingoid bases. *J. Biol. Chem.* 284:26322–30
99. Han G, Gupta SD, Gable K, Niranjanakumari S, Moitra P, et al. 2009. Identification of small subunits of mammalian serine palmitoyltransferase that confer distinct acyl-CoA substrate specificities. *Proc. Natl. Acad. Sci. USA* 106:8186–91
100. Eichler FS, Hornemann T, McCampbell A, Kuljis D, Penno A, et al. 2009. Overexpression of the wild-type SPT1 subunit lowers desoxysphingolipid levels and rescues the phenotype of HSAN1. *J. Neurosci.* 29:14646–51
101. Pruett ST, Bushnev A, Hagedorn K, Adiga M, Haynes CA, et al. 2008. Biodiversity of sphingoid bases ("sphingosines") and related amino alcohols. *J. Lipid Res.* 49:1621–39
102. Menaldino DS, Bushnev A, Sun A, Liotta DC, Symolon H, et al. 2003. Sphingoid bases and de novo ceramide synthesis: enzymes involved, pharmacology and mechanisms of action. *Pharmacol. Res.* 47:373–81
103. Albi E, Cataldi S, Rossi G, Magni MV. 2003. A possible role of cholesterol-sphingomyelin/phosphatidylcholine in nuclear matrix during rat liver regeneration. *J. Hepatol.* 38:623–28
104. Albi E, Pieroni S, Viola Magni MP, Sartori C. 2003. Chromatin sphingomyelin changes in cell proliferation and/or apoptosis induced by ciprofibrate. *J. Cell. Physiol.* 196:354–61
105. Ledeen RW, Wu G. 2006. Gangliosides of the nuclear membrane: a crucial locus of cytoprotective modulation. *J. Cell Biochem.* 97:893–903
106. Albi E, Cataldi S, Villani M, Perrella G. 2009. Nuclear phosphatidylcholine and sphingomyelin metabolism of thyroid cells changes during stratospheric balloon flight. *J. Biomed. Biotechnol.* 2009:125412
107. Albi E, La Porta CA, Cataldi S, Magni MV. 2005. Nuclear sphingomyelin-synthase and protein kinase Cδ in melanoma cells. *Arch. Biochem. Biophys.* 438:156–61
108. Kleuser B, Maceyka M, Milstien S, Spiegel S. 2001. Stimulation of nuclear sphingosine kinase activity by platelet-derived growth factor. *FEBS Lett.* 503:85–90
109. Cocco L, Maraldi NM, Manzoli FA, Gilmour RS, Lang A. 1980. Phospholipid interactions in rat liver nuclear matrix. *Biochem. Biophys. Res. Commun.* 96:890–98
110. Shiraishi T, Imai S, Uda Y. 2003. The presence of ceramidase activity in liver nuclear membrane. *Biol. Pharm. Bull.* 26:775–79
111. Haines DS, Strauss KI, Gillespie DH. 1991. Cellular response to double-stranded RNA. *J. Cell Biochem.* 46:9–20
112. Albi E, Magni MP. 1997. Chromatin neutral sphingomyelinase and its role in hepatic regeneration. *Biochem. Biophys. Res. Commun.* 236:29–33
113. Albi E, Magni MV. 1999. Sphingomyelin synthase in rat liver nuclear membrane and chromatin. *FEBS Lett.* 460:369–72

114. Albi E, Lazzarini R, Magni MV. 2003. Reverse sphingomyelin-synthase in rat liver chromatin. *FEBS Lett.* 549:152–56
115. Wang J, Wu G, Miyagi T, Lu ZH, Ledeen RW. 2009. Sialidase occurs in both membranes of the nuclear envelope and hydrolyzes endogenous GD1a. *J. Neurochem.* 111:547–54
116. Wu G, Lu ZH, Xie X, Ledeen RW. 2004. Susceptibility of cerebellar granule neurons from GM2/GD2 synthase-null mice to apoptosis induced by glutamate excitotoxicity and elevated KCl: rescue by GM1 and LIGA20. *Glycoconj. J.* 21:305–13
117. James JL, Clawson GA, Chan CH, Smuckler EA. 1981. Analysis of the phospholipid of the nuclear envelope and endoplasmic reticulum of liver cells by high pressure liquid chromatography. *Lipids* 16:541–45
118. Keenan TW, Berezney R, Crane FL. 1972. Lipid composition of further purified bovine liver nuclear membranes. *Lipids* 7:212–15
119. Albi E, Magni MV. 2002. The presence and the role of chromatin cholesterol in rat liver regeneration. *J. Hepatol.* 36:395–400
120. Lein M, Stibane I, Mansour R, Hege C, Roigas J, et al. 2006. Complications, urinary continence, and oncologic outcome of 1000 laparoscopic transperitoneal radical prostatectomies—experience at the Charite Hospital Berlin, Campus Mitte. *Eur. Urol.* 50:1278–82
121. Schmidt GW, Stibane H, Ulbrich F, Griesse H. 1967. Edwards syndrome with double trisomy D and E (mosaicism). *Klin. Wochenschr.* 45:634–38
122. Thevissen K, Francois IE, Winderickx J, Pannecouque C, Cammue BP. 2006. Ceramide involvement in apoptosis and apoptotic diseases. *Mini. Rev. Med. Chem.* 6:699–709
123. Fasutino R, Cheung P, Richard M, Dibrov E, Kneesch A, et al. 2008. Ceramide regulation of nuclear protein import. *J. Lipid Res.* 49:654–62
124. Martelli AM, Evangelisti C, Nyakern M, Manzoli FA. 2006. Nuclear protein kinase C. *Biochim. Biophys. Acta* 1761:542–51
125. Evangelisti C, Bortul R, Fala F, Tabellini G, Goto K, Martelli AM. 2007. Nuclear diacylglycerol kinases: emerging downstream regulators in cell signaling networks. *Histol. Histopathol.* 22:573–79
126. Sun B, Murray NR, Fields AP. 1997. A role for nuclear phosphatidylinositol-specific phospholipase C in the G2/M phase transition. *J. Biol. Chem.* 272:26313–17
127. Deacon EM, Pettitt TR, Webb P, Cross T, Chahal H, et al. 2002. Generation of diacylglycerol molecular species through the cell cycle: a role for 1-stearoyl, 2-arachidonyl glycerol in the activation of nuclear protein kinase C-βII at G2/M. *J. Cell Sci.* 115:983–89
128. Abboushi N, El-Hed A, El-Assaad W, Kozhaya L, El-Sabban ME, et al. 2004. Ceramide inhibits IL-2 production by preventing protein kinase C-dependent NF-κB activation: possible role in protein kinase Cθ regulation. *J. Immunol.* 173:3193–200
129. Albi E, Micheli M, Viola Magni MP. 1996. Phospholipids and nuclear RNA. *Cell Biol. Int.* 20:407–12
130. Chocian G, Chabowski A, Zendzian-Piotrowska M, Harasim E, Lukaszuk B, Gorski J. 2010. High fat diet induces ceramide and sphingomyelin formation in rat's liver nuclei. *Mol. Cell. Biochem.* 340:125–31
131. Adams JM 2nd, Pratipanawatr T, Berria R, Wang E, DeFronzo RA, et al. 2004. Ceramide content is increased in skeletal muscle from obese insulin-resistant humans. *Diabetes* 53:25–31
132. Chavez JA, Holland WL, Bar J, Sandhoff K, Summers SA. 2005. Acid ceramidase overexpression prevents the inhibitory effects of saturated fatty acids on insulin signaling. *J. Biol. Chem.* 280:20148–53
133. Andreyev AY, Fahy E, Guan Z, Kelly S, Li X, et al. 2010. Subcellular organelle lipidomics in TLR-4-activated macrophages. *J. Lipid Res.* 51:2785–97
134. Schimazu T, Horinouchi S, Yoshida M. 2007. Multiple histone deacetylases and the CREB-binding protein regulate pre-mRNA 3′-end processing. *J. Biol. Chem.* 282:4470–78
135. Pettus BJ, Bielawska A, Spiegel S, Roddy P, Hannun YA, Chalfant CE. 2003. Ceramide kinase mediates cytokine- and calcium ionophore-induced arachidonic acid release. *J. Biol. Chem.* 278:38206–13
136. Lamour NF, Subramanian P, Wijesinghe DS, Stahelin RV, Bonventre JV, Chalfant CE. 2009. Ceramide 1-phosphate is required for the translocation of group IVA cytosolic phospholipase A2 and prostaglandin synthesis. *J. Biol. Chem.* 284:26897–907

137. Urs AN, Dammer E, Sewer M. 2006. Sphingosine regulates the transcription of CYP17 by binding to steroidogenic factor-1. *Endocrinology* 147:5249–58
138. Sewer MB, Waterman MR. 2003. ACTH modulation of transcription factors responsible for steroid hydroxylase gene expression in the adrenal cortex. *Microsc. Res. Tech.* 61:300–7
139. Sewer MB, Waterman MR. 2002. cAMP-dependent transcription of steroidogenic genes in the human adrenal cortex requires a dual-specificity phosphatase in addition to protein kinase A. *J. Mol. Endocrinol.* 29:163–74
140. Maceyka M, Sankala H, Hait NC, Le Stunff H, Liu H, et al. 2005. SphK1 and SphK2, sphingosine kinase isoenzymes with opposing functions in sphingolipid metabolism. *J. Biol. Chem.* 280:37118–29
141. Keenan TW, Morre DJ, Huang CM. 1972. Distribution of gangliosides among subcellular fractions from rat liver and bovine mammary gland. *FEBS Lett.* 24:204–8
142. Matyas GR, Morre DJ. 1987. Subcellular distribution and biosynthesis of rat liver gangliosides. *Biochim. Biophys. Acta* 921:599–614
143. Katoh N, Kira T, Yuasa A. 1993. Protein kinase C substrates and ganglioside inhibitors in bovine mammary nuclei. *J. Dairy Sci.* 76:3400–9
144. Wu G, Lu ZH, Ledeen RW. 1995. Induced and spontaneous neuritogenesis are associated with enhanced expression of ganglioside GM1 in the nuclear membrane. *J. Neurosci.* 15:3739–46
145. Saito M, Sugiyama K. 2002. Characterization of nuclear gangliosides in rat brain: concentration, composition, and developmental changes. *Arch. Biochem. Biophys.* 398:153–59
146. Humbert JP, Matter N, Artault JC, Koppler P, Malviya AN. 1996. Inositol 1,4,5-trisphosphate receptor is located to the inner nuclear membrane vindicating regulation of nuclear calcium signaling by inositol 1,4,5-trisphosphate. Discrete distribution of inositol phosphate receptors to inner and outer nuclear membranes. *J. Biol. Chem.* 271:478–85
147. Gilchrist JS, Pierce GN. 1993. Identification and purification of a calcium-binding protein in hepatic nuclear membranes. *J. Biol. Chem.* 268:4291–99
148. Saito M, Hagita H, Ito M, Ando S, Yu RK. 2002. Age-dependent reduction in sialidase activity of nuclear membranes from mouse brain. *Exp. Gerontol.* 37:937–41
149. Wang J, Wu G, Miyagi T, Lu ZH, Ledeen RW. 2009. Sialidase occurs in both membranes of the nuclear envelope and hydrolyzes endogenous GD1a. *J. Neurochem.* 111:547–54
150. Parkinson ME, Smith CG, Garland PB, van Heyningen S. 1989. Identification of cholera toxin-binding sites in the nucleus of intestinal epithelial cells. *FEBS Lett.* 242:309–13
151. Copani A, Melchiorri D, Caricasole A, Martini F, Sale P, et al. 2002. β-Amyloid-induced synthesis of the ganglioside GD3 is a requisite for cell cycle reactivation and apoptosis in neurons. *J. Neurosci.* 22:3963–68
152. Philipson KD, Nicoll DA. 2000. Sodium-calcium exchange: a molecular perspective. *Annu. Rev. Physiol.* 62:111–33
153. Xie X, Wu G, Lu ZH, Rohowsky-Kochan C, Ledeen RW. 2004. Presence of sodium-calcium exchanger/GM1 complex in the nuclear envelope of non-neural cells: nature of exchanger-GM1 interaction. *Neurochem. Res.* 29:2135–46
154. Ledeen R, Wu G. 2011. New findings on nuclear gangliosides: overview on metabolism and function. *J. Neurochem.* 116:714–20
155. Kofuji P, Lederer WJ, Schulze DH. 1994. Mutually exclusive and cassette exons underlie alternatively spliced isoforms of the Na/Ca exchanger. *J. Biol. Chem.* 269:5145–49
156. He S, Ruknudin A, Bambrick LL, Lederer WJ, Schulze DH. 1998. Isoform-specific regulation of the Na^+/Ca^{2+} exchanger in rat astrocytes and neurons by PKA. *J. Neurosci.* 18:4833–41
157. Wu G, Xie X, Lu ZH, Ledeen RW. 2009. Sodium-calcium exchanger complexed with GM1 ganglioside in nuclear membrane transfers calcium from nucleoplasm to endoplasmic reticulum. *Proc. Natl. Acad. Sci. USA* 106:10829–34
158. Xie X, Wu G, Ledeen RW. 2004. C6 cells express a sodium-calcium exchanger/GM1 complex in the nuclear envelope but have no exchanger in the plasma membrane: comparison to astrocytes. *J. Neurosci. Res.* 76:363–75
159. Wu G, Lu ZH, Kulkarni N, Amin R, Ledeen RW. 2011. Mice lacking major brain gangliosides develop Parkinsonism. *Neurochem. Res.* 36:1706–14

160. Ledeen R, Wu G. 2011. New findings on nuclear gangliosides: overview on metabolism and function. *J. Neurochem.* 116:714–20
161. Liu Y, Wada R, Kawai H, Sango K, Deng C, et al. 1999. A genetic model of substrate deprivation therapy for a glycosphingolipid storage disorder. *J. Clin. Investig.* 103:497–505
162. Wu G, Xie X, Lu ZH, Ledeen RW. 2001. Cerebellar neurons lacking complex gangliosides degenerate in the presence of depolarizing levels of potassium. *Proc. Natl. Acad. Sci. USA* 98:307–12
163. Fujiwara Y. 2008. Cyclic phosphatidic acid—a unique phospholipid. *Biochim. Biophys. Acta* 1781(9):519–24

Adenosine and Hypoxia-Inducible Factor Signaling in Intestinal Injury and Recovery

Sean P. Colgan and Holger K. Eltzschig

Departments of Medicine and Anesthesiology and the Mucosal Inflammation Program, University of Colorado School of Medicine, Aurora, Colorado 80045; email: Sean.Colgan@UCDenver.edu

Keywords

metabolism, inflammation, nucleotide, nucleoside, nucleotidase, mucosa, colitis, ischemia, neutrophil, epithelium, endothelium, murine model

Abstract

The gastrointestinal mucosa has proven to be an interesting tissue in which to investigate disease-related metabolism. In this review, we outline some of the evidence that implicates hypoxia-mediated adenosine signaling as an important signature within both healthy and diseased mucosa. Studies derived from cultured cell systems, animal models, and human patients have revealed that hypoxia is a significant component of the inflammatory microenvironment. These studies have revealed a prominent role for hypoxia-induced factor (HIF) and hypoxia signaling at several steps along the adenine nucleotide metabolism and adenosine receptor signaling pathways. Likewise, studies to date in animal models of intestinal inflammation have demonstrated an almost uniformly beneficial influence of HIF stabilization on disease outcomes. Ongoing studies to define potential similarities with and differences between innate and adaptive immune responses will continue to teach us important lessons about the complexity of the gastrointestinal tract. Such information has provided new insights into disease pathogenesis and, importantly, will provide insights into new therapeutic targets.

INTRODUCTION

The primary functions of the gastrointestinal tract are the processing of ingested nutrients, waste removal, fluid homeostasis, and the development of oral tolerance to nonpathogenic antigens. These dynamic processes occur in conjunction with the constant flux of new antigenic material and require that the mucosal immune system appropriately dampen inflammatory and immunological reactions to harmless ingested antigens.

The intestinal epithelium lies juxtaposed to the mucosal immune system and lines the entire gastrointestinal tract. Covering a surface area of approximately 300 m² in the adult human, the intestinal epithelium consists of a monolayer of cells with intercellular tight junctions, a complex three-dimensional structure, and a thick mucous gel layer and provides a dynamic and regulated barrier to the flux of the luminal contents to the lamina propria (1, 2). The gastrointestinal tract exists in a state of low-grade inflammation. Such a state results from the constant processing of luminal antigenic material during the development of oral tolerance and the priming of the mucosal immune system for rapid and effective responses to antigens or microbes that may penetrate the barrier. The anatomy and function of the intestine provide a fascinating oxygenation profile, whereby even under physiological conditions, the intestinal mucosa experiences profound fluctuations in blood flow, oxygenation, and metabolic shifts. For example, less than 5% of total blood volume resides in the intestine (small and large combined) during fasting. This proportion can increase to as much as 30% of the cardiac output following ingestion of a large meal. Such changes in blood flow result in marked shifts in local partial pressure of oxygen (pO_2), and it is perhaps not surprising that the epithelium has evolved a number of features to cope with such large metabolic fluxes. Studies comparing functional responses between epithelial cells from different tissues have revealed that intestinal epithelial cells (IECs) seem to be uniquely resistant to hypoxia and that low pO_2 within the normal intestinal epithelial barrier (so-called physiological hypoxia) may be a regulatory adaptation mechanism in response to the steep O_2 gradient (3). Thus, the availability of O_2 in both health and disease regulates both the absorptive and barrier properties of the intestinal epithelium (4). Here, we discuss the signaling pathways involved in adaptation to hypoxia, with a particular focus on adenine nucleotide metabolism and signaling.

HYPOXIA AND THE MUCOSAL IMMUNE RESPONSE

Sites of mucosal inflammation are characterized by profound changes in tissue metabolism, including local depletion of nutrients, imbalances in tissue O_2 supply and demand, and the generation of large amounts of adenine nucleotide metabolites (5, 6). As shown in **Table 1**, the local generation of these adenine nucleotide metabolites is driven largely by hypoxia through mechanisms involving the transcriptional regulator hypoxia-inducible factor (HIF) (see below).

These inflammation-associated changes in metabolism can be attributed, at least in part, to the initial recruitment of cells of the innate immune system, including myeloid cells such as neutrophils [polymorphonuclear leukocytes (PMNs)] and monocytes. PMNs are recruited by chemical signals generated at sites of active inflammation as part of the innate host immune response to microorganisms. In transit, these cells expend tremendous amounts of energy. Large amounts of ATP, for example, are needed for the high actin turnover required for cell migration (7). Once PMNs reach inflammation sites, the nutrient, energy, and O_2 demands of the PMNs increase to accomplish the processes of phagocytosis and microbial killing. PMNs are primarily glycolytic cells, with few mitochondria and little energy produced from respiration (8). A predominantly glycolytic metabolism ensures that PMNs can function at the low O_2 concentrations (even anoxia) associated with inflammatory lesions.

IECs: intestinal epithelial cells

HIF: hypoxia-inducible factor

PMNs: polymorphonuclear leukocytes

Table 1 Hypoxia and HIF targets in adenine nucleotide metabolism and signaling[a]

Gene product	Function	Regulation	Reference
CD39	ATP/ADP → AMP	Sp1	101
CD73	AMP → Ado	HIF-1	78
A_1AR	Signaling	Unknown	146
$A_{2A}AR$	Signaling	HIF-2	114
$A_{2B}AR$	Signaling	HIF-1	113
ENT-1	Ado transport	HIF-1	147
Ado kinase	Ado → AMP	HIF-1	83
Ado deaminase	Ado → inosine	Unknown	82

[a]Abbreviations: Ado, adenosine; A_1AR, adenosine A_1 receptor; $A_{2A}AR$, adenosine A_{2A} receptor; $A_{2B}AR$, adenosine A_{2B} receptor; CD39, ecto-nucleoside triphosphate diphosphohydrolase 1; CD73, ecto-5'-nucleotidase; ENT, equilibrative nucleoside transporter; HIF, hypoxia-inducible factor.

Once at the inflammatory site, PMNs recognize and engulf pathogens and activate the release of antibacterial peptides, proteases, and reactive oxygen species (ROS) (e.g., superoxide anion, hydrogen peroxide, hydroxyl radical, and hypochlorous acid) into the vacuole, which together kill the invading microbes (9). ROS are produced by phagocytes in a powerful oxidative burst that is driven by a rapid increase in O_2 uptake and glucose consumption, which in turn triggers further generation of ROS. When activated, PMNs can consume up to 10 times more O_2 than can any other cell in the body. Notably, the PMN oxidative burst is not hindered by even relatively low O_2 (as low as 4.5% O_2) (10), which is important in that ROS can be generated in the relatively low O_2 environments of inflamed intestinal mucosa (4).

In contrast to cells of the myeloid lineage, T and B cells utilize glucose, amino acids, and lipids as energy sources during oxidative phosphorylation. Mitogenic stimulation of thymocytes and naive T cells is a highly energy-demanding process. As lymphocytes proliferate, they become more and more dependent on glucose uptake. Stimulated proliferation of thymocytes can result in nearly 20-fold increases in glucose uptake, which is accomplished by high expression of glucose transporter-1 (11) and is tightly controlled by HIF (see below). For example, IL-7- and IL-4-dependent pathways instruct nutrient uptake in naive T cells (12). During periods of high proliferation, even in the presence of adequate O_2 concentration, lymphocytes shift toward aerobic glycolysis for ATP synthesis, and lactate production from glycolysis can increase by as much as 40-fold in mitogen-stimulated T cells (6). When glucose becomes limiting, as it often does at sites of high immune activity, T cells can utilize alternative energy sources, such as glutamine, within the TCA cycle (13). In the past 10 years we have begun to understand the nature of interactions between microenvironmental metabolic changes and the generation of recruitment signals and molecular mechanisms of leukocyte migration into these areas. The metabolic changes that occur as a result of the recruitment and activation of leukocytes during inflammation provide information about the potential sources of hypoxia at the intestinal epithelial barrier.

INFLAMMATORY BOWEL DISEASE AS A MODEL DISEASE FOR INFLAMMATORY HYPOXIA AND NUCLEOTIDE METABOLISM

Inflammatory bowel disease (IBD), such as Crohn's disease and ulcerative colitis, is an interesting disease in which to study the metabolic events and metabolism of adenine nucleotides associated with inflammation, particularly the development of severe hypoxia within inflammatory lesions (4).

IBD: inflammatory bowel disease

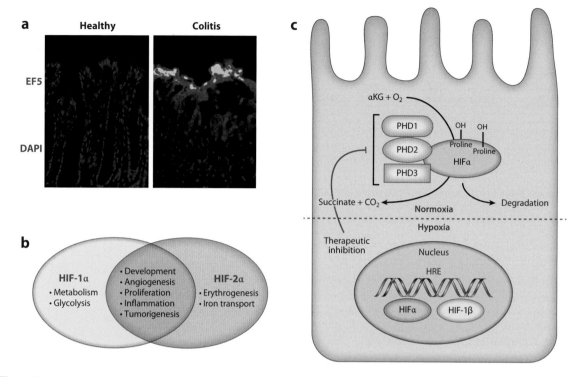

Figure 1

Localization of hypoxia and mechanism of hypoxia-inducible factor (HIF) stabilization. (*a*) Tissue sections from healthy control or trinitrobenzene sulfonic acid (TNBS) colitis were examined for localization of the 2-nitroimidazole compound 2-(2-nitro-1H-imidazol-1-yl)-*N*-(2,2,3,3,-pentafluoropropyl) acetamide (EF5) (*red*). Nuclear counterstaining with 4′,6-diamidino-2-phenylindole (DAPI) is shown in blue. Hematoxylin and eosin staining of an adjacent section is shown for orientation. (*b*) General features of HIF. Overlapping functions of HIF-1α and HIF-2α include the regulation of tissue development, vascular angiogenesis, cell proliferation, and multiple inflammation targets and genes that promote or suppress tumorigenesis. Specific target genes for HIF-1α include those genes involved primarily in metabolism, including genes within the glycolytic pathways. HIF-2α-specific targets include erythropoietin and gene targets involved in duodenal iron transport. (*c*) The biochemical pathway of HIF hydroxylation by the combination of α-ketoglutarate (αKG), molecular oxygen (O_2), and the prolyl hydroxylase (PHD) enzymes in normoxia. When O_2 or prolyl hydroxylation becomes limiting (due to hypoxia or PHD inhibition), the HIF-1α subunit and/or HIF-2α subunit stabilize and bind to the HIF-1β subunit within the nucleus, where the dimer becomes transcriptionally active upon binding to the hypoxia-response element (HRE) DNA consensus sequence.

Microvascular abnormalities that describe IBD patients have been associated with abnormal blood flow to the intestine, including increased production of tissue vasoconstrictor molecules and the reduced generation of nitric oxide by endothelial cells (14), as well as vascular endothelial growth factor–dependent pathological angiogenesis (15). In addition, studies of active inflammation in mouse models of IBD have shown the intestinal epithelial cell to be a primary target of hypoxia (16).

Prominent epithelial hypoxia in murine models of IBD was revealed using 2-nitroimidazole dyes, a class of compounds that undergo intracellular metabolism, depending on tissue oxygenation (**Figure 1**) (17). Tissue staining with these nitroimidazole dyes revealed two profound observations. First, in the healthy colon, physiological hypoxia predominates (**Figure 1**). These dyes adduct with proteins, peptides, and amino acids at a pO_2 of less than 10 mm Hg, and therefore such low O_2 levels may regulate basal gene expression in otherwise healthy IECs (5). Second, the inflammatory lesions seen in these mouse models are profoundly hypoxic or even anoxic, with levels of oxygen similar to those seen in some large tumors, and penetrate deep into the mucosal tissue. Multiple contributing factors, such as vasculitis, vasoconstriction, edema, and increased O_2 consumption, likely predispose the inflamed intestinal epithelia to decreased O_2 delivery and hypoxia (16). Although these 2-nitroimidazole compounds have not been used clinically to image inflammatory lesions, they have shown significant clinical utility in tumor imaging and in the identification of stroke regions within the brains of patients (18). As opposed to other imaging techniques, these compounds have the advantages that they image only viable tissue and are not active within necrotic or apoptotic regions of the tissue (19). Likewise, studies are under way to use these compounds as adjunct radiosensitizers for enhancing chemotherapy targeting (20).

CD73: ecto-5′-nucleotidase

$A_{2B}AR$: adenosine A_{2B} receptor

HYPOXIA-INDUCIBLE FACTOR

The studies that identified inflammation-related hypoxia also revealed stabilization of HIF-1α within the inflammatory lesions (16). Many cell types, including IECs (21), express both HIF-1α and HIF-2α, and murine genetic studies suggest that these proteins have nonredundant roles (22). Some studies suggest that distinct transcriptional responses mediated by HIF-1α and HIF-2α may be integrated in ways that support particular adaptations to hypoxia. For example, the transcriptional responses that coordinate the glycolytic pathways include more than 11 target genes and seem to be more selective for the HIF-1α isoform than for the HIF-2α isoform (**Figure 1**) (22). Likewise, studies addressing the selectivity of the two isoforms for erythropoietin induction suggest a more prominent role for HIF-2α (22). Currently, this specificity is not well understood. Some studies indicate that binding of HIF-1α or HIF-2α to other transcription factors at the site of DNA binding may determine such specificity (22), but findings are not conclusive.

Several studies demonstrate that HIF triggers the transcription of a number of genes that enable IECs to act as an effective barrier. Originally guided by microarray analysis of hypoxic IECs (23), these studies were validated in animal models of intestinal inflammation (16, 24–28) and in human intestinal inflammation tissues (29–31). Interestingly, the functional proteins encoded by a number of uniquely hypoxia-inducible genes in intestinal epithelia localize primarily to the most luminal aspect of polarized epithelia, providing significant support for the hypothesis that hypoxia supports a barrier-protective phenotype. Molecular studies of this hypoxia-elicited pathway(s) show a dependency on HIF-mediated transcriptional responses. Notably, epithelial barrier–protective pathways driven by HIF tend not to be the classical regulators of barrier function, such as the tight junction proteins occludin or claudins. Rather, the HIF-regulated molecules include molecules that support overall tissue integrity and promote increased mucin production (32), that modify mucin (e.g., intestinal trefoil factor) (3), that promote xenobiotic clearance via P-glycoprotein (33), that enhance nucleotide metabolism [by ecto-5′-nucleotidase (CD73)] (23, 34), and that drive nucleotide signaling [e.g., adenosine A_{2B} receptor ($A_{2B}AR$)] (34).

As an extension of the original studies identifying HIF stabilization within the intestinal mucosa, Karhausen et al. (16) generated transgenic mice expressing either mutant Hif1-α (causing constitutive repression of *Hif1-α*) or mutant von Hippel–Lindau (causing constitutive overexpression of HIF) targeted to IECs. Loss of epithelial HIF-1α resulted in a more severe colitic phenotype than in wild-type animals, including increased epithelial permeability, enhanced loss of body weight, and decreased colon length. Constitutively active intestinal epithelial HIF (mutant von Hippel–Lindau) was protective for each of these individual parameters. These findings may be somewhat model dependent because another study found that epithelial HIF–based signaling also promotes inflammation (28). Nonetheless, the findings of Karhausen et al. (16) confirmed that IECs can adapt to hypoxia and that HIF may contribute to this adaptation.

PHD: prolyl hydroxylase

Nonepithelial cell types within the gastrointestinal mucosa have also been studied for HIF expression and response to hypoxia. Activated T cells show increased expression of HIF-1α, which prevents them from undergoing activation-induced cell death in hypoxic settings. T cell survival in hypoxia is, at least in part, mediated by the vasoactive peptide adrenomedullin (35). Other studies using chimeric mice bearing HIF-1α-deficient T and B cells have revealed lineage-specific defects that result in increased autoimmunity, including autoantibodies, increased rheumatoid factor, and kidney damage (36). HIF function has also been studied in some detail in myeloid cells. Cre-*LoxP*-based elimination of HIF-1α in cells of the myeloid lineage (lysozyme M promoter) has revealed multiple features that importantly implicate metabolic control of myeloid function (37). In particular, these studies show that PMN and macrophage bacterial killing capacities are severely limited in the absence of HIF-1α, as HIF-1α is central to the production of antimicrobial peptides and granule proteases. These findings are explained, at least in part, by the inability of myeloid cells to mount appropriate metabolic responses to diminished O_2 characteristic of infectious sites (37). Finally, compelling evidence has revealed that HIF-1α transcriptionally controls the critical integrin important in all myeloid cell adhesion and transmigration, namely, the β2 integrin (CD18) (38). Such findings are important for our current understanding of the role of functional PMNs in IBD. A recent study, for example, used PMN depletion techniques to document a central role for PMNs in the resolution of inflammation in several murine IBD models (39).

OXYGEN-SENSING PROLYL HYDROXYLASES AND HYPOXIA-INDUCIBLE FACTOR EXPRESSION

The molecular mechanisms of HIF stabilization have been clarified over the past decade. Three prolyl hydroxylases (PHDs) termed PHD1, PHD2, and PHD3 as well as factor-inhibiting HIF (FIH) contribute to HIF pathway regulation (40). As depicted in **Figure 1**, these hydroxylases are encoded by different genes, and their gene product enzymes demonstrate tissue-specific expression patterns (40). All three PHDs and FIH are found in the intestinal epithelium (24, 27, 41). Genetic studies have revealed significantly different phenotypes in mice lacking the individual isoforms of the PHDs. For example, studies in $Phd1^{-/-}$ mice have revealed differences in basal metabolic profiles and decreased exercise performance (42). Disease models have demonstrated that these animals are protected against acute liver ischemia, muscle ischemia, and dextran sodium sulfate (DSS)-induced colitis (42–44). Homozygous PHD2 deletion is embryonic lethal due to developmental angiogenesis defects (45, 46). PHD2 heterozygous knockout animals show enhanced tumor angiogenesis but decreased metastasis (45). $Phd3^{-/-}$ mice manifest reduced neuronal apoptosis, abnormal sympathoadrenal development, and reduced blood pressure (47). These diverse phenotypes strongly suggest distinct isoform-specific functions in vivo.

In the presence of 2-oxoglutarate, Fe^{2+}, and molecular O_2, PHDs hydroxylate the α subunit of HIF and lead to subsequent ubiquitination and degradation via the proteasome. Hypoxia or pharmacological agents (such as DMOG) inhibit HIF hydroxylases that lead to HIF stabilization. The impact of HIF hydroxylase inhibitors on epithelial cell gene expression is not restricted to regulation via HIF. For example, HIF hydroxylases can regulate nuclear factor-κB (NF-κB) (40, 48). The transcriptional targets of HIF hydroxylases can impact epithelial barrier function in a number of ways. For example, HIF regulates the expression of a family of barrier-protective factors, including intestinal trefoil factor (3), the mucins (32), and actin cytoskeletal cross-linkers (49). Likewise, NF-κB is thought to be largely protective in the intestinal epithelium via the inhibition of enterocyte apoptosis (24).

OXYGEN-SENSING PROLYL HYDROXYLASES IN THE CONVERGENCE OF HYPOXIA AND INFLAMMATION

The O_2-dependent regulatory role of PHDs is not restricted to HIF stabilization. For example, NF-κB is activated during inflammation and may interact in fundamental ways with the HIF pathway. NF-κB consists of either homodimers or heterodimers that, on activation, translocate to the nucleus and bind with the transcriptional coactivator CBP/p300 to begin transcription or repression of various genes. The inhibitory IκB proteins regulate NF-κB activity (50). The best-studied complex is IκBα bound to the NF-κB p50-p65 dimer (50). This interaction with IκBα inhibits NF-κB from binding to DNA and maintains the complex in the cytoplasm. On activation by various extracellular signals, IκB kinase (IKK) is activated, resulting in phosphorylation (51) and polyubiquitylation of IκBα (52). The S26 proteasome then selectively degrades polyubiquitinated IκBα. Once dissociated from IκBα, NF-κB rapidly enters the nucleus and activates gene expression.

Recent studies have indicated that PHDs also regulate NF-κB-dependent pathways and that PHD inhibitors for murine colitis also target the NF-κB pathway. Indeed, hypoxia activates NF-κB, and such activation is, at least in part, dependent on PHD-mediated hydroxylation of IKKβ (41, 53). In normoxia, IKKβ activity is held in check through LXXLAP-dependent hydroxylation by PHD1 and PHD2 (41). Conditional deletion of the NF-κB pathway in IECs in mice leads to an increased susceptibility to colitis (54), a phenotype similar to that of the mice expressing homozygous mutant HIF-1α (16). This implicates epithelial NF-κB in a prominently protective role in colitis, probably through the expression of antiapoptotic genes in IECs and through enhanced epithelial barrier function. Some studies suggest that both the HIF and NF-κB pathways may also be influenced by mediators found within inflammatory sites, including microbial products, cytokines, and even intact bacteria (37). NF-κB is a classic transcriptional regulator activated by a spectrum of agonists, the activation of which drives a complex series of receptor-mediated signaling pathways. Recent studies indicate that NF-κB-mediated signaling activates HIF-1α transcription (55). Inflammation-associated upregulation of HIF-1α mRNA occurs in an NF-κB-dependent manner (55). HIF-1 may also promote increased NF-κB activity in hypoxia (48). Thus, a cross-regulatory loop may exist between the HIF and NF-κB pathways and may involve other transcriptional regulators that bear nonredundant PHD sensitivity, including activating transcription factor-4 and Notch (56, 57), both critical regulators of cell fate. Given that IECs are in an environment with constant exposure to potentially inflammatory stimuli, the cross-regulation of HIF and NF-κB may have profound implications for intestinal epithelial cell function and survival under both homeostatic and disease conditions.

The identification of HIF-selective PHDs has provided unique opportunities for the development of PHD-based therapies (58, 59). Although there is wide interest in developing HIF-1 inhibitors as potential cancer therapies, opportunities also exist to selectively stabilize HIF in an attempt to promote inflammatory resolution in IBD. For example, 2-OG analogs effectively stabilize HIF-α (58). Although this approach is not selective for particular PHD isoforms, some in vitro studies suggest that marked differences in substrate specificity may exist and could be harnessed for selectivity. For example, all PHDs hydroxylate the C-ODD domain more efficiently than the N-ODD domain, and PHD2 hydroxylates the N-ODD domain less efficiently on HIF-2α than on HIF-1α. In addition, PHD3 does not hydroxylate the N-ODD domain of HIF-1α (60, 61). Additionally, the protection afforded by PHD inhibitors (e.g., decreased tissue inflammatory cytokines, increased barrier function, decreased epithelial apoptosis) may involve both HIF and NF-κB activity.

A₁AR: adenosine A₁ receptor

Given the central role of HIF-mediated signaling in erythropoietin production, PHD inhibitors have been developed and are in clinical trials for the treatment of anemia (62). Investigators have developed several PHD inhibitors, including direct inhibitors, analogs of naturally occurring cyclic hydroxamates, and antagonists of α-ketoglutarate (5). Each of these molecules serves as a competitive PHD inhibitor through the substitution for α-ketoglutarate in the hydroxylation reaction shown in **Figure 1**. Within the gastrointestinal tract, the PHD inhibitors DMOG and FG-4497 have been used effectively to reduce symptoms in at least two mouse models of colitis (24, 27). Indeed, these studies show that both DMOG and FG-4497 have an overall beneficial influence on multiple parameters studied in chemically induced trinitrobenzene sulfonic acid (TNBS) or DSS mouse models of colitis. In these mouse models, the drugs were well tolerated, with no significant adverse side effects. In our experience, both FG-4497 and DMOG can be delivered by multiple routes of administration (intraperitoneal, oral, and intravenous), and both FG-4497 and DMOG are absorbed orally, with only a slight loss of efficacy compared with intraperitoneal administration. To date, although no human trials have been initiated for the treatment of IBD, a recently published proof-of-principle study in end-stage renal disease demonstrates the efficacy of PHD inhibitors in elevating the HIF target erythropoietin (63).

ADENOSINE: A BIOMEDICAL DISCOVERY STORY DATING BACK TO THE 1920s

Although this review focuses on the contribution of hypoxia and adenine nucleotide metabolism to intestinal disease, the original discovery of adenosine as a signaling molecule stems from a completely different mindset. In fact, the first description of adenosine signaling dates back almost 90 years. In 1927, two scientists from the University of Cambridge, United Kingdom, were the first to observe specific signaling by adenosine. In their studies, Drury & Szent-Gyorgyi (64) performed intravascular injections of an extract derived from cardiac tissues in an intact animal. They were somewhat surprised to observe a robust and transient slowing of the heart rate upon intravascular injection of this tissue extract (64). Utilizing what were then the state-of-the-art chemical purification methods of the early twentieth century, the authors identified the biological activity within the cardiac tissue extract as an "adenine compound" (64). Following this groundbreaking discovery, almost 50 years passed until the heart rate–slowing influences of extracellular adenosine were translated from bench to bedside. In the 1980s, the heart rate–slowing influence of adenosine was considered for the treatment of patients with supraventricular tachycardia (a disturbance of cardiac rhythm) (65–68). Intravenous treatment with a bolus of adenosine results in a transient complete heart block, leading to a complete standstill of the heart for typically 5 to 10 s. When adenosine is cleared, the heart rate recovers, and if treatment is successful, a normal sinus rhythm prevails (66). Today, studies in gene-targeted mice for individual adenosine receptors (ARs) provide convincing evidence that the transient slowing of the heart rate is due to the activation of adenosine A_1 receptors (A_1ARs) expressed on cardiac tissues (69–71). Other clinical applications of the direct or indirect influences of adenosine include its role as an arterial vasodilator during pharmacologically induced stress echocardiography (72) or as an inhibitor of platelet aggregation (73, 74). Moreover, the nonspecific AR antagonist caffeine has been used to treat headaches, whereas the AR antagonist theophylline has been used to treat obstructive airway disease (75). Thus, adenosine-based medical therapy plays an important role in the treatment of medical or surgical patients. The more recent discoveries of adenosine generation and signaling as potential therapeutic targets for the treatment of inflammatory diseases are less well developed as therapies for human diseases.

STRUCTURAL AND BIOLOGICAL RELATIONSHIP BETWEEN EXTRACELLULAR NUCLEOTIDES AND ADENOSINE

Adenosine belongs to the chemical group of nucleosides and is structurally composed of the purine-based nucleobase adenine bound to a ribose sugar moiety via a β-N_9-glycosidic bond. In the extracellular compartment, adenosine is generated predominantly through the phosphohydrolysis of extracellular nucleotides, particularly ATP or ADP. Enzymatic conversion of ATP or ADP by the ecto-nucleoside triphosphate diphosphohydrolase (E-NTPDase) 1 (CD39) to AMP and subsequent conversion of AMP to adenosine by CD73 represent the major pathway for extracellular adenosine generation (76–78). Extracellular adenosine can activate any of four distinct ARs (A_1AR, A_{2A}AR, A_{2B}AR, or A_3AR), and signaling is terminated by the relatively short half-life of extracellular adenosine. Passive transport from the extracellular compartment into the intracellular space through adenosine transporters is responsible for the short half-life of adenosine in circulation (68, 79–81). Within the cytosol, adenosine is deaminated to inosine by adenosine deaminase or is rephosphorylated to AMP by adenosine kinase (82, 83). As alluded to above and as depicted in **Table 1**, hypoxia directly influences many aspects of this pathway.

E-NTPDase: ecto-nucleoside triphosphate diphosphohydrolase

CD39: ecto-nucleoside triphosphate diphosphohydrolase 1

A_{2A}AR: adenosine A_{2A} receptor

A_3AR: adenosine A_3 receptor

SOURCES OF NUCLEOTIDES DURING INJURY

During inflammation or hypoxia, a number of cell types actively release adenine nucleotides, particularly in the form of ATP or ADP (68, 80, 84). Likewise, as the intracellular concentrations of ATP are high (approximately 5–7 mM), cellular necrosis, lysis, or programmed cell death (apoptosis) is associated with the liberation of large amounts of ATP. For example, recent studies investigated the role of ATP released by apoptotic cells as a so-called find-me signal for promoting phagocytic clearance (85, 86). In this context, a very elegant study provided several lines of evidence for extracellular nucleotides as a critical apoptotic cell find-me signal. Enzymatic removal of ATP (by apyrase or the expression of ectopic CD39) abrogated the ability of apoptotic cell supernatants to recruit monocytes in vitro and in vivo. Subsequent studies identified the ATP receptor $P2Y_2$ as a critical sensor of nucleotides released by apoptotic cells. The results from this study pinpointed nucleotides as a critical find-me cue released by apoptotic cells to promote $P2Y_2$-dependent recruitment of phagocytes (86). Interestingly, an additional study identified (*a*) a specific mechanism of nucleotide release from apoptotic cells, (*b*) the plasma membrane channel pannexin 1 (PANX1) as a mediator of find-me signal and nucleotide release from apoptotic cells and selective plasma membrane permeability during apoptosis, and (*c*) a new mechanism of PANX1 activation by caspases (85).

Other studies have investigated the contributions of inflammatory cells to extracellular nucleotide release. Given the association of neutrophils (PMNs) with adenine nucleotide/nucleoside signaling in the inflammatory milieu, we hypothesized that PMNs may represent an important source of extracellular ATP (87, 88). Initial studies using high-performance liquid chromatography (HPLC) and luminometric ATP detection assays revealed that PMNs release ATP through activation-dependent pathways. After excluding lytic ATP release, pharmacological strategies revealed a mechanism involved in PMN ATP release via connexin 43 (Cx43) hemichannels. Cx43 molecules assemble as hexadimers (so-called connexons) that form junctional complexes between different cell types. More recently, and in addition to their role as gap-junctional proteins, studies implicate Cx43 connexons as intercellular signaling channels via ATP release (89, 90). In the above studies defining ATP release from human PMNs (87, 88), the authors confirmed their findings in PMNs derived from induced $Cx43^{-/-}$ mice, whereby activated PMNs released less than 15% of ATP relative to littermate controls and Cx43 heterozygote PMNs were intermediate in their capacity for ATP release. This study implicated Cx43 in activated PMN ATP release, therein contributing to the innate metabolic control of the inflammatory milieu (87). Subsequent studies by others revealed that human neutrophils release ATP predominantly from the leading edge of their cell surface as a mechanism to amplify chemotactic signals and direct cell orientation by feedback through $P2Y_2$ nucleotide receptors (91, 92).

An additional source of extracellular ATP release comes from platelets, which release nucleotides at high concentrations upon activation by ADP or collagen via dense granule release (93). In this context, a recent study highlighted the interaction between PMNs and platelets in regulating intestinal inflammation and fluid transport via nucleotide release (94). Mucosal diseases are often characterized by a mixed inflammatory infiltrate that includes PMNs and platelets. These studies showed that platelets migrate across intestinal epithelia in a PMN-dependent manner. Furthermore, platelet-PMN comigration occurred in intestinal tissue derived from human patients with IBD. The translocated platelets release large quantities of ATP, which is metabolized to adenosine via a two-step enzymatic reaction involving CD73 and CD39-like molecules expressed on IECs. Subsequent studies revealed a mechanism involving adenosine-mediated activation of electrogenic chloride secretion, with concomitant water movement into the intestinal lumen, originally described by Madara et al. (95). Together, these studies demonstrated that E-NTPDases are expressed on IECs and interact with platelet-derived nucleotides through a mechanism involving platelets that piggy back across mucosal barriers while attached to the surface of PMNs (**Figure 2**) (94).

HYPOXIA PROMOTES THE PHOSPHOHYDROLYSIS OF EXTRACELLULAR ATP/ADP TO AMP

As discussed above, there is strong evidence that mucosal inflammation is characterized by the release of extracellular nucleotides from multiple cell types. Extracellular nucleotides either activate extracellular ATP receptors or are rapidly converted to adenosine in a two-step enzymatic process, including phosphohydrolysis to AMP and the conversion of AMP to adenosine. This metabolic pathway can be readily detected through the use of non-native etheno-nucleotide derivatives using HPLC (**Figure 3**). AMP generation from ATP/ADP is achieved mainly enzymatically by the E-NTPDases, a recently described family of ubiquitously expressed membrane-bound enzymes (76, 96). The catalytic sites of plasma membrane–expressed E-NTPDases 1–3 and 8 are exposed to the extracellular milieu, whereas the other E-NTPDases are intracellular (76). The presumptive biological role of plasma membrane–bound E-NTPDases is to fine-tune extracellular nucleotide levels. For example, E-NTPDase 1 (CD39) plays an important role in vascular endothelial function by blocking platelet aggregation via the phosphohydrolysis of ATP and ADP from the blood to maintain vascular integrity (97–99). At the same time, E-NTPDase 1 is also important in the maintenance of platelet functionality by preventing platelet $P2Y_1$ receptor desensitization. As such, mice gene targeted for E-NTPDase 1 ($Cd39^{-/-}$ mice) show prolonged bleeding time with minimally perturbed coagulation parameters (100) and increased vascular permeability as measured by Evan's blue dye extravasation (**Figure 4**) (3, 23, 33, 34).

Several studies have provided evidence that the CD39 transcript, protein, and function are under the direct control of hypoxia-dependent signaling pathways. The first evidence comes from two studies that subjected vascular endothelia or intestinal epithelia to hypoxia and observed robust increases in the expression and function of CD39 (and CD39-like molecules) (78, 88). To define these molecular principles, a recent study determined whether the human CD39 (hCD39) promoter was hypoxia responsive (101). In view of the likelihood of transcription-mediated induction of CD39 during hypoxia, attention was directed to the 5′-UTR for potential hypoxia-regulated

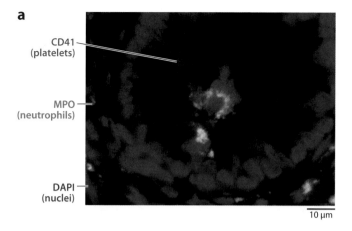

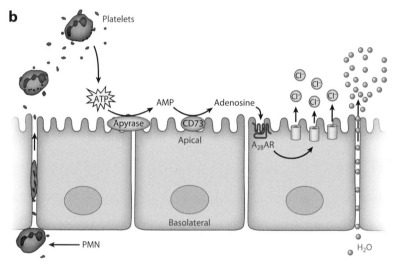

Figure 2

Platelet–polymorphonuclear leukocyte (PMN) cotransmigration in crypt abscesses from human inflammatory bowel disease. (*a*) A merged fluorescence image localizing PMNs in green [anti-myeloperoxidase (MPO)], platelets in red (anti-CD41), and nuclei 4′,6-diamidino-2-phenylindole (DAPI) stain in blue from a human ulcerative colitis crypt abscess. (*b*) A model of facilitated platelet translocation and activation of epithelial electrogenic Cl⁻ secretion during PMN transmigration. During active inflammation, platelets are caught in the flow of PMN transmigration. PMN- and platelet-derived ATP is selectively metabolized to adenosine by a two-step enzymatic reaction involving CD39 and CD73. Adenosine binding to apical adenosine A_{2B} receptors (A_{2B}AR) results in activation of electrogenic Cl⁻ secretion and the paracellular movement of water.

transcription factor sequences. 5′-RACE (5′-rapid amplification of cDNA ends) results confirmed the findings of Maliszewski et al. (102), who identified the transcription start site at 81 bp upstream of the start codon. Sequence comparison of this region in the hCD39 and mouse CD39 promoter regions revealed only a single-base-pair difference and an identical expression profile for both specificity protein-1 (Sp1) and GATA3 binding sites, suggesting that this region is highly conserved. Subsequent studies with site-directed mutagenesis of the central transcription factor

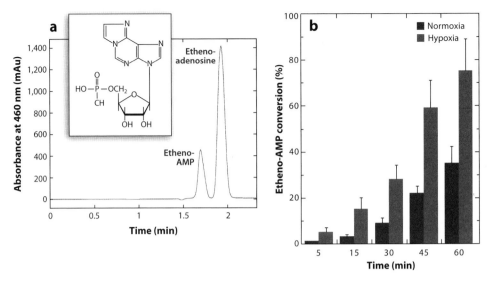

Figure 3

Biochemical analysis of intestinal epithelial CD73 activity. (*a*) A representative high-performance liquid chromatography (HPLC) tracing demonstrating peak resolution between etheno-AMP (structure shown in *inset*) and etheno-adenosine. (*b*) T84 intestinal epithelial monolayers were exposed to normoxia (pO$_2$ 147 torr) or hypoxia (pO$_2$ 20 torr) for 36 h and washed. Surface CD73 activity was determined by HPLC analysis of etheno-AMP conversion to etheno-adenosine at the indicated sampling times.

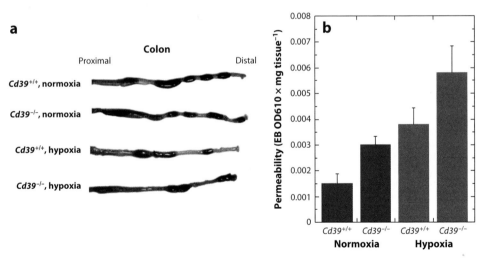

Figure 4

Analysis of colonic vascular leak in CD39-deficient (*Cd39*$^{-/-}$) mice. (*a*) Wild-type (*Cd39*$^{+/+}$) and *Cd39*$^{-/-}$ mice were administered intravenous Evans blue (EB) solution (0.2 ml of 0.5% in phosphate-buffered solution) and were exposed to normobaric hypoxia (8% O$_2$, 92% N$_2$) or room air for 4 h. Animals were sacrificed, and their colons were harvested and imaged. (*b*) A plot of extracted EB, expressed as optical density (OD) at 610 nm mg^{-1} tissue.

binding site in Sp1 or GATA-3 in the hCD39 promoter and analysis of hypoxia inducibility revealed that only Sp1 contributes to this response (101).

As CD39 hydrolyzes nucleotides to generate adenosine, another pathway examined the role of hypoxia-induced CD39 in inflammatory bowel disease. Here, Friedman et al. (103) hypothesized that CD39 might protect against IBD. They studied these possibilities in a mouse model of colitis using mice with global CD39 deletion and tested whether human genetic polymorphisms in the CD39 gene might influence susceptibility to Crohn's disease. $Cd39^{-/-}$ mice were highly susceptible to chemically induced colitis; heterozygote mice showed an intermediate phenotype. Moreover, these researcherss identified a common single-nucleotide polymorphism (SNP) that tags CD39 mRNA expression levels in humans. The SNP tagging low levels of CD39 expression was associated with increased susceptibility to Crohn's disease in a case-control cohort composed of 1,748 Crohn's patients and 2,936 controls. These data indicate that CD39 deficiency exacerbates murine colitis and suggest that CD39 polymorphisms are associated with IBD in humans (103). Other studies identified CD39 as a specific marker for regulatory T cells and implicate CD39-dependent ATP/ADP breakdown in autocrine enhancement of the anti-inflammatory functions of this group of T cells (104).

HYPOXIA-INDUCIBLE FACTOR CONTROLS ADENOSINE GENERATION BY REGULATING ECTO-5′-NUCLEOTIDASE (CD73)

In the extracellular compartment, AMP phosphohydrolysis to adenosine is achieved primarily by the ecto-enzyme CD73, which is the pacemaker enzyme for extracellular adenosine production (77). CD73 is bound to the extracellular compartment of the membrane via a glycosylphosphatidylinositol anchor (77). This anatomic localization within the extracellular membrane and orientation toward the extracellular compartment would allow CD73 to be released from the cell membrane during injurious conditions. However, the function of circulating CD73 and its potential as a biomarker of human disease have yet to be established. CD73 is expressed ubiquitously, with the highest expression levels in the intestine (105). Consistent with this notion, several studies have implicated CD73 in dampening hypoxia-elicited inflammation of the intestine. For instance, pharmacological inhibition or gene-targeted deletion of CD73 is associated with intestinal or vascular permeability dysfunction upon exposure of mice to ambient hypoxia (8% over 4 h) (78, 105). During experimental colitis induced by the hapten TNBS, $Cd73^{-/-}$ mice developed a more severe phenotype (106). Cytokine profiling revealed similar increases in both interferon (IFN)-γ and tumor necrosis factor-α mRNA in colitic animals, independent of genotype. However, IL-10 mRNA increased in wild-type mice on day 3 after TNBS administration, whereas $Cd73^{-/-}$ mice mounted no IL-10 response. This IL-10 response was restored in the $Cd73^{-/-}$ mice by exogenous IFN-αA. Further cytokine profiling revealed that a transient IFN-αA induction precedes IL-10 induction. Together, these studies indicate a critical regulatory role for CD73-modulated IFN-αA in the acute inflammatory phase of TNBS colitis, thereby implicating IFN-αA as a protective element of adenosine signaling during mucosal inflammation (106). Other studies have revealed a protective role for CD73-dependent adenosine generation during intestinal ischemia-reperfusion injury (107) and in hypoxic preconditioning. These studies highlight an important role of CD73 in dampening inflammatory responses in the context of tissue hypoxia.

To dissect the direct consequences of hypoxia on CD73 expression and function, a study exposed IECs to hypoxia and observed robust induction of CD73 transcript, protein, and enzymatic activity (78). Examination of the CD73 gene promoter identified at least one binding site for HIF-1, and the inhibition of HIF-1α expression by antisense oligonucleotides resulted in significant inhibition of hypoxia-inducible CD73 expression. Studies using luciferase reporter constructs revealed a significant increase in activity in cells subjected to hypoxia; no such increase was seen in truncated and mutated constructs lacking a functional HIF binding site. In vivo studies in a murine hypoxia model revealed that hypoxia-induced CD73 may protect the epithelial barrier because the CD73 inhibitor αβ-methylene ADP promoted increased intestinal permeability. These results identify a HIF-1-dependent regulatory pathway for CD73 and indicate that CD39 and CD73 protect epithelial barrier function during hypoxia. Studies of intestinal ischemia-reperfusion injury demonstrate that gene-targeted mice for HIF-1α suffer from a more severe phenotype that is associated with attenuated CD73 levels (108). Conversely, treatment with the pharmacological HIF activator DMOG provides potent protection from intestinal ischemia-reperfusion injury in wild-type mice but is ineffective in $Cd73^{-/-}$ mice (108).

A recent study identified mutations in the gene encoding CD73 (*5NTE*) that cause human disease (109). This study revealed a severe vascular phenotype wherein mutations of CD73 result in arterial calcification. This study identified nine persons with calcifications of the lower-extremity arteries and hand and foot joint capsules: all five siblings in one family, three siblings in another family, and one patient in a third family. All mutations resulted in nonfunctional CD73. Genetic rescue experiments normalized the CD73 and alkaline phosphatase activity in patients' cells, and adenosine treatment reduced the levels of alkaline phosphatase and calcification. The authors conclude that mutations in the *NT5E* gene are associated with symptomatic arterial and joint calcifications, supporting a role for this metabolic pathway in inhibiting ectopic tissue calcification. In this study (109), the authors developed a complex model proposing that vascular cells produce adenosine via the conversion of ATP to AMP and pyrophosphate by ecto-nucleotide pyrophosphatase phosphodiesterase 1 (ENPP1), with subsequent hydrolysis of AMP to adenosine by CD73. Decreased levels of adenosine secondary to lower CD73 activity boost alkaline phosphatase activity, which clears pyrophosphate. Although deletion of *Enpp1* in mice recapitulates disease in infancy with arterial calcifications and ectopic osteochondral differentiation (110), there is no evidence that CD73 deletion in mice is associated with arterial calcifications (77, 105). This work raises many interesting issues and provides an important basis for further study (111).

ADENOSINE RECEPTORS AND HYPOXIA

Extracellular adenosine exerts its biological signaling actions through the activation of any of four ARs. Whereas activation of the A_1AR or the A_3AR leads to attenuation of intracellular cAMP levels, activation of the high-affinity $A_{2A}AR$ or the low-affinity $A_{2B}AR$ is associated with elevation of cAMP levels (68). The crystal structure of agonist- and antagonist-bound $A_{2A}AR$ was recently solved (112).

As we discuss in the paragraph above, hypoxia shifts the balance to nucleotide signaling by enhancing the phosphohydrolysis of adenosine precursor nucleotides. This shift from ATP toward adenosine signaling involves Sp1-dependent induction of CD39 and HIF-dependent induction of CD73 during conditions of hypoxia. In addition, hypoxia directly influences adenosine signaling events. In fact, investigators have described two hypoxia-elicited, transcriptionally regulated pathways for ARs, including HIF-1α-dependent induction of $A_{2B}AR$ (113) and HIF-2α-dependent induction of $A_{2A}AR$ (114). $A_{2B}AR$ has the lowest affinity of the ARs. However, extracellular adenosine levels that are sufficient to activate $A_{2B}AR$ can be achieved, particularly during conditions of hypoxia or ischemia (84, 115).

The first evidence for hypoxia-dependent enhancement of $A_{2B}AR$ signaling comes from a study that examined expressional and transcriptional responses of ARs during hypoxia (88). The authors profiled the relative expression of ARs in normoxic or hypoxic endothelial cells by microarray analysis. These experiments identified selective induction of $A_{2B}AR$. Several studies validated these microarray results, and subsequent analysis revealed a mechanism involving direct HIF-1α-dependent regulation of the $A_{2B}AR$ promoter (113). Induction of $A_{2B}AR$ in hypoxia has translated to a strong anti-inflammatory phenotype that, at least in part, includes barrier protection in several different tissues. Consistent with these findings, other studies demonstrated induction of $A_{2B}AR$ in intestinal ischemia-reperfusion injury (108, 116) and experimental colitis (117) in conjunction with attenuated inflammation (118) and improved organ function (68, 79, 84, 119).

ENT: equilibrative nucleoside transporter

A second transcriptional pathway that is under the control of hypoxia signaling involves $A_{2A}AR$. Genetic and pharmacological studies strongly implicated $A_{2A}AR$ in attenuation of inflammatory responses in a wide range of models (36, 104, 120–122). Studies that examined the transcriptional control of $A_{2A}AR$ during conditions of hypoxia focused on HIF-2α (128). Unlike the case for HIF-1α, which regulates a unique set of genes, most genes regulated by HIF-2α overlap with those induced by HIF-1α. Thus, the unique contribution of HIF-2α remains largely obscure. By using adenoviral mutant HIF-1α or adenoviral mutant HIF-2α constructs, in which the HIFs are transcriptionally active under normoxic conditions, a study from the White laboratory (114) demonstrated that HIF-2α, but not HIF-1α, regulates $A_{2A}AR$ expression in primary cultures of human lung endothelial cells, suggesting nonredundant, tissue-specific roles for HIF-1α and HIF-2α in the regulation of ARs.

In addition to directly influencing AR expression, hypoxia was recently implicated in a pathway that examined mechanisms of indirect AR amplification during conditions of hypoxia. A previous study had demonstrated that the neuronal guidance molecule netrin-1 requires interactions with $A_{2B}AR$ for axon outgrowth and cAMP production (123). On the basis of other studies showing that signaling events involving netrin-1 can attenuate acute inflammatory responses (124), a subsequent study investigated the contribution of netrin-1 signaling to hypoxia-induced inflammation (125). The authors detected HIF-1α-dependent induction of expression of the gene encoding netrin-1 (*Ntn1*) in hypoxic epithelia. Neutrophil transepithelial migration studies showed that by engaging $A_{2B}ARs$ on neutrophils, netrin-1 attenuated neutrophil transmigration. Moreover, exogenous netrin-1 suppressed hypoxia-elicited inflammation in wild-type but not in $A_{2B}AR$-deficient mice, and inflammatory hypoxia was enhanced in $Ntn1^{+/-}$ mice relative to that in $Ntn1^{+/+}$ mice. These studies demonstrate that HIF-dependent induction of netrin-1 attenuates hypoxia-elicited inflammation at mucosal surfaces by enhancing signaling events through $A_{2B}AR$ (125).

OTHER INFLUENCES OF HYPOXIA ON ADENOSINE METABOLISM

As discussed above, hypoxia elicits a coordinated response that results in increased enzymatic production and signaling events of adenosine. Other studies have implicated hypoxia signaling in the regulation of the extracellular half-life of adenosine through transcriptional control of adenosine transporters (26, 81, 126). The short half-life of extracellular adenosine is attributable to its uptake via nucleoside transporters. Two group of nucleoside transporters have been described: the equilibrative nucleoside transporters (ENTs), which allow for passive diffusion of adenosine following its gradient across the cell membrane, and the active concentrative nucleoside transporters, which transport adenosine in exchange for Na^+ (81, 127–129). The ENTs have been implicated in the functional regulation of adenosine signaling during conditions of hypoxia (26, 81, 126, 130–134).

Hypoxia also strongly influences intracellular adenosine metabolism. Within the intracellular space, adenosine can undergo deamination by adenosine deaminase to inosine (82) or phosphorylation to AMP by adenosine kinase (83, 135). Despite its intracellular location, adenosine kinase regulates extracellular adenosine signaling, most likely through hypoxia-mediated inhibition of adenosine kinase (135–138); such inhibition involves HIF-1α-regulated repression of adenosine kinase (83).

MECHANISMS OF ANTI-INFLAMMATION MEDIATED BY ADENOSINE

Surprisingly little is known about the actual mechanisms of adenosine-mediated anti-inflammation. Although signal transduction through the various ARs is well characterized, less is known about postreceptor events. One particularly intriguing mechanism suggests that adenosine inhibits NF-κB through actions on proteasomal degradation of IκB proteins (139). These findings were based on previous work suggesting that commensal bacteria inhibit NF-κB through Cullin-1 (Cul-1) deneddylation (140). Studies addressing adenosine signaling mechanisms revealed that adenosine and adenosine analogs display a dose-dependent deneddylation of Cul-1, with rank order of receptor potencies as follows: $A_{2B}AR > A_1AR \gg A_{2A}AR = A_3AR$ (139). Regulated protein degradation is an essential feature of cell signaling for many adaptive processes. The proteasomal degradation of IκB proteins that inhibit NF-κB is one such example of a rapid response by the cell to signal for cell growth, differentiation, apoptosis, or inflammation. The E3 SCF ubiquitin ligase, which is specific to IκB-family members and comprises SKP1, Cul-1, and the F-box domain of β-TrCP, is responsible for the polyubiquitination of IκB (141). E3 SCF requires the COP9 signalosome (CSN) to bind Nedd8 to Cul-1 in order to be active, and deneddylated Cul-1 is incapable of IκB ubiquitination and, hence, of the inactivation of NF-κB (142). Deneddylation reactions on Cullin targets via CSN-associated proteolysis are increasingly implicated as a central point for Cullin-mediated E3 ubiquitylation (143). Notably, other pathways for deneddylation have been reported. For example, the identification of the Nedd8-specific proteases NEDP1 and DEN1 has provided new insights into this emerging field. NEDP1/DEN1 have isopeptidase activity capable of directly deneddylating Cullin targets (144, 145). How adenosine influences NEDP1/DEN1 activity is not known.

CONCLUSION

The interplay of metabolic pathways in health and disease defines an elegant lesson in biology. Studies in model disease systems and human patients have allowed for the identification of metabolic changes that have proven fundamental to our understanding of disease pathogenesis.

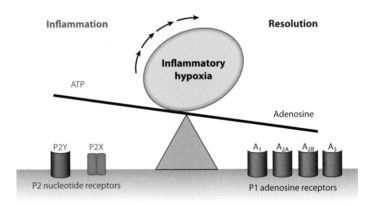

Figure 5

Model of cooperation between nucleotide and nucleoside receptors in inflammation. In areas of ongoing inflammation, diminished O_2 supply (inflammatory hypoxia) coordinates the metabolism of nucleotides to adenosine and subsequent signaling via P1 adenosine receptors. Much of the inflammatory nucleotide signaling occurs via P2 receptors, whereas P1 receptors contribute significantly to the resolution of ongoing inflammation.

The interdependence of HIF and adenosine shows how these biochemical and physiological pathways yield insights both to better understand tissue function and to define new targets as templates for the development of novel therapies that promote the resolution of inflammatory disease (**Figure 5**).

> SUMMARY POINTS
>
> 1. The microenvironment of an inflammatory lesion is depleted of O_2 (e.g., is hypoxic).
> 2. Hypoxia-inducible factor (HIF) functions as one of the master regulators of O_2 homeostasis.
> 3. HIF prolyl hydroxylases are a primary sensor of O_2 in both health and disease.
> 4. Adenine nucleotides represent a dominant metabolic fingerprint of hypoxia.
> 5. HIF coordinates adenine nucleotide metabolism and regulates adenosine signaling.

DISCLOSURE STATEMENT

The authors are not aware of any affiliations, memberships, funding, or financial holdings that might be perceived as influencing the objectivity of this review.

ACKNOWLEDGMENTS

This work was supported by National Institutes of Health grants DK50189, HL60569, HL92188, HL098294, and DK83385 and by funding from the Crohn's and Colitis Foundation of America.

LITERATURE CITED

1. Laukoetter MG, Bruewer M, Nusrat A. 2006. Regulation of the intestinal epithelial barrier by the apical junctional complex. *Curr. Opin. Gastroenterol.* 22:85–89
2. Turner JR. 2009. Intestinal mucosal barrier function in health and disease. *Nat. Rev. Immunol.* 9:799–809
3. Furuta GT, Turner JR, Taylor CT, Hershberg RM, Comerford KM, et al. 2001. Hypoxia-inducible factor 1-dependent induction of intestinal trefoil factor protects barrier function during hypoxia. *J. Ex. Med.* 193:1027–34
4. Taylor CT, Colgan SP. 2008. Hypoxia and gastrointestinal disease. *J. Mol. Med.* 85:1295–300
5. Colgan SP, Taylor CT. 2010. Hypoxia: an alarm signal during intestinal inflammation. *Nat. Rev. Gastroenterol. Hepatol.* 7:281–87
6. Kominsky DJ, Campbell EL, Colgan SP. 2010. Metabolic shifts in immunity and inflammation. *J. Immunol.* 184:4062–68
7. Pollard TD, Borisy GG. 2003. Cellular motility driven by assembly and disassembly of actin filaments. *Cell* 112:453–65
8. Borregaard N, Herlin T. 1982. Energy metabolism of human neutrophils during phagocytosis. *J. Clin. Investig.* 70:550–57
9. El-Benna J, Dang PM, Gougerot-Pocidalo MA. 2008. Priming of the neutrophil NADPH oxidase activation: role of $p47^{phox}$ phosphorylation and NOX2 mobilization to the plasma membrane. *Semin. Immunopathol.* 30:279–89
10. Gabig TG, Bearman SI, Babior BM. 1979. Effects of oxygen tension and pH on the respiratory burst of human neutrophils. *Blood* 53:1133–39
11. Greiner EF, Guppy M, Brand K. 1994. Glucose is essential for proliferation and the glycolytic enzyme induction that provokes a transition to glycolytic energy production. *J. Biol. Chem.* 269:31484–90

12. Plas DR, Rathmell JC, Thompson CB. 2002. Homeostatic control of lymphocyte survival: potential origins and implications. *Nat. Immunol.* 3:515–21
13. Fox CJ, Hammerman PS, Thompson CB. 2005. Fuel feeds function: energy metabolism and the T-cell response. *Nat. Rev. Immunol.* 5:844–52
14. Hatoum OA, Heidemann J, Binion DG. 2006. The intestinal microvasculature as a therapeutic target in inflammatory bowel disease. *Ann. N. Y. Acad. Sci.* 1072:78–97
15. Danese S, Dejana E, Fiocchi C. 2007. Immune regulation by microvascular endothelial cells: directing innate and adaptive immunity, coagulation, and inflammation. *J. Immunol.* 178:6017–22
16. Karhausen JO, Furuta GT, Tomaszewski JE, Johnson RS, Colgan SP, Haase VH. 2004. Epithelial hypoxia-inducible factor-1 is protective in murine experimental colitis. *J. Clin. Investig.* 114:1098–106
17. Evans SM, Hahn S, Pook DR, Jenkins WT, Chalian AA, et al. 2000. Detection of hypoxia in human squamous cell carcinoma by EF5 binding. *Cancer Res.* 60:2018–24
18. Takasawa M, Moustafa RR, Baron JC. 2008. Applications of nitroimidazole in vivo hypoxia imaging in ischemic stroke. *Stroke* 39:1629–37
19. Kizaka-Kondoh S, Konse-Nagasawa H. 2009. Significance of nitroimidazole compounds and hypoxia-inducible factor-1 for imaging tumor hypoxia. *Cancer Sci.* 100:1366–73
20. Overgaard J. 2007. Hypoxic radiosensitization: adored and ignored. *J. Clin. Oncol.* 25:4066–74
21. Mastrogiannaki M, Matak P, Keith B, Simon MC, Vaulont S, Peyssonnaux C. 2009. HIF-2α, but not HIF-1α, promotes iron absorption in mice. *J. Clin. Investig.* 119:1159–66
22. Ratcliffe PJ. 2007. HIF-1 and HIF-2: working alone or together in hypoxia? *J. Clin. Investig.* 117:862–65
23. Synnestvedt K, Furuta GT, Comerford KM, Louis N, Karhausen J, et al. 2002. Ecto-5′-nucleotidase (CD73) regulation by hypoxia-inducible factor-1 (HIF-1) mediates permeability changes in intestinal epithelia. *J. Clin. Investig.* 110:993–1002
24. Cummins EP, Seeballuck F, Keely SJ, Mangan NE, Callanan JJ, et al. 2008. The hydroxylase inhibitor dimethyloxalylglycine is protective in a murine model of colitis. *Gastroenterology* 134:156–65
25. Han IO, Kim HS, Kim HC, Joe EH, Kim WK. 2003. Synergistic expression of inducible nitric oxide synthase by phorbol ester and interferon-γ is mediated through NF-κB and ERK in microglial cells. *J. Neurosci. Res.* 73:659–69
26. Morote-Garcia JC, Rosenberger P, Nivillac NM, Coe IR, Eltzschig HK. 2009. Hypoxia-inducible factor-dependent repression of equilibrative nucleoside transporter 2 attenuates mucosal inflammation during intestinal hypoxia. *Gastroenterology* 136:607–18
27. Robinson A, Keely S, Karhausen J, Gerich ME, Furuta GT, Colgan SP. 2008. Mucosal protection by hypoxia-inducible factor prolyl hydroxylase inhibition. *Gastroenterology* 134:145–55
28. Shah YM, Ito S, Morimura K, Chen C, Yim SH, et al. 2008. Hypoxia-inducible factor augments experimental colitis through an MIF-dependent inflammatory signaling cascade. *Gastroenterology* 134:2036–48
29. Giatromanolaki A, Sivridis E, Maltezos E, Papazoglou D, Simopoulos C, et al. 2003. Hypoxia inducible factor 1α and 2α overexpression in inflammatory bowel disease. *J. Clin. Pathol.* 56:209–13
30. Mariani F, Sena P, Marzona L, Riccio M, Fano R, et al. 2009. Cyclooxygenase-2 and Hypoxia-Inducible Factor-1α protein expression is related to inflammation, and up-regulated since the early steps of colorectal carcinogenesis. *Cancer Lett.* 279:221–29
31. Matthijsen RA, Derikx JP, Kuipers D, van Dam RM, Dejong CH, Buurman WA. 2009. Enterocyte shedding and epithelial lining repair following ischemia of the human small intestine attenuate inflammation. *PLoS ONE* 4:e7045
32. Louis NA, Hamilton KE, Canny G, Shekels LL, Ho SB, Colgan SP. 2006. Selective induction of mucin-3 by hypoxia in intestinal epithelia. *J. Cell Biochem.* 99:1616–27
33. Comerford KM, Wallace TJ, Karhausen J, Louis NA, Montalto MC, Colgan SP. 2002. Hypoxia-inducible factor-1-dependent regulation of the multidrug resistance (*MDR1*) gene. *Cancer Res.* 62:3387–94
34. Eltzschig HK, Ibla JC, Furuta GT, Leonard MO, Jacobson KA, et al. 2003. Coordinated adenine nucleotide phosphohydrolysis and nucleoside signaling in posthypoxic endothelium: role of ectonucleotidases and adenosine A_{2B} receptors. *J. Ex. Med.* 198:783–96
35. Makino Y, Nakamura H, Ikeda E, Ohnuma K, Yamauchi K, et al. 2003. Hypoxia-inducible factor regulates survival of antigen receptor-driven T cells. *J. Immunol.* 171:6534–40

36. Sitkovsky M, Lukashev D. 2005. Regulation of immune cells by local-tissue oxygen tension: HIF1α and adenosine receptors. *Nat. Rev. Immunol.* 5:712–21
37. Nizet V, Johnson RS. 2009. Interdependence of hypoxic and innate immune responses. *Nat. Rev. Immunol.* 9:609–17
38. Kong T, Eltzschig HK, Karhausen J, Colgan SP, Shelley CS. 2004. Leukocyte adhesion during hypoxia is mediated by HIF-1-dependent induction of β2 integrin gene expression. *Proc. Natl. Acad. Sci. USA* 101:10440–45
39. Kuhl AA, Kakirman H, Janotta M, Dreher S, Cremer P, et al. 2007. Aggravation of different types of experimental colitis by depletion or adhesion blockade of neutrophils. *Gastroenterology* 133:1882–92
40. Kaelin WG Jr, Ratcliffe PJ. 2008. Oxygen sensing by metazoans: the central role of the HIF hydroxylase pathway. *Mol. Cell* 30:393–402
41. Cummins EP, Berra E, Comerford KM, Ginouves A, Fitzgerald KT, et al. 2006. Prolyl hydroxylase-1 negatively regulates IκB kinase-β, giving insight into hypoxia-induced NFκB activity. *Proc. Natl. Acad. Sci. USA* 103:18154–59
42. Aragones J, Schneider M, Van Geyte K, Fraisl P, Dresselaers T, et al. 2008. Deficiency or inhibition of oxygen sensor Phd1 induces hypoxia tolerance by reprogramming basal metabolism. *Nat. Genet.* 40:170–80
43. Schneider M, Van Geyte K, Fraisl P, Kiss J, Aragones J, et al. 2009. Loss or silencing of the PHD1 prolyl hydroxylase protects livers of mice against ischemia/reperfusion injury. *Gastroenterology* 138:1143–54
44. Tambuwala MM, Cummins EP, Lenihan CR, Kiss J, Stauch M, et al. 2010. Loss of prolyl hydroxylase-1 protects against colitis through reduced epithelial cell apoptosis and increased barrier function. *Gastroenterology* 139:2093–101
45. Mazzone M, Dettori D, Leite de Oliveira R, Loges S, Schmidt T, et al. 2009. Heterozygous deficiency of PHD2 restores tumor oxygenation and inhibits metastasis via endothelial normalization. *Cell* 136:839–51
46. Ozolins TR, Fisher TS, Nadeau DM, Stock JL, Klein AS, et al. 2009. Defects in embryonic development of EGLN1/PHD2 knockdown transgenic mice are associated with induction of Igfbp in the placenta. *Biochem. Biophys. Res. Commun.* 390:372–76
47. Bishop T, Gallagher D, Pascual A, Lygate CA, de Bono JP, et al. 2008. Abnormal sympathoadrenal development and systemic hypotension in $PHD3^{-/-}$ mice. *Mol. Cell. Biol.* 28:3386–400
48. Taylor CT. 2008. Interdependent roles for hypoxia inducible factor and nuclear factor-κB in hypoxic inflammation. *J. Physiol.* 586:4055–59
49. Rosenberger P, Khoury J, Kong T, Weissmuller T, Robinson AM, Colgan SP. 2007. Identification of vasodilator-stimulated phosphoprotein (VASP) as an HIF-regulated tissue permeability factor during hypoxia. *FASEB J.* 21:2613–21
50. Chen Z, Hagler J, Palombella VJ, Melandri F, Scherer D, et al. 1995. Signal-induced site-specific phosphorylation targets IκBα to the ubiquitin-proteasome pathway. *Genes Dev.* 9:1586–97
51. Luo JL, Kamata H, Karin M. 2005. IKK/NF-κB signaling: balancing life and death—a new approach to cancer therapy. *J. Clin. Investig.* 115:2625–32
52. Chen ZJ. 2005. Ubiquitin signalling in the NF-κB pathway. *Nat. Cell Biol.* 7:758–65
53. Cockman ME, Lancaster DE, Stolze IP, Hewitson KS, McDonough MA, et al. 2006. Posttranslational hydroxylation of ankyrin repeats in IκB proteins by the hypoxia-inducible factor (HIF) asparaginyl hydroxylase, factor inhibiting HIF (FIH). *Proc. Natl. Acad. Sci. USA* 103:14767–72
54. Zaph C, Troy AE, Taylor BC, Berman-Booty LD, Guild KJ, et al. 2007. Epithelial-cell-intrinsic IKK-β expression regulates intestinal immune homeostasis. *Nature* 446:552–56
55. Rius J, Guma M, Schachtrup C, Akassoglou K, Zinkernagel AS, et al. 2008. NF-κB links innate immunity to the hypoxic response through transcriptional regulation of HIF-1α. *Nature* 453:807–11
56. Coleman ML, McDonough MA, Hewitson KS, Coles C, Mecinovic J, et al. 2007. Asparaginyl hydroxylation of the Notch ankyrin repeat domain by factor inhibiting hypoxia-inducible factor. *J. Biol. Chem.* 282:24027–38
57. Koditz J, Nesper J, Wottawa M, Stiehl DP, Camenisch G, et al. 2007. Oxygen-dependent ATF-4 stability is mediated by the PHD3 oxygen sensor. *Blood* 110:3610–17
58. Mole DR, Schlemminger I, McNeill LA, Hewitson KS, Pugh CW, et al. 2003. 2-Oxoglutarate analogue inhibitors of HIF prolyl hydroxylase. *Bioorg. Med. Chem. Lett.* 13:2677–80

59. Masson N, Ratcliffe PJ. 2003. HIF prolyl and asparaginyl hydroxylases in the biological response to intracellular O_2 levels. *J. Cell Sci.* 116:3041–49
60. Schofield CJ, Ratcliffe PJ. 2004. Oxygen sensing by HIF hydroxylases. *Nat. Rev. Mol. Cell Biol.* 5:343–54
61. Bruick RK. 2003. Oxygen sensing in the hypoxic response pathway: regulation of the hypoxia-inducible transcription factor. *Genes. Dev.* 17:2614–23
62. Jelkmann W. 2007. Control of erythropoietin gene expression and its use in medicine. *Methods Enzymol.* 435:179–97
63. Bernhardt WM, Wiesener MS, Scigalla P, Chou J, Schmieder RE, et al. 2010. Inhibition of prolyl hydroxylases increases erythropoietin production in ESRD. *J. Am. Soc. Nephrol.* 21:2151–56
64. Drury AN, Szent-Gyorgyi A. 1929. The physiological activity of adenine compounds with especial reference to their action upon the mammalian heart. *J. Physiol.* 68:213–37
65. Belhassen B, Pelleg A. 1984. Electrophysiologic effects of adenosine triphosphate and adenosine on the mammalian heart: clinical and experimental aspects. *J. Am. Coll. Cardiol.* 4:414–24
66. Delacretaz E. 2006. Clinical practice. Supraventricular tachycardia. *N. Engl. J. Med.* 354:1039–51
67. Blomstrom-Lundqvist C, Scheinman MM, Aliot EM, Alpert JS, Calkins H, et al. 2003. ACC/AHA/ESC guidelines for the management of patients with supraventricular arrhythmias—executive summary: a report of the American College of Cardiology/American Heart Association Task Force on Practice Guidelines and the European Society of Cardiology Committee for Practice Guidelines (Writing Committee to Develop Guidelines for the Management of Patients with Supraventricular Arrhythmias). *Circulation* 108:1871–909
68. Eltzschig HK. 2009. Adenosine: an old drug newly discovered. *Anesthesiology* 111:904–15
69. Koeppen M, Eckle T, Eltzschig HK. 2009. Selective deletion of the A1 adenosine receptor abolishes heart-rate slowing effects of intravascular adenosine in vivo. *PLoS ONE* 4:e6784
70. Matherne GP, Linden J, Byford AM, Gauthier NS, Headrick JP. 1997. Transgenic A1 adenosine receptor overexpression increases myocardial resistance to ischemia. *Proc. Natl. Acad. Sci. USA* 94:6541–46
71. Yang JN, Chen JF, Fredholm BB. 2009. Physiological roles of A1 and A2A adenosine receptors in regulating heart rate, body temperature, and locomotion as revealed using knockout mice and caffeine. *Am. J. Physiol. Heart Circ. Physiol.* 296:1141–49
72. Picano E, Trivieri MG. 1999. Pharmacologic stress echocardiography in the assessment of coronary artery disease. *Curr. Opin. Cardiol.* 14:464–70
73. Hart ML, Kohler D, Eckle T, Kloor D, Stahl GL, Eltzschig HK. 2008. Direct treatment of mouse or human blood with soluble 5′-nucleotidase inhibits platelet aggregation. *Arterioscler. Thromb. Vasc. Biol.* 28:1477–83
74. Sacco RL, Diener HC, Yusuf S, Cotton D, Ounpuu S, et al. 2008. Aspirin and extended-release dipyridamole versus clopidogrel for recurrent stroke. *N. Engl. J. Med.* 359:1238–51
75. Fanta CH. 2009. Asthma. *N. Engl. J. Med.* 360:1002–14
76. Robson SC, Wu Y, Sun X, Knosalla C, Dwyer K, Enjyoji K. 2005. Ectonucleotidases of CD39 family modulate vascular inflammation and thrombosis in transplantation. *Semin. Thromb. Hemost.* 31:217–33
77. Colgan SP, Eltzschig HK, Eckle T, Thompson LF. 2006. Physiological roles of 5′-ectonucleotidase (CD73). *Purinergic Signal.* 2:351–60
78. Synnestvedt K, Furuta GT, Comerford KM, Louis N, Karhausen J, et al. 2002. Ecto-5′-nucleotidase (CD73) regulation by hypoxia-inducible factor-1 mediates permeability changes in intestinal epithelia. *J. Clin. Investig.* 110:993–1002
79. Bauerle JD, Grenz A, Kim JH, Lee HT, Eltzschig HK. 2011. Adenosine generation and signaling during acute kidney injury. *J. Am. Soc. Nephrol.* 22:14–20
80. Eckle T, Koeppen M, Eltzschig HK. 2009. Role of extracellular adenosine in acute lung injury. *Physiology* 24:298–306
81. Loffler M, Morote-Garcia JC, Eltzschig SA, Coe IR, Eltzschig HK. 2007. Physiological roles of vascular nucleoside transporters. *Arterioscler. Thromb. Vasc. Biol.* 27:1004–13
82. Eltzschig HK, Faigle M, Knapp S, Karhausen J, Ibla J, et al. 2006. Endothelial catabolism of extracellular adenosine during hypoxia: the role of surface adenosine deaminase and CD26. *Blood* 108:1602–10
83. Morote-Garcia JC, Rosenberger P, Kuhlicke J, Eltzschig HK. 2008. HIF-1-dependent repression of adenosine kinase attenuates hypoxia-induced vascular leak. *Blood* 111:5571–80

84. Aherne CM, Kewley EM, Eltzschig HK. 2010. The resurgence of A2B adenosine receptor signaling. *Biochim. Biophys. Acta* 1808:1329–39
85. Chekeni FB, Elliott MR, Sandilos JK, Walk SF, Kinchen JM, et al. 2010. Pannexin 1 channels mediate "find-me" signal release and membrane permeability during apoptosis. *Nature* 467:863–67
86. Elliott MR, Chekeni FB, Trampont PC, Lazarowski ER, Kadl A, et al. 2009. Nucleotides released by apoptotic cells act as a find-me signal to promote phagocytic clearance. *Nature* 461:282–86
87. Eltzschig HK, Eckle T, Mager A, Kuper N, Karcher C, et al. 2006. ATP release from activated neutrophils occurs via connexin 43 and modulates adenosine-dependent endothelial cell function. *Circ. Res.* 99:1100–8
88. Eltzschig HK, Ibla JC, Furuta GT, Leonard MO, Jacobson KA, et al. 2003. Coordinated adenine nucleotide phosphohydrolysis and nucleoside signaling in posthypoxic endothelium: role of ectonucleotidases and adenosine A2B receptors. *J. Exp. Med.* 198:783–96
89. Novak I. 2003. ATP as a signaling molecule: the exocrine focus. *News Physiol. Sci.* 18:12–17
90. Goodenough DA, Paul DL. 2003. Beyond the gap: functions of unpaired connexon channels. *Nat. Rev. Mol. Cell Biol.* 4:285–94
91. Chen Y, Corriden R, Inoue Y, Yip L, Hashiguchi N, et al. 2006. ATP release guides neutrophil chemotaxis via P2Y2 and A3 receptors. *Science* 314:1792–95
92. Linden J. 2006. Purinergic chemotaxis. *Science* 314:1689–90
93. Gordon JL. 1986. Extracellular ATP: effects, sources and fate. *Biochem. J.* 233:309–19
94. Weissmuller T, Campbell EL, Rosenberger P, Scully M, Beck PL, et al. 2008. PMNs facilitate translocation of platelets across human and mouse epithelium and together alter fluid homeostasis via epithelial cell-expressed ecto-NTPDases. *J. Clin. Investig.* 118:3682–92
95. Madara JL, Patapoff TW, Gillece-Castro B, Colgan SP, Parkos CA, et al. 1993. 5′-Adenosine monophosphate is the neutrophil-derived paracrine factor that elicits chloride secretion from T84 intestinal epithelial cell monolayers. *J. Clin. Investig.* 91:2320–25
96. Zimmermann H. 2000. Extracellular metabolism of ATP and other nucleotides. *Naunyn Schmiedebergs Arch. Pharmacol.* 362:299–309
97. Pinsky DJ, Broekman MJ, Peschon JJ, Stocking KL, Fujita T, et al. 2002. Elucidation of the thromboregulatory role of CD39/ectoapyrase in the ischemic brain. *J. Clin. Investig.* 109:1031–40
98. Marcus AJ, Broekman MJ, Drosopoulos JHF, Islam N, Alyonycheva TN, et al. 1997. The endothelial cell ecto-ADPase responsible for inhibition of platelet function is CD39. *J. Clin. Investig.* 99:1351–60
99. Kohler D, Eckle T, Faigle M, Grenz A, Mittelbronn M, et al. 2007. CD39/ectonucleoside triphosphate diphosphohydrolase 1 provides myocardial protection during cardiac ischemia/reperfusion injury. *Circulation* 116:1784–94
100. Enjyoji K, Sevigny J, Lin Y, Frenette PS, Christie PD, Esch JS, et al. 1999. Targeted disruption of CD39/ATP diphosphohydrolase results in disordered hemostasis and thromboregulation. *Nat. Med.* 5:1010–17
101. Eltzschig HK, Kohler D, Eckle T, Kong T, Robson SC, Colgan SP. 2009. Central role of Sp1-regulated CD39 in hypoxia/ischemia protection. *Blood* 113:224–32
102. Maliszewski CR, Delespesse GJ, Schoenborn MA, Armitage RJ, Fanslow WC, et al. 1994. The CD39 lymphoid cell activation antigen. Molecular cloning and structural characterization. *J. Immunol.* 153:3574–83
103. Friedman DJ, Künzli BM, A-Rahim YI, Sevigny J, Berberat PO, et al. 2009. CD39 deletion exacerbates experimental murine colitis and human polymorphisms increase susceptibility to inflammatory bowel disease. *Proc. Natl. Acad. Sci. USA* 106:16788–93
104. Deaglio S, Dwyer KM, Gao W, Friedman D, Usheva A, et al. 2007. Adenosine generation catalyzed by CD39 and CD73 expressed on regulatory T cells mediates immune suppression. *J. Exp. Med.* 204:1257–65
105. Thompson LF, Eltzschig HK, Ibla JC, Van De Wiele CJ, Resta R, et al. 2004. Crucial role for ecto-5′-nucleotidase (CD73) in vascular leakage during hypoxia. *J. Exp. Med.* 200:1395–405
106. Louis NA, Robinson AM, MacManus CF, Karhausen J, Scully M, Colgan SP. 2008. Control of IFN-αA by CD73: implications for mucosal inflammation. *J. Immunol.* 180:4246–55

107. Hart ML, Henn M, Kohler D, Kloor D, Mittelbronn M, et al. 2008. Role of extracellular nucleotide phosphohydrolysis in intestinal ischemia-reperfusion injury. *FASEB J.* 22:2784–97
108. Hart ML, Grenz A, Gorzolla IC, Schittenhelm J, Dalton JH, Eltzschig HK. 2011. Hypoxia-inducible factor-1α-dependent protection from intestinal ischemia/reperfusion injury involves ecto-5′-nucleotidase (CD73) and the A2B adenosine receptor. *J. Immunol.* 186:4367–74
109. St Hilaire C, Ziegler SG, Markello TC, Brusco A, Groden C, et al. 2011. NT5E mutations and arterial calcifications. *N. Engl. J. Med.* 364:432–42
110. Terkeltaub R. 2006. Physiologic and pathologic functions of the NPP nucleotide pyrophosphatase/phosphodiesterase family focusing on NPP1 in calcification. *Purinergic Signal.* 2:371–77
111. Eltzschig HK, Robson SC. 2011. NT5E mutations and arterial calcifications. *N. Engl. J. Med.* 364:1577–78
112. Xu F, Wu H, Katritch V, Han GW, Jacobson KA, et al. 2011. Structure of an agonist-bound human A2A adenosine receptor. *Science* 332:322–27
113. Kong T, Westerman KA, Faigle M, Eltzschig HK, Colgan SP. 2006. HIF-dependent induction of adenosine A2B receptor in hypoxia. *FASEB J.* 20:2242–50
114. Ahmad A, Ahmad S, Glover LE, Miller S, Shannon JM, et al. 2009. Adenosine A_{2A} receptor is a unique angiogenic target of HIF-2α in pulmonary endothelial cells. *Proc. Natl. Acad. Sci. USA* 106:10684–89
115. Fredholm BB. 2007. Adenosine, an endogenous distress signal, modulates tissue damage and repair. *Cell Death Differ.* 14:1315–23
116. Hart ML, Jacobi B, Schittenhelm J, Henn M, Eltzschig HK. 2009. Cutting edge: A2B adenosine receptor signaling provides potent protection during intestinal ischemia/reperfusion injury. *J. Immunol.* 182:3965–68
117. Frick JS, MacManus CF, Scully M, Glover LE, Eltzschig HK, Colgan SP. 2009. Contribution of adenosine A2B receptors to inflammatory parameters of experimental colitis. *J. Immunol.* 182:4957–64
118. Eltzschig HK, Thompson LF, Karhausen J, Cotta RJ, Ibla JC, et al. 2004. Endogenous adenosine produced during hypoxia attenuates neutrophil accumulation: coordination by extracellular nucleotide metabolism. *Blood* 104:3986–92
119. Grenz A, Homann D, Eltzschig HK. 2011. Extracellular adenosine: a "safety signal" that dampens hypoxia-induced inflammation during ischemia. *Antioxid. Redox Signal.* 15(8):2221–34
120. Ohta A, Sitkovsky M. 2001. Role of G-protein-coupled adenosine receptors in downregulation of inflammation and protection from tissue damage. *Nature* 414:916–20
121. Sitkovsky MV, Lukashev D, Apasov S, Kojima H, Koshiba M, et al. 2004. Physiological control of immune response and inflammatory tissue damage by hypoxia-inducible factors and adenosine A_{2A} receptors. *Annu. Rev. Immunol.* 22:657–82
122. Lappas CM, Day YJ, Marshall MA, Engelhard VH, Linden J. 2006. Adenosine A_{2A} receptor activation reduces hepatic ischemia reperfusion injury by inhibiting CD1d-dependent NKT cell activation. *J. Exp. Med.* 203:2639–48
123. Corset V, Nguyen-Ba-Charvet KT, Forcet C, Moyse E, Chedotal A, Mehlen P. 2000. Netrin-1-mediated axon outgrowth and cAMP production requires interaction with adenosine A2b receptor. *Nature* 407:747–50
124. Ly NP, Komatsuzaki K, Fraser IP, Tseng AA, Prodhan P, et al. 2005. Netrin-1 inhibits leukocyte migration in vitro and in vivo. *Proc. Natl. Acad. Sci. USA* 102:14729–34
125. Rosenberger P, Schwab JM, Mirakaj V, Masekowsky E, Mager A, et al. 2009. Hypoxia-inducible factor-dependent induction of netrin-1 dampens inflammation caused by hypoxia. *Nat. Immunol.* 10:195–202
126. Eltzschig HK, Abdulla P, Hoffman E, Hamilton KE, Daniels D, et al. 2005. HIF-1-dependent repression of equilibrative nucleoside transporter (ENT) in hypoxia. *J. Exp. Med.* 202:1493–505
127. Baldwin SA, Beal PR, Yao SY, King AE, Cass CE, Young JD. 2004. The equilibrative nucleoside transporter family, SLC29. *Pflüg. Arch.* 447:735–43
128. Griffiths M, Beaumont N, Yao SY, Sundaram M, Boumah CE, et al. 1997. Cloning of a human nucleoside transporter implicated in the cellular uptake of adenosine and chemotherapeutic drugs. *Nat. Med.* 3:89–93
129. Coe IR, Griffiths M, Young JD, Baldwin SA, Cass CE. 1997. Assignment of the human equilibrative nucleoside transporter (hENT1) to 6p21.1-p21.2. *Genomics* 45:459–60

130. Casanello P, Torres A, Sanhueza F, Gonzalez M, Farias M, et al. 2005. Equilibrative nucleoside transporter 1 expression is downregulated by hypoxia in human umbilical vein endothelium. *Circ. Res.* 97:16–24
131. Chaudary N, Naydenova Z, Shuralyova I, Coe IR. 2004. Hypoxia regulates the adenosine transporter, mENT1, in the murine cardiomyocyte cell line, HL-1. *Cardiovasc. Res.* 61:780–88
132. Rose JB, Naydenova Z, Bang A, Eguchi M, Sweeney G, et al. 2009. Equilibrative nucleoside transporter 1 plays an essential role in cardioprotection. *Am. J. Physiol. Heart Circ. Physiol.* 298:771–77
133. Hart ML, Much C, Kohler D, Schittenhelm J, Gorzolla IC, et al. 2008. Use of a hanging-weight system for liver ischemic preconditioning in mice. *Am. J. Physiol. Gastrointest. Liver Physiol.* 294:1431–40
134. Eckle T, Grenz A, Kohler D, Redel A, Falk M, et al. 2006. Systematic evaluation of a novel model for cardiac ischemic preconditioning in mice. *Am. J. Physiol. Heart Circ. Physiol.* 291:2533–40
135. Decking UKM, Schlieper G, Kroll K, Schrader J. 1997. Hypoxia-induced inhibition of adenosine kinase potentiates cardiac adenosine release. *Circ. Res.* 81:154–64
136. Peart JN, Gross GJ. 2005. Cardioprotection following adenosine kinase inhibition in rat hearts. *Basic Res. Cardiol.* 100:328–36
137. Ely SW, Matherne GP, Coleman SD, Berne RM. 1992. Inhibition of adenosine metabolism increases myocardial interstitial adenosine concentrations and coronary flow. *J. Mol. Cell Cardiol.* 24:1321–32
138. Pak MA, Haas HL, Decking UK, Schrader J. 1994. Inhibition of adenosine kinase increases endogenous adenosine and depresses neuronal activity in hippocampal slices. *Neuropharmacology* 33:1049–53
139. Khoury J, Ibla JC, Neish AS, Colgan SP. 2007. Antiinflammatory adaptation to hypoxia through adenosine-mediated cullin-1 deneddylation. *J. Clin. Investig.* 117:703–11
140. Neish AS, Gewirtz AT, Zeng H, Young AN, Hobert ME, et al. 2000. Prokaryotic regulation of epithelial responses by inhibition of IκB-α ubiquitination. *Science* 289:1560–63
141. Cope GA, Deshaies RJ. 2003. COP9 signalosome: a multifunctional regulator of SCF and other cullin-based ubiquitin ligases. *Cell* 114:663–71
142. Cardozo T, Pagano M. 2004. The SCF ubiquitin ligase: insights into a molecular machine. *Nat. Rev. Mol. Cell Biol.* 5:739–51
143. Parry G, Estelle M. 2004. Regulation of cullin-based ubiquitin ligases by the Nedd8/RUB ubiquitin-like proteins. *Semin. Cell Dev. Biol.* 15:221–29
144. Mendoza HM, Shen LN, Botting C, Lewis A, Chen J, et al. 2003. NEDP1, a highly conserved cysteine protease that deNEDDylates Cullins. *J. Biol. Chem.* 278:25637–43
145. Wu K, Yamoah K, Dolios G, Gan-Erdene T, Tan P, et al. 2003. DEN1 is a dual function protease capable of processing the C terminus of Nedd8 and deconjugating hyper-neddylated CUL1. *J. Biol. Chem.* 278:28882–91
146. Castillo CA, Leon D, Ruiz MA, Albasanz JL, Martin M. 2008. Modulation of adenosine A_1 and A_{2A} receptors in C6 glioma cells during hypoxia: involvement of endogenous adenosine. *J. Neurochem.* 105:2315–29
147. Eltzschig HK, Abdulla P, Hoffman E, Hamilton KE, Daniels D, et al. 2005. HIF-1-dependent repression of equilibrative nucleoside transporter (ENT) in hypoxia. *J. Exp. Med.* 202:1493–505

Toll-Like Receptor–Gut Microbiota Interactions: Perturb at Your Own Risk!

Frederic A. Carvalho,[1,2,*] Jesse D. Aitken,[3,*] Matam Vijay-Kumar,[3] and Andrew T. Gewirtz[3]

[1]Pharmacologie Fondamentale et Clinique de la Douleur, Clermont Université, Université d'Auvergne, F-63000 Clermont-Ferrand, France

[2]Inserm, U 766, F-63001 Clermont-Ferrand, France

[3]Center for Inflammation, Immunity and Infection, Department of Biology, Georgia State University, Atlanta, Georgia 30303; email: agewirtz@gsu.edu

*These authors contributed equally.

Keywords

inflammatory bowel disease, cancer, metabolic syndrome, MyD88, intestinal homeostasis

Abstract

The well-being of the intestine and its host requires that this organ execute its complex function amid colonization by a large and diverse microbial community referred to as the gut microbiota. A myriad of interacting mechanisms of mucosal immunity permit the gut to corral the microbiota in such a way as to maximize the benefits and to minimize the danger of living in close proximity to this large microbial biomass. Toll-like receptors and Nod-like receptors, collectively referred to as pattern recognition receptors (PRRs), recognize a variety of microbial components and, hence, play a central role in governing the interface between host and microbiota. This review examines mechanisms by which PRR-microbiota interactions are regulated so as to allow activation of host defense when necessary while preventing excessive inflammation, which can have a myriad of negative consequences for the host. Analysis of published studies performed in human subjects and a variety of murine disease models reveals the central theme that PRRs play a key role in maintaining a healthful stable relationship between the intestine and its microbiota. In contrast, although select genetic ablations of PRR signaling may protect against some chronic diseases, the overriding theme of studies performed to date is that perturbations of PRR-microbiota interactions are more likely to promote disease states associated with inflammation.

INTRODUCTION. GUT MICROBIOTA: YOUR BEST FRENEMY FOREVER

OTU: operational taxonomic unit

BFF: best frenemy forever

The mammalian intestine is inhabited by a large diverse community of microbes collectively referred to as the gut microbiota. The human microbiota consists of approximately 100 trillion (10^{14}) bacteria that weigh in sum 1–2 kg and are composed of 6–10 major phyla and approximately 5,000 distinct species, sometimes referred to as operational taxonomic units (OTUs) to reflect that they are defined by the sequence of their 16S rRNA (1). Importantly, given that the 16S rRNA sequence is often insufficient to describe key aspects of microbial genotype (for example, it cannot distinguish pathogenic and harmless *Escherichia coli* strains), the true genetic and phenotypic diversity of the gut microbiota is underestimated by OTU cataloging and, at present, remains incompletely defined. Nonetheless, increasingly detailed descriptions of the genetic composition of the microbiota have been reported and continue to shed light on the various functions provided by the gut microbiota, particularly in terms of its metabolism (2, 3). The composition of the microbiota, which is normally acquired during and shortly after birth, is relatively plastic during early development but, once stabilized, is thought to maintain many of its characteristics throughout the life of the host. Although this review does not discuss the composition of the gut microbiota in detail, we note that the most prevalent bacterial phyla in the microbiota are the Bacteroidetes and Firmicutes, strict anaerobes that are often very difficult to culture, and Proteobacteria, which include bacteria, such as *E. coli* and *Helicobacter pylori*, that have been associated with inflammation in a variety of scenarios (4).

The improved understanding of the gut microbiota indicates that the traditional dichotomy of commensal versus pathogen is clearly too simplistic. Rather, some bacteria such as *Lactobacillus* and *Bifidobacteria* are highly associated with good health, and animal model studies support the concept that they provide far more benefit than harm to their hosts and thus should be viewed as symbiotic (5). Conversely, some bacteria, such as select *E. coli* species, cause no harm to healthy hosts but are associated with, and have the capacity to exacerbate, disease and are now generally viewed as opportunistic pathogens (6, 7). Thus, it seems reasonable to view a healthy gut microbiota as a continuum of symbiotic to potentially pathogenic bacteria. When viewed holistically, the gut microbiota clearly has an overall benefit to the host in that germ-free (also referred as gnotobiotic) mice, which lack a microbiota, have considerable immune and metabolic defects; the latter defects result in a requirement for greater caloric consumption relative to body mass (8). Another major benefit of the microbiota is that it serves as an effective line of defense. Specifically, during attempted colonization by pathogens, the microbiota functions as an entrenched competitor for food, space, and adhesion molecules while also producing metabolites, notably peroxides, biosurfactants, and organic acids, with direct antimicrobial properties (9, 10). By-products of commensal metabolism, including butyrate and lithocholic acid, induce the synthesis of mucus and antimicrobial factors (11), and some probiotic bacteria promote epithelial barrier function (12). Host detection of circulating microbiota-derived peptidoglycan by Nod1 is essential to the maintenance of basal immune competence (13). However, that multiple mouse models of inflammation and autoimmune disease, from colitis to arthritis to experimental autoimmune encephalomyelitis (14), require a gut microbiota and that the composition of the microbiota is a determinant of disease indicate that the microbiota can also constitute a major threat to its host. Thus, given the great benefit the microbiota confers on key areas of host biology and its potential to harm its host if not managed properly, and considering that its composition is long lasting, one could refer to the gut microbiota as the host's best frenemy forever (BFF).

The complex nature of the gut microbiota–host relationship presents a substantial challenge for the intestinal mucosal immune system. Specifically, this immune system must police a large active "port" (an appropriate way to view the intestine) so that it can expediently detect and clear pathogens and keep opportunists in check while avoiding harm to beneficial microbes and host tissues. This review discusses the mechanisms by which mucosal immunity fulfills this function, focusing primarily on the important role played by pattern recognition receptors (PRRs) of the innate immune system. Discussion is limited to two major classes of PRRs, namely, the Toll-like receptors (TLRs) and the nucleotide-binding oligomerization domain (Nod)-like receptors (NLRs). Both TLRs and NLRs recognize a variety of broadly conserved microbial components and allow the innate immune system to sense a wide variety of bacteria, viruses, fungi, and parasites as well as endogenously produced danger-associated molecular patterns (DAMPs) like extranuclear self-DNA, free ATP, HMGB1, and uric acid. The basic immunobiology of PRR recognition of microbial ligands and mechanisms of signal transduction have been extensively reviewed elsewhere. Rather, this review focuses on how PRRs allow the intestinal immune system to carry out its function, with a particular focus on the panoply of negative consequences that result when alteration of the host and/or the gut microbiota perturbs PRR–gut microbiota interactions.

PRR: pattern recognition receptor
TLR: Toll-like receptor
NLR: Nod-like receptor

USING COMPARTMENTALIZATION AND SIGNAL REGULATION TO AVOID EXCESSIVE PRR ACTIVATION IN THE GUT

When one considers the enormous microbial biomass in the gut and the fact that several TLRs can sense their cognate agonists at picomolar concentrations, it is apparent that mechanisms must exist to prevent excessive TLR activation in the gut. Indeed, the host has evolved a suite of redundant overlapping processes to prevent constant PRR activation while maintaining the ability to activate PRRs when needed and enjoying the benefits conferred by microbiota stability (summarized in **Figure 1**). Among the most important of such mechanisms is the mucus layer, which separates the fecal stream from the intestinal epithelia. Consequently, bacteria rarely have direct contact with gut epithelial cells. The extent to which the mucus layer impedes TLR agonists released by bacteria is less clear, but its gel-like properties likely enable it to serve as a physical barrier to PRR activation. Goblet cells are responsible for the synthesis of mucins, most notably MUC2 but also others (MUC5, MUC6, MUC19, and MUC7), whose high levels of O-glycosylation and oligomerization (although MUC7 does not oligomerize) confer the physical properties necessary to demarcate the boundary between lumen and epithelia (15, 16). Not surprisingly, MUC2-deficient mice fail to produce mucus and thus are more easily colonized by pathogens and suffer exacerbated dextran sulfate sodium (DSS) colitis that is likely driven, at least in part, by increased TLR activation (17–19). Under normal genetic conditions, intestinal mucus is composed of two distinct layers, a thin sterile inner layer and a partially colonized outer layer whose thickness increases with adjacent bacterial load (20). Goblet cells also secrete trefoil factor 3, which modulates the viscosity of the mucus layer and plays a role in epithelial repair (21), and resistin-like molecule β (Relmβ), which promotes mucin secretion and limits bacterial and parasitic feeding at the epithelia (22).

The production of antibacterial factors also mediates physical exclusion of bacteria in the gut. Paneth cells produce the majority of secreted antimicrobials. Two Paneth cell enzymes, lysozyme and secretory phospholipase A_2, are constitutively excreted even in the absence of gut bacteria and are generally antimicrobial, as both enzymes hydrolyze lipopolysaccharide (LPS) and thereby lyse walled bacteria (23). Another class of antimicrobials is the defensins, small (2–6-kDa) pore-forming peptides that are also referred to as cryptidins when produced by Paneth cells. Defensins are divided into the α, β, and θ groups (θ-defensins are not produced by humans) on the basis of the positions of cysteine residues and differences in tertiary structure (24). α-Defensins are constitutively secreted and exert a selective pressure on the microbiota, as mice deficient in matrix metalloproteinase 7 (MMP7), required for processing of pro-α-defensin, exhibit expansion of the

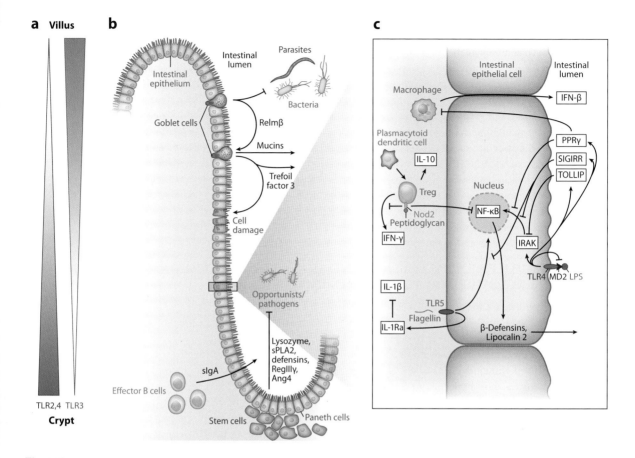

Figure 1

Compartmentalization and gut-customized signaling prevent excessive inflammation in the gut. (*a*) Some TLRs are differentially expressed on the basis of distance from the crypt to better protect against pertinent pathogen types. (*b*) The intestine uses physical barriers and antimicrobial peptides to keep the majority of bacteria at a close, yet safe, distance. (*c*) Many other immune signaling pathways are calibrated to block or avoid excessive inflammatory signaling (see text for details).

phyla Bacteroidetes and contraction of the phyla Firmicutes, hallmarks of microbiotal dysregulation (25). Members of the cathelicidin family, of which humans and mice each express only a single example, are also broadly antimicrobial and can induce chemotaxis in T cells and macrophages (26). RegIIIγ, a member of the C-type lectin family referred to as HIP/PAP in humans, preferentially binds to gram-positive bacterial peptidoglycan, resulting in lysis. RegIIIγ is constitutively expressed in conventional, but not in germ-free, mice (27). Angiogenin 4, originally characterized as a vascular factor, also exerts species-specific antimicrobial activity, which may depend in part on its ability to hydrolyze RNA, and its expression is virtually nonexistent in germ-free mice (23, 28). Other proteins, including lipocalin 2 and calprotectin, limit microbial invasion by removing essential metals, specifically, iron, calcium, and zinc, although upregulation of these compounds generally occurs in inflammatory states (29, 30). Enteric mucus also contains nonspecific secretory immunoglobulin A (sIgA), which limits bacterial adhesion and invasion by preventing epithelial adhesion and by forming clumps of bacteria physically incapable of individual translocation beyond the epithelia (31). This clump-forming property of immunoglobulin A (IgA) also reduces transmission to other potential hosts, a property that contributes to the establishment and maintenance of herd immunity. Generally, no single antimicrobial compound reaches an inhibitory concentration in vivo (32), but rather the sum of all secreted antimicrobial compounds shapes microbiota composition and constitutively primes the innate immune response against invasive pathogens. Whereas Paneth cells constantly secrete some antimicrobial factors, accelerated degranulation occurs in response to TLR9 activation by bacterial deoxy-cytidylate-phosphate-deoxy-guanylate (CpG) DNA (33). Moreover, Paneth cell α-defensin secretion is accelerated in response to ligation of TLRs 2, 3, and 4 by lipotechoic acid, CpG, and LPS, respectively (24). In the gut, β-defensins, which are also expressed by epithelial cells, are induced in response to nuclear factor-κB (NF-κB), which is induced by PRR activation, particularly TLR5 (34). Thus, proper PRR activation in the gut may be viewed as preventing the excessive PRR activation that might result if the microbiota were not properly managed.

Ig: immunoglobulin

Despite the abundance and efficacy of the physical obstacles to PRR accessibility, additional mechanisms are utilized to prevent aberrant PRR activation in the gut. One such mechanism is for the intestine to be selective about the cell types and conditions in which TLRs are expressed. TLRs 2 and 4 are barely expressed in healthy intestinal epithelial cells (IECs) but are upregulated in conditions associated with immune activation such as inflammatory bowel disease (IBD) (35). Aberrant activation of TLR4, the most proinflammatory of the PRRs, is also avoided by limiting the availability of coreceptor MD-2 in IECs (36). Two other factors essential to TLR4 activation, CD14 and LPS-binding protein (LBP), represent additional regulatory targets. TLRs 2 and 4 are also expressed at greater levels by IECs that have yet to migrate up the villus, ensuring that robust activation of these PRRs occurs only if the crypt, which is not normally colonized, is threatened (37). TLR5, the receptor for flagellin, is expressed basolaterally by IECs, a strategy that allows the host to generate a response only to invasive microbes (38). Some pathogens exploit TLR5's capacity to attract the immune cells required for systemic translocation by releasing flagellin monomers in response to contact with host lysophospholipids (39). Constant exposure to flagellin can result in tolerization to TLR5 signaling in IECs, a mechanism that may protect against opportunistic flagellin expression (40). Similarly, macrophages are rendered tolerant to LPS upon repeated exposure, a property that may limit inflammation and damage resulting from recurrent infection (41). TLR3 is expressed immediately beneath the luminal surface of IECs; expression increases as cells move toward the villus, presumably to detect invasive pathogens before they menace the crypt (37). TLR9 is unique among the TLRs in that it provides a brake for signaling through all TLRs. Whereas basolateral activation of TLR9 elicits a classic TLR inflammatory response, apical TLR9 attenuates such a response via an alternative signaling pathway that blunts interleukin (IL)-8 activity and that can be induced by commensal DNA (42, 43). Constant basolateral exposure to unmethylated bacterial DNA leads to TLR9 hyporesponsiveness, another mechanism by which aberrant inflammation, and even autoimmunity, is avoided (44).

Another means of reducing excessive PRR activation is through TLR signaling, which activates negative regulators of both TLR and NLR signaling. For example, TLR5 signaling in IECs does not result in IL-1β production but rather induces robust production of secretory IL-1Ra, which acts to competitively inhibit IL-1β (45). The Th2 cytokines IL-4 and IL-13 induce hyporesponsiveness to TLR4 signaling in IECs (46). Some negative regulators of TLR signaling include single immunoglobulin IL-1R-related molecule, an Ig-like protein that inhibits epithelial TLR and IL-1β signaling (47). Toll-interacting protein, which is upregulated in the presence of LPS, inhibits the signaling of TLRs 2 and 4 through the IL-1R-associated kinase pathway (48). That these two molecules, as well as peroxisome proliferator–activated receptor-γ, which also inhibits interferon (IFN)-β production in macrophages (49), inhibit NF-κB activation and are upregulated following ligation of TLR4 speaks to the need to carefully modulate the inflammatory response to invasive bacteria. Plasmacytoid dendritic cells differentially promote the development of IL-10-producing regulatory T cells and, upon ligation of Nod2, limit IFN-γ production and inhibit NF-κB (50), a mechanism that limits the autoimmune potential of these cells upon TLR9 ligation of self-DNA bound to antimicrobial cathelicidin (51). Thus, despite the enormous number of PRR ligands in the intestine, PRRs can be utilized to protect the gut without activating dangerous levels of inflammation.

PRR-MICROBIOTA INTERACTIONS IN THE GUT PROMOTE HOST DEFENSE DEVELOPMENT

Much of our understanding of PRR–gut microbiota interactions comes from studies involving germ-free mice and/or mice deficient in specific PRRs or molecules by which PRRs signal. Many studies have used MyD88-deficient mice to represent mice broadly deficient in TLR signaling, reflecting that MyD88 is required for signaling by all TLRs except TLR3. However, because MyD88-deficient mice have defective signaling in response to the inflammasome cytokines IL-1β and IL-18, which are major products of NLR signaling, they effectively lack signaling through TLRs and NLRs. Germ-free mice exhibit reduced epithelial proliferation and increased permeability (52), although observations that mice lacking MyD88 preserve relatively normal barrier function argue against a major role for TLRs and inflammasome cytokines in gut barrier development (**Figure 2**) (53). However, interaction between bacterial components and TLRs is crucial for IEC proliferation, as macrophages isolated from germ-free mice exhibit decreased expression of trophic factors due to defective TLR signaling (54). As efficacious expression of several TLRs depends on the presence of the intestinal microbiota, germ-free mice are more susceptible to pathogenic infection (55). Germ-free mice exhibit developmental defects (postbirth) in both innate and adaptive immunity, suggesting that recognition of the microbiota by PRRs is essential for normal immune function (56). Germ-free mice fail to make Paneth cell antimicrobials such as Relmβ and RegIIIγ (27). A similar phenotype is exhibited in conventional mice (i.e., those having a microbiota) that lack MyD88 (57). Such MyD88-dependent expression of antimicrobials reflects, at least in part, a direct role for MyD88 in Paneth cells. Lack of Paneth cell products was not observed in mice lacking any individual TLR, suggesting redundancy in TLR and/or NLR signaling for this aspect of development. Lack of MyD88-mediated antimicrobial proteins is associated with increased bacterial penetration in the intestinal mucosa and may be corrected by engineering Paneth cells to express these proteins. Thus, PRR signaling by Paneth cells helps confine the microbiota to the lumen. Although the overall effect of the absence of MyD88-dependent PRR signaling on the gut microbiota is relatively moderate in the basal state, this scenario largely reflects the efficacy of upregulated mucosal adaptive immunity as a compensatory mechanism (53). Germ-free mice are also deficient in mucosal adaptive immunity. Specifically, germ-free mice

---

Figure 2

Mucosal immune development relies on PRR-microbiota interactions. (*a*) In the normal state, PRR-microbiota interactions result in the secretion of antimicrobial peptides, growth factors that promote epithelial proliferation, and the development of gut-associated lymphoid tissue (GALT). (*b*) In the absence of gut microbiota, there is a dramatic reduction in antimicrobial peptides, reduced epithelial proliferation, and complete loss of GALT. (*c*) Loss of MyD88 recapitulates the loss of AMP and moreover results in increased bacterial translocation.

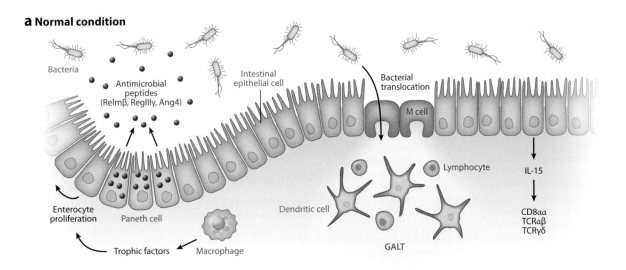
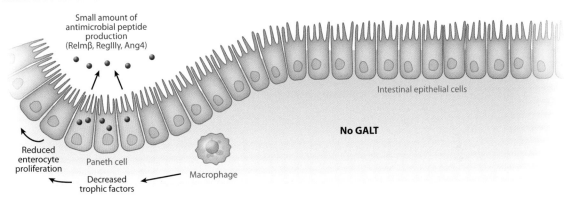
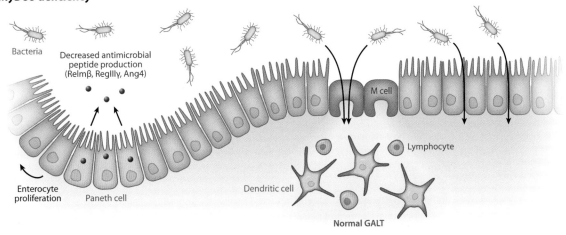

almost completely lack gut-associated lymphoid tissue (GALT), including Peyer's patches and isolated lymphoid follicles (58). Such lack of GALT alters mucosal immune responses in response to pathogens and allergens. However, the extent of GALT is relatively normal in MyD88-deficient mice, arguing against the necessity of TLR signaling per se (59). TLR signaling does play a key role in generating the basal population of intestinal lymphocytes, as CD8αα, TCRαβ, and TCRγδ intestinal intraepithelial lymphocytes were significantly decreased in MyD88-deficient mice (60). Bone marrow chimera experiments indicated that MyD88 signaling in IECs that results in IL-15 production is important for driving this cell population. Together, these results indicate that, even in the absence of overt challenges, PRR-microbiota interactions play a role in developing and maintaining host defense in the gut.

PRR-MICROBIOTA INTERACTIONS MAINTAIN HOMEOSTASIS AND RESTORE THE GUT UPON CHALLENGE: PUTTING OUT THE FIRE QUICKLY USES LESS WATER

Janeway's (61) prescient prediction of the existence of PRRs and the notion that they would exist to detect foreign microbes that posed danger to the host held that such PRRs would typically exist in a quiescent state in the absence of pathogens. Observations that intestinal epithelial cells were generally unresponsive to LPS and responded only to flagellin that reached the basolateral surface were in accord with this view and presented a means by which the host might use PRRs to discriminate between commensal and pathogenic bacteria (38, 62). Thus, the 2004 report by Medzhitov and colleagues (63) that demonstrated microbiota-induced PRR signaling under basal conditions initiated a paradigm shift in this field. Specifically, they observed that loss of MyD88 altered IEC survival, thus requiring greater proliferation to maintain the intestinal mucosa. Moreover, they observed MyD88-dependent basal production of cytokines, including keratinocyte-derived chemokine, IL-6, and tumor necrosis factor α (TNFα). Although these studies remain subject to the above-mentioned caveats that apply to MyD88 mice, they highly suggest ongoing PRR activation in the colon in the absence of an overt challenge. Such PRR activation may take the form of a tonic uniform activation driven by the low concentration of PRR ligands. In accord, apical stimulation of IECs promotes homeostatic signaling (44). Alternatively, basal PRR-mediated cytokine production may reflect localized activation of PRRs on immune cells and/or on IECs as a result of local microabrasions of the epithelium that occur and are resolved regularly.

Although lack of basal MyD88-mediated PRR signaling does not consistently result in an obvious basal phenotype, it is in accord with widespread anecdotal observations that MyD88 mice can be quite difficult to breed in some vivariums and tend to be somewhat runted relative to wild-type littermates. Moreover, that there is ongoing PRR recognition of the microbiota helps explain the widely observed phenomenon that rederivation of mouse strains, which can be viewed as exchanging mouse microbiota for that of the surrogate mother, can have a clear effect on phenotype, even when the original mice lacked any known pathogens. The lack of basal MyD88 signaling combined with the inability to activate such signaling under insult make clear that, although MyD88 activates classic proinflammatory signaling, the outcome of such signaling helps the host manage a variety of challenges. Most strikingly, loss of MyD88 renders mice highly susceptible to intestinal damage and death induced by the chemical colitigen DSS (63, 64). The severe gut pathology of MyD88-deficient mice in response to DSS correlates with failure to upregulate typical stress responses such as heat shock proteins and cytokines/chemokines. This finding suggests that, although these responses rapidly induce inflammation, they play a pivotal role in helping the host manage the challenge, ultimately promoting resolution and restoration of basal phenotype. Work by Abreu and colleagues (65) demonstrating that increased gut damage is preceded by markedly fewer inflammatory infiltrates supports the notion that these inflammatory cells indeed help resolve inflammation. The effect of MyD88 ablation on DSS colitis was largely mimicked in wild-type mice by placing them on broad-spectrum antibiotics, arguing that the absence of MyD88 signaling reflects loss of PRR signaling rather than a role for IL-1β or IL-18 signaling (63). The effect of MyD88 deficiency on DSS colitis was partially phenocopied by loss of TLRs 2 and 4 (63, 65), suggesting that these receptors were activated upon DSS treatment. TLR5- and TLR9-deficient mice also exhibit more severe colitis, supporting the logical notion that TLR-mediated tolerogenic signaling is optimized in wild-type mice (66, 67). Steppenback and colleagues (64) showed that DSS-induced, MyD88-dependent signaling activated the proliferation of epithelial early progenitor cells that mediate epithelial restitution of mucosal injury induced by either DSS or radiation.

The above-described studies utilized mice wholly deficient in TLRs and/or MyD88 mice and thus used the cell types in the gut in which PRR signaling remains poorly defined. MyD88 expression in leukocytes is sufficient to drive epithelial proliferation in response to DSS (68). However, another experiment showed that epithelial cell MyD88 is required for proper restitution in response to DSS colitis (68). Again, all experiments involving MyD88-deficient mice must be interpreted with the notion that any observed phenotypes may reflect failure to respond to IL-1β and IL-18. Accordingly, given that IECs may lack the capacity to produce inflammasome cytokines but are responsive to these cytokines, many epithelial defects resulting from MyD88 deficiency may reflect loss of response to inflammasome cytokines rather than loss of TLR signaling.

A unifying theme of many experiments using MyD88-deficient mice in acute models of inflammation in various systems is that absence of MyD88 reduced inflammation but resulted ultimately in more extensive injury. A similar theme emerged when these mice were infected with *Citrobacter rodentium*, which causes transient inflammation and hyperplasia in wild-type mice. Specifically, such mice exhibited reduced inflammation but ultimately died from failing to clear the pathogen and/or from resulting tissue damage (69, 70). However, this phenotype was largely mimicked by loss of IL-1R, reflecting a broad block of PRR signaling rather than loss of TLR deficiency (71). Somewhat analogously, MyD88-deficient mice subjected to the *Salmonella*/streptomycin model of induced gastroenteritis, used to approximate the gut-predominant salmonellosis seen in humans, exhibited no inflammation 48 h following challenge but ultimately greater mortality (72). The strain or environment may influence the extent to which *Salmonella*/streptomycin gastroenteritis is MyD88 dependent (73). Nonetheless, in many models the broad block of PRR signaling that results from loss of MyD88 results in a substantial reduction in inflammatory response at the cost of greater net damage. In contrast, loss of an individual TLR, namely, the flagellin receptor TLR5, resulted in only transient delay in inflammation, resulting in more severe inflammation (72). Importantly, such delayed but enhanced inflammation phenocopied previous observations in wild-type mice that aflagellate *Salmonella* induced greater inflammatory pathology than did an aflagellate *Salmonella* mutant (74). The difference in these results likely reflects that, in the absence of the rapid neutrophil recruitment observed in wild-type mice infected with wild-type *Salmonella*, the bacteria persists until it activates other TLRs and NLRs (**Figure 3**). Thus, overall, one can make the case that rapid activation of an inflammatory response upon pathogen detection is analogous to a highly responsive fire-fighting department: Activating the alarm quickly allows the fire to be put out quickly, thus minimizing collateral damage.

CHRONIC COLITIS AND PRR-MICROBIOTA INTERACTIONS

IBD is well associated with an elevated immune response to the gut microbiota (75, 76). The IBD-associated immune response is not merely a marker of disease but rather a defining mediator

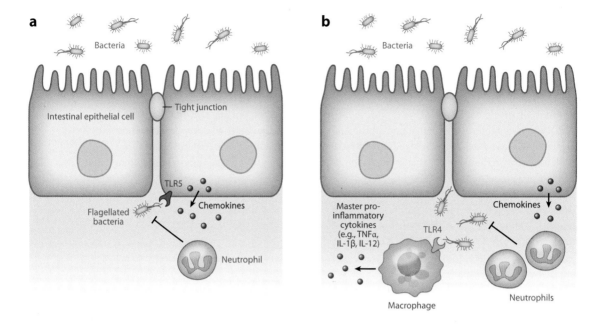

Figure 3
Putting out the fire quickly minimizes inflammation. (*a*) TLR5 quickly detects the presence of flagellated bacteria, resulting in the timely recruitment of neutrophils that efficiently neutralize the perturbing bacteria. (*b*) In the absence of TLR5, the bacteria persist and proliferate before activating the macrophage TLR4, which triggers robust inflammation, including the production of master proinflammatory cytokines.

that drives the clinical manifestations of this disorder. The extent to which the elevated immune response originates from an aberrant microbiota or immune dysregulation or results from an underlying immune deficiency is less clear. Indeed, animal models of colitis and human studies lend some support to all three possibilities, and each of these distinct possibilities may underlie a subset of IBD. The primary evidence that IBD can result from an alteration in PRR-microbiota interactions is the association of mutations in the gene encoding the NLR Nod2, which recognizes the bacterial cell wall component muramyl dipeptide (77, 78). The mechanism by which alterations in Nod2 function may result in IBD remains under investigation but includes the possibilities that *Nod2* mutations result in gain of function (79) and/or failure to downregulate signaling by other TLRs (80–82). These mechanisms may prove to play a role in IBD, but the simplest explanation involves the *Nod2* frameshift mutation. This loss-of-function mutation results in innate immune deficiency, specifically, an inability to efficiently respond to the occasional constituent of the gut microbiota that breaches the mucous layer and is internalized by host cells of the mucosa (intestinal macrophages seem to be a good candidate). In this scenario, other PRRs would eventually detect the bacteria, resulting in clearance and likely adaptive immunity; both events would take place at a higher frequency than if the microbe had been detected and cleared by an efficient, functional Nod2. The notion that Nod2 is directly involved in the clearance of bacteria in vivo is supported by the observation that Nod2-deficient mice exhibit a modest delay in clearing orally administered *Listeria monocytogenes* but that loss of Nod2 has not been shown to result in or to promote development of colitis or secondary immune dysregulation (83). Two mouse studies have demonstrated that an innate immune deficiency can result in chronic colitis. Nenci et al. (84) showed that defective NF-κB signaling in epithelial cells resulted in robust colitis driven by commensal microbiota. Such colitis was associated with TNFα production by macrophages, and genetic ablation of *TNFα* ameliorated the colitis (84). We observed paradigmatically analogous results in TLR5-deficient mice. Specifically, some (∼30%) of mice lacking TLR5 developed spontaneous colitis (85). The colitis was driven by altered PRR-microbiota interactions in that rederivation resulting in microbiotal exchange reduced the severity and incidence of such colitis (86). Furthermore, genetic deletion of *TLR4* prevented such colitis. TLR5KO (where KO denotes knockout) colitis was associated with elevated adaptive immunity, breeding the mice onto a RAG1-deficient background ameliorated disease severity, and transfer of total splenocytes from colitic TLR5KO mice also transferred colitis to RAG1KO mice (A.T. Gewirtz & M. Vijay-Kumar, unpublished observations). Thus, at least in mouse models, a primary innate immune deficiency can drive chronic inflammation that can then be perpetuated and exacerbated by adaptive immunity, supporting the notion that altered mucosa-PRR interactions (or altered innate-adaptive cross talk) can be an initiating event in IBD.

Both murine models of colitis and human IBD are also associated with an elevated adaptive immune response to the gut microbiota. Mouse models of IBD clearly indicate that such an elevated adaptive immune response can be either the result of a primary immune dysregulation (e.g., IL-10-deficient mice) or a consequence of ongoing unresolved acute inflammation. An example of the latter would be the chronic model of DSS colitis, wherein repeated exposure to DSS results in self-perpetuating disease (87). TLR5KO colitis appears to be another example of this concept: Absence of this innate immune receptor results in an increased elevated adaptive immune response to the microbiota. Importantly, regardless of the underlying cause of the adaptive immune response, that transfer of splenocytes to immunodeficient mice typically transfers colitis indicates that such a response plays a role in driving disease. In models of colitis driven by adaptive immunity, several studies have observed that disease is ameliorated on ablation of PRR signaling. For example, MyD88-deficient mice are protected from the spontaneous colitis that normally results from the loss of IL-10 signaling due to loss of the *IL-10* gene or as a result of antibody-mediated neutralization of the IL-10 receptor (68, 88). A similar pattern occurs regardless of whether the colitis was spontaneous (89) or was induced via colonization of *Helicobacter hepaticus*. However, analogous to the case for acute inflammation, loss of just TLR4 on an IL-10-deficient background increased the severity of immune dysregulation and resulted in greater disease (90). In further support of the role of TLR4 in protecting against T cell–mediated colitis, Raz and colleagues (91) observed that loss of TLR4 exacerbated the severity of colitis both in the IL-10KO model and in T cell–mediated transfer of disease. In contrast, loss of TLR9 signaling, which was strongly protective in acute DSS colitis (92, 93), ameliorated disease in multiple models of chronic T cell–mediated colitis (94, 95). There are limited data on the role of TLRs in human disease, but we note that, although in mice loss of TLR5 exacerbates disease in multiple models of acute inflammation, humans carrying a dominant negative TLR5 that reduces function by roughly 75% display reduced levels of antibodies to flagellin and, in some genetic backgrounds, are protected against developing IBD (96). Thus, the overarching theme gleaned from these observations is that loss of PRR-microbiota interactions affects the signaling processes that regulate the T cells responsible for mediating chronic inflammation. The consequences of such dysregulation can vary greatly depending on the specific pathway altered as well as on a variety of genetic and environmental factors. Hence, pharmacological intervention may be risky or at least require a level of personalization that is currently not possible but may prove to be the future of medical treatment for IBD and other chronic inflammatory disorders.

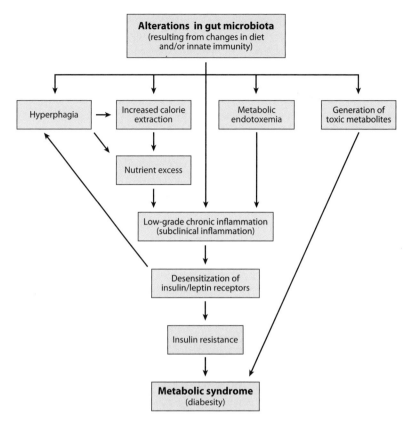

Figure 4

Ways by which alterations in gut microbiota can result in metabolic disease.

METABOLIC DISEASES AND ALTERED PRR–GUT MICROBIOTA INTERACTIONS

It is increasingly appreciated that the intestine's role in metabolism extends well beyond food digestion and nutrient absorption and that rather the intestine exhibits considerable regulatory influence over much of the brain and endocrine system. Thus, considering the complexity of gut microbiota and its interactions with the immune system and the host, one can imagine a myriad of ways by which PRR-microbiota interactions might influence metabolism. Moreover, the composition of a society's microbiota can be envisaged to change quickly relative to that society's genome due to the substantial influence of the modern diet, hygienic practices, and antibiotic use on microbiota composition. Thus, societal alterations in microbiota, with *H. pylori* likely being one of many examples of this phenomenon (97), are a plausible factor in the epidemic increase in metabolic and other associated diseases. Below, we discuss some of the mechanisms by which PRR-microbiota interactions influence metabolic disease. These mechanisms can be broadly categorized as direct consequences of microbial metabolism, wherein PRRs may regulate the composition and activity of the microbiota or may result from an altered microbiota activating PRRs or other components of the immune system (**Figure 4**).

Pioneering studies from Gordon and colleagues (98) observed differences in the composition of the microbiota, specifically, changes in the Bacteroidetes/Firmicutes ratios in obese humans.

Similar changes in microbiota composition were observed in mice fed obesogenic high-fat diets (99). Such changes in microbiota composition are unlikely to be a consequence of obesity because similar changes were observed in mice that were resistant to obesity due to Relmβ deficiency (100). Rather, such changes may play a role in obesity development, as the transfer of aberrant microbial communities into wild-type germ-free mice resulted in increased fat mass (98). This increase in fat mass occurred without increased food consumption and was associated with reduced energy content in the feces, indicating that the observed obesity was the result of increased caloric extraction. Although increased efficiency of energy extraction has obvious benefits to the host, it may be detrimental in a society in which food is plentiful. Another means by which microbial metabolism may negatively influence the host was recently reported by Wang et al. (101), who demonstrated that the microbiota plays a role in generating harmful metabolites from dietary lipids that are linked to heart disease. Specifically, their study demonstrates that the gut microbiota metabolizes dietary phosphatidylcholine (PC), also known as lecithin, to the noxious odoriferous gas trimethylamine (TMA), which is eventually metabolized in the host liver to trimethyl N-oxide (TMAO). Furthermore, microbiota ablation in atherosclerosis-prone ApoEKO mice protected against increased plasma TMAO, aortic macrophage content, and atherosclerosis. More importantly, Wang et al. (101) also found that levels of PC metabolites predict coronary vascular disease risk: Serum from humans with coronary artery disease, peripheral artery disease, and a history of myocardial infarction contains elevated levels of PC metabolites, suggesting PC's relevance to human disease.

Researchers recently observed that changes in the Bacteroidetes/Firmicutes ratio are not specific to obesity or to a high-fat diet but also occur in response to various proinflammatory agents, including *C. rodentium*, DSS, and a neutralizing antibody to the IL-10 receptor (45, 102). The increasing appreciation that a significant amount of metabolic disease can be viewed as low-grade chronic inflammation and observations that altered PRR-microbiota interactions promote inflammation suggest that such inflammation may be a major means by which the microbiota can negatively affect metabolism. As Hotamisligil & Spiegelman (103) showed, proinflammatory signaling cross-desensitizes insulin receptor signaling, resulting in insulin resistance, which is central to the development of metabolic syndrome. Metabolic syndrome denotes abnormalities in parameters such as hypertension, hyperlipidemia, and dysglycemia, which lead to the development of type II diabetes and cardiovascular disease. By showing that a high-fat diet can induce metabolic endotoxemia, Cani et al. (104) demonstrated that the microbiota can promote metabolic disease via inflammation. Specifically, their studies demonstrate that high-fat feeding increases intestinal gram-negative bacterial load, thus increasing the amount of luminal LPS that leaks into systemic circulation and drives metabolic endotoxemia. Such high-fat feeding also resulted in classical symptoms of metabolic inflammation such as elevated macrophage infiltration in adipose tissue, body weight gain, and diabetes risk. To study mechanistically the PRRs involved in causing such metabolic inflammation, Cani et al. (104) challenged CD14-deficient mice with high-fat diets and found that these mice resisted most of the high-fat diet–induced indices of metabolic disease. Another study from Cani et al. (105) demonstrates that a high-fat diet given for 4 weeks significantly reduced *Lactobacillus* spp. and *Bacteroides* spp. Such a change in microbiota composition in high-fat diet–fed mice was accompanied by increased cecal and plasma LPS and increased intestinal permeability due to decreased expression of the tight junction protein ZO-1. Moreover, the administration of broad-spectrum antibiotics for 4 weeks substantially attenuated high-fat diet–induced effects. Interestingly, metabolic endotoxemia was positively associated with inflammation, oxidative stress, and macrophage infiltration markers. In addition, antibiotic administration to *ob/ob* mice reduced plasma LPS load and inflammatory markers and improved insulin sensitivity without decreasing body weights, thus strengthening the connection to inflammation rather than to obesity. In addition, gut microbiota–derived LPS signaling via TLR4 also generates other systemic inflammatory markers such as SAA3 and HMGB1, which serve as endogenous ligands for TLR4, amplifying the inflammatory response. Further evidence of the involvement of LPS in metabolic disease comes from studies using mice lacking functional TLR4. A number of studies have shown that TLR4-deficient mice are protected against high-fat diet–induced adipose tissue inflammation and metabolic disorders (106). Consistent with these data, Cani et al. (107) showed a partial reversion of inflammatory markers in *ob/ob* mice lacking the LPS coreceptor CD14. Two mechanisms have been proposed for luminal LPS absorption in the gut: the absorption of LPS via lipid-rich chylomicrons (108) and the leakage of LPS through tight junctions of the epithelial monolayer. These mechanisms are not mutually exclusive but may function in parallel and may be linked to metabolic endotoxemia.

Our studies observed a paradigmatically similar association of metabolic syndrome with microbiota-mediated, PRR-driven low-grade inflammation that was independent of TLR4 (86). Specifically, we observed that noncolitic TLR5KO mice (noncolitic mice in the original colony or mice from a rederived colony that lacked colitis) exhibited 15–20% more body weight than did wild-type mice, as well as epididymal fat pads that were 2.5-fold larger than those of wild-type littermates at 20 weeks of age. This increase in fat mass correlated with substantial increases in serum triglycerides, cholesterol, and blood pressure. Furthermore, TLR5KO mice exhibited mild elevations in overnight fasting blood glucose, mild loss of glycemic control, hyperinsulinemia, and insulin resistance. In accordance with their hyperinsulinemia, TLR5KO mice had both more and larger functional pancreatic islets. Thus, TLR5KO mice have a mild loss of glycemic control that is likely driven by insulin resistance and that is compensated for by increased insulin production, thus mimicking the conditions typically seen in humans with metabolic syndrome. When TLR5KO mice were challenged with a high-fat diet, they developed more severe hyperglycemia (>120 mg dl^{-1}); exhibited inflammatory infiltrates in the pancreatic islets (insulitis); and displayed one of the more severe manifestations of metabolic syndrome, hepatic steatosis. TLR5KO mice exhibited hyperphagia, and restricting their food consumption to that of wild-type littermates prevented most aspects of their metabolic syndrome. Insulin resistance persisted, indicating that this aspect of the TLR5KO phenotype is not a consequence of increased food consumption or adiposity but rather may result from inflammation. The development of metabolic syndrome in TLR5KO mice is independent of TLR4 and TLR2 but is dependent upon MyD88, suggesting the involvement of another PRR and/or inflammasome-generated IL-1β or IL-18. That ablation of microbiota by broad-spectrum antibiotics corrected, at least in part, most aspects of metabolic syndrome further supports the notion that changes in gut microbiota play a role in the metabolic syndrome exhibited by TLR5KO mice. Microbiotas of wild-type and TLR5KO mice exhibited species-level differences, and transplantation of TLR5KO microbiota into germ-free wild-type recipients recapitulated most aspects of metabolic syndrome, suggesting that alterations in gut microbiota play a role in their metabolic syndrome.

Recent work also implicates PRR-microbiota interactions in the pathogenesis of type I diabetes. Although type I diabetes is an autoimmune disease rather than an inflammatory disorder, inflammation may connect the microbiota with this disorder. A study in rats revealed that reducing gut microbiota by antibiotics and thus lowering antigenic and PRR ligand load in the gut reduced inflammation and protected against β-cell destruction (109). However, lack of PRR signaling may also promote type I diabetes. A recent study by Chevonsky and colleagues (110) observed that nonobese diabetic (NOD) mice deficient in MyD88 are protected against the development of type I diabetes, whereas germ-free MyD88KO/NOD mice develop robust diabetes. Colonization of germ-free MyD88KO/NOD mice with altered Schaedler bacteria significantly reduced the incidence of diabetes, and broad-spectrum antibiotic administration increased diabetes incidence in these mice. Although the mechanism remains to be determined, these results may reflect that some degree of innate immune activation is necessary to develop regulatory T cells that can prevent autoimmunity.

PRR-MICROBIOTA INTERACTIONS AND CANCER

Colorectal cancer (CRC) is a leading cause of morbidity in developed countries (111). Chronic inflammation promotes the development of a variety of cancers, including CRC, in part as a result of the antiapoptotic signaling induced by the proinflammatory transcription factor NF-κB and various growth factors (e.g., epidermal growth factor) produced to resolve inflammation. In accordance, CRC risk is considerably elevated in patients with IBD, in whom this disease is referred to as colitis-associated cancer (CAC) (111, 113). Given that altered PRR-microbiota interactions result in inflammation in general and in IBD in particular, it would be logical to suspect that PRR-microbiota interactions may play a role in CAC. A common mouse model of CAC utilizes a combination of the carcinogen azoxymethane (AOM) and the colitigen DSS. The great sensitivity of MyD88-deficient mice makes it difficult to use these mice to broadly examine the role of PRR signaling in this model. However, that TLR4-deficient mice are protected against neoplasia in this model indicates that TLR4 signaling can promote carcinogenesis (114, 115). Surprisingly, loss of TLR4 protected against tumor development without a marked effect on colitis. This result may reflect that TLR4 modulates tumor development independent of inflammation or that the acute nature of the experimental system made it difficult to observe a pronounced difference in inflammation. The idea that TLR signaling promotes cancer development was also seen in a more physiological model of CAC developed by Jobin and colleagues (116). Specifically, these investigators observed that IL-10-deficient mice given AOM developed tumors in a manner that depended upon the presence of a gut microbiota (i.e., germ-free mice were protected) and TLR signaling (i.e., MyD88-deficient mice were protected), indicating that TLR-microbiota interactions promote tumor development. In this model, seemingly more physiological than the DSS/AOM model, tumor development correlated very well with the extent of colitis in these IL-10-deficient mice, arguing strongly that neoplasia is largely a consequence of inflammation. The role of NLRs in CAC has not been extensively investigated, but initial studies indicate that at least some NLRs may protect against cancer: NLRP3-deficient mice exhibited altered severity of both colitis (117, 118) and tumor frequency in the AOM/DSS model (119). Thus, interfering with PRR-microbiota interactions can affect the development of cancer, likely via the microbiota's effects on inflammation. Another inflammasome-related protein, NLRC4, may play a central role in CAC through the regulation of epithelial response to injury rather than through induced inflammation (120).

The most widely used model of spontaneous cancer not dependent upon inflammation involves mice carrying a mutation in the adenomatous polyposis coli gene (*Apc*), commonly referred to as $Apc^{Min/+}$ mice (121). Two caveats of this model are that (*a*) tumors occur largely in the small intestines of these mice and that (*b*) such tumors, although not directly dependent upon overt inflammation, can nonetheless be promoted by inflammation. Germ-free $Apc^{Min/+}$ mice develop slightly fewer tumors in comparison to conventionally housed $Apc^{Min/+}$ mice, suggesting that PRR-microbiota interactions may promote disease in this model (121). That $Apc^{Min/+}$ mice lacking MyD88 are protected from tumor development in this model supports this notion (122). Such protection requires that MyD88 be absent in IECs, although the absence of MyD88 in bone marrow–derived cells did not affect tumor development (123). IEC MyD88 signaling promotes tumors via activating ERK-dependent stabilization of the *c-myc* oncogene. Such MyD88 signaling may reflect that the microbiota directly activates growth, thus promoting TLR signaling in IECs, but may also reflect a role for IL-1β activating these events in IECs. Thus, overall, although additional work is needed to better define the role and mechanisms of PRR-microbiota interactions in CRC, the scenario may prove to be somewhat analogous to that of colitis, wherein select targeted manipulations of PRR-microbiota interactions may protect against cancer but many manipulations make disease worse.

CONCLUSIONS AND SPECULATIONS

This article discusses the interactions between the PRRs of the innate immune system and the intestinal microbiota. One theme we seek to convey above is that, although the direct consequence of PRR-microbiota interactions can generally be viewed as eliciting proinflammatory gene expression, blocking the interactions between PRRs and microbial products in many in vivo scenarios often causes greater inflammation. This concept has been revealed in various models of inflammation, including that induced by pathogens, chemicals, and innate immune deficiencies. That blockade of receptors mediating inflammation begets greater inflammatory pathology reflects that, although the immediate consequence of PRR activation is typically to activate inflammation, such activation is typically an appropriate response to the prevailing conditions and challenges and initiates signaling to resolve the inflammation and to repair the tissue. Another general theme of this article is that intestinal inflammation can take various forms, including overt flares of inflammation, chronic inflammation, metabolic disease, and cancer. Consequently, altered PRR-microbiota interactions may play a role in a variety of these disease states. On the one hand, this interface between host and microbiota seems to present a target for developing therapies to potentially treat a wide variety of disorders. On the other hand, many manipulations of PRR-microbiota interactions make things worse, suggesting that this strategy may be attempting to manipulate an already optimized system of host defense. Thus, it might be more practical to manipulate the microbiota side of this relationship, but we suggest such strategies should proceed with caution. PRR-microbiota interactions reflect a very highly evolved relationship. Observational studies may allow the identification of specific components of that relationship that seem amiss. Although it may be possible to eventually perform a targeted correction of the relationship, our current lack of understanding of PRR-microbiota interactions would seem to make manipulations on the microbial side as prone to failure as those performed on the host. Greater study of the fascinating interactions between the microbiota and host PRRs is warranted. However, those researchers seeking to perturb PRR-microbiota interactions should do so with the appreciation that such perturbation may serve to exacerbate, rather than treat, disease.

SUMMARY POINTS

1. Gut microbiota provides benefits and potential detriments to the host, although it needs to be well managed to be beneficial to the host.
2. Gut microbiota is a large source of innate immune activators. A variety of mechanisms prevent inappropriate innate immune activation in the gut.
3. PRR recognition of gut microbiota drives the development of innate and adaptive immunity in the gut.
4. A deficiency in PRR recognition of gut microbiota manifests as excessive inflammation, especially in response to challenge.
5. A deficiency in PRR recognition of gut microbiota may result in chronic low-grade inflammation and/or may protect against chronic inflammation.

6. Altered PRR-microbiota interactions can drive low-grade inflammation that can lead to cancer.
7. Altered PRR-microbiota interactions can result in metabolic disease by several different mechanisms.

DISCLOSURE STATEMENT

The authors are not aware of any affiliations, memberships, funding, or financial holdings that might be perceived as affecting the objectivity of this review.

LITERATURE CITED

1. Hooper LV, Midtvedt T, Gordon JI. 2002. How host-microbial interactions shape the nutrient environment of the mammalian intestine. *Annu. Rev. Nutr.* 22:283–307
2. Martin FP, Sprenger N, Yap IK, Wang Y, Bibiloni R, et al. 2009. Panorganismal gut microbiome-host metabolic crosstalk. *J. Proteome Res.* 8:2090–105
3. Turnbaugh PJ, Gordon JI. 2008. An invitation to the marriage of metagenomics and metabolomics. *Cell* 134:708–13
4. Carvalho FA, Barnich N, Sauvanet P, Darcha C, Gelot A, Darfeuille-Michaud A. 2008. Crohn's disease–associated *Escherichia coli* LF82 aggravates colitis in injured mouse colon via signaling by flagellin. *Inflamm. Bowel Dis.* 14:1051–60
5. Klijn A, Mercenier A, Arigoni F. 2005. Lessons from the genomes of bifidobacteria. *FEMS Microbiol. Rev.* 29:491–509
6. Barnich N, Carvalho FA, Glasser AL, Darcha C, Jantscheff P, et al. 2007. CEACAM6 acts as a receptor for adherent-invasive *E. coli*, supporting ileal mucosa colonization in Crohn disease. *J. Clin. Investig.* 117:1566–74
7. Carvalho FA, Barnich N, Sivignon A, Darcha C, Chan CH, et al. 2009. Crohn's disease adherent-invasive *Escherichia coli* colonize and induce strong gut inflammation in transgenic mice expressing human CEACAM. *J. Exp. Med.* 206:2179–89
8. Wostmann BS, Larkin C, Moriarty A, Bruckner-Kardoss E. 1983. Dietary intake, energy metabolism, and excretory losses of adult male germfree Wistar rats. *Lab. Anim. Sci.* 33:46–50
9. Alakomi HL, Skytta E, Saarela M, Mattila-Sandholm T, Latva-Kala K, Helander IM. 2000. Lactic acid permeabilizes gram-negative bacteria by disrupting the outer membrane. *Appl. Environ. Microbiol.* 66:2001–5
10. Searle LE, Best A, Nunez A, Salguero FJ, Johnson L, et al. 2009. A mixture containing galactooligosaccharide, produced by the enzymic activity of *Bifidobacterium bifidum*, reduces *Salmonella enterica* serovar Typhimurium infection in mice. *J. Med. Microbiol.* 58:37–48
11. Gudmundsson GH, Bergman P, Andersson J, Raqib R, Agerberth B. 2010. Battle and balance at mucosal surfaces—the story of *Shigella* and antimicrobial peptides. *Biochem. Biophys. Res. Commun.* 396:116–19
12. Ewaschuk JB, Diaz H, Meddings L, Diederichs B, Dmytrash A, et al. 2008. Secreted bioactive factors from *Bifidobacterium infantis* enhance epithelial cell barrier function. *Am. J. Physiol. Gastrointest. Liver Physiol.* 295:1025–34
13. Clarke TB, Davis KM, Lysenko ES, Zhou AY, Yu Y, Weiser JN. 2010. Recognition of peptidoglycan from the microbiota by Nod1 enhances systemic innate immunity. *Nat. Med.* 16(2):228–31
14. Lee YK, Menezes JS, Umesaki Y, Mazmanian SK. 2011. Proinflammatory T-cell responses to gut microbiota promote experimental autoimmune encephalomyelitis. *Proc. Natl. Acad. Sci. USA* 108(Suppl. 1):4615–22
15. McGuckin MA, Linden SK, Sutton P, Florin TH. 2011. Mucin dynamics and enteric pathogens. *Nat. Rev. Microbiol.* 9:265–78

16. Moran AP, Gupta A, Joshi L. 2011. Sweet-talk: role of host glycosylation in bacterial pathogenesis of the gastrointestinal tract. *Gut* 60:1412–25
17. Bergstrom KS, Kissoon-Singh V, Gibson DL, Ma C, Montero M, et al. 2010. Muc2 protects against lethal infectious colitis by disassociating pathogenic and commensal bacteria from the colonic mucosa. *PLoS Pathog.* 6:e1000902
18. Heazlewood CK, Cook MC, Eri R, Price GR, Tauro SB, et al. 2008. Aberrant mucin assembly in mice causes endoplasmic reticulum stress and spontaneous inflammation resembling ulcerative colitis. *PLoS Med.* 5:e54
19. Johansson ME, Gustafsson JK, Sjoberg KE, Petersson J, Holm L, et al. 2010. Bacteria penetrate the inner mucus layer before inflammation in the dextran sulfate colitis model. *PLoS ONE* 5:e12238
20. Johansson ME, Larsson JM, Hansson GC. 2011. The two mucus layers of colon are organized by the MUC2 mucin, whereas the outer layer is a legislator of host-microbial interactions. *Proc. Natl. Acad. Sci. USA* 108(Suppl. 1):4659–65
21. Madsen J, Nielsen O, Tornoe I, Thim L, Holmskov U. 2007. Tissue localization of human trefoil factors 1, 2, and 3. *J. Histochem. Cytochem.* 55:505–13
22. Herbert DR, Yang JQ, Hogan SP, Groschwitz K, Khodoun M, et al. 2009. Intestinal epithelial cell secretion of RELM-β protects against gastrointestinal worm infection. *J. Exp. Med.* 206:2947–57
23. Mukherjee S, Vaishnava S, Hooper LV. 2008. Multi-layered regulation of intestinal antimicrobial defense. *Cell Mol. Life Sci.* 65:3019–27
24. Selsted ME, Ouellette AJ. 2005. Mammalian defensins in the antimicrobial immune response. *Nat. Immunol.* 6:551–57
25. Salzman NH, Hung K, Haribhai D, Chu H, Karlsson-Sjoberg J, et al. 2010. Enteric defensins are essential regulators of intestinal microbial ecology. *Nat. Immunol.* 11:76–83
26. Iimura M, Gallo RL, Hase K, Miyamoto Y, Eckmann L, Kagnoff MF. 2005. Cathelicidin mediates innate intestinal defense against colonization with epithelial adherent bacterial pathogens. *J. Immunol.* 174:4901–7
27. Cash HL, Whitham CV, Behrendt CL, Hooper LV. 2006. Symbiotic bacteria direct expression of an intestinal bactericidal lectin. *Science* 313:1126–30
28. Hooper LV, Stappenbeck TS, Hong CV, Gordon JI. 2003. Angiogenins: a new class of microbicidal proteins involved in innate immunity. *Nat. Immunol.* 4:269–73
29. Bachman MA, Miller VL, Weiser JN. 2009. Mucosal lipocalin 2 has pro-inflammatory and iron-sequestering effects in response to bacterial enterobactin. *PLoS Pathog.* 5:e1000622
30. Urban CF, Ermert D, Schmid M, Abu-Abed U, Goosmann C, et al. 2009. Neutrophil extracellular traps contain calprotectin, a cytosolic protein complex involved in host defense against *Candida albicans*. *PLoS Pathog.* 5:e1000639
31. Mestecky J, Russell MW. 2009. Specific antibody activity, glycan heterogeneity and polyreactivity contribute to the protective activity of S-IgA at mucosal surfaces. *Immunol. Lett.* 124:57–62
32. Mendez-Samperio P. 2010. The human cathelicidin hCAP18/LL-37: a multifunctional peptide involved in mycobacterial infections. *Peptides* 31:1791–98
33. Rumio C, Besusso D, Palazzo M, Selleri S, Sfondrini L, et al. 2004. Degranulation of Paneth cells via Toll-like receptor 9. *Am. J. Pathol.* 165:373–81
34. Uehara A, Fujimoto Y, Fukase K, Takada H. 2007. Various human epithelial cells express functional Toll-like receptors, NOD1 and NOD2 to produce anti-microbial peptides, but not proinflammatory cytokines. *Mol. Immunol.* 44:3100–11
35. Cario E, Podolsky DK. 2000. Differential alteration in intestinal epithelial cell expression of Toll-like receptor 3 (TLR3) and TLR4 in inflammatory bowel disease. *Infect. Immun.* 68:7010–17
36. Vamadevan AS, Fukata M, Arnold ET, Thomas LS, Hsu D, Abreu MT. 2010. Regulation of Toll-like receptor 4-associated MD-2 in intestinal epithelial cells: a comprehensive analysis. *Innate Immun.* 16:93–103
37. Furrie E, Macfarlane S, Thomson G, Macfarlane GT. 2005. Toll-like receptors-2, -3 and -4 expression patterns on human colon and their regulation by mucosal-associated bacteria. *Immunology* 115:565–74

38. Gewirtz AT, Navas TA, Lyons S, Godowski PJ, Madara JL. 2001. Cutting edge: Bacterial flagellin activates basolaterally expressed TLR5 to induce epithelial proinflammatory gene expression. *J. Immunol.* 167:1882–85
39. Subramanian N, Qadri A. 2006. Lysophospholipid sensing triggers secretion of flagellin from pathogenic salmonella. *Nat. Immunol.* 7(6):583–89
40. Sun J, Fegan PE, Desai AS, Madara JL, Hobert ME. 2007. Flagellin-induced tolerance of the Toll-like receptor 5 signaling pathway in polarized intestinal epithelial cells. *Am. J. Physiol. Gastrointest. Liver Physiol.* 292:767–78
41. Biswas SK, Lopez-Collazo E. 2009. Endotoxin tolerance: new mechanisms, molecules and clinical significance. *Trends Immunol.* 30(10):475–87
42. Ghadimi D, Vrese M, Heller KJ, Schrezenmeir J. 2010. Effect of natural commensal-origin DNA on Toll-like receptor 9 (TLR9) signaling cascade, chemokine IL-8 expression, and barrier integrity of polarized intestinal epithelial cells. *Inflamm. Bowel Dis.* 16:410–27
43. Vilaysane A, Muruve DA. 2009. The innate immune response to DNA. *Semin. Immunol.* 21:208–14
44. Lee J, Gonzales-Navajas JM, Raz E. 2008. The "polarizing-tolerizing" mechanism of intestinal epithelium: its relevance to colonic homeostasis. *Semin. Immunopathol.* 30:3–9
45. Carvalho FA, Aitken JD, Gewirtz AT, Vijay-Kumar M. 2011. TLR5 activation induces secretory interleukin-1 receptor antagonist (sIL-1Ra) and reduces inflammasome-associated tissue damage. *Mucosal Immunol.* 4:102–11
46. Mueller T, Terada T, Rosenberg IM, Shibolet O, Podolsky DK. 2006. Th2 cytokines down-regulate TLR expression and function in human intestinal epithelial cells. *J. Immunol.* 176:5805–14
47. Khan MA, Steiner TS, Sham HP, Bergstrom KS, Huang JT, et al. 2010. The single IgG IL-1-related receptor controls TLR responses in differentiated human intestinal epithelial cells. *J. Immunol.* 184:2305–13
48. Zhang G, Ghosh S. 2002. Negative regulation of Toll-like receptor-mediated signaling by Tollip. *J. Biol. Chem.* 277:7059–65
49. Zhao W, Wang L, Zhang M, Wang P, Zhang L, et al. 2011. Peroxisome proliferator-activated receptor gamma negatively regulates IFN-β production in Toll-like receptor (TLR) 3- and TLR4-stimulated macrophages by preventing interferon regulatory factor 3 binding to the IFN-β promoter. *J. Biol. Chem.* 286:5519–28
50. Castellaneta A, Sumpter TL, Chen L, Tokita D, Thomson AW. 2009. NOD2 ligation subverts IFN-α production by liver plasmacytoid dendritic cells and inhibits their T cell allostimulatory activity via B7-H1 up-regulation. *J. Immunol.* 183:6922–32
51. Tian J, Avalos AM, Mao SY, Chen B, Senthil K, et al. 2007. Toll-like receptor 9-dependent activation by DNA-containing immune complexes is mediated by HMGB1 and RAGE. *Nat. Immunol.* 8:487–96
52. Abrams GD, Bauer H, Sprinz H. 1963. Influence of the normal flora on mucosal morphology and cellular renewal in the ileum. A comparison of germ-free and conventional mice. *Lab. Investig.* 12:355–64
53. Slack E, Hapfelmeier S, Stecher B, Velykoredko Y, Stoel M, et al. 2009. Innate and adaptive immunity cooperate flexibly to maintain host-microbiota mutualism. *Science* 325:617–20
54. Pull SL, Doherty JM, Mills JC, Gordon JI, Stappenbeck TS. 2005. Activated macrophages are an adaptive element of the colonic epithelial progenitor niche necessary for regenerative responses to injury. *Proc. Natl. Acad. Sci. USA* 102:99–104
55. Lundin A, Bok CM, Aronsson L, Bjorkholm B, Gustafsson JA, et al. 2008. Gut flora, Toll-like receptors and nuclear receptors: a tripartite communication that tunes innate immunity in large intestine. *Cell Microbiol.* 10:1093–103
56. Cebra JJ. 1999. Influences of microbiota on intestinal immune system development. *Am. J. Clin. Nutr.* 69:S1046–51
57. Vaishnava S, Behrendt CL, Ismail AS, Eckmann L, Hooper LV. 2008. Paneth cells directly sense gut commensals and maintain homeostasis at the intestinal host-microbial interface. *Proc. Natl. Acad. Sci. USA* 105:20858–63
58. Rhee KJ, Sethupathi P, Driks A, Lanning DK, Knight KL. 2004. Role of commensal bacteria in development of gut-associated lymphoid tissues and preimmune antibody repertoire. *J. Immunol.* 172:1118–24

59. Iiyama R, Kanai T, Uraushihara K, Ishikura T, Makita S, et al. 2003. Normal development of the gut-associated lymphoid tissue except Peyer's patch in MyD88-deficient mice. *Scand. J. Immunol.* 58:620–27
60. Yu Q, Tang C, Xun S, Yajima T, Takeda K, Yoshikai Y. 2006. MyD88-dependent signaling for IL-15 production plays an important role in maintenance of CD8αα TCRαβ and TCRγδ intestinal intraepithelial lymphocytes. *J. Immunol.* 176:6180–85
61. Janeway CA Jr. 1989. Approaching the asymptote? Evolution and revolution in immunology. *Cold Spring Harb. Symp. Quant. Biol.* 54(Pt. 1):1–13
62. Gewirtz AT, Simon PO Jr, Schmitt CK, Taylor LJ, Hagedorn CH, et al. 2001. *Salmonella typhimurium* translocates flagellin across intestinal epithelia, inducing a proinflammatory response. *J. Clin. Investig.* 107:99–109
63. Rakoff-Nahoum S, Paglino J, Eslami-Varzaneh F, Edberg S, Medzhitov R. 2004. Recognition of commensal microflora by Toll-like receptors is required for intestinal homeostasis. *Cell* 118:229–41
64. Brown SL, Riehl TE, Walker MR, Geske MJ, Doherty JM, et al. 2007. Myd88-dependent positioning of Ptgs2-expressing stromal cells maintains colonic epithelial proliferation during injury. *J. Clin. Investig.* 117:258–69
65. Fukata M, Michelsen KS, Eri R, Thomas LS, Hu B, et al. 2005. Toll-like receptor-4 is required for intestinal response to epithelial injury and limiting bacterial translocation in a murine model of acute colitis. *Am. J. Physiol. Gastrointest. Liver Physiol.* 288:1055–65
66. Ivison SM, Himmel ME, Hardenberg G, Wark PA, Kifayet A, et al. 2010. TLR5 is not required for flagellin-mediated exacerbation of DSS colitis. *Inflamm. Bowel Dis.* 16:401–9
67. Lee J, Mo JH, Katakura K, Alkalay I, Rucker AN, et al. 2006. Maintenance of colonic homeostasis by distinctive apical TLR9 signalling in intestinal epithelial cells. *Nat. Cell Biol.* 8:1327–36
68. Asquith MJ, Boulard O, Powrie F, Maloy KJ. 2010. Pathogenic and protective roles of MyD88 in leukocytes and epithelial cells in mouse models of inflammatory bowel disease. *Gastroenterology* 139:519–29
69. Gibson DL, Ma C, Bergstrom KS, Huang JT, Man C, Vallance BA. 2008. MyD88 signalling plays a critical role in host defence by controlling pathogen burden and promoting epithelial cell homeostasis during *Citrobacter rodentium*-induced colitis. *Cell Microbiol.* 10:618–31
70. Lebeis SL, Bommarius B, Parkos CA, Sherman MA, Kalman D. 2007. TLR signaling mediated by MyD88 is required for a protective innate immune response by neutrophils to *Citrobacter rodentium*. *J. Immunol.* 179:566–77
71. Lebeis SL, Powell KR, Merlin D, Sherman MA, Kalman D. 2009. Interleukin-1 receptor signaling protects mice from lethal intestinal damage caused by the attaching and effacing pathogen *Citrobacter rodentium*. *Infect. Immun.* 77:604–14
72. Vijay-Kumar M, Aitken JD, Kumar A, Neish AS, Uematsu S, et al. 2008. Toll-like receptor 5-deficient mice have dysregulated intestinal gene expression and nonspecific resistance to *Salmonella*-induced typhoid-like disease. *Infect. Immun.* 76:1276–81
73. Hapfelmeier S, Stecher B, Barthel M, Kremer M, Muller AJ, et al. 2005. The *Salmonella* pathogenicity island (SPI)-2 and SPI-1 type III secretion systems allow *Salmonella* serovar *typhimurium* to trigger colitis via MyD88-dependent and MyD88-independent mechanisms. *J. Immunol.* 174:1675–85
74. Vijay-Kumar M, Wu H, Jones R, Grant G, Babbin B, et al. 2006. Flagellin suppresses epithelial apoptosis and limits disease during enteric infection. *Am. J. Pathol.* 169:1686–700
75. Rakoff-Nahoum S, Hao L, Medzhitov R. 2006. Role of Toll-like receptors in spontaneous commensal-dependent colitis. *Immunity* 25:319–29
76. Sartor RB. 2006. Mechanisms of disease: pathogenesis of Crohn's disease and ulcerative colitis. *Nat. Clin. Pract. Gastroenterol. Hepatol.* 3:390–407
77. Hugot JP, Chamaillard M, Zouali H, Lesage S, Cezard JP, et al. 2001. Association of NOD2 leucine-rich repeat variants with susceptibility to Crohn's disease. *Nature* 411:599–603
78. Ogura Y, Bonen DK, Inohara N, Nicolae DL, Chen FF, et al. 2001. A frameshift mutation in *NOD2* associated with susceptibility to Crohn's disease. *Nature* 411:603–6
79. Maeda S, Hsu LC, Liu H, Bankston LA, Iimura M, et al. 2005. *Nod2* mutation in Crohn's disease potentiates NF-κB activity and IL-1β processing. *Science* 307:734–38

80. Hedl M, Li J, Cho JH, Abraham C. 2007. Chronic stimulation of Nod2 mediates tolerance to bacterial products. *Proc. Natl. Acad. Sci. USA* 104:19440–45
81. Watanabe T, Kitani A, Murray PJ, Strober W. 2004. NOD2 is a negative regulator of Toll-like receptor 2-mediated T helper type 1 responses. *Nat. Immunol.* 5:800–8
82. Watanabe T, Kitani A, Murray PJ, Wakatsuki Y, Fuss IJ, Strober W. 2006. Nucleotide binding oligomerization domain 2 deficiency leads to dysregulated TLR2 signaling and induction of antigen-specific colitis. *Immunity* 25:473–85
83. Kobayashi KS, Chamaillard M, Ogura Y, Henegariu O, Inohara N, et al. 2005. Nod2-dependent regulation of innate and adaptive immunity in the intestinal tract. *Science* 307:731–34
84. Nenci A, Becker C, Wullaert A, Gareus R, van Loo G, et al. 2007. Epithelial NEMO links innate immunity to chronic intestinal inflammation. *Nature* 446:557–61
85. Vijay-Kumar M, Sanders CJ, Taylor RT, Kumar A, Aitken JD, et al. 2007. Deletion of TLR5 results in spontaneous colitis in mice. *J. Clin. Investig.* 117:3909–21
86. Vijay-Kumar M, Aitken JD, Carvalho FA, Cullender TC, Mwangi S, et al. 2010. Metabolic syndrome and altered gut microbiota in mice lacking Toll-like receptor 5. *Science* 328:228–31
87. Okayasu I, Hatakeyama S, Yamada M, Ohkusa T, Inagaki Y, Nakaya R. 1990. A novel method in the induction of reliable experimental acute and chronic ulcerative colitis in mice. *Gastroenterology* 98:694–702
88. Karrasch T, Kim JS, Muhlbauer M, Magness ST, Jobin C. 2007. Gnotobiotic IL-10$^{-/-}$;NF-κB^{EGFP} mice reveal the critical role of TLR/NF-κB signaling in commensal bacteria-induced colitis. *J. Immunol.* 178:6522–32
89. Carvalho FA, Nalbantoglu I, Ortega-Fernandez S, Aitken JD, Su Y, et al. 2011. IL-1β promotes susceptibility of Toll-like receptor 5 KO mice to colitis. *Gut* gut.2011.240556
90. Matharu KS, Mizoguchi E, Cotoner CA, Nguyen DD, Mingle B, et al. 2009. Toll-like receptor 4-mediated regulation of spontaneous *Helicobacter*-dependent colitis in IL-10-deficient mice. *Gastroenterology* 137:1380–90
91. Gonzalez-Navajas JM, Fine S, Law J, Datta SK, Nguyen KP, et al. 2010. TLR4 signaling in effector CD4$^+$ T cells regulates TCR activation and experimental colitis in mice. *J. Clin. Investig.* 120:570–81
92. Obermeier F, Dunger N, Strauch UG, Hofmann C, Bleich A, et al. 2005. CpG motifs of bacterial DNA essentially contribute to the perpetuation of chronic intestinal inflammation. *Gastroenterology* 129:913–27
93. Rachmilewitz D, Katakura K, Karmeli F, Hayashi T, Reinus C, et al. 2004. Toll-like receptor 9 signaling mediates the anti-inflammatory effects of probiotics in murine experimental colitis. *Gastroenterology* 126:520–28
94. Fukata M, Breglio K, Chen A, Vamadevan AS, Goo T, et al. 2008. The myeloid differentiation factor 88 (MyD88) is required for CD4$^+$ T cell effector function in a murine model of inflammatory bowel disease. *J. Immunol.* 180:1886–94
95. Heimesaat MM, Nogai A, Bereswill S, Plickert R, Fischer A, et al. 2010. MyD88/TLR9 mediated immunopathology and gut microbiota dynamics in a novel murine model of intestinal graft-versus-host disease. *Gut* 59:1079–87
96. Gewirtz AT, Vijay-Kumar M, Brant SR, Duerr RH, Nicolae DL, Cho JH. 2006. Dominant-negative TLR5 polymorphism reduces adaptive immune response to flagellin and negatively associates with Crohn's disease. *Am. J. Physiol. Gastrointest. Liver Physiol.* 290:1157–63
97. Cover TL, Blaser MJ. 2009. *Helicobacter pylori* in health and disease. *Gastroenterology* 136:1863–73
98. Turnbaugh PJ, Ley RE, Mahowald MA, Magrini V, Mardis ER, Gordon JI. 2006. An obesity-associated gut microbiome with increased capacity for energy harvest. *Nature* 444:1027–31
99. Ley RE, Backhed F, Turnbaugh P, Lozupone CA, Knight RD, Gordon JI. 2005. Obesity alters gut microbial ecology. *Proc. Natl. Acad. Sci. USA* 102:11070–75
100. Hildebrandt MA, Hoffman C, Sherrill-Mix SA, Keilbaugh SA, Hamady M, et al. 2009. High-fat diet determines the composition of the murine gut microbiome independently of obesity. *Gastroenterology* 137(5):1716–24
101. Wang Z, Klipfell E, Bennett BJ, Koeth R, Levison BS, et al. 2011. Gut flora metabolism of phosphatidylcholine promotes cardiovascular disease. *Nature* 472:57–63

102. Lupp C, Robertson ML, Wickham ME, Sekirov I, Champion OL, et al. 2007. Host-mediated inflammation disrupts the intestinal microbiota and promotes the overgrowth of Enterobacteriaceae. *Cell Host Microbe* 2:119–29
103. Hotamisligil GS, Spiegelman BM. 1994. Tumor necrosis factor α: a key component of the obesity-diabetes link. *Diabetes* 43:1271–78
104. Cani PD, Amar J, Iglesias MA, Poggi M, Knauf C, et al. 2007. Metabolic endotoxemia initiates obesity and insulin resistance. *Diabetes* 56:1761–72
105. Cani PD, Possemiers S, Van de Wiele T, Guiot Y, Everard A, et al. 2009. Changes in gut microbiota control inflammation in obese mice through a mechanism involving GLP-2-driven improvement of gut permeability. *Gut* 58:1091–103
106. Poggi M, Bastelica D, Gual P, Iglesias MA, Gremeaux T, et al. 2007. C3H/HeJ mice carrying a toll-like receptor 4 mutation are protected against the development of insulin resistance in white adipose tissue in response to a high-fat diet. *Diabetologia* 50:1267–76
107. Cani PD, Bibiloni R, Knauf C, Waget A, Neyrinck AM, Delzenne NM, Burcelin R. 2008. Changes in gut microbiota control metabolic endotoxemia-induced inflammation in high-fat diet-induced obesity and diabetes in mice. *Diabetes* 57(6):1470–81
108. Vreugdenhil AC, Rousseau CH, Hartung T, Greve JW, van't Veer C, Buurman WA. 2003. Lipopolysaccharide (LPS)-binding protein mediates LPS detoxification by chylomicrons. *J. Immunol.* 170:1399–405
109. Brugman S, Klatter FA, Visser JT, Wildeboer-Veloo AC, Harmsen HJ, et al. 2006. Antibiotic treatment partially protects against type 1 diabetes in the Bio-Breeding diabetes-prone rat. Is the gut flora involved in the development of type 1 diabetes? *Diabetologia* 49:2105–8
110. Wen L, Ley RE, Volchkov PY, Stranges PB, Avanesyan L, et al. 2008. Innate immunity and intestinal microbiota in the development of Type 1 diabetes. *Nature* 455:1109–13
111. Center MM, Jemal A, Smith RA, Ward E. 2009. Worldwide variations in colorectal cancer. *CA Cancer J. Clin.* 59:366–78
112. Eaden JA, Abrams KR, Mayberry JF. 2001. The risk of colorectal cancer in ulcerative colitis: a meta-analysis. *Gut* 48:526–35
113. Rutter M, Saunders B, Wilkinson K, Rumbles S, Schofield G, et al. 2004. Severity of inflammation is a risk factor for colorectal neoplasia in ulcerative colitis. *Gastroenterology* 126:451–59
114. Fukata M, Chen A, Vamadevan AS, Cohen J, Breglio K, et al. 2007. Toll-like receptor-4 promotes the development of colitis-associated colorectal tumors. *Gastroenterology* 133:1869–81
115. Fukata M, Hernandez Y, Conduah D, Cohen J, Chen A, et al. 2009. Innate immune signaling by Toll-like receptor-4 (TLR4) shapes the inflammatory microenvironment in colitis-associated tumors. *Inflamm. Bowel Dis.* 15:997–1006
116. Uronis JM, Muhlbauer M, Herfarth HH, Rubinas TC, Jones GS, Jobin C. 2009. Modulation of the intestinal microbiota alters colitis-associated colorectal cancer susceptibility. *PLoS ONE* 4:e6026
117. Bauer C, Duewell P, Mayer C, Lehr HA, Fitzgerald KA, et al. 2010. Colitis induced in mice with dextran sulfate sodium (DSS) is mediated by the NLRP3 inflammasome. *Gut* 59:1192–99
118. Zaki MH, Boyd KL, Vogel P, Kastan MB, Lamkanfi M, Kanneganti TD. 2010. The NLRP3 inflammasome protects against loss of epithelial integrity and mortality during experimental colitis. *Immunity* 32:379–91
119. Allen IC, TeKippe EM, Woodford RM, Uronis JM, Holl EK, et al. 2010. The NLRP3 inflammasome functions as a negative regulator of tumorigenesis during colitis-associated cancer. *J. Exp. Med.* 207:1045–56
120. Hu B, Elinav E, Huber S, Booth CJ, Strowig T, et al. 2010. Inflammation-induced tumorigenesis in the colon is regulated by caspase-1 and NLRC4. *Proc. Natl. Acad. Sci. USA* 107:21635–40
121. Dove WF, Clipson L, Gould KA, Luongo C, Marshall DJ, et al. 1997. Intestinal neoplasia in the Apc^{Min} mouse: independence from the microbial and natural killer (*beige* locus) status. *Cancer Res.* 57:812–14
122. Rakoff-Nahoum S, Medzhitov R. 2007. Regulation of spontaneous intestinal tumorigenesis through the adaptor protein MyD88. *Science* 317:124–27
123. Lee SH, Hu LL, Gonzalez-Navajas J, Seo GS, Shen C, et al. 2010. ERK activation drives intestinal tumorigenesis in $Apc^{min/+}$ mice. *Nat. Med.* 16:665–70

The Calyx of Held Synapse: From Model Synapse to Auditory Relay

J. Gerard G. Borst[*] and John Soria van Hoeve

Department of Neuroscience, Erasmus MC, University Medical Center, 3015 GE Rotterdam, The Netherlands; email: g.borst@erasmusmc.nl, j.soriavanhoeve@erasmusmc.nl

Keywords

synaptic maturation, medial nucleus of the trapezoid body, transmitter release, vesicle pools, short-term plasticity

Abstract

The calyx of Held is an axosomatic terminal in the auditory brainstem that has attracted anatomists because of its giant size and physiologists because of its accessibility to patch-clamp recordings. The calyx allows the principal neurons in the medial nucleus of the trapezoid body (MNTB) to provide inhibition that is both well timed and sustained to many other auditory nuclei. The special adaptations that allow the calyx to drive its principal neuron even when frequencies are high include a large number of release sites with low release probability, a large readily releasable pool, fast presynaptic calcium clearance and little delayed release, a large quantal size, and fast AMPA-type glutamate receptors. The transformation from a synapse that is unremarkable except for its giant size into a fast and reliable auditory relay happens in just a few days. In rodents this transformation is essentially ready when hearing starts.

[*]Corresponding author.

INTRODUCTION

Relay synapse: a synapse that faithfully passes along presynaptic signals independently of presynaptic firing frequency

The calyx of Held synapse is a giant axosomatic synapse in the auditory brainstem that functions as a fast, inverting, relay synapse. Because of its giant size, it harbors many release sites, allowing it to drive the principal neurons of the medial nucleus of the trapezoid body (MNTB) at high frequencies. The principal neurons can thus provide both well-timed and sustained inhibition to many other brainstem nuclei. The calyx of Held has long fascinated anatomists. It was first described by Hans Held using the newly developed Golgi technique (1). At approximately the turn of the nineteenth century, many of the main neuroanatomists provided descriptions of the calyx's basic anatomy (reviewed in Reference 2). For technical reasons, these studies were often performed on young animals because the Golgi technique works less well on myelinated fibers. A century later, the calyx became popular again, this time with cell physiologists, following the discovery that this terminal and its postsynaptic partner are accessible to patch-clamp recordings in brain slices (3, 4). Again, much of the work has been done on immature synapses, mainly because of the poor visibility in mature slices owing to the presence of myelin. Both because of this technical bias and because the calyx of Held synapse is easily identifiable at all developmental changes, its maturation has been well studied. In this review we argue that the large body of work on the developing calyx synapse allows a detailed answer to the question of what it takes to become an auditory relay synapse. An ideal relay synapse faithfully passes along presynaptic signals independently of presynaptic firing frequency. We first discuss the projections of the principal neurons because the impact the principal neurons have on their targets shows that not only speed but especially endurance are essential requirements for the function of the calyx of Held synapse. We then summarize what is known about the formation and early maturation of this synapse. Finally, we discuss which changes enable the calyx synapse to transform in a few days from a model synapse, which has been used to elucidate basic mechanisms of synaptic transmission and whose properties resemble those of many other synapses in the brain, into the adult relay synapse.

WHAT IS THE FUNCTION OF THE CALYX OF HELD SYNAPSE?

The Calyx Is the Main Synaptic Input of Principal Neurons

The MNTB lies within the axons of the trapezoid body, ventromedial to the medial superior olive (MSO) (**Figure 1**). Each principal neuron is contacted by a single, giant, axosomatic terminal, the calyx of Held (**Figure 2**) (2). The principal neurons receive few other inputs, and the distal dendrites are only sparsely innervated (5). Their firing behavior is dominated by this calyceal input, which can evoke large, glutamatergic, typically suprathreshold excitatory postsynaptic potentials (EPSPs) (6, 7). The calyx of Held synapse is formed by fibers originating from the globular bushy cells (GBCs) of the anteroventral cochlear nucleus (AVCN) and, to a lesser extent, the posteroventral cochlear nucleus (PVCN) (**Figure 1**) (8). Principal neurons of the MNTB are characterized by their globular or oval perikaryon of 15–20 μm containing an eccentric nucleus; they have relatively short dendrites, which branch extensively near the cell (5, 9, 10). Apart from the large calyceal input, they also receive somatic inhibitory inputs whose origin and function are unknown (11) and noncalyceal excitatory inputs, which may partially represent collaterals from the calyceal inputs (12–15).

Response to Sound of Principal Neurons

Almost all cells in the MNTB are excited only by contralateral sounds, as expected from the dominant calyceal input to the principal neurons (16). Owing to the thick, heavily myelinated

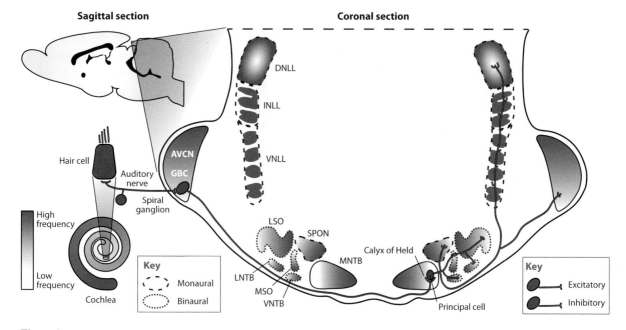

Figure 1

Anatomical connections of principal neurons. Principal neurons in the medial nucleus of the trapezoid body (MNTB) receive a single large calyceal input, which originates predominantly from globular bushy cells (GBC) in the anteroventral cochlear nucleus (AVCN). The GBC receive direct inputs from the auditory nerve. Principal neurons innervate the lateral superior olive (LSO); the superior paraolivary nucleus (SPON), which is termed the dorsal medial periolivary nucleus in nonrodents; the medial superior olive (MSO); and several other ipsilateral periolivary regions, of which only the lateral nucleus of the trapezoid body (LNTB) and ventral nucleus of the trapezoid body (VNTB) are shown (8). Projections outside the superior olivary complex include the ventral, intermediate, and dorsal nucleus of the lateral lemniscus (VNLL, INLL, and DNLL, respectively). The principal neurons project to both monaural and binaural nuclei, as indicated by the outlines of the target nuclei. The projections are typically tonotopic (7–10). The light-dark gradient inside the nuclei indicates the low-frequency (LF) to high-frequency (HF) tonotopic organization of the different nuclei (8, 16, 218).

axons of the GBCs, the minimum sound response latency of principal neurons is only 3 to 5 ms, only slightly longer than the delay of GBCs and shorter than for many other superior olivary complex (SOC) nuclei (5, 10, 17). The MNTB is tonotopically organized. High frequencies are represented medially and low frequencies laterally (10, 16). The principal neurons typically have monotonic rate-level functions at the characteristic frequency (17). During tone presentations, sustained rates typically go up to ∼300 Hz (17–20). Because the calyx synapse rarely fails, the response to sounds of principal neurons strongly resembles the responses of GBCs. This similarity includes the presence of a V-shaped frequency response area in most principal neurons (16); a predominance of high-frequency units (16); primary-like, primary-like-with-notch, or phase-locked responses to tones (5, 10, 16–18, 20, 21); and a highly variable spontaneous activity frequency, which can range from less than 1 Hz to more than 100 Hz. On average the spontaneous frequency is ∼30 Hz in rodents (17). A difference with the GBC is that the excitatory input to the principal neurons is dominated by a single input, whereas the GBCs receive many inputs. Phase locking, which is a measure for temporal precision, is therefore similar in the principal neurons and in the GBCs, although the GBCs can show better phase locking than the auditory nerve fibers (22–24).

Frequency response area: the strength of response of a neuron to tones as a function of frequency and sound pressure level

Primary like: response pattern of an auditory neuron that is characterized by a poststimulus time histogram that is similar to that of the (primary) auditory nerve fibers

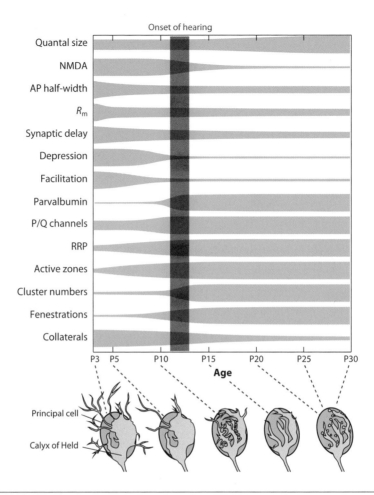

Figure 2

Timeline of major developmental steps in the rodent calyx of Held synapse. Active zones denotes number of active zones (71–73); age, age in postnatal days (P); AP half-width, full width at half-maximum of the postsynaptic action potential during slice whole-cell recordings (40, 48, 50) or estimated on the basis of juxtacellular in vivo recordings (51, 52); cluster numbers, number of vesicle clusters (74); collaterals, total length of calycine collaterals (2, 14, 45, 46); depression, amount of short-term depression during in vivo spontaneous activity (51, 52) [this measure is analogous to measurements of release probability in slices (73, 94, 139)]; facilitation, short-term facilitation decay time constant (52); fenestrations, appearance of fenestrations in the calyx of Held (45, 46); NMDA, size of synaptic NMDA-type glutamate receptor conductance (50, 92, 99); P/Q channels, fraction of presynaptic calcium channels that are P/Q type (68, 200); parvalbumin, presynaptic parvalbumin concentration (219) [a somewhat earlier onset has also been reported (203)]; quantal size, miniature excitatory postsynaptic current (EPSC) amplitude (18, 48, 49, 94, 95, 139, 220, 221); R_m, postsynaptic membrane resistance (40, 48, 106); RRP, size of the readily releasable pool (73, 94, 95, 139, 197) [for the earliest time points only the size of the EPSC has been measured (49)]; synaptic delay, delay between in vivo pre- and postsynaptic action potentials (51, 52).

Projections of the Principal Neurons

The principal neurons of the MNTB are glycinergic (reviewed in Reference 8); they can inhibit many other ipsilateral nuclei both within and outside of the SOC. The MNTB is the major source of intrinsic afferents of neurons within the SOC (8). **Figure 1** illustrates the main projections of the principal neurons.

The function of the principal neuron projections has been well studied in their main target, the lateral superior olive (LSO), and also in the superior paraolivary nucleus (SPON), whereas it is more controversial in the MSO. Neurons in the LSO are strongly inhibited by sound stimuli at the contralateral ear via the principal neurons of the MNTB, but these neurons also receive a strong excitatory input from the ipsilateral ear, directly from the cochlear nucleus. Despite the presence of an extra synapse in the contralateral pathway, the large caliber of the GBC axon and the short synaptic delay at the calyx of Held guarantee that the inhibition from the contralateral ear arrives almost at the same time as the excitation from the ipsilateral ear (reviewed in Reference 25). LSO neurons can compute the position in azimuth for complex sounds over a wide range of intensities. Although timing information is important during sound onset and in LSO neurons with low characteristic frequencies, cells in the LSO foremost make use of intensity differences between both ears to localize sounds (reviewed in Reference 25).

Similar to the situation in the LSO, the MNTB input to the SPON also has a clear tonic component. Most of the cells in the rat SPON display little or no spontaneous activity and fail to discharge during tone stimulation. These cells are tonically inhibited by the principal neurons of the MNTB, even in the absence of sound inputs, but fire at the offset of tones, when the MNTB neurons are silent (26). The cells in the SPON provide a major, monaural, inhibitory input to the inferior colliculus. The reciprocal firing behavior between the MNTB and the SPON can sharpen contrasts in amplitude-modulated sounds, improve gap detection, and play a role in sound duration coding in the rat (27). The input from the MNTB probably has a similar function in the cat and the rabbit DMPO/SPON (16, 21, 28), whereas responses in the gerbil SPON are more heterogeneous (29, 30).

The cellular function of the MNTB inputs to the LSO and SPON is thus remarkably similar. In both cases, the MNTB neurons normally provide a sustained inhibitory input by virtue of their high spontaneous rates, which reduce firing frequency or even abolish firing altogether in these targets in the absence of sound (26, 31). During sound stimulation, the input from the MNTB dominates in the case of the SPON and overrides other, excitatory inputs. In the case of the LSO, neurons tend to be completely silenced when sound intensities are higher at the contralateral ear (25).

The role of the MNTB input to the MSO is more controversial. Neurons in the MSO are excited by inputs from both ears and are specialized in detecting small differences in arrival times of sounds at both ears (reviewed in Reference 25). The inhibitory input again has a tonic function because spontaneous firing rates increase when inhibition is blocked pharmacologically (32, 33), similar to the situation in the SPON and the LSO. Blocking inhibition pharmacologically typically shifted the best delay from contralateral leading toward 0 ms (32, 33). One interpretation of this finding is that the inhibition leads the excitation from the contralateral ear. However, direct experimental support for this theory is still lacking, and this theory has been challenged on both theoretical and experimental grounds (34). Alternatively, inhibition may extend the dynamic range of responses to sound by preventing saturation of spike rates, or it may narrow the interaction window between the EPSPs from both ears, analogous to its proposed functions in the nucleus laminaris, the bird homolog of the MSO (35).

In conclusion, the main function of the principal neurons of the MNTB is to provide inhibition to many other nuclei. This inhibition has two, somewhat conflicting characteristics. It has a clear tonic component, as exemplified by the effect on the spontaneous firing frequencies of neurons in the SPON, LSO, and MSO. At the same time, this inhibition can be well timed, as exemplified by the ability of SPON neurons to detect small gaps in sounds (27) and by the ability of low-frequency LSO neurons to detect small differences in interaural arrival times (5, 23). The solution that has evolved to these two requirements is the calyx of Held synapse. In the remainder of this review, we

Primary like with notch: response pattern of an auditory neuron that is characterized by a primary-like response but with a brief notch in its poststimulus time histogram

Phase locking: the ability to respond preferentially at a particular phase of a tone

Best delay: the delay between the arrival times of inputs originating from both ears to which a neuron responds best

discuss the specializations that allow the calyx synapse to be both fast and reliable, with a special emphasis on the developmental transition from the immature model synapse that has been used extensively to characterize the basic biophysical mechanisms of transmitter release to the adult relay synapse that faithfully transmits information even at high frequencies.

WHAT MAKES THE CALYX SYNAPSE FAST?

The highly specialized anatomy of the calyx immediately suggests that speed was important in its evolution. Its axosomatic location minimizes the time needed to trigger an action potential in the axon. The large number of vesicles that can be released following a presynaptic action potential allows a single input to trigger a spike, thus minimizing jitter in the arrival time compared with the situation with multiple inputs.

Formation and Early Maturation of the Calyx Synapse

The presence of the calyx is of great importance for the proper development of the principal neuron. Its adult size (36, 37) and even its survival during early development (38, 39) may depend on the presence of the calyx.

The mechanisms that ensure that a principal neuron receives a single calyx of Held are still relatively unexplored, but some progress has been made in recent years. The earliest synaptic contacts between the GBCs and the principal neurons of the MNTB are formed in the mouse at embryonic day 17 (E17) (40). The initial innervation is divergent, and many of these contacts are dendritic (13, 14, 40). In rodents, at approximately postnatal days 2–3 (P2–3) a presumably preexisting somatic contact expands to form a protocalyx (**Figure 3**). The early calyx has many collaterals, which can contact nearby principal neurons (14). A single GBC axon can form multiple calyces (1, 41, 42), but conversely, even though there is both electrophysiological evidence and anatomical evidence suggesting that multiple calyces can contact a principal neuron (13, 43, 44), there is good evidence that the large majority of principal neurons receive only a single calyceal input.

As extreme as the adult morphology is, the formation of the calyx synapse parallels the situation generally observed elsewhere in the brain. The initial innervation is exuberant, allowing sufficient choice for the dynamic GBC axons to find the appropriate partner (14). Following the establishment of the calyx synapse, most of these superfluous connections disappear (**Figure 2**) (2, 14, 45, 46). The gradual reduction of the total area innervated by a single axon within a target nucleus is seen elsewhere as well (reviewed in Reference 47). However, in the case of the calyx formation, the end result is very extreme, as the innervated area is eventually almost fully restricted to the soma of the principal neuron.

Owing mainly to its high input resistance, the postsynaptic cell is very excitable before calyx formation (14). In the first postnatal week, its input resistance decreases (40, 48) while at the same time the calyceal synaptic currents dramatically increase in size (**Figure 2**) (13, 49). The concurrent decrease in excitability and increase in calyceal currents strongly diminish the relative impact of noncalyceal excitatory inputs.

The accessibility of both the presynaptic terminal and its postsynaptic partner to patch-clamp recordings in slices has allowed a relatively detailed dissection of the different steps of the signal transduction across the synapse. Until shortly before hearing onset, the physiology of the calyx synapse exhibits mostly gradual changes. Examples are the decrease in the duration of the presynaptic action potentials, the postsynaptic currents, or the synaptic delay (50–52). However, more dramatic changes occur shortly before hearing onset. **Figure 2** gives a schematic overview

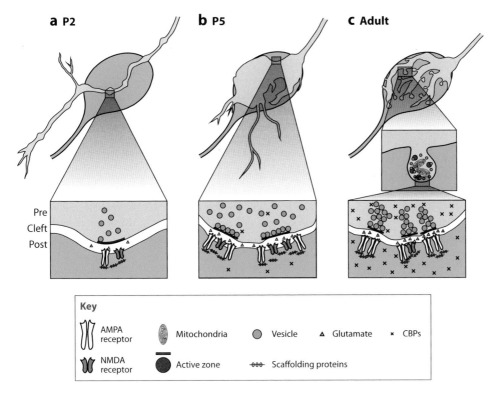

Figure 3

Schematic view of the three main stages in the development of the rodent calyx of Held synapse. (*a*) Before the calyx forms, principal neurons are contacted at the soma and dendrites by a divergent projection. (*b*) The immature calyx of Held synapse. (*c*) The mature calyx of Held synapse. Abbreviations: CBPs, calcium-binding proteins; P, postnatal day; post, postsynaptic; pre, presynaptic.

of the time course of the developmental changes discussed in the next sections, whereas **Figure 3** illustrates some of the differences between the precursor of the calyx synapse, the immature calyx synapse, and the adult relay synapse.

Presynaptic Action Potential

The calyx fires a brief, overshooting action potential (3), which is typically followed by a longer-lasting depolarizing afterpotential (4). The large size of the calyx presents a considerable capacitive load for the action potential (53, 54). To overcome this capacitive load, the last axonal heminode is unusually long and contains a high density of Na^+ channels (55). The terminal does not contain Na^+ channels. The absence of Na^+ channels not only reduces the workload of Na^+ carriers but also may produce a waveform with a shorter half-width (55). Some of the Na^+ channels have a very negative activation voltage. This sustained current contributes to the resting membrane potential and lowers the resting conductance (56). In addition, some of the Na^+ channels preferably open during repolarization. These resurgent currents contribute to the depolarizing afterpotential (57).

Both low-threshold K^+ channels and high-threshold delayed rectifier-type K^+ channels are present presynaptically. The low-threshold K^+ channels include Kv1.1 and Kv1.2 channels, which are concentrated at the transition zone between the axon and the terminal (58), and Kv1.3 and

Kv7.5 (KCNQ5) channels, which are located in the calyx (59, 60). The Kv7.5 channels have a very negative activation threshold, which makes them important for setting the resting potential (61). The Kv1 channels may reduce excitability and prevent firing on the depolarizing afterpotential (62). Deletion of Kv1.1 leads to a loss of temporal fidelity and to the inability to follow high-frequency, amplitude-modulated sound stimulation in vivo (63).

The high-threshold K^+ current is composed mostly of Kv3.1, which gates rapidly but inactivates slowly (58). It is largely excluded from the release face (64). The calyx also expresses the hyperpolarization-activated mixed cationic current I_h, which contributes little to transmitter release but can influence the resting membrane potential (65) and can contribute to a posttetanic afterhyperpolarization (66).

Changes in both Na^+ and K^+ channels cause the developmental increase in the action potential speed (50). Mature presynaptic Na^+ channels recover more rapidly from inactivation (55). The density of both Kv1 and Kv3 presynaptic K^+ currents becomes larger between P7 and P14, with little change afterward (67). The increase in Kv3.1 (64) contributes to the decrease in the width of the presynaptic action potentials.

Presynaptic Calcium Influx

In immature calyces the presynaptic action potential opens a large fraction of the high-voltage-activated Ca^{2+} currents during the repolarization phase (53). Up to hearing onset, the size of the presynaptic Ca^{2+} currents increases (49). In the immature calyx, the channels consist of a mixture of N-, R-, and P/Q-type Ca^{2+} channels. The P/Q-type channels couple more efficiently to release than the other types do, suggesting that they are more concentrated at active zones (68, 69). Around hearing onset, transmission becomes entirely dependent on P/Q-type Ca^{2+} channels (**Figure 2**) (70).

Presynaptic Transmitter Release

The most striking property of transmission at the calyx synapse is the large number of release sites. The calyx has approximately 300–700 active zones (71–74) and can release more than 100 vesicles with a single action potential (75).

Transmitter release depends critically on the amount of Ca^{2+} influx (75, 76). The relation between the Ca^{2+} concentration in the terminal and the release rates is highly nonlinear. At Ca^{2+} concentrations of between approximately 2 and 5 μM, the release scales with the fourth or fifth power of the Ca^{2+} concentration (77, 78). The Ca^{2+} sensor that triggers rapid vesicle fusion in the calyx is synaptotagmin-2 (79), an isoform with fast kinetics (80), in agreement with studies in which fast, uniform Ca^{2+} transients were used to trigger transmitter release (81).

A small elevation of the intraterminal Ca^{2+} concentration above the resting value leads to increased transmitter release (77, 78, 82, 83). Synaptotagmin-2 is unlikely to trigger this increase because in the absence of synaptotagmin-2 spontaneous release is larger instead of smaller. This finding suggests that at these low Ca^{2+} concentrations release is driven by a different Ca^{2+} sensor (79, 84), possibly one of the other synaptotagmins that are present presynaptically (85).

Postsynaptic Potentials

Release of glutamate from the calyx activates both AMPA- and NMDA-type glutamate receptors in the principal neuron (86). The AMPA receptors show fast gating owing to the relative abundance of the *flop* splice variant of the GluA4 subunit (87), which further increases its expression after

P8 (88–90), resulting in even faster gating (50, 91). AMPA-type excitatory postsynaptic currents (EPSCs) become larger during maturation, with little increase after hearing onset (92, 93). In contrast, miniature EPSCs show little change in amplitude until hearing onset (49, 94), but their amplitude becomes considerably larger afterward (**Figure 2**) (18, 95). The large quantal size is in agreement with the relatively large size of the presynaptic vesicles (71, 73). A consequence of this increase in quantal size is that fewer vesicles are needed to trigger the same conductance change in the adult principal neuron. AMPA receptors also become more concentrated at postsynaptic densities (96, 97), and scaffolding proteins such as Homer and PSD-95 may contribute to this change (98).

The NMDA-EPSC is large in the immature principal neurons (86) but is strongly reduced after hearing onset (50), even though it remains present in the adult synapse (**Figure 2**) (90, 99, 100). This developmental downregulation is postponed by cochlear ablation before hearing onset (92) or in a mouse mutant that lacks spontaneous inputs (101). The time course of the NMDA-EPSCs becomes faster during maturation (50); a switch from NR2B to NR2A subunits contributes to this speedup (99). Both the developmental decline in the NMDAR currents and their faster kinetics allow greater fidelity of transmission (50, 92, 93). One of the functions of the NMDA receptors in the adult synapse is to stimulate, in the principal neurons, the production of nitric oxide, which can regulate synaptic strength (99, 100).

Even though glutamate is thought to briefly reach millimolar concentrations in the synaptic cleft following vesicle fusion (102), a single vesicle does not saturate the postsynaptic AMPA and NMDA receptors opposite to a release site (103). Because the calyx releases many vesicles following a single action potential, the EPSP is generally suprathreshold. The soma of principal neurons has a low Na^+ channel density, and therefore the postsynaptic action potential is most likely always triggered in the axon, after which it back propagates to the soma (55, 104).

Miniature excitatory postsynaptic currents (EPSCs): EPSCs that have been recorded in the absence of presynaptic action potentials, presumably reflecting the currents evoked by the release of a single vesicle

Coincidence detector: an auditory neuron that is specialized in detecting the coincident arrival of inputs originating from both ears

Cell Physiology of the Principal Neurons

The basic physiological properties of the principal neurons are in agreement with the idea that they provide well-timed inhibition, but their passive properties indicate that they are not pure timing specialists. Timing specialists such as the principal neurons of the MSO or the octopus cells of the cochlear nucleus excel in the detection of the exact arrival times of different synaptic inputs (105), whereas synaptic integration is unimportant in the principal neurons of the MNTB, which are dominated by the single calyceal input. MSO neurons and octopus cells have very strong low-threshold K^+ conductance and I_h, allowing both a depolarized membrane potential and a very short time constant (105). Input resistance (∼100 MΩ) and membrane time constant (∼4 ms) of adult MNTB neurons are at least tenfold higher than those of, for example, MSO neurons, mostly because of a decrease after hearing onset in the latter (106). For the proper function of MSO neurons as coincidence detectors, the spontaneous inputs must remain subthreshold, whereas for the relay function of the principal neurons of the MNTB, the opposite is true. If the principal neurons of the MNTB had a membrane resistance comparable to that of the MSO neurons, this would put much greater strain on the synaptic conductance needed to reach threshold. In addition, summation of EPSPs may counteract postsynaptic spike depression (18). The input resistance of principal neurons in the MNTB is also higher than that of bushy cells of the AVCN or that of principal neurons in the LSO, in part because of a lower I_h conductance in the MNTB (107). The precision of first-spike latencies in response to tones in principal neurons of the MNTB is lower than that of, for example, the LSO neurons (108). Phase locking of rat principal neurons has a timing jitter of ∼0.15 ms (24).

Other currents that contribute to the resting conductance in the MNTB principal neurons are K$^+$ leak currents (109) and low-threshold, dendrotoxin-sensitive voltage-activated K$^+$ channels (110), which consist of Kv1.1/Kv1.2 heteromers that are located at the initial segment (111). Kv1 channels contribute to the threshold and to the amplitude of the postsynaptic action potentials (111, 112). These channels ensure that the calyceal EPSP typically triggers only a single action potential (3, 113). The axon initial segment also contains high-threshold Kv2 channels, which have slow gating, making them important during high-frequency signaling (114). In addition, the principal neurons express Kv3 channels, which have a high activation threshold and show rapid gating. These channels contribute to the brief duration of the postsynaptic action potentials (110, 115). An increased expression after hearing onset (62) further decreases the duration of the postsynaptic action potential (50), especially in medial, high-frequency cells, which have higher expression of Kv3.1 (116) and briefer postsynaptic action potentials than in the lateral, low-frequency part of the MNTB (117). This channel is regulated in a complex way by activity. Its medial-to-lateral gradient depends on hearing (118, 119) and on the presence of the fragile X mental retardation protein (120). Intense electrical activity in the MNTB leads to dephosphorylation of this channel, which in turn increases its activity, facilitating high-frequency spiking (121). In contrast, intense activity can also stimulate nitric oxide production in the principal neuron, whereas nitric oxide inhibits the postsynaptic Kv3 channels (100).

Finally, the principal neurons have a high density of Na$^+$-activated K$^+$ channels, which also contribute to temporal fidelity (122). These channels are activated by the fragile X mental retardation protein, but in contrast to the Kv3.1 channels they are especially prominent in the lateral part of the MNTB (123).

WHAT GIVES THE CALYX SYNAPSE STAMINA?

To act as a proper relay synapse, the calyx synapse must be not only fast but reliable, even during prolonged high-frequency inputs. Its many release sites share the workload, and in addition, a set of changes happening around hearing onset conspire to make the release from the calyx largely independent of firing frequency (**Figure 2**). At the same time, changes in the postsynaptic neurons ensure that, even at very high frequencies, almost all presynaptic action potentials are followed by a postsynaptic action potential.

Firing Pattern Before Hearing Onset

Neurons in the peripheral auditory system fire in bursts before hearing onset (124). The principal neurons are no exception (20, 125). This activity is caused by ATP-driven Ca^{2+} action potentials in cochlear hair cells, leading to a characteristic firing pattern, which is already present at P4 (125).

Reliability of the Calyx Synapse

At high spike rates, the synaptic delay of the calyx synapse can increase considerably (12, 64, 126–128). Nevertheless, the adult synapse typically shows few transmission failures in vivo. In the cat, postsynaptic failures have been observed only during very high frequency electrical stimulation (12), but not otherwise (127). One mouse study observed failures in approximately half of the cells (18), but another mouse study observed no failures (51). In the rat, failures were observed in approximately half of the cells (52). Failures were observed only rarely in the gerbil (129). Hence, across species, failures exist in the mature animal, but they are typically rare events. Even in the mouse, the impact of failures on tone adaptation is not very large compared with the decrease in firing rate caused by adaptation in the cochlea or cochlear nucleus (130). Failures are caused by spike depression and by stochastic fluctuations in the size of EPSPs (18). Although EPSPs are likely to be tonically depressed owing to the high spontaneous firing frequencies (131), short-term synaptic depression does not appear to make a large contribution to the fluctuations in the size of the EPSPs (18).

Before hearing onset the synapse also fails (51, 52). A difference is that before hearing onset, short-term depression plays a much more important role in the failures, even though it is counteracted by short-term facilitation at short intervals (52).

Residual Ca^{2+}: the global increase in the presynaptic Ca^{2+} concentration following an action potential

Short-Term Depression

Marked short-term depression is observed in every cell during in vivo recordings before hearing onset (51, 52). The mechanisms underlying short-term depression have been well studied in slices. The three main mechanisms that can contribute to short-term depression are depletion of the pool of vesicles with high release probability (132, 133), inactivation of presynaptic Ca^{2+} channels (134, 135), and postsynaptic receptor desensitization (73, 136, 137). Their relative contribution depends on stimulus pattern and developmental stage. At low stimulus frequencies, short-term depression is mainly presynaptic. Postsynaptic receptor desensitization becomes especially important during prolonged high-frequency stimulation in immature synapses (73, 93, 138, 139), whereas in vivo at that time, the synapse fires only a few action potentials at high frequency (51, 52), suggesting that its impact is limited in vivo.

Short-Term Facilitation and Posttetanic Potentiation

Short-term facilitation can be observed in almost every cell in vivo before hearing onset (52), and in the cases in which it is not observed, it is probably still present but obscured by synaptic depression (4). It manifests itself as an increase in the size of the responses that decays in tens to hundreds of milliseconds. Similar to the underlying mechanism at other synapses (140), residual Ca^{2+} triggers an increase in release probability (141). Small elevations of the Ca^{2+} concentration above rest can increase transmitter release appreciably (82, 83, 142–144). Residual Ca^{2+} increases the release probability in several different ways. It increases Ca^{2+} influx by speeding up the activation of P/Q-type Ca^{2+} channels (145–148). Calmodulin mediates this facilitation of the Ca^{2+} channels (149). Furthermore, residual Ca^{2+} can sum with the Ca^{2+} entering during the next action potential, leading to a somewhat higher Ca^{2+} concentration at the release sensor (150). These two mechanisms have been estimated to account for approximately half of the facilitation. In the third mechanism, residual Ca^{2+} binds to endogenous Ca^{2+} buffers, thus preventing them from competing with the Ca^{2+} entering during the next action potential (151). These three mechanisms—all of which increase the Ca^{2+} concentration at the phasic sensor for transmitter release, synaptotagmin (79)—may not fully account for facilitation. Multiple other mechanisms have been postulated (152).

Facilitation can be observed after single stimuli and decays with a time constant of less than 1 s. Ca^{2+} accumulates in the calyx during high-frequency trains, and the final decay of the residual Ca^{2+} to resting values following high-frequency trains can be much slower (143, 144) because of the slow release of Ca^{2+} from mitochondria (153, 154). If the time course of the decay of residual Ca^{2+} takes seconds or even minutes, synaptic facilitation gives way to posttetanic potentiation (143, 144).

Readily Releasable Pool

At any synapse, the number of vesicles that are released following an action potential is determined by the product of the total number of vesicles that can be released, which is generally termed the

Readily releasable pool (RRP): the pool of vesicles that can be released by a very large stimulus

readily releasable pool (RRP), and by the average release probability of these vesicles. Investigators have made great efforts to estimate the relative contribution of a reduction of size and the average release probability of the RRP to short-term depression at the calyx of Held synapse. To do this, both the size of the RRP and the release probability must be measured accurately, and much effort has gone into improving the quantification of both (77, 132, 133, 136, 155–157). To measure RRP size, a depleting stimulus is given, the total number of vesicles is estimated from total release, and replenishment is corrected for. RRP size has been estimated at between 700 and 5,000 vesicles, depending on the measurement method (77, 133, 136, 155–157). The total number of docked vesicles at the calyx of Held falls within this range, suggesting that most docked vesicles are immediately releasable (71, 73). Following pool depletion, the recovery of the RRP can be studied by giving another depleting stimulus at variable intervals. These experiments indicate that the vesicles that recover most rapidly have a relatively low release probability (132). These experiments thus indicate that the release probability of the RRP is heterogeneous, and more detailed studies indicate that at rest the RRP can be subdivided into a fast-releasing pool and a slow-releasing pool (156). Even though a lower Ca^{2+} sensitivity can contribute to the lower release probability of the slow-releasing pool (155), the difference is insufficient to explain the observed differences in release probability, indicating that a larger distance from open Ca^{2+} channels makes an important contribution (158). Recovery from short-term depression is typically quite slow. Synaptic transmission recovers with a time constant of approximately 1 s in vivo (52). In slice studies, a similar slow recovery was observed following 10-Hz stimuli (159). This slow time course is hard to reconcile with the challenge to sustain transmission at the high frequencies that the calyx synapse operates at in vivo. However, the recovery from short-term depression represents both replenishment of the RRP and an increase in the release probability, and the existence of vesicles showing rapid replenishment provides a plausible mechanism to sustain transmission at high frequencies. Various explanations for the discrepancy between the apparent recovery speed during and after a stimulus train have been provided. Interestingly, following longer trains at high frequency that lead to strong pool depletion, the recovery from short-term depression becomes biphasic (160). A model containing two pools, (*a*) a pool with high release probability but slow replenishment and (*b*) a pool with low release probability but rapid replenishment, captures many aspects of the response to trains of different frequencies at the cerebellar mossy fiber–to–granule cell synapse, including the biphasic recovery (161). Further refinement may come by taking synaptic facilitation into account (132). For example, adding facilitation to the model helps to explain the observation in auditory synapses in the avian brainstem that recovery can be faster following high-frequency trains than following low-frequency trains (162). The rapidly replenished, reluctant vesicles will be facilitated (132, 163) because release of these vesicles depends relatively strongly on the buildup of residual Ca^{2+} (156). However, to what extent these vesicles contribute to phasic release is still under debate (164, 165). An additional explanation for the use dependence of the RRP replenishment is provided by evidence that accumulation of Ca^{2+} can directly speed up replenishment of the fast-releasing pool in the calyx (160, 166) and in many other terminals (reviewed in Reference 167). Replenishment depends linearly on the presynaptic Ca^{2+} concentration (163), presumably via binding to the Ca^{2+}-binding protein calmodulin (166, 168). Replenishment of the fast-releasing pool depends on the presence of cAMP (169), which acts via the cAMP-dependent guanine nucleotide exchange factor (170, 171). Additional proteins involved in regulating the vesicular release probability include protein kinase C and munc-13, which can increase transmitter release by increasing the sensitivity to Ca^{2+} (83, 172, 173); VAMP-2 (158, 174); and synaptotagmin-2 (175). An increase in release probability and both stimulation and inhibition of vesicle replenishment have been reported for myosin light-chain kinase (153, 168, 176, 177).

The mechanisms that regulate the size of the high-release-probability pool are of great interest because these mechanisms control to a large extent the strength of individual synapses (reviewed in Reference 178). The ability to measure and control the size of different vesicle pools at the calyx of Held will therefore continue to be valuable tools in the efforts to elucidate the biochemical mechanisms that control these pools.

Bulk endocytosis: a form of endocytosis during which large endosome-like structures directly bud off from the plasma membrane

Endocytosis

The net changes in the surface area of the calyx that result from exo- and endocytosis can be estimated from its capacitance as measured under voltage clamp (54, 157). Even individual fusion events have been measured by patching the release face of the calyx after removing the principal neuron (179). Two kinetically different forms of endocytosis can be discriminated at the calyx: rapid endocytosis, which takes a few seconds, and slow endocytosis, which takes tens of seconds or longer (157). The rapid form is observed only during intense activity (180). Which biochemical mechanisms are responsible for these differences is not clear, but the available evidence suggests that both dynamin and clathrin, which characterize the classical endocytosis pathway, are involved in both forms (181–185). Evidence for the presence of rapid and slow endocytosis has also been obtained in other terminals, including the ribbon-type synapse of retinal bipolar cells, in which the underlying biochemical mechanisms are also still debated (186). Following endocytosis, vesicles become part of the recycling pool of vesicles, which has been estimated to be only 10–20% of all vesicles (182, 187).

Following very strong stimuli, vesicles may fuse with each other, leading to a mode of fusion termed compound fusion (188). A third form of endocytosis occurs under these conditions; exocytosis can overwhelm the normal endocytic capacity, and the calyx resorts to bulk endocytosis (187, 189). However, this form of endocytosis is unlikely to contribute during physiological stimuli.

Similar to its action regarding the replenishment of the RRP, Ca^{2+} can also speed up the time course of all three forms of endocytosis (190); relatively high Ca^{2+} concentrations are needed (185, 191). In immature terminals Ca^{2+} probably exerts its effects via calmodulin (190), which in turn acts via calcineurin (185). Other similarities to the RRP replenishment include the involvement of VAMP and GTP hydrolysis (182, 191). These similarities between the RRP replenishment and endocytosis lead to the obvious question to what extent the two are connected. Available evidence suggests that endocytosed vesicles are not immediately reused for the replenishment, but especially after large stimuli, release sites should be cleared before RRP replenishment can take place (184, 191, 192). To what extent the apparent Ca^{2+} dependence of the RRP replenishment can be attributed to release site clearance is not yet known.

An important point is that—apart from the large number of release sites—there is little that is exceptional about the vesicle pools in the calyx, including the way they are replenished or the way that vesicles are endocytosed. When it was possible to do similar experiments at other synapses, for example, using optical techniques, similar results were obtained (167, 178, 193, 194). The studies in the immature calyx provide information about the strategies that allow it to limit short-term depression. In vivo, the immature calyx fires at bursts of a few action potentials at 100 Hz but also has long periods of rest. In its maximum firing frequencies or the ability to sustain them, it is thus not remarkable compared with many nonauditory synapses, whereas the adult calyx is specialized in high-frequency signaling. Moreover, in contrast to the immature situation, the adult calyx shows little evidence for depression in vivo, whereas firing rates are much higher. In this respect it behaves in a manner similar to that of other relay-type synapses such as the lateral geniculate nucleus or cerebellar mossy fiber–to–granule cell synapses, which also seem able to minimize the impact of short-term depression, even at high firing rates (reviewed in Reference 195). The next section discusses in more detail the developmental changes that transform the calyx into a relay synapse.

Developmental Changes in Short-Term Depression

Just before hearing onset, the bursting firing pattern of the immature calyx makes way for the primary-like firing pattern of the adult synapse (20, 51, 52). On average, the firing rates are much higher following hearing onset, but remarkably the impact of short-term plasticity is strongly reduced (**Figure 2**) (51, 52). The synaptic changes that accompany the maturation of the calyx synapse into an auditory relay synapse have been well documented. Remarkably, many of these changes occur before hearing onset (**Figure 2**) (52) and do not seem to depend on afferent activity (101).

As discussed above, the main three causes for synaptic depression are depletion of vesicles with high release probability, Ca^{2+} channel inactivation, and postsynaptic receptor desensitization. The impact of each of these three factors is strongly reduced after hearing onset. Whereas release sites of the immature calyx of Held synapse have a release probability that is similar to that of most other synapses (178), one of the most important developmental changes occurring just before hearing onset is a strong decrease in release probability (50, 73, 94). The main cause for this decrease is the smaller duration of the presynaptic action potentials after hearing onset (50). As a result, the influx of Ca^{2+} becomes much briefer and smaller in amplitude (196). In addition, a small decrease in the Ca^{2+} affinity of the phasic sensor of transmitter release may contribute to the reduction of release probability after hearing onset (95, 197). Around hearing onset, the number of active zones increases, and their size decreases. However, the density of docked vesicles does not change, and therefore the number of docked vesicles per active zone decreases (**Figure 3**) (73). Several other developmental changes counteract the decrease in the release probability. The size of the RRP becomes larger (**Figure 2**) (50, 73, 94), effectively compensating for the decrease in release probability.

The opening of the presynaptic Ca^{2+} channel creates a steep gradient of increased intracellular Ca^{2+} concentration, which is termed a Ca^{2+} domain or nanodomain. If open channels are sufficiently close, their nanodomains will overlap and create a so-called microdomain of elevated Ca^{2+}. Several experiments suggest that in the immature calyx most released vesicles face microdomains instead of nanodomains, suggesting relatively loose coupling between Ca^{2+} channels and vesicles (53, 68, 75, 150, 198). Presynaptic RIM scaffolding proteins are important both for the localization of Ca^{2+} channels to the active zone and for vesicle docking (199).

After hearing onset, the coupling between Ca^{2+} influx and transmitter release becomes more efficient (73). A developmental replacement of N- and R-type Ca^{2+} channels, which couple less efficiently to transmitter release (68), with P/Q-type Ca^{2+} channels (200) contributes to this increase in exocytic efficiency. The diffusional distance between Ca^{2+} channels and vesicles decreases (95, 196); the developmental reorganization of the presynaptic localization of the filamentous protein septin 5 contributes to this phenomenon (201). This tighter coupling counteracts the reduction of the presynaptic influx of Ca^{2+} due to the developmental decrease in the width of the presynaptic action potential (95, 196).

Even though the time course of the slow phase of the recovery from synaptic depression is not changed (94), the effect of these changes in vesicle pools and Ca^{2+}-secretion coupling is that depletion of the RRP is strongly reduced during high-frequency signaling (50, 73, 94). These same factors reduce the contribution of activation of the presynaptic metabotropic glutamate receptor to short-term depression (94).

A second important developmental change is that clearance of Ca^{2+} becomes faster with maturation (202), which is accompanied by a strong increase in the expression of Ca^{2+}-binding proteins

(46, 203). As a result, Ca^{2+} channel inactivation during high-frequency trains is strongly reduced (73, 149).

The impact of the third main mechanism contributing to short-term depression, postsynaptic receptor desensitization, is also strongly reduced. In addition to the lower release probability, which leads to less accumulation of glutamate, glutamate is cleared more rapidly from the synaptic cleft, and a subunit switch ensures that AMPA receptors recover more rapidly from desensitization (73, 93, 138, 139). Morphological changes in the calyx probably facilitate glutamate clearance. At approximately P8 in rodents, parts of the calyx become thinner, and by P14 it becomes fenestrated, assuming its mature, floral-like structure (**Figures 2** and **3**) (46, 204). Glial processes containing glutamate transporters can occupy the newly created windows in the calyx (45). Astrocytes are entirely responsible for glutamate clearance from the calyx of Held synapse (205). Their transporters are not present in the synaptic cleft but are localized in processes next to the region of the principal cell not contacted by the calyx and in processes next to the calyx (45, 102, 206). As a result, the contribution of postsynaptic receptor desensitization to short-term depression becomes negligible after hearing onset.

Delayed release: the release following the brief, phasic release period. Also termed asynchronous release

Other morphological changes probably also contribute to an increased resistance to synaptic depression (but see Reference 207). As the calyx becomes more fenestrated, the branching of the calyx becomes more elaborate, and the second- and third-order branches typically consist of ellipsoid swellings. These swellings are often linked to the parent branches by narrow necks. These swellings vary in size from bouton-like to much larger and may constitute distinct biochemical compartments (74, 208, 209). The synaptic vesicle clusters are located predominantly in the swellings within the higher-order branches of the terminal (74). The calyx contains many mitochondria, especially in the central regions (204, 210, 211); a subset of these are attached to puncta adherentia via a cytoskeletal network termed the mitochondria-associated adherens complex. This network can be found as early as P3 (13). It places the mitochondria in the vicinity of the active zone, which may allow them to help the calyx meet metabolic demands, buffer Ca^{2+}, or synthesize glutamate (208, 212). The synaptic vesicles typically form donut-like assemblies around this central cluster of interconnected mitochondria (74).

Developmental Changes in Short-Term Facilitation

By itself, the reduction of the release probability occurring around hearing onset is expected to uncover synaptic facilitation and to transform this synapse from a predominantly depressing synapse into a facilitating synapse (4). However, in vivo observations do not support this scenario (52). After hearing onset, the amount of facilitation decreases, and the decay of facilitation becomes much faster (**Figure 2**) (52). The main cause for these events is probably a faster decay of residual Ca^{2+}. Before hearing onset, Ca^{2+} clearance speeds up during maturation (202). The concentration of the slow mobile Ca^{2+}-binding protein parvalbumin shows a strong developmental increase (**Figure 2**) (46, 203). The Ca^{2+}-binding protein calretinin also increases, but its expression is more variable than that of parvalbumin (203). The increase in presynaptic Ca^{2+} buffers contributes to a much faster decay of the residual Ca^{2+}. The faster decay of the Ca^{2+} concentration following an action potential has several consequences. As discussed above, synaptic facilitation critically depends on residual Ca^{2+}, providing a plausible explanation for the much faster decay of synaptic facilitation. The amount of delayed release decreases (49, 50, 140). The faster decay of the Ca^{2+} concentration is also expected to limit the amount of Ca^{2+} channel facilitation, thus reducing the contribution of changes in Ca^{2+} influx to short-term plasticity. The reduced buildup of residual Ca^{2+} during a high-frequency train can be expected to increase temporal precision (128). Similar mechanisms are probably responsible for the reduction of posttetanic potentiation, which depends on a long-lasting increase in residual Ca^{2+} at more mature synapses (144).

These changes are accompanied by a downregulation of the effects of group II/III metabotropic glutamate receptors (102, 213), adenosine (214), serotonin (215), and noradrenalin (216) on transmitter release. The developmental decline in the stimulatory effects of cAMP on transmitter release is consistent with the reduction of the effects of these modulators (171).

The Throughput at the Mature Calyx

With an estimated quantal content as low as 20 vesicles (18), assuming ~600 active zones (71–74), release probability is only ~3% per active zone in the adult calyx. Even during loud tones, when the calyx synapse fires at ~300 Hz (17–20), on average only ~10 vesicles per active zone per second are released. Similar estimates come from slice recordings (139, 192). This means that, even under these conditions, endocytosis, which is probably the rate-limiting step in the vesicle cycle, can keep up with exocytosis because the maximal endocytic rate has been estimated to be approximately threefold higher (190), although this rate is lower following smaller stimuli (182, 191, 217). At a spontaneous frequency of 30 Hz, ~1 vesicle per active zone per second thus has to be recycled, a challenge that is not especially remarkable compared with what other synapses can accomplish (reviewed in References 193 and 194).

The above calculations, even though they are somewhat speculative, suggest that in the calyx of Held synapse or other synapses that are specialized in high-frequency signaling, the same constraints apply to vesicle recycling as in other, less ambitious synapses. The giant size of the calyx has thus evolved as an adaptation to avoid an adult life of permanent exhaustion. In the adult calyx the many release sites in combination with the other adaptations discussed in this review allow the calyx to function as a relay synapse, even at the high firing frequencies that are common in the auditory pathway.

SUMMARY POINTS

1. Principal neurons of the medial nucleus of the trapezoid body provide inhibition to many target nuclei that not only is well timed but also has a clear tonic component.

2. Each principal neuron is contacted by a single, giant axosomatic terminal, the calyx of Held, which can drive its postsynaptic cell reliably in the adult situation, even when firing rates are high.

3. The cell physiological properties of the principal neurons are in agreement with the idea that they are not pure timing specialists.

4. Principal neurons project to monaural and binaural nuclei both within and outside of the superior olivary complex, indicating that the role of the calyx of Held synapse is not restricted to a function in sound localization.

5. The accessibility of the calyx of Held to patch-clamp recordings has allowed a detailed dissection of the role of the different metabotropic receptors, ion channels, and ion transporters that control transmitter release and of the different roles of calcium ions in controlling phasic transmitter release, short-term plasticity, and vesicle recycling.

6. The readily releasable pool of the calyx of Held is heterogeneous with regard to release probability, and different kinetic phases can be distinguished in the time course of the pool's replenishment following depletion.

7. The most important property of the calyx that allows it to function as a relay synapse is that it has so many active zones; the many active zones allow the vesicle throughput per active zone to be kept to levels that are similar to those at many other synapses.

8. Around hearing onset a change in firing pattern from bursting to primary like of the calyx of Held synapse is accompanied by many pre- and postsynaptic changes contributing to its relay function, including a decrease in release probability, changes in calcium-secretion coupling, an increase in the size of the readily releasable pool, an increase in quantal size, a faster decay of short-term facilitation, increased potassium channel conductance, faster gating of glutamate receptors, and a downregulation of NMDA receptors.

DISCLOSURE STATEMENT

The authors are not aware of any affiliations, memberships, funding, or financial holdings that might be perceived as affecting the objectivity of this review.

ACKNOWLEDGMENTS

We thank Drs. Erwin Neher, Holger Taschenberger, Takeshi Sakaba, Christopher Kushmerick, and Thomas Kuner for their helpful comments on an earlier version of this manuscript.

LITERATURE CITED

1. Held H. 1893. Die centrale Gehörleitung. *Arch. Anat. Physiol. Anat. Abt.* 201–48
2. Morest DK. 1968. The growth of synaptic endings in the mammalian brain: a study of the calyces of the trapezoid body. *Z. Anat. Entwicklungsgesch.* 127:201–20
3. Forsythe ID. 1994. Direct patch recording from identified presynaptic terminals mediating glutamatergic EPSCs in the rat CNS, in vitro. *J. Physiol.* 479:381–87
4. Borst JGG, Helmchen F, Sakmann B. 1995. Pre- and postsynaptic whole-cell recordings in the medial nucleus of the trapezoid body of the rat. *J. Physiol.* 489:825–40
5. Smith PH, Joris PX, Yin TCT. 1998. Anatomy and physiology of principal cells of the medial nucleus of the trapezoid body (MNTB) of the cat. *J. Neurophysiol.* 79:3127–42
6. Forsythe ID, Barnes-Davies M. 1993. The binaural auditory pathway: membrane currents limiting multiple action potential generation in the rat medial nucleus of the trapezoid body. *Proc. R. Soc. Lond. Ser. B* 251:143–50
7. Banks MI, Smith PH. 1992. Intracellular recordings from neurobiotin-labeled cells in brain slices of the rat medial nucleus of the trapezoid body. *J. Neurosci.* 12:2819–37
8. Thompson AM, Schofield BR. 2000. Afferent projections of the superior olivary complex. *Microsc. Res. Tech.* 51:330–54
9. Morest DK. 1968. The collateral system of the medial nucleus of the trapezoid body of the cat, its neuronal architecture and relation to the olivo-cochlear bundle. *Brain Res.* 9:288–311
10. Sommer I, Lingenhöhl K, Friauf E. 1993. Principal cells of the rat medial nucleus of the trapezoid body: an intracellular in vivo study of their physiology and morphology. *Exp. Brain Res.* 95:223–39
11. Awatramani GB, Turecek R, Trussell LO. 2004. Inhibitory control at a synaptic relay. *J. Neurosci.* 24:2643–47
12. Guinan JJ Jr, Li RY-S. 1990. Signal processing in brainstem auditory neurons which receive giant endings (calyces of Held) in the medial nucleus of the trapezoid body of the cat. *Hear. Res.* 49:321–34
13. Hoffpauir BK, Grimes JL, Mathers PH, Spirou GA. 2006. Synaptogenesis of the calyx of Held: rapid onset of function and one-to-one morphological innervation. *J. Neurosci.* 26:5511–23

14. Rodriguez-Contreras A, van Hoeve JS, Habets RLP, Locher H, Borst JGG. 2008. Dynamic development of the calyx of Held synapse. *Proc. Natl. Acad. Sci. USA* 105:5603–8
15. Hamann M, Billups B, Forsythe ID. 2003. Non-calyceal excitatory inputs mediate low fidelity synaptic transmission in rat auditory brainstem slices. *Eur. J. Neurosci.* 18:2899–902
16. Guinan JJ Jr, Norris BE, Guinan SS. 1972. Single auditory units in the superior olive complex II: tonotopic organization and locations of unit categories. *Int. J. Neurosci.* 4:147–66
17. Kopp-Scheinpflug C, Tolnai S, Malmierca MS, Rübsamen R. 2008. The medial nucleus of the trapezoid body: comparative physiology. *Neuroscience* 154:160–70
18. Lorteije JAM, Rusu SI, Kushmerick C, Borst JGG. 2009. Reliability and precision of the mouse calyx of Held synapse. *J. Neurosci.* 29:13770–84
19. Spirou GA, Brownell WE, Zidanic M. 1990. Recordings from cat trapezoid body and HRP labeling of globular bushy cell axons. *J. Neurophysiol.* 63:1169–90
20. Sonntag M, Englitz B, Kopp-Scheinpflug C, Rübsamen R. 2009. Early postnatal development of spontaneous and acoustically evoked discharge activity of principal cells of the medial nucleus of the trapezoid body: an in vivo study in mice. *J. Neurosci.* 29:9510–20
21. Guinan JJ Jr, Guinan SS, Norris BE. 1972. Single auditory units in the superior olive complex I: responses to sounds and classifications based on physiological properties. *Int. J. Neurosci.* 4:101–20
22. Joris PX, Smith PH. 2008. The volley theory and the spherical cell puzzle. *Neuroscience* 154:65–76
23. Tollin DJ, Yin TCT. 2005. Interaural phase and level difference sensitivity in low-frequency neurons in the lateral superior olive. *J. Neurosci.* 25:10648–57
24. Paolini AG, FitzGerald JV, Burkitt AN, Clark GM. 2001. Temporal processing from the auditory nerve to the medial nucleus of the trapezoid body in the rat. *Hear. Res.* 159:101–16
25. Grothe B, Pecka M, McAlpine D. 2010. Mechanisms of sound localization in mammals. *Physiol. Rev.* 90:983–1012
26. Kulesza RJ Jr, Kadner A, Berrebi AS. 2007. Distinct roles for glycine and GABA in shaping the response properties of neurons in the superior paraolivary nucleus of the rat. *J. Neurophysiol.* 97:1610–20
27. Kadner A, Berrebi AS. 2008. Encoding of temporal features of auditory stimuli in the medial nucleus of the trapezoid body and superior paraolivary nucleus of the rat. *Neuroscience* 151:868–87
28. Kuwada S, Batra R. 1999. Coding of sound envelopes by inhibitory rebound in neurons of the superior olivary complex in the unanesthetized rabbit. *J. Neurosci.* 19:2273–87
29. Behrend O, Brand A, Kapfer C, Grothe B. 2002. Auditory response properties in the superior paraolivary nucleus of the gerbil. *J. Neurophysiol.* 87:2915–28
30. Dehmel S, Kopp-Scheinpflug C, Dörrscheidt GJ, Rübsamen R. 2002. Electrophysiological characterization of the superior paraolivary nucleus in the Mongolian gerbil. *Hear. Res.* 172:18–36
31. Boudreau JC, Tsuchitani C. 1968. Binaural interaction in the cat superior olive S segment. *J. Neurophysiol.* 31:442–54
32. Brand A, Behrend O, Marquardt T, McAlpine D, Grothe B. 2002. Precise inhibition is essential for microsecond interaural time difference coding. *Nature* 417:543–47
33. Pecka M, Brand A, Behrend O, Grothe B. 2008. Interaural time difference processing in the mammalian medial superior olive: the role of glycinergic inhibition. *J. Neurosci.* 28:6914–25
34. Joris P, Yin TCT. 2007. A matter of time: internal delays in binaural processing. *Trends Neurosci.* 30:70–78
35. Burger RM, Fukui I, Ohmori H, Rubel EW. 2011. Inhibition in the balance: binaurally coupled inhibitory feedback in sound localization circuitry. *J. Neurophysiol.* 106:4–14
36. Pasic TR, Moore DR, Rubel EW. 1994. Effect of altered neuronal activity on cell size in the medial nucleus of the trapezoid body and ventral cochlear nucleus of the gerbil. *J. Comp. Neurol.* 348:111–20
37. Jean-Baptiste M, Morest DK. 1975. Transneuronal changes of synaptic endings and nuclear chromatin in the trapezoid body following cochlear ablations in cats. *J. Comp. Neurol.* 162:111–34
38. Toyoshima M, Sakurai K, Shimazaki K, Takeda Y, Shimoda Y, Watanabe K. 2009. Deficiency of neural recognition molecule NB-2 affects the development of glutamatergic auditory pathways from the ventral cochlear nucleus to the superior olivary complex in mouse. *Dev. Biol.* 336:192–200
39. Maricich SM, Xia A, Mathes EL, Wang VY, Oghalai JS, et al. 2009. *Atoh1*-lineal neurons are required for hearing and for the survival of neurons in the spiral ganglion and brainstem accessory auditory nuclei. *J. Neurosci.* 29:11123–33

40. Hoffpauir BK, Kolson DR, Mathers PH, Spirou GA. 2010. Maturation of synaptic partners: functional phenotype and synaptic organization tuned in synchrony. *J. Physiol.* 588:4365–85
41. Rodriguez-Contreras A, de Lange RPJ, Lucassen PJ, Borst JGG. 2006. Branching of calyceal afferents during postnatal development in the rat auditory brainstem. *J. Comp. Neurol.* 496:214–28
42. Kuwabara N, DiCaprio RA, Zook JM. 1991. Afferents to the medial nucleus of the trapezoid body and their collateral projections. *J. Comp. Neurol.* 314:684–706
43. Bergsman JB, De Camilli P, McCormick DA. 2004. Multiple large inputs to principal cells in the mouse medial nucleus of the trapezoid body. *J. Neurophysiol.* 92:545–52
44. Wimmer VC, Nevian T, Kuner T. 2004. Targeted in vivo expression of proteins in the calyx of Held. *Pflüg. Arch.* 449:319–33
45. Ford MC, Grothe B, Klug A. 2009. Fenestration of the calyx of Held occurs sequentially along the tonotopic axis, is influenced by afferent activity, and facilitates glutamate clearance. *J. Comp. Neurol.* 514:92–106
46. Kandler K, Friauf E. 1993. Pre- and postnatal development of efferent connections of the cochlear nucleus in the rat. *J. Comp. Neurol.* 328:161–84
47. Sanes JR, Yamagata M. 2009. Many paths to synaptic specificity. *Annu. Rev. Cell Dev. Biol.* 25:161–95
48. Rusu SI, Borst JGG. 2011. Developmental changes in intrinsic excitability of principal neurons in the rat medial nucleus of the trapezoid body. *Dev. Neurobiol.* 71:284–95
49. Chuhma N, Ohmori H. 1998. Postnatal development of phase-locked high-fidelity synaptic transmission in the medial nucleus of the trapezoid body of the rat. *J. Neurosci.* 18:512–20
50. Taschenberger H, von Gersdorff H. 2000. Fine-tuning an auditory synapse for speed and fidelity: developmental changes in presynaptic waveform, EPSC kinetics, and synaptic plasticity. *J. Neurosci.* 20:9162–73
51. Sonntag M, Englitz B, Typlt M, Rübsamen R. 2011. The calyx of Held develops adult-like dynamics and reliability by hearing onset in the mouse in vivo. *J. Neurosci.* 31:6699–709
52. Crins TTH, Rusu SI, Rodríguez-Contreras A, Borst JGG. 2011. Developmental changes in short-term plasticity at the rat calyx of Held synapse. *J. Neurosci.* 31:11706–17
53. Borst JGG, Sakmann B. 1998. Calcium current during a single action potential in a large presynaptic terminal of the rat brainstem. *J. Physiol.* 506:143–57
54. Sun J-Y, Wu XS, Wu W, Jin SX, Dondzillo A, Wu L-G. 2004. Capacitance measurements at the calyx of Held in the medial nucleus of the trapezoid body. *J. Neurosci. Methods* 134:121–31
55. Leão RM, Kushmerick C, Pinaud R, Renden R, Li G-L, et al. 2005. Presynaptic Na^+ channels: locus, development, and recovery from inactivation at a high-fidelity synapse. *J. Neurosci.* 25:3724–38
56. Huang H, Trussell LO. 2008. Control of presynaptic function by a persistent Na^+ current. *Neuron* 60:975–79
57. Kim JH, Kushmerick C, von Gersdorff H. 2010. Presynaptic resurgent Na^+ currents sculpt the action potential waveform and increase firing reliability at a CNS nerve terminal. *J. Neurosci.* 30:15479–90
58. Dodson PD, Billups B, Rusznák Z, Szûcs G, Barker MC, Forsythe ID. 2003. Presynaptic rat Kv1.2 channels suppress synaptic terminal hyperexcitability following action potential invasion. *J. Physiol.* 550:27–33
59. Gazula V-R, Strumbos JG, Mei X, Chen H, Rahner C, Kaczmarek LK. 2010. Localization of Kv1.3 channels in presynaptic terminals of brainstem auditory neurons. *J. Comp. Neurol.* 518:3205–20
60. Caminos E, Garcia-Pino E, Martinez-Galan JR, Juiz JM. 2007. The potassium channel KCNQ5/Kv7.5 is localized in synaptic endings of auditory brainstem nuclei of the rat. *J. Comp. Neurol.* 505:363–78
61. Huang H, Trussell LO. 2011. KCNQ5 channels control resting properties and release probability of a synapse. *Nat. Neurosci.* 14:840–47
62. Ishikawa T, Nakamura Y, Saitoh N, Li W-B, Iwasaki S, Takahashi T. 2003. Distinct roles of Kv1 and Kv3 potassium channels at the calyx of Held presynaptic terminal. *J. Neurosci.* 23:10445–53
63. Kopp-Scheinpflug C, Fuchs K, Lippe WR, Tempel BL, Rübsamen R. 2003. Decreased temporal precision of auditory signaling in *Kcna1*-null mice: an electrophysiological study in vivo. *J. Neurosci.* 23:9199–207
64. Elezgarai I, Díez J, Puente N, Azkue JJ, Benítez R, et al. 2003. Subcellular localization of the voltage-dependent potassium channel Kv3.1b in postnatal and adult rat medial nucleus of the trapezoid body. *Neuroscience* 118:889–98

65. Cuttle MF, Rusznák Z, Wong AY, Owens S, Forsythe ID. 2001. Modulation of a presynaptic hyperpolarization-activated cationic current (I_h) at an excitatory synaptic terminal in the rat auditory brainstem. *J. Physiol.* 534:733–44
66. Kim JH, Sizov I, Dobretsov M, von Gersdorff H. 2007. Presynaptic Ca^{2+} buffers control the strength of a fast post-tetanic hyperpolarization mediated by the α3 Na^+/K^+-ATPase. *Nat. Neurosci.* 10:196–205
67. Nakamura Y, Takahashi T. 2007. Developmental changes in potassium currents at the rat calyx of Held presynaptic terminal. *J. Physiol.* 581:1101–12
68. Wu L-G, Westenbroek RE, Borst JGG, Catterall WE, Sakmann B. 1999. Calcium channel types with distinct presynaptic localization couple differentially to transmitter release in single calyx-type synapses. *J. Neurosci.* 19:726–36
69. Inchauspe CG, Forsythe ID, Uchitel OD. 2007. Changes in synaptic transmission properties due to the expression of N-type calcium channels at the calyx of Held synapse of mice lacking P/Q-type calcium channels. *J. Physiol.* 584:835–51
70. Iwasaki S, Momiyama A, Uchitel OD, Takahashi T. 2000. Developmental changes in calcium channel types mediating central synaptic transmission. *J. Neurosci.* 20:59–65
71. Sätzler K, Söhl LF, Bollmann JH, Borst JGG, Frotscher M, et al. 2002. Three-dimensional reconstruction of a calyx of Held and its postsynaptic principal neuron in the medial nucleus of the trapezoid body. *J. Neurosci.* 22:10567–79
72. Dondzillo A, Sätzler K, Horstmann H, Altrock WD, Gundelfinger ED, Kuner T. 2010. Targeted three-dimensional immunohistochemistry reveals localization of presynaptic proteins Bassoon and Piccolo in the rat calyx of Held before and after the onset of hearing. *J. Comp. Neurol.* 518:1008–29
73. Taschenberger H, Leão RM, Rowland KC, Spirou GA, von Gersdorff H. 2002. Optimizing synaptic architecture and efficiency for high-frequency transmission. *Neuron* 36:1127–43
74. Wimmer VC, Horstmann H, Groh A, Kuner T. 2006. Donut-like topology of synaptic vesicles with a central cluster of mitochondria wrapped into membrane protrusions: a novel structure-function module of the adult calyx of Held. *J. Neurosci.* 26:109–16
75. Borst JGG, Sakmann B. 1996. Calcium influx and transmitter release in a fast CNS synapse. *Nature* 383:431–34
76. Barnes-Davies M, Forsythe ID. 1995. Pre- and postsynaptic glutamate receptors at a giant excitatory synapse in rat auditory brainstem slices. *J. Physiol.* 488:387–406
77. Schneggenburger R, Neher E. 2000. Intracellular calcium dependence of transmitter release rates at a fast central synapse. *Nature* 406:889–93
78. Bollmann JH, Sakmann B, Borst JGG. 2000. Calcium sensitivity of glutamate release in a calyx-type terminal. *Science* 289:953–57
79. Sun J, Pang ZP, Qin D, Fahim AT, Adachi R, Südhof TC. 2007. A dual-Ca^{2+}-sensor model for neurotransmitter release in a central synapse. *Nature* 450:676–82
80. Xu J, Mashimo T, Südhof TC. 2007. Synaptotagmin-1, -2, and -9: Ca^{2+} sensors for fast release that specify distinct presynaptic properties in subsets of neurons. *Neuron* 54:567–81
81. Bollmann JH, Sakmann B. 2005. Control of synaptic strength and timing by the release-site Ca^{2+} signal. *Nat. Neurosci.* 8:426–34
82. Awatramani GB, Price GD, Trussell LO. 2005. Modulation of transmitter release by presynaptic resting potential and background calcium levels. *Neuron* 48:109–21
83. Lou X, Scheuss V, Schneggenburger R. 2005. Allosteric modulation of the presynaptic Ca^{2+} sensor for vesicle fusion. *Nature* 435:497–501
84. Kochubey O, Schneggenburger R. 2011. Synaptotagmin increases the dynamic range of synapses by driving Ca^{2+}-evoked release and by clamping a near-linear remaining Ca^{2+} sensor. *Neuron* 69:736–48
85. Xiao L, Han Y, Runne H, Murray H, Kochubey O, et al. 2010. Developmental expression of Synaptotagmin isoforms in single calyx of Held-generating neurons. *Mol. Cell. Neurosci.* 44:374–85
86. Forsythe ID, Barnes-Davies M. 1993. The binaural auditory pathway: Excitatory amino acid receptors mediate dual timecourse excitatory postsynaptic currents in the rat medial nucleus of the trapezoid body. *Proc. R. Soc. Lond. Ser. B* 251:151–57

87. Geiger JRP, Melcher T, Koh D-S, Sakmann B, Seeburg PH, et al. 1995. Relative abundance of subunit mRNAs determines gating and Ca^{2+} permeability of AMPA receptors in principal neurons and interneurons in rat CNS. *Neuron* 15:193–204
88. Joshi I, Shokralla S, Titis P, Wang L-Y. 2004. The role of AMPA receptor gating in the development of high-fidelity neurotransmission at the calyx of Held synapse. *J. Neurosci.* 24:183–96
89. Koike-Tani M, Saitoh N, Takahashi T. 2005. Mechanisms underlying developmental speeding in AMPA-EPSC decay time at the calyx of Held. *J. Neurosci.* 25:199–207
90. Caicedo A, Eybalin M. 1999. Glutamate receptor phenotypes in the auditory brainstem and mid-brain of the developing rat. *Eur. J. Neurosci.* 11:51–74
91. Yang Y-M, Aitoubah J, Lauer AM, Nuriya M, Takamiya K, et al. 2011. GluA4 is indispensable for driving fast neurotransmission across a high-fidelity central synapse. *J. Physiol.* 589:4209–27
92. Futai K, Okada M, Matsuyama K, Takahashi T. 2001. High-fidelity transmission acquired via a developmental decrease in NMDA receptor expression at an auditory synapse. *J. Neurosci.* 21:3342–49
93. Joshi I, Wang L-Y. 2002. Developmental profiles of glutamate receptors and synaptic transmission at a single synapse in the mouse auditory brainstem. *J. Physiol.* 540:861–73
94. Iwasaki S, Takahashi T. 2001. Developmental regulation of transmitter release at the calyx of Held in rat auditory brainstem. *J. Physiol.* 534:861–71
95. Wang L-Y, Neher E, Taschenberger H. 2008. Synaptic vesicles in mature calyx of Held synapses sense higher nanodomain calcium concentrations during action potential-evoked glutamate release. *J. Neurosci.* 28:14450–58
96. Hermida D, Mateos JM, Elezgarai I, Puente N, Bilbao A, et al. 2010. Spatial compartmentalization of AMPA glutamate receptor subunits at the calyx of Held synapse. *J. Comp. Neurol.* 518:163–74
97. Hermida D, Elezgarai I, Puente N, Alonso V, Anabitarte N, et al. 2006. Developmental increase in postsynaptic α-amino-3-hydroxy-5-methyl-4 isoxazolepropionic acid receptor compartmentalization at the calyx of Held synapse. *J. Comp. Neurol.* 495:624–34
98. Soria Van Hoeve J, Borst JGG. 2010. Delayed appearance of the scaffolding proteins PSD-95 and Homer-1 at the developing rat calyx of Held synapse. *J. Comp. Neurol.* 518:4581–90
99. Steinert JR, Postlethwaite M, Jordan MD, Chernova T, Robinson SW, Forsythe ID. 2010. NMDAR-mediated EPSCs are maintained and accelerate in time course during maturation of mouse and rat auditory brainstem in vitro. *J. Physiol.* 588:447–63
100. Steinert JR, Kopp-Scheinpflug C, Baker C, Challiss RAJ, Mistry R, et al. 2008. Nitric oxide is a volume transmitter regulating postsynaptic excitability at a glutamatergic synapse. *Neuron* 60:642–56
101. Erazo-Fischer E, Striessnig J, Taschenberger H. 2007. The role of physiological afferent nerve activity during in vivo maturation of the calyx of Held synapse. *J. Neurosci.* 27:1725–37
102. Renden R, Taschenberger H, Puente N, Rusakov DA, Duvoisin R, et al. 2005. Glutamate transporter studies reveal the pruning of metabotropic glutamate receptors and absence of AMPA receptor desensitization at mature calyx of Held synapses. *J. Neurosci.* 25:8482–97
103. Ishikawa T, Sahara Y, Takahashi T. 2002. A single packet of transmitter does not saturate postsynaptic glutamate receptors. *Neuron* 34:613–21
104. Leão RN, Leão RM, da Costa LF, Rock Levinson S, Walmsley B. 2008. A novel role for MNTB neuron dendrites in regulating action potential amplitude and cell excitability during repetitive firing. *Eur. J. Neurosci.* 27:3095–108
105. Trussell LO. 1999. Synaptic mechanisms for coding timing in auditory neurons. *Annu. Rev. Physiol.* 61:477–96
106. Scott LL, Mathews PJ, Golding NL. 2005. Posthearing developmental refinement of temporal processing in principal neurons of the medial superior olive. *J. Neurosci.* 25:7887–95
107. Leao KE, Leao RN, Sun H, Fyffe REW, Walmsley B. 2006. Hyperpolarization-activated currents are differentially expressed in brainstem auditory nuclei. *J. Physiol.* 576(Pt. 3):849–64
108. Tsuchitani C. 1997. Input from the medial nucleus of trapezoid body to an interaural level detector. *Hear. Res.* 105:211–24
109. Berntson AK, Walmsley B. 2008. Characterization of a potassium-based leak conductance in the medial nucleus of the trapezoid body. *Hear. Res.* 244:98–106

110. Brew HM, Forsythe ID. 1995. Two voltage-dependent K^+ conductances with complementary functions in postsynaptic integration at a central auditory synapse. *J. Neurosci.* 15:8011–22

111. Dodson PD, Barker MC, Forsythe ID. 2002. Two heteromeric Kv1 potassium channels differentially regulate action potential firing. *J. Neurosci.* 22:6953–61

112. Klug A, Trussell LO. 2006. Activation and deactivation of voltage-dependent K^+ channels during synaptically driven action potentials in the MNTB. *J. Neurophysiol.* 96:1547–55

113. Wu SH, Kelly JB. 1991. Physiological properties of neurons in the mouse superior olive: membrane characteristics and postsynaptic responses studied in vitro. *J. Neurophysiol.* 65:230–46

114. Johnston J, Griffin SJ, Baker C, Skrzypiec A, Chernova T, Forsythe ID. 2008. Initial segment Kv2.2 channels mediate a slow delayed rectifier and maintain high frequency action potential firing in medial nucleus of the trapezoid body neurons. *J. Physiol.* 586:3493–509

115. Wang L-Y, Gan L, Forsythe ID, Kaczmarek LK. 1998. Contribution of the Kv3.1 potassium channel to high-frequency firing in mouse auditory neurons. *J. Physiol.* 509:183–94

116. Li W, Kaczmarek LK, Perney TM. 2001. Localization of two high-threshold potassium channel subunits in the rat central auditory system. *J. Comp. Neurol.* 437:196–218

117. Brew HM, Forsythe ID. 2005. Systematic variation of potassium current amplitudes across the tonotopic axis of the rat medial nucleus of the trapezoid body. *Hear. Res.* 206:116–32

118. von Hehn CA, Bhattacharjee A, Kaczmarek LK. 2004. Loss of Kv3.1 tonotopicity and alterations in cAMP response element-binding protein signaling in central auditory neurons of hearing impaired mice. *J. Neurosci.* 24:1936–40

119. Strumbos JG, Polley DB, Kaczmarek LK. 2010. Specific and rapid effects of acoustic stimulation on the tonotopic distribution of Kv3.1b potassium channels in the adult rat. *Neuroscience* 167:567–72

120. Strumbos JG, Brown MR, Kronengold J, Polley DB, Kaczmarek LK. 2010. Fragile X mental retardation protein is required for rapid experience-dependent regulation of the potassium channel Kv3.1b. *J. Neurosci.* 30:10263–71

121. Song P, Yang Y, Barnes-Davies M, Bhattacharjee A, Hamann M, et al. 2005. Acoustic environment determines phosphorylation state of the Kv3.1 potassium channel in auditory neurons. *Nat. Neurosci.* 8:1335–42

122. Yang B, Desai R, Kaczmarek LK. 2007. Slack and Slick K_{Na} channels regulate the accuracy of timing of auditory neurons. *J. Neurosci.* 27:2617–27

123. Brown MR, Kronengold J, Gazula V-R, Chen Y, Strumbos JG, et al. 2010. Fragile X mental retardation protein controls gating of the sodium-activated potassium channel Slack. *Nat. Neurosci.* 13:819–21

124. Blankenship AG, Feller MB. 2010. Mechanisms underlying spontaneous patterned activity in developing neural circuits. *Nat. Rev. Neurosci.* 11:18–29

125. Tritsch NX, Rodríguez-Contreras A, Crins TTH, Wang HC, Borst JGG, Bergles DE. 2010. Calcium action potentials in hair cells pattern auditory neuron activity before hearing onset. *Nat. Neurosci.* 13:1050–52

126. Tolnai S, Englitz B, Scholbach J, Jost J, Rübsamen R. 2009. Spike transmission delay at the calyx of Held in vivo: rate dependence, phenomenological modeling, and relevance for sound localization. *J. Neurophysiol.* 102:1206–17

127. Mc Laughlin M, van der Heijden M, Joris PX. 2008. How secure is in vivo synaptic transmission at the calyx of Held? *J. Neurosci.* 28:10206–19

128. Fedchyshyn MJ, Wang L-Y. 2007. Activity-dependent changes in temporal components of neurotransmission at the juvenile mouse calyx of Held synapse. *J. Physiol.* 581:581–602

129. Englitz B, Tolnai S, Typlt M, Jost J, Rübsamen R. 2009. Reliability of synaptic transmission at the synapses of Held in vivo under acoustic stimulation. *PLoS ONE* 4:e7014

130. Lorteije JAM, Borst JGG. 2011. Contribution of the mouse calyx of Held synapse to tone adaptation. *Eur. J. Neurosci.* 33:251–58

131. Hermann J, Pecka M, von Gersdorff H, Grothe B, Klug A. 2007. Synaptic transmission at the calyx of Held under in vivo-like activity levels. *J. Neurophysiol.* 98:807–20

132. Wu LG, Borst JGG. 1999. The reduced release probability of releasable vesicles during recovery from short-term synaptic depression. *Neuron* 23:821–32

133. Schneggenburger R, Meyer AC, Neher E. 1999. Released fraction and total size of a pool of immediately available transmitter quanta at a calyx synapse. *Neuron* 23:399–409
134. Forsythe ID, Tsujimoto T, Barnes-Davies M, Cuttle MF, Takahashi T. 1998. Inactivation of presynaptic calcium current contributes to synaptic depression at a fast central synapse. *Neuron* 20:797–807
135. Xu J, Wu L-G. 2005. The decrease in the presynaptic calcium current is a major cause of short-term depression at a calyx-type synapse. *Neuron* 46:633–45
136. Neher E, Sakaba T. 2001. Combining deconvolution and noise analysis for the estimation of transmitter release rates at the calyx of Held. *J. Neurosci.* 21:444–61
137. Wong AY, Graham BP, Billups B, Forsythe ID. 2003. Distinguishing between presynaptic and postsynaptic mechanisms of short-term depression during action potential trains. *J. Neurosci.* 23:4868–77
138. Koike-Tani M, Kanda T, Saitoh N, Yamashita T, Takahashi T. 2008. Involvement of AMPA receptor desensitization in short-term synaptic depression at the calyx of Held in developing rats. *J. Physiol.* 586:2263–75
139. Taschenberger H, Scheuss V, Neher E. 2005. Release kinetics, quantal parameters and their modulation during short-term depression at a developing synapse in the rat CNS. *J. Physiol.* 568:513–37
140. Zucker RS, Regehr WG. 2002. Short-term synaptic plasticity. *Annu. Rev. Physiol.* 64:355–405
141. Müller M, Felmy F, Schwaller B, Schneggenburger R. 2007. Parvalbumin is a mobile presynaptic Ca^{2+} buffer in the calyx of Held that accelerates the decay of Ca^{2+} and short-term facilitation. *J. Neurosci.* 27:2261–71
142. Hori T, Takahashi T. 2009. Mechanisms underlying short-term modulation of transmitter release by presynaptic depolarization. *J. Physiol.* 587:2987–3000
143. Habets RLP, Borst JGG. 2005. Post-tetanic potentiation in the rat calyx of Held synapse. *J. Physiol.* 564:173–87
144. Korogod N, Lou X, Schneggenburger R. 2005. Presynaptic Ca^{2+} requirements and developmental regulation of posttetanic potentiation at the calyx of Held. *J. Neurosci.* 25:5127–37
145. Borst JGG, Sakmann B. 1998. Facilitation of presynaptic calcium currents in the rat brainstem. *J. Physiol.* 513:149–55
146. Cuttle MF, Tsujimoto T, Forsythe ID, Takahashi T. 1998. Facilitation of the presynaptic calcium current at an auditory synapse in rat brainstem. *J. Physiol.* 512:723–29
147. Inchauspe CG, Martini FJ, Forsythe ID, Uchitel OD. 2004. Functional compensation of P/Q by N-type channels blocks short-term plasticity at the calyx of Held presynaptic terminal. *J. Neurosci.* 24:10379–83
148. Ishikawa T, Kaneko M, Shin HS, Takahashi T. 2005. Presynaptic N-type and P/Q-type Ca^{2+} channels mediating synaptic transmission at the calyx of Held of mice. *J. Physiol.* 568:199–209
149. Nakamura T, Yamashita T, Saitoh N, Takahashi T. 2008. Developmental changes in calcium/calmodulin-dependent inactivation of calcium currents at the rat calyx of Held. *J. Physiol.* 586:2253–61
150. Meinrenken CJ, Borst JGG, Sakmann B. 2002. Calcium secretion coupling at calyx of Held governed by nonuniform channel-vesicle topography. *J. Neurosci.* 22:1648–67
151. Felmy F, Neher E, Schneggenburger R. 2003. Probing the intracellular calcium sensitivity of transmitter release during synaptic facilitation. *Neuron* 37:801–11
152. Müller M, Felmy F, Schneggenburger R. 2008. A limited contribution of Ca^{2+} current facilitation to paired-pulse facilitation of transmitter release at the rat calyx of Held. *J. Physiol.* 586:5503–20
153. Lee JS, Kim M-H, Ho W-K, Lee S-H. 2008. Presynaptic release probability and readily releasable pool size are regulated by two independent mechanisms during posttetanic potentiation at the calyx of Held synapse. *J. Neurosci.* 28:7945–53
154. Billups B, Forsythe ID. 2002. Presynaptic mitochondrial calcium sequestration influences transmission at mammalian central synapses. *J. Neurosci.* 22:5840–47
155. Wölfel M, Lou X, Schneggenburger R. 2007. A mechanism intrinsic to the vesicle fusion machinery determines fast and slow transmitter release at a large CNS synapse. *J. Neurosci.* 27:3198–210
156. Sakaba T, Neher E. 2001. Quantitative relationship between transmitter release and calcium current at the calyx of Held synapse. *J. Neurosci.* 21:462–76
157. Sun J-Y, Wu L-G. 2001. Fast kinetics of exocytosis revealed by simultaneous measurements of presynaptic capacitance and postsynaptic currents at a central synapse. *Neuron* 30:171–82

158. Wadel K, Neher E, Sakaba T. 2007. The coupling between synaptic vesicles and Ca^{2+} channels determines fast neurotransmitter release. *Neuron* 53:563–75
159. von Gersdorff H, Schneggenburger R, Weis S, Neher E. 1997. Presynaptic depression at a calyx synapse: the small contribution of metabotropic glutamate receptors. *J. Neurosci.* 17:8137–46
160. Wang L-Y, Kaczmarek LK. 1998. High-frequency firing helps replenish the readily releasable pool of synaptic vesicles. *Nature* 394:384–88
161. Hallermann S, Fejtova A, Schmidt H, Weyhersmüller A, Silver RA, et al. 2010. Bassoon speeds vesicle reloading at a central excitatory synapse. *Neuron* 68:710–23
162. Macleod KM, Horiuchi TK. 2011. A rapid form of activity-dependent recovery from short-term synaptic depression in the intensity pathway of the auditory brainstem. *Biol. Cybern.* 104:209–23
163. Hosoi N, Sakaba T, Neher E. 2007. Quantitative analysis of calcium-dependent vesicle recruitment and its functional role at the calyx of Held synapse. *J. Neurosci.* 27:14286–98
164. Sakaba T. 2006. Roles of the fast-releasing and the slowly releasing vesicles in synaptic transmission at the calyx of Held. *J. Neurosci.* 26:5863–71
165. Müller M, Goutman JD, Kochubey O, Schneggenburger R. 2010. Interaction between facilitation and depression at a large CNS synapse reveals mechanisms of short-term plasticity. *J. Neurosci.* 30:2007–16
166. Sakaba T, Neher E. 2001. Calmodulin mediates rapid recruitment of fast-releasing synaptic vesicles at a calyx-type synapse. *Neuron* 32:1–13
167. Fioravante D, Regehr WG. 2011. Short-term forms of presynaptic plasticity. *Curr. Opin. Neurobiol.* 21:269–74
168. Lee JS, Ho W-K, Lee S-H. 2010. Post-tetanic increase in the fast-releasing synaptic vesicle pool at the expense of the slowly releasing pool. *J. Gen. Physiol.* 136:259–72
169. Sakaba T, Neher E. 2001. Preferential potentiation of fast-releasing synaptic vesicles by cAMP at the calyx of Held. *Proc. Natl. Acad. Sci. USA* 98:331–36
170. Sakaba T, Neher E. 2003. Direct modulation of synaptic vesicle priming by $GABA_B$ receptor activation at a glutamatergic synapse. *Nature* 424:775–78
171. Kaneko M, Takahashi T. 2004. Presynaptic mechanism underlying cAMP-dependent synaptic potentiation. *J. Neurosci.* 24:5202–8
172. Hori T, Takai Y, Takahashi T. 1999. Presynaptic mechanism for phorbol ester-induced synaptic potentiation. *J. Neurosci.* 19:7262–67
173. Lou X, Korogod N, Brose N, Schneggenburger R. 2008. Phorbol esters modulate spontaneous and Ca^{2+}-evoked transmitter release via acting on both Munc13 and protein kinase C. *J. Neurosci.* 28:8257–67
174. Sakaba T, Stein A, Jahn R, Neher E. 2005. Distinct kinetic changes in neurotransmitter release after SNARE protein cleavage. *Science* 309:491–94
175. Young SM Jr, Neher E. 2009. Synaptotagmin has an essential function in synaptic vesicle positioning for synchronous release in addition to its role as a calcium sensor. *Neuron* 63:482–96
176. Srinivasan G, Kim JH, von Gersdorff H. 2008. The pool of fast releasing vesicles is augmented by myosin light chain kinase inhibition at the calyx of Held synapse. *J. Neurophysiol.* 99:1810–24
177. Fioravante D, Chu Y, Myoga MH, Leitges M, Regehr WG. 2011. Calcium-dependent isoforms of protein kinase C mediate posttetanic potentiation at the calyx of Held. *Neuron* 70:1005–19
178. Branco T, Staras K. 2009. The probability of neurotransmitter release: variability and feedback control at single synapses. *Nat. Rev. Neurosci.* 10:373–83
179. He L, Wu X-S, Mohan R, Wu L-G. 2006. Two modes of fusion pore opening revealed by cell-attached recordings at a synapse. *Nature* 444:102–5
180. Wu W, Xu J, Wu X-S, Wu L-G. 2005. Activity-dependent acceleration of endocytosis at a central synapse. *J. Neurosci.* 25:11676–83
181. Xu J, McNeil B, Wu W, Nees D, Bai L, Wu L-G. 2008. GTP-independent rapid and slow endocytosis at a central synapse. *Nat. Neurosci.* 11:45–53
182. Yamashita T, Hige T, Takahashi T. 2005. Vesicle endocytosis requires dynamin-dependent GTP hydrolysis at a fast CNS synapse. *Science* 307:124–27
183. Lou X, Paradise S, Ferguson SM, De Camilli P. 2008. Selective saturation of slow endocytosis at a giant glutamatergic central synapse lacking dynamin 1. *Proc. Natl. Acad. Sci. USA* 105:17555–60

184. Wu X-S, Wu L-G. 2009. Rapid endocytosis does not recycle vesicles within the readily releasable pool. *J. Neurosci.* 29:11038–42
185. Yamashita T, Eguchi K, Saitoh N, von Gersdorff H, Takahashi T. 2010. Developmental shift to a mechanism of synaptic vesicle endocytosis requiring nanodomain Ca^{2+}. *Nat. Neurosci.* 13:838–44
186. Wu L-G, Ryan TA, Lagnado L. 2007. Modes of vesicle retrieval at ribbon synapses, calyx-type synapses, and small central synapses. *J. Neurosci.* 27:11793–802
187. de Lange RPJ, de Roos ADG, Borst JGG. 2003. Two modes of vesicle recycling in the rat calyx of Held. *J. Neurosci.* 23:10164–73
188. He L, Xue L, Xu J, McNeil BD, Bai L, et al. 2009. Compound vesicle fusion increases quantal size and potentiates synaptic transmission. *Nature* 459:93–97
189. Wu W, Wu L-G. 2007. Rapid bulk endocytosis and its kinetics of fission pore closure at a central synapse. *Proc. Natl. Acad. Sci. USA* 104:10234–39
190. Wu XS, McNeil BD, Xu J, Fan J, Xue L, et al. 2009. Ca^{2+} and calmodulin initiate all forms of endocytosis during depolarization at a nerve terminal. *Nat. Neurosci.* 12:1003–10
191. Hosoi N, Holt M, Sakaba T. 2009. Calcium dependence of exo- and endocytotic coupling at a glutamatergic synapse. *Neuron* 63:216–29
192. Neher E. 2010. What is rate-limiting during sustained synaptic activity: vesicle supply or the availability of release sites. *Front. Synaptic Neurosci.* 2:144
193. Dittman J, Ryan TA. 2009. Molecular circuitry of endocytosis at nerve terminals. *Annu. Rev. Cell Dev. Biol.* 25:133–60
194. Denker A, Rizzoli SO. 2010. Synaptic vesicle pools: an update. *Front. Synaptic Neurosci.* 2:135
195. Borst JGG. 2010. The low synaptic release probability in vivo. *Trends Neurosci.* 33:259–66
196. Fedchyshyn MJ, Wang L-Y. 2005. Developmental transformation of the release modality at the calyx of Held synapse. *J. Neurosci.* 25:4131–40
197. Kochubey O, Han Y, Schneggenburger R. 2009. Developmental regulation of the intracellular Ca^{2+} sensitivity of vesicle fusion and Ca^{2+}-secretion coupling at the rat calyx of Held. *J. Physiol.* 587:3009–23
198. Borst JGG, Sakmann B. 1999. Effect of changes in action potential shape on calcium currents and transmitter release in a calyx-type synapse of the rat auditory brainstem. *Philos. Trans. R. Soc. Lond. Ser. B* 354:347–55
199. Han Y, Kaeser PS, Südhof TC, Schneggenburger R. 2011. RIM determines Ca^{2+} channel density and vesicle docking at the presynaptic active zone. *Neuron* 69:304–16
200. Iwasaki S, Takahashi T. 1998. Developmental changes in calcium channel types mediating synaptic transmission in rat auditory brainstem. *J. Physiol.* 509:419–23
201. Yang Y-M, Fedchyshyn MJ, Grande G, Aitoubah J, Tsang CW, et al. 2010. Septins regulate developmental switching from microdomain to nanodomain coupling of Ca^{2+} influx to neurotransmitter release at a central synapse. *Neuron* 67:100–15
202. Chuhma N, Ohmori H. 2001. Differential development of Ca^{2+} dynamics in presynaptic terminal and postsynaptic neuron of the rat auditory synapse. *Brain Res.* 904:341–44
203. Felmy F, Schneggenburger R. 2004. Developmental expression of the Ca^{2+}-binding proteins calretinin and parvalbumin at the calyx of Held of rats and mice. *Eur. J. Neurosci.* 20:1473–82
204. Kil J, Kageyama GH, Semple MN, Kitzes LM. 1995. Development of ventral cochlear nucleus projections to the superior olivary complex in gerbil. *J. Comp. Neurol.* 353:317–40
205. Palmer MJ, Taschenberger H, Hull C, Tremere L, von Gersdorff H. 2003. Synaptic activation of presynaptic glutamate transporter currents in nerve terminals. *J. Neurosci.* 23:4831–41
206. Elezgarai I, Bilbao A, Mateos JM, Azkue JJ, Benítez R, et al. 2001. Group II metabotropic glutamate receptors are differentially expressed in the medial nucleus of the trapezoid body in the developing and adult rat. *Neuroscience* 104:487–98
207. Schwenger DB, Kuner T. 2010. Acute genetic perturbation of exocyst function in the rat calyx of Held impedes structural maturation, but spares synaptic transmission. *Eur. J. Neurosci.* 32:974–84
208. Rowland KC, Irby NK, Spirou GA. 2000. Specialized synapse-associated structures within the calyx of Held. *J. Neurosci.* 20:9135–44
209. Spirou GA, Chirila FV, von Gersdorff H, Manis PB. 2008. Heterogeneous Ca^{2+} influx along the adult calyx of Held: a structural and computational study. *Neuroscience* 154:171–85

210. Lenn NJ, Reese TS. 1966. The fine structure of nerve endings in the nucleus of the trapezoid body and the ventral cochlear nucleus. *Am. J. Anat.* 118:375–90
211. Nakajima Y. 1971. Fine structure of the medial nucleus of the trapezoid body of the bat with special reference to two types of synaptic endings. *J. Cell Biol.* 50:121–34
212. Perkins GA, Tjong J, Brown JM, Poquiz PH, Scott RT, et al. 2010. The micro-architecture of mitochondria at active zones: electron tomography reveals novel anchoring scaffolds and cristae structured for high-rate metabolism. *J. Neurosci.* 30:1015–26
213. Elezgarai I, Benítez R, Mateos JM, Lázaro E, Osorio A, et al. 1999. Developmental expression of the group III metabotropic glutamate receptor mGluR4a in the medial nucleus of the trapezoid body of the rat. *J. Comp. Neurol.* 411:431–40
214. Kimura M, Saitoh N, Takahashi T. 2003. Adenosine A_1 receptor-mediated presynaptic inhibition at the calyx of Held of immature rats. *J. Physiol.* 553:415–26
215. Mizutani H, Hori T, Takahashi T. 2006. 5-HT_{1B} receptor-mediated presynaptic inhibition at the calyx of Held of immature rats. *Eur. J. Neurosci.* 24:1946–54
216. Leão RM, von Gersdorff H. 2002. Noradrenaline increases high-frequency firing at the calyx of Held synapse during development by inhibiting glutamate release. *J. Neurophysiol.* 87:2297–306
217. Sun J-Y, Wu XS, Wu L-G. 2002. Single and multiple vesicle fusion induce different rates of endocytosis at a central synapse. *Nature* 417:555–59
218. Kelly JB, Liscum A, van Adel B, Ito M. 1998. Projections from the superior olive and lateral lemniscus to tonotopic regions of the rat's inferior colliculus. *Hear. Res.* 116:43–54
219. Lohmann C, Friauf E. 1996. Distribution of the calcium-binding proteins parvalbumin and calretinin in the auditory brainstem of adult and developing rats. *J. Comp. Neurol.* 367:90–109
220. Postlethwaite M, Hennig MH, Steinert JR, Graham BP, Forsythe ID. 2007. Acceleration of AMPA receptor kinetics underlies temperature-dependent changes in synaptic strength at the rat calyx of Held. *J. Physiol.* 579:69–84
221. Kushmerick C, Renden R, von Gersdorff H. 2006. Physiological temperatures reduce the rate of vesicle pool depletion and short-term depression via an acceleration of vesicle recruitment. *J. Neurosci.* 26:1366–77

Neurotransmitter Corelease: Mechanism and Physiological Role

Thomas S. Hnasko and Robert H. Edwards

Departments of Physiology & Neurology, University of California, San Francisco, California 94158-2517; email: thomas.hnasko@ucsf.edu, robert.edwards@ucsf.edu

Keywords

glutamate corelease, neurotransmitter costorage, synaptic vesicle pools, vesicular neurotransmitter transporters

Abstract

Neurotransmitter identity is a defining feature of all neurons because it constrains the type of information they convey, but many neurons release multiple transmitters. Although the physiological role for corelease has remained poorly understood, the vesicular uptake of one transmitter can regulate filling with the other by influencing expression of the H^+ electrochemical driving force. In addition, the sorting of vesicular neurotransmitter transporters and other synaptic vesicle proteins into different vesicle pools suggests the potential for distinct modes of release. Corelease thus serves multiple roles in synaptic transmission.

INTRODUCTION TO THE NEUROTRANSMITTER CYCLE

Exocytosis: the fusion of a vesicle with the plasma membrane

LDCV: large dense-core vesicle

Classical neurotransmitters: small molecules synthesized or recycled locally, transported into vesicles, and released to convey an extracellular signal

GABA: γ-aminobutyric acid

ACh: acetylcholine

VGAT: vesicular GABA (and glycine) transporter

VMAT: vesicular monoamine transporter

Plasma membrane neurotransmitter transporter: active at the plasma membrane, this protein transports neurotransmitter from the extracellular space back into the presynaptic neuron or glia

SERT: serotonin transporter

Chemical neurotransmission depends on the regulated synthesis and release of a range of soluble mediators. In the case of lipophilic or gaseous molecules such as endocannabinoids and nitric oxide, which readily penetrate biological membranes, release is regulated at the level of synthesis. However, the hydrophilic compounds that mediate most forms of both synaptic transmission and neuromodulation are packaged into vesicles that undergo regulated release by exocytosis. For neural peptides, synthesis and translocation into the secretory pathway occur at the endoplasmic reticulum, with subsequent packaging into large dense-core vesicles (LDCVs) at the *trans*-Golgi network. LDCVs then translocate to release sites in the axon or dendrites and undergo regulated release in response to the appropriate physiological stimulus. However, the time required for passage through the secretory pathway and along neuronal processes limits the capacity for sustained release and hence high-frequency transmission. Fast synaptic transmission is thus mediated by classical neurotransmitters that undergo local synthesis and recycling. Indeed, synaptic vesicles recycle locally, at the nerve terminal, through a carefully orchestrated process of exo- and endocytosis known as the synaptic vesicle cycle (1). In addition, release from rapidly recycling synaptic vesicles depends on their capacity to refill with transmitter at the nerve terminal, and presynaptic boutons have developed mechanisms to recapture released transmitter as well as to synthesize it de novo as part of a parallel, integrated process known as the neurotransmitter cycle. The expression of specialized biosynthetic enzymes and transporters required for the neurotransmitter cycle thus defines transmitter phenotype. A recent proteomic analysis indeed shows that glutamatergic and GABAergic synaptic vesicles differ primarily in the expression of vesicular transporters for glutamate and GABA (γ-aminobutyric acid) (2).

NEUROTRANSMITTER CORELEASE

Although it has generally been assumed that neurons release only one classical neurotransmitter, exceptions continue to accumulate. The first demonstration of corelease involved ATP and acetylcholine (ACh) in the electric organ of *Torpedo californica* (3, 4). Subsequent work showed that ATP is frequently stored and released with other, often cationic classical transmitters in the central and peripheral nervous systems of both invertebrates and vertebrates (for review see Reference 5). Because the vesicular GABA transporter (VGAT, also known as vesicular inhibitory amino acid transporter) also transports glycine (6), that some neurons release both inhibitory transmitters is not surprising (7–9). Similarly, the vesicular monoamine transporter VMAT2 recognizes serotonin and histamine as well as catecholamines and is expressed by essentially all monoamine neurons. The biosynthetic enzymes for different monoamines are expressed by specific subpopulations, but the plasma membrane monoamine transporters show only modest substrate selectivity, indicating the potential for uptake, storage, and release of one monoamine by a neuron that does not produce that particular transmitter. For example, the antidepressant drug fluoxetine, which selectively inhibits the plasma membrane serotonin transporter (SERT), redistributes serotonin from serotonergic to dopaminergic terminals, where serotonin also undergoes release, and this redistribution may contribute to the antidepressant action of fluoxetine (10). In addition, glutamate-releasing thalamocortical neurons (as well as some retinal ganglion cells) express SERT and VMAT2 transiently during development, conferring the ability to take up and release serotonin during the critical period for maturation of this projection (11). Conversely, many monoamine neurons corelease glutamate when grown in culture (12, 13), and dopaminergic periglomerular cells in the olfactory bulb also corelease GABA (14). Even motor neurons thought to release only ACh may corelease glutamate from collateral synapses within the spinal cord (15).

Although the evidence for the corelease of classical neurotransmitters in vivo is clear, and the occurrence more widespread than originally anticipated, the physiological significance remains largely unknown. In this review, we therefore focus on the consequences of corelease for vesicle filling, neurotransmission, synaptic plasticity, and behavior.

VESICLE FILLING

Proton Electrochemical Driving Force

The filling of synaptic vesicles with neurotransmitter depends on the energy stored in a H^+ electrochemical gradient ($\Delta\mu_{H^+}$) produced by the vacuolar-type H^+-ATPase. The vacuolar H^+ pump resembles the F0/F1 ATPase (ATP synthase) of mitochondria in structure and function. However, rather than using H^+ flux to produce ATP, the vacuolar H^+ pump uses ATP hydrolysis to drive H^+ transport into membranes of the secretory pathway, including endosomes, lysosomes, synaptic vesicles, and LDCVs (16). $\Delta\mu_{H^+}$ in turn comprises both a chemical gradient (ΔpH) and a membrane potential ($\Delta\psi$), and the transport of all classical transmitters into synaptic vesicles depends on both components (**Figure 1**). However, classical studies have shown that the different transport activities depend to differing extents on ΔpH and $\Delta\psi$ due to the charge on the substrate and the stoichiometry of coupling to H^+.

$\Delta\mu_{H^+}$: H^+ electrochemical gradient

Vacuolar-type H^+-ATPase (or H^+ pump): a complex of V0 and V1 subunits homologous to the F0 and F1 subunits of mitochondrial ATP synthase; uses ATP hydrolysis to pump H^+ into organelles

ΔpH: pH gradient

$\Delta\psi$: organelle membrane potential

VAChT: vesicular ACh transporter

VGLUT: vesicular glutamate transporter

Vesicular Transporters

The vesicular transporters for monoamines (VMAT) and ACh (VAChT) exchange two luminal H^+ for each molecule of cytosolic transmitter (17–19). However, only the charged monoamine is recognized, and ACh is permanently protonated, so each transport cycle results in a net loss from the lumen of 2 H^+ but only +1 charge, accounting for the greater dependence of these activities on ΔpH than on $\Delta\psi$. The greater consumption of ΔpH than of $\Delta\psi$ in turn requires the replacement of more H^+ than charge by the H^+ pump. Because the number of charges pumped by the H^+-ATPase must equal the number of H^+, the regeneration of the gradients dissipated by vesicular monoamine and ACh transport thus requires an additional mechanism that can restore the necessary balance.

VGAT recognizes glycine as well as GABA. GABA and glycine exchange for an unknown number of H^+, and as zwitterions, their uptake depends equally on ΔpH and on $\Delta\psi$ (20, 21). Despite the clear role for ΔpH in vesicular GABA transport, recent work using functional reconstitution of purified mammalian VGAT has suggested that the activity requires cotransport of 2 Cl^- and hence relies predominantly if not exclusively on $\Delta\psi$ (22). Previous work had not identified a requirement for Cl^-, but the apparent affinity of VGAT for Cl^- appears high, suggesting that only low concentrations may be required (22). However, the assays used may reflect only kinetics, and determining the stoichiometry using thermodynamic measurements at equilibrium will be important.

In contrast to VMAT and VAChT, the vesicular glutamate transporters (VGLUTs) depend primarily on $\Delta\psi$. The three isoforms (VGLUT1–3) exhibit generally complementary patterns of expression in the brain but very similar transport activity (reviewed in References 23–25). Despite the primary reliance on $\Delta\psi$, VGLUT activity retains some dependence on ΔpH even after dissipation of $\Delta\psi$ (26, 27), suggesting that the mechanism involves H^+ exchange. Independently of H^+ coupling, however, glutamate uptake depends more on $\Delta\psi$ because at neutral pH, glutamate is anionic. If exchanged for nH^+ (and the stoichiometry of coupling remains unknown), glutamate influx results in an efflux of $n + 1$ charge. The VGLUTs thus produce an imbalance between

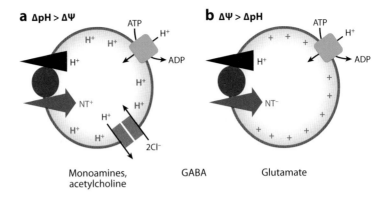

Figure 1
Vesicular neurotransmitter transporters depend differentially on the chemical and electrical components of the H⁺ electrochemical gradient ($\Delta\mu_{H+}$). The vacuolar-type H⁺-ATPase generates the $\Delta\mu_{H+}$ required for transport of all classical neurotransmitters into synaptic vesicles. However, different vesicular neurotransmitter transporters rely to differing extents on the two components of $\Delta\mu_{H+}$: the chemical gradient (ΔpH) and the electrical gradient ($\Delta\psi$). (*a*) The vesicular accumulation of monoamines and acetylcholine (ACh) involves the exchange of protonated cytosolic transmitter for two luminal H⁺. The resulting movement of more H⁺ than charge dictates a greater dependence on ΔpH than on $\Delta\psi$ for both vesicular ACh transport and vesicular monoamine transport. (*b*) Vesicular glutamate transport may not involve H⁺ translocation. In the absence of $\Delta\psi$, however, disruption of ΔpH inhibits uptake, suggesting that the transport of anionic glutamate involves exchange for nH⁺, resulting in the movement of $n + 1$ charge and hence greater dependence on $\Delta\psi$ than on ΔpH. Transport of the neutral zwitterion GABA (and glycine) involves the movement of an equal number of H⁺ and charge, consistent with the similar dependence of vesicular GABA transporter on ΔpH and $\Delta\psi$. These differences suggest that vesicles storing monoamines or ACh may have mechanisms to favor the accumulation of ΔpH at the expense of $\Delta\psi$, whereas those storing glutamate may promote a larger $\Delta\psi$. The extent to which vesicles differ in their expression of these two components remains unknown, but intracellular Cl⁻ carriers such as the synaptic vesicle–associated ClC-3 promote vesicle acidification by dissipating the positive $\Delta\psi$ developed by the vacuolar H⁺ pump, thereby disinhibiting the pump to make larger ΔpH. The vesicular glutamate transporters (VGLUTs) can also contribute to ΔpH formation because entry of glutamate as an anion similarly dissipates $\Delta\psi$ to promote ΔpH. Interestingly, a Cl⁻ conductance associated with the VGLUTs may also promote acidification by Cl⁻ (41).

Vesicular neurotransmitter transporter: located on secretory vesicles, this class of transporters uses the electrochemical gradient produced by the vacuolar proton pump to fill vesicles with neurotransmitter

charge and H⁺ similar to but opposite that created by VMAT and VAChT, implicating additional mechanisms to balance the two components of $\Delta\mu_{H+}$ so that the H⁺ pump can continue to function.

It is widely assumed that the expression of a vesicular neurotransmitter transporter confers the potential for regulated release of available substrate. Indeed, all known transporters contain signals that target them to endocytic vesicles, even in non-neural cells (28), and the expression of $\Delta\mu_{H+}$ by endosomes should in principle drive their activity. Heterologous expression of the VMATs by a range of cell lines indeed confers robust monoamine uptake by endosomes. However, measuring the activity of other vesicular transporters after heterologous expression has been extremely difficult, perhaps because they have a much lower apparent affinity (low millimolar K_m) for substrate than do the VMATs ($K_m < 1$ μM), but at least in some cases perhaps because the endosomes of non-neural cells lack essential components such as factors that regulate the expression of $\Delta\mu_{H+}$ as ΔpH or $\Delta\psi$.

THE REGULATION OF ΔpH BY ANION FLUX

More attention has focused on the factors that promote the formation of ΔpH than on those promoting Δψ because organelle ΔpH is easier to measure than Δψ and because it is presumed to have a more important biological role: in ligand dissociation from receptors within the endocytic pathway, in the processing of propeptides within the biosynthetic pathway, and in proteolytic degradation within lysosomes, as well as in vesicular neurotransmitter transport. Importantly, in vitro studies have repeatedly shown that the simple addition of ATP to activate the H^+ pump does not suffice to produce substantial ΔpH. With activation of the H^+ pump, Δψ accumulates before the bulk concentration of H^+ increases, arresting the activity of the pump before the development of ΔpH. Dissipation of Δψ, generally considered to involve anion entry, allows the ATPase to continue pumping H^+ and to produce ΔpH.

ClC: a family of Cl^- carriers including the intracellular members that mediate Cl^-/H^+ exchange

Chloride

The principal anion involved in vesicle acidification is presumed to be Cl^-. In the absence of Cl^-, synaptic vesicles and other isolated organelles show only a small acidification upon the addition of ATP. The addition of Cl^- then leads to a concentration-dependent increase in ΔpH, presumably by dissipating Δψ (26, 27, 29). Intracellular members of the ClC Cl^- channel family are considered to mediate the Cl^- permeability of acidic vesicles, with ClC-3 the predominant but probably not the only isoform on synaptic vesicles (30). Interestingly, work on the related ClCs 4–7 as well as on a bacterial homolog shows that these proteins do not function as channels but rather as Cl^-/H^+ exchangers with a stoichiometry of $2Cl^-:1H^+$ (31–35). In this case, Cl^- entry is coupled to H^+ efflux, which seems counterproductive because Cl^- entry acts primarily to increase ΔpH. In the case of ClCs, however, the loss of 1 H^+ is accompanied by the loss of +3 charge, dissipating Δψ more than ΔpH and thus stimulating the H^+-ATPase to replenish these gradients. For an equivalent $[Cl^-]$ gradient, $2Cl^-:1H^+$ exchange would thus produce a larger ΔpH than a simple Cl^- channel would (29, 36, 37). For $2Cl^-:1H^+$ exchange, the concentration gradient of Cl^- at equilibrium is predicted by the equation

$$2\log_{10}([Cl^-]_i/[Cl^-]_o) = \log_{10}([H^+]_i/[H^+]_o) + 3\Delta\psi/(2.3RT/F), \qquad 1.$$

where R is the gas constant, T is the absolute temperature, F is Faraday's constant, and the vATPase determines ΔpH and Δψ. Estimating that the proton pump can generate a total $\Delta\mu_{H+} \sim 3$ (i.e., ΔpH $\sim$ 3 pH units, Δψ $\sim$180 mV, or a combination of both) (38, 39),

$$3 = \log_{10}([H^+]_i/[H^+]_o) + \Delta\psi/(2.3RT/F). \qquad 2.$$

Replacing Δψ in Equation 1 with $(2.3RT/F)(3 - \log_{10}([H^+]_i/[H^+]_o))$ predicts

$$\log_{10}([H^+]_i/[H^+]_o) = 4.5 - \log_{10}([Cl^-]_i/[Cl^-]_o). \qquad 3.$$

In contrast, if the ClC or another protein present on synaptic vesicles functioned as a simple Cl^- channel, the concentration gradient of Cl^- at equilibrium would be predicted by the Nernst equation:

$$\log_{10}([Cl^-]_i/[Cl^-]_o) = \Delta\psi/(2.3RT/F).$$

Replacing Δψ with $(2.3RT/F)(3 - \log_{10}([H^+]_i/[H^+]_o))$, as above,

$$\log_{10}([H^+]_i/[H^+]_o) = 3 - \log_{10}([Cl^-]_i/[Cl^-]_o). \qquad 4.$$

For an equivalent concentration gradient of anion, the H^+ exchange mechanism thus counterintuitively produces a substantially larger ΔpH (by 1.5 pH units) than does a simple ion channel.

KO: knockout

Conversion of two ClCs into Cl⁻ channels in knockin mice indeed impaired the function of the endocytic pathway (36, 37). However, no change in acidification was observed, raising the possibility that the two mechanisms differ primarily in the luminal concentration of Cl⁻. It is unclear why changes in luminal Cl⁻ would affect the function of the endocytic pathway if not through a change in ΔpH, but the anion gradients likely differ between the two mechanisms.

Considering the established role of ClCs in endosome/lysosome acidification, it is surprising that recent work has suggested a primary role for the VGLUTs in Cl⁻ flux by synaptic vesicles. Originally, the analysis of ClC-3 knockout (KO) mice had suggested a role for that isoform in the acidification of synaptic vesicles, but the analysis was complicated by severe degeneration of the hippocampus and the retina (30). In younger ClC-3 KO mice, the defect appeared much less significant (40, 41). In contrast, synaptic vesicles from VGLUT1 KO mice showed a more profound defect in acidification due to Cl⁻ (41), suggesting that the VGLUTs mediate Cl⁻ flux by synaptic vesicles. Indeed, the expression of other so-called type I phosphate transporters of the VGLUT family confers a Cl⁻ conductance (42), and the VGLUTs also promote acidification of synaptic vesicles by Cl⁻ (41, 43). In addition to their essential role in packaging glutamate, the VGLUTs may thus exhibit Cl⁻ channel activity, and vesicular glutamate transport shows a clear biphasic dependence on Cl⁻ (26, 27). In addition, the Cl⁻ dependence of glutamate transport may reflect allosteric activation rather than effects on the driving force (44).

However, it remains unclear how a Cl⁻ conductance might contribute to the kinetic properties of glutamate transport. Recent work has indeed failed to detect any Cl⁻ flux after functional reconstitution of purified VGLUT2 (45), and the analysis involved direct measurement of flux rather than indirect effects on acidification. Thus, whether the VGLUTs and/or ClCs mediate Cl⁻ entry into glutamatergic synaptic vesicles remains uncertain. Taken together, however, the data suggest that synaptic vesicles storing glutamate, which are the most abundant in brain, express more VGLUT than ClC; Cl⁻ entry would indeed dissipate the $\Delta\psi$ required for vesicular glutamate transport, and previous work has suggested that substrates can inhibit the Cl⁻ conductance associated with VGLUTs and related proteins (42, 43).

What then would be the role for a Cl⁻ conductance associated with glutamatergic vesicles? Recent work in reconstituted proteoliposomes has suggested that Cl⁻ efflux can promote glutamate uptake (41). Immediately after endocytosis, synaptic vesicles should contain large amounts of Cl⁻ captured from the extracellular space. Although luminal Cl⁻ may exchange directly for cytosolic glutamate, Cl⁻ efflux more likely generates the $\Delta\psi$ required for vesicular glutamate transport, and this possibility requires direct testing. A high priority is to determine whether luminal Cl⁻ influences the filling of native synaptic vesicles, rather than simply the filling of artificial membranes whose much larger size may confer new properties. In any case, the acidification of nonglutamatergic synaptic vesicles presumably depends on ClCs, and ClC-3 may be only one of several isoforms involved. Indeed, recent work using ClC-3 KO mice has shown major defects in GABA release, apparently due to the impaired acidification of GABAergic synaptic vesicles (40).

Previous work has also demonstrated the synergistic effect of ATP (also an anion) on serotonin uptake by chromaffin granules (46). Although ATP is present in all cells, this effect presumably requires vesicular nucleotide transport, which may occur only in cells that release ATP. Manipulation of the recently described vesicular nucleotide transporter VNUT (47) will therefore be required to assess the physiological role of ATP.

Glutamate

Independently of the Cl⁻ flux that may be mediated by VGLUTs, vesicular glutamate transport itself has profound effects on ΔpH (26, 27). As an anion, glutamate, like Cl⁻, dissipates $\Delta\psi$

and hence promotes ΔpH. Indeed, glutamate alone acidifies synaptic vesicles in the presence of ATP to activate the H^+ pump, presumably reflecting the abundance of glutamatergic vesicles in the mammalian brain. We do not know the stoichiometry of ionic coupling by the VGLUTs, but the sensitivity to ΔpH (27, 43, 48) supports a H^+ exchange mechanism despite the primary dependence of VGLUTs on $\Delta\psi$. Assuming the exchange of 1 H^+ for 1 glutamate and hence the movement of +2 charge,

VTA: ventral tegmental area

$$\log_{10}([glu^-]_i/[glu^-]_o) = \log_{10}([H^+]_i/[H^+]_o) + 2\Delta\psi/(2.3RT/F). \qquad 5.$$

Again replacing $\Delta\psi$ with $(2.3RT/F)(3 - \log_{10}([H^+]_i/[H^+]_o))$,

$$\log_{10}([H^+]_i/[H^+]_o) = 6 - \log_{10}([glu^-]_i/[glu^-]_o). \qquad 6.$$

For a given anion gradient, glutamate flux through the VGLUTs (Equation 6) is therefore predicted to generate ΔpH 1.5 units greater than Cl^- flux does through even an intracellular ClC (Equation 3) and 3 units greater than Cl^- flux does through a channel (Equation 4).

Consistent with these predictions, we found that different anions have nonredundant effects on vesicle filling with transmitter (29), presumably by producing different ΔpH. Although Cl^- suffices to promote ΔpH and to stimulate the ΔpH-dependent storage of cationic transmitters (29, 49), we and others found that glutamate can also increase the packaging of monoamines (29, 50, 51) and ACh (52) into isolated synaptic vesicles. Indeed, a subset of monoamine and cholinergic neurons express VGLUTs: A number of catecholamine populations including midbrain dopamine neurons in the ventral tegmental area (VTA) express VGLUT2 (53, 54), whereas serotonergic neurons in the dorsal raphe and cholinergic interneurons in the striatum express VGLUT3 (55–57; reviewed in Reference 25). However, it is unclear how glutamate promotes vesicle filling in the presence of substantially higher cytosolic Cl^- concentrations, and most previous work showing stimulation of vesicle filling by glutamate has relied on very low Cl^- (50–52). We recently found that the effects of glutamate on monoamine filling persist even at physiological Cl^- (20 mM) (29), indicating that the two anions do not have redundant roles. Surprisingly, glutamate produces larger synaptic vesicle pH gradients than does Cl^- at concentrations up to ~12 mM. The acidification by glutamate saturates at concentrations greater than 2–4 mM, consistent with the known VGLUT K_m (1–3 mM). In addition, the acidification produced by glutamate is more stable than that produced by Cl^-: After inhibition of the H^+ pump, ΔpH collapses immediately in vesicles acidified with Cl^-, but much more slowly in those acidified with glutamate (29). Although glutamate has a much higher pK_a than does Cl^- and can thus serve as a better buffer, the increased stability of ΔpH is in large part attributable to the mechanism of anion flux. In the absence of an electrical shunt, synaptic vesicle ΔpH is quite stable because the efflux of H^+ will create a negative $\Delta\psi$ that opposes further efflux. In the case of vesicles acidified with Cl^-, however, H^+ can leave the vesicle because Cl^- efflux through a channel-like mechanism dissipates $\Delta\psi$. In the case of vesicles acidified with glutamate, glutamate cannot leave the vesicle because the H^+ exchange mechanism opposes coupled H^+ influx into acidic vesicles. In contrast to a channel, the H^+ exchange mechanism thus serves to lock H^+ inside synaptic vesicles, stabilizing ΔpH and promoting vesicular uptake of monoamines and ACh (**Figure 2**) (29).

Using KO mice, recent work has demonstrated the physiological significance of VGLUT coexpression with VMAT2 or VAChT on synaptic vesicles in vivo. Originally, there was some concern that adult dopamine neurons did not express VGLUT2 (58, 59), and expression does appear to be highest early in development or after injury (60–63). However, mature conditional knockout (cKO) mice lacking VGLUT2 selectively in dopamine neurons clearly show a reduction in both dopamine storage and evoked dopamine release (29) that presumably accounts for their reduced response to psychostimulants (29, 64). The reduction is anatomically restricted to the ventral

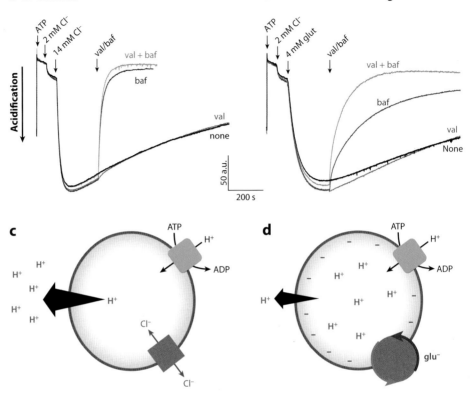

Figure 2

Glutamate flux produces larger and more stable changes in vesicular ΔpH than does Cl^-. Changes in ΔpH of isolated synaptic vesicles were monitored using acridine orange (5 µM) in 140 mM choline gluconate, 10 mM K^+ gluconate, 10 mM HEPES, pH 7.4. Acidification was triggered by the sequential addition of 1 mM ATP and 2 mM Cl^- followed by either (*a*) 14 mM Cl^- or (*b*) 4 mM glutamate; more Cl^- is required to produce an equivalent initial change in ΔpH. The traces in black indicate vesicles without any further addition. Where indicated by the arrows, the K^+ ionophore valinomycin (val) (50 nM; *gray*), the proton pump inhibitor bafilomycin (baf) (250 nM; *dark blue/red*), or both (*light blue/pink*) were added. The rate of alkalinization immediately after bafilomycin addition (*dark blue/red*) is much faster in the vesicles acidified with Cl^-, indicating that vesicles acidified with glutamate maintain a more stable ΔpH. Although increased buffering may contribute to the stabilization of ΔpH by glutamate, valinomycin accelerates the bafilomycin-induced collapse in ΔpH across membranes acidified with glutamate (*pink*), but not across membranes acidified with Cl^- (*light blue*), indicating an important role for negative $\Delta\psi$ in the stability of ΔpH in glutamate-acidified vesicles. We hypothesize that the negative $\Delta\psi$ developing upon H^+ efflux impedes further dissipation of ΔpH. In the case of vesicles acidified with Cl^-, anion efflux through a channel (*c*) would shunt the developing negative $\Delta\psi$, allowing the continued efflux of H^+ and rapid collapse of ΔpH. In the case of vesicles acidified with glutamate, a H^+/anion exchange mechanism (*d*) would impede glutamate efflux because it would be coupled to the uphill movement of H^+ into acidic vesicles. Because glutamate efflux is disfavored, H^+ efflux is slow and ΔpH more stable. Thus, the differences in mechanism of anion flux (channel versus H^+ exchange) confer differences in the stability of ΔpH. Glutamate thus serves to lock H^+, and hence cationic transmitters such as acetylcholine and monoamines, inside secretory vesicles. Panels *a* and *b* reproduced from Reference 29 with permission from Elsevier.

striatum, consistent with the expression of VGLUT2 by VTA dopamine neurons projecting to the ventral striatum but not by their neighbors in the substantia nigra pars compacta (SNc) that innervate the dorsal striatum (53, 54). These data are also consistent with the presence of TH^+ asymmetric (presumably excitatory) synapses in the ventral but not the dorsal striatum (65, 66).

Because VGLUT proteins usually localize exclusively to axon terminals, identification of $VGLUT^+$ cell populations has generally required quantitative polymerase chain reaction, in situ hybridization, or alternatively immunoelectron or confocal microscopy to examine nerve terminals directly. However, the low levels of VGLUT2 in mature dopamine neurons have sometimes eluded detection with the less sensitive of these methods, leading to conflicting conclusions about the expression of VGLUT2 by midbrain dopamine neurons (53, 54, 58–64, 67, 68). Using transgenic mice expressing GFP (green fluorescent protein) under the control of VGLUT2-regulatory elements, we observed clear colocalization of GFP with tyrosine hydroxylase in a medial subset of VTA neurons (29), consistent with a recent comprehensive report using in situ hybridization (54). Because the coexpressing neurons compose only a fraction of all dopamine neurons in the VTA, the effect of the KO on dopamine stores in vivo may greatly underestimate the effect on this subset. Thus, midbrain dopamine neurons may differ dramatically in the storage and release of dopamine, due to the heterogeneous expression of VGLUT2.

SNc: substantia nigra pars compacta

Quantal size: the postsynaptic response to the release of a single secretory vesicle

A KO of VGLUT3 has also been used to assess the role of glutamate storage and release by cholinergic interneurons of the striatum, which along with serotonin neurons in the raphe express high levels of VGLUT3. Constitutive disruption of VGLUT3 produces increased locomotor activity that can be reversed by the inhibition of acetylcholinesterase, and the animals show a reduction in vesicular ACh (and serotonin) uptake and release (50, 52). In contrast to wild-type animals, these animals also show no stimulation of vesicular ACh or serotonin transport by glutamate. However, the expression of VGLUT3 by a number of neuronal populations and the unconditional inactivation of VGLUT3 in these animals make it difficult to conclude that the behavioral abnormalities reflect a specific alteration in ACh release by striatal interneurons. The biochemical effect of glutamate on ACh and monoamine costorage thus seems clear, but the conditional inactivation of VGLUT3 or even of VAChT in genetically defined cell populations will be required to address the role of this phenomenon in behavior.

Although the dissipation of $\Delta\psi$ required for vesicle acidification has generally been attributed to anion entry, recent observations from non-neural cells suggest a role for cation efflux in lysosome ΔpH (69). Cl^- clearly promotes lysosome acidification in vitro, but this report suggests a smaller role in intact (or at least permeabilized) cells, with the efflux of luminal cation (apparently K^+) responsible in vivo. Nonetheless, the considerable data from ClC KO mice documenting effects on acidification within the endosome/lysosome pathway make it very difficult to exclude a role for Cl^- and these proteins in ΔpH formation.

THE REGULATION OF $\Delta\psi$ BY CATION FLUX

Do endocytic vesicles have a specific mechanism to promote formation of $\Delta\psi$? Or does $\Delta\psi$ result simply from the absence of a counterion such as Cl^- or glutamate? In general, $\Delta\psi$ has received little attention for an independent role in the secretory pathway, but vesicular glutamate transport clearly depends on $\Delta\psi$. Although recent attention has focused on the expression of VGLUTs as a presynaptic determinant of quantal size (70–72), the number of transporters per vesicle will change primarily the kinetics of transport, not the thermodynamic equilibrium reached at steady state (1). However, changes in the driving force should have dramatic effects on the extent as well as on the rate of vesicle filling, so the regulation of $\Delta\psi$ has important implications for transmitter release.

NHE: Na$^+$/H$^+$ exchanger

SAC: starburst amacrine cell

DSGC: direction-selective ganglion cell

LSO: lateral superior olive

Although very little is known about the factors that promote the formation of $\Delta\psi$, recent work has identified intracellular members of the Na$^+$/H$^+$ exchanger (NHE) family that could serve this function. NHEs catalyze the electroneutral exchange of monovalent cation for H$^+$, and plasma membrane isoforms have an important role in the regulation of cytosolic pH (73). Intracellular isoforms recognize K$^+$ as well as Na$^+$, and several isoforms localize to endosomes (74), where they should dissipate ΔpH and thus enable the H$^+$ pump to increase $\Delta\psi$. Interestingly, recent human genetic studies have implicated intracellular isoform NHE6 in Angelman syndrome (75) and NHE9 in autism (76).

Cation channels may also influence the formation of $\Delta\psi$. In this case, K$^+$ entry would promote the formation of $\Delta\psi$ independently of the H$^+$ pump. Interestingly, the TRPM7 (transient receptor potential cation channel, subfamily M, member 7) channel localizes to synaptic vesicles and influences quantal size, although it also interacts with proteins involved in fusion and affects the frequency of release (77, 78). However, the work on TRPM7 has involved cholinergic neurons, whereas the presence of an active K$^+$ conductance on synaptic vesicles may shunt the $\Delta\psi$ required for vesicular glutamate transport.

INDEPENDENT ROLES FOR CORELEASED NEUROTRANSMITTERS

In addition to the presynaptic consequences for vesicle filling, corelease has implications for the activation of postsynaptic receptors. Both coreleased transmitters may activate receptors, with the potential for distinct modes of signaling, and recent work has begun to elucidate the physiological role of corelease.

Corelease of GABA and Acetylcholine from Starburst Amacrine Cells

Starburst amacrine cells (SACs) contribute to direction-selective motion sensing by the vertebrate retina. SACs have a radially symmetric dendritic morphology that overlaps with dendrites from neighboring SACs as well as direction-selective ganglion cells (DSGCs) in the inner plexiform layer. Dual recordings show that SACs release more GABA onto DSGCs in response to light moving in the nonpreferred direction than in response to light moving in the preferred direction. Indeed, GABA release, presumably from SACs, appears to be essential for direction selectivity (79).

In addition to inhibitory GABA, SACs release ACh, activating nicotinic (nACh) receptors on DSGCs. However, the activation of nACh receptors is not required for direction selectivity (79). To characterize the release of both transmitters, a recent study using paired recordings demonstrated that, whereas GABA release by SACs is selective for movement in the null direction, the cholinergic response is greater with movement in the preferred direction (80). Both GABA and ACh currents depend on external Ca^{2+}, supporting a vesicular release mechanism, but ACh release shows much less sensitivity to Ca^{2+} than does GABA release, providing physiological evidence that different vesicle populations mediate release of the two transmitters. These observations are consistent with a proposed dual role for SACs as encoding direction selectivity through GABA release and encoding motion sensitivity through ACh release.

GABA and Glutamate Corelease from the Medial Nucleus of the Trapezoid Body

Neurons in the lateral superior olive (LSO) function as interaural coincidence detectors essential for sound localization. They accomplish this by integrating tonotopically precise excitatory input from the ipsilateral cochlear nucleus with inhibitory GABAergic and glycinergic inputs from the contralateral medial nucleus of the trapezoid body (MNTB). During development, however, MNTB neurons transiently express VGLUT3 and corelease glutamate between postnatal day (P)0 and P12 (81). In VGLUT3 KO mice, MNTB cells still form synapses onto LSO neurons that are indistinguishable from those in control animals at P1–2; however, the strengthening of these inhibitory synapses that normally occurs by P10–12 fails to occur in VGLUT3-null mice (82). Furthermore, tonotopic projections from the MNTB that project diffusely within the LSO at P1 fail to sharpen normally in the absence of VGLUT3. But why is glutamate release important when GABA is excitatory [due to a shift in E_{Cl} (the equilibrium potential for Cl^-)] during the same time frame? Presumably, the specific activation of NMDA receptors confers the plasticity required for normal development (81). The results thus support a role for glutamate corelease in synapse refinement that underlies sound localization in the auditory system.

MNTB: medial nucleus of the trapezoid body

GABA and Glutamate Corelease from Hippocampal Mossy Fibers

In the hippocampus, mossy fibers derived from granule cells in the dentate gyrus form glutamatergic synapses onto CA3 pyramidal neurons, where they also corelease GABA. Early in development, pyramidal neurons express VGAT and glutamic acid decarboxylase, the enzyme responsible for GABA biosynthesis, but the genes involved subsequently downregulate (83–85). For the first 3 weeks after birth, stimulation of mossy fiber inputs produces GABA-mediated currents in pyramidal neurons (86). However, the significance of this transient GABA corelease remains unknown, and at this time, GABA currents are still excitatory due to the shift in Cl^- reversal potential.

Interestingly, epileptic activity rekindles expression of the GABAergic phenotype in adult granule cells (87–90). At this point, GABA transmission is inhibitory and may thus serve a distinct, homeostatic role to restrain the excitability responsible for epilepsy.

Monoamine and Glutamate Corelease

The first clear evidence that monoamine neurons corelease glutamate derived from dissociated neurons grown in isolation so that they could form synapses onto only themselves. Stimulation of both serotonin (12) and dopamine (13) neurons produced fast excitatory currents blocked by glutamate receptor antagonists, indicating the potential for glutamate corelease to activate postsynaptic receptors. However, the postnatal decline in VGLUT2 expression by midbrain dopamine neurons (60, 63) raised the possibility that VGLUT2 expression in vitro (67) might simply reflect dedifferentiation. The low level of VGLUT2 expression by midbrain dopamine neurons in the adult raised further questions about the physiological relevance of these in vitro observations. The phenotype of mice lacking VGLUT2 specifically in dopamine neurons and the anatomical evidence for VGLUT2 expression by a medial subset of VTA neurons have provided clear evidence for the effects of glutamate on costored dopamine but have not directly addressed the role of glutamate as an independent signal.

In 2004, the Rayport laboratory (91) published a landmark study that used an acute, horizontal slice preparation to demonstrate the presence of a monosynaptic glutamatergic projection from VTA to nucleus accumbens at both P10 and P21. The next year, the Seamans laboratory (92) showed that VTA stimulation in vivo rapidly leads to glutamate release in the prefrontal cortex (PFC). Although both of these studies supported an independent role for the glutamate released by dopamine neurons, questions remained about the specificity of stimulation, particularly after the identification of purely glutamatergic neurons in the ventral midbrain (53, 59) that we now know also project to both the ventral striatum and the PFC (54, 93).

In contrast, genetic approaches have recently provided definitive physiological evidence that glutamate released by at least a subset of dopamine neurons in adult mice activates ionotropic glutamate receptors on postsynaptic medium spiny neurons in the striatum. Using cre recombinase selectively expressed by dopamine neurons to activate a conditional allele of the light-activated cation channel channelrhodopsin-2, we and others observed glutamate responses evoked by direct illumination of the striatum (94, 95). In addition to the increased specificity, the ability to stimulate glutamate release directly at presynaptic boutons circumvented the unavoidable transection of mesolimbic projections in horizontal slices, resulting in larger postsynaptic responses. Robust glutamate-mediated AMPA receptor currents were observed in the ventral striatum but not in the dorsal striatum, even though light evoked dopamine release at both sites (94), consistent with the restricted expression of VGLUT2 by dopamine neurons in the VTA but not in the SNc (53, 54). Furthermore, the cKO of VGLUT2 in dopamine neurons completely abolished these responses (94).

What then is the role of this glutamate signal? The most robust phenotype observed in cKO mice that lack glutamate corelease from dopamine neurons is a reduction in psychostimulant-induced locomotion (29, 64). This may be most easily explained by the reduction in dopamine release that we attribute to a reduction in vesicular dopamine storage (29). However, the activation of postsynaptic ionotropic receptors by the glutamate released from dopamine neurons likely encodes distinct information.

One possibility is that the glutamate released by dopamine terminals contributes to the prediction-error signal encoded in the firing rates of dopamine neurons (96, 97). A subset of tonically active midbrain (presumably dopamine) neurons burst fire in response to unexpected rewards or to rewards better than predicted by a conditioned cue. Conversely, they slow or pause firing in response to rewards worse than predicted (98). Consistent with these changes in firing, extracellular dopamine measured by fast-scan cyclic voltammetry changes as predicted in rodents performing goal-directed tasks (99). However, we do not know how dopamine signaling by metabotropic G protein–coupled dopamine receptors can maintain the fidelity of synaptic transmission required for learning tasks dependent on subsecond cue discrimination. As a neuromodulator activating G protein–coupled receptors, dopamine presumably acts on slower timescales (i.e., seconds to minutes). In contrast, the glutamate coreleased by dopamine neurons produces a rapid, transient postsynaptic response more tightly coupled to dopamine neuron firing and is thus well positioned to convey temporally precise information about reward (for excellent reviews see References 100 and 101). This hypothesis predicts deficits in reward learning by cKO mice lacking VGLUT2 in dopamine neurons, but initial assessment using conditioned place preference (CPP) showed no such deficits (29). However, mice can also learn CPP in the absence of dopamine (102, 103), and the cue-reward pairing involved in CPP occurs continuously over the course of 20 min and may therefore not depend on transient subsecond bursts in dopamine neuron firing.

The expression of channelrhodopsin in raphe nuclei has also revealed an optically evoked glutamate-mediated response in the hippocampus, presumably from the population of serotonergic neurons expressing VGLUT3 (104). However, these experiments did not use genetic manipulation to limit channelrhodopsin expression to serotonergic neurons, so the responses may derive from neighboring nonserotonergic neurons in the raphe. Indeed, despite the strong expression of VGLUT3 mRNA in raphe nuclei, to what extent VGLUT3 and serotonergic markers are coexpressed or compose separate neuronal populations, similar to the nondopaminergic VGLUT2$^+$ population of neurons in the medial midbrain, remains unclear. However, the anatomical evidence supports VGLUT3 expression by at least a subset of serotonergic neurons (105–109).

Acetylcholine and Glutamate Corelease

Channelrhodopsin was also used recently to demonstrate that, in addition to the role of glutamate costorage in promoting vesicular ACh filling in striatal interneurons (52), the released glutamate activates ionotropic receptors on medium spiny neurons. Consistent with VGLUT3 expression by these cells, the response was abolished in VGLUT3 KO mice (110). Recent work has also identified corelease of ACh and glutamate by neurons of the medial habenula. Expressed in cholinergic neurons, channelrhodopsin confers light-evoked release of glutamate as well as of ACh within the interpeduncular nucleus of the midbrain (111). However, brief illumination evokes primarily the glutamate response, with the ACh response requiring more sustained stimulation. Released from the same neuron, the two transmitters may thus subserve distinct roles in signaling, perhaps due to differences in the distance between release site and postsynaptic receptors (i.e., between synaptic and volume transmission) or perhaps as a function of release from different vesicle populations.

DISTINCT AND OVERLAPPING POOLS OF SYNAPTIC VESICLES

The ability of one transmitter to affect the storage of another through changes in the H^+ electrochemical driving force requires localization of the two vesicular transporters to the same secretory vesicle, but several recent observations suggest that release can also occur from distinct vesicle populations. In retinal SACs, GABA release and ACh release respond differently to Ca^{2+} (80), providing unequivocal evidence for release from different vesicles. Immunolabeling for endogenously expressed proteins also suggests that dopaminergic release sites are heterogeneous in their capacity to store glutamate (13, 53, 60–62, 67, 68). In midbrain dopamine neurons, heterologous expression of differentially tagged vesicular glutamate and monoamine transporters shows colocalization at most boutons, but a significant fraction express only one or the other (112), consistent with the original suggestion that catecholamine and glutamate markers may segregate to distinct synapses both in vitro and in vivo (13, 65). However, in contrast to the VGLUTs, which generally reside only at presynaptic boutons, VMAT2 localizes to dendrites as well as to axons, but the segregation occurs even with the analysis restricted to axonal sites. The segregation of monoamine and glutamate markers to different release sites may indeed contribute to the failure to detect VGLUT expression in tyrosine hydroxylase–positive striatal projections by immunoelectron microscopy (60, 68). Hippocampal neurons show no evidence of such segregation, indicating mechanisms specific to dopamine neurons. In addition, optical imaging with a pHluorin-based reporter shows that field stimulation evokes release of a greater proportion of VGLUT1 than VMAT2 at boutons (112), suggesting that the two proteins exhibit overlapping but differential localization to synaptic vesicle pools.

Considerable previous work has shown that only a fraction of the synaptic vesicles in a presynaptic bouton are available for evoked release, even after prolonged stimulation (113). This so-called recycling pool can be only a small fraction of all the vesicles present, with the remaining, so-called resting pool of uncertain physiological role. Because the proportion of several synaptic vesicle proteins in this recycling pool is generally the same (~50–60%), it has been assumed that they will all exhibit the same distribution between recycling and resting (unresponsive) pools. However, the relatively small recycling pool size of VMAT2 (20–30%) indicates that in addition to the segregation of dopamine and glutamate vesicles at different boutons, dopamine and glutamate vesicles also segregate to at least some extent within individual boutons where they both reside. Interestingly, the differential exocytosis of VMAT2 and VGLUT occurs in hippocampal as well as in midbrain dopamine neurons, indicating the potential for differential corelease of classical transmitters by many if not all neuronal populations.

Recent work has suggested that the VGLUTs may control the probability of transmitter release, perhaps accounting for the differential release of two transmitters by the same neuron. The distribution of VGLUT1 and -2 originally suggested a correlation of VGLUT1 with synapses having a low probability of release (such as hippocampal synapses and parallel fiber synapses in the cerebellum) and VGLUT2 with synapses having a high probability of release (114). Although it has been difficult to understand how the transporter might control fusion, recent work has indeed suggested that the known interaction of VGLUT1 with the endocytic protein endophilin (115) may also influence exocytosis (116). Alternatively, the two transporters may simply recycle through slightly different mechanisms, consistent with the role of endophilin in endocytosis, and these mechanisms may generate vesicles with different release probability. Rather than influencing the release machinery, the transporter may thus simply target to vesicles with different properties. The difference between VMAT2 and the VGLUTs in overall recycling pool size supports this possibility, but it may be more difficult to assess directly the targeting of VGLUT1 and -2 to distinct subsets within the recycling pool. Because synaptic vesicles have generally been considered homogeneous in terms of biochemical composition, considerable basic work will be required to characterize the properties of these subsets and to identify the proteins responsible for their properties, as well as the mechanisms responsible for sorting these proteins into functionally distinct vesicle pools.

SUMMARY POINTS

1. The filling of synaptic vesicles with different transmitters relies on different components of $\Delta\mu_{H+}$.
2. $\Delta\mu_{H+}$ can be expressed as ΔpH, $\Delta\psi$, or a combination of both.
3. The entry of Cl^- and other anions promotes the formation of ΔpH by dissipating $\Delta\psi$, thereby disinhibiting the H^+ pump.
4. Cation flux may promote the formation of $\Delta\psi$.
5. Many neuronal populations corelease two classical transmitters.
6. Costorage with glutamate promotes the vesicular transport of monoamines and ACh.
7. Coreleased neurotransmitters can activate their cognate postsynaptic receptors.
8. Corelease of two transmitters can also occur from independent vesicle populations.

DISCLOSURE STATEMENT

The authors are not aware of any affiliations, memberships, funding, or financial holdings that might be perceived as affecting the objectivity of this review.

LITERATURE CITED

1. Edwards RH. 2007. The neurotransmitter cycle and quantal size. *Neuron* 55:835–58
2. Gronborg M, Pavlos NJ, Brunk I, Chua JJ, Munster-Wandowski A, et al. 2010. Quantitative comparison of glutamatergic and GABAergic synaptic vesicles unveils selectivity for few proteins including MAL2, a novel synaptic vesicle protein. *J. Neurosci.* 30:2–12
3. Whittaker VP, Dowdall MJ, Boyne AF. 1972. The storage and release of acetylcholine by cholinergic nerve terminals: recent results with non-mammalian preparations. *Biochem. Soc. Symp.* 1972:49–68

4. Silinsky EM. 1975. On the association between transmitter secretion and the release of adenine nucleotides from mammalian motor nerve terminals. *J. Physiol.* 247:145–62
5. Burnstock G. 2004. Cotransmission. *Curr. Opin. Pharmacol.* 4:47–52
6. Wojcik SM, Katsurabayashi S, Guillemin I, Friauf E, Rosenmund C, et al. 2006. A shared vesicular carrier allows synaptic corelease of GABA and glycine. *Neuron* 50:575–87
7. Awatramani GB, Turecek R, Trussell LO. 2005. Staggered development of GABAergic and glycinergic transmission in the MNTB. *J. Neurophysiol.* 93:819–28
8. Jonas P, Bischofberger J, Sandkuhler J. 1998. Corelease of two fast neurotransmitters at a central synapse. *Science* 281:419–24
9. Nabekura J, Katsurabayashi S, Kakazu Y, Shibata S, Matsubara A, et al. 2004. Developmental switch from GABA to glycine release in single central synaptic terminals. *Nat. Neurosci.* 7:17–23
10. Zhou FM, Liang Y, Salas R, Zhang L, De Biasi M, Dani JA. 2005. Corelease of dopamine and serotonin from striatal dopamine terminals. *Neuron* 46:65–74
11. Lebrand C, Cases O, Adelbrecht C, Doye A, Alvarez C, et al. 1996. Transient uptake and storage of serotonin in developing thalamic neurons. *Neuron* 17:823–35
12. Johnson MD. 1994. Synaptic glutamate release by postnatal rat serotonergic neurons in microculture. *Neuron* 12:433–42
13. Sulzer D, Joyce MP, Lin L, Geldwert D, Haber SN, et al. 1998. Dopamine neurons make glutamatergic synapses in vitro. *J. Neurosci.* 18:4588–602
14. Maher BJ, Westbrook GL. 2008. Co-transmission of dopamine and GABA in periglomerular cells. *J. Neurophysiol.* 99:1559–64
15. Nishimaru H, Restrepo CE, Ryge J, Yanagawa Y, Kiehn O. 2005. Mammalian motor neurons corelease glutamate and acetylcholine at central synapses. *Proc. Natl. Acad. Sci. USA* 102:5245–49
16. Forgac M. 2007. Vacuolar ATPases: rotary proton pumps in physiology and pathophysiology. *Nat. Rev. Mol. Cell Biol.* 8:917–29
17. Johnson RG, Carty SE, Scarpa A. 1981. Proton: substrate stoichiometries during active transport of biogenic amines in chromaffin ghosts. *J. Biol. Chem.* 256:5773–80
18. Knoth J, Zallakian M, Njus D. 1981. Stoichiometry of H^+-linked dopamine transport in chromaffin granule ghosts. *Biochemistry* 20:6625–29
19. Nguyen ML, Cox GD, Parsons SM. 1998. Kinetic parameters for the vesicular acetylcholine transporter: Two protons are exchanged for one acetylcholine. *Biochemistry* 37:13400–10
20. Hell JW, Maycox PR, Jahn R. 1990. Energy dependence and functional reconstitution of the γ-aminobutyric acid carrier from synaptic vesicles. *J. Biol. Chem.* 265:2111–17
21. Kish PE, Fischer-Bovenkerk C, Ueda T. 1989. Active transport of γ-aminobutyric acid and glycine into synaptic vesicles. *Proc. Natl. Acad. Sci. USA* 86:3877–81
22. Juge N, Muroyama A, Hiasa M, Omote H, Moriyama Y. 2009. Vesicular inhibitory amino acid transporter is a Cl^-/γ-aminobutyrate co-transporter. *J. Biol. Chem.* 284:35073–78
23. Fremeau RT Jr, Voglmaier S, Seal RP, Edwards RH. 2004. VGLUTs define subsets of excitatory neurons and suggest novel roles for glutamate. *Trends Neurosci.* 27:98–103
24. Takamori S. 2006. VGLUTs: "exciting" times for glutamatergic research? *Neurosci. Res.* 55:343–51
25. El Mestikawy S, Wallen-Mackenzie A, Fortin GM, Descarries L, Trudeau LE. 2011. From glutamate co-release to vesicular synergy: vesicular glutamate transporters. *Nat. Rev. Neurosci.* 12:204–16
26. Maycox PR, Deckwerth T, Hell JW, Jahn R. 1988. Glutamate uptake by brain synaptic vesicles. Energy dependence of transport and functional reconstitution in proteoliposomes. *J. Biol. Chem.* 263:15423–28
27. Tabb JS, Kish PE, Van Dyke R, Ueda T. 1992. Glutamate transport into synaptic vesicles. Roles of membrane potential, pH gradient, and intravesicular pH. *J. Biol. Chem.* 267:15412–18
28. Tan PK, Waites C, Liu Y, Krantz DE, Edwards RH. 1998. A leucine-based motif mediates the endocytosis of vesicular monoamine and acetylcholine transporters. *J. Biol. Chem.* 273:17351–60
29. **Hnasko TS, Chuhma N, Zhang H, Goh GY, Sulzer D, et al. 2010. Vesicular glutamate transport promotes dopamine storage and glutamate corelease in vivo. *Neuron* 65:643–56**
30. Stobrawa SM, Breiderhoff T, Takamori S, Engel D, Schweizer M, et al. 2001. Disruption of ClC-3, a chloride channel expressed on synaptic vesicles, leads to a loss of the hippocampus. *Neuron* 29:185–96

29. Using conditional knockout mice lacking VGLUT2 specifically in dopamine neurons, this paper demonstrates the costorage and corelease of glutamate by VTA dopamine neurons. It also elucidates the nonredundant roles of glutamate and Cl^- in formation and stabilization of ΔpH.

31. Accardi A, Miller C. 2004. Secondary active transport mediated by a prokaryotic homologue of ClC Cl⁻ channels. *Nature* 427:803–7

32. Graves AR, Curran PK, Smith CL, Mindell JA. 2008. The Cl⁻/H⁺ antiporter ClC-7 is the primary chloride permeation pathway in lysosomes. *Nature* 453:788–92

33. Picollo A, Pusch M. 2005. Chloride/proton antiporter activity of mammalian CLC proteins ClC-4 and ClC-5. *Nature* 436:420–23

34. Scheel O, Zdebik AA, Lourdel S, Jentsch TJ. 2005. Voltage-dependent electrogenic chloride/proton exchange by endosomal CLC proteins. *Nature* 436:424–27

35. Neagoe I, Stauber T, Fidzinski P, Bergsdorf EY, Jentsch TJ. 2010. The late endosomal ClC-6 mediates proton/chloride countertransport in heterologous plasma membrane expression. *J. Biol. Chem.* 285:21689–97

36. Novarino G, Weinert S, Rickheit G, Jentsch TJ. 2010. Endosomal chloride-proton exchange rather than chloride conductance is crucial for renal endocytosis. *Science* 328:1398–401

37. Weinert S, Jabs S, Supanchart C, Schweizer M, Gimber N, et al. 2010. Lysosomal pathology and osteopetrosis upon loss of H⁺-driven lysosomal Cl⁻ accumulation. *Science* 328:1401–3

38. Johnson RG, Carty SE, Scarpa A. 1985. Coupling of H⁺ gradients to catecholamine transport in chromaffin granules. *Ann. N. Y. Acad. Sci.* 456:254–67

39. Johnson RG, Scarpa A. 1979. Protonmotive force and catecholamine transport in isolated chromaffin granules. *J. Biol. Chem.* 254:3750–60

40. Riazanski V, Deriy LV, Shevchenko PD, Le B, Gomez EA, Nelson DJ. 2011. Presynaptic CLC-3 determines quantal size of inhibitory transmission in the hippocampus. *Nat. Neurosci.* 14:487–94

41. **Schenck S, Wojcik SM, Brose N, Takamori S. 2009. A chloride conductance in VGLUT1 underlies maximal glutamate loading into synaptic vesicles. *Nat. Neurosci.* 12:156–62**

42. Busch AE, Schuster A, Waldegger S, Wagner CA, Zempel G, et al. 1996. Expression of a renal type I sodium/phosphate transporter (NaPi-1) induces a conductance in *Xenopus* oocytes permeable for organic and inorganic anions. *Proc. Natl. Acad. Sci. USA* 93:5347–51

43. Bellocchio EE, Reimer RJ, Fremeau RT Jr, Edwards RH. 2000. Uptake of glutamate into synaptic vesicles by an inorganic phosphate transporter. *Science* 289:957–60

44. Wolosker H, de Souza DO, de Meis L. 1996. Regulation of glutamate transport into synaptic vesicles by chloride and proton gradient. *J. Biol. Chem.* 271:11726–31

45. **Juge N, Gray JA, Omote H, Miyaji T, Inoue T, et al. 2010. Metabolic control of vesicular glutamate transport and release. *Neuron* 68:99–112**

46. Bankston LA, Guidotti G. 1996. Characterization of ATP transport into chromaffin granule ghosts. Synergy of ATP and serotonin accumulation in chromaffin granule ghosts. *J. Biol. Chem.* 271:17132–38

47. Sawada K, Echigo N, Juge N, Miyaji T, Otsuka M, et al. 2008. Identification of a vesicular nucleotide transporter. *Proc. Natl. Acad. Sci. USA* 105:5683–86

48. Takamori S, Rhee JS, Rosenmund C, Jahn R. 2000. Identification of a vesicular glutamate transporter that defines a glutamatergic phenotype in neurons. *Nature* 407:189–94

49. Erickson JD, Masserano JM, Barnes EM, Ruth JA, Weiner N. 1990. Chloride ion increases [³H]dopamine accumulation by synaptic vesicles purified from rat striatum: inhibition by thiocyanate ion. *Brain Res.* 516:155–60

50. Amilhon B, Lepicard E, Renoir T, Mongeau R, Popa D, et al. 2010. VGLUT3 (vesicular glutamate transporter type 3) contribution to the regulation of serotonergic transmission and anxiety. *J. Neurosci.* 30:2198–210

51. Zander JF, Munster-Wandowski A, Brunk I, Pahner I, Gomez-Lira G, et al. 2010. Synaptic and vesicular coexistence of VGLUT and VGAT in selected excitatory and inhibitory synapses. *J. Neurosci.* 30:7634–45

52. **Gras C, Amilhon B, Lepicard EM, Poirel O, Vinatier J, et al. 2008. The vesicular glutamate transporter VGLUT3 synergizes striatal acetylcholine tone. *Nat. Neurosci.* 11:292–300**

53. Kawano M, Kawasaki A, Sakata-Haga H, Fukui Y, Kawano H, et al. 2006. Particular subpopulations of midbrain and hypothalamic dopamine neurons express vesicular glutamate transporter 2 in the rat brain. *J. Comp. Neurol.* 498:581–92

---

41. Uses purified, reconstituted VGLUT1 and vesicle acidification to suggest that Cl⁻ efflux promotes vesicular glutamate uptake.

45. Along with showing that ketone bodies influence vesicular glutamate transport by competing with Cl⁻ at an allosteric site, this work could not detect a VGLUT-mediated Cl⁻ conductance by directly measuring Cl⁻ flux.

52. Focusing on cholinergic interneurons of the striatum, this paper demonstrates an important role for glutamate entry through VGLUT3 on synaptic vesicle filling with ACh.

54. Yamaguchi T, Wang HL, Li X, Ng TH, Morales M. 2011. Mesocorticolimbic glutamatergic pathway. *J. Neurosci.* 31:8476–90
55. Fremeau RT Jr, Burman J, Qureshi T, Tran CH, Proctor J, et al. 2002. The identification of vesicular glutamate transporter 3 suggests novel modes of signaling by glutamate. *Proc. Natl. Acad. Sci. USA* 99:14488–93
56. Gras C, Herzog E, Bellenchi GC, Bernard V, Ravassard P, et al. 2002. A third vesicular glutamate transporter expressed by cholinergic and serotoninergic neurons. *J. Neurosci.* 22:5442–51
57. Schafer MK, Varoqui H, Defamie N, Weihe E, Erickson JD. 2002. Molecular cloning and functional identification of mouse vesicular glutamate transporter 3 and its expression in subsets of novel excitatory neurons. *J. Biol. Chem.* 277:50734–48
58. Nair-Roberts RG, Chatelain-Badie SD, Benson E, White-Cooper H, Bolam JP, Ungless MA. 2008. Stereological estimates of dopaminergic, GABAergic and glutamatergic neurons in the ventral tegmental area, substantia nigra and retrorubral field in the rat. *Neuroscience* 152:1024–31
59. Yamaguchi T, Sheen W, Morales M. 2007. Glutamatergic neurons are present in the rat ventral tegmental area. *Eur. J. Neurosci.* 25:106–18
60. Berube-Carriere N, Riad M, Dal Bo G, Levesque D, Trudeau LE, Descarries L. 2009. The dual dopamine-glutamate phenotype of growing mesencephalic neurons regresses in mature rat brain. *J. Comp. Neurol.* 517:873–91
61. Dal Bo G, Berube-Carriere N, Mendez JA, Leo D, Riad M, et al. 2008. Enhanced glutamatergic phenotype of mesencephalic dopamine neurons after neonatal 6-hydroxydopamine lesion. *Neuroscience* 156:59–70
62. Descarries L, Berube-Carriere N, Riad M, Dal Bo G, Mendez JA, Trudeau LE. 2008. Glutamate in dopamine neurons: synaptic versus diffuse transmission. *Brain Res. Rev.* 58:290–302
63. Mendez JA, Bourque MJ, Dal Bo G, Bourdeau ML, Danik M, et al. 2008. Developmental and target-dependent regulation of vesicular glutamate transporter expression by dopamine neurons. *J. Neurosci.* 28:6309–18
64. Birgner C, Nordenankar K, Lundblad M, Mendez JA, Smith C, et al. 2010. VGLUT2 in dopamine neurons is required for psychostimulant-induced behavioral activation. *Proc. Natl. Acad. Sci. USA* 107:389–94
65. Hattori T, Takada M, Moriizumi T, Van der Kooy D. 1991. Single dopaminergic nigrostriatal neurons form two chemically distinct synaptic types: possible transmitter segregation within neurons. *J. Comp. Neurol.* 309:391–401
66. Meredith GE, Wouterlood FG. 1993. Identification of synaptic interactions of intracellularly injected neurons in fixed brain slices by means of dual-label electron microscopy. *Microsc. Res. Tech.* 24:31–42
67. Dal Bo G, St-Gelais F, Danik M, Williams S, Cotton M, Trudeau LE. 2004. Dopamine neurons in culture express VGLUT2 explaining their capacity to release glutamate at synapses in addition to dopamine. *J. Neurochem.* 88:1398–405
68. Moss J, Ungless MA, Bolam JP. 2011. Dopaminergic axons in different divisions of the adult rat striatal complex do not express vesicular glutamate transporters. *Eur. J. Neurosci.* 33:1205–11
69. Steinberg BE, Huynh KK, Brodovitch A, Jabs S, Stauber T, et al. 2010. A cation counterflux supports lysosomal acidification. *J. Cell Biol.* 189:1171–86
70. Wojcik SM, Rhee JS, Herzog E, Sigler A, Jahn R, et al. 2004. An essential role for vesicular glutamate transporter 1 (VGLUT1) in postnatal development and control of quantal size. *Proc. Natl. Acad. Sci. USA* 101:7158–63
71. Wilson NR, Kang J, Hueske EV, Leung T, Varoqui H, et al. 2005. Presynaptic regulation of quantal size by the vesicular glutamate transporter VGLUT1. *J. Neurosci.* 25:6221–34
72. De Gois S, Jeanclos E, Morris M, Grewal S, Varoqui H, Erickson JD. 2006. Identification of endophilins 1 and 3 as selective binding partners for VGLUT1 and their co-localization in neocortical glutamatergic synapses: implications for vesicular glutamate transporter trafficking and excitatory vesicle formation. *Cell Mol. Neurobiol.* 26:679–93
73. Orlowski J, Grinstein S. 2004. Diversity of the mammalian sodium/proton exchanger SLC9 gene family. *Pflüg. Arch.* 447:549–65

74. Nakamura N, Tanaka S, Teko Y, Mitsui K, Kanazawa H. 2005. Four Na$^+$/H$^+$ exchanger isoforms are distributed to Golgi and post-Golgi compartments and are involved in organelle pH regulation. *J. Biol. Chem.* 280:1561–72

75. Gilfillan GD, Selmer KK, Roxrud I, Smith R, Kyllerman M, et al. 2008. SLC9A6 mutations cause X-linked mental retardation, microcephaly, epilepsy, and ataxia, a phenotype mimicking Angelman syndrome. *Am. J. Hum. Genet.* 82:1003–10

76. Morrow EM, Yoo SY, Flavell SW, Kim TK, Lin Y, et al. 2008. Identifying autism loci and genes by tracing recent shared ancestry. *Science* 321:218–23

77. Brauchi S, Krapivinsky G, Krapivinsky L, Clapham DE. 2008. TRPM7 facilitates cholinergic vesicle fusion with the plasma membrane. *Proc. Natl. Acad. Sci. USA* 105:8304–8

78. Krapivinsky G, Mochida S, Krapivinsky L, Cibulsky SM, Clapham DE. 2006. The TRPM7 ion channel functions in cholinergic synaptic vesicles and affects transmitter release. *Neuron* 52:485–96

79. Demb JB. 2007. Cellular mechanisms for direction selectivity in the retina. *Neuron* 55:179–86

80. **Lee S, Kim K, Zhou ZJ. 2010. Role of ACh-GABA cotransmission in detecting image motion and motion direction. *Neuron* 68:1159–72**

81. Gillespie DC, Kim G, Kandler K. 2005. Inhibitory synapses in the developing auditory system are glutamatergic. *Nat. Neurosci.* 8:332–38

82. **Noh J, Seal RP, Garver JA, Edwards RH, Kandler K. 2010. Glutamate co-release at GABA/glycinergic synapses is crucial for the refinement of an inhibitory map. *Nat. Neurosci.* 13:232–38**

83. Gutierrez R, Romo-Parra H, Maqueda J, Vivar C, Ramirez M, et al. 2003. Plasticity of the GABAergic phenotype of the "glutamatergic" granule cells of the rat dentate gyrus. *J. Neurosci.* 23:5594–98

84. Maqueda J, Ramirez M, Lamas M, Gutierrez R. 2003. Glutamic acid decarboxylase (GAD)67, but not GAD65, is constitutively expressed during development and transiently overexpressed by activity in the granule cells of the rat. *Neurosci. Lett.* 353:69–71

85. Walker MC, Ruiz A, Kullmann DM. 2001. Monosynaptic GABAergic signaling from dentate to CA3 with a pharmacological and physiological profile typical of mossy fiber synapses. *Neuron* 29:703–15

86. Trudeau LE, Gutierrez R. 2007. On cotransmission & neurotransmitter phenotype plasticity. *Mol. Interv.* 7:138–46

87. Gutierrez R. 2000. Seizures induce simultaneous GABAergic and glutamatergic transmission in the dentate gyrus-CA3 system. *J. Neurophysiol.* 84:3088–90

88. Lehmann H, Ebert U, Loscher W. 1996. Immunocytochemical localization of GABA immunoreactivity in dentate granule cells of normal and kindled rats. *Neurosci. Lett.* 212:41–44

89. Nadler JV. 2003. The recurrent mossy fiber pathway of the epileptic brain. *Neurochem. Res.* 28:1649–58

90. Romo-Parra H, Vivar C, Maqueda J, Morales MA, Gutierrez R. 2003. Activity-dependent induction of multitransmitter signaling onto pyramidal cells and interneurons of hippocampal area CA3. *J. Neurophysiol.* 89:3155–67

91. Chuhma N, Zhang H, Masson J, Zhuang X, Sulzer D, et al. 2004. Dopamine neurons mediate a fast excitatory signal via their glutamatergic synapses. *J. Neurosci.* 24:972–81

92. Lavin A, Nogueira L, Lapish CC, Wightman RM, Phillips PE, Seamans JK. 2005. Mesocortical dopamine neurons operate in distinct temporal domains using multimodal signaling. *J. Neurosci.* 25:5013–23

93. Gorelova N, Mulholland PJ, Chandler LJ, Seamans JK. 2011. The glutamatergic component of the mesocortical pathway emanating from different subregions of the ventral midbrain. *Cereb. Cortex.* In press

94. **Stuber GD, Hnasko TS, Britt JP, Edwards RH, Bonci A. 2010. Dopaminergic terminals in the nucleus accumbens but not in the dorsal striatum corelease glutamate. *J. Neurosci.* 30:8229–33**

95. **Tecuapetla F, Patel JC, Xenias H, English D, Tadros I, et al. 2010. Glutamatergic signaling by mesolimbic dopamine neurons in the nucleus accumbens. *J. Neurosci.* 30:7105–10**

96. Hollerman JR, Schultz W. 1998. Dopamine neurons report an error in the temporal prediction of reward during learning. *Nat. Neurosci.* 1:304–9

97. Schultz W. 2002. Getting formal with dopamine and reward. *Neuron* 36:241–63

98. Schultz W, Dayan P, Montague PR. 1997. A neural substrate of prediction and reward. *Science* 275:1593–99

---

80. Elegantly shows that starburst amacrine cells of the retina release GABA and ACh from distinct vesicle pools and that the differential release contributes to direction selectivity in retinal ganglion cells.

82. Demonstrates a developmental role for VGLUT3 in the corelease of glutamate by inhibitory MNTB neurons and its importance for synaptic refinement within the lateral superior olive that contributes to sound localization.

95. Along with Reference 94, this work uses optogenetics to show that glutamate released by dopamine neurons activates ionotropic glutamate receptors on postsynaptic medium spiny neurons in the nucleus accumbens of adult mice.

99. Phillips PE, Stuber GD, Heien ML, Wightman RM, Carelli RM. 2003. Subsecond dopamine release promotes cocaine seeking. *Nature* 422:614–18
100. Lapish CC, Kroener S, Durstewitz D, Lavin A, Seamans JK. 2007. The ability of the mesocortical dopamine system to operate in distinct temporal modes. *Psychopharmacology* 191:609–25
101. Lapish CC, Seamans JK, Judson Chandler L. 2006. Glutamate-dopamine cotransmission and reward processing in addiction. *Alcohol. Clin. Exp. Res.* 30:1451–65
102. Hnasko TS, Sotak BN, Palmiter RD. 2005. Morphine reward in dopamine-deficient mice. *Nature* 438:854–57
103. Hnasko TS, Sotak BN, Palmiter RD. 2007. Cocaine-conditioned place preference by dopamine-deficient mice is mediated by serotonin. *J. Neurosci.* 27:12484–88
104. **Varga V, Losonczy A, Zemelman BV, Borhegyi Z, Nyiri G, et al. 2009. Fast synaptic subcortical control of hippocampal circuits. *Science* 326:449–53**
105. Hioki H, Fujiyama F, Nakamura K, Wu SX, Matsuda W, Kaneko T. 2004. Chemically specific circuit composed of vesicular glutamate transporter 3- and preprotachykinin B-producing interneurons in the rat neocortex. *Cereb. Cortex* 14:1266–75
106. Hioki H, Nakamura H, Ma YF, Konno M, Hayakawa T, et al. 2010. Vesicular glutamate transporter 3-expressing nonserotonergic projection neurons constitute a subregion in the rat midbrain raphe nuclei. *J. Comp. Neurol.* 518:668–86
107. Jackson J, Bland BH, Antle MC. 2009. Nonserotonergic projection neurons in the midbrain raphe nuclei contain the vesicular glutamate transporter VGLUT3. *Synapse* 63:31–41
108. Mintz EM, Scott TJ. 2006. Colocalization of serotonin and vesicular glutamate transporter 3-like immunoreactivity in the midbrain raphe of Syrian hamsters (*Mesocricetus auratus*). *Neurosci. Lett.* 394:97–100
109. Shutoh F, Ina A, Yoshida S, Konno J, Hisano S. 2008. Two distinct subtypes of serotonergic fibers classified by co-expression with vesicular glutamate transporter 3 in rat forebrain. *Neurosci. Lett.* 432:132–36
110. Higley MJ, Gittis AH, Oldenburg IA, Balthasar N, Seal RP, et al. 2011. Cholinergic interneurons mediate fast VGluT3-dependent glutamatergic transmission in the striatum. *PLoS ONE* 6:e19155
111. **Ren J, Qin C, Hu F, Tan J, Qiu L, et al. 2011. Habenula "cholinergic" neurons co-release glutamate and acetylcholine and activate postsynaptic neurons via distinct transmission modes. *Neuron* 69:445–52**
112. **Onoa B, Li H, Gagnon-Bartsch JA, Elias LA, Edwards RH. 2010. Vesicular monoamine and glutamate transporters select distinct synaptic vesicle recycling pathways. *J. Neurosci.* 30:7917–27**
113. Rizzoli SO, Betz WJ. 2005. Synaptic vesicle pools. *Nat. Rev. Neurosci.* 6:57–69
114. Fremeau RT Jr, Troyer MD, Pahner I, Nygaard GO, Tran CH, et al. 2001. The expression of vesicular glutamate transporters defines two classes of excitatory synapse. *Neuron* 31:247–60
115. Voglmaier SM, Kam K, Yang H, Fortin DL, Hua Z, et al. 2006. Distinct endocytic pathways control the rate and extent of synaptic vesicle protein recycling. *Neuron* 51:71–84
116. Weston MC, Nehring RB, Wojcik SM, Rosenmund C. 2011. Interplay between VGLUT isoforms and endophilin A1 regulates neurotransmitter release and short-term plasticity. *Neuron* 69:1147–59

---

104. Uses channelrhodopsin to demonstrate the release of glutamate from hippocampal projections of serotonergic raphe nuclei.

111. Relies on optogenetics to demonstrate a glutamatergic response in the interpeduncular nucleus elicited by optical stimulation of cholinergic projections from the medial habenula.

112. Using heterologous expression in primary dissociated culture, this study shows that VGLUT2 and VMAT2 traffic to overlapping but distinct synaptic boutons and respond differently to stimulation when expressed in midbrain dopamine neurons.

Small-Conductance Ca^{2+}-Activated K$^+$ Channels: Form and Function

John P. Adelman,[1,*] James Maylie,[2] and Pankaj Sah[3,*]

[1]Vollum Institute, Oregon Health & Science University, Portland, Oregon 97239; email: adelman@ohsu.edu

[2]Department of Obstetrics and Gynecology, Oregon Health & Science University, Portland, Oregon 97239; email: mayliej@ohsu.edu

[3]The Queensland Brain Institute, The University of Queensland, Brisbane, Queensland, 4072, Australia; email: pankaj.sah@uq.edu.au

*Corresponding authors.

Keywords

Ca^{2+} gating, intrinsic excitability, synaptic transmission, synaptic plasticity, learning and memory

Abstract

Small-conductance Ca^{2+}-activated K$^+$ channels (SK channels) are widely expressed throughout the central nervous system. These channels are activated solely by increases in intracellular Ca^{2+}. SK channels are stable macromolecular complexes of the ion pore–forming subunits with calmodulin, which serves as the intrinsic Ca^{2+} gating subunit, as well as with protein kinase CK2 and protein phosphatase 2A, which modulate Ca^{2+} sensitivity. Well-known for their roles in regulating somatic excitability in central neurons, SK channels are also expressed in the postsynaptic membrane of glutamatergic synapses, where their activation and regulated trafficking modulate synaptic transmission and the induction and expression of synaptic plasticity, thereby affecting learning and memory. In this review we discuss the molecular and functional properties of SK channels and their physiological roles in central neurons.

INTRODUCTION

Afterhyperpolarization (AHP): the hyperpolarization phase of a neuron's action potential in which the membrane potential falls below the normal resting potential

Calcium (Ca^{2+}) is a ubiquitous second messenger; cytosolic Ca^{2+} concentration is tightly controlled by a combination of buffer proteins, Ca^{2+} pumps, and transporters (1). Rises in cytosolic Ca^{2+} are used to signal diverse actions (2). The activation of Ca^{2+}-dependent ion channels is perhaps the most rapid response and leads to immediate as well as long-term physiological changes. Several types of ion channels that respond to increases in cytosolic Ca^{2+} have been described (3); of these, Ca^{2+}-activated K^+ channels have the widest distribution and are the most understood.

Increased K^+ permeability subsequent to elevated cytosolic Ca^{2+} was initially described in red blood cells more than 50 years ago (4). Twenty years later, pharmacological manipulation of cytosolic Ca^{2+} led to the identification of Ca^{2+}-dependent K^+ channels in molluscan neurons (5). Subsequently, a variety of Ca^{2+}-activated K^+ channels that can be separated on the basis of their molecular and pharmacological properties were described. Big-conductance K^+ channels (BK channels) have single-channel conductances of 100–200 pS and are blocked by low concentrations of tetraethyl ammonium as well as by a number of specific peptide antagonists (6). Although BK channels are formally voltage-dependent K^+ channels that are modulated by Ca^{2+} (7), under physiological conditions their activation requires a rise in cytosolic Ca^{2+} (8) such as that which occurs during action potentials (APs). The activity of these channels contributes to the repolarization phase of the spike and to the fast component of the afterhyperpolarization (fAHP) (9). Small-conductance Ca^{2+}-activated K^+ channels (SK channels) have lower single-channel conductance (10–20 pS) and are voltage independent, as they are gated directly by submicromolar concentrations of intracellular Ca^{2+}. These latter channels are widely distributed and play multiple roles in a variety of cell types (10). Within the central nervous system, SK channels are expressed in many regions, and their activity has rapid effects on intrinsic excitability and synaptic transmission, as well as on long-term changes that affect learning and memory formation. A third type, the intermediate-conductance Ca^{2+}-activated K^+ channel (IK channel), is structurally and functionally similar to an SK channel and is part of the same gene family but is largely absent in central neurons.

IDENTIFICATION OF SK CHANNELS

Researchers first showed more than 70 years ago that injection of bee venom has toxic actions on the central nervous system (11). The active neurotoxin apamin, an 18-amino-acid peptide with two disulfide bridges and a C-terminal amide (12), was subsequently isolated from *Apis mellifera*. Sublethal intravenous injections of apamin into rodents induce extreme motor coordination deficits and hyperactivity; lethal doses result in tonic convulsions and respiratory failure. Early studies using visceral smooth muscle and guinea pig liver showed that apamin could block the hyperpolarizing effects of adrenergic agonists or ATP and that this effect was due to blockade of "Ca^{2+}-mediated increases in potassium permeability" (13). Electrophysiological studies then established that apamin blocks a voltage-insensitive Ca^{2+}-activated K^+ current (14). The underlying channels were subsequently identified as SK channels (15). Apamin is a remarkably selective SK channel blocker, and after more than 70 years, SK channels are the only known targets for apamin. Thus, the effects of apamin at the molecular, cellular, and even behavioral levels may be ascribed to SK channel blockade.

SK CHANNEL CLONING AND STRUCTURE

The cDNA clones encoding SK channels were isolated using a degenerate oligonucleotide probe designed upon the consensus sequence for K^+-selective pores that identified three clones: SK1,

SK2, and SK3 (16). Subsequent sequence homology screens identified the fourth member of the family, IK1 (17, 18). SK1, SK2, and SK3 mRNAs are expressed throughout the brain in overlapping yet distinct patterns (19), whereas IK1 expression in neurons is much more limited. The patterns of SK mRNA expression are largely mirrored by the respective SK protein distributions (20). The International Union of Pharmacology has now placed all Ca^{2+}-activated K^+ channels into one gene family, in which they are identified as $K_{Ca}1.1$ (BK channels); $K_{Ca}2.1$, $K_{Ca}2.2$, $K_{Ca}2.3$ (SK1, SK2, and SK3 channels); and $K_{Ca}3.1$ (IK1 channels) (21).

SK channel subunits share the serpentine transmembrane (TM) topology of voltage-activated K^+ channels, with six TM domains and cytosolic N and C termini. The fourth TM domain, which in voltage-activated K^+ channels is decorated with positively charged residues on one face of the predicted α helix and composes the voltage sensor (22, 23), contains three positively charged residues in SK subunits. However, cloned SK channels reflect their native counterparts in lacking any voltage dependency (15, 24), possessing a relatively small unitary conductance ($\sim$10 pS in symmetrical K^+), and being gated solely by submicromolar concentrations of intracellular Ca^{2+} ions (apparent $K_D \sim 0.5$ μM) (16, 18, 25). The TM core of the SK channel subunits contains the canonical K^+-selective signature sequence in the pore loop between TM domains 5 and 6. The reversal potential (26) and ionic selectivity of the apamin-sensitive Ca^{2+}-activated K^+ current are consistent with this structure. Interestingly, studies of homomeric rat SK2 channels expressed in *Xenopus* oocytes suggest a significant Na^+ permeability of these channels (27).

Despite their overall similar architecture, SK channels share only $\sim$40% conservation with voltage-activated K^+ channels. The most conserved domain among the SK channel subunits does not, however, reside in the TM core of the protein. Rather, it is the part of the channel immediately C-terminal to the sixth TM domain, in the intracellular C terminus of the subunits (see below). The remainder of the C-terminal domains and the intracellular N-terminal domains is less related between the different SK subunits, suggesting that these regions may impart functional and physiological distinctions to this family of channels that are otherwise not markedly different in either their single-channel conductance or Ca^{2+} sensitivity. The SK channels are highly conserved among mammalian species, and SK channel genes are recognizable in many organisms, including *Drosophila*, which has one gene, and *Caenorhabditis elegans*, which has four SK paralogs.

Functional SK channels assemble as homomeric tetramers (16). However, different subunits can also coassemble into heteromeric channels both in heterologous expression systems (28) and in native tissue (29). Alternative exon splicing generates further diversity in the primary structures of the SK subunits; the SK1 gene has up to 32 potential mRNA splice variants (30). Several splice variants of SK2 and SK3, including one that lacks apamin sensitivity, have also been described (31). In addition, the SK2 gene uses two promoters to generate different SK2 isoforms that differ in the length of the intracellular N-terminal domain (29) (see below). The physiological significance of heteromeric channel assembly or splice diversity is not yet understood.

SK CHANNEL PHARMACOLOGY

Expression of human SK1 or rat or mouse SK2/SK3 produces functional, homomeric channels (16, 32). In contrast, rat or mouse SK1 cDNA does not produce functional plasma membrane channels (33), but SK1 can coassemble with SK2 (28, 33). Expression of homomeric SK channels produces Ca^{2+}-activated K^+ currents that are potently but differentially blocked by apamin: SK2 is the most sensitive ($EC_{50} \sim 40$ pM), hSK1 channels are the least sensitive ($EC_{50} \sim 10$ nM), and SK3 demonstrates an intermediate sensitivity to apamin ($EC_{50} \sim 1$ nM) (16, 34). Apamin sensitivity is endowed in large part by several amino acids in the outer vestibule of the pore (35, 36). However, the interactions may be more complex and involve external residues outside of the pore (37).

1-Ethyl-2-benzimidazolinone (1-EBIO), 6,7-dichloro-1H-indole-2,3-dione 3-oxime (NS309), and N-cyclohexyl-N-[2-(3,5-dimethyl-pyrazol-1-yl)-6-methyl-4-pyrimidinamine] (CyPPA): SK channel modulators that increase the apparent Ca^{2+} sensitivity of the channels

Calmodulin (CaM): a ubiquitously expressed Ca^{2+}-binding protein that transduces Ca^{2+} signals by modulating the activity of binding protein partners

CaM-binding domain (CaMBD): the region of SK channels that constitutively binds CaM

Indeed, recent work shows that apamin binding has higher affinity than the effective blocking potency, suggesting that apamin may inhibit SK channel activity not by physically occluding the pore but rather by acting as an allosteric inhibitor (36).

Other, less selective compounds such as bicuculline, D-tubocurarine, and scyllatoxin (38) also block SK channels. The majority of peptide toxins described to date—including apamin and scyllatoxin, as well as organic compounds such as D-tubocurarine and quaternary salts of bicuculline, dequalinium, and cyclophane derivatives such as UCL1684 and UCL1848 (39)—are non-subtype-specific SK channel blockers. The only known exceptions so far are tamapin, a scorpion toxin, and leiurotoxin-Dab, an analog of scyllatoxin, which preferentially block SK2 (38). Several compounds that enhance SK channel activity have also been described. These include a series of structurally similar compounds, including 1-ethyl-2-benzimidazolinone (1-EBIO) and 6,7-dichloro-1H-indole-2,3-dione 3-oxime (NS309) (40) as well as the structurally distinct trisubstituted pyrimidine N-cyclohexyl-N-[2-(3,5-dimethyl-pyrazol-1-yl)-6-methyl-4-pyrimidinamine] (CyPPA) (41). All these compounds act by enhancing the Ca^{2+} sensitivity of SK channels.

Ca^{2+} GATING OF SK CHANNELS

Ca^{2+} gates SK channels, but the pore-forming subunits do not contain an intrinsic Ca^{2+}-binding domain. Rather, Ca^{2+} gating is endowed by a constitutive interaction between the pore-forming subunits and calmodulin (CaM) (**Figure 1a**). CaM binds to the intracellular domain immediately following the sixth TM domain, the CaM-binding domain (CaMBD), which is highly conserved across the SK family. Each subunit of the tetrameric channel is thought to bind one CaM, and the binding and unbinding of Ca^{2+} ions to the N-lobe E-F hands of CaM are transduced via conformational changes into channel opening and closure, respectively (25). The structure of the isolated CaMBD in complex with Ca^{2+}-CaM reveals a dimeric organization of two CaMBDs and two CaMs, with the N-lobe E-F hands occupied by Ca^{2+} ions. Each Ca^{2+}-loaded CaM interacts with each of the CaMBDs, but the CaMBDs do not make direct contact with each

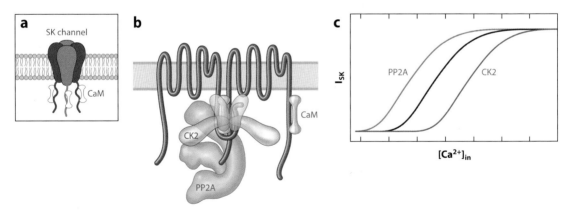

Figure 1

SK channels are macromolecular signaling complexes. (*a*) The core components of SK channels are tetrameric assemblies of the pore-forming subunits, which can be homomeric or heteromeric and have constitutively bound calmodulin (CaM) that mediates Ca^{2+} gating. (*b*) Schematic of two of the four subunits of an SK channel. Protein kinase CK2 and protein phosphatase 2A (PP2A) are also constitutive components of the channels. CK2 and PP2A alter the Ca^{2+} sensitivity of the SK channels by phosphorylating or dephosphorylating SK-associated CaM. (*c*) Ca^{2+} dose-response relationships. The resting Ca^{2+} sensitivity of channel open probability is shown in black. CK2 acts on closed SK channels to decrease Ca^{2+} sensitivity, whereas PP2A acts on open SK channels to increase Ca^{2+} sensitivity. Panel *b* is from Reference 51.

other. The interaction of CaM with the CaMBD alters the geometry of the C-lobe E-F hands on CaM, rendering them incapable of chelating Ca^{2+} ions (42). In contrast to the dimeric structure of the complex in the presence of Ca^{2+}, the structure of the Ca^{2+}-free complex is monomeric and extended (43). Together, the structures suggest that Ca^{2+} gating of SK channels involves a transition from an extended tetramer of monomers to a folded dimer of dimers that rotates a region of the CaMBD, thereby translating Ca^{2+} binding into mechanical force to open the channel gate. Interestingly, in addition to the role of Ca^{2+} in gating, the association of SK channels with CaM is critical for plasma membrane expression (44).

> **Long-term potentiation (LTP):** an activity-dependent change in synaptic strength thought to underlie some forms of learning

The SK channel family members show very similar Ca^{2+} gating profiles. They respond rapidly to Ca^{2+}, and the dose-response relationships show a Hill coefficient of 3–5, indicating cooperative, Ca^{2+}-dependent gating. Fast application of saturating Ca^{2+} (10 µM) to inside-out patches shows that SK channels have activation time constants of 5–15 ms and deactivation time constants of ∼50 ms (25). Consistent with the macroscopic currents, single-channel analysis shows that SK channels are voltage independent and that their gating is well described by a model with four closed states and two open states, with Ca^{2+}-dependent transitions between the closed states (24).

The SK channels show inward rectification that was initially attributed to pore block by internal divalent cations at positive membrane potentials. However, a recent report shows that inward rectification in SK channels is an intrinsic property that is mediated by three charged residues in the sixth TM domain. Importantly, these residues also contribute to setting the intrinsic open probability of SK channels in the absence of Ca^{2+}, affecting the apparent Ca^{2+} affinity for activation (45). The activation gate in SK channels appears to be different from that in voltage-activated K^+ channels, in which the gate is located at the intracellular end of the pore. On the basis of a methanethiosulfonate reagent and Cd^{2+} modifications of introduced cysteine residues, as well as Ba^{2+} trapping, the activation gate for SK channels likely resides deep in the inner vestibule, perhaps in the selectivity filter (46).

TRAFFICKING AND MODULATION OF SK CHANNELS

There are multiple predicted protein phosphorylation sites on the SK channels (16), and some of these sites modulate SK channel trafficking. For example, protein kinase A (PKA) phosphorylation of serine residues within the C-terminal domain of SK2 reduces plasma membrane expression in COS cells (47). PKA phosphorylation of SK channels in amygdala pyramidal neurons reduces delivery of these channels to the plasma membrane (48) (see below), and although the exact sites have not been identified in these cells, such residues are likely to be the same as those identified in hippocampal neurons. In CA1 pyramidal neurons, these PKA sites regulate long-term potentiation (LTP)-dependent trafficking (49) (see below).

SK2 and SK3 channels are multiprotein complexes and, in addition to the pore-forming subunits and CaM, have constitutively bound protein kinase CK2 and protein phosphatase 2A (PP2A) (**Figure 1b**) (50, 51). Results from structure-function studies are consistent with a model in which CK2 contacts multiple domains on the intracellular termini of the pore-forming subunits, whereas PP2A is bound by a discrete domain just C-terminal to the CaMBD (**Figure 1b**) (51). CK2 does not appear to phosphorylate the SK2 subunit. Rather, CK2 phosphorylates SK2-bound CaM at threonine 80 [CaM(T80)], but only when the channels are closed, which reduces the apparent Ca^{2+} sensitivity of the channels by approximately fivefold. This state dependency of CK2 activity is engendered by the relative disposition of a single lysine, K121, in the N-terminal domain of the channels. In the open state, PP2A activity dephosphorylates CaM(T80), increasing the Ca^{2+} sensitivity of the channels (**Figure 1c**) (51). Thus, native SK2 and SK3 channels do not have a defined Ca^{2+} sensitivity. Rather, Ca^{2+} sensitivity is Ca^{2+} dependent, as reflected by the state dependency of modulation. In neurons, this mechanism endows an activity dependency to SK2 channel activity.

CK2/PP2A modulation of native SK channels has been demonstrated in a number of systems. In superior cervical ganglion neurons and dorsal root ganglion neurons, noradrenaline and somatostatin decrease SK channel activity independently of their effects on voltage-dependent Ca^{2+} (Cav) channels that provide the Ca^{2+} for SK channel gating. The inhibition by these neurotransmitters occurs in a CK2-dependent manner, resulting in phosphorylation of SK-associated CaM and thereby reducing SK channel activity that underlies the afterhyperpolarization (AHP) in these neurons (52). In hippocampal pyramidal neurons, activation of M1 muscarinic acetylcholine receptors (mAChRs) increases excitatory postsynaptic potentials (EPSPs), N-methyl-D-aspartate-type glutamate receptor (NMDAR) synaptic currents, and spine Ca^{2+} transients, enhancing NMDAR-dependent LTP (see Reference 53). However, two recent reports show that these actions of acetylcholine (ACh) are not mediated by direct modulation of synaptic glutamate receptors but rather that they are transduced through SK channel modulation. One study found that ACh works through SK2-bound CK2 to reduce the Ca^{2+} sensitivity of spine SK channels that are activated by synaptically evoked Ca^{2+} entry and that normally function to reduce EPSPs (54) (see below). However, another study did not see effects of CK2 and instead pointed to protein kinase C modulation of SK channels as underlying the effects of M1 mAChR activation on synaptic transmission (55). Whether ACh works through CK2 or PKC, the result is decreased SK channel activity that reduces the repolarizing influence of synaptic SK channels (see below).

N-Methyl-D-aspartate receptors (NMDARs): a major subtype of glutamate receptor that mediates excitatory neurotransmission

Nicotinic acetylcholine receptors (nAChRs): nicotine-sensitive ionotropic acetylcholine receptors that usually mediate an excitatory postsynaptic response

Ca^{2+} SOURCES FOR SK CHANNEL ACTIVATION

SK channels are activated by elevations in cytosolic Ca^{2+} from several different sources: Ca^{2+} influx via Cav channels (**Figure 2a**); Ca^{2+} influx via Ca^{2+}-permeable agonist-gated ion channels, such as NMDARs and nicotinic acetylcholine receptors (nAChRs); and Ca^{2+} released from intracellular Ca^{2+} stores by generation of inositol trisphosphate (IP3) via G protein–coupled receptors, Ca^{2+}-induced Ca^{2+} release (CICR), or a combination of the two. Cytosolic Ca^{2+} then returns to resting levels due to a combination of diffusion and buffering, uptake into organelles, and extrusion by exchangers in the plasma membrane (56). These mechanisms have different cellular distributions and kinetic properties, and as SK channel activity largely follows that of free Ca^{2+} near the channel, the kinetics of the macroscopic current thus depends on the source and location of Ca^{2+}.

Ca^{2+} entering the cytosol via Cav channels leads to local and large (greater than micromolar) increases in the immediate vicinity of the Cav pore but rapidly equilibrates due to a combination

---

Figure 2

Ca^{2+} sources for SK channel activation. (*a*) Somatodendritic SK channels can be activated by Ca^{2+} influx through voltage-dependent Ca^{2+}(Cav) channels that are activated by action potentials (APs). Ca^{2+} influx during the APs activates somatic SK channels, and their activity contributes to the afterhyperpolarization (AHP), affecting AP frequency, as revealed by the effects of apamin. (*b*) SK channels can also be activated by Ca^{2+} release from internal stores, such as that which occurs following G protein–coupled metabotropic glutamate receptor (mGluR) activation, leading to the generation of inositol trisphosphate (IP3), or by Ca^{2+}-induced Ca^{2+} release (CICR) subsequent to Ca^{2+} influx through Cav channels and the activation of ryanodine receptors or IP3 receptors (IP3R). (*c*) An example of dendritic Ca^{2+} influx and the associated AHP before and after activation of G protein–coupled receptors and the generation of IP3. An AP (*left*; *lower trace*) leads to a rise in dendritic Ca^{2+} (*left*; *upper trace*), which is shown as a change in fluorescence with Oregon green. In the presence of IP3, Ca^{2+} influx during the AP is amplified by CICR (*right*; *upper trace*), activating SK channels and thus enhancing the AHP (*right*; *lower trace*). V_m denotes membrane voltage.

of diffusion and buffering (57). Cav channels are activated largely during the upstroke of APs. Activation of SK channels by inflowing Ca^{2+} is determined by their proximity to Cav channels (58), and their activity tracks the time course of changes in free Ca^{2+} in their vicinity (26, 58).

In contrast, Ca^{2+} influx via transmitter-gated channels is slower due to the intrinsically slow kinetics of transmitter-gated channels. Two Ca^{2+}-permeable ionotropic receptors that couple to SK channels are NMDARs (59, 60) and α9-containing nAChRs (61). In neurons of the hippocampus and amygdala, NMDARs are expressed in dendritic spines, and SK2-containing channels are found in close apposition. The synaptically activated Ca^{2+} transient is large enough (62) to activate

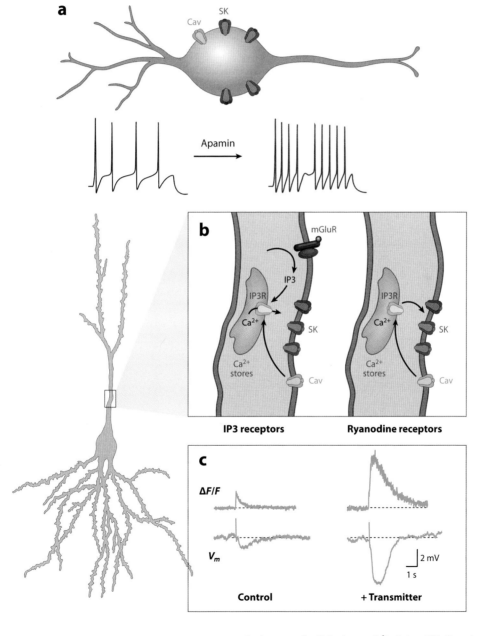

Long-term depression (LTD): an activity-dependent change in synaptic strength thought to underlie some forms of learning

SK2-containing channels. Because of the restricted geometry of the spine head and the intrinsically slow kinetics of the NMDAR, spine Ca^{2+} transients largely track the kinetics of the excitatory postsynaptic current (62), and the time course of SK channel activity follows (59). Interestingly, in spines, Ca^{2+} influx through NMDARs may not be the principal source of Ca^{2+} that activates spine SK channels (see below). In contrast, Ca^{2+} influx through α9-containing nAChRs directly and rapidly activates nearby SK2-containing channels (see below).

Ca^{2+} is also stored in the smooth endoplasmic reticulum and can be released by activation of IP3 receptors subsequent to IP3 generation by metabotropic receptor activation. Ca^{2+} ions act cooperatively to activate IP3 receptors, and IP3 generation can cooperatively interact with AP-induced Ca^{2+} rises to amplify Ca^{2+} release from intracellular stores (**Figure 2b**). This form of Ca^{2+} release activates SK channels (63, 64). Activation of IP3 receptors and the subsequent release of Ca^{2+} from intracellular stores are intrinsically slow, and the Ca^{2+} transients evoked by such release are in general an order of magnitude slower than those due to influx from extracellular sources. Thus, in midbrain dopamine neurons, both synaptically released glutamate (65) and ACh (66) evoke a slow inhibitory postsynaptic potential (IPSP) that results from mobilization of Ca^{2+} from IP3-sensitive intracellular stores and subsequent activation of SK channels. A similar cholinergic-mediated hyperpolarization is also present in cortical (63) and amygdala (48) pyramidal neurons. Interestingly, when coupled to Ca^{2+} influx via APs, the actions of IP3 on Ca^{2+} release can be amplified by CICR from IP3 receptors, and store-released Ca^{2+} can trigger an AHP that greatly inhibits AP generation (64).

Ca^{2+} release from intracellular stores can also occur via ryanodine receptors (RyRs) (**Figure 2b**). Thus, in Meynert cells of the nucleus basalis (67), ventral midbrain dopamine neurons from young animals (68), and neurons in the medial preoptic hypothalamus (69), SK channels underlie slow spontaneous miniature outward currents that are activated by RyR-mediated Ca^{2+} release from intracellular stores. Little is known about the function of the spontaneous miniature outward currents, but they may influence both spontaneous firing and transitions to burst firing. The spontaneous release of Ca^{2+} from internal stores and the activation of SK channels have been implicated in the generation of spontaneous hyperpolarizations in dopamine neurons from neonatal rats (70).

P/Q-type Ca^{2+} channel activity is necessary for SK2 channel activation in Purkinje neurons, but the SK2 channels may not be directly gated by Ca^{2+} ions entering the cytosol through P/Q-type channels. Junctophilins are components of junctional complexes between the plasma membrane and the endoplasmic reticulum that mediate cross talk between cell-surface and intracellular RyRs and IP3Rs. In response to climbing fiber stimulation, Purkinje cells fire a complex spike, followed by an apamin-sensitive AHP that is absent in junctophilin 3, 4 double-null mice (JP-DKO). Moreover, in control mice, blocking RyRs but not IP3Rs abolishes the SK-mediated AHP, and immunogold electron microscopy has shown colocalization of RyRs and SK2 channels in Purkinje cell dendrites. Finally, protocols that induce long-term depression (LTD) at parallel fiber–to–Purkinje cell synapses, considered a cellular basis for cerebellar motor learning (71), result in LTP in JP-DKO mice or after apamin application to control slices. Thus, in Purkinje cells, Ca^{2+} influx through P/Q-type channels is required to induce CICR, and Ca^{2+} ions released from ryanodine-sensitive stores finally gate the SK channels. The strict coupling of P/Q-type Ca^{2+} channels and SK channel activity actually reflects a structural platform, including junctophilins, that is specialized for Ca^{2+}-mediated communication between P/Q-type channels, RyRs, and SK channels, which cooperate to generate the AHP (72).

P/Q-type Ca^{2+} channel mutant mice are models for human episodic ataxia type 2 and highlight the importance of the strict coupling between P/Q-type channels and SK channels in Purkinje cells. In *leaner* and *ducky* mice, each with decreased P/Q-type Ca^{2+} currents, the firing precision in Purkinje cells is markedly reduced. Presynaptic P/Q-type channels are often important for neurotransmitter release, and it was believed that the ataxia was caused by impaired synaptic transmission within the cerebellum. However, closer examination has revealed that the firing precision deficit is due to the reduced activation of postsynaptic P/Q-coupled SK2 channels in Purkinje cells, evidenced as a reduced medium afterhyperpolarization (mAHP). The AHP together with the precision deficit were normalized by the application of the SK channel agonist 1-EBIO. Remarkably, in vivo perfusion of 1-EBIO into the cerebellum significantly improved the motor coordination deficits seen in *ducky* mice as well as the dyskinesia and ataxia in *tottering* mice (73). Similarly, the structurally related, FDA-approved compound chlorzoxanone also normalized Purkinje cell precision, and orally administered chlorzoxanone ameliorated motor deficits and reduced the severity, frequency, and duration of dyskinesia episodes without any obvious adverse effects (74).

SK CHANNELS AND INTRINSIC EXCITABILITY

In many neurons a single AP or bursts of APs are followed by a prolonged AHP that may depend on Ca^{2+} influx (75). The AHP influences the intrinsic excitability of neurons, the firing of APs from a given input signal, in two ways. First, the increased K^+ conductance that underlies the AHP that follows each AP influences the voltage trajectory between APs, thus setting the frequency of cell firing. Second, bursts of APs result in summation of the AHP that ultimately blocks AP firing, a phenomenon known as spike frequency adaptation (76).

The AHP may display several overlapping kinetic components, the fast (f), medium (m), and slow (s) AHPs (77). The fAHP, which overlaps the falling phase of the AP and contributes to spike repolarization and the initial component of the AHP, typically lasts for 10–20 ms (78, 79). The fAHP is frequently due to the activity of BK channels (78).

The mAHP activates rapidly and decays over several hundred milliseconds. In many, but not all, cases the mAHP is blocked by apamin, which increases excitability and identifies this component as being due to SK channel activity (**Figure 2a**). An apamin-sensitive component of the AHP has been demonstrated in a wide variety of neurons, including spinal motoneurons (80), neurosecretory neurons in the supraoptic area of the hypothalamus (81), vagal motoneurons (82), pyramidal neurons in the sensory cortex (83) and the lateral and basolateral amygdala (64), interneurons in the nucleus reticularis of the thalamus (nRt) (84), striatal cholinergic interneurons (85), hippocampal interneurons in the stratum oriens-alveus (86) and the stratum radiatum (87), cholinergic nucleus basalis neurons (88), paraventricular neurons (89), rat subthalamic neurons (90), cerebellar Purkinje neurons (91; but see Reference 92), noradrenergic neurons of the locus coeruleus, serotonergic neurons of the dorsal raphe, midbrain dopamine neurons (93), circadian clock neurons in the suprachiasmatic nucleus of the hypothalamus (94), and mitral cells of the olfactory bulb (95).

The sAHP typically lasts several seconds. The underlying current, IsAHP, has a prominent rising phase lasting several hundred milliseconds and decays over several seconds. The IsAHP was identified as a K^+ current that is activated by the influx of extracellular Ca^{2+}, as this current is blocked by inhibiting Ca^{2+} influx or by buffering cytosolic Ca^{2+} (96). Even though apamin does not block the sAHP, the sAHP was thought to be due to an apamin-insensitive type of SK channel. However, it now seems unlikely that SK channels underlie the slow AHP. First, all SK channels open rapidly to exogenously applied Ca^{2+}, whereas the IsAHP has a slow rising phase that outlasts the immediate rise in cytosolic Ca^{2+} following APs or step depolarization. Second, rapidly buffering cytosolic Ca^{2+} does not rapidly decrease the IsAHP (97). Finally, the IsAHP in CA1 pyramidal neurons remains intact in SK transgenic mice lacking SK1, SK2, or SK3 channels

(98). The molecular identities of the sAHP channels, and their mechanisms of activation, remain unknown.

The contribution of SK channels to the AHP is due largely to their activation by Ca^{2+} influx via Cav channels. However, in most cases the identity of the Cav channels that provide the Ca^{2+} and the subtype of the SK channel that contributes to the AHP are not known. Moreover, in those cases in which these are known, the linkage of specific Cav channels to specific types of SK channels does not appear to be absolutely specific. For example, in cerebellar Purkinje neurons, P/Q-type Ca^{2+} channels are selectively coupled to SK2 channels (92), whereas in their downstream target neurons of the deep cerebellar nucleus, N-type Ca^{2+} channels provide the Ca^{2+} source for SK channel activity (99). Indeed, some neurons have more than one functional population of SK channels, which can be coupled to different Ca^{2+} sources. Thus, in midbrain dopamine neurons, a slow oscillation of the membrane potential drives intrinsic pacemaker activity (100). The depolarizing phase has been attributed to L-type Ca^{2+} channels (101), whereas the repolarization phase results from the activation of SK channels. In contrast, the AHP in these neurons, also mediated by SK activity, is driven by Ca^{2+} influx via T-type Ca^{2+} channels (102). Similarly, in spinal motoneurons there are two functionally distinct SK channel populations: one that mediates the mAHP and is activated by Ca^{2+} influx through N- and P/Q-type Ca^{2+} channels and another population that forms a feedback loop with persistent L-type Ca^{2+} channel currents (103).

Despite the variability in the coupling of Cav channels and SK channels, in the same neuron there is discrete and specific coupling of Ca^{2+} channels to different Ca^{2+}-activated K^+ channels. This is highlighted by the exquisite selectivity shown for BK and SK channel activation in acutely dissociated hippocampal neurons. In somatic cell–attached patch recordings, Ca^{2+} influx through L-type channels activates SK channels without activating BK channels that are present in the same patch. The latency between L-type channel and SK channel openings suggests that these channels are 50–150 nm apart. In contrast, Ca^{2+} influx through N-type channels activates only BK channels; opening of the two channel types is almost coincident (104). The situation is reversed in superior cervical neurons (105) and in vagal motoneurons (106), in which N-type Ca^{2+} channels provide the Ca^{2+} for SK channels that are responsible for the mAHP whereas different Cav channels mediate the BK channel activation that underlies the fAHP. This absolute segregation of coupling between channels reflects the strict delineation of submembrane Ca^{2+} microdomains.

When one is assessing the contribution of different channels to the AHP, the recording technique used is an important contributor to data interpretation. The amplitude of the medium and slow components of the AHP is generally studied using somatic intracellular recordings in which depolarizing current injections generate a single AP or trains of APs. These APs are generated at the axon's initial segment and then back propagate into the dendritic tree with different levels of attenuation. Consequently, different populations of Ca^{2+} channels are activated in different neuronal compartments. The AHP is then measured as the membrane potential returns to baseline, and the contribution of different channels is determined by pharmacological means.

To identify the currents that may underlie the AHP, depolarizing voltage steps generate an unclamped Ca^{2+} action current. Upon repolarization, the activity of Ca^{2+}-dependent K^+ currents is measured as a tail current. The identity of the underlying channels is determined by kinetic and pharmacological analysis. However, there may not be a strict correspondence between the AHP measured in current clamp mode and the IAHP measured in voltage clamp mode. For example, the contribution of SK channels to the mAHP in hippocampal CA1 pyramidal neurons has been controversial. One study reports that apamin reduced the mAHP (107), whereas other studies find no apamin sensitivity to the mAHP. These latter studies indicate that the mAHP in CA1 pyramidal neurons is due to H and M channels (108), despite the fact that tail currents consistently show a pronounced apamin-sensitive component (98, 107, 108). Pyramidal neurons of the lateral amygdala reflect a similar situation (109).

SK CHANNEL EFFECTS ON NETWORK ACTIVITY

SK channels are not usually active at rest, requiring elevated levels of cytosolic Ca^{2+} for activation. Hypothalamic gonadotropin-releasing hormone (GnRH) neurons provide a noteworthy exception in showing tonic SK channel activity (110), and there appear to be two apamin-sensitive components of the AHP that may reflect differently coupled SK channel populations. The pulsatile release of GnRH is crucial to fertility and reflects burst firing in GnRH neurons. Burst firing induces time-locked Ca^{2+} transients that are generated by Ca^{2+} influx through L-type Ca^{2+} channels and are amplified by IP3-mediated Ca^{2+} release from internal stores. These Ca^{2+} transients rapidly activate SK channels that underlie a component of the AHP and regulate both intraburst and interburst dynamics, which are critical for GnRH neuron burst firing (111). The Ca^{2+} transients also activate an apamin-sensitive sAHP (112).

> **Gonadotropin-releasing hormone (GnRH):** a tropic peptide hormone that stimulates the release of luteinizing hormone and follicle-stimulating hormone from pituitary gonadotrophs

A distinctly different type of coupling exists in the nRt, a thin inhibitory network interposed between thalamocortical projection neurons and the cortex that is crucial for information transfer and arousal control. Neurons in the thalamocortical system cooperate to produce synchronized, rhythmic network activity, which underlies the slow waves characteristic of sleep electroencephalograms. Rhythmogenesis is accompanied by low-threshold burst discharges in thalamic neurons, which require T-type Ca^{2+} channels. Intrinsic to nRt neurons is the ability to generate oscillatory activity comprising an alternating sequence of low-threshold Ca^{2+} spikes separated by SK channel–mediated AHPs (113). In the dendrites of nRt neurons, Ca^{2+} influx through T-type channels activates competing targets, SK2 channels and sarco/endoplasmic reticulum Ca^{2+}-ATPases, to generate and regulate the strength of nRt oscillations. The key role of SK2 channels in mediating these oscillations is highlighted by the fact that SK2-null mice have fragmented sleep patterns (114).

In thalamic brain slices, robust epileptiform oscillations underlying absence epilepsy can be induced by the application of bicuculline methiodide, which blocks both $GABA_A$ receptors and SK2 channels; more specific blockers of each channel (picrotoxin and apamin, respectively) are insufficient on their own. Blocking SK channels with apamin enhances the intrinsic excitability of nRt neurons (115) but has no effect on relay neurons, showing that changes in the intrinsic properties of individual neurons and changes at the circuit level can robustly modulate these oscillations and suggesting that SK channels may be therapeutic targets for absence epilepsy (116).

Thalamic circuits are also important targets for general anesthetics. In nRt neurons, the application of propofol increases intrinsic excitability, temporal summation, and spike firing rate by reducing the mAHP. Apamin mimics and occludes this action. Consistent with increased excitability of nRt neurons, propofol increases GABAergic transmission, and the activation of nRt neurons is essential for propofol to elicit thalamocortical suppression, indicating that propofol-mediated blockade of SK channels may play a critical role in gating spike firing in thalamocortical relay neurons (117).

SK CHANNELS AND LEARNING AND MEMORY

Apamin crosses the blood-brain barrier (118), and early studies showed that systemic apamin administration facilitates learning and memory. The first such study, using appetitive learning paradigms, reported that systemic apamin injections accelerate acquisition of the bar-pressing response and also accelerate bar-pressing rates (119). Apamin facilitates the encoding of memory, as assessed by habituation of exploratory activity, and improves performance on the novel object recognition task (120). Apamin-treated mice exhibit faster learning of the platform location during the initial training trials in the Morris water maze (**Figure 3a,b**) (121). Thus, blockade of SK channels facilitates an early stage of hippocampus-dependent spatial memory encoding. Apamin also improves task acquisition in a learned extinction operant behavior protocol (122) and enhances working memory in a medial prefrontal cortex–dependent spatial delayed alternation task (123).

Systemic apamin facilitates the encoding of contextual fear memory (124). The hippocampus is critical for encoding the association between the context and footshock (125), whereas the amygdala is essential for encoding the associations of footshock with both the context and the tone (126). Apamin-treated mice exhibit strong contextual fear memory, which is typically seen after more extensive conditioning. Interestingly, apamin does not affect the strength of amygdala-dependent cued fear memory. These findings suggest that systemic apamin facilitates the encoding of hippocampus-dependent contextual fear memory and that SK channels limit memory encoding. However, some aspects of working/short-term memory are disrupted in SK3-deficient mice (127).

In contrast to the effects of blocking SK channels with apamin, increasing SK channel activity impairs learning. In the Morris water maze, SK2-overexpressing mice cannot learn the platform location, despite an extensive training paradigm (128). Similarly, the SK channel agonists 1-EBIO and CyPPA impair the encoding, but not the retrieval, of object memory in a spontaneous object recognition task (124).

Experimental models of amnesia show that SK channel activity is also implicated in memory retention, such as that which may occur in normal aging or after brain damage or is associated with Alzheimer's disease. Apamin dose-dependently alleviates deficits in spatial reference and working memory induced by partial electrolytic hippocampal lesion, as assessed with the radial arm maze and the water maze (129). Medial-septal lesions cause a cholinergic deficit in the hippocampus, and the degeneration of cholinergic cells is associated with the development of cognitive disorders in Alzheimer's disease and during normal brain aging (130). Systemic apamin administration reverses the navigation failure presented in medial-septal-lesioned mice during the initial and reversal learning stages of the Morris water maze task (129). Apamin also attenuates the memory deficits caused by scopolamine, a cholinergic antagonist that produces amnesia, most likely by affecting hippocampal and cortical activity (131).

In addition to influencing the acquisition phase, changes in SK channel activity may contribute to the expression of learning and memory. Hippocampus-dependent learning is impaired in older animals. Hippocampal SK3 expression increases with age, and decreasing SK3 expression by the infusion of antisense oligonucleotides improves memory in aged mice (132). One recent study reported that SK2 and SK3 mRNA levels in hippocampus, assessed by in situ hybridization, were transiently downregulated in the early stages of a spatial learning task (133). Another study reported that the enhanced AP firing seen in the nucleus accumbens core that occurs following protracted abstinence from alcohol is associated with reduced SK3 expression, thus reducing the AHP. Moreover, direct infusion of 1-EBIO into the nucleus accumbens core specifically reduces the motivation to obtain alcohol after abstinence (134).

Taken together, these studies underscore the importance of SK channels in information processing and storage at the systems level. Such studies suggest that SK channels may be suitable targets for therapeutic intervention in learning deficits associated with trauma, pathologies, normal aging, and alcohol addiction.

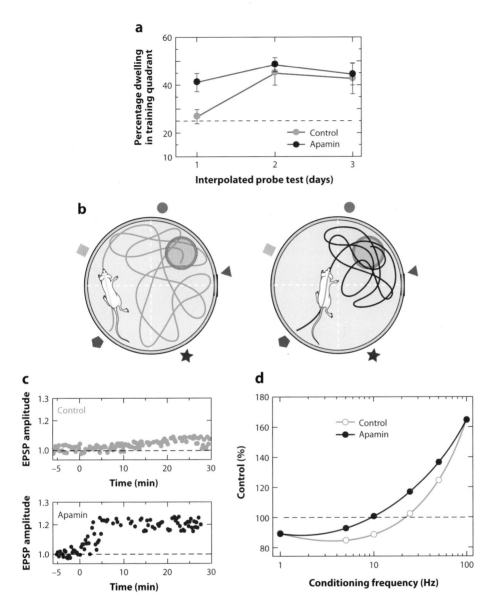

Figure 3

SK channel activity affects learning and the induction of synaptic plasticity. (*a*) Apamin administration facilitates acquisition of the Morris water maze task. Apamin-treated mice (*red*) perform significantly better than saline-treated control mice (*blue*) after minimal training, although the former do not learn to any greater final extent. (*b*) Search paths for control (*left*) and apamin-treated (*right*) mice during the first probe test. (*c*) Apamin increases the amount of long-term potentiation (LTP). In the control condition, a 25-Hz tetanus delivered to the stratum radiatum does not result in a change in synaptic strength in area CA1 of the hippocampus but results in LTP in the presence of apamin. EPSP denotes excitatory postsynaptic potential. (*d*) Blocking SK channel activity with apamin facilitates the induction of synaptic plasticity by shifting the frequency-response relationship to reduced conditioning frequencies. Panels *a*, *b*, and *d* are after Reference 121.

SK CHANNELS, SYNAPTIC TRANSMISSION, AND PLASTICITY

> **2-Amino-3-(5-methyl-3-oxo-1,2-oxazol-4-yl)propanoic acid receptors (AMPARs):** a major subtype of glutamate receptor that mediates excitatory neurotransmission

Synaptic plasticity is a leading model for activity-dependent cellular changes underlying certain forms of memory. Two early papers provided the first direct links between SK channel activity and synaptic plasticity. One study showed that blocking SK channels with apamin during a 5-Hz tetanus facilitated the induction of LTP in area CA1 of the hippocampus (135). A second study demonstrated that apamin increases the magnitude of LTP induced by a single 100-Hz tetanus but has no effect on a maximal induction protocol (three repeated 100-Hz stimulations) (136). This work was later expanded to examine a range of conditioning frequencies and revealed that apamin facilitates the induction of synaptic plasticity in area CA1 by shifting the frequency-response relationship to lower conditioning frequencies (**Figure 3c,d**) (121). Brain-derived neurotrophic factor (BDNF) has a similar effect as does apamin on LTP induction in area CA1 induced by theta-pattern stimulations, and the effects of BDNF and apamin are not additive, suggesting that BDNF may facilitate LTP induction by downregulating SK channel activity (137). Although presaging later studies that revealed the likely cellular mechanisms behind these effects of apamin (see below), all these studies concluded that apamin affects the induction of synaptic plasticity by reducing the AHP and increasing somatic excitability, either directly in CA1 pyramidal neurons (121, 135, 137) or indirectly through CA1 interneurons (136).

Two studies published simultaneously revealed the likely cellular mechanism by which SK channel activity modulates the induction of synaptic plasticity. These papers showed that, in pyramidal neurons of the hippocampus and amygdala, SK channels in spines are activated by synaptically driven Ca^{2+} influx and that their activity modulates synaptic responses. The activation of synaptic SK channels provides a repolarizing conductance that counters 2-amino-3-(5-methyl-3-oxo-1,2-oxazol-4-yl)propanoic acid (AMPA)-mediated depolarization, reducing EPSPs (**Figure 4a,b**) (59, 60). Thus, blocking synaptic SK channels with apamin increases EPSPs, and this effect can be occluded or reversed by blocking NMDARs before or after apamin application, respectively.

Interestingly, most of the spine Ca^{2+} transient is due to Ca^{2+} influx through NMDARs and in CA1 pyramidal neurons, and SK2 and NMDARs are closely colocalized in the synaptic membrane (**Figure 4a**) (49). Yet the source of Ca^{2+} that activates the spine SK channels may not be influx through NMDARs. One study showed that blocking R-type Cav channels with the peptide toxin SNX-482 is sufficient to block spine SK channel activation while NMDARs are active, leading to the suggestion that NMDAR activity boosts the depolarization of the spine membrane potential; such depolarization adds to the 2-amino-3-(5-methyl-3-oxo-1,2-oxazol-4-yl)propanoic acid receptor (AMPAR)-mediated depolarization and effectively activates R-type Ca^{2+} channels that fuel SK channel activation (138). In addition, a recent report showed that in prefrontal layer V pyramidal neurons, Ca^{2+} influx through NMDARs and R-type and L-type Ca^{2+} channels as well as Ca^{2+} ions released from IP3-sensitive stores activate spine SK channels (139). These results suggest the presence of very restricted Ca^{2+} domains within the spine head that may couple to distinct transduction mechanisms.

Similar to hippocampal CA1 pyramidal neurons, in pyramidal neurons of the amygdala, tetanic stimulation of cortical afferents leads to LTP of the synaptic input, and LTP is significantly enhanced in the presence of apamin (59). Thus, in both of these brain areas, SK channel activity limits synaptic plasticity (**Figure 4**). The contributions of SK channels to synaptic responses provide a mechanism by which SK channel activity modulates synaptic plasticity and perhaps learning (see above). The repolarizing influence of synaptic SK channel activity on the spine membrane potential contributes to reinstating the block of NMDARs by external Mg^{2+} ions, thus reducing Ca^{2+} influx through NMDARs and reducing the spine Ca^{2+} transient (59, 60). The duration and direction of synaptic plasticity are determined by the degree of activation of NMDARs. The

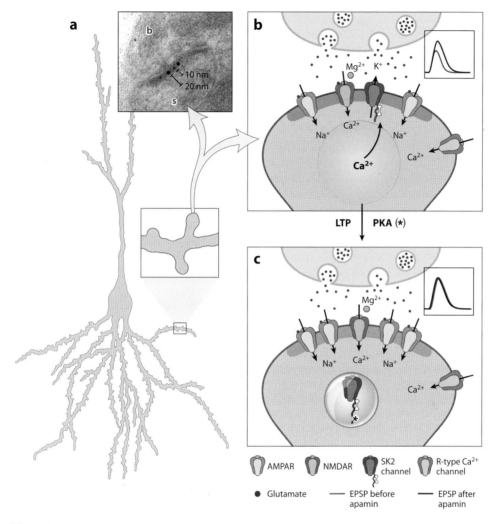

Figure 4

SK2 channels modulate synaptic responses, and SK channel trafficking contributes to long-term potentiation (LTP). (*a*) Schematic of a pyramidal neuron. (*Top inset*) Immunoelectron micrograph shows that SK2 (10-nm particles) is expressed in the synaptic membrane, specifically the postsynaptic density, close to *N*-methyl-D-aspartate-type glutamate receptors (NMDARs) (20-nm particles) in dendritic spines (s). b denotes bouton. (*Lower inset*) A dendritic segment with spines. From Reference 49. (*b*) Schematic shows a presynaptic terminal releasing glutamate onto a postsynaptic spine of a pyramidal neuron. Postsynaptic SK2 channels are activated by synaptically evoked Ca^{2+} influx and provide a repolarizing conductance that opposes the 2-amino-3-(5-methyl-3-oxo-1,2-oxazol-4-yl) propanoic acid receptor (AMPAR)-mediated depolarization, reducing the excitatory postsynaptic potential (EPSP). (*Inset*) The EPSP before apamin (*gray trace*) and after apamin (*red trace*). Apamin increases the EPSP. (*c*) Phosphorylation of SK2 channels by protein kinase A (PKA) either during LTP induction or due to receptor activation leads to endocytosis of postsynaptic SK channels. (*Inset*) Following phosphorylation of SK channels, apamin does not affect the EPSP. Gray trace, before apamin; red trace, after apamin.

magnitude of the Ca^{2+} influx through NMDARs in dendritic spines likely determines whether the synapse undergoes LTP or LTD: Large Ca^{2+} influxes lead to LTP, and modest Ca^{2+} influxes lead to LTD (140, 141). Thus, synaptic SK channel activity effectively shifts the frequency-response relationship to stronger conditioning stimulations; blocking SK channels with apamin increases Ca^{2+} influx through NMDARs and facilitates the induction of synaptic plasticity.

The activation of SK channels by Ca^{2+} influx via transmitter-gated ion channels can also result in unexpected consequences for neurotransmission. In cochlear outer hair cells, SK2 channels are tightly coupled to α9-containing Ca^{2+}-permeable nAChRs. Ca^{2+} influx through the normally excitatory nAChR rapidly activates closely positioned SK2 channels that mediate a paradoxical inhibitory postsynaptic current (61).

Spine SK channel trafficking also contributes to LTP expression. At Shaffer collateral to CA1 synapses, LTP induction recruits additional GluA1-containing AMPARs to the synaptic membrane (142), effectively increasing the depolarizing drive upon excitatory synaptic transmission. In addition, endocytosis removes SK2-containing channels from the synaptic membrane, decreasing the natural shunting of the EPSP by these channels. This process requires PKA phosphorylation of serine residues in the C-terminal domain of SK2 (**Figure 4c**). Thus, LTP at these synapses reflects the sum of increased AMPARs and decreased SK2-containing channels (49). The activity-dependent exocytosis of GluA1-containing AMPARs and the endocytosis of synaptic SK2-containing channels are mechanistically and unidirectionally coupled, as selectively inhibiting GluA1-containing AMPAR exocytosis blocks SK2-containing channel endocytosis, but blocking SK2-containing channel endocytosis does not interfere with GluA1-containing AMPAR exocytosis (143). In amygdala pyramidal neurons, β-adrenoreceptor stimulation regulates synaptic SK2 expression through PKA activation. Here, different from the situation in CA1 pyramidal neurons, SK2 channels undergo constitutive recycling into and out of the synaptic membrane, and PKA selectively disrupts exocytotic delivery of SK channels, thus enhancing synaptic transmission and plasticity (48).

These studies emphasize the importance of synaptic SK2-containing channel localization. The SK2 gene (*KCNN2*) encodes two isoforms that differ only in the length of the intracellular N-terminal domain; SK2-L has 207 additional N-terminal amino acids compared with SK2-S (29). A recent study showed that SK2-L expression is essential for synaptic SK2-containing channel localization and function. In transgenic mice that selectively lack SK2-L, i.e., SK2-S-only mice, SK2-S-containing channels are expressed in the extrasynaptic spine plasma membrane but are specifically excluded from the synaptic membrane of dendritic spines on CA1 pyramidal neurons. Consequently, apamin does not affect EPSPs or LTP in these mice. In addition, SK2-S-only mice phenocopy apamin-treated mice in the novel object recognition task. Thus, SK2-L expression directs synaptic SK2-containing channel localization that is necessary for normal synaptic responses, plasticity, and learning (144).

Activity- and experience-dependent changes in synaptic transmission are cellular models for some forms of learning. The studies discussed above demonstrate a central role of synaptic SK channels in this process. Importantly, synaptic gain, the modification of neuronal responsiveness to synaptic activity, may be influenced not only by alteration in synaptic transmission (e.g., LTP or LTD) but also by modulation of intrinsic excitability. Indeed, SK channel activity that contributes to the AHP and shapes intrinsic excitability is also subject to activity-dependent plasticity. An example is layer V pyramidal neurons from the rat sensorimotor cortex. Afferent stimulations in the presence of AMPAR and NMDAR blockers result in an activity-dependent increase in intrinsic excitability (LTP-IE); more APs are evoked for a given current injection after stimulation than before stimulation. The LTP-IE requires mGluR5 receptor activation and is largely occluded by apamin. Consistent with this, the mAHP is reduced. In addition, after LTP-IE, spike trains showed fewer failures and greater temporal fidelity, similar to the effects of apamin in the absence of stimulation. In contrast, enhancing SK channel activity with 1-EBIO has largely opposite effects as does apamin (145).

In the cerebellum, stimulation of parallel fiber inputs to Purkinje cells results in LTP as well as in increased IE and a decreased AHP. Depolarizing current injections, independent of synaptic stimulation, can also induce the intrinsic plasticity that is associated with a decrease in the spike threshold, and following each AP the rate of depolarization toward spike threshold is enhanced. These changes in IE are also seen following apamin application and require activation of protein phosphatases 1, 2A, and 2B as well as of protein kinase CK2. The increased IE does not alter the input-output function, but the impact of parallel fiber signaling is reduced despite increased excitability. Finally, intrinsic plasticity lowers the probability of subsequent LTP induction at parallel fiber–to–Purkinje cell synapses (146).

The firing rate of oxytocin neurosecretory cells in the paraventricular nucleus and supraoptic hypothalamus determines the timing and quantity of peripheral hormone release. Under resting conditions, oxytocin neurons fire in a slow, irregular pattern (147). However, during lactation, these neurons display brief, high-frequency bursts of activity preceding each milk ejection (148). In late pregnancy and during lactation, the amplitudes of the mAHP and the sAHP increase, contributing to the short, synchronous bursts that release a bolus of oxytocin required for maximal mammary contraction and milk ejection. Centrally applied oxytocin can induce neuronal plasticity, and the administration of an oxytocin antagonist blocks the plasticity of AHPs observed in pregnant rats (149).

The functional output of a neuron reflects the integration of synaptic responses and intrinsic membrane excitability. Homeostatic plasticity refers to the process by which neurons adjust their output in response to activity. In nucleus accumbens medium spiny neurons, IE is decreased in response to a prolonged increase in NMDAR-mediated excitatory synaptic responses. This is accomplished by the induction of an apamin-sensitive component of the mAHP and is abolished in rats pretreated with cocaine (150).

CONCLUSIONS

In the past 15 years we have witnessed a remarkable expansion of our understanding of SK channels in central neurons. The identification of SK channel clones allowed the discovery that CaM mediates Ca^{2+} gating, and proteomics identified protein kinase CK2 and PP2A as stable components of SK channels in the brain. Although SK channels have long been appreciated for their roles in regulating somatic excitability, the synaptic localization and function of SK channels have only recently been discovered. Such findings have led to an understanding that synaptic SK channels influence neurotransmission as well as the induction and expression of synaptic plasticity and may explain the effects of apamin on learning and memory. Transgenic mouse models and apamin administration have clearly shown that SK channels influence learning, sleep, and addictive behaviors. This list will undoubtedly continue to grow, thereby expanding the possible therapeutic applications for drugs that target SK channels.

SUMMARY POINTS

1. SK channels are Ca^{2+}-activated K^+ channels that are widely expressed in the brain.
2. SK channels are multiprotein complexes of the pore-forming subunits in complex with CaM (which mediates Ca^{2+} gating), CK2, and PP2A, which modulate the Ca^{2+} sensitivity of the SK channels.

3. SK channels are found in different subcellular compartments, even within the same neuron, and are coupled to a variety of Ca^{2+} sources.

4. In many neurons, SK channel activity contributes to the AHP and modulates intrinsic excitability.

5. In many principal neurons, SK channels are expressed in the postsynaptic membrane, where they modulate synaptic responses, influence the induction of synaptic plasticity, and affect learning.

6. Spine SK channel trafficking is regulated by PKA and contributes to the expression of LTP.

FUTURE ISSUES

1. There are likely additional proteins that are stable components of SK channels, and the different molecular building blocks may associate differently in distinct cell types and subcellular compartments. If so, how are the different macromolecular SK channel complexes segregated and trafficked to their final destinations, and how do the SK-associated proteins influence specific physiological roles?

2. How and why are different populations of SK channels, even within the same neuron, coupled to different Ca^{2+} sources?

3. An emerging theme is that there are cell-type-specific differences in the synaptic functions of SK channels, and recent findings raise new questions about their dynamic trafficking within dendritic spines. How does SK2-L direct synaptic localization, how is synaptic SK channel recycling accomplished under basal conditions, and how are SK channel endocytosis and AMPAR exocytosis mechanistically coupled upon the induction of LTP?

4. SK channels are also expressed in dendrites. Are SK channels important for dendritic integration, and if so, how do they complement the other ion channel types that also contribute to this role?

5. The SK subunits form heteromeric channels, but little is known about the relative abundance of homomeric and heteromeric SK channels or about the specific roles of heteromeric SK channels and what each subtype engenders.

6. SK1 mRNAs are also expressed in the brain, yet there is virtually nothing known about the roles of SK1 or the multiple splice variants that are conserved across mammalian species.

DISCLOSURE STATEMENT

The authors are not aware of any affiliations, memberships, funding, or financial holdings that might be perceived as affecting the objectivity of this review.

ACKNOWLEDGMENTS

We are grateful to past and present members of our research groups, our collaborators, and many others in the field for the insights they have provided into SK channels. We also apologize to colleagues whose outstanding work on SK channels was not cited due to length restrictions. We thank Ms. Lori Vaskalis for graphic art. Our work is supported by the National Institutes of Health (NS038880, NS065855, and NS071314 to J.P.A. and MH081860 to J.M.) and by the National Health and Medical Research Council (P.S.).

LITERATURE CITED

1. Carafoli E. 1987. Intracellular calcium homeostasis. *Annu. Rev. Biochem.* 56:395–433
2. Ghosh A, Greenberg ME. 1995. Calcium signaling in neurons: molecular mechanisms and cellular consequences. *Science* 268:239–47
3. Hille B. 2001. *Ionic Channels of Excitable Membranes*. Sunderland: Sinauer
4. Gárdos G. 1958. The function of calcium in the potassium permeability of human erythrocytes. *Biochim. Biophys. Acta* 30:653–54
5. Meech RW. 1978. Calcium-dependent potassium activation in nervous tissues. *Annu. Rev. Biophys. Bioeng.* 7:1–18
6. Coetzee WA, Amarillo Y, Chiu J, Chow A, Lau D, et al. 1999. Molecular diversity of K^+ channels. *Ann. N. Y. Acad. Sci.* 868:233–85
7. Cui J, Cox DH, Aldrich RW. 1997. Intrinsic voltage dependence and Ca^{2+} regulation of *mslo* large conductance Ca-activated K^+ channels. *J. Gen. Physiol.* 109:647–73
8. Marty A. 1989. The physiological role of calcium-dependent channels. *Trends Neurosci.* 12:420–24
9. Faber ES, Sah P. 2003. Ca^{2+}-activated K^+ (BK) channel inactivation contributes to spike broadening during repetitive firing in the rat lateral amygdala. *J. Physiol.* 552:483–97
10. Faber ES, Sah P. 2003. Calcium-activated potassium channels: multiple contributions to neuronal function. *Neuroscientist* 9:181–94
11. Habermann E. 1984. Apamin. *Pharmacol. Ther.* 25:255–70
12. Habermann E. 1963. *Recent Advances in the Pharmacology of Toxins*. Oxford, UK: Pergamon
13. Vladimirova IA, Shuba MF. 1978. Effect of strychnine, hydrastine and apamin on synaptic transmission in smooth muscle cells. *Neirofiziologiia* 10:295–99
14. Hugues M, Schmid H, Lazdunski M. 1982. Identification of a protein component of the Ca^{2+}-dependent K^+ channel by affinity labelling with apamin. *Biochem. Biophys. Res. Commun.* 107:1577–82
15. Blatz AL, Magleby KL. 1986. Single apamin-blocked Ca-activated K^+ channels of small conductance in cultured rat skeletal muscle. *Nature* 323:718–20
16. Kohler M, Hirschberg B, Bond CT, Kinzie JM, Marrion NV, et al. 1996. Small-conductance, calcium-activated potassium channels from mammalian brain. *Science* 273:1709–14
17. Ishii TM, Silvia C, Hirschberg B, Bond CT, Adelman JP, Maylie J. 1997. A human intermediate conductance calcium-activated potassium channel. *Proc. Natl. Acad. Sci. USA* 94:11651–56
18. Joiner WJ, Wang L-Y, Tang MD, Kaczmarek LK. 1997. hSK4, a member of a novel subfamily of calcium-activated potassium channels. *Proc. Natl. Acad. Sci. USA* 94:11013–18
19. Stocker M, Pedarzani P. 2000. Differential distribution of three Ca^{2+}-activated K^+ channel subunits, SK1, SK2, and SK3, in the adult rat central nervous system. *Mol. Cell. Neurosci.* 15:476–93
20. Sailer CA, Hu H, Kaufmann WA, Trieb M, Schwarzer C, et al. 2002. Regional differences in distribution and functional expression of small-conductance Ca^{2+}-activated K^+ channels in rat brain. *J. Neurosci.* 22:9698–707
21. Wei AD, Gutman GA, Aldrich R, Chandy KG, Grissmer S, Wulff H. 2005. International Union of Pharmacology. LII. Nomenclature and molecular relationships of calcium-activated potassium channels. *Pharmacol. Rev.* 57:463–72
22. Bezanilla F. 2000. The voltage sensor in voltage-dependent ion channels. *Physiol. Rev.* 80:555–92
23. Catterall WA. 2010. Ion channel voltage sensors: structure, function, and pathophysiology. *Neuron* 67:915–28
24. Hirschberg B, Maylie J, Adelman JP, Marrion NV. 1998. Gating of recombinant small-conductance Ca-activated K^+ channels by calcium. *J. Gen. Physiol.* 111:565–81

25. Xia XM, Fakler B, Rivard A, Wayman G, Johnson-Pais T, et al. 1998. Mechanism of calcium gating in small-conductance calcium-activated potassium channels. *Nature* 395:503–7
26. Sah P. 1992. Role of calcium influx and buffering in the kinetics of Ca^{2+}-activated K^+ current in rat vagal motoneurons. *J. Neurophysiol.* 68:2237–47
27. Shin N, Soh H, Chang S, Kim DH, Park CS. 2005. Sodium permeability of a cloned small-conductance calcium-activated potassium channel. *Biophys. J.* 89:3111–19
28. Monaghan AS, Benton DC, Bahia PK, Hosseini R, Shah YA, et al. 2004. The SK3 subunit of small conductance Ca^{2+}-activated K^+ channels interacts with both SK1 and SK2 subunits in a heterologous expression system. *J. Biol. Chem.* 279:1003–9
29. Strassmaier T, Bond CT, Sailer CA, Knaus HG, Maylie J, Adelman JP. 2005. A novel isoform of SK2 assembles with other SK subunits in mouse brain. *J. Biol. Chem.* 280:21231–36
30. Shmukler BE, Bond CT, Wilhelm S, Bruening-Wright A, Maylie J, et al. 2001. Structure and complex transcription pattern of the mouse SK1 K_{Ca} channel gene, *KCNN1*. *Biochim. Biophys. Acta* 1518:36–46
31. Wittekindt OH, Visan V, Tomita H, Imtiaz F, Gargus JJ, et al. 2004. An apamin- and scyllatoxin-insensitive isoform of the human SK3 channel. *Mol. Pharmacol.* 65:788–801
32. Shah M, Haylett DG. 2000. The pharmacology of hSK1 Ca^{2+}-activated K^+ channels expressed in mammalian cell lines. *Br. J. Pharmacol.* 129:627–30
33. Benton DC, Monaghan AS, Hosseini R, Bahia PK, Haylett DG, Moss GW. 2003. Small conductance Ca^{2+}-activated K^+ channels formed by the expression of rat SK1 and SK2 genes in HEK 293 cells. *J. Physiol.* 553:13–19
34. Grunnet M, Jensen BS, Olesen SP, Klaerke DA. 2001. Apamin interacts with all subtypes of cloned small-conductance Ca^{2+}-activated K^+ channels. *Pflüg. Arch.* 441:544–50
35. Ishii TM, Maylie J, Adelman JP. 1997. Determinants of apamin and *d*-tubocurarine block in SK potassium channels. *J. Biol. Chem.* 272:23195–200
36. Lamy C, Goodchild SJ, Weatherall KL, Jane DE, Liegeois JF, et al. 2010. Allosteric block of $K_{Ca}2$ channels by apamin. *J. Biol. Chem.* 285:27067–77
37. Nolting A, Ferraro T, D'Hoedt D, Stocker M. 2007. An amino acid outside the pore region influences apamin sensitivity in small conductance Ca^{2+}-activated K^+ channels. *J. Biol. Chem.* 282:3478–86
38. Weatherall KL, Goodchild SJ, Jane DE, Marrion NV. 2010. Small conductance calcium-activated potassium channels: from structure to function. *Prog. Neurobiol.* 91:242–55
39. Conejo-Garcia A, Campos JM. 2008. Bis-quinoliniumcyclophanes: highly potent and selective non-peptidic blockers of the apamin-sensitive Ca^{2+}-activated K^+ channel. *Curr. Med. Chem.* 15:1305–15
40. Cao Y, Dreixler JC, Roizen JD, Roberts MT, Houamed KM. 2001. Modulation of recombinant small-conductance Ca^{2+}-activated K^+ channels by the muscle relaxant chlorzoxazone and structurally related compounds. *J. Pharmacol. Exp. Ther.* 296:683–89
41. Hougaard C, Eriksen BL, Jorgensen S, Johansen TH, Dyhring T, et al. 2007. Selective positive modulation of the SK3 and SK2 subtypes of small conductance Ca^{2+}-activated K^+ channels. *Br. J. Pharmacol.* 151:655–65
42. Schumacher MA, Rivard AF, Bachinger HP, Adelman JP. 2001. Structure of the gating domain of a Ca^{2+}-activated K^+ channel complexed with Ca^{2+}/calmodulin. *Nature* 410:1120–24
43. Schumacher MA, Crum M, Miller MC. 2004. Crystal structures of apocalmodulin and an apocalmodulin/SK potassium channel gating domain complex. *Structure* 12:849–60
44. Lee WS, Ngo-Anh TJ, Bruening-Wright A, Maylie J, Adelman JP. 2003. Small conductance Ca^{2+}-activated K^+ channels and calmodulin: cell surface expression and gating. *J. Biol. Chem.* 278:25940–46
45. Li W, Aldrich RW. 2011. Electrostatic influences of charged inner pore residues on the conductance and gating of small conductance Ca^{2+}-activated K^+ channels. *Proc. Natl. Acad. Sci. USA* 108:5946–53
46. Bruening-Wright A, Lee WS, Adelman JP, Maylie J. 2007. Evidence for a deep pore activation gate in small conductance Ca^{2+}-activated K^+ channels. *J. Gen. Physiol.* 130:601–10
47. Ren Y, Barnwell LF, Alexander JC, Lubin FD, Adelman JP, et al. 2006. Regulation of surface localization of the small conductance Ca^{2+}-activated potassium channel, Sk2, through direct phosphorylation by cAMP-dependent protein kinase. *J. Biol. Chem.* 281:11769–79

48. Faber ES, Delaney AJ, Power JM, Sedlak PL, Crane JW, Sah P. 2008. Modulation of SK channel trafficking by beta adrenoceptors enhances excitatory synaptic transmission and plasticity in the amygdala. *J. Neurosci.* 28:10803–13
49. Lin MT, Lujan R, Watanabe M, Adelman JP, Maylie J. 2008. SK2 channel plasticity contributes to LTP at Schaffer collateral-CA1 synapses. *Nat. Neurosci.* 11:170–77
50. Bildl W, Strassmaier T, Thurm H, Andersen J, Eble S, et al. 2004. Protein kinase CK2 is coassembled with small conductance Ca^{2+}-activated K^+ channels and regulates channel gating. *Neuron* 43:847–58
51. Allen D, Fakler B, Maylie J, Adelman JP. 2007. Organization and regulation of small conductance Ca^{2+}-activated K^+ channel multiprotein complexes. *J. Neurosci.* 27:2369–76
52. Maingret F, Coste B, Hao J, Giamarchi A, Allen D, et al. 2008. Neurotransmitter modulation of small-conductance Ca^{2+}-activated K^+ channels by regulation of Ca^{2+} gating. *Neuron* 59:439–49
53. Maylie J, Adelman JP. 2010. Cholinergic signaling through synaptic SK channels: It's a protein kinase but which one? *Neuron* 68:809–11
54. Giessel AJ, Sabatini BL. 2010. M1 muscarinic receptors boost synaptic potentials and calcium influx in dendritic spines by inhibiting postsynaptic SK channels. *Neuron* 68:936–47
55. Buchanan KA, Petrovic MM, Chamberlain SE, Marrion NV, Mellor JR. 2010. Facilitation of long-term potentiation by muscarinic M_1 receptors is mediated by inhibition of SK channels. *Neuron* 68:948–63
56. De Shutter E, Smolen P. 1998. Calcium dynamics in large neuronal networks. In *Methods in Neuronal Modelling*, ed. C Koch, I Segev, pp. 211–51. Cambridge, MA: Mass. Inst. Technol. Press. 2nd ed.
57. Blaustein MP. 1988. Calcium transport and buffering in neurons. *Trends Neurosci.* 11:438–43
58. Fakler B, Adelman JP. 2008. Control of K_{Ca} channels by calcium nano/microdomains. *Neuron* 59:873–81
59. Faber ES, Delaney AJ, Sah P. 2005. SK channels regulate excitatory synaptic transmission and plasticity in the lateral amygdala. *Nat. Neurosci.* 8:635–41
60. Ngo-Anh TJ, Bloodgood BL, Lin M, Sabatini BL, Maylie J, Adelman JP. 2005. SK channels and NMDA receptors form a Ca^{2+}-mediated feedback loop in dendritic spines. *Nat. Neurosci.* 8:642–49
61. Oliver D, Klocker N, Schuck J, Baukrowitz T, Ruppersberg JP, Fakler B. 2000. Gating of Ca^{2+}-activated K^+ channels controls fast inhibitory synaptic transmission at auditory outer hair cells. *Neuron* 26:595–601
62. Sabatini BL, Oertner TG, Svoboda K. 2002. The life cycle of Ca^{2+} ions in dendritic spines. *Neuron* 33:439–52
63. Gulledge AT, Stuart GJ. 2005. Cholinergic inhibition of neocortical pyramidal neurons. *J. Neurosci.* 25:10308–20
64. Power JM, Sah P. 2008. Competition between calcium-activated K^+ channels determines cholinergic action on firing properties of basolateral amygdala projection neurons. *J. Neurosci.* 28:3209–20
65. Fiorillo CD, Williams JT. 1998. Glutamate mediates an inhibitory postsynaptic potential in dopamine neurons. *Nature* 394:78–82
66. Morikawa H, Imani F, Khodakhah K, Williams JT. 2000. Inositol 1,4,5-triphosphate-evoked responses in midbrain dopamine neurons. *J. Neurosci.* 20:RC103
67. Arima J, Matsumoto N, Kishimoto K, Akaike N. 2001. Spontaneous miniature outward currents in mechanically dissociated rat Meynert neurons. *J. Physiol.* 534:99–107
68. Cui G, Okamoto T, Morikawa H. 2004. Spontaneous opening of T-type Ca^{2+} channels contributes to the irregular firing of dopamine neurons in neonatal rats. *J. Neurosci.* 24:11079–87
69. Klement G, Druzin M, Haage D, Malinina E, Arhem P, Johansson S. 2010. Spontaneous ryanodine-receptor-dependent Ca^{2+}-activated K^+ currents and hyperpolarizations in rat medial preoptic neurons. *J. Neurophysiol.* 103:2900–11
70. Seutin V, Mkahli F, Massotte L, Dresse A. 2000. Calcium release from internal stores is required for the generation of spontaneous hyperpolarizations in dopaminergic neurons of neonatal rats. *J. Neurophysiol.* 83:192–97
71. Linden DJ, Connor JA. 1995. Long-term synaptic depression. *Annu. Rev. Neurosci.* 18:319–57
72. Kakizawa S, Kishimoto Y, Hashimoto K, Miyazaki T, Furutani K, et al. 2007. Junctophilin-mediated channel crosstalk essential for cerebellar synaptic plasticity. *EMBO J.* 26:1924–33
73. Walter JT, Alviña K, Womack MD, Chevez C, Khodakah K. 2006. Decreases in the precision of Purkinje cell pacemaking cause cerebellar dysfunction and ataxia. *Nat. Neurosci.* 9:389–97

74. Alviña K, Khodakhah K. 2010. K_{Ca} channels as therapeutic targets in episodic ataxia type-2. *J. Neurosci.* 30:7249–57
75. Barrett EF, Barret JN. 1976. Separation of two voltage-sensitive potassium currents, and demonstration of a tetrodotoxin-resistant calcium current in frog motoneurones. *J. Physiol.* 255:737–74
76. Madison DV, Nicoll RA. 1982. Noradrenaline blocks accommodation of pyramidal cell discharge in the hippocampus. *Nature* 299:636–38
77. Sah P. 1996. Ca^{2+}-activated K^+ currents in neurones: types, physiological roles and modulation. *Trends Neurosci.* 19:150–54
78. Storm JF. 1987. Action potential repolarization and a fast after-hyperpolarization in rat hippocampal pyramidal cells. *J. Physiol.* 385:733–59
79. Gu N, Vervaeke K, Storm JF. 2007. BK potassium channels facilitate high-frequency firing and cause early spike frequency adaptation in rat CA1 hippocampal pyramidal cells. *J. Physiol.* 580:859–82
80. Zhang L, Krnjevic K. 1987. Apamin depresses selectively the after-hyperpolarization of cat spinal motoneurons. *Neurosci. Lett.* 74:58–62
81. Bourque CW, Brown DA. 1987. Apamin and d-tubocurarine block the afterhyperpolarization of rat supraoptic neurosecretory neurons. *Neurosci. Lett.* 82:185–90
82. Sah P, McLachlan EM. 1992. Potassium currents contributing to action potential repolarization and the afterhyperpolarization in rat vagal motoneurons. *J. Neurophysiol.* 68:1834–41
83. Schwindt PC, Spain WJ, Foehring RC, Chubb MC, Crill WE. 1988. Slow conductances in neurons from cat sensorimotor cortex in vitro and their role in slow excitability changes. *J. Neurophysiol.* 59:450–67
84. Avanzini G, de Curtis M, Panzica F, Spreafico R. 1989. Intrinsic properties of nucleus reticularis thalami neurones of the rat studied in vitro. *J. Physiol.* 416:111–22
85. Goldberg JA, Wilson CJ. 2005. Control of spontaneous firing patterns by the selective coupling of calcium currents to calcium-activated potassium currents in striatal cholinergic interneurons. *J. Neurosci.* 25:10230–38
86. Zhang L, McBain CJ. 1995. Potassium conductances underlying repolarization and afterhyperpolarization in rat CA1 hippocampal interneurones. *J. Physiol.* 488(Pt. 3):661–72
87. Savic N, Pedarzani P, Sciancalepore M. 2001. Medium afterhyperpolarization and firing pattern modulation in interneurons of stratum radiatum in the CA3 hippocampal region. *J. Neurophysiol.* 85:1986–97
88. Williams S, Serafin M, Muhlethaler M, Bernheim L. 1997. Distinct contributions of high- and low-voltage-activated calcium currents to afterhyperpolarizations in cholinergic nucleus basalis neurons of the guinea pig. *J. Neurosci.* 17:7307–15
89. Chen QH, Toney GM. 2009. Excitability of paraventricular nucleus neurones that project to the rostral ventrolateral medulla is regulated by small-conductance Ca^{2+}-activated K^+ channels. *J. Physiol.* 587:4235–47
90. Hallworth NE, Wilson CJ, Bevan MD. 2003. Apamin-sensitive small conductance calcium-activated potassium channels, through their selective coupling to voltage-gated calcium channels, are critical determinants of the precision, pace, and pattern of action potential generation in rat subthalamic nucleus neurons in vitro. *J. Neurosci.* 23:7525–42
91. Womack MD, Khodakhah K. 2003. Somatic and dendritic small-conductance calcium-activated potassium channels regulate the output of cerebellar Purkinje neurons. *J. Neurosci.* 23:2600–7
92. Edgerton JR, Reinhart PH. 2003. Distinct contributions of small and large conductance Ca^{2+}-activated K^+ channels to rat Purkinje neuron function. *J. Physiol.* 548:53–69
93. Shepard PD, Stump D. 1999. Nifedipine blocks apamin-induced bursting activity in nigral dopamine-containing neurons. *Brain Res.* 817:104–9
94. Teshima K, Kim SH, Allen CN. 2003. Characterization of an apamin-sensitive potassium current in suprachiasmatic nucleus neurons. *Neuroscience* 120:65–73
95. Maher BJ, Westbrook GL. 2005. SK channel regulation of dendritic excitability and dendrodendritic inhibition in the olfactory bulb. *J. Neurophysiol.* 94:3743–50
96. Alger BE, Nicoll RA. 1980. Epileptiform burst after hyperpolarization: calcium-dependent potassium potential in hippocampal CA1 pyramidal cells. *Science* 210:1122–24
97. Sah P, Clements JD. 1999. Photolytic manipulation of $[Ca^{2+}]_i$ reveals slow kinetics of potassium channels underlying the afterhyperpolarization in hippocampal pyramidal neurons. *J. Neurosci.* 19:3657–64

98. Bond CT, Herson PS, Strassmaier T, Hammond R, Stackman R, et al. 2004. Small conductance Ca^{2+}-activated K^+ channel knock-out mice reveal the identity of calcium-dependent afterhyperpolarization currents. *J. Neurosci.* 24:5301–6
99. Alviña K, Ellis-Davies G, Khodakhah K. 2009. T-type calcium channels mediate rebound firing in intact deep cerebellar neurons. *Neuroscience* 158:635–41
100. Fujimura K, Matsuda Y. 1989. Autogenous oscillatory potentials in neurons of the guinea pig substantia nigra pars compacta in vitro. *Neurosci. Lett.* 104:53–57
101. Chan CS, Guzman JN, Ilijic E, Mercer JN, Rick C, et al. 2007. "Rejuvenation" protects neurons in mouse models of Parkinson's disease. *Nature* 447:1081–86
102. Wolfart J, Roeper J. 2002. Selective coupling of T-type calcium channels to SK potassium channels prevents intrinsic bursting in dopaminergic midbrain neurons. *J. Neurosci.* 22:3404–13
103. Li X, Bennett DJ. 2007. Apamin-sensitive calcium-activated potassium currents (SK) are activated by persistent calcium currents in rat motoneurons. *J. Neurophysiol.* 97:3314–30
104. Marrion NV, Tavalin SJ. 1998. Selective activation of Ca^{2+}-activated K^+ channels by colocalized Ca^{2+} channels in hippocampal neurons. *Nature* 395:900–5
105. Davies PJ, Ireland DR, McLachlan EM. 1996. Sources of Ca^{2+} for different Ca^{2+}-activated K^+ conductances in neurones of the rat superior cervical ganglion. *J. Physiol.* 495(Pt. 2):353–66
106. Sah P. 1995. Different calcium channels are coupled to potassium channels with distinct physiological roles in vagal neurons. *Proc. R. Soc. Lond. Ser. B* 260:105–11
107. Stocker M, Krause M, Pedarzani P. 1999. An apamin-sensitive Ca^{2+}-activated K^+ current in hippocampal pyramidal neurons. *Proc. Natl. Acad. Sci. USA* 96:4662–67
108. Gu N, Vervaeke K, Hu H, Storm JF. 2005. Kv7/KCNQ/M and HCN/h, but not $K_{Ca}2$/SK channels, contribute to the somatic medium after-hyperpolarization and excitability control in CA1 hippocampal pyramidal cells. *J. Physiol.* 566:689–715
109. Faber ES, Sah P. 2002. Physiological role of calcium-activated potassium currents in the rat lateral amygdala. *J. Neurosci.* 22:1618–28
110. Liu X, Herbison AE. 2008. Small-conductance calcium-activated potassium channels control excitability and firing dynamics in gonadotropin-releasing hormone (GnRH) neurons. *Endocrinology* 149:3598–604
111. Lee K, Duan W, Sneyd J, Herbison AE. 2010. Two slow calcium-activated afterhyperpolarization currents control burst firing dynamics in gonadotropin-releasing hormone neurons. *J. Neurosci.* 30:6214–24
112. Kato M, Tanaka N, Usui S, Sakuma Y. 2006. The SK channel blocker apamin inhibits slow afterhyperpolarization currents in rat gonadotropin-releasing hormone neurones. *J. Physiol.* 574:431–42
113. Bal T, McCormick DA. 1993. Mechanisms of oscillatory activity in guinea-pig nucleus reticularis thalami in vitro: a mammalian pacemaker. *J. Physiol.* 468:669–91
114. Cueni L, Canepari M, Lujan R, Emmenegger Y, Watanabe M, et al. 2008. T-type Ca^{2+} channels, SK2 channels and SERCAs gate sleep-related oscillations in thalamic dendrites. *Nat. Neurosci.* 11:683–92
115. Kasten MR, Rudy B, Anderson MP. 2007. Differential regulation of action potential firing in adult murine thalamocortical neurons by Kv3.2, Kv1, and SK potassium and N-type calcium channels. *J. Physiol.* 584:565–82
116. Kleiman-Weiner M, Beenhakker MP, Segal WA, Huguenard JR. 2009. Synergistic roles of $GABA_A$ receptors and SK channels in regulating thalamocortical oscillations. *J. Neurophysiol.* 102:203–13
117. Ying SW, Goldstein PA. 2005. Propofol-block of SK channels in reticular thalamic neurons enhances GABAergic inhibition in relay neurons. *J. Neurophysiol.* 93:1935–48
118. Cheng-Raude D, Treloar M, Habermann E. 1976. Preparation and pharmacokinetics of labeled derivatives of apamin. *Toxicon* 14:467–76
119. Messier C, Mourre C, Bontempi B, Sif J, Lazdunski M, Destrade C. 1991. Effect of apamin, a toxin that inhibits Ca^{2+}-dependent K^+ channels, on learning and memory processes. *Brain Res.* 551:322–26
120. Deschaux O, Bizot JC, Goyffon M. 1997. Apamin improves learning in an object recognition task in rats. *Neurosci. Lett.* 222:159–62
121. Stackman RW, Hammond RS, Linardatos E, Gerlach A, Maylie J, et al. 2002. Small conductance Ca^{2+}-activated K^+ channels modulate synaptic plasticity and memory encoding. *J. Neurosci.* 22:10163–71
122. Deschaux O, Bizot JC. 2005. Apamin produces selective improvements of learning in rats. *Neurosci. Lett.* 386:5–8

123. Brennan AR, Dolinsky B, Vu MA, Stanley M, Yeckel MF, Arnsten AF. 2008. Blockade of IP_3-mediated SK channel signaling in the rat medial prefrontal cortex improves spatial working memory. *Learn. Mem.* 15:93–96
124. Vick KA, Guidi M, Stackman RW Jr. 2010. In vivo pharmacological manipulation of small conductance Ca^{2+}-activated K^+ channels influences motor behavior, object memory and fear conditioning. *Neuropharmacology* 58:650–59
125. Matus-Amat P, Higgins EA, Barrientos RM, Rudy JW. 2004. The role of the dorsal hippocampus in the acquisition and retrieval of context memory representations. *J. Neurosci.* 24:2431–39
126. Phillips RG, LeDoux JE. 1992. Differential contribution of amygdala and hippocampus to cued and contextual fear conditioning. *Behav. Neurosci.* 106:274–85
127. Jacobsen JP, Redrobe JP, Hansen HH, Petersen S, Bond CT, et al. 2009. Selective cognitive deficits and reduced hippocampal brain-derived neurotrophic factor mRNA expression in small-conductance calcium-activated K^+ channel deficient mice. *Neuroscience* 163:73–81
128. Hammond RS, Bond CT, Strassmaier T, Ngo-Anh TJ, Adelman JP, et al. 2006. Small-conductance Ca^{2+}-activated K^+ channel type 2 (SK2) modulates hippocampal learning, memory, and synaptic plasticity. *J. Neurosci.* 26:1844–53
129. Ikonen S, Riekkinen P Jr. 1999. Effects of apamin on memory processing of hippocampal-lesioned mice. *Eur. J. Pharmacol.* 382:151–56
130. Kasa P, Rakonczay Z, Gulya K. 1997. The cholinergic system in Alzheimer's disease. *Prog. Neurobiol.* 52:511–35
131. Inan SY, Aksu F, Baysal F. 2000. The effects of some K^+ channel blockers on scopolamine- or electroconvulsive shock-induced amnesia in mice. *Eur. J. Pharmacol.* 407:159–64
132. Blank T, Nijholt I, Kye MJ, Radulovic J, Spiess J. 2003. Small-conductance, Ca^{2+}-activated K^+ channel SK3 generates age-related memory and LTP deficits. *Nat. Neurosci.* 6:911–12
133. Mpari B, Sreng L, Manrique C, Mourre C. 2010. $K_{Ca}2$ channels transiently downregulated during spatial learning and memory in rats. *Hippocampus* 20:352–63
134. Hopf FW, Bowers MS, Chang SJ, Chen BT, Martin M, et al. 2010. Reduced nucleus accumbens SK channel activity enhances alcohol seeking during abstinence. *Neuron* 65:682–94
135. Norris CM, Halpain S, Foster TC. 1998. Reversal of age-related alterations in synaptic plasticity by blockade of L-type Ca^{2+} channels. *J. Neurosci.* 18:3171–79
136. Behnisch T, Reymann KG. 1998. Inhibition of apamin-sensitive calcium dependent potassium channels facilitate the induction of long-term potentiation in the CA1 region of rat hippocampus in vitro. *Neurosci. Lett.* 253:91–94
137. Kramar EA, Lin B, Lin CY, Arai AC, Gall CM, Lynch G. 2004. A novel mechanism for the facilitation of theta-induced long-term potentiation by brain-derived neurotrophic factor. *J. Neurosci.* 24:5151–61
138. Bloodgood BL, Sabatini BL. 2007. Nonlinear regulation of unitary synaptic signals by $CaV_{2.3}$ voltage-sensitive calcium channels located in dendritic spines. *Neuron* 53:249–60
139. Faber ES. 2010. Functional interplay between NMDA receptors, SK channels and voltage-gated Ca^{2+} channels regulates synaptic excitability in the medial prefrontal cortex. *J. Physiol.* 588:1281–92
140. Lisman J. 1989. A mechanism for the Hebb and the anti-Hebb processes underlying learning and memory. *Proc. Natl. Acad. Sci. USA* 86:9574–78
141. Malenka RC, Nicoll RA. 1993. NMDA-receptor-dependent synaptic plasticity: multiple forms and mechanisms. *Trends Neurosci.* 16:521–27
142. Hayashi Y, Shi SH, Esteban JA, Piccini A, Poncer JC, Malinow R. 2000. Driving AMPA receptors into synapses by LTP and CaMKII: requirement for GluR1 and PDZ domain interaction. *Science* 287:2262–67
143. Lin MT, Lujan R, Watanabe M, Frerking M, Maylie J, Adelman JP. 2010. Coupled activity-dependent trafficking of synaptic SK2 channels and AMPA receptors. *J. Neurosci.* 30:11726–34
144. Allen D, Bond CT, Lujan R, Ballesteros-Merino C, Lin MT, et al. 2011. The SK2-long isoform directs synaptic localization and function of SK2-containing channels. *Nature Neurosci.* 14:744–49
145. Sourdet V, Russier M, Daoudal G, Ankri N, Debanne D. 2003. Long-term enhancement of neuronal excitability and temporal fidelity mediated by metabotropic glutamate receptor subtype 5. *J. Neurosci.* 23:10238–48

146. Belmeguenai A, Hosy E, Bengtsson F, Pedroarena CM, Piochon C, et al. 2010. Intrinsic plasticity complements long-term potentiation in parallel fiber input gain control in cerebellar Purkinje cells. *J. Neurosci.* 30:13630–43
147. Poulain DA, Wakerley JB. 1982. Electrophysiology of hypothalamic magnocellular neurones secreting oxytocin and vasopressin. *Neuroscience* 7:773–808
148. Wakerley JB, Lincoln DW. 1973. The milk-ejection reflex of the rat: a 20- to 40-fold acceleration in the firing of paraventricular neurones during oxytocin release. *J. Endocrinol.* 57:477–93
149. Teruyama R, Lipschitz DL, Wang L, Ramoz GR, Crowley WR, et al. 2008. Central blockade of oxytocin receptors during mid-late gestation reduces amplitude of slow afterhyperpolarization in supraoptic oxytocin neurons. *Am. J. Physiol. Endocrinol. Metab.* 295:1167–71
150. Ishikawa M, Mu P, Moyer JT, Wolf JA, Quock RM, et al. 2009. Homeostatic synapse-driven membrane plasticity in nucleus accumbens neurons. *J. Neurosci.* 29:5820–31

The Calcium-Sensing Receptor Beyond Extracellular Calcium Homeostasis: Conception, Development, Adult Physiology, and Disease

Daniela Riccardi and Paul J. Kemp

Division of Pathophysiology and Repair, School of Biosciences, Cardiff University, Cardiff, CF10 3AX, United Kingdom; email: riccardi@cf.ac.uk, kemp@cf.ac.uk

Keywords

G protein–coupled receptor, GPCR, extracellular, free, ionized Ca^{2+} concentration, Ca^{2+}, extracellular Ca^{2+}-sensing receptor, CaSR, embryonic development, calcimimetics, calcilytics

Abstract

The extracellular calcium-sensing receptor (CaSR) is the first identified G protein–coupled receptor to be activated by an ion, extracellular calcium (Ca^{2+}). Since the identification of the CaSR in 1993, genetic mutations in the CaSR gene, and murine models in which CaSR expression has been manipulated, have clearly demonstrated the importance of this receptor in the maintenance of stable, free, ionized Ca^{2+} concentration in the extracellular fluids. These functions have been extensively reviewed elsewhere. However, the distribution pattern and expression of the CaSR in lower vertebrates strongly suggest that the CaSR must play a role that is independent of mineral cation metabolism. This review addresses the involvement of the CaSR in nutrient sensing; its putative and demonstrated functions during conception, embryonic development, and birth; and its contributions to adult physiology and disease, with reference to CaSR-based therapeutics. Recent ongoing developments concerning the role of the CaSR in stem cell differentiation are also reviewed.

CaSR: calcium-sensing receptor

EVOLUTIONARY ASPECTS OF CALCIUM BIOLOGY

The calcium ion (Ca^{2+}) is the fifth most common ion in the Earth's crust and oceans. It has been employed in a wide variety of biological functions and has therefore significantly contributed to the success and diversification in the animal kingdom. Ca^{2+} is particularly important in humans, in whom Ca^{2+} represents, by mass, the most abundant mineral in the body. Such widespread use of Ca^{2+} has led to the evolution of many intracellular proteins involved in Ca^{2+} transport and storage and to the development of molecules that can detect its abundance in the extracellular milieu.

Ca^{2+} played a pivotal role during the evolution from prokaryotes to eukaryotes, a process that began approximately 2 billion years ago. Owing to the progressive exhaustion of the Fe^{2+}/H_2S buffering system in the sea, oxygen (O_2) supply increased, and oxidative processes became prevalent (1). Accordingly, the ambient redox potential increased from -0.2 V (which, incidentally, is very similar to the cytosolic potential of any cell) to $+0.8$ V, a value that closely resembles the current redox potential of the sea (2). Such dramatic effects in the ecosystem exerted a significant pressure on prokaryotic life, which, out of necessity, underwent major changes to adapt and survive. Ca^{2+} is versatile, is highly reactive, is readily available, promotes plasma membrane stability, and acts as a cofactor in many enzymatic processes (3). As a consequence, Ca^{2+} played a major role in facilitating the necessary adaptive changes of prokaryotic organisms and contributed to the evolution of more complex organisms and their land colonization (1). Indeed, colonization of the main land mass could not have occurred without the evolution of several Ca^{2+}-regulated processes (4). Such processes include (*a*) the evolution of a cytoskeleton, which allowed cells to migrate and to change shape and size; (*b*) nutrient seeking and danger avoidance, which required the development of gravity-detecting devices; (*c*) the formation of an exoskeleton or an endoskeleton, which became important as organisms increased in size and complexity; and (*d*) the evolution of hard-shelled eggs and the ability to supply milk, which facilitated the survival of immature embryos on land. Because the common denominator of all these activities is the requirement for Ca^{2+} (4), the development of specialized sensors designed to monitor changes in the ambient content of this ion became a necessity for most living organisms. Although single-membrane-spanning, Ca^{2+}-sensing molecules have been identified in plants (5), Ca^{2+} sensors in the animal kingdom are generally G protein–coupled receptors (GPCRs) (6).

EVOLUTION OF THE CALCIUM-SENSING RECEPTOR (CaSR)

Advanced computer-generated homology models have shown that proteins have evolved as a composite of several modular components, each of which carries a specific functional domain of the molecule (7). At the level of the primary sequence, individual modular components of homologous genes are remarkably different. However, 3-D modeling has revealed striking similarities in the tertiary structure of these modules (8). The ability to combine these modular structures has allowed organisms to adapt and survive in an ever-changing environment. GPCRs are examples of how the modular organization has been employed successfully, and their intrinsic multimodular structures have allowed them to be extensively utilized, from yeast to humans.

The Ca^{2+}-sensing receptor (CaSR) was first identified in 1993 (9). Functional and bioinformatics observations unanimously confirmed that this first ion-sensing, intrinsic membrane protein to be discovered was a GPCR. Comparative observations promptly suggested that the CaSR evolved in the broader context of proteins that sense nutrients—such as ions, amino acids, and sugars—from a common ancestor that belonged to the bacterial nutrient periplasmic binding protein family (10, 11). The function of this family of proteins is to transport bacterial nutrients, namely amino acids and vitamins (12). These proteins can differ substantially in their molecular weight (which can vary between 26 and 60 kDa). However, crystallographic data show that they bear a remarkable resemblance to each other in that they all share a bilobed structure reminiscent of a Venus flytrap: a structure containing two protomers that are separated by a cleft region (12). Ligand binding evokes conformational changes that bring the two protomers closer to each other, a step now deemed essential for subsequent G protein activation. The Venus flytrap constitutes the first module, or sensor, and the transmitter is represented by a seven-membrane-spanning region, which connects the cell's exterior to the cell's interior (13, 14) and represents the second modular component of GPCRs. As for all GPCRs, the CaSR's ability to transduce the signal does not depend upon ion/nutrient uptake. Following the heptahelical domain is the third module, represented by the intracellular C-terminal region, which is involved in signal transduction and receptor desensitization (13).

Phylogenic observations of the transmembrane region of GPCRs have resulted in categorization of GPCRs into groups A, B, and C. The CaSR is a member of the group C GPCRs (see Reference 15), which also includes metabotropic glutamate receptors (mGluRs) 1–8, basic amino acid receptors [i.e., the GPCR family C, group 6, subtype A (GPRC6A)], sweet and umami taste receptors (e.g., T1R1 and -3), pheromone receptors [e.g., vomeronasal organ (V2R)], and γ-amino-butyric acid (i.e., $GABA_BR1$ and -2). Although members of group C GPCRs are molecularly distinct, their pharmacological profiles overlap. For instance, Ca^{2+} is the physiological ligand at the CaSR, but several amino acids allosterically modulate receptor sensitivity for this ion. Similarly, amino acids, GABA, and certain nutrients act as orthosteric agonists at mGluRs, GPRC6A, $GABA_BRs$, and T1Rs, and Ca^{2+} modulates these responses. Group C GPCRs function as obligate dimers (either homo- or heterodimers). Homodimerization has been demonstrated for mGluRs (16), CaSR (17), and GPRC6A (18), whereas $GABA_B$ (19) and T1Rs (20) are constitutive heterodimers. In addition, CaSR, mGluR, and $GABA_B$ can also heterodimerize, giving rise to novel functional units with pharmacological profiles and plasma membrane expression different from that of their native homodimeric assemblies (recently reviewed in Reference 21). Specifically, CaSR heterodimers include (*a*) CaSR/mGluR1a and CaSR/mGluR5 (22) and (*b*) $CaSR/GABA_BR1$ and $CaSR/GABA_BR2$ (23). Crystallographic data obtained from the extracellular domain of mGluR show that the two lobes of the Venus flytrap exist in open and closed conformations. Agonist binding occurs in the pocket of lobe I, which subsequently promotes the closure of lobe II, a process that brings about the conformational change necessary for signal transmission (24, 25).

Although Ca^{2+} is the physiological stimulus at the CaSR, it can also be activated by many di-, tri-, and polyvalent cations, which act in an orthosteric fashion [i.e., their ability to activate the CaSR is independent of the presence of extracellular Ca^{2+} (reviewed in Reference 26)]. Furthermore, CaSR function can also be modulated by several physiological stimuli, including ionic strength, extracellular pH, L-aromatic amino acids, and naturally occurring polyamines (27). Although these stimuli are ineffective in the absence of extracellular Ca^{2+}, they can profoundly affect the EC_{50} (effective concentration necessary to induce a 50% effect) values for Ca^{2+} binding at the receptor. Known as allosteric modulators of the CaSR, such naturally occurring stimuli exert physiologically relevant effects on parathyroid hormone (PTH) secretion (28, 29), mineral ion absorption and water balance by the kidney (reviewed in References 29 and 30 and mentioned below), and gut epithelium differentiation and secretion (31). These functions are reminiscent of the role of the CaSR in lower vertebrates. For example, in elasmobranch and teleost fish—in which, in the absence of parathyroid glands, the CaSR is highly expressed in organs involved in osmoregulation (gills, kidney, rectal glands, and intestine)—the CaSR acts as a salinity sensor (32).

In addition to responding to physiological stimuli, the CaSR is also a useful drug target. Indeed, positive allosteric CaSR modulators (or calcimimetics) are now employed for the treatment of many diseases in which PTH levels are altered; such modulators are the first drugs of this kind to reach the market. Similarly, negative allosteric CaSR modulators (also known as calcilytics) are also being explored by the pharmaceutical industry and are currently in trials for the treatment of age-related osteoporosis. The development of CaSR therapeutics has been extensively reviewed elsewhere (e.g., Reference 33) but is further discussed below in the context of the role of the CaSR in disease states (see also **Table 1**).

> **Calcimimetics:** positive allosteric CaSR modulators
>
> **Calcilytics:** negative allosteric CaSR modulators
>
> $[Ca^{2+}]$: calcium ion concentration

ROLES FOR THE CaSR IN REPRODUCTION, DEVELOPMENT, AND BIRTH

A large body of evidence gathered over the past decade supports a role of the CaSR in reproductive biology. Indeed, the CaSR is expressed in egg and sperm, where it is involved in maturation, activation, and fertilization. The CaSR is essential for implantation, development, and maturation of the embryo and regulates placental Ca^{2+} transfer from mother to fetus. Postnatally, the coordinated actions of the CaSR in the neonatal parathyroid glands and in the milk-producing breast alveoli of the lactating mother ensure optimal Ca^{2+} supply to the newborn infant.

The CaSR in Reproduction

In Japanese quail, the CaSR is expressed in the granulosa cells and in follicles undergoing ovulation, but not in immature or smaller follicles (34). The CaSR is also expressed in human and equine oocytes and cumulus cells, where receptor expression appears to be greatest in the metaphase I stage (35). Furthermore, the CaSR may play a role in oocyte development and maturation (36).

Sperm maturation and survival are exquisitely Ca^{2+}-sensitive processes, as evidenced in mice in which transient receptor potential vanilloid 6 (TRPV6) has been genetically inactivated. The distal part of the epididymal ducts of these mice exhibits a luminal Ca^{2+} concentration ($[Ca^{2+}]$) tenfold greater than the physiological concentration. This high concentration results in reduced sperm viability and functionality (37). CaSR mRNA and protein are found in rat testis and sperm. In both the rat and pig, activation of the sperm CaSR by means of the positive allosteric modulator AMG 641 increased sperm mobility but did not affect sperm survival (38).

Rat studies provide evidence that CaSR signaling may also be involved in blastocyst implantation and decidualization (39). These studies demonstrate that CaSR mRNA levels in the lumen of the uterus vary during the estrus cycle: These levels are at their zenith before ovulation and are at their nadir when the corpus luteum begins to form and then regresses. Following fertilization, CaSR expression increases (between day 1 and day 3 of postconception); expression is highest first in the luminal epithelium and subsequently increases in the stromal cells. This estrogen- and progesterone-dependent expression temporarily decreases at day 4 (even in the face of high progesterone levels) and then rises again at days 6 through 9 in the implanting blastocysts and in the subluminal stromal endometrial cells undergoing decidualization (39).

The human placenta actively transports ∼30 g of Ca^{2+} from the mother to the fetus throughout pregnancy (40). Most such transport occurs during the third trimester (41, 42). Fetal plasma $[Ca^{2+}]$ is kept above that of the adult (42). This reset is independent of maternal $[Ca^{2+}]$ and has been ascribed to the effects of parathyroid hormone–related peptide (PTHrP), which is highly expressed in the placenta and in the fetal circulation (42). PTHrP also regulates placental Ca^{2+} transport, bone resorption, and renal Ca^{2+} excretion (42). Using murine models of constitutive CaSR deletion and of fetal parathyroid gland ablation, researchers demonstrated that PTH is

Table 1 Diseases arising from altered CaSR expression and/or function in different cells/tissues of the body and the biological consequences of CaSR dysfunction[a]

Putative disease/pathology	Target cell/tissue	Consequences	References
Infertility	Egg	⇓ Maturation	36
	Sperm	⇓ Mobility	38
	Uterus	⇓ Embryo implantation	39
Impaired embryonic growth/survival	Placenta	⇓ Ca^{2+} transport	43, 44
Lung hypo/hyperplasia	Fetal lung	⇓ Fluid secretion ⇑ Branching morphogenesis	50
Impaired sympathetic innervation	Fetal PNS	⇓ Neurite outgrowth	54
Impaired LTP	Perinatal hippocampus	⇓ Neurite outgrowth	
Delayed growth plate development	Fetal cartilage	⇓ Chondrocyte function	59
Impaired skeletal development and postnatal growth	Fetal bone	⇓ Osteoblast function	59
Low milk production	Breast alveolus	⇓ Milk production	75
Rickets	Kidney proximal tubule	⇓ 1-Hydroxylase activity	83, 84
Bartter syndrome type V	Thick ascending limb of Henle's loop	⇑ Na^+, Ca^{2+}, Mg^{2+} wasting ⇑ Plasma renin	85
Kidney stones (nephrocalcinosis)	Collecting duct principal cells	⇓ Water excretion	89, 90
	Collecting duct intercalated cells	⇓ Urine acidification	91
Hypertension	Juxtaglomerular apparatus	⇑ Renin secretion	92
Proteinuria	Kidney glomerular podocytes	⇓ Cytoskeletal integrity ⇑ Apoptosis	93
Pancreatic dysfunction, impaired gut secretion, obesity	Duodenum	⇓ CCK secretion	104
Secretory diarrhea Colon cancer	Colon	⇓ Fluid secretion ⇓ Cell differentiation, ⇑ cell proliferation	109
Pancreatitis	Exocrine pancreas	⇓ Fluid secretion	110
Diabetes	Endocrine pancreas	⇓ Insulin secretion	116
Primary hyperparathyroidism	Parathyroid, kidney	⇑ PTH secretion	127
Secondary hyperparathyroidism	Parathyroid, kidney	⇓ $1,25(OH)_2D_3$ synthesis, ⇓ VDR, ⇑ P_i, PTH	133
Cancer	Colon, prostate	⇑ PTHrP secretion	142, 143
Osteoporosis	Osteoblast	⇓ Osteoblast activity	147
	Osteoclast	⇓ Osteoclast apoptosis	148–150
Alzheimer's disease	CNS	⇑ Neuronal death	156
Epilepsy	CNS	⇑ Neuronal excitability	157, 158

[a]Abbreviations: CCK, cholecystokinin; CNS, central nervous system; LTP, long-term potentiation; PNS, peripheral nervous system; PTH, parathyroid hormone; PTHrP, parathyroid hormone–related peptide; VDR, vitamin D receptor.

present in both the fetus and the placenta, where it plays a role in fetal Ca^{2+} homeostasis and mineralization and where the CaSR maintains levels similar to those measured in the adult despite the relative fetal hypercalcemia (43, 44). Furthermore, there is functional expression of the CaSR in both the villous and extravillous regions of the human placenta (45). CaSR distribution at the apical side of syncytiotrophoblast cells points to a role for the receptor in the control of materno-fetal Ca^{2+} (and possibly Mg^{2+}) exchange, whereas extravillous CaSR expression in cytotrophoblast cells suggests a role for the CaSR in the process of uterine invasion and/or in the development of placental immunity (46).

The CaSR in Development

As discussed above, in the mammalian fetus, both total $[Ca^{2+}]$ and free, ionized serum $[Ca^{2+}]$ significantly exceed those of the adult. Indeed, human fetal free, ionized $[Ca^{2+}]$ is approximately 1.6 mM, whereas in the adult it is between 1.1 and 1.3 mM (47–49). This relative hypercalcemia is maintained largely by PTHrP, with a minor role played by fetal PTH or calcitriol $[1,25(OH)_2D_3]$. The reason for this reset of fetal Ca^{2+} levels is currently unknown, but the reset appears to be unrelated to maternal Ca^{2+} and $1,25(OH)_2D_3$ levels because maternal deficiencies of either Ca^{2+} or $1,25(OH)_2D_3$ do not affect fetal divalent mineral skeleton content (reviewed by Reference 42). This relative fetal hypercalcemia may protect skeletal mineralization in the face of adverse conditions in the womb, such as low O_2 availability or low pH, and/or may help the fetus with the transition at birth, when the maternal supply of Ca^{2+} is no longer available. Owing to the widespread distribution of the CaSR in the developing mouse embryo (**Figure 1a,b**), we hypothesized that higher Ca^{2+} concentration, acting through the fetal CaSR, is an important regulator of many developmental processes. Indeed, CaSR immunoreactivity is present in the developing mouse [embryonic day (E)11.5 to E16.5] (50). The strongest expression of the CaSR is in several branching organs (namely the lung, the kidney, and the intestine), the central and peripheral nervous systems (CNS and PNS, respectively), cartilage, and bone (see **Figure 1a** and Reference 21).

The CaSR in the developing viscera. During embryonic life, cells of the lung epithelium secrete fluid into the developing lung lumen (51). Such Cl^--driven fluid secretion is vital to provide the mechanical stimulus for branching and a template around which branching can proceed. The overall effect is an appropriately distended lung, matched by an adequate number of branches, both of which ensure that optimal gaseous exchange occurs at birth. A mismatch in this process, such as that evoked by lung fluid drainage or tracheal occlusion, yields hypo- or hyperplastic lungs, respectively, and is associated with long-term morbidity well into adulthood (see Reference 52 for a recent review). We recently demonstrated that CaSR expression is developmentally regulated in the fetal mouse (50) and in the human lung (**Figure 1c**). Such expression is confined to the pseudoglandular stage—E11.5 to E16.5 in mice and weeks 9 to 16 in humans—the period during which branching morphogenesis takes place (52). Using mouse lung explants in culture, we manipulated extracellular $[Ca^{2+}]$ ($[Ca^{2+}]_o$) and showed that $[Ca^{2+}]$ mimicking that present in the fetus (i.e., 1.7 mM) evokes CaSR activation. Such activation results in the stimulation of Cl^--driven lung fluid secretion and the suppression of lung-branching morphogenenesis (50). These effects cannot be observed in the presence of adult $[Ca^{2+}]$ (i.e., $\sim$1.2 mM). Thus, CaSR activation in the developing lung ensures that the fluid secretion rate matches the extent of branching morphogenesis.

The CaSR is also expressed in the large tubules and the branching ureteric buds of the perinatal rat nephron (53). Our recent observations show that CaSR immunoreactivity can be detected in embryonic mouse metanephroi at E13.5 (**Figure 1**). Together, these observations point to roles for the CaSR in ureteric bud branching, as well as in renal handling of divalent cations, in the perinatal period.

As in the mouse fetal lung and kidney, the developing mouse intestine (at E13.5) also expresses the CaSR in the epithelium (**Figure 1**). Given that impaired fetal development sets the basis for adult disease, CaSR expression in both the developing kidney and the gut warrants further investigation.

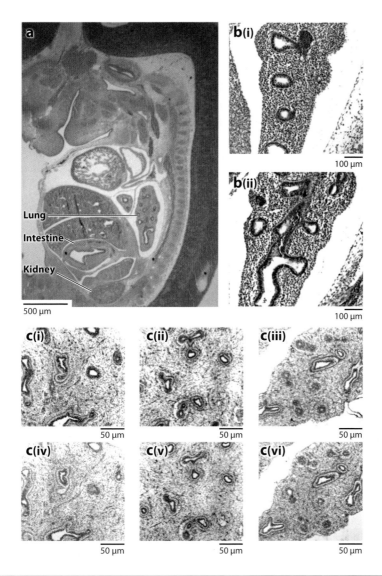

Figure 1

Embryonic expression of the Ca^{2+}-sensing receptor (CaSR). (*a*) CaSR immunoreactivity in a 5-μm-thick section through a whole mouse embryo (13.5 days postconception) showing extensive expression of the receptor in many branching organs, including the lung, the intestine, and the kidney. With kind permission from Springer Science+Business Media (see Reference 21). (*b*) Higher-magnification photomicrographs showing (*i*) specificity and (*ii*) localization of CaSR immunoreactivity in embryonic mouse lung (13.5 days postconception). (*i*) An image from a 5-μm section cut from the same embryo as that shown in panel *a*. Here, an irrelevant rabbit IgG was used instead of the CaSR primary antibody to demonstrate the specificity of the immunohistochemical labeling. (*ii*) An image from the same slide as that shown in panel *a* reveals that CaSR immunoreactivity at this embryonic stage is confined to the lung epithelium (*i*). (*c*) CaSR immunoreactivity in a 5-μm-thick section of the developing human lung showing CaSR epithelial and mesenchymal expression at (*i*) 8, (*ii*) 10, and (*iii*) 11 weeks postconception. Immunoreactivity was absent from serial sections incubated with an irrelevant rabbit IgG (*iv–vi*). All tissues were paraformaldehyde fixed and paraffin embedded. The primary antibody used for all panels of this figure was a rabbit polyclonal directed against an epitope in the N terminus of the CaSR protein. Binding of the primary antibody was detected using diaminobenzidine (*brown pigment*), and the counterstain was Harris's hematoxylin (*blue pigment*).

The CaSR in the developing nervous system. During development of the PNS, neuronal processes migrate until they reach their final target organs. This process is exquisitely sensitive to environmental cues and is regulated by neurotrophins. We recently showed that the CaSR is highly expressed in sympathetic neurons of the developing mouse superior cervical ganglion, where, as seen in the developing lung, CaSR expression is developmentally regulated (54). CaSR expression occurs only when sympathetic axons innervate their peripheral tissues, with a peak between E18 and postnatal day 0. Pharmacological (via CaSR allosteric modulators), genetic (via CaSR knockout mice), or molecular (via a CaSR dominant negative construct) modulation of CaSR function in vitro or in vivo significantly affects both neurite outgrowth and target field innervation exclusively at these developmental time points but is completely ineffective outside this developmental window (54). Because of the relative fetal hypercalcemia (see Reference 42), the CaSR is maximally activated during embryonic life. Within this context, CaSR regulation of the developmental programs in the lung and the PNS is achieved by transient upregulation of receptor expression at specific developmental time points, rather than by the canonical agonist activation of the receptor more commonly observed postnatally (33).

In the CNS, the CaSR has been detected in many regions of the developing and adult rat brain and spinal cord, and the receptor distribution pattern is consistent with a role in myelinization and oligodendrocyte maturation (55, 56). In the early postnatal rat hippocampus, CaSR expression appears to parallel the ontogeny of long-term potentiation (57), suggesting a role in learning and memory. Postnatal CaSR expression is highest in neurons of the subfornical organ, a thirst center. Some neurons of this organ project to the supraoptic and paraventricular nuclei, where they regulate the activity of vasopressin-secreting neurons. These observations strongly suggest that the CaSR is involved in the central integration of fluid and electrolyte balance (58).

The CaSR in the developing skeleton. Murine models of selective ablation of the CaSR from the cartilage and bone-forming cells, the osteoblasts, have elegantly demonstrated that the CaSR plays a major role in skeletal development (59). Constitutive CaSR gene deletion from chondrocytes was embryonically lethal (59). When the death of these mice was averted by using conditional, tamoxifen-induced knockout later in development, a key role for the receptor in the regulation of the growth plate was revealed. The porcine developing tooth organ also expresses CaSR protein, mostly in the predentine and ameloblasts of the developing molars (60). Furthermore, observations in two-week-old mice, in which the CaSR was constitutively knocked out, demonstrated impaired tooth formation and mandible development. However, when the investigators rescued the hypercalcemia of these mice by crossing them with vitamin D or PTH knockout animals, these researchers discovered that these effects were due to altered Ca^{2+}, phosphorus, and PTHrP levels, rather than to a direct effect of CaSR ablation in this organ (61).

The CaSR at Birth

Preeclampsia and premature delivery constitute major clinical problems for both the mother and the fetus and are associated with significant morbidity and mortality (62). A Cochrane systematic review and meta-analysis study of double-blind, placebo-controlled mothers—either at high risk or on low dietary Ca^{2+} intake—found that dietary Ca^{2+} supplementation significantly reduces maternal blood pressure, the risk of preeclampsia, and the incidence of premature birth (63–66). Furthermore, dietary Ca^{2+} intake in rats and humans is inversely related with blood pressure in offspring, an effect that is independent of vitamin D levels (67–70). The role of the CaSR in mediating any of the effects of dietary Ca^{2+} on maternal preeclampsia, offspring blood pressure, or birth prematurity needs thorough investigation.

After birth, fetal hypocalcemia develops in 20–30% of neonates and is accentuated in preterm infants born before the thirty-seventh week of gestation (71), in newborns from diabetic mothers (72), and in babies delivered through Cesarean section (73). This hypocalcemia has been ascribed to a combination of (a) the loss of maternal Ca^{2+} supply through the placenta and (b) a switch from largely PTHrP-regulated Ca^{2+} homeostasis to activation of the PTH and $1,25(OH)_2D_3$ axis in the newborn, which normally plays only a limited role during fetal life (reviewed in Reference 42). Recent evidence, garnered from a mouse model of constitutive CaSR ablation in the context of targeted deletion of PTH, shows that in neonates of dams fed on different dietary Ca^{2+} intakes, CaSR ablation reduces bone formation markers and activates osteoclast-dependent bone resorption (74). In lactating mothers, dietary Ca^{2+} intake affects both milk production and the total Ca^{2+} content of milk. These effects are mediated though PTHrP production. Studies carried out in CaSR knockout mice in vivo and in a 3-D culture model of the milk alveolus show that CaSR activation stimulates transcellular Ca^{2+} transport into maternal milk and inhibits PTHrP production (75).

Overall, these observations suggest that the CaSR plays a pivotal role in development. Why fetal hypercalcemia develops in the context of suppressed PTH secretion remains to be understood. We hypothesize that other contributing factors, such as fetal pH, O_2 availability, and/or serum protein content, may reset receptor sensitivity for Ca^{2+}, thereby resulting in the excessive suppression of fetal plasma PTH. In addition, given the aforementioned role for the receptor in the perinatal period, investigators should test whether the different degrees of fetal hypocalcemia affect CaSR-regulated processes, particularly in the CNS, the PNS, and the developing kidney.

The CaSR IN PHYSIOLOGY

Parathyroid glands and circulating PTH appeared first in evolution in terrestrial vertebrates. Parathyroid glands are considered to be evolutionarily related to fish gills owing to (a) their similar embryological derivation from the pharyngeal pouch endoderm (76); (b) the presence of a parathyroid-specific transcription factor, Gcm-2; and (c) the expression of Gcm-2 targets, PTH, and the CaSR (77). Because in fish the CaSR is a salinity sensor and is involved in osmoregulation (32), the parathyroid glands may represent a modification of the gills and may have arisen as a consequence of land colonization. In humans, genetic mutations in the CaSR demonstrate that its main physiological role is in extracellular Ca^{2+} (Ca^{2+}_o) homeostasis. Furthermore, accumulating evidence gathered over the past 15 years points to significant roles of the CaSR outside the Ca^{2+} homeostatic system.

The CaSR and the Extracellular Calcium Homeostatic System

The CaSR maintains a stable $[Ca^{2+}]_o$ in two ways: (a) through the control of PTH secretion by the parathyroid glands and (b) through both PTH-dependent and PTH-independent regulation of the target organs involved in Ca^{2+} homeostasis, namely the kidney and the bone. Although the role of the CaSR within the Ca^{2+}_o homeostatic system has been exhaustively reviewed elsewhere, what follows is a brief discussion of the role of the CaSR in the kidney and bone.

The CaSR in the parathyroid glands. Within the parathyroid glands, PTH is produced and stored inside vesicles in the chief cells. Ca^{2+}_o is a key regulator of parathyroid gland function, with ~50% of this activity already suppressed at physiological plasma $[Ca^{2+}]_o$ (78). Acutely, the CaSR exerts a physiological brake on PTH secretion. Pharmacological studies using positive allosteric modulators of the CaSR and targeted disruption of the CaSR in the parathyroid glands have shown that the CaSR also suppresses parathyroid gene expression (through posttranslational mechanisms) (79) as well as parathyroid cell proliferation (80).

The CaSR in the kidney. The CaSR is highly expressed throughout the kidney, where its cellular distribution can be apical, basolateral, or both, depending upon the segment (81). Immunohistochemical and functional observations have shown that the CaSR is present apically in the proximal tubule and the medullary collecting duct, basolaterally in the thick ascending limb, and apically/basolaterally in both the distal tubule and the cortical collecting duct (81). This distribution pattern indicates that the CaSR is capable of monitoring the Ca^{2+} content in both the pro-urine and the blood (reviewed in Reference 29). In vitro and in vivo studies have shown that the main functions of the CaSR in the kidney involve (*a*) divalent mineral cation homeostasis, (*b*) water and NaCl reabsorption, (*c*) modulation of the actions of calciotropic hormone effects, and (*d*) renin secretion. These have been recently reviewed (29), and species differences notwithstanding, the tubular functions are summarized in **Figures 2–5** (see below). In the proximal tubule, where the CaSR is expressed apically, activation of the receptor blunts the phosphaturic action of PTH (82) and inhibits 1α-hydroxylase activity (**Figure 2**) (83, 84). In the thick ascending limb of Henle's loop, the CaSR monitors serum $[Ca^{2+}]$ and is expressed basolaterally. At this tubular site, approximately 20–25% of the filtered Ca^{2+} is reabsorbed, largely through the paracellular route.

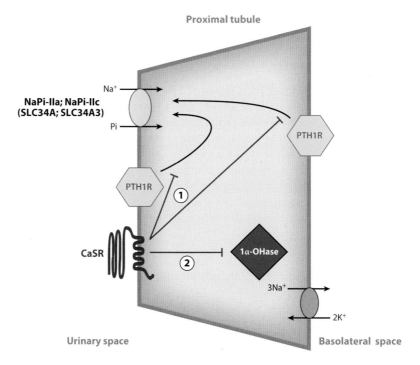

Figure 2

Functions of the CaSR in the proximal tubule. In mouse kidney proximal tubules, parathyroid hormone (PTH), acting on the type 1 PTH receptor (PTH1R), inhibits Na^+-dependent phosphate reabsorption via the Na^+-dependent inorganic phosphate transporter NaPi-IIa and NaPi-IIc. ① Activation of the apical CaSR reverses the inhibitory effects of PTH on phosphate reabsorption (82). ② In HKC8 cells, an SV40-transformed, human proximal tubule cell line, high Ca^{2+} inhibits the mitochondrial enzyme activity of cytochrome P450 25-hydroxyvitamin D3-1α-hydroxylase (1α-OHase) (83, 84).

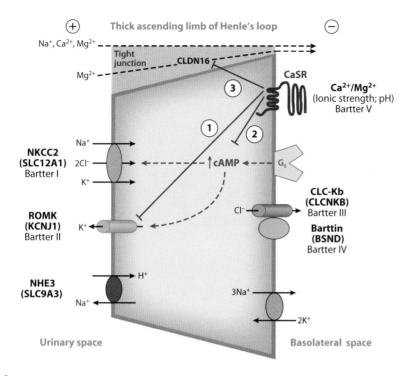

Figure 3

Functions of the CaSR in the thick ascending limb of Henle's loop. Under baseline conditions, the concerted activities of the apical electroneutral NKCC2 transporter and ROMK channel (which is the rate-limiting step for apical K^+ recycling) with the basolateral CLC-Kb contribute to the generation of a lumen-positive transepithelial potential difference. This creates the driving force for nonselective, paracellular reabsorption of monovalent (Na^+) and divalent (Ca^{2+}, Mg^{2+}) cations. In addition, the activity of the apical transporters can be increased by the second messenger cAMP, whose level is elevated as a consequence of G_s-coupled GPCR activation (such as, for instance, calcitonin and vasopressin in the medullary thick ascending limb and parathyroid hormone in the cortical thick ascending limb). When hypercalcemia/hypermagnesemia activate the basolateral CaSR, this dampens the activity of the apical transporters, both under baseline conditions, by inhibiting apical K^+ recycling ① or, under hormone-stimulated conditions (*dotted blue lines*), by reducing intracellular cAMP levels ②. This effect is reminiscent of the phenotype observed in patients with certain activating mutations of the CaSR who exhibit a Bartter's-like phenotype (Bartter's type V variant) (85). Additionally, mutations in NKCC2 (SLC12A1), ROMK (KCNJ1), CLC-Kb (CLCNKB), and Barttin (BSND) result in Bartter's type syndromes I, II, III, and IV, respectively. In MDCK cells, the CaSR agonists Ca^{2+} and Mg^{2+} induce lysosomal translocation of the tight junctional protein claudin-16 (CLDN16), which creates a Mg^{2+}-selective paracellular permeability, thereby reducing paracellular Mg^{2+} transport in this nephron segment ③. In the medullary thick ascending limb, where water permeability is low, basolateral NaCl as well as Ca^{2+} and Mg^{2+} are rapidly accumulated. In addition, the CaSR is allosterically modulated in a negative fashion by both ionic strength and acidification (29).

In this nephron segment, hypercalcemia causes a loop diuretic–like effect with a reduction in transcellular NaCl transport, as well as a decrease in the driving force for paracellular Ca^{2+} and Mg^{2+} reabsorption. Such a phenotype is recapitulated by activating mutations in the CaSR gene, which cause one of the Bartter variants (Bartter type V). This variant is characterized by NaCl and Mg^{2+} wasting, hypercalciuria, and hyperreninemia (**Figure 3**) (85). Additional Bartter variants include mutations in NKCC2 (Bartter type I); in the ROMK channel (Bartter type II); in the basolateral Cl^- channel, CLC-Kb (Bartter type III); and in its modifying subunit, Barttin (BSND, Bartter

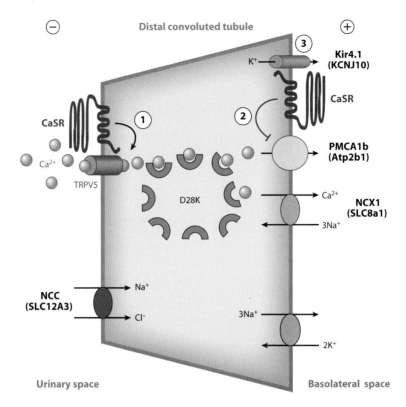

Figure 4

Functions of the CaSR in the distal convoluted tubule. The apical CaSR is biochemically linked to transient receptor potential vanilloid member 5 (TRPV5) (86), and receptor activation results in increased TRPV5 channel activity. The physiological significance of this observation is that NaCl and Ca^{2+} handling can be uncoupled in the distal convoluted tubule. Therefore, impairment of NaCl reabsorption by the thick ascending limb leads to increased NaCl and Ca^{2+} delivery to the distal convoluted tubule. Because of the CaSR-dependent activation of TRPV5, sufficient Ca^{2+} can still be reabsorbed in the distal convoluted tubule, thereby avoiding urinary Ca^{2+} wasting while eliminating excess NaCl ①. In addition, CaSR activation in the distal tubule–derived cell line MDCK inhibits the activity of the plasma membrane Ca^{2+}-ATPase 1b (PMCA1b) ② (87). Finally, in HEK293 cells heterologously expressing the CaSR and Kir4.1, receptor activation reduces membrane expression of this K^+ channel (88). In the kidney, this would result in an inhibition of basolateral K^+ recycling (with an attendant reduction in NaCl reabsorption) through Kir4.1 ③. Other abbreviations: D28K, calbindin D28K; NCC, thiazide-sensitive Na^+/Cl^- cotransporter; NCX1, Na^+/Ca^{2+} exchanger.

type IV). In the distal tubule, where CaSR expression has been detected apically and in intracellular vesicle-like structures of the human kidney, observations gathered from recombinant systems suggest that receptor activation increases apical Ca^{2+} entry through the calcitriol-regulated transient receptor potential vanilloid member 5 (TRPV5) (86). The CaSR appears to regulate basolateral Ca^{2+} exit via the plasma membrane Ca^{2+}-ATPase (**Figure 4**) (87).

The CaSR is present in both principal and intercalated cells of the collecting ducts (81). In the medullary collecting duct, hypercalcemia or abnormal Ca^{2+} handling can lead to hypercalciuria, which, if severe, can result in kidney stone formation. Studies carried out in isolated rat collecting ducts (89), in cell lines derived from rabbit collecting ducts (90), and in collecting ducts isolated from transgenic mice lacking both TRPV5 and the B1 subunit of the H^+-ATPase have shown

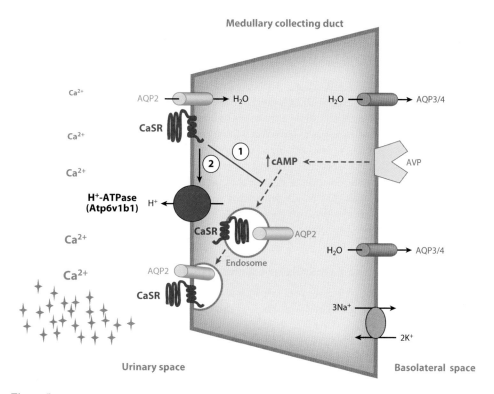

Figure 5

Documented functions of the CaSR in the kidney medullary collecting duct. The antidiuretic hormone vasopressin (AVP), acting through its basolateral G protein–coupled receptor, increases water permeability by promoting the trafficking of vesicles containing aquaporin 2 (AQP2) water channels to the apical membrane. This process is mediated via an increase in intracellular cAMP content (*dotted blue lines*) (89, 90). ① In the concomitant presence of a hypercalciuric stimulus, activation of the apical CaSR reduces the osmotic water permeability by inhibiting the AVP-mediated apical insertion of AQP2 water channels into principal cells. ② In addition, CaSR activation increases the activity of the H^+-ATPase in intercalated cells (91). Together, these maneuvers lead to the production of less concentrated, more acidic urine, thereby reducing the risk of calcium stone formation. Only AVP-regulated water permeability is under the control of the CaSR because basolateral aquaporin 3 (AQP3) and aquaporin 4 (AQP4) water channel expression is unaffected.

that CaSR activation in this part of the nephron promotes polyuria (by inhibiting the vasopressin-mediated apical insertion of aquaporin 2 water channels) and urine acidification (by promoting activity of the B1 subunit of the vacuolar H^+-ATPase in intercalated cells of the collecting ducts) (**Figure 5**). Together, these maneuvers aim at reducing the incidence of Ca^{2+} stone formation, particularly the incidence of phosphate-containing crystals (91).

In addition to the role of the CaSR in mineral ion absorption, its intrarenal distribution clearly suggests non-Ca^{2+} homeostatic functions of this receptor. These include a role for the CaSR in the inhibition of renin secretion by the juxtaglomerular apparatus (92). Additionally, CaSR expression has been detected in murine podocytes (which form the glomerular filter), where, in an experimental model of glomerulosclerosis, calcimimetics have been shown to improve cell viability and functional integrity of the cytoskeleton. Such functional integrity is essential to the maintenance of the foot processes of these cells and, therefore, to filter integrity (93). Given the rising incidence of chronic kidney disease in the population worldwide, these observations suggest the possibility of using pharmacological modulators of the CaSR to evoke blood pressure–lowering effects and/or to reduce glomerular proteinuria, both of which are strong predictors of loss of renal function (94).

The CaSR in bone. CaSR mRNA and protein are expressed in rat and mouse femora (59, 95). In rat frozen sections, this expression is localized to the osteoblasts (bone-forming cells), the osteocytes (mechanosensing cells), and the osteoclasts (bone-resorbing cells) (95). The CaSR is also expressed in a plethora of osteoblast-derived cell lines (reviewed in Reference 96). Experiments carried out in vitro (95) and in vivo (97) show that activation of the CaSR is intimately involved in the regulation of many aspects of osteoblast function from proliferation to differentiation to mineralization. Cell-specific ablation of the CaSR in osteoblasts demonstrates an important role for the CaSR in these cells that is independent of any possible effects of the CaSR on PTH secretion (59). The CaSR also plays an essential role in osteoclast differentiation and apoptosis (98).

The CaSR Outside the Extracellular Calcium Homeostatic System

As discussed above, the CaSR appeared early in evolution, and it is therefore not surprising that many of its functions are associated with homeostatic mechanisms beyond divalent mineral ion metabolism. For brevity, here we focus on the proposed roles of the CaSR in the gastrointestinal system, the CNS, and the PNS, for which significant literature exists.

The CaSR in the gastrointestinal system. Functional expression of the CaSR has been detected all along the gastrointestinal system, from the taste buds in the tongue to the colon. In all these organs, the CaSR is involved in nutrient sensing, in the regulation of hormone and fluid secretion, and in cell differentiation.

In human taste buds, CaSR activators act as flavor enhancers, suggestive of a role for the receptor in taste perception (99). The human esophagus and esophageal epithelial cell lines express the CaSR; in these cell lines, receptor activation has been linked to proliferative signals and secretion of the chemokine interleukin 8 (100). Furthermore, studies carried out in mice and humans in vivo indicate that, in the stomach, CaSR activation regulates gastrin and proton secretion (101, 102), as well as cell proliferation (102), supporting the idea that this receptor acts as a multimodal chemosensor.

Cholecystokinin (CCK) secretion is crucial to many aspects of gastrointestinal physiology, including gallbladder contraction; pancreatic enzyme secretion; bile production; gastric acid secretion; and, because of CCK's ability to interact with orexogenic peptides, appetite control (103). The CaSR is expressed in enteroendocrine cells of the proximal duodenum (104). These cells secrete CCK in response to aromatic amino acids (L-tryptophan and L-phenylalanine), and calcilytics inhibit this effect (104). These observations provide a compelling explanation for how amino acid sensing is linked to hormone secretion in the small intestine.

Numerous observations in the colon support that the CaSR suppresses fluid secretion evoked by secretagogues such as bacterial toxins (105, 106). In addition, dietary Ca^{2+} supplementation reduces the risk of colon cancer, whereas low Ca^{2+} promotes epithelial cell proliferation (107). Loss of CaSR expression, possibly occurring as a consequence of epigenetic mechanisms, is also associated with neoplastic transformation (108). Together, these observations indicate the possibility of employing CaSR-based therapeutics in developing countries to prevent the secretory diarrhea that is induced by bacterial toxins and/or to suppress malignant transformation in colorectal carcinogenesis (109).

CaSR transcripts and protein are expressed in acinar cells, ducts, and islets of Langerhans of rat (110) and human (111) pancreata. In isolated pancreatic ducts, CaSR agonists administered from the luminal, but not the basolateral, side evoked fluid secretion, possibly reducing the risk of intraductal precipitates, which may lead to pancreatic stone formation and predispose one to pancreatitis (112–114). Indeed, patients with inactivating CaSR mutations exhibit an increased risk of acute pancreatitis (115).

In the endocrine pancreas, CaSR activation evokes insulin secretion in isolated human islets and in insulinoma-derived cell lines (116). Recent evidence suggests that interaction with L-type voltage-gated Ca^{2+} channels may be a possible mechanism by which such an effect may occur (117).

CaSR expression has been reported in rat liver hepatocytes, where pharmacological receptor activation stimulates bile flow (118). CaSR expression is absent from nonparenchymal cells of the liver, i.e., stellate, Kupffer, and endothelial cells.

The CaSR in the nervous system. The CaSR is expressed throughout the nervous system (119), but for most cell types we do not know how it contributes to physiology and, perhaps more importantly, to disease. However, we know that the CaSR plays key roles in controlling neuronal excitability via cell-specific regulation of several classes of ion channels. Here, we discuss some of the emerging evidence for CaSR-dependent ion channel regulation and focus principally on pathways and mechanisms that have been clearly defined, from receptor to effector to physiological response. In so doing, we highlight some emerging themes in neuronal physiology in which $[Ca^{2+}]_o$ is a demonstrable and crucial determinant of function and touch on evidence for a role of the CaSR in the pathophysiology of diseases such as Alzheimer's disease and epilepsy.

High neuronal activity can result in decreased cleft $[Ca^{2+}]$, which may negatively impact the efficiency of synaptic transmission during repetitive firing (120). Cortical neurons express the CaSR (121). During high synaptic activity, the decrease in cleft $[Ca^{2+}]$ results in an increase in excitability via activation of a voltage-dependent, nonselective cation channel; this process depends on the expression of the CaSR (122). Thus, decreased cleft $[Ca^{2+}]$ reduces CaSR receptor occupancy, and this depression of CaSR activity leads to the activation of cation channels. This striking effect of the CaSR on the regulation of neuronal excitability was elegantly demonstrated in paired neocortical neuronal recordings in which excitatory postsynaptic currents were almost doubled in neurons from $CaSR^{-/-}$ mice compared with those from wild-type mice (122). This ability of the CaSR to orchestrate the increase in neuronal excitability in the face of the reduced cleft $[Ca^{2+}]$ appears to be a physiological compensatory mechanism crucial to maintain neuronal function during periods of medium to intense excitation. Furthermore, CaSR activation in neurons is linked to tonic glutamate release (123), further reinforcing the idea that Ca^{2+}-dependent optimization of neuronal function depends on the CaSR.

The idea that CaSR-linked ion channels are critical components controlling neuronal excitability has also been demonstrated directly in hippocampal cultures. However, in this case, the ion channel in question is a voltage-independent Na^+ leak conductance that is activated upon a reduction in Ca^{2+}_o, thereby exciting hippocampal neurons (124). Interestingly, although functional coupling of $[Ca^{2+}]_o$ to ion channel activity depends on the CaSR and its cognate G proteins, physical association of the channel and the receptor with two further proteins (UNC79 and UNC80) within a protein complex is required for the signal transduction process. Indeed, knockout of either the Na^+ leak channel or UNC79 uncouples neuronal excitability from changes in $[Ca^{2+}]_o$ (124). Thus, where fast cellular adaptation to acute changes in Ca^{2+}_o is a physiological necessity, the CaSR may be associated with ion channels in an efficient and advantageous manner.

In addition to protein-protein interactions, several other ion channel activities have been linked to the CaSR. For instance, $[Ca^{2+}]_o$ is important in tooth dentin formation, and CaSR expression has been demonstrated in trigeminal ganglia, sensory axons, and tooth dental pulp (125). Increased Ca^{2+}_o or treatment with high concentrations of the calcimimetic compound NPS R-467 evokes robust augmentation of blood flow in the tooth pulp. Such an increase is presumably a stimulus that promotes dentin formation. This response is completely blocked by pretreatment with the specific BK_{Ca} (large-conductance, voltage-dependent, Ca^{2+}-activated K^+ channel) channel blocker iberiotoxin. These data suggest that the CaSR can cause vasodilatation via Ca^{2+}_o-induced, K^+ channel–dependent smooth muscle cell hyperpolarization, a mechanism not dissimilar, at least in principle, to that seen in mesenteric arterioles (126). The details of how each component of the dental pulp system contributes to the overall physiological response remain conjecture. However, cross talk between neuronal CaSR and smooth muscle K^+ channels—rather than channel/receptor colocalization within a single cell type, as is seen in the CNS—is likely required for this particular Ca^{2+}_o-dependent modulation of cell excitability (122, 124).

THE CaSR IN PATHOLOGY AND CaSR THERAPEUTICS

Genetic CaSR Diseases

Several diseases of Ca^{2+} metabolism are a consequence of altered CaSR expression or function. Genetic CaSR loss-of-function mutations in humans (127) and in mice (59, 128) yield hypercalcemia, which occurs in the face of inappropriately low urinary Ca^{2+} excretion. Whereas heterozygous inactivation produces a relatively asymptomatic phenotype, a mutation in both alleles leads to profound symptomatic hypercalcemia with strongly elevated plasma PTH levels. Indeed, neonatal severe hyperparathyroidism in humans is associated with significant morbidity and mortality and requires parathyroidectomy early in life (129). In contrast, activating CaSR mutations in humans (130) are characterized by inappropriately low plasma Ca^{2+} levels in the face of excessive urinary Ca^{2+} wasting with inappropriately suppressed plasma PTH levels as a result of overactivation of parathyroid CaSR. Analysis of the phenotype both in humans and in mice underscores an important role of the CaSR in the regulation of mineral ion metabolism through its direct effects not only on the parathyroid glands but also on the kidney (85) and bone (97). A growing list of CaSR identified mutations can be found at http://www.casrdb.mcgill.ca/. Furthermore, activating and inactivating autoantibodies against the CaSR have also been identified, and they exhibit the biochemical and clinical hallmarks of patients with genetic CaSR mutations (reviewed in Reference 29).

Acquired Diseases Linked to CaSR Disturbances

Several studies carried out in humans and rats have demonstrated loss of CaSR expression in parathyroid glands during chronic kidney disease (e.g., References 131 and 132). Such loss of expression is likely due to an increase in plasma inorganic phosphate (P_i) levels, which arise as a consequence of the failing kidney. Excessive plasma P_i drives PTH secretion. Such secretion, together with the loss of the CaSR, results in a vicious circle that leads to an even greater loss of renal function and to secondary hyperparathyroidism (133). This increase in plasma PTH further exacerbates the pathological changes to Ca^{2+} and P_i metabolism, with profound pathological alterations in the renal and skeletal functions that are associated with significant morbidity and mortality. Furthermore, advanced chronic kidney disease is also accompanied by vascular and ectopic calcifications, which increase the risk of fatal cardiovascular events (134). The CaSR is expressed in noncalcified areas of human femoral arteries, although its expression is lost in calcified areas of the same blood vessel (135). Additionally, loss of receptor expression in vascular smooth muscle cells in vitro, induced by infecting cells with a dominant negative form of the CaSR, exacerbates the mineralizing effects of high Ca^{2+} (135). Owing to their ability to increase the sensitivity of the CaSR to Ca^{2+}, calcimimetics have recently proven to be very valuable for the treatment of patients with advanced chronic kidney disease. Calcimimetics reduce parathyroid gland hyperplasia and the gland's elevated plasma PTH levels (see, for instance, Reference 136). In vitro studies suggest that calcimimetics may also be beneficial in reducing the calcification of vascular smooth muscle cells that occurs as a consequence of deranged Ca^{2+} and P_i homeostasis (135, 137). Recent evidence also suggests that calcimimetics reduce all-cause and cardiovascular mortality in advanced secondary hyperparathyroidism patients receiving hemodialysis (138).

The CaSR and cancer. Recent evidence suggests a dichotomous role for CaSR signaling in either suppressing or promoting several types of benign or malignant transformation [reviewed by Mentaverri and colleagues (139)]. For instance, although normal colon and parathyroid glands express the CaSR, its expression is greatly reduced, if not completely absent, in colon carcinoma or parathyroid adenoma (140, 141). In contrast, CaSR signaling induces PTHrP secretion in breast and prostate cancer cells (142, 143). Such an effect has been ascribed to the alternate usage of G proteins in normal versus malignant transformation. Thus, pharmacological CaSR modulators could be used either to enhance CaSR function where receptor expression is lost—for instance, during parathyroid adenoma and colon carcinoma—or to suppress overactive CaSR function in breast or prostate cancer cells (144).

The CaSR and osteoporosis. With increasing age, the balance between bone formation and bone resorption is lost, which results in significant and progressive damage to bone strength and microarchitecture to the point that osteopenia and even osteoporosis develop. Owing to its ability to prevent bone breakdown while enhancing bone formation, strontium ranelate has become a novel therapeutic agent used for the treatment of age-related osteoporosis (145). Strontium ranelate can activate the CaSR in vitro (146), and the CaSR may play a role in both the anabolic and anticatabolic effects of this cation. Indeed, CaSR activation increases osteoblast proliferation and osteoclast apoptosis (147) while inhibiting osteoclast maturation (148–150). Furthermore, although sustained increases in plasma PTH exert a bone catabolic effect, transient bursts in this hormone stimulate new bone formation (151, 152). Recent studies have shown that orally bioavailable calcilytic compounds can improve bone mineral density in rats and can produce short-lived bursts in plasma PTH in humans (153). Whether such compounds can produce the much desired bone anabolic effect in postmenopausal women remains to be determined (154).

The CaSR and Alzheimer's disease. β-Amyloid peptides are potent agonists at the CaSR, and in hippocampal neurons, the CaSR can couple β-amyloid presentation to the activation of nonselective cation channels and to a consequent chronic elevation in intracellular Ca^{2+} (155). Although the CaSR is widely distributed among neuronal populations of the brain (119), it is also expressed at relatively high levels in astrocytes (156). Expression in this cell type appears key to the neuronal cell death seen in late-onset Alzheimer's disease. As $Ab_{(1-42)}$—the most toxic product of Alzheimer's precursor protein cleavage—builds up in the brain of an Alzheimer's patient, astrocyte CaSR is activated. This receptor occupancy by $Ab_{(1-42)}$ results in G protein–dependent mitogen-activated protein kinase activation and nitric oxide synthesis. In the presence of O_2 free radicals, nitric oxide synthesis generates concentrations of peroxynitrite sufficient to induce neuron cell death (156). Again, cross talk between neighboring cell types seems important to the response, a factor that might be usefully exploited in future clinical treatments.

The CaSR and epilepsy. At least two reports published in the past 4 years have given credence to the idea originally posited by Brown and colleagues (119) that the CaSR may be involved in controlling neuronal excitability and may thereby contribute to the generation of epileptic seizures. A rare CaSR missense mutation was identified, by linkage analysis, in five patients with idiopathic generalized epilepsy (157). CaSR endoplasmic reticulum retention mutations have been identified in patients with chronic pancreatitis and idiopathic epilepsy syndrome. These mutants lead to increased plasma membrane CaSR expression, thereby leading to an activating phenotype that, if occurring in neurons or glia, may contribute to the altered neuronal excitability characteristic of epileptic seizures (158).

NOVEL ASPECTS OF CaSR BIOLOGY: STEM CELLS

Although specification of stem cells to different cell populations depends on the cells' response to extrinsic factors, there is a dearth of information regarding possible roles for either Ca^{2+}_o or the CaSR in embryonic stem cell proliferation and subsequent differentiation. In contrast, CaSR activation has been implicated in several adult stem cell populations, in which the CaSR is intimately involved in proliferation, specification, lodgment, and engraftment.

The CaSR in Lodgment and Engraftment in the Ostosteal Niche

The CaSR has been identified in several populations of bone marrow–derived cells of both murine and human origin, including hematopoietic stem cells (HSCs) (159, 160). Furthermore, neonatal CaSR$^{-/-}$ mice demonstrate significantly fewer HSCs in their bone marrow than do wild-type mice, even though both genotypes have comparable HSC counts in other sites such as the spleen. This finding suggests that CaSR activation may be important for the migration of HSCs to that particular stem cell niche (161). Such an idea was reinforced by the demonstration that HSCs from CaSR$^{-/-}$ donors migrated poorly to the bone marrow of irradiated wild-type recipients, whereas engraftment of wild-type HSCs was normal (161). Importantly, the reduced lodgment of HSC in this niche was due, in part, to impaired binding to collagen I, a major component of the endosteal niche. These data suggest strongly that CaSR activation drives the migration of HSCs to the bone marrow during development and the consequent switch of hematopoiesis from fetal liver to postnatal bone marrow (162). Such activation is facilitated by the high $[Ca^{2+}]_o$ thought to be present in this particular microenvironment (163). Recently, such enhancement of HSC lodgment was mimicked in vivo using a commercially available calcimimetic (164). This development thus reinforces the idea that HSC engraftment depends heavily on CaSR activation and further strengthens the notion that calcimimetic or calcilytic treatment may be beneficial in HSC engraftment or mobilization, respectively (161), as discussed succinctly in Reference 162.

The evidence from the HSC/endosteal niche studies suggests that such CaSR-evoked signaling is also important for stem cell migration and retention in other niches. In this regard, preliminary evidence suggests that CaSR activation in a subpopulation of bone marrow–derived cells leads to improved engraftment in injured/ischemic muscle, where such cells participate in the neovascularization process (165).

The CaSR in Stem Cell Proliferation in the Central Nervous System

During neurogenesis, the differentiation of oligodendrocyte progenitor cells leads to the specification of oligodendrocytes. In vitro, neural stem cells directed down an oligodendrocyte lineage express higher levels of CaSR mRNA than do those specified to either neuronal or astrocyte lineages, and these differential CaSR expression levels are maintained to cell maturity (56). During oligodendrocyte specification, the CaSR is expressed most robustly in the oligodendrocyte precursors, and CaSR expression diminishes from the premyelination stage to maturity. Importantly, CaSR activation is proproliferative at the precursor stage and, later in specification, induces transcription of myelin basic protein mRNA in preoligodendrocytes. Together with the observation that myelin basic protein is diminished in cerebella of $CaSR^{-/-}$ mice during development, these data demonstrate that the CaSR is crucial for oligodendrocyte expansion and specification both in vivo and in vitro.

The CaSR in Umbilical Cord Stem Cells

Similarly, sieved umbilical cord matrix mesenchymal stem cells from the horse express the CaSR. Population doubling times of a subpopulation of these stem cells are reduced by both high Ca^{2+}_o and the calcimimetic compound NPS R-467 (albeit only when used at a very high concentration). This CaSR-dependent increase in proliferation is associated with a depression in the receptor's cell surface expression over time. Because these effects are antagonized by treatment with the calcilytic compound NPS-2,390, the CaSR may once again be linked to stem cell expansion (166). Taken together, these recent data open the door to a new era of CaSR research, with the exciting possibility that CaSR pharmaceutics may offer great benefits in the design and implementation of stem cell–based therapies and disease modeling.

SUMMARY POINTS

1. The CaSR is a G protein–coupled receptor (GPCR) for inorganic ions, and its main physiological roles are in divalent mineral ion and water homeostasis.

2. The CaSR is an evolutionarily old molecule that has evolved within the broader context of nutrient sensing. As such, it has maintained its ability (*a*) to work in a polymodal mode and (*b*) to respond to diverse physiological agonists in different parts of the body.

3. The pharmacological profile of the CaSR overlaps somewhat with the profiles of other group C GPCRs.

4. Many genetic and acquired disturbances in Ca^{2+} metabolism are due to abnormal CaSR expression and/or function.

5. Because fetal plasma Ca^{2+} levels in mammals are maintained above those of the adult, and fetal hypocalcemia often develops immediately after birth, emerging roles for the CaSR include both fetal development and perinatal Ca^{2+} homeostasis.

6. The CaSR is expressed outside of the Ca^{2+} homeostatic system, in which its roles include the regulation of hormone and fluid secretion, the regulation of neuronal signaling, and the regulation of cell fate.

7. Novel, emerging roles for the CaSR in stem cell biology include the control of lodgment and engraftment in particular stem cell niches.

8. Calcimimetics, the first allosteric modulators to reach the market, are widely used to treat hyperphosphatemia/hyperparathyroid disorders. Calcilytics are currently undergoing clinical trials for the treatment of age-related osteoporosis.

DISCLOSURE STATEMENT

The authors are not aware of any affiliations, memberships, funding, or financial holdings that might be perceived as affecting the objectivity of this review.

ACKNOWLEDGMENTS

The authors wish to apologize to the many scientists working in the CaSR field whose work could not be cited due to space limitations. In addition, the authors would like to thank Professors Eniko Kallay and Romuald Mentaverri for their critical input, Dr. Brenda Finney for the immunohistochemical localization of the CaSR in fetal tissue, the Biotechnology and Biological Sciences Research Council (grant BB/D01591X) for funding the embryonic development studies, and the FP7 Marie Curie ITN grant ("Multifaceted CaSR," grant 264663; **http://www.multifaceted-casr.org/**) for our current funding.

LITERATURE CITED

1. Williams RJ. 2006. The evolution of calcium biochemistry. *Biochim. Biophys. Acta* 1763:1139–46
2. Krebs J, Heizmann CW. 2007. *Calcium: A Matter of Life or Death*. Amsterdam: Elsevier
3. Ehrstrom M, Eriksson LE, Israelachvili J, Ehrenberg A. 1973. The effects of some cations and anions on spin labeled cytoplasmic membranes of *Bacillus subtilis*. *Biochem. Biophys. Res. Commun.* 55:396–402
4. Ross MD. 1984. The influence of gravity on structure and function of animals. *Adv. Space Res.* 4:305–14
5. Han S, Tang R, Anderson LK, Woerner TE, Pei ZM. 2003. A cell surface receptor mediates extracellular Ca^{2+} sensing in guard cells. *Nature* 425:196–200
6. Bockaert J, Pin JP. 1999. Molecular tinkering of G protein-coupled receptors: an evolutionary success. *EMBO J.* 18:1723–29
7. Bork P, Downing AK, Kieffer B, Campbell ID. 1996. Structure and distribution of modules in extracellular proteins. *Q. Rev. Biophys.* 29:119–67
8. Felder CB, Graul RC, Lee AY, Merkle HP, Sadee W. 1999. The Venus flytrap of periplasmic binding proteins: an ancient protein module present in multiple drug receptors. *AAPS PharmSci.* 1:E2
9. Brown EM, Gamba G, Riccardi D, Lombardi M, Butters R, et al. 1993. Cloning and characterization of an extracellular Ca^{2+}-sensing receptor from bovine parathyroid. *Nature* 366:575–80
10. Conklin BR, Bourne HR. 1994. Homeostatic signals. Marriage of the flytrap and the serpent. *Nature* 367:22
11. Quiocho FA, Ledvina PS. 1996. Atomic structure and specificity of bacterial periplasmic receptors for active transport and chemotaxis: variation of common themes. *Mol. Microbiol.* 20:17–25
12. Cao J, Huang S, Qian J, Huang J, Jin L, et al. 2009. Evolution of the class C GPCR Venus flytrap modules involved positive selected functional divergence. *BMC Evol. Biol.* 9:67
13. Kniazeff J, Prezeau L, Rondard P, Pin JP, Goudet C. 2011. Dimers and beyond: the functional puzzles of class C GPCRs. *Pharmacol. Ther.* 130:9–25
14. Rondard P, Goudet C, Kniazeff J, Pin JP, Prezeau L. 2011. The complexity of their activation mechanism opens new possibilities for the modulation of mGlu and GABAB class C G protein-coupled receptors. *Neuropharmacology* 60:82–92
15. Alexander SP, Mathie A, Peters JA. 2008. Guide to Receptors and Channels (GRAC), 3rd edition. *Br. J. Pharmacol.* 153(Suppl. 2):1–209

16. Romano C, Yang WL, O'Malley KL. 1996. Metabotropic glutamate receptor 5 is a disulfide-linked dimer. *J. Biol. Chem.* 271:28612–16
17. Ward DT, Brown EM, Harris HW. 1998. Disulfide bonds in the extracellular calcium-polyvalent cation-sensing receptor correlate with dimer formation and its response to divalent cations in vitro. *J. Biol. Chem.* 273:14476–83
18. Pi M, Faber P, Ekema G, Jackson PD, Ting A, et al. 2005. Identification of a novel extracellular cation-sensing G-protein-coupled receptor. *J. Biol. Chem.* 280:40201–9
19. White JH, Wise A, Main MJ, Green A, Fraser NJ, et al. 1998. Heterodimerization is required for the formation of a functional GABA$_B$ receptor. *Nature* 396:679–82
20. Zhao GQ, Zhang Y, Hoon MA, Chandrashekar J, Erlenbach I, et al. 2003. The receptors for mammalian sweet and umami taste. *Cell* 115:255–66
21. Riccardi D, Finney BA, Wilkinson WJ, Kemp PJ. 2009. Novel regulatory aspects of the extracellular Ca^{2+}-sensing receptor, CaR. *Pflüg. Arch.* 458:1007–22
22. Gama L, Wilt SG, Breitwieser GE. 2001. Heterodimerization of calcium sensing receptors with metabotropic glutamate receptors in neurons. *J. Biol. Chem.* 276:39053–59
23. Cheng Z, Tu C, Rodriguez L, Chen TH, Dvorak MM, et al. 2007. Type B γ-aminobutyric acid receptors modulate the function of the extracellular Ca^{2+}-sensing receptor and cell differentiation in murine growth plate chondrocytes. *Endocrinology* 148:4984–92
24. Bessis AS, Bertrand HO, Galvez T, De Colle C, Pin JP, Acher F. 2000. Three-dimensional model of the extracellular domain of the type 4a metabotropic glutamate receptor: new insights into the activation process. *Protein Sci.* 9:2200–9
25. Galvez T, Prezeau L, Milioti G, Franek M, Joly C, et al. 2000. Mapping the agonist-binding site of GABA$_B$ type 1 subunit sheds light on the activation process of GABA$_B$ receptors. *J. Biol. Chem.* 275:41166–74
26. Nemeth EF. 2004. Calcimimetic and calcilytic drugs: just for parathyroid cells? *Cell Calcium* 35:283–89
27. Brown EM, Macleod RJ. 2001. Extracellular calcium sensing and extracellular calcium signaling. *Physiol. Rev.* 81:239–97
28. Conigrave AD, Mun HC, Delbridge L, Quinn SJ, Wilkinson M, Brown EM. 2004. L-Amino acids regulate parathyroid hormone secretion. *J. Biol. Chem.* 279:38151–59
29. Riccardi D, Brown EM. 2010. Physiology and pathophysiology of the calcium-sensing receptor in the kidney. *Am. J. Physiol. Ren. Physiol.* 298:485–99
30. Hebert SC, Brown EM, Harris HW. 1997. Role of the Ca^{2+}-sensing receptor in divalent mineral ion homeostasis. *J. Exp. Biol.* 200:295–302
31. Hebert SC, Cheng S, Geibel J. 2004. Functions and roles of the extracellular Ca^{2+}-sensing receptor in the gastrointestinal tract. *Cell Calcium* 35:239–47
32. Nearing J, Betka M, Quinn S, Hentschel H, Elger M, et al. 2002. Polyvalent cation receptor proteins (CaRs) are salinity sensors in fish. *Proc. Natl. Acad. Sci. USA* 99:9231–36
33. Ward DT, Riccardi D. 2011. New concepts in calcium-sensing receptor pharmacology and signalling. *Br. J. Pharmacol.* In press; doi: 10.1111/j.1476-5381.2011.01511.x
34. Diez-Fraile A, Mussche S, Vanden Berghe T, Espeel M, Vandenabeele P, D'Herde KG. 2010. Expression of calcium-sensing receptor in quail granulosa explants: a key to survival during folliculogenesis. *Anat. Rec.* 293:890–99
35. Dell'Aquila ME, De Santis T, Cho YS, Reshkin SJ, Caroli AM, et al. 2006. Localization and quantitative expression of the calcium-sensing receptor protein in human oocytes. *Fertil. Steril.* 85(Suppl. 1):1240–47
36. De Santis T, Casavola V, Reshkin SJ, Guerra L, Ambruosi B, et al. 2009. The extracellular calcium-sensing receptor is expressed in the cumulus-oocyte complex in mammals and modulates oocyte meiotic maturation. *Reproduction* 138:439–52
37. Weissgerber P, Kriebs U, Tsvilovskyy V, Olausson J, Kretz O, et al. 2011. Male fertility depends on Ca^{2+} absorption by TRPV6 in epididymal epithelia. *Sci. Signal.* 4:ra27
38. Mendoza FJ, Perez-Marin CC, Garcia-Marin L, Madueno JA, Henley C, et al. 2011. Localization, distribution and function of the calcium-sensing receptor in sperm. *J. Androl.* In press
39. Xiao LJ, Yuan JX, Li YC, Wang R, Hu ZY, Liu YX. 2005. Extracellular Ca^{2+}-sensing receptor expression and hormonal regulation in rat uterus during the peri-implantation period. *Reproduction* 129:779–88

40. Lafond J, Simoneau L. 2006. Calcium homeostasis in human placenta: role of calcium-handling proteins. *Int. Rev. Cytol.* 250:109–74
41. Trotter M, Hixon BB. 1974. Sequential changes in weight, density, and percentage ash weight of human skeletons from an early fetal period through old age. *Anat. Rec.* 179:1–18
42. Kovacs CS, Kronenberg HM. 1997. Maternal-fetal calcium and bone metabolism during pregnancy, puerperium, and lactation. *Endocr. Rev.* 18:832–72
43. Kovacs CS, Ho-Pao CL, Hunzelman JL, Lanske B, Fox J, et al. 1998. Regulation of murine fetal-placental calcium metabolism by the calcium-sensing receptor. *J. Clin. Investig.* 101:2812–20
44. Kovacs CS, Manley NR, Moseley JM, Martin TJ, Kronenberg HM. 2001. Fetal parathyroids are not required to maintain placental calcium transport. *J. Clin. Investig.* 107:1007–15
45. Bradbury RA, Sunn KL, Crossley M, Bai M, Brown EM, et al. 1998. Expression of the parathyroid Ca^{2+}-sensing receptor in cytotrophoblasts from human term placenta. *J. Endocrinol.* 156:425–30
46. Bradbury RA, Cropley J, Kifor O, Lovicu FJ, de Iongh RU, et al. 2002. Localization of the extracellular Ca^{2+}-sensing receptor in the human placenta. *Placenta* 23:192–200
47. Delivoria-Papadopoulos M, Battaglia FC, Bruns PD, Meschia G. 1967. Total, protein-bound, and ultrafilterable calcium in maternal and fetal plasmas. *Am. J. Physiol.* 213:363–66
48. Schauberger CW, Pitkin RM. 1979. Maternal-perinatal calcium relationships. *Obstet. Gynecol.* 53:74–76
49. Wadsworth JC, Kronfeld DS, Ramberg CF Jr. 1982. Parathyrin and calcium homeostasis in the fetus. *Biol. Neonate* 41:101–9
50. Finney BA, del Moral PM, Wilkinson WJ, Cayzac S, Cole M, et al. 2008. Regulation of mouse lung development by the extracellular calcium-sensing receptor, CaR. *J. Physiol.* 586:6007–19
51. Olver RE, Strang LB. 1974. Ion fluxes across the pulmonary epithelium and the secretion of lung liquid in the foetal lamb. *J. Physiol.* 241:327–57
52. Warburton D, El-Hashash A, Carraro G, Tiozzo C, Sala F, et al. 2010. Lung organogenesis. *Curr. Top. Dev. Biol.* 90:73–158
53. Chattopadhyay N, Baum M, Bai M, Riccardi D, Hebert SC, et al. 1996. Ontogeny of the extracellular calcium-sensing receptor in rat kidney. *Am. J. Physiol. Ren. Physiol.* 271:736–43
54. Vizard TN, O'Keeffe GW, Gutierrez H, Kos CH, Riccardi D, Davies AM. 2008. Regulation of axonal and dendritic growth by the extracellular calcium-sensing receptor. *Nat. Neurosci.* 11:285–91
55. Ferry S, Traiffort E, Stinnakre J, Ruat M. 2000. Developmental and adult expression of rat calcium-sensing receptor transcripts in neurons and oligodendrocytes. *Eur. J. Neurosci.* 12:872–84
56. Chattopadhyay N, Espinosa-Jeffrey A, Tfelt-Hansen J, Yano S, Bandyopadhyay S, et al. 2008. Calcium receptor expression and function in oligodendrocyte commitment and lineage progression: potential impact on reduced myelin basic protein in CaR-null mice. *J. Neurosci. Res.* 86:2159–67
57. Chattopadhyay N, Legradi G, Bai M, Kifor O, Ye C, et al. 1997. Calcium-sensing receptor in the rat hippocampus: a developmental study. *Brain Res. Dev. Brain Res.* 100:13–21
58. Bandyopadhyay S, Tfelt-Hansen J, Chattopadhyay N. 2010. Diverse roles of extracellular calcium-sensing receptor in the central nervous system. *J. Neurosci. Res.* 88:2073–82
59. Chang W, Tu C, Chen TH, Bikle D, Shoback D. 2008. The extracellular calcium-sensing receptor (CaSR) is a critical modulator of skeletal development. *Sci. Signal.* 1:ra1
60. Mathias RS, Mathews CH, Machule C, Gao D, Li W, Denbesten PK. 2001. Identification of the calcium-sensing receptor in the developing tooth organ. *J. Bone Miner. Res.* 16:2238–44
61. Sun W, Liu J, Zhou X, Xiao Y, Karaplis A, et al. 2010. Alterations in phosphorus, calcium and PTHrP contribute to defects in dental and dental alveolar bone formation in calcium-sensing receptor-deficient mice. *Development* 137:985–92
62. Steegers EA, von Dadelszen P, Duvekot JJ, Pijnenborg R. 2010. Pre-eclampsia. *Lancet* 376:631–44
63. Hofmeyr GJ, Lawrie TA, Atallah AN, Duley L. 2010. Calcium supplementation during pregnancy for preventing hypertensive disorders and related problems. *Cochrane Database Syst. Rev.* 8:CD001059
64. Astrup A. 2010. Calcium reduces risk of pre-eclampsia. *Lancet* 376:1986
65. Sibai BM. 2010. Calcium supplementation during pregnancy reduces risk of high blood pressure, pre-eclampsia and premature birth compared with placebo? *Evid. Based Med.* 16:40–41
66. Tan PC. 2008. Review: calcium supplementation during pregnancy reduces the risk of pre-eclampsia. *Evid. Based Med.* 13:83

67. Bergel E, Belizán JM. 2002. A deficient maternal calcium intake during pregnancy increases blood pressure of the offspring in adult rats. *BJOG* 109:540–45
68. Gillman MW, Rifas-Shiman SL, Kleinman KP, Rich-Edwards JW, Lipshultz SE. 2004. Maternal calcium intake and offspring blood pressure. *Circulation* 110:1990–95
69. Johnson LE, DeLuca HF. 2002. Reproductive defects are corrected in vitamin D-deficient female rats fed a high calcium, phosphorus and lactose diet. *J. Nutr.* 132:2270–73
70. Johnson LE, DeLuca HF. 2001. Vitamin D receptor null mutant mice fed high levels of calcium are fertile. *J. Nutr.* 131:1787–91
71. Tsang RC, Chen IW, Friedman MA, Chen I. 1973. Neonatal parathyroid function: role of gestational age and postnatal age. *J. Pediatr.* 83:728–38
72. Tsang RC, Kleinman LI, Sutherland JM, Light IJ. 1972. Hypocalcemia in infants of diabetic mothers. Studies in calcium, phosphorus, and magnesium metabolism and parathormone responsiveness. *J. Pediatr.* 80:384–95
73. Bagnoli F, Bruchi S, Garosi G, Pecciarini L, Bracci R. 1990. Relationship between mode of delivery and neonatal calcium homeostasis. *Eur. J. Pediatr.* 149:800–3
74. Shu L, Ji J, Zhu Q, Cao G, Karaplis A, et al. 2011. The calcium-sensing receptor mediates bone turnover induced by dietary calcium and parathyroid hormone in neonates. *J. Bone Miner. Res.* 26:1057–71
75. VanHouten J, Dann P, McGeoch G, Brown EM, Krapcho K, et al. 2004. The calcium-sensing receptor regulates mammary gland parathyroid hormone-related protein production and calcium transport. *J. Clin. Investig.* 113:598–608
76. Okabe M, Graham A. 2004. The origin of the parathyroid gland. *Proc. Natl. Acad. Sci. USA* 101:17716–19
77. Mizobuchi M, Ritter CS, Krits I, Slatopolsky E, Sicard G, Brown AJ. 2009. Calcium-sensing receptor expression is regulated by glial cells missing-2 in human parathyroid cells. *J. Bone Miner. Res.* 24:1173–79
78. Brown EM. 1991. Extracellular Ca^{2+} sensing, regulation of parathyroid cell function, and role of Ca^{2+} and other ions as extracellular (first) messengers. *Physiol. Rev.* 71:371–411
79. Levi R, Ben-Dov IZ, Lavi-Moshayoff V, Dinur M, Martin D, et al. 2006. Increased parathyroid hormone gene expression in secondary hyperparathyroidism of experimental uremia is reversed by calcimimetics: correlation with posttranslational modification of the *trans* acting factor AUF1. *J. Am. Soc. Nephrol.* 17:107–12
80. Colloton M, Shatzen E, Miller G, Stehman-Breen C, Wada M, et al. 2005. Cinacalcet HCl attenuates parathyroid hyperplasia in a rat model of secondary hyperparathyroidism. *Kidney Int.* 67:467–76
81. Riccardi D, Hall AE, Chattopadhyay N, Xu JZ, Brown EM, Hebert SC. 1998. Localization of the extracellular Ca^{2+}/polyvalent cation-sensing protein in rat kidney. *Am. J. Physiol. Ren. Physiol.* 274:611–22
82. Ba J, Brown D, Friedman PA. 2003. Calcium-sensing receptor regulation of PTH-inhibitable proximal tubule phosphate transport. *Am. J. Physiol. Ren. Physiol.* 285:1233–43
83. Bland R, Walker EA, Hughes SV, Stewart PM, Hewison M. 1999. Constitutive expression of 25-hydroxyvitamin D3-1α-hydroxylase in a transformed human proximal tubule cell line: evidence for direct regulation of vitamin D metabolism by calcium. *Endocrinology* 140:2027–34
84. Zehnder D, Bland R, Walker EA, Bradwell AR, Howie AJ, et al. 1999. Expression of 25-hydroxyvitamin D3-1α-hydroxylase in the human kidney. *J. Am. Soc. Nephrol.* 10:2465–73
85. Watanabe S, Fukumoto S, Chang H, Takeuchi Y, Hasegawa Y, et al. 2002. Association between activating mutations of calcium-sensing receptor and Bartter's syndrome. *Lancet* 360:692–94
86. Topala CN, Schoeber JP, Searchfield LE, Riccardi D, Hoenderop JG, Bindels RJ. 2009. Activation of the Ca^{2+}-sensing receptor stimulates the activity of the epithelial Ca^{2+} channel TRPV5. *Cell Calcium* 45:331–39
87. Blankenship KA, Williams JJ, Lawrence MS, McLeish KR, Dean WL, Arthur JM. 2001. The calcium-sensing receptor regulates calcium absorption in MDCK cells by inhibition of PMCA. *Am. J. Physiol. Ren. Physiol.* 280:815–22
88. Cha SK, Huang C, Ding Y, Qi X, Huang CL, Miller RT. 2011. Calcium-sensing receptor decreases cell surface expression of the inwardly rectifying K^+ channel Kir4.1. *J. Biol. Chem.* 286(3):1828–35
89. Sands JM, Flores FX, Kato A, Baum MA, Brown EM, et al. 1998. Vasopressin-elicited water and urea permeabilities are altered in IMCD in hypercalcemic rats. *Am. J. Physiol. Ren. Physiol.* 274:978–85

90. Procino G, Carmosino M, Tamma G, Gouraud S, Laera A, et al. 2004. Extracellular calcium antagonizes forskolin-induced aquaporin 2 trafficking in collecting duct cells. *Kidney Int.* 66:2245–55
91. Renkema KY, Alexander RT, Bindels RJ, Hoenderop JG. 2008. Calcium and phosphate homeostasis: concerted interplay of new regulators. *Ann. Med.* 40:82–91
92. Atchison DK, Ortiz-Capisano MC, Beierwaltes WH. 2010. Acute activation of the calcium-sensing receptor inhibits plasma renin activity in vivo. *Am. J. Physiol. Regul. Integr. Comp. Physiol.* 299:1020–26
93. Oh J, Beckmann J, Bloch J, Hettgen V, Mueller J, et al. 2011. Stimulation of the calcium-sensing receptor stabilizes the podocyte cytoskeleton, improves cell survival, and reduces toxin-induced glomerulosclerosis. *Kidney Int.* 80(5):483–92
94. Ruggenenti P, Cravedi P, Remuzzi G. 2010. The RAAS in the pathogenesis and treatment of diabetic nephropathy. *Nat. Rev. Nephrol.* 6:319–30
95. Dvorak MM, Siddiqua A, Ward DT, Carter DH, Dallas SL, et al. 2004. Physiological changes in extracellular calcium concentration directly control osteoblast function in the absence of calciotropic hormones. *Proc. Natl. Acad. Sci. USA* 101:5140–45
96. Yamaguchi T. 2008. The calcium-sensing receptor in bone. *J. Bone Miner. Metab.* 26:301–11
97. Dvorak MM, Chen TH, Orwoll B, Garvey C, Chang W, et al. 2007. Constitutive activity of the osteoblast Ca^{2+}-sensing receptor promotes loss of cancellous bone. *Endocrinology* 148:3156–63
98. Mentaverri R, Yano S, Chattopadhyay N, Petit L, Kifor O, et al. 2006. The calcium sensing receptor is directly involved in both osteoclast differentiation and apoptosis. *FASEB J.* 20:2562–64
99. Ohsu T, Amino Y, Nagasaki H, Yamanaka T, Takeshita S, et al. 2010. Involvement of the calcium-sensing receptor in human taste perception. *J. Biol. Chem.* 285:1016–22
100. Justinich CJ, Mak N, Pacheco I, Mulder D, Wells RW, et al. 2008. The extracellular calcium-sensing receptor (CaSR) on human esophagus and evidence of expression of the CaSR on the esophageal epithelial cell line (HET-1A). *Am. J. Physiol. Gastrointest. Liver Physiol.* 294:120–29
101. Ceglia L, Harris SS, Rasmussen HM, Dawson-Hughes B. 2009. Activation of the calcium sensing receptor stimulates gastrin and gastric acid secretion in healthy participants. *Osteoporos. Int.* 20:71–78
102. Feng J, Petersen CD, Coy DH, Jiang JK, Thomas CJ, et al. 2010. Calcium-sensing receptor is a physiologic multimodal chemosensor regulating gastric G-cell growth and gastrin secretion. *Proc. Natl. Acad. Sci. USA* 107:17791–96
103. Chandra R, Liddle RA. 2007. Cholecystokinin. *Curr. Opin. Endocrinol. Diabetes Obes.* 14:63–67
104. Wang Y, Chandra R, Samsa LA, Gooch B, Fee BE, et al. 2011. Amino acids stimulate cholecystokinin release through the Ca^{2+}-sensing receptor. *Am. J. Physiol. Gastrointest. Liver Physiol.* 300:528–37
105. Cheng SX, Geibel JP, Hebert SC. 2004. Extracellular polyamines regulate fluid secretion in rat colonic crypts via the extracellular calcium-sensing receptor. *Gastroenterology* 126:148–58
106. Geibel J, Sritharan K, Geibel R, Geibel P, Persing JS, et al. 2006. Calcium-sensing receptor abrogates secretagogue-induced increases in intestinal net fluid secretion by enhancing cyclic nucleotide destruction. *Proc. Natl. Acad. Sci. USA* 103:9390–97
107. Kallay E, Kifor O, Chattopadhyay N, Brown EM, Bischof MG, et al. 1997. Calcium-dependent c-*myc* proto-oncogene expression and proliferation of Caco-2 cells: a role for a luminal extracellular calcium-sensing receptor. *Biochem. Biophys. Res. Commun.* 232:80–83
108. Hizaki K, Yamamoto H, Taniguchi H, Adachi Y, Nakazawa M, et al. 2011. Epigenetic inactivation of calcium-sensing receptor in colorectal carcinogenesis. *Mod. Pathol.* 24:876–84
109. Geibel JP, Hebert SC. 2009. The functions and roles of the extracellular Ca^{2+}-sensing receptor along the gastrointestinal tract. *Annu. Rev. Physiol.* 71:205–17
110. Bruce JI, Yang X, Ferguson CJ, Elliott AC, Steward MC, et al. 1999. Molecular and functional identification of a Ca^{2+} (polyvalent cation)-sensing receptor in rat pancreas. *J. Biol. Chem.* 274:20561–68
111. Racz GZ, Kittel A, Riccardi D, Case RM, Elliott AC, Varga G. 2002. Extracellular calcium sensing receptor in human pancreatic cells. *Gut* 51:705–11
112. Layer P, Hotz J, Goebell H. 1986. Stimulatory effect of hypercalcemia on pancreatic secretion is prevented by pretreatment with cholecystokinin and cholinergic agonists. *Pancreas* 1:478–82
113. Frick TW, Hailemariam S, Heitz PU, Largiader F, Goodale RL. 1990. Acute hypercalcemia induces acinar cell necrosis and intraductal protein precipitates in the pancreas of cats and guinea pigs. *Gastroenterology* 98:1675–81

114. Frick TW, Wiegand D, Bimmler D, Fernandez-del Castillo C, Rattner DW, Warshaw AL. 1994. A rat model to study hypercalcemia-induced acute pancreatitis. *Int. J. Pancreatol.* 15:91–96
115. Whitcomb DC. 2010. Genetic aspects of pancreatitis. *Annu. Rev. Med.* 61:413–24
116. Gray E, Muller D, Squires PE, Asare-Anane H, Huang GC, et al. 2006. Activation of the extracellular calcium-sensing receptor initiates insulin secretion from human islets of Langerhans: involvement of protein kinases. *J. Endocrinol.* 190:703–10
117. Parkash J. 2011. Glucose-mediated spatial interactions of voltage dependent calcium channels and calcium sensing receptor in insulin producing beta-cells. *Life Sci.* 88:257–64
118. Canaff L, Petit JL, Kisiel M, Watson PH, Gascon-Barre M, Hendy GN. 2001. Extracellular calcium-sensing receptor is expressed in rat hepatocytes. Coupling to intracellular calcium mobilization and stimulation of bile flow. *J. Biol. Chem.* 276:4070–79
119. Yano S, Brown EM, Chattopadhyay N. 2004. Calcium-sensing receptor in the brain. *Cell Calcium* 35:257–64
120. Rusakov DA, Fine A. 2003. Extracellular Ca^{2+} depletion contributes to fast activity-dependent modulation of synaptic transmission in the brain. *Neuron* 37:287–97
121. Ruat M, Molliver ME, Snowman AM, Snyder SH. 1995. Calcium sensing receptor: molecular cloning in rat and localization to nerve terminals. *Proc. Natl. Acad. Sci. USA* 92:3161–65
122. Phillips CG, Harnett MT, Chen W, Smith SM. 2008. Calcium-sensing receptor activation depresses synaptic transmission. *J. Neurosci.* 28:12062–70
123. Vyleta NP, Smith SM. 2011. Spontaneous glutamate release is independent of calcium influx and tonically activated by the calcium-sensing receptor. *J. Neurosci.* 31:4593–606
124. Lu B, Zhang Q, Wang H, Wang Y, Nakayama M, Ren D. 2010. Extracellular calcium controls background current and neuronal excitability via an UNC79-UNC80-NALCN cation channel complex. *Neuron* 68:488–99
125. Heyeraas KJ, Haug SR, Bukoski RD, Awumey EM. 2008. Identification of a Ca^{2+}-sensing receptor in rat trigeminal ganglia, sensory axons, and tooth dental pulp. *Calcif. Tissue Int.* 82:57–65
126. Weston AH, Absi M, Ward DT, Ohanian J, Dodd RH, et al. 2005. Evidence in favor of a calcium-sensing receptor in arterial endothelial cells: studies with calindol and Calhex 231. *Circ. Res.* 97:391–98
127. Pollak MR, Brown EM, Chou YH, Hebert SC, Marx SJ, et al. 1993. Mutations in the human Ca^{2+}-sensing receptor gene cause familial hypocalciuric hypercalcemia and neonatal severe hyperparathyroidism. *Cell* 75:1297–303
128. Ho C, Conner DA, Pollak MR, Ladd DJ, Kifor O, et al. 1995. A mouse model of human familial hypocalciuric hypercalcemia and neonatal severe hyperparathyroidism. *Nat. Genet.* 11:389–94
129. Thakker RV. 2004. Diseases associated with the extracellular calcium-sensing receptor. *Cell Calcium* 35:275–82
130. Pollak MR, Brown EM, Estep HL, McLaine PN, Kifor O, et al. 1994. Autosomal dominant hypocalcaemia caused by a Ca^{2+}-sensing receptor gene mutation. *Nat. Genet.* 8:303–7
131. Kifor O, Moore FD Jr, Wang P, Goldstein M, Vassilev P, et al. 1996. Reduced immunostaining for the extracellular Ca^{2+}-sensing receptor in primary and uremic secondary hyperparathyroidism. *J. Clin. Endocrinol. Metab.* 81:1598–606
132. Yano S, Sugimoto T, Tsukamoto T, Chihara K, Kobayashi A, et al. 2000. Association of decreased calcium-sensing receptor expression with proliferation of parathyroid cells in secondary hyperparathyroidism. *Kidney Int.* 58:1980–86
133. Goodman WG, Quarles LD. 2008. Development and progression of secondary hyperparathyroidism in chronic kidney disease: lessons from molecular genetics. *Kidney Int.* 74:276–88
134. London GM, Marchais SJ, Guerin AP, Metivier F. 2005. Arteriosclerosis, vascular calcifications and cardiovascular disease in uremia. *Curr. Opin. Nephrol. Hypertens.* 14:525–31
135. Alam MU, Kirton JP, Wilkinson FL, Towers E, Sinha S, et al. 2009. Calcification is associated with loss of functional calcium-sensing receptor in vascular smooth muscle cells. *Cardiovasc. Res.* 81:260–68
136. Drueke TB, Ritz E. 2009. Treatment of secondary hyperparathyroidism in CKD patients with cinacalcet and/or vitamin D derivatives. *Clin. J. Am. Soc. Nephrol.* 4:234–41
137. Koleganova N, Piecha G, Ritz E, Schirmacher P, Muller A, et al. 2009. Arterial calcification in patients with chronic kidney disease. *Nephrol. Dial. Transplant.* 24:2488–96

138. Block GA, Zaun D, Smits G, Persky M, Brillhart S, et al. 2010. Cinacalcet hydrochloride treatment significantly improves all-cause and cardiovascular survival in a large cohort of hemodialysis patients. *Kidney Int.* 78:578–89
139. Saidak Z, Mentaverri R, Brown EM. 2009. The role of the calcium-sensing receptor in the development and progression of cancer. *Endocr. Rev.* 30:178–95
140. Chakrabarty S, Wang H, Canaff L, Hendy GN, Appelman H, Varani J. 2005. Calcium sensing receptor in human colon carcinoma: interaction with Ca^{2+} and 1,25-dihydroxyvitamin D_3. *Cancer Res.* 65:493–98
141. Corbetta S, Eller-Vainicher C, Vicentini L, Lania A, Mantovani G, et al. 2007. Modulation of cyclin D1 expression in human tumoral parathyroid cells: effects of growth factors and calcium sensing receptor activation. *Cancer Lett.* 255:34–41
142. Mamillapalli R, VanHouten J, Zawalich W, Wysolmerski J. 2008. Switching of G-protein usage by the calcium-sensing receptor reverses its effect on parathyroid hormone-related protein secretion in normal versus malignant breast cells. *J. Biol. Chem.* 283:24435–47
143. Yano S, Macleod RJ, Chattopadhyay N, Tfelt-Hansen J, Kifor O, et al. 2004. Calcium-sensing receptor activation stimulates parathyroid hormone-related protein secretion in prostate cancer cells: role of epidermal growth factor receptor transactivation. *Bone* 35:664–72
144. Chakravarti B, Dwivedi SK, Mithal A, Chattopadhyay N. 2009. Calcium-sensing receptor in cancer: good cop or bad cop? *Endocrine* 35:271–84
145. Marie PJ, Felsenberg D, Brandi ML. 2010. How strontium ranelate, via opposite effects on bone resorption and formation, prevents osteoporosis. *Osteoporos. Int.* 22:1659–67
146. Coulombe J, Faure H, Robin B, Ruat M. 2004. In vitro effects of strontium ranelate on the extracellular calcium-sensing receptor. *Biochem. Biophys. Res. Commun.* 323:1184–90
147. Chattopadhyay N, Quinn SJ, Kifor O, Ye C, Brown EM. 2007. The calcium-sensing receptor (CaR) is involved in strontium ranelate-induced osteoblast proliferation. *Biochem. Pharmacol.* 74:438–47
148. Rybchyn MS, Slater M, Conigrave AD, Mason RS. 2011. An Akt-dependent increase in canonical Wnt signaling and a decrease in sclerostin protein levels are involved in strontium ranelate-induced osteogenic effects in human osteoblasts. *J. Biol. Chem.* 286:23771–79
149. Caudrillier A, Hurtel-Lemaire AS, Wattel A, Cournarie F, Godin C, et al. 2010. Strontium ranelate decreases receptor activator of nuclear factor-κB ligand-induced osteoclastic differentiation in vitro: involvement of the calcium-sensing receptor. *Mol. Pharmacol.* 78:569–76
150. Fromigue O, Hay E, Barbara A, Petrel C, Traiffort E, et al. 2009. Calcium sensing receptor-dependent and receptor-independent activation of osteoblast replication and survival by strontium ranelate. *J. Cell Mol. Med.* 13:2189–99
151. Hock JM, Gera I. 1992. Effects of continuous and intermittent administration and inhibition of resorption on the anabolic response of bone to parathyroid hormone. *J. Bone Miner. Res.* 7:65–72
152. Dobnig H, Turner RT. 1997. The effects of programmed administration of human parathyroid hormone fragment (1–34) on bone histomorphometry and serum chemistry in rats. *Endocrinology* 138:4607–12
153. John MR, Widler L, Gamse R, Buhl T, Seuwen K, et al. 2011. ATF936, a novel oral calcilytic, increases bone mineral density in rats and transiently releases parathyroid hormone in humans. *Bone* 49:233–41
154. Widler L, Altmann E, Beerli R, Breitenstein W, Bouhelal R, et al. 2011. 1-Alkyl-4-phenyl-6-alkoxy-1H-quinazolin-2-ones: a novel series of potent calcium-sensing receptor antagonists. *J. Med. Chem.* 53:2250–63
155. Ye C, Ho-Pao CL, Kanazirska M, Quinn S, Rogers K, et al. 1997. Amyloid-β proteins activate Ca^{2+}-permeable channels through calcium-sensing receptors. *J. Neurosci. Res.* 47:547–54
156. Chiarini A, Dal Pra I, Marconi M, Chakravarthy B, Whitfield JF, Armato U. 2009. Calcium-sensing receptor (CaSR) in human brain's pathophysiology: roles in late-onset Alzheimer's disease (LOAD). *Curr. Pharm. Biotechnol.* 10:317–26
157. Kapoor A, Satishchandra P, Ratnapriya R, Reddy R, Kadandale J, et al. 2008. An idiopathic epilepsy syndrome linked to 3q13.3-q21 and missense mutations in the extracellular calcium sensing receptor gene. *Ann. Neurol.* 64:158–67
158. Stepanchick A, McKenna J, McGovern O, Huang Y, Breitwieser GE. 2011. Calcium sensing receptor mutations implicated in pancreatitis and idiopathic epilepsy syndrome disrupt an arginine-rich retention motif. *Cell Physiol. Biochem.* 26:363–74

159. Yamaguchi T, Chattopadhyay N, Kifor O, Brown EM. 1998. Extracellular calcium (Ca^{2+}_o)-sensing receptor in a murine bone marrow-derived stromal cell line (ST2): potential mediator of the actions of Ca^{2+}_o on the function of ST2 cells. *Endocrinology* 139:3561–68
160. House MG, Kohlmeier L, Chattopadhyay N, Kifor O, Yamaguchi T, et al. 1997. Expression of an extracellular calcium-sensing receptor in human and mouse bone marrow cells. *J. Bone Miner. Res.* 12:1959–70
161. Adams GB, Chabner KT, Alley IR, Olson DP, Szczepiorkowski ZM, et al. 2006. Stem cell engraftment at the endosteal niche is specified by the calcium-sensing receptor. *Nature* 439:599–603
162. Drueke TB. 2006. Haematopoietic stem cells—role of calcium-sensing receptor in bone marrow homing. *Nephrol. Dial. Transplant.* 21:2072–74
163. Silver IA, Murrills RJ, Etherington DJ. 1988. Microelectrode studies on the acid microenvironment beneath adherent macrophages and osteoclasts. *Exp. Cell Res.* 175:266–76
164. Lam BS, Cunningham C, Adams GB. 2011. Pharmacologic modulation of the calcium-sensing receptor enhances hematopoietic stem cell lodgment in the adult bone marrow. *Blood* 117:1167–75
165. Wu Q, Shao H, Darwin ED, Li J, Yang B, et al. 2009. Extracellular calcium increases CXCR4 expression on bone marrow-derived cells and enhances pro-angiogenesis therapy. *J. Cell Mol. Med.* 13:3764–73
166. Martino NA, Lange-Consiglio A, Cremonesi F, Valentini L, Caira M, et al. 2011. Functional expression of the extracellular calcium sensing receptor (CaSR) in equine umbilical cord matrix size-sieved stem cells. *PLoS ONE* 6:e17714

RELATED RESOURCES

1. Tu CL, Chang W, Bikle DD. 2011. The calcium-sensing receptor-dependent regulation of cell-cell adhesion and keratinocyte differentiation requires Rho and filamin A. *J. Investig. Dermatol.* 131:1119–28
2. Tu CL, Chang W, Xie Z, Bikle DD. 2008. Inactivation of the calcium sensing receptor inhibits E-cadherin-mediated cell-cell adhesion and calcium-induced differentiation in human epidermal keratinocytes. *J. Biol. Chem.* 283:3519–28

Cell Biology and Pathology of Podocytes

Anna Greka and Peter Mundel

Department of Medicine, Massachusetts General Hospital and Harvard Medical School, Charlestown, Massachusetts 02129; email: greka.anna@mgh.harvard.edu, mundel.peter@mgh.harvard.edu

Keywords

glomerulus, foot process, slit diaphragm, actin cytoskeleton, calcium, angiotensin, synaptopodin, TRPC channels, Rac1, RhoA, integrins, mTOR, autophagy

Abstract

As an integral member of the filtration barrier in the kidney glomerulus, the podocyte is in a unique geographical position: It is exposed to chemical signals from the urinary space (Bowman's capsule), it receives and transmits chemical and mechanical signals to/from the glomerular basement membrane upon which it elaborates, and it receives chemical and mechanical signals from the vascular space with which it also communicates. As with every cell, the ability of the podocyte to receive signals from the surrounding environment and to translate them to the intracellular milieu is dependent largely on molecules residing on the cell membrane. These molecules are the first-line soldiers in the ongoing battle to sense the environment, to respond to friendly signals, and to defend against injurious foes. In this review, we take a membrane biologist's view of the podocyte, examining the many membrane receptors, channels, and other signaling molecules that have been implicated in podocyte biology. Although we attempt to be comprehensive, our goal is not to capture every membrane-mediated pathway but rather to emphasize that this approach may be fruitful in understanding the podocyte and its unique properties.

INTRODUCTION

With every cardiac cycle, the kidney glomerulus filters blood into an ultrafiltrate that will ultimately become urine. Architecturally, the glomerulus or renal corpuscle consists of a glomerular tuft and Bowman's capsule. The basic unit of the glomerular tuft is a single capillary. The glomerular basement membrane (GBM) provides the primary structural scaffold for the glomerular tuft. Endothelial and smooth muscle–like mesangial cells providing capillary support are located inside the GBM, whereas podocytes are attached to the outer part of the GBM (**Figure 1**). There are therefore four resident cell types in the glomerulus: endothelial cells, mesangial cells, parietal epithelial cells of Bowman's capsule, and podocytes (**Figure 1*a***). Podocytes are pericyte-like cells with a complex cellular organization consisting of a cell body, major processes, and foot processes (FPs). Podocyte FPs elaborate into a characteristic interdigitating pattern with FPs of neighboring podocytes, forming in between the filtration slits that are bridged by the glomerular slit diaphragm (SD) (**Figure 1*b*,*c***). Podocyte FPs and the interposed SD cover the outer part of the GBM and play a major role in establishing the selective permeability of the glomerular filtration barrier, which explains why podocyte injury is typically associated with marked albuminuria. Podocytes are highly differentiated cells with limited capability to undergo cell division in the adult, and the loss of podocytes is a hallmark of progressive kidney disease. The function of podocytes is based largely on the dynamic regulation of their complex cell architecture, in particular the FP structure. Over the past decade, there has been a growing understanding of the role of membrane proteins in transducing extracellular cues into the podocyte intracellular milieu to effect critical architectural changes. Membrane proteins, e.g., ion channels, integrins, and growth factor receptors, are taking center stage in podocyte biology research. Here we provide insight into the physiological and pathophysiological pathways that regulate podocyte structure and function in health and disease.

PODOCYTE ARCHITECTURE: THE CENTRAL ROLE OF FOOT PROCESSES

Mature podocytes consist of three morphologically distinct segments: a cell body, major processes, and FPs (1). From the cell body, microtubule-rich major processes split into FPs (**Figure 1*b***) containing an actin-based cytoskeleton (**Figure 1*c*,*d***). Podocyte FPs elaborate into a highly branched interdigitating network with FPs of neighboring podocytes. The SD (**Figure 1*c***) is a multiprotein complex similar to an adherens junction. The SD bridges the filtration slits between

---

Figure 1

The function of podocytes is based on their intricate cell architecture. (*a*) A glomerulus contains a capillary tuft that receives primary structural support from the glomerular basement membrane (GBM). Glomerular endothelial cells (E) lining the capillary lumen (CL) and mesangial cells (M) are located on the blood side of the GBM, whereas podocyte foot processes (FP) cover the outer part of the GBM. Podocyte cell bodies (CB) and major processes (MP) float in Bowman's space (BS) in primary urine. Along its route from the CL to BS (*blue arrow*), the plasma ultrafiltrate passes through the fenestrated endothelium, the GBM, and the filtration slits that are covered by the slit diaphragm (SD). AA, afferent arteriole; DT, distal tubule; EE, efferent arteriole; PE, parietal epithelium; PT, proximal tubule. (*b*) Scanning electron microscopy view from BS highlighting the intricate shape of podocytes. MP link the CB to FP, which interdigitate with FP of neighboring podocytes, thereby forming the filtration slits. (*c*) Transmission electron microscopy image of the filtration barrier consisting of fenestrated endothelium, GBM, and podocyte FP with the SD covering the filtration slits. (*d*) Podocyte FP are defined by three membrane domains: the apical membrane domain (AMD) (*blue*), the basal membrane domain (BMD) (*red*), and the SD (*black*). All three domains are connected to the underpinning actin cytoskeleton (*gray*) and with each other. PAB denotes a single parallel, contractile actin bundle containing α-actinin-4, myosin 9, and synaptopodin. (*e*) Under conditions of proteinuria, FP lose their normal interdigitating pattern and instead show effacement. A continuous sheet of cytoplasm filled with reorganized actin filaments is clearly visible.

opposing podocyte FPs (1), thereby establishing the final barrier to urinary protein loss (2). FPs are characterized by a podosome-like cortical network of short, branched actin filaments and by the presence of highly ordered, parallel contractile actin filament bundles (**Figure 1***d*) (3), which are thought to modulate the permeability of the filtration barrier through changes in FP morphology. The function of podocytes is based largely on their complex architecture, in particular on the maintenance of highly ordered, parallel, contractile actin filament bundles in FPs (3). In fact, similar to (pseudo)unipolar sensory neurons, the podocyte cell body and primary processes may simply play a trophic, supporting role to the critically important FPs. FPs are functionally defined by three membrane domains: the apical membrane domain, the SD, and the basal membrane domain

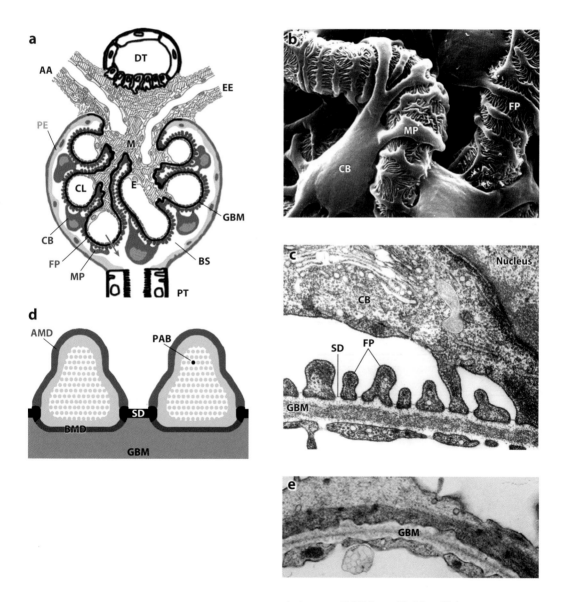

associated with the GBM (**Figure 1d**). All three domains are physically and functionally linked to the FP actin cytoskeleton, thus lending a central role to actin for both podocyte function and dysfunction. Interference with any of the three FP domains changes the actin cytoskeleton from parallel, contractile bundles (3) into a dense network. The results are (*a*) FP effacement reflected by the simplification of the FP structure and loss of the normal interdigitating pattern (**Figure 1e**) and (*b*) proteinuria (4). FP effacement requires the active reorganization of actin filaments (5, 6), a process regulated at the molecular level by a multitude of signaling events involving, but not limited to, integrin activation, G protein–coupled receptor (GPCR) and growth factor receptor activity, and calcium (Ca^{2+}) influx pathways as upstream modulators of the actin cytoskeleton. In vivo studies have shown that FP effacement is always temporally related to the emergence of proteinuria (4, 7, 8) and thus suggest a causative link between these two events, which is a widely held belief in our field. However, the molecular steps leading to FP effacement are so hierarchical and precise that they lend credence to what remains a tantalizing hypothesis: FP effacement may be the podocyte's desperate attempt to adapt to toxic cues to prevent proteinuria (W. Kriz, personal communication).

PODOCYTE ORIGINS: PODOCYTE MATURATION IS ASSOCIATED WITH A PHYSIOLOGICAL EPITHELIAL-TO-MESENCHYMAL TRANSITION

During renal development, epithelial precursors give rise to mature podocytes, which are mesenchyme-like cells. Glomerular development proceeds in four stages: the renal vesicle stage, the S-shaped body stage, the capillary loop stage, and the maturing-glomeruli stage (9, 10). The transition from the S-shaped body stage to the capillary loop stage is critical for podocyte differentiation (1). During the early developmental stages, simple undifferentiated epithelial cells with apically located tight junctions form the immature podocyte precursor cell population (11). At this stage, the expression of podocalyxin (12) and of the tight junction protein ZO-1 (13) commences, and expression of the podocyte-specific transcription factor Wilm's tumor protein 1 (WT-1) is highest (14). As these immature podocytes enter the capillary loop stage, they lose their mitotic activity and begin to establish their characteristic complex cell architecture, including the appearance of FPs and the reorganization of cell-cell junctions into SDs (11), which are essentially modified adherens junctions (15). At this stage, ZO-1 migrates from its apical location down to the level of the slit membrane, where it is observed in a punctate pattern along the filtration slits (13). A similar apical-to-basal redistribution pattern occurs in the case of the atypical protein kinase C (aPKC), an important cell polarity protein (16–18). The phenotypic conversion occurring during the S-shaped body stage is accompanied by the expression of synaptopodin (19) and by the reappearance of vimentin (20), a phenotypic marker of mesenchymal cells.

The canonical developmental pathway leading to the formation of podocyte FPs is governed by the coordinated activities of numerous membrane or membrane-associated proteins or the transcription factors that regulate the expression of such membrane proteins. Mouse genetic studies have shown a central role for $\alpha 3$ integrin in the formation of mature FPs (21), a phenotype mirrored by the podocyte-specific deletion of $\beta 1$ integrin (22, 23). The deletion of the nonclassical protocadherin FAT1 revealed the importance of this molecule for proper FP formation (24). Disruption of the cytoskeletal adaptor proteins Nck1/2, which are linked to the critical SD protein nephrin (25), resulted in complete failure to develop FPs (26), thereby phenocopying $\alpha 3$ integrin knockout mice. This result is not surprising, as Nck proteins are critically involved in integrin signaling (27). Furthermore, mice lacking podocalyxin also fail to form FPs and die soon after birth from renal failure (28). The podocyte transcription factor Kreisler (**Table 1**), whose targets include the SD proteins nephrin (25) and podocin (29), is also essential for FP formation (30).

Table 1 Transcription factors in podocyte development, physiology, and pathology

Name	Putative target(s)	References
Wilm's tumor protein 1 (WT1)	Podxl (podocalyxin), Nphs1 (nephrin)	177–180
LMX1B	α3 and α4 collagens, Nphs2 (podocin)	115, 181, 182
FOXC2 (Mfh2)	Nphs2, α3 and α4 collagens, MafB	183
Kreisler (MafB)	Nphs1, Nphs2, CD2AP	30, 184
Pax2	WT1?	185, 186
ZHX1/ZHX2	WT1?	187, 188
Pod1 (epicardin)	Nphs2, α3 collagen, MafB?	31, 189
ZEB2	P/E cadherins	190
HIFα	CXCR4	191–193
Rbpj (Notch pathway)	?	95
NFAT	TRPC6	98, 138, 139
β-Catenin	?	194, 195

Moreover, disruption of Pod1 (also known as epicardin) leads to aberrant FP formation (**Table 1**) (31). These genetic studies reveal the importance of these molecules in podocyte development and, more specifically, in FP development. **Table 1** summarizes transcription factors involved in the regulation of podocyte structure and function during development as well as under physiological and pathological conditions.

Podocyte development is relevant to mechanisms of podocyte injury in the adult. Mature podocytes are unable to undergo cell division in vivo. The evidence for this hypothesis comes from studies showing that the number of podocyte nuclei does not increase during postnatal and hypertrophic kidney growth (32–36). The transition of podocytes from the S-shaped body stage to the capillary loop stage and their branching into specialized cell architecture involve a neuron-like complex cell differentiation process, which is mutually exclusive with cell division. Interestingly, in response to certain stimuli [e.g., fibroblast growth factor (FGF)-2], podocytes may reenter the cell cycle and undergo nuclear division but cannot complete cell division (33). The only exception to this phenomenon is noted in HIV nephropathy, in which podocytes undergo a tumor-like proliferation, albeit without reparative capacity (37–40). Although the underlying molecular mechanisms leading to the effective arrest of cytokinesis in vivo remain to be established, cyclins and cyclin-dependent kinases and their inhibitors have been implicated (reviewed in detail in Reference 41).

The inability of mature podocytes to undergo cell division and to replenish their population renders the glomerulus vulnerable to toxic cues, leading to significant podocyte loss, which is a hallmark in the development of chronic kidney failure (42). The search for a podocyte stem cell to replenish a diminishing podocyte pool has led to the notion that a population of cells in the parietal epithelium of the glomerulus may migrate into the glomerular tuft and differentiate into mature podocytes (43, 44). However, more recent evidence suggests that migrating parietal epithelial cells may be maladaptive for glomerular function (45). Further studies are likely to reveal if there is a population of cells capable of replenishing lost podocytes.

THE SLIT DIAPHRAGM: A MULTIPROTEIN SIGNALING HUB

The SD is a complex signal transduction unit with characteristics of a modified adherens junction that spans the 30–50-nm-wide filtration slits (15). The extracellular portion of the SD is made

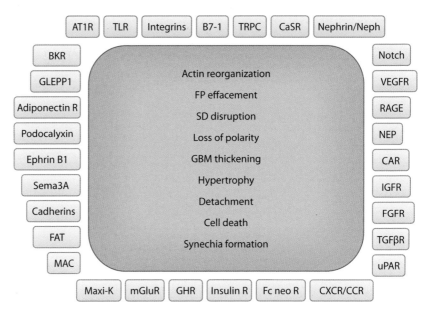

Figure 2

Podocyte plasma membrane proteins and a canonical pattern of injury. Shown is an (incomplete) list of membrane proteins that have been implicated in the regulation of podocyte function in health and disease. Injured podocytes respond with a finite repertoire of changes, as depicted here. Our ability to repair the pathways initiated by the molecules on the podocyte plasma membrane to the precise cellular phenotypes listed here will provide not only novel insight but enormous opportunities for successful therapeutic interventions. Abbreviations: adiponectin R, adiponectin receptor; AT1R, angiotensin type 1 receptor; BKR, bradykinin receptor; CAR, Coxsackie and adenovirus receptor; CaSR, calcium-sensing receptor; CXCR/CCR, C-X-C/C-C chemokine receptor; FAT, protocadherin FAT1; Fc neo R, neonatal Fc receptor; FGFR, fibroblast growth factor receptor; FP, foot process; GBM, glomerular basement membrane; GHR, growth hormone receptor; GLEPP1, glomerular epithelial (podocyte) protein 1; IGFR, insulin growth factor receptor; insulin R, insulin receptor; MAC, membrane attack complex; Maxi-K, large-conductance calcium-activated potassium channel (also known as BK channel); mGluR, metabotropic glutamate receptor; NEP, neutral endopeptidase; RAGE, receptor for advanced glycation end products; SD, slit diaphragm; Sema3A, semaphorin-3A; TGFβR, transforming growth factor β receptor; TLR, Toll-like receptor; TRPC, transient receptor potential canonical; uPAR, urokinase receptor; VEGFR, vascular endothelial growth factor receptor.

up of rod-like units that are thought to be composed of the extracellular domains of various transmembrane proteins such as nephrin and FAT (1). These rods are connected by a linear bar, forming a zipper-like pattern, which leaves pores the same size as or smaller than albumin (1). The SD's cytoplasmic portion contains a region of Triton-X-100-resistant (1), electron-dense material, which is reminiscent of the highly insoluble specialization of the submembranous actin cytoskeleton in neurons known as the postsynaptic density (PSD) (46). The PSD contains multiple receptors and ion channels linked through a multitude of adaptor proteins to the cytoskeletal core, forming a large protein network (47). Similarly, at the SD, ion channels such as TRPCs (where TRPC denotes transient receptor potential canonical) (48), receptors, integrins, and other membrane proteins (**Figure 2**) are connected to actin via a variety of adaptor and effector proteins.

The disruption of SD structure or function is a common theme in many kidney diseases arising at the podocyte level (49). The SD is thought to function as a key sensor for and as a regulator of adaptations in FP shape and length (7, 8). For example, the movement of each FP needs to be precisely coordinated with that of the FPs of neighboring podocytes to ensure the integrity of the filtration barrier. Such coordination is likely achieved through functional coupling of opposed FPs, which generates signaling cascades on both ends of the SD. This multiprotein network likely serves a far more complex role as a signaling platform and is not simply a physical sieve.

Nephrin is a well-known and widely studied SD membrane protein (8). Mutations in the *NPHS1* gene encoding for nephrin have been identified as the cause of congenital nephrotic syndrome of the Finnish type (25). Nephrin has a single transmembrane domain; a short intracellular tail; and a long, immunoglobulin-like extracellular moiety, which is thought to align parallel to the extracellular domains of neighboring nephrin molecules. Through its intracellular domain, nephrin is connected to the actin cytoskeleton by several adapter proteins and plays a pivotal part in the regulation of podocyte actin dynamics (7, 50). Among other pathways (reviewed in detail in Reference 7), a recently discovered signaling pathway couples nephrin to the actin cytoskeleton via the adapter protein Nck (26, 51). After nephrin phosphorylation by Fyn (52), Nck binds to phospho-nephrin and to N-WASP (26, 51), activating the Arp2/3 complex, a major regulator of actin dynamics (7, 26, 50, 51). Recent work has also shown that Fyn phosphorylation of nephrin promotes activation of phosphoinositide 3 kinase and Rac1 activity (53; please also see below).

Notably, a large body of human genetic, animal, and physiological studies, including micropuncture experiments, over more than four decades had established the podocyte FPs with the interposed SD as the final barrier to albumin (2, 8). However, some years ago, on the basis of live-imaging two-photon microscopy results, researchers developed the alternative hypothesis that nephrotic levels of albumin pass across the normal glomerulus filters and are subsequently retrieved by the proximal tubule (54). Through the use of the same two-photon microscopy approach, two recent independent studies convincingly refuted this leaky glomerular barrier hypothesis (55–57). Both Tanner (55) and Peti-Peterdi (57) directly confirmed the classical view that the glomerular filter is the primary barrier for albumin and that the glomerular sieving coefficient for albumin is extremely low (55–57), thereby corroborating the size and charge selectivity of the glomerular barrier (55, 58–60).

PODOCYTE INJURY IS THE HALLMARK OF PROTEINURIA AND GLOMERULAR DISEASE

Podocyte injury is the common denominator in many forms of human and experimental glomerular disease such as minimal change disease, focal segmental glomerulosclerosis (FSGS), membranous glomerulopathy, diabetic nephropathy (DN), and lupus nephritis (2, 4). The best-characterized pattern of injury involves a reorganization of the FP actin cytoskeleton that leads to FP effacement and to SD disruption (61, 62). The disruption of any of the three FP domains (**Figure 1d**), with the concomitant transformation of the actin cytoskeleton from parallel contractile bundles (3) into a dense network and loss of the normal interdigitating pattern (**Figure 1e**), leads to proteinuria (4).

There is a growing understanding of the sequence of events that mediates FP effacement and proteinuria (**Figure 3a**), followed over time by further phenotypic changes such as podocyte hypertrophy and ultimately podocyte detachment and loss (**Figure 3b**). The initial patterns of injury include (*a*) changes in SD structure or function (52, 63, 64), (*b*) interference with the GBM or the podocyte:GBM interaction (21, 65–70), (*c*) dysregulation of podocyte Ca^{2+} homeostasis (71, 72), (*d*) dysfunction of the podocyte actin cytoskeleton (50, 51, 62, 73–76), (*e*) modulation of the negative surface charge of podocytes (5, 77, 78), (*f*) activation of innate immunity pathways such as B7-1 signaling (79–81), (*g*) upregulation of CatL-mediated proteolysis (82–86), and (*h*) disturbances in the transcriptional regulation of podocyte function (87).

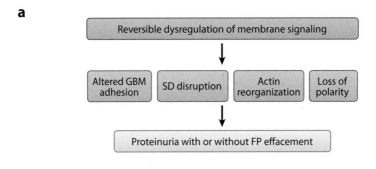

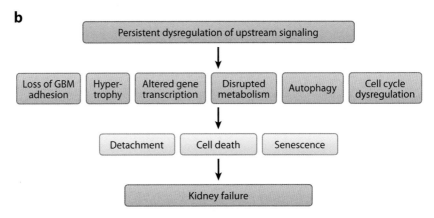

Figure 3

Reversible and irreversible consequences of dysregulated podocyte signaling. (*a*) Dysregulated signaling at the plasma membrane may lead to reversible morphological changes. Therefore, proteinuria may arise, with or without foot process (FP) effacement (see text for details). However, if the upstream injurious signals are reversed, the cell morphology can revert back to physiological patterns. SD denotes slit diaphragm.
(*b*) Persistence of podocyte injury is manifest in the activation of cellular processes that lead to irreversible changes such as loss of adhesion to the glomerular basement membrane (GBM), cell hypertrophy, transcriptional changes, disrupted metabolic pathways, autophagy, and cell cycle dysregulation. These irreversible changes can cause podocyte cell death or the detachment of podocytes from the GBM. Podocyte senescence may also be a manifestation of persistent injury, although less is known about this mechanism. The resulting loss of podocytes ultimately leads to irreversible glomerulosclerosis and kidney failure.

Importantly, a common theme to recent advances is the ever-growing significance of membrane proteins in the emergence of FP effacement and progressive podocyte injury (**Figures 2** and **3**). Growth factor receptors such as vascular endothelial growth factor (88, 89) and transforming growth factor β (90), GPCRs such as the angiotensin type 1 receptor (AT1R) (91–93), signaling through Notch (94, 95) or integrins (21–23, 96, 97), ion channels such as the TRPCs (48, 71, 98–100), and many other molecules (summarized in **Figure 2**) have been implicated in early podocyte injury. Although such molecules work in concert under physiological conditions, they become functionally uncoupled under disease conditions, leading to a disrupted cytoskeleton, the best known podocyte injury pattern to date.

Early podocyte injury is reversible if the actin cytoskeleton is repaired, allowing FPs to branch once again into their interdigitating pattern (**Figure 3a**). Sustained chronic podocyte injury can lead to the loss of glomerular and ultimately entire kidney function through three principal mechanisms: (*a*) the dysregulated pathway, (*b*) the inflammatory pathway, and (*c*) the degenerative pathway (101). First, in the dysregulated pathway, dedifferentiation of podocytes leads to podocyte proliferation within Bowman's space and the collapse of the glomerular tuft, with GBM wrinkling and capillary loss. Thus, so-called collapsing glomerulopathy occurs, for example, in HIV-associated nephropathy. Second, inflammatory mechanisms can lead to the fixation of podocytes to the parietal basement membrane followed by the establishment of tuft adhesions to Bowman's capsule (101). Further proliferation of podocytes and parietal cells results in the formation of cellular crescents. When the lesion heals by fibrosis, segmental glomerulosclerosis occurs (101). Finally, in the degenerative form, which is most commonly observed, the persistence of podocyte injury can cause cell body attenuation, podocyte hypertrophy, detachment from the GBM, and podocyte death followed by the formation of synechiae by the attachment of parietal epithelial cells to nude GBM (**Figure 3*b***). This attachment results in misdirected filtration toward the interstitium (101, 102). Through a series of ensuing changes (reviewed in detail in Reference 101), the loss of podocytes ultimately leads to glomerulosclerosis and to kidney failure (101). The contribution of proteinuria to the progression of kidney disease is a matter of debate (103). Some investigators believe that proteinuria can induce podocyte damage (104) or tubulointerstitial inflammation and progressive injury (105). However, severe experimental protein loss across the glomerular filter caused by repeated injection of rats with the antinephrin antibody 5-1-6 did not lead to progressive renal failure (106). The role of proteinuria in the progression of kidney failure may depend on the route of protein loss. Significant podocyte loss per glomerulus, and thus misdirected filtration into the periglomerular interstitium, leads to tubular destruction and the progression of kidney failure (101, 107). In contrast, protein loss across the filtration barrier without significant podocyte loss does not lead to disease progression and may be reversible (**Figure 3*a***) (101, 108).

GENETIC CAUSES OF PODOCYTE INJURY

Human genetics have fueled our progress toward a molecular understanding of the SD and the modulators of FP architecture. In the past 15 years, human genetic studies revealed that mutations in the genes encoding nephrin (25), podocin (29), phospholipase C ε (109), and coenzyme Q10 biosynthesis mono-oxygenase 6 (110) give rise to early-onset proteinuria. Additionally, adult-onset proteinuria such as FSGS is associated with mutations in the genes encoding α-actinin 4 (62), CD2AP (111), INF2 (112), TRPC6 (48, 100, 113), and synaptopodin (114). Mutations in *LMX1B*, which encodes a transcription factor for collagen, result in podocyte abnormalities due to impaired cell adhesion to the abnormal GBM (115). Similarly, mutations in the gene encoding laminin $\beta 2$, another component of the GBM, lead to podocyte injury and proteinuria (116). Finally, recent exome sequencing as well as a whole-genome linkage analysis revealed *MYO1E* mutations in childhood proteinuric disease and FSGS; *MYO1E* encodes a mutant form of nonmuscle class I myosin (117, 118). **Table 2** summarizes human mutations affecting podocyte structure and function. Mouse genetic studies have revealed that additional proteins regulating the plasticity of the podocyte actin cytoskeleton such as Rho GDP dissociation inhibitor α (Rho GDIα) (119), podocalyxin (6), FAT1 (120), Nck1/2 (26, 51), synaptopodin (121, 122), and cofilin (123) are also of critical importance for sustained function of the glomerular filtration barrier. Taken together, human and mouse genetics have revealed that the dysregulation of the highly specialized podocyte actin cytoskeleton is closely associated with disease phenotypes. At present, the regulation of the actin cytoskeleton is probably the best understood aspect of podocyte function (see below).

With the advent of genomics, large population studies have revealed common variations in a number of genes that predispose or confer susceptibility to acquired proteinuric kidney disease.

Table 2 Human mutations affecting podocyte structure and function

Protein	Gene	Associated disease	Mode of inheritance	Clinical description	Reference
Nephrin	NPHS1	Congenital nephrotic syndrome of the Finnish type (CNF)	Autosomal recessive (AR)	Often massive proteinuria in utero and nephrotic syndrome postnatally; resistant to treatment	25
Podocin	NPHS2	Corticosteroid-resistant nephrotic syndrome (SRNS)	AR	Variable onset and severity of nephropathy; resistance to corticosteroid therapy	29
Coenzyme Q10 biosynthesis mono-oxygenase 6	COQ6	Corticosteroid-resistant nephrotic syndrome (SRNS)	AR	Early-onset SNRS with sensorineural deafness	110
PLCε1	PLCE1	Inherited nephrotic syndrome	AR	Nephrosis, diffuse mesangial sclerosis, and end-stage kidney disease (ESKD); may be reversible after early treatment	109
Laminin β2	LAMB2	Pierson's syndrome Occasionally oligosymptomatic disease variants	AR	Onset of nephrosis postnatally, mesangial sclerosis, microcoria	116
α-Actinin-4	ACTN4	Focal segmental glomerulosclerosis (FSGS)	Autosomal dominant (AD)	Mild proteinuria in adolescence, slow progression to FSGS and ESKD in adulthood	62
TRPC6	TRPC6	FSGS	AD	Proteinuria in adolescence and early adulthood; progression to FSGS and ESKD; pediatric/sporadic cases reported	48, 100
MYH9	MYH9	Epstein syndrome Fechtner syndrome	AD	Thrombocytopenia, hearing defects, and progressive proteinuria	126
LMX1B	LMX1B	Nail-patella syndrome	AD	Variable penetrance, nephrotic syndrome, and skeletal/nail abnormalities in children	115, 196
WT1	WT1	Denys-Drash syndrome (DDS) Frasier's syndrome (FS)	AD	Male pseudohermaphroditism with progressive nephropathy and ESKD; development of FSGS by 3 years (DDS) or later (FS)	197, 198
CD2AP	CD2AP	Sporadic FSGS	N/A	FSGS in African-American patients	111
Synaptopodin	SYNPO	Sporadic FSGS	N/A	FSGS in Chinese patients	114
Myosin 1E	MYO1E	Childhood FSGS	N/A	Progressive, steroid-resistant FSGS	117, 118
Apolipoprotein L-1	APOL1	Sporadic FSGS	N/A	Progressive proteinuria, FSGS, and ESKD in African-American patients	127
Glypican 5	GPC5	Acquired nephrotic syndrome	N/A	Progressive proteinuria and ESKD	128

These gene polymorphisms are not directly linked to podocyte-specific defects, but this is an area of active research at this time. A large locus containing numerous genes was recently identified in African-American populations (124, 125). Initial studies revealed *MYH9* as a likely gene candidate in this locus. This was an attractive hypothesis, given previous work showing that *MYH9* is responsible for two genetic causes of proteinuria: Epstein and Fechtner syndromes (**Table 2**) (126). Interestingly, further work revealed that the likely candidate gene conferring risk for kidney disease is *APOL1*. This gene encodes apolipoprotein L1, a molecule that is known for its trypanolytic properties and that confers an evolutionary advantage in African-Americans (127). Furthermore, common variations in *GPC5* (which encodes glypican 5) are also associated with acquired nephrotic syndrome (128).

MOLECULAR REGULATORS OF PODOCYTE FUNCTION IN HEALTH AND DISEASE

The past decade has brought significant new knowledge about membrane proteins that initiate signaling pathways important for podocyte structure and function (**Figure 2**). These molecules act as sensors of the podocyte's complex extracellular environment, receiving cues from the urinary (Bowman's) space, the vascular (capillary) space, and the GBM. These signals are subsequently transduced to the intracellular environment, modulating the actin cytoskeleton, gene transcription, and cellular metabolism pathways, among many others (**Figures 2** and **3**). Here we focus on a few signaling cascades that multiple scientists have investigated: Ca^{2+} signaling and TRPC channels, angiotensin signaling, integrins, and a mammalian target of rapamycin (mTOR) cascade intersecting with autophagy pathways. The roles of numerous other molecules are not clearly understood. **Figure 2**, a schematic of a podocyte plasma membrane decorated by membrane proteins, represents our concerted effort to be inclusive of the proteins studied in podocytes to date.

CALCIUM SIGNALING IN PODOCYTES

In conjunction with mounting evidence that proteinuria and podocyte FP effacement are mediated by rearrangements of the actin cytoskeleton (7, 51, 129), disrupted Ca^{2+} signaling and homeostasis were postulated as early events in podocyte injury (130). Complement C5b-9 complex–mediated podocyte damage is associated with an increase in intracellular Ca^{2+} concentration ($[Ca^{2+}]_i$) (131). Protamine sulfate, which can cause FP effacement in vivo (132, 133), also increases $[Ca^{2+}]_i$ in vitro in both cultured cells (134) and isolated glomeruli (72). Bradykinin and angiotensin II (Ang II) application on differentiated mouse and rat podocytes, respectively, increases $[Ca^{2+}]_i$ (135, 136). A recent intriguing study also revealed that the Ca^{2+}-sensing receptor has prosurvival and actin-stabilizing effects in podocytes (137). Importantly, Ang II evokes a nonselective cationic current recorded from podocytes of isolated rat glomeruli (93). Once TRPC6 channel mutations were found in patients with familial FSGS (48, 100), TRP channels emerged as prime candidates for this as-yet-uncharacterized conductance. Recently, detailed electrophysiology unveiled TRPC5 and TRPC6 as the channels downstream of the Ang II–evoked nonselective cationic conductance (71), initially identified more than a decade ago (93).

Activation of the Ca^{2+}-dependent phosphatase calcineurin leads to cathepsin L–mediated cleavage of synaptopodin and to proteinuria (83). The calcineurin inhibitor cyclosporine A (CsA) prevents synaptopodin degradation in vitro, and mice resistant to cathepsin-mediated synaptopodin degradation are protected from proteinuria in vivo (83). Conversely, the activation of calcineurin in podocytes is sufficient to cause degradation of synaptopodin and proteinuria (83). Thus, the preservation of synaptopodin and the podocyte actin cytoskeleton provides an intriguing podocyte-specific, T cell– and NFAT-independent mechanism for the long-known antiproteinuric effect of calcineurin inhibition (83). However, NFAT-dependent mechanisms are also important in podocytes. Ca^{2+}-conducting and FSGS-causing TRPC6 mutations, but not wild-type TRPC6, induce constitutive activation of calcineurin- and NFAT-dependent gene transcription (138). Importantly, the activation of NFAT signaling in podocytes is sufficient to cause glomerulosclerosis in mice (139). A recent study supports the notion of a positive feedback loop, whereby NFAT-mediated increases in TRPC6 channel transcription lead to proteinuria (98). Although not experimentally addressed in this study, the proposed mechanism of injury relates to excessive Ca^{2+} influx due to aberrantly high numbers of active TRPC6 channels on the podocyte plasma membrane. A similar feed forward cycle was demonstrated in a mouse model of cardiac hypertrophy: Calcineurin- and NFAT-mediated increases in TRPC6 transcription promote pathological cardiac remodeling (140). More recently, however, investigators showed that other TRPC channels mediate calcineurin-NFAT activation in cardiac myocytes (140–142), strongly suggesting that this pathway is not specific to TRPC6.

ANGIOTENSIN SIGNALING IN PODOCYTES

The podocyte-specific overexpression of the AT1R in rats is sufficient to cause proteinuria and FSGS-type lesions. This model may be one of the best rodent models of FSGS to date because the model mimics the human disease by exhibiting hyperlipidemia, hypoalbuminemia, and progressive kidney failure (91). At the cellular level, the activation of the AT1R transactivates the epidermal growth factor receptor (EGFR) in tubular epithelia (143) and in podocytes (144). Following an increase in cytosolic $[Ca^{2+}]$, AT1R-EGFR interactions also activate downstream serine/threonine kinases such as the mitogen-activated protein kinase pathway (145). Ang II induces membrane ruffling and the loss of stress fibers (92), similar to the depletion of synaptopodin (83, 122) or of TRPC6 (71; see section below). Intriguingly, the AT1R signals to TRPC5 and TRPC6 channels in podocytes (71; see section below). The most compelling evidence to study the effects of angiotensin in podocytes comes from human trials in which the inhibition of AT1R signaling through angiotensin-converting enzyme inhibitors or angiotensin receptor blockers delayed disease progression in patients with diabetic kidney disease (146, 147).

CALCIUM, TRPC CHANNELS, RHO GTPASES, AND THE REGULATION OF THE ACTIN CYTOSKELETON

Previous studies in many cell types, including fibroblasts and neurons, had established an intimate association between Ca^{2+} influx and the activation of the Rho GTPases, which are cytoskeleton master regulators (148, 149). Given the central role played by Ca^{2+} and the actin cytoskeleton in podocyte biology, it is not surprising that the Rho GTPases have been the subject of numerous recent studies in podocytes. At the cellular level, a stationary podocyte phenotype, which suggests intact FPs, is thought to be due to the relative predominance of RhoA activity. In contrast, a predominance of Cdc42/Rac1 activity mediates a disease-associated motile phenotype, suggesting unstable or retracted FPs (7). Synaptopodin has emerged as an important regulator of Rho GTPases in podocytes. Synaptopodin promotes RhoA signaling by preventing its ubiquitination and proteasomal degradation. This results in the preservation of stress fibers in vitro and in safeguarding against proteinuria in vivo (83, 122). Synaptopodin-depleted podocytes, in which RhoA is ubiquitinated, display loss of stress fiber formation and aberrant filopodia (121). Synaptopodin also suppresses Cdc42 signaling through the inhibition of Cdc42:IRSp53:Mena complexes (150).

Recent studies thus support the notion that synaptopodin stabilizes the kidney filter by promoting RhoA and inhibiting Cdc42, thereby preventing the reorganization of the podocyte FP cytoskeleton into a migratory phenotype (83). The promotility GTPase Rac1 was also implicated in podocyte biology: Rho GDIα–null mice develop heavy albuminuria (119) due to increased, constitutively active Rac1 signaling in podocytes (75). Rac1 promotes the accumulation of mineralocorticoid receptor into the podocyte nucleus through p21-activated kinase phosphorylation (75). Pharmacological intervention with a Rac1-specific small-molecule inhibitor (NSC23766) diminishes mineralocorticoid receptor hyperactivity and ameliorates proteinuria and renal damage in this mouse model of proteinuria (75). Taken together, these studies suggest an important role for the RhoGTPases in podocyte health and disease.

Since the discovery of gain-of-function *TRPC6* mutations in familial FSGS (48, 100), the molecular mechanisms involving TRPC channels have become a central question in podocyte biology. Recently, AT1R-activated TRPC5 and TRPC6 channels were unveiled as antagonistic regulators of actin dynamics and cell motility in podocytes through the regulation of Rac1 and RhoA, respectively (**Figure 4**) (71). The latter study showed that TRPC6 depletion results in the loss of stress fibers, Rac1 activation, and increased motility, all of which are rescued by constitutively active RhoA (71). Conversely, TRPC5 depletion leads to enhanced stress fiber formation, RhoA activation, and decreased motility, all of which are reversed by constitutively active Rac1 (**Figure 4**) (71). Two distinct signaling microdomains emerged: TRPC5 specifically interacts with and activates Rac1, whereas TRPC6 specifically interacts with and activates RhoA (**Figure 4**) (71). Consistent with previous studies (83), CsA restored synaptopodin expression in TRPC6-depleted cells (71). In contrast, synaptopodin expression was preserved in TRPC5-depleted podocytes (71). These results significantly extended our mechanistic understanding of TRPC channelopathies and

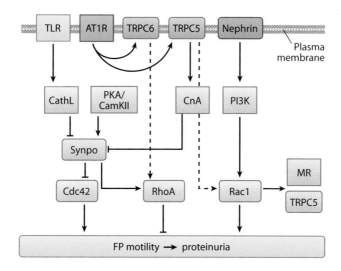

Figure 4

An example of pathways at the intersection of AT1R and TRPC signaling at the plasma membrane as they converge on synaptopodin and Rho GTPase signaling in the cytosol to effect critical cytoskeletal remodeling. Dashed arrows indicate indirect effects between molecules. Abbreviations: TLR, Toll-like receptor; AT1R, angiotensin type 1 receptor; TRPC6/5, transient receptor potential canonical channels 6 and 5; CathL, cathepsin L; PKA/CamKII, protein kinase A/calcium-calmodulin-dependent protein kinase II; CnA, calcineurin; PI3K, phosphoinositide 3 kinase; Synpo, synaptopodin; MR, mineralocorticoid receptor; FP, foot process.

provided a mechanistic link between AT1R activation, Ca^{2+} signaling, TRPC channels, and Rho GTPase activity in podocytes.

Mice overexpressing wild-type *TRPC6* and *TRPC6* gain-of-function mutants develop albuminuria and FSGS-type lesions (151). Although proteinuria in these mice was modest with low and variable penetrance and structural abnormalities were not observed before 6–8 months of age (151), these results are consistent with Ca^{2+} overload–mediated cellular injury and death, ultimately leading to FSGS. In keeping with cell culture data showing RhoA activation downstream of TRPC6 activity (71), recent work revealed that mice overexpressing constitutively active RhoA in a podocyte-specific manner also developed proteinuria and FSGS-type lesions (152). Taken together, these studies suggest that an unopposed or overactive TRPC6-RhoA pathway may cause irreversible injury, podocyte loss, and kidney failure. In keeping with this, a recent study reported that *TRPC6*-null mice were significantly protected from the proteinuric effects of Ang II (99). The observed protective role of *TRPC6* deletion is surprising, given that *TRPC6*-null mice were initially reported to be hypertensive at baseline due to (over)compensation by TRPC3 for the loss of TRPC6 (153). In the light of the antagonistic effects of TRPC5 and TRPC6 on podocyte actin dynamics (71), future studies will be needed to address possible compensatory or antagonistic relationships between TRPC channels in TRPC6 or other TRPC knockout mice. The development of podocyte-specific, inducible TRPC knockout mice is also likely to illuminate our understanding of in vivo TRPC channel function in podocytes.

INTEGRIN SIGNALING IN PODOCYTES

Integrins are heterodimeric cell adhesion receptors whose function is central to inflammation, immunity, tumor progression, development, and the maintenance of normal tissue architecture in mature organs (154, 155). Cell adhesion and spreading as well as remodeling of the extracellular matrix (ECM) involve bidirectional signaling and physical linkage between the ECM, integrins, and the cytoskeleton (154, 155). The actin cytoskeleton of podocyte FPs is linked to the GBM by α3β1 integrin (23, 96, 156), αvβ3 integrin (64, 157), and α-/β-dystroglycans (65, 158). Podocytes also express β4 integrin, and mutations in the gene encoding β4 integrin are associated with congenital FSGS and skin disease (159). Genetic inactivation of α3 (21) or β1 (22, 23) integrin causes podocyte FP effacement and kidney failure in newborn mice, thereby underscoring the critical role of α3β1 integrin in the development and maintenance of the glomerular filter. A recent study (160) suggested that soluble uPAR (suPAR) may be a factor responsible for the development of recurrent FSGS (161). According to this study, suPAR leads to recurrent FSGS through the activation of β3 integrin in podocytes (160). Of note, these data are in contrast with previous studies identifying β3 integrin as a mediator of protective osteopontin signaling in podocytes (157). Future studies will be required to address the precise function of β3 integrin signaling in podocytes. In addition, the relative contribution of β3 integrin activation versus loss of sphingomyelin phosphodiesterase acid–like 3b protein expression (162) to the pathogenesis of recurrent FSGS remains to be resolved.

The β1 integrin–binding protein integrin-linked kinase (ILK) is another mediator of progressive podocyte damage (163). Podocyte-specific deletion of the ILK gene in mice causes progressive FSGS and renal failure (22, 164, 165). Activation of ILK in podocytes induces Wnt signaling, which in turn leads to the reduction of CD2AP and P-cadherin expression, podocyte detachment, and proliferation (166). Moreover, overexpression of ILK causes the rearrangement of the podocyte actin cytoskeleton, presumably via ILK-mediated phosphorylation of α-actinin (163). ILK serves as an adaptor that biochemically and functionally connects the GBM with the SD: It interacts with nephrin, α-actinin, PINCH, and α-parvin, playing a crucial role in podocyte adhesion, morphology, and survival (165, 167). A recent study also brought to light the importance of focal adhesion kinase (FAK) in podocytes, showing that its inhibition protects against proteinuria and FP effacement (168). This study concluded that podocytes isolated from conditional, podocyte-specific FAK knockout mice demonstrated reduced spreading and migration. How the FAK, the ILK, and other integrin-mediated pathways intersect to mediate adaptive or maladaptive podocyte adhesion to the GBM will undoubtedly be the subject of many future inquiries.

The role of B7-1 as a bidirectional regulator of T cell activation and tolerance is well established and involves the binding of B7-1 to its cognate receptors CD28, CTLA, or PD-L1 (169). Most interestingly, however, B7-1 is also an inducible mediator of podocyte injury and proteinuria (81). Initially identified due to its significant upregulation in α3 integrin knockout mice, it is also upregulated in podocytes in human and experimental lupus nephritis and in nephrin knockout mice (81). The clinical significance of these results was underscored by the observation that podocyte expression of B7-1 correlates with the severity of human lupus nephritis. The latter study (81) also found that in vivo exposure to low-dose lipopolysaccharide (LPS) rapidly upregulates B7-1 in podocytes of wild-type and SCID mice, thereby leading to proteinuria. In contrast, mice lacking B7-1 are protected from LPS-induced proteinuria, demonstrating a functional link between podocyte B7-1 expression and proteinuria (81). LPS signaling through Toll-like receptor-4 reorganizes the podocyte actin cytoskeleton. Moreover, the activation of B7-1 in cultured podocytes leads to the reorganization of the vital SD proteins nephrin and CD2AP (81). Thus, the upregulation of B7-1 in podocytes by LPS contributes to the pathogenesis of proteinuria by disrupting the glomerular filter (81). Given the initial observation that B7-1 is upregulated in the absence of α3 integrin in vivo, future studies will be required to address how B7-1 might intersect with integrin-mediated signaling in podocytes.

PODOCYTE MAMMALIAN TARGET OF RAPAMYCIN SIGNALING AND AUTOPHAGY

The mTOR signaling cascade regulates a wide array of cellular processes, including cell growth, proliferation, and autophagy, in response to nutrients such as glucose, amino acids, and growth factors. mTOR is an evolutionarily conserved protein kinase and forms two functional complexes termed mTOR complex 1 (mTORC1) and mTORC2. mTORC1 is a rapamycin-sensitive protein kinase that senses nutrient availability (170). mTOR signaling in podocytes has come to the forefront for two reasons. First, the mTOR inhibitor rapamycin, which is clinically used for immunosuppression after organ transplantation, can induce proteinuria, although the underlying molecular mechanisms are poorly understood. The idea that rapamycin can impair podocyte function is supported by the observation that its application on cultured podocytes decreases the protein abundance of nephrin, TRPC6, and Nck and reduces cell adhesion and motility (171). Nephrin expression is also reduced in glomeruli of kidney-transplanted patients undergoing mTOR inhibition therapy (172). However, rapamycin can prevent the protamine sulfate–induced and Ca^{2+}-dependent disruption of the podocyte actin cytoskeleton, suggesting a potential beneficial effect of rapamycin on proteinuria (72).

In a second independent line of research, two recent studies revealed the importance of mTOR signaling in podocyte function and diabetic nephropathy in humans and mice. Genetic deletion of mTORC1 in mouse podocytes induced proteinuria and progressive glomerulosclerosis, whereas simultaneous deletion of both mTORC1 and mTORC2 from mouse podocytes aggravated the glomerular lesions (173). In contrast, increased mTOR activity accompanied human diabetic nephropathy, characterized by early glomerular hypertrophy and hyperfiltration. Curtailing mTORC1 signaling in mice by genetically reducing mTORC1 copy number in podocytes prevented glomerulosclerosis and ameliorated glomerular disease progression in diabetic nephropathy (173). Similarly, the activity of the mTOR complex was enhanced in podocytes of diabetic animals (174). Furthermore, podocyte-specific mTORC1 activation induced by ablation of an upstream negative regulator (PcKOTsc1) recapitulated many DN features, including podocyte loss, GBM thickening, mesangial expansion, and proteinuria, in nondiabetic young and adult mice (174). Abnormal mTORC1 activation caused mislocalization of SD proteins and endoplasmic reticulum (ER) stress in podocytes. Conversely, reduction of ER stress with a chemical chaperone significantly protected against both the podocyte phenotypic switch and podocyte loss in PcKOTsc1 mice. Finally, genetic reduction of podocyte-specific mTORC1 in diabetic animals suppressed the development of DN (174). Although protein synthesis and autophagic degradation are regulated in an opposite manner by mTOR (170), such regulation could be beneficial under certain conditions if these two events occurred in unison to handle rapid protein turnover. In keeping with this notion, a recent study revealed that spatial coupling of mTOR and autophagy augments the secretory phenotype of podocytes and macrophages (175). Narita and colleagues (175) observed a distinct cellular compartment at the *trans* side of the Golgi apparatus, the TOR-autophagy spatial coupling compartment (TASCC), where (auto)lysosomes and mTOR accumulated during Ras-induced senescence. mTOR recruitment to the TASCC was amino acid dependent and Rag guanosine triphosphatase dependent, and disruption of mTOR localization to the TASCC suppressed interleukin-6/8 synthesis. TASCC formation was observed during macrophage differentiation and in podocytes; both macrophages and podocytes displayed increased protein secretion. The spatial coupling of cells' catabolic and anabolic machinery could augment their respective functions and facilitate the mass synthesis of secretory proteins (175). Collectively, these data underscore the critical role of mTOR signaling in podocytes for glomerular function in health and disease. These results demonstrate the requirement for tightly balanced mTOR activity in podocyte homeostasis and suggest that mTOR inhibition may protect podocytes and prevent progressive diabetic nephropathy (173). Given the detrimental effect of prolonged rapamycin treatment on podocyte function, it will be important to test if reduction of podocyte mTOR activity can be harnessed as a potential therapeutic strategy to treat DN.

OUTLOOK AND FUTURE DIRECTIONS

In recent decades, the podocyte has been firmly established as the cell responsible for proteinuria and kidney damage. Yet much remains to be learned about this complex cell. In this review, we take a membrane biologist's view, examining membrane receptors, ion channels, and other signaling molecules implicated in podocyte biology. This approach may be fruitful in understanding the podocyte and its unique geographical properties because it brings a new perspective to the many challenges that lie ahead. Although we have made significant inroads in identifying the molecular components of this complex cell, many central questions remain. One challenge is to gain a further understanding of how these molecules fit together and work in unison under physiological conditions and how the system is perturbed after a noxious insult. A membrane biology perspective may help in this regard if we consider the plasma membrane as the beginning of Ariadne's thread, which can help us navigate our way through what is truly a signaling labyrinth. Although disrupted cytoskeletal dynamics have emerged as a central mechanism for podocyte injury, another challenge is to go beyond the cytoskeleton in an effort to assign a molecular signature to pathways that may mediate the propagation of injury, such as metabolic dysregulation and autophagy. Indeed, little is known about the molecular determinants of pathological progression in podocytes. For example, does podocyte cell death precede detachment or vice versa? The future study of appropriate animal models of disease progression is also likely to be instrumental in this area. Finally, transmembrane receptors and ion channels have proven to be successful drug targets in other areas (176) and therefore should garner significant attention as putative drug targets for antiproteinuric therapies. The challenge will be to discern which of the multitude of podocyte membrane proteins, many of which we discuss here, will be the critical target(s) of choice.

DISCLOSURE STATEMENT

The authors are not aware of any affiliations, memberships, funding, or financial holdings that might be perceived as affecting the objectivity of this review.

ACKNOWLEDGMENTS

We apologize to our colleagues whose work we were not able to cite in this review due to space limitations. Reviews were often quoted at the expense of original work. We thank W. Kriz and M.A. Arnaout for helpful discussions. A.G. is funded by a NephCure Young Investigator Grant, the ASN Gottschalk Award, and NIH grant DK083511; P.M., by NIH grants DK57683 and DK062472.

LITERATURE CITED

1. Mundel P, Kriz W. 1995. Structure and function of podocytes: an update. *Anat. Embryol.* 192:385–97
2. Somlo S, Mundel P. 2000. Getting a foothold in nephrotic syndrome. *Nat. Genet.* 24:333–35
3. Drenckhahn D, Franke RP. 1988. Ultrastructural organization of contractile and cytoskeletal proteins in glomerular podocytes of chicken, rat, and man. *Lab. Investig.* 59:673–82
4. Kerjaschki D. 2001. Caught flat-footed: podocyte damage and the molecular bases of focal glomerulosclerosis. *J. Clin. Investig.* 108:1583–87
5. Takeda T, McQuistan T, Orlando RA, Farquhar MG. 2001. Loss of glomerular foot processes is associated with uncoupling of podocalyxin from the actin cytoskeleton. *J. Clin. Investig.* 108:289–301
6. Schmieder S, Nagai M, Orlando RA, Takeda T, Farquhar MG. 2004. Podocalyxin activates RhoA and induces actin reorganization through NHERF1 and Ezrin in MDCK cells. *J. Am. Soc. Nephrol.* 15:2289–98
7. Faul C, Asanuma K, Yanagida-Asanuma E, Kim K, Mundel P. 2007. Actin up: regulation of podocyte structure and function by components of the actin cytoskeleton. *Trends Cell Biol.* 17:428–37
8. Tryggvason K, Patrakka J, Wartiovaara J. 2006. Hereditary proteinuria syndromes and mechanisms of proteinuria. *N. Engl. J. Med.* 354:1387–401
9. Saxen L, Sariola H. 1987. Early organogenesis of the kidney. *Pediatr. Nephrol.* 1:385–92
10. Sorokin L, Ekblom P. 1992. Development of tubular and glomerular cells of the kidney. *Kidney Int.* 41:657–64
11. Reeves W, Caulfield JP, Farquhar MG. 1978. Differentiation of epithelial foot processes and filtration slits: sequential appearance of occluding junctions, epithelial polyanion, and slit membranes in developing glomeruli. *Lab. Investig.* 39:90–100
12. Schnabel E, Dekan G, Miettinen A, Farquhar MG. 1989. Biogenesis of podocalyxin—the major glomerular sialoglycoprotein—in the newborn rat kidney. *Eur. J. Cell Biol.* 48:313–26
13. Schnabel E, Anderson JM, Farquhar MG. 1990. The tight junction protein ZO-1 is concentrated along slit diaphragms of the glomerular epithelium. *J. Cell Biol.* 111:1255–63
14. Mundlos S, Pelletier J, Darveau A, Bachmann M, Winterpacht A, Zabel B. 1993. Nuclear localization of the protein encoded by the Wims' tumor gene *WT1* in embryonic and adult tissues. *Development* 119:1329–41
15. Reiser J, Kriz W, Kretzler M, Mundel P. 2000. The glomerular slit diaphragm is a modified adherens junction. *J. Am. Soc. Nephrol.* 11:1–8

16. Hartleben B, Schweizer H, Lubben P, Bartram MP, Moller CC, et al. 2008. Neph-Nephrin proteins bind the Par3-Par6-atypical protein kinase C (aPKC) complex to regulate podocyte cell polarity. *J. Biol. Chem.* 283:23033–38
17. Huber TB, Hartleben B, Winkelmann K, Schneider L, Becker JU, et al. 2009. Loss of podocyte aPKCλ/ι causes polarity defects and nephrotic syndrome. *J. Am. Soc. Nephrol.* 20:798–806
18. Hirose T, Satoh D, Kurihara H, Kusaka C, Hirose H, et al. 2009. An essential role of the universal polarity protein, aPKCλ, on the maintenance of podocyte slit diaphragms. *PLoS ONE* 4:e4194
19. Mundel P, Gilbert P, Kriz W. 1991. Podocytes in glomerulus of rat kidney express a characteristic 44 KD protein. *J. Histochem. Cytochem.* 39:1047–56
20. Holthofer H, Miettinen A, Lehto VP, Lehtonen E, Virtanen I. 1984. Expression of vimentin and cytokeratin types of intermediate filament proteins in developing and adult human kidneys. *Lab. Investig.* 50:552–29
21. Kreidberg JA, Donovan MJ, Goldstein SL, Rennke H, Shepherd K, et al. 1996. Alpha 3 beta 1 integrin has a crucial role in kidney and lung organogenesis. *Development* 122:3537–47
22. Kanasaki K, Kanda Y, Palmsten K, Tanjore H, Lee SB, et al. 2008. Integrin β1-mediated matrix assembly and signaling are critical for the normal development and function of the kidney glomerulus. *Dev. Biol.* 313:584–93
23. Pozzi A, Jarad G, Moeckel GW, Coffa S, Zhang X, et al. 2008. β1 integrin expression by podocytes is required to maintain glomerular structural integrity. *Dev. Biol.* 316:288–301
24. Ciani L, Patel A, Allen ND, Ffrench-Constant C. 2003. Mice lacking the giant protocadherin mFAT1 exhibit renal slit junction abnormalities and a partially penetrant cyclopia and anophthalmia phenotype. *Mol. Cell. Biol.* 23:3575–82
25. Kestila M, Lenkkeri U, Mannikko M, Lamerdin J, McCready P, et al. 1998. Positionally cloned gene for a novel glomerular protein—nephrin—is mutated in congenital nephrotic syndrome. *Mol. Cell* 1:575–82
26. Jones N, Blasutig IM, Eremina V, Ruston JM, Bladt F, et al. 2006. Nck adaptor proteins link nephrin to the actin cytoskeleton of kidney podocytes. *Nature* 440:818–23
27. Brakebusch C, Fassler R. 2003. The integrin-actin connection, an eternal love affair. *EMBO J.* 22:2324–33
28. Doyonnas R, Kershaw DB, Duhme C, Merkens H, Chelliah S, et al. 2001. Anuria, omphalocele, and perinatal lethality in mice lacking the CD34-related protein podocalyxin. *J. Exp. Med.* 194:13–27
29. Boute N, Gribouval O, Roselli S, Benessy F, Lee H, et al. 2000. NPHS2, encoding the glomerular protein podocin, is mutated in autosomal recessive steroid-resistant nephrotic syndrome. *Nat. Genet.* 24:349–54
30. Sadl V, Jin F, Yu J, Cui S, Holmyard D, et al. 2002. The mouse *Kreisler (Krml1/MafB)* segmentation gene is required for differentiation of glomerular visceral epithelial cells. *Dev. Biol.* 249:16–29
31. Quaggin SE, Schwartz L, Cui S, Igarashi P, Deimling J, et al. 1999. The basic-helix-loop-helix protein Pod1 is critically important for kidney and lung organogenesis. *Development* 126:5771–83
32. Pabst R, Sterzl RB. 1983. Cell renewal of glomerular cell types in normal rats. An autoradiographic analysis. *Kidney Int.* 24:626–31
33. Kriz W, Elger M, Kretzler M, Uiker S, Koeppen-Hagemann I, et al. 1994. The role of podocytes in the development of glomerular sclerosis. *Kidney Int.* 45:S64–72
34. Nagata M, Yamaguchi Y, Komatsu Y, Ito K. 1995. Mitosis and the presence of binucleate cells among glomerular podocytes in diseased human kidneys. *Nephron* 70:68–71
35. Rasch R, Norgaard JO. 1983. Renal enlargement: comparative autoradiographic studies of ^{3}H-thymidine uptake in diabetic and uninephrectomized rats. *Diabetologia* 25:280–87
36. Fries JW, Sandstrom DJ, Meyer TW, Rennke HG. 1989. Glomerular hypertrophy and epithelial cell injury modulate progressive glomerulosclerosis in the rat. *Lab. Investig.* 60:205–18
37. Barisoni L, Kriz W, Mundel P, D'Agati V. 1999. The dysregulated podocyte phenotype: a novel concept in the pathogenesis of collapsing idiopathic focal segmental glomerulosclerosis and HIV-associated nephropathy. *J. Am. Soc. Nephrol.* 10:51–61
38. Barisoni L, Mokrzycki M, Sablay L, Nagata M, Yamase H, Mundel P. 2000. Podocyte cell cycle regulation and proliferation in collapsing glomerulopathies. *Kidney Int.* 58:137–43
39. Husain M, Gusella GL, Klotman ME, Gelman IH, Ross MD, et al. 2002. HIV-1 Nef induces proliferation and anchorage-independent growth in podocytes. *J. Am. Soc. Nephrol.* 13:1806–15

40. Shankland SJ, Eitner F, Hudkins KL, Goodpaster T, D'Agati V, Alpers CE. 2000. Differential expression of cyclin-dependent kinase inhibitors in human glomerular disease: role in podocyte proliferation and maturation. *Kidney Int.* 58:674–83
41. Shankland SJ, Wolf G. 2000. Cell cycle regulatory proteins in renal disease: role in hypertrophy, proliferation, and apoptosis. *Am. J. Physiol. Ren. Physiol.* 278:515–29
42. Kriz W, Gretz N, Lemley KV. 1998. Progression of glomerular diseases: Is the podocyte the culprit? *Kidney Int.* 54:687–97
43. Smeets B, Uhlig S, Fuss A, Mooren F, Wetzels JF, et al. 2009. Tracing the origin of glomerular extracapillary lesions from parietal epithelial cells. *J. Am. Soc. Nephrol.* 20:2604–15
44. Appel D, Kershaw DB, Smeets B, Yuan G, Fuss A, et al. 2009. Recruitment of podocytes from glomerular parietal epithelial cells. *J. Am. Soc. Nephrol.* 20:333–43
45. Smeets B, Kuppe C, Sicking EM, Fuss A, Jirak P, et al. 2011. Parietal epithelial cells participate in the formation of sclerotic lesions in focal segmental glomerulosclerosis. *J. Am. Soc. Nephrol.* 22:1262–74
46. Kennedy MB. 1993. The postsynaptic density. *Curr. Opin. Neurobiol.* 3:732–37
47. Kim E, Sheng M. 2004. PDZ domain proteins of synapses. *Nat. Rev. Neurosci.* 5:771–81
48. Reiser J, Polu KR, Moller CC, Kenlan P, Altintas MM, et al. 2005. TRPC6 is a glomerular slit diaphragm-associated channel required for normal renal function. *Nat. Genet.* 37:739–44
49. Durvasula RV, Shankland SJ. 2006. Podocyte injury and targeting therapy: an update. *Curr. Opin. Nephrol. Hypertens.* 15:1–7
50. Tryggvason K, Pikkarainen T, Patrakka J. 2006. Nck links nephrin to actin in kidney podocytes. *Cell* 125:221–24
51. Verma R, Kovari I, Soofi A, Nihalani D, Patrie K, Holzman LB. 2006. Nephrin ectodomain engagement results in Src kinase activation, nephrin phosphorylation, Nck recruitment, and actin polymerization. *J. Clin. Investig.* 116:1346–59
52. Verma R, Wharram B, Kovari I, Kunkel R, Nihalani D, et al. 2003. Fyn binds to and phosphorylates the kidney slit diaphragm component Nephrin. *J. Biol. Chem.* 278:20716–23
53. Zhu J, Sun N, Aoudjit L, Li H, Kawachi H, et al. 2008. Nephrin mediates actin reorganization via phosphoinositide 3-kinase in podocytes. *Kidney Int.* 73:556–66
54. Russo LM, Sandoval RM, McKee M, Osicka TM, Collins AB, et al. 2007. The normal kidney filters nephrotic levels of albumin retrieved by proximal tubule cells: Retrieval is disrupted in nephrotic states. *Kidney Int.* 71:504–13
55. Tanner GA. 2009. Glomerular sieving coefficient of serum albumin in the rat: a two-photon microscopy study. *Am. J. Physiol. Ren. Physiol.* 296:1258–65
56. Tanner GA, Rippe C, Shao Y, Evan AP, Williams JC Jr. 2009. Glomerular permeability to macromolecules in the *Necturus* kidney. *Am. J. Physiol. Ren. Physiol.* 296:1269–78
57. Peti-Peterdi J. 2009. Independent two-photon measurements of albumin GSC give low values. *Am. J. Physiol. Ren. Physiol.* 296:1255–57
58. Haraldsson B, Jeansson M. 2009. Glomerular filtration barrier. *Curr. Opin. Nephrol. Hypertens.* 18:331–35
59. Haraldsson B, Nystrom J, Deen WM. 2008. Properties of the glomerular barrier and mechanisms of proteinuria. *Physiol. Rev.* 88:451–87
60. Navar LG. 2009. Glomerular permeability: a never-ending saga. *Am. J. Physiol. Ren. Physiol.* 296:1266–68
61. Tryggvason K, Wartiovaara J. 2001. Molecular basis of glomerular permselectivity. *Curr. Opin. Nephrol. Hypertens.* 10:543–49
62. Kaplan JM, Kim SH, North KN, Rennke H, Correia LA, et al. 2000. Mutations in *ACTN4*, encoding α-actinin-4, cause familial focal segmental glomerulosclerosis. *Nat. Genet.* 24:251–56
63. Simons M, Schwarz K, Kriz W, Miettinen A, Reiser J, et al. 2001. Involvement of lipid rafts in nephrin phosphorylation and organization of the glomerular slit diaphragm. *Am. J. Pathol.* 159:1069–77
64. Wei C, Moller CC, Altintas MM, Li J, Schwarz K, et al. 2008. Modification of kidney barrier function by the urokinase receptor. *Nat. Med.* 14:55–63
65. Regele HM, Fillipovic E, Langer B, Poczewki H, Kraxberger I, et al. 2000. Glomerular expression of dystroglycans is reduced in minimal change nephrosis but not in focal segmental glomerulosclerosis. *J. Am. Soc. Nephrol.* 11:403–12

66. Raats CJ, Bakker MA, van den Born J, Berden JH. 1997. Hydroxyl radicals depolymerize glomerular heparan sulfate in vitro and in experimental nephrotic syndrome. *J. Biol. Chem.* 272:26734–41
67. Noakes PG, Miner JH, Gautam M, Cunningham JM, Sanes JR, Merlie JP. 1995. The renal glomerulus of mice lacking s-laminin/laminin β2: nephrosis despite molecular compensation by laminin β1. *Nat. Genet.* 10:400–6
68. Kretzler M, Teixeira VP, Unschuld PG, Cohen CD, Wanke R, et al. 2001. Integrin-linked kinase as a candidate downstream effector in proteinuria. *FASEB J.* 15:1843–45
69. Lorenzen J, Shah R, Biser A, Staicu SA, Niranjan T, et al. 2008. The role of osteopontin in the development of albuminuria. *J. Am. Soc. Nephrol.* 19:884–90
70. Sachs N, Kreft M, van den Bergh Weerman MA, Beynon AJ, Peters TA, et al. 2006. Kidney failure in mice lacking the tetraspanin CD151. *J. Cell Biol.* 175:33–39
71. Tian D, Jacobo SM, Billing D, Rozkalne A, Gage SD, et al. 2010. Antagonistic regulation of actin dynamics and cell motility by TRPC5 and TRPC6 channels. *Sci. Signal.* 3:ra77
72. Vassiliadis J, Bracken C, Matthews D, O'Brien S, Schiavi S, Wawersik S. 2011. Calcium mediates glomerular filtration through calcineurin and mTORC2/Akt signaling. *J. Am. Soc. Nephrol.* 22:1453–61
73. Smoyer WE, Mundel P. 1998. Regulation of podocyte structure during the development of nephrotic syndrome. *J. Mol. Med.* 76:172–83
74. Kos CH, Le TC, Sinha S, Henderson JM, Kim SH, et al. 2003. Mice deficient in α-actinin-4 have severe glomerular disease. *J. Clin. Investig.* 111:1683–90
75. Shibata S, Nagase M, Yoshida S, Kawarazaki W, Kurihara H, et al. 2008. Modification of mineralocorticoid receptor function by Rac1 GTPase: implication in proteinuric kidney disease. *Nat. Med.* 14:1370–76
76. Lu TC, He JC, Wang ZH, Feng X, Fukumi-Tominaga T, et al. 2008. HIV-1 Nef disrupts the podocyte actin cytoskeleton by interacting with diaphanous interacting protein. *J. Biol. Chem.* 283:8173–82
77. Orlando RA, Takeda T, Zak B, Schmieder S, Benoit VM, et al. 2001. The glomerular epithelial cell anti-adhesin podocalyxin associates with the actin cytoskeleton through interactions with ezrin. *J. Am. Soc. Nephrol.* 12:1589–98
78. Galeano B, Klootwijk R, Manoli I, Sun M, Ciccone C, et al. 2007. Mutation in the key enzyme of sialic acid biosynthesis causes severe glomerular proteinuria and is rescued by N-acetylmannosamine. *J. Clin. Investig.* 117:1585–94
79. Nagase M, Shibata S, Yoshida S, Nagase T, Gotoda T, Fujita T. 2006. Podocyte injury underlies the glomerulopathy of Dahl salt-hypertensive rats and is reversed by aldosterone blocker. *Hypertension* 47:1084–93
80. Navarro-Munoz M, Ibernon M, Perez V, Ara J, Espinal A, et al. 2011. Messenger RNA expression of B7-1 and NPHS1 in urinary sediment could be useful to differentiate between minimal change disease and focal segmental glomerulosclerosis in adult patients. *Nephrol. Dial. Transplant.* In press
81. Reiser J, von Gersdorff G, Loos M, Oh J, Asanuma K, et al. 2004. Induction of B7-1 in podocytes is associated with nephrotic syndrome. *J. Clin. Investig.* 113:1390–97
82. Asanuma K, Shirato I, Ishidoh K, Kominami E, Tomino Y. 2002. Selective modulation of the secretion of proteinases and their inhibitors by growth factors in cultured differentiated podocytes. *Kidney Int.* 62:822–31
83. Faul C, Donnelly M, Merscher-Gomez S, Chang YH, Franz S, et al. 2008. The actin cytoskeleton of kidney podocytes is a direct target of the antiproteinuric effect of cyclosporine A. *Nat. Med.* 14:931–38
84. Reiser J, Oh J, Shirato I, Asanuma K, Hug A, et al. 2004. Podocyte migration during nephrotic syndrome requires a coordinated interplay between cathepsin L and α3 integrin. *J. Biol. Chem.* 279:34827–32
85. Ronco P. 2007. Proteinuria: Is it all in the foot? *J. Clin. Investig.* 117:2079–82
86. Sever S, Altintas MM, Nankoe SR, Moller CC, Ko D, et al. 2007. Proteolytic processing of dynamin by cytoplasmic cathepsin L is a mechanism for proteinuric kidney disease. *J. Clin. Investig.* 117:2095–104
87. Quaggin SE. 2002. Transcriptional regulation of podocyte specification and differentiation. *Microsc. Res. Tech.* 57:208–11
88. Eremina V, Sood M, Haigh J, Nagy A, Lajoie G, et al. 2003. Glomerular-specific alterations of VEGF-A expression lead to distinct congenital and acquired renal diseases. *J. Clin. Investig.* 111:707–16
89. Eremina V, Jefferson JA, Kowalewska J, Hochster H, Haas M, et al. 2008. VEGF inhibition and renal thrombotic microangiopathy. *N. Engl. J. Med.* 358:1129–36

90. Böttinger EP, Bitzer M. 2002. TGF-β signaling in renal disease. *J. Am. Soc. Nephrol.* 13:2600–10
91. Hoffmann S, Podlich D, Hahnel B, Kriz W, Gretz N. 2004. Angiotensin II type 1 receptor overexpression in podocytes induces glomerulosclerosis in transgenic rats. *J. Am. Soc. Nephrol.* 15:1475–87
92. Hsu HH, Hoffmann S, Endlich N, Velic A, Schwab A, et al. 2008. Mechanisms of angiotensin II signaling on cytoskeleton of podocytes. *J. Mol. Med.* 86:1379–94
93. Gloy J, Henger A, Fischer KG, Nitschke R, Mundel P, et al. 1997. Angiotensin II depolarizes podocytes in the intact glomerulus of the rat. *J. Clin. Investig.* 99:2772–81
94. Waters AM, Wu MY, Onay T, Scutaru J, Liu J, et al. 2008. Ectopic Notch activation in developing podocytes causes glomerulosclerosis. *J. Am. Soc. Nephrol.* 19:1139–57
95. Niranjan T, Bielesz B, Gruenwald A, Ponda MP, Kopp JB, et al. 2008. The Notch pathway in podocytes plays a role in the development of glomerular disease. *Nat. Med.* 14:290–98
96. Kreidberg JA. 2000. Functions of α3 β1 integrin. *Curr. Opin. Cell Biol.* 12:548–53
97. Clement LC, Avila-Casado C, Mace C, Soria E, Bakker WW, et al. 2011. Podocyte-secreted angiopoietin-like-4 mediates proteinuria in glucocorticoid-sensitive nephrotic syndrome. *Nat. Med.* 17:117–22
98. Nijenhuis T, Sloan AJ, Hoenderop JG, Flesche J, van Goor H, et al. 2011. Angiotensin II contributes to podocyte injury by increasing TRPC6 expression via an NFAT-mediated positive feedback signaling pathway. *Am. J. Pathol.* 179:1719–32
99. Eckel J, Lavin PJ, Finch EA, Mukerji N, Burch J, et al. 2011. TRPC6 enhances angiotensin II-induced albuminuria. *J. Am. Soc. Nephrol.* 22:526–35
100. Winn MP, Conlon PJ, Lynn KL, Farrington MK, Creazzo T, et al. 2005. A mutation in the TRPC6 cation channel causes familial focal segmental glomerulosclerosis. *Science* 308:1801–4
101. Kriz W, LeHir M. 2005. Pathways to nephron loss starting from glomerular diseases—insights from animal models. *Kidney Int.* 67:404–19
102. Mundel P, Shankland SJ. 2002. Podocyte biology and response to injury. *J. Am. Soc. Nephrol.* 13:3005–15
103. Zandi-Nejad K, Eddy AA, Glassock RJ, Brenner BM. 2004. Why is proteinuria an ominous biomarker of progressive kidney disease? *Kidney Int.* 66:S76–89
104. Morigi M, Buelli S, Angioletti S, Zanchi C, Longaretti L, et al. 2005. In response to protein load podocytes reorganize cytoskeleton and modulate endothelin-1 gene: implication for permselective dysfunction of chronic nephropathies. *Am. J. Pathol.* 166:1309–20
105. Wilmer WA, Rovin BH, Hebert CJ, Rao SV, Kumor K, Hebert LA. 2003. Management of glomerular proteinuria: a commentary. *J. Am. Soc. Nephrol.* 14:3217–32
106. Kikuchi H, Kawachi H, Ito Y, Matsui K, Nosaka H, et al. 2000. Severe proteinuria, sustained for 6 months, induces tubular epithelial cell injury and cell infiltration in rats but not progressive interstitial fibrosis. *Nephrol. Dial. Transplant.* 15:799–810
107. Kriz W, Hartmann I, Hosser H, Hahnel B, Kranzlin B, et al. 2001. Tracer studies in the rat demonstrate misdirected filtration and peritubular filtrate spreading in nephrons with segmental glomerulosclerosis. *J. Am. Soc. Nephrol.* 12:496–506
108. Gross ML, Hanke W, Koch A, Ziebart H, Amann K, Ritz E. 2002. Intraperitoneal protein injection in the axolotl: the amphibian kidney as a novel model to study tubulointerstitial activation. *Kidney Int.* 62:51–59
109. Hinkes B, Wiggins RC, Gbadegesin R, Vlangos CN, Seelow D, et al. 2006. Positional cloning uncovers mutations in *PLCE1* responsible for a nephrotic syndrome variant that may be reversible. *Nat. Genet.* 38:1397–405
110. Heeringa SF, Chernin G, Chaki M, Zhou W, Sloan AJ, et al. 2011. *COQ6* mutations in human patients produce nephrotic syndrome with sensorineural deafness. *J. Clin. Investig.* 121:2013–24
111. Kim JM, Wu H, Green G, Winkler CA, Kopp JB, et al. 2003. CD2-associated protein haploinsufficiency is linked to glomerular disease susceptibility. *Science* 300:1298–300
112. Brown EJ, Schlondorff JS, Becker DJ, Tsukaguchi H, Uscinski AL, et al. 2010. Mutations in the formin gene *INF2* cause focal segmental glomerulosclerosis. *Nat. Genet.* 42:72–76
113. Heeringa SF, Moller CC, Du J, Yue L, Hinkes B, et al. 2009. A novel *TRPC6* mutation that causes childhood FSGS. *PLoS ONE* 4:e7771

114. Dai S, Wang Z, Pan X, Wang W, Chen X, et al. 2010. Functional analysis of promoter mutations in the *ACTN4* and *SYNPO* genes in focal segmental glomerulosclerosis. *Nephrol. Dial. Transplant.* 25:824–35
115. Morello R, Zhou G, Dreyer SD, Harvey SJ, Ninomiya Y, et al. 2001. Regulation of glomerular basement membrane collagen expression by LMX1B contributes to renal disease in nail patella syndrome. *Nat. Genet.* 27:205–8
116. Zenker M, Aigner T, Wendler O, Tralau T, Muntefering H, et al. 2004. Human laminin β2 deficiency causes congenital nephrosis with mesangial sclerosis and distinct eye abnormalities. *Hum. Mol. Genet.* 13:2625–32
117. Mele C, Iatropoulos P, Donadelli R, Calabria A, Maranta R, et al. 2011. *MYO1E* mutations and childhood familial focal segmental glomerulosclerosis. *N. Engl. J. Med.* 365:295–306
118. Sanna-Cherchi S, Burgess KE, Nees SN, Caridi G, Weng PL, et al. 2011. Exome sequencing identified *MYO1E* and *NEIL1* as candidate genes for human autosomal recessive steroid-resistant nephrotic syndrome. *Kidney Int.* 80:389–96
119. Togawa A, Miyoshi J, Ishizaki H, Tanaka M, Takakura A, et al. 1999. Progressive impairment of kidneys and reproductive organs in mice lacking Rho GDIα. *Oncogene* 18:5373–80
120. Moeller MJ, Soofi A, Braun GS, Li X, Watzl C, et al. 2004. Protocadherin FAT1 binds Ena/VASP proteins and is necessary for actin dynamics and cell polarization. *EMBO J.* 23:3769–79
121. Asanuma K, Kim K, Oh J, Giardino L, Chabanis S, et al. 2005. Synaptopodin regulates the actin-bundling activity of α-actinin in an isoform-specific manner. *J. Clin. Investig.* 115:1188–98
122. Asanuma K, Yanagida-Asanuma E, Faul C, Tomino Y, Kim K, Mundel P. 2006. Synaptopodin orchestrates actin organization and cell motility via regulation of RhoA signalling. *Nat. Cell Biol.* 8:485–91
123. Garg P, Verma R, Cook L, Soofi A, Venkatareddy M, et al. 2010. Actin-depolymerizing factor cofilin-1 is necessary in maintaining mature podocyte architecture. *J. Biol. Chem.* 285:22676–88
124. Kao WH, Klag MJ, Meoni LA, Reich D, Berthier-Schaad Y, et al. 2008. *MYH9* is associated with nondiabetic end-stage renal disease in African Americans. *Nat. Genet.* 40:1185–92
125. Kopp JB, Smith MW, Nelson GW, Johnson RC, Freedman BI, et al. 2008. *MYH9* is a major-effect risk gene for focal segmental glomerulosclerosis. *Nat. Genet.* 40:1175–84
126. Arrondel C, Vodovar N, Knebelmann B, Grunfeld JP, Gubler MC, et al. 2002. Expression of the nonmuscle myosin heavy chain IIA in the human kidney and screening for *MYH9* mutations in Epstein and Fechtner syndromes. *J. Am. Soc. Nephrol.* 13:65–74
127. Genovese G, Friedman DJ, Ross MD, Lecordier L, Uzureau P, et al. 2010. Association of trypanolytic ApoL1 variants with kidney disease in African Americans. *Science* 329:841–45
128. Okamoto K, Tokunaga K, Doi K, Fujita T, Suzuki H, et al. 2011. Common variation in *GPC5* is associated with acquired nephrotic syndrome. *Nat. Genet.* 43:459–63
129. Garg P, Verma R, Nihalani D, Johnstone DB, Holzman LB. 2007. Neph1 cooperates with nephrin to transduce a signal that induces actin polymerization. *Mol. Cell. Biol.* 27:8698–712
130. Kerjaschki D. 1978. Polycation-induced dislocation of slit diaphragms and formation of cell junctions in rat kidney glomeruli. The effects of low temperature, divalent cations, colchicine, and cytochalasin B. *Lab. Investig.* 39:430–40
131. Cybulsky AV, Bonventre JV, Quigg RJ, Lieberthal W, Salant DJ. 1990. Cytosolic calcium and protein kinase C reduce complement-mediated glomerular epithelial injury. *Kidney Int.* 38:803–11
132. Seiler MW, Rennke HG, Venkatachalam MA, Cotran RS. 1977. Pathogenesis of polycation-induced alterations ("fusion") of glomerular epithelium. *Lab. Investig.* 36:48–61
133. Seiler MW, Venkatachalam MA, Cotran RS. 1975. Glomerular epithelium: structural alterations induced by polycations. *Science* 189:390–93
134. Rudiger F, Greger R, Nitschke R, Henger A, Mundel P, Pavenstadt H. 1999. Polycations induce calcium signaling in glomerular podocytes. *Kidney Int.* 56:1700–9
135. Mundel P, Reiser J, Kriz W. 1997. Induction of differentiation in cultured rat and human podocytes. *J. Am. Soc. Nephrol.* 8:697–705
136. Henger A, Huber T, Fischer KG, Nitschke R, Mundel P, et al. 1997. Angiotensin II increases the cytosolic calcium activity in rat podocytes in culture. *Kidney Int.* 52:687–93

137. Oh J, Beckmann J, Bloch J, Hettgen V, Mueller J, et al. 2011. Stimulation of the calcium-sensing receptor stabilizes the podocyte cytoskeleton, improves cell survival, and reduces toxin-induced glomerulosclerosis. *Kidney Int.* 80:483–92
138. Schlondorff J, Del Camino D, Carrasquillo R, Lacey V, Pollak MR. 2009. TRPC6 mutations associated with focal segmental glomerulosclerosis cause constitutive activation of NFAT-dependent transcription. *Am. J. Physiol. Cell Physiol.* 296:58–69
139. Wang Y, Jarad G, Tripathi P, Pan M, Cunningham J, et al. 2010. Activation of NFAT signaling in podocytes causes glomerulosclerosis. *J. Am. Soc. Nephrol.* 21:1657–66
140. Kuwahara K, Wang Y, McAnally J, Richardson JA, Bassel-Duby R, et al. 2006. TRPC6 fulfills a calcineurin signaling circuit during pathologic cardiac remodeling. *J. Clin. Investig.* 116:3114–26
141. Onohara N, Nishida M, Inoue R, Kobayashi H, Sumimoto H, et al. 2006. TRPC3 and TRPC6 are essential for angiotensin II-induced cardiac hypertrophy. *EMBO J.* 25:5305–16
142. Wu X, Eder P, Chang B, Molkentin JD. 2010. TRPC channels are necessary mediators of pathologic cardiac hypertrophy. *Proc. Natl. Acad. Sci. USA* 107:7000–5
143. Chen J, Chen JK, Neilson EG, Harris RC. 2006. Role of EGF receptor activation in angiotensin II-induced renal epithelial cell hypertrophy. *J. Am. Soc. Nephrol.* 17:1615–23
144. Flannery PJ, Spurney RF. 2006. Transactivation of the epidermal growth factor receptor by angiotensin II in glomerular podocytes. *Nephron Exp. Nephrol.* 103:e109–18
145. Inagami T, Eguchi S, Numaguchi K, Motley ED, Tang H, et al. 1999. Cross-talk between angiotensin II receptors and the tyrosine kinases and phosphatases. *J. Am. Soc. Nephrol.* 10(Suppl. 11):57–61
146. Brenner BM, Cooper ME, de Zeeuw D, Keane WF, Mitch WE, et al. 2001. Effects of losartan on renal and cardiovascular outcomes in patients with type 2 diabetes and nephropathy. *N. Engl. J. Med.* 345:861–69
147. Lewis EJ, Hunsicker LG, Clarke WR, Berl T, Pohl MA, et al. 2001. Renoprotective effect of the angiotensin-receptor antagonist irbesartan in patients with nephropathy due to type 2 diabetes. *N. Engl. J. Med.* 345:851–60
148. Aspenstrom P, Fransson A, Saras J. 2004. Rho GTPases have diverse effects on the organization of the actin filament system. *Biochem. J.* 377:327–37
149. Etienne-Manneville S, Hall A. 2002. Rho GTPases in cell biology. *Nature* 420:629–35
150. Yanagida-Asanuma E, Asanuma K, Kim K, Donnelly M, Young Choi H, et al. 2007. Synaptopodin protects against proteinuria by disrupting Cdc42:IRSp53:Mena signaling complexes in kidney podocytes. *Am. J. Pathol.* 171:415–27
151. Krall P, Canales CP, Kairath P, Carmona-Mora P, Molina J, et al. 2010. Podocyte-specific overexpression of wild type or mutant Trpc6 in mice is sufficient to cause glomerular disease. *PLoS ONE* 5:e12859
152. Zhu L, Jiang R, Aoudjit L, Jones N, Takano T. 2011. Activation of RhoA in podocytes induces focal segmental glomerulosclerosis. *J. Am. Soc. Nephrol.* 22:1621–30
153. Dietrich A, Mederos YSM, Gollasch M, Gross V, Storch U, et al. 2005. Increased vascular smooth muscle contractility in $TRPC6^{-/-}$ mice. *Mol. Cell. Biol.* 25:6980–89
154. Hynes RO. 2002. Integrins: bidirectional, allosteric signaling machines. *Cell* 110:673–87
155. Arnaout MA, Mahalingam B, Xiong JP. 2005. Integrin structure, allostery, and bidirectional signaling. *Annu. Rev. Cell Dev. Biol.* 21:381–410
156. Kerjaschki D, Ojha PP, Susani M, Horvat R, Binder S, et al. 1989. A β1-integrin receptor for fibronectin in human kidney glomeruli. *Am. J. Pathol.* 134:481–89
157. Schordan S, Schordan E, Endlich K, Endlich N. 2011. αv-Integrins mediate the mechanoprotective action of osteopontin in podocytes. *Am. J. Physiol. Ren. Physiol.* 300:119–32
158. Raats CJ, van Den Born J, Bakker MA, Oppers-Walgreen B, Pisa BJ, et al. 2000. Expression of agrin, dystroglycan, and utrophin in normal renal tissue and in experimental glomerulopathies. *Am. J. Pathol.* 156:1749–65
159. Kambham N, Tanji N, Seigle RL, Markowitz GS, Pulkkinen L, et al. 2000. Congenital focal segmental glomerulosclerosis associated with β4 integrin mutation and epidermolysis bullosa. *Am. J. Kidney Dis.* 36:190–96
160. Wei C, El Hindi S, Li J, Fornoni A, Goes N, et al. 2011. Circulating urokinase receptor as a cause of focal segmental glomerulosclerosis. *Nat. Med.* 17:952–60

161. Savin VJ, Sharma R, Sharma M, McCarthy ET, Swan SK, et al. 1996. Circulating factor associated with increased glomerular permeability to albumin in recurrent focal segmental glomerulosclerosis. *N. Engl. J. Med.* 334:878–83
162. Fornoni A, Sageshima J, Wei C, Merscher-Gomez S, Aguillon-Prada R, et al. 2011. Rituximab targets podocytes in recurrent focal segmental glomerulosclerosis. *Sci. Transl. Med.* 3:85ra46
163. Blattner SM, Kretzler M. 2005. Integrin-linked kinase in renal disease: connecting cell-matrix interaction to the cytoskeleton. *Curr. Opin. Nephrol. Hypertens.* 14:404–10
164. El-Aouni C, Herbach N, Blattner SM, Henger A, Rastaldi MP, et al. 2006. Podocyte-specific deletion of integrin-linked kinase results in severe glomerular basement membrane alterations and progressive glomerulosclerosis. *J. Am. Soc. Nephrol.* 17:1334–44
165. Dai C, Stolz DB, Bastacky SI, St-Arnaud R, Wu C, et al. 2006. Essential role of integrin-linked kinase in podocyte biology: bridging the integrin and slit diaphragm signaling. *J. Am. Soc. Nephrol.* 17:2164–75
166. de Paulo V, Teixeira C, Blattner SM, Li M, Anders HJ, et al. 2005. Functional consequences of integrin-linked kinase activation in podocyte damage. *Kidney Int.* 67:514–23
167. Yang Y, Guo L, Blattner SM, Mundel P, Kretzler M, Wu C. 2005. Formation and phosphorylation of the PINCH-1-integrin linked kinase-α-parvin complex are important for regulation of renal glomerular podocyte adhesion, architecture, and survival. *J. Am. Soc. Nephrol.* 16:1966–76
168. Ma H, Togawa A, Soda K, Zhang J, Lee S, et al. 2010. Inhibition of podocyte FAK protects against proteinuria and foot process effacement. *J. Am. Soc. Nephrol.* 21:1145–56
169. Keir ME, Butte MJ, Freeman GJ, Sharpe AH. 2008. PD-1 and its ligands in tolerance and immunity. *Annu. Rev. Immunol.* 26:677–704
170. Wullschleger S, Loewith R, Hall MN. 2006. TOR signaling in growth and metabolism. *Cell* 124:471–84
171. Vollenbroker B, George B, Wolfgart M, Saleem MA, Pavenstadt H, Weide T. 2009. mTOR regulates expression of slit diaphragm proteins and cytoskeleton structure in podocytes. *Am. J. Physiol. Ren. Physiol.* 296:418–26
172. Biancone L, Bussolati B, Mazzucco G, Barreca A, Gallo E, et al. 2010. Loss of nephrin expression in glomeruli of kidney-transplanted patients under m-TOR inhibitor therapy. *Am. J. Transplant.* 10:2270–78
173. Godel M, Hartleben B, Herbach N, Liu S, Zschiedrich S, et al. 2011. Role of mTOR in podocyte function and diabetic nephropathy in humans and mice. *J. Clin. Investig.* 121:2197–209
174. Inoki K, Mori H, Wang J, Suzuki T, Hong S, et al. 2011. mTORC1 activation in podocytes is a critical step in the development of diabetic nephropathy in mice. *J. Clin. Investig.* 121:2181–96
175. Narita M, Young AR, Arakawa S, Samarajiwa SA, Nakashima T, et al. 2011. Spatial coupling of mTOR and autophagy augments secretory phenotypes. *Science* 332:966–70
176. Ashcroft FM. 2006. From molecule to malady. *Nature* 440:440–47
177. Call KM, Glaser T, Ito CY, Buckler AJ, Pelletier J, et al. 1990. Isolation and characterization of a zinc finger polypeptide gene at the human chromosome 11 Wilms' tumor locus. *Cell* 60:509–20
178. Gessler M, Poustka A, Cavenee W, Neve RL, Orkin SH, Bruns GA. 1990. Homozygous deletion in Wilms tumours of a zinc-finger gene identified by chromosome jumping. *Nature* 343:774–78
179. Huang A, Campbell CE, Bonetta L, McAndrews-Hill MS, Chilton-MacNeill S, et al. 1990. Tissue, developmental, and tumor-specific expression of divergent transcripts in Wilms tumor. *Science* 250:991–94
180. Bonetta L, Kuehn SE, Huang A, Law DJ, Kalikin LM, et al. 1990. Wilms tumor locus on 11p13 defined by multiple CpG island-associated transcripts. *Science* 250:994–97
181. Miner JH, Morello R, Andrews KL, Li C, Antignac C, et al. 2002. Transcriptional induction of slit diaphragm genes by Lmx1b is required in podocyte differentiation. *J. Clin. Investig.* 109:1065–72
182. Rohr C, Prestel J, Heidet L, Hosser H, Kriz W, et al. 2002. The LIM-homeodomain transcription factor Lmx1b plays a crucial role in podocytes. *J. Clin. Investig.* 109:1073–82
183. Takemoto M, He L, Norlin J, Patrakka J, Xiao Z, et al. 2006. Large-scale identification of genes implicated in kidney glomerulus development and function. *EMBO J.* 25:1160–74
184. Moriguchi T, Hamada M, Morito N, Terunuma T, Hasegawa K, et al. 2006. MafB is essential for renal development and F4/80 expression in macrophages. *Mol. Cell. Biol.* 26:5715–27

185. Dehbi M, Ghahremani M, Lechner M, Dressler G, Pelletier J. 1996. The paired-box transcription factor, PAX2, positively modulates expression of the Wilms' tumor suppressor gene (*WT1*). *Oncogene* 13:447–53
186. Ryan G, Steele-Perkins V, Morris JF, Rauscher FJ 3rd, Dressler GR. 1995. Repression of Pax-2 by WT1 during normal kidney development. *Development* 121:867–75
187. Clement LC, Liu G, Perez-Torres I, Kanwar YS, Avila-Casado C, Chugh SS. 2007. Early changes in gene expression that influence the course of primary glomerular disease. *Kidney Int.* 72:337–47
188. Liu G, Clement LC, Kanwar YS, Avila-Casado C, Chugh SS. 2006. ZHX proteins regulate podocyte gene expression during the development of nephrotic syndrome. *J. Biol. Chem.* 281:39681–92
189. Cui S, Li C, Ema M, Weinstein J, Quaggin SE. 2005. Rapid isolation of glomeruli coupled with gene expression profiling identifies downstream targets in Pod1 knockout mice. *J. Am. Soc. Nephrol.* 16:3247–55
190. Kumar PA, Kotlyarevska K, Dejkhmaron P, Reddy GR, Lu C, et al. 2010. Growth hormone (GH)-dependent expression of a natural antisense transcript induces zinc finger E-box-binding homeobox 2 (ZEB2) in the glomerular podocyte: a novel action of GH with implications for the pathogenesis of diabetic nephropathy. *J. Biol. Chem.* 285:31148–56
191. Ding M, Cui S, Li C, Jothy S, Haase V, et al. 2006. Loss of the tumor suppressor Vhlh leads to upregulation of Cxcr4 and rapidly progressive glomerulonephritis in mice. *Nat. Med.* 12:1081–87
192. Brukamp K, Jim B, Moeller MJ, Haase VH. 2007. Hypoxia and podocyte-specific *Vhlh* deletion confer risk of glomerular disease. *Am. J. Physiol. Ren. Physiol.* 293:1397–407
193. Freeburg PB, Robert B, St John PL, Abrahamson DR. 2003. Podocyte expression of hypoxia-inducible factor (HIF)-1 and HIF-2 during glomerular development. *J. Am. Soc. Nephrol.* 14:927–38
194. Heikkila E, Juhila J, Lassila M, Messing M, Perala N, et al. 2010. β-Catenin mediates adriamycin-induced albuminuria and podocyte injury in adult mouse kidneys. *Nephrol. Dial. Transplant.* 25:2437–46
195. He W, Kang YS, Dai C, Liu Y. 2011. Blockade of Wnt/β-catenin signaling by paricalcitol ameliorates proteinuria and kidney injury. *J. Am. Soc. Nephrol.* 22:90–103
196. Dreyer SD, Zhou G, Baldini A, Winterpacht A, Zabel B, et al. 1998. Mutations in *LMX1B* cause abnormal skeletal patterning and renal dysplasia in nail patella syndrome. *Nat. Genet.* 19:47–50
197. Pelletier J, Bruening W, Kashtan CE, Mauer SM, Manivel JC, et al. 1991. Germline mutations in the Wilm's tumor suppressor gene are associated with abnormal urogential development in Denys-Drash syndrome. *Cell* 67:437–47
198. Pelletier J, Bruening W, Li FP, Haber DA, Glaser T, Housman DE. 1991. *WT1* mutations contribute to abnormal genital system development and hereditary Wilms' tumour. *Nature* 353:431–34

A New Look at Electrolyte Transport in the Distal Tubule

Dominique Eladari,[1,2,3] Régine Chambrey,[1,2] and Janos Peti-Peterdi[4]

[1]Centre de Recherche des Cordeliers, Université Paris Descartes, INSERM UMRS 872, Equipe 3, F-75006, Paris, France; email: dominique.eladari@crc.jussieu.fr, regine.chambrey@crc.jussieu.fr

[2]Université Pierre et Marie Curie, CNRS ERL7226, F-75006, Paris, France

[3]Département de Physiologie, Hôpital Européen Georges Pompidou, Assistance Publique-Hôpitaux de Paris, F-75015, Paris, France

[4]Department of Physiology and Biophysics, Keck School of Medicine, Zilkha Neurogenetic Institute, University of Southern California, Los Angeles, California 90033; email: petipete@usc.edu

Keywords

pendrin, purinergic signaling, serine protease, hypertension, potassium, sodium, acid-base

Abstract

The distal nephron plays a critical role in the renal control of homeostasis. Until very recently most studies focused on the control of Na^+, K^+, and water balance by principal cells of the collecting duct and the regulation of solute and water by hormones from the renin-angiotensin-aldosterone system and by antidiuretic hormone. However, recent studies have revealed the unexpected importance of renal intercalated cells, a subtype of cells present in the connecting tubule and collecting ducts. Such cells were thought initially to be involved exclusively in acid-base regulation. However, it is clear now that intercalated cells absorb NaCl and K^+ and hence may participate in the regulation of blood pressure and potassium balance. The second paradigm-challenging concept we highlight is the emerging importance of local paracrine factors that play a critical role in the renal control of water and electrolyte balance.

INTRODUCTION

The composition of the body's different compartments is constant as a result of the dynamic and controlled exchange of solute and water between body cells and our surrounding environment. The kidney plays a central role in this function, termed electrolyte and water homeostasis. This equilibrium is achieved by controlling the amount of water and ions excreted into urine to exactly match the daily input of various substances due either to dietary intake or to metabolic production. Renal homeostasis is the result of the combination of two primary processes: glomerular filtration and tubular ion transport. The glomeruli filter daily very large amounts (~180 liters) of water and of ions (e.g., 25,000 mmol of Na^+, 4,300 mmol of HCO_3^-, 720 mmol of K^+). These quantities easily exceed those in the body. Accordingly, the renal tubule must absorb most substances filtered at the glomerulus to avoid their loss into urine. However, as indicated above, a small fraction of water and ions that exactly matches the daily input from diet and metabolism must also be excreted.

The nephron can be functionally divided into three different portions: the proximal part, which corresponds approximately to the proximal tubule; the loop of Henle; and the distal nephron. The first portion achieves massive absorption of water and solute. The second portion creates the cortico-papillary gradient of solute and osmoles in the interstitial space and hence plays a critical role in the control of transport of water and ions by the distal nephron. The distal tubule and the collecting ducts (CDs) are the final nephron segments that are sensitive to systemic hormones like aldosterone and vasopressin (AVP). Depending on these hormones' actions, these final nephron segments can reabsorb 10–15% of filtered salt and water. These tubule segments enable fine-tuning of systemic salt, water, K^+, and acid-base balance.

The distal nephron features cellular heterogeneity. The distal nephron begins after the macula densa and is composed of five successive segments: the distal convoluted tubule (DCT), the connecting tubule (CNT), the cortical collecting duct (CCD), the outer medullary collecting duct (OMCD), and the inner medullary collecting duct (IMCD) (1). Even though the CNT, the CCD, and the OMCD exhibit some functional differences and hence might be viewed as different nephron segments, they also share many characteristics because they are composed of very similar cell types (2–4). Thus, for the sake of simplicity, we hereafter refer to these nephron segments as the distal nephron. Several different cell types are identified along the distal nephron: DCT cells, principal cells (PCs), intercalated cells (ICs), and IMCD cells. Additional cell subtypes exist. For example, CNT cells and PCs differ with regard to Ca^{2+} transport mechanisms, which are present in the former but are absent in the latter. In mice, rats, and humans, but not in rabbits, a cell type with features in common with both DCT and CNT cells exists in the latter portion of the DCT. However, because all these three cell subtypes are characterized by the presence of the epithelial Na^+ channel ENaC, they are considered to be PCs in this review. ICs undergo transdifferentiation from an acid-secreting phenotype termed α-intercalated cells (α-ICs) to a base-secreting phenotype termed β-intercalated cells (β-ICs) (5–7). The mechanisms underlying this functional plasticity have recently been reviewed in detail (8). α-ICs secrete H^+ through an apical vacuolar H^+-ATPase (vH^+-ATPase) and reclaim HCO_3^- via the basolateral Cl^-/HCO_3^- exchanger AE1 (SLC4A1). β-ICs have the opposite polarity: They extrude H^+ through a basolateral vH^+-ATPase and secrete HCO_3^- into urine via the apical Cl^-/HCO_3^- exchanger pendrin (9–13). However, even if they represent a unique cell type, α-ICs and β-ICs represent two states of differentiation with completely opposite functions and express different $Cl^-/HCO3^-$ exchangers. Therefore, in this review we consider α-ICs and β-ICs to be two different cell types. There is at least one other subtype of ICs, termed the non-α, non-β IC, which expresses both the H^+ pump and the Cl^-/HCO_3^- exchanger pendrin at the apical membrane (11, 14). Whether these cells represent a transitional form of β-ICs to α-ICs or have a specific function is not known. For simplicity, we here use the term β-IC to refer to both canonical β-ICs and non-α, non-β ICs.

Our classical view of the distal nephron assumes that this cellular heterogeneity reflects a very high degree of functional specialization. For example, DCT cells mediate net reabsorption of NaCl and are impermeable to water. Therefore, DCT cells are critical for extracellular fluid volume and blood pressure regulation, as well as for urine dilution. In contrast, PCs mediate Na^+/K^+ exchange and have highly regulated water absorption. Thus, they are important for K^+ homeostasis and urine concentration. Finally, ICs can excrete either acid or base, depending on the acid-base status. Recent evidence challenges the view that hormone-regulatory loops activate or inhibit only one specific cell type and thus one specific function. For example, the role of ICs is not restricted to acid-base regulation because they can also transport NaCl and absorb K^+ and thus participate in both blood pressure and plasma K^+ regulation. Also, a large number of studies have examined the regulation of homeostasis by the classical hormones AVP and aldosterone, whereas important local autocrine and paracrine regulatory mechanisms that are based on the special tissue environment, diverse cell types, and unique cell metabolism of distal nephron epithelia have received much less attention.

In this article we highlight recently identified, emerging functions of ICs and paracrine mechanisms in the distal nephron that may play paradigm-shifting roles in renal physiology, body electrolyte and water homeostasis, and blood pressure control. Our mission is to depict the central role of ICs and to give the distal nephron a new face, depicting it as a complex, environmentally sensitive, semiautonomous nephron segment fully loaded with local sensory and regulatory machinery for the ultimate control of kidney functions and body homeostasis.

INTERCALATED CELLS ABSORB CHLORIDE AND PARTICIPATE IN BLOOD PRESSURE REGULATION

Two different molecular mechanisms of NaCl transport have been identified along the distal nephron. In the DCT, Na^+ and Cl^- are reabsorbed via the NaCl cotransporter NCC. First cloned by Gamba et al. (15) from the flounder bladder, NCC belongs to the superfamily of cation Cl^- cotransporters (16) and is encoded by the gene *SLC12A3* (17, 18). The importance of NCC in blood pressure regulation is attested by observations that inactivating mutations of *SLC12A3* in humans (19), or targeted disruption of *Slc12a3* in mice (20), cause Gitelman's syndrome, an inherited recessive disease characterized by low blood pressure. More recently, researchers showed that even a mild decrease in NCC activity observed in individuals harboring heterozygous inactivating mutations in the *SLC12A3* locus can significantly lower blood pressure (21). Moreover, NCC is the canonical target of thiazide diuretics, the oldest (22) but still one of the most effective therapies for human hypertension (23, 24).

In the CNT and the CD, NCC is not expressed, and Na^+ reabsorption occurs through ENaC, which is located in the PCs (3, 25). ENaC is a heteromultimeric channel (26) that belongs to the degenerin gene superfamily, which also includes channels involved in mechanotransduction or neurotransmission and the H^+-gated cation channel ASIC (27). In the mammalian kidney, ENaC has been characterized as a target for the diuretic amiloride (28, 29). Because ENaC is electrogenic, Na^+ absorption by PCs leads to the development of a lumen-negative transepithelial voltage, which is a critical determinant of K^+ secretion. Thus, the function of PCs is generally viewed as that of Na^+/K^+ exchange, even though both ion fluxes are not directly coupled by a single membrane transport protein (30–32). Because significant permeability of the paracellular pathway for Cl^- has been measured in isolated CCDs (32–35), until recently researchers believed (36) that Cl^- absorption by the CNT and CD is passive, occurring through the paracellular pathway, and is driven by the lumen-negative transepithelial voltage. However, several studies indicate that significant transcellular Cl^- absorption also exists (33). The primary mechanism for

Cl$^-$ transport in the CCD involves an electroneutral anion exchanger that can operate in either a Cl$^-$/Cl$^-$ self-exchange or a Cl$^-$/HCO$_3^-$ exchange mode (37–39). This transport activity has been detected exclusively in β-ICs (37, 40, 41) and appears to be directly linked to HCO$_3^-$ secretion (38, 42). The transepithelial movement of Cl$^-$ is effected by an apical Cl$^-$/HCO$_3^-$ exchanger, operating in series with basolateral Cl$^-$ conductance, that exhibits the same functional properties as does the voltage-gated chloride channel ClC-Kb (43).

A considerable advance in our understanding of Cl$^-$ transport by the CD originated from the identification of pendrin (SLC26A4) as the apical Cl$^-$/HCO$_3^-$ exchanger in β-ICs (12). Pendrin is not restricted to the kidney but is also expressed in the inner ear and thyroid. Pendrin is responsible for Pendred's syndrome, which is characterized by deafness associated with goiter (44). Initially described as a sulfate transporter on the basis of sequence homologies with SAT-1, a prototype of sulfate transporter (45), pendrin is not a sulfate transporter, but rather a Cl$^-$/anion exchanger (46). Pendrin has a very broad substrate specificity and can accept iodide, formate, or HCO$_3^-$ (47–49). In the inner ear, pendrin mediates Cl$^-$/HCO$_3^-$ exchange. Pendrin mediates HCO$_3^-$ secretion into the lumen of the cochlea, the vestibular labyrinth, and the endolymphatic sac of the inner ear (50–53). During embryonic development of the inner ear, pendrin participates in the net absorption of fluid in the endolymphatic sac (53). Loss of pendrin during the embryonic phase of development leads to an enlargement of the inner ear, which in the cochlea causes failure to develop a robust hearing phenotype, leading to deafness in children and in mouse models (52, 54, 55). The role of pendrin in the thyroid is less well understood. However, because pendrin is expressed in the apical membrane of thyrocytes and can exchange iodide for Cl$^-$, it may be important for the secretion of iodide into the colloid and for thyroid hormone organification (56, 57).

A study by Royaux et al. (12) elucidated the role of pendrin in the kidney. These authors showed that pendrin is confined to the apical domain of β-ICs. They further clarified that pendrin represents the long-sought molecular basis for apical Cl$^-$/HCO$_3^-$ exchange in β-ICs. Thus, disruption of *Slc26a4* abolishes HCO$_3^-$ secretion in CDs isolated from alkalotic animals (12). In addition, Wall et al. (58) subsequently reported that pendrin disruption not only impairs HCO$_3^-$ secretion but also totally abolishes Cl$^-$ reabsorption. They demonstrated that pendrin-mediated Cl$^-$ transport is the most important, if not the only, pathway for Cl$^-$ absorption in the collecting system (58).

That Cl absorption is not passive but rather occurs through β-ICs and involves a protein-mediated mechanism raises several important questions. (*a*) Is Cl$^-$ transport specifically regulated by changes in extracellular fluid volume or in Cl$^-$ balance? (*b*) Is Cl$^-$ transport an independent determinant of vascular fluid and blood pressure? (*c*) How are Cl$^-$ and Na$^+$ fluxes synchronized to yield net NaCl absorption because they are not operated by the same cell type?

We designed a study in which rats were treated by different challenges to manipulate the amount of Cl$^-$ excreted into urine, and we measured renal pendrin protein abundance by semiquantitative Western blot analyses. Rats administered 0.28 M NaCl, NH$_4$Cl, NaHCO$_3$, KCl, or KHCO$_3$ solution for 6 days or fed a low-NaCl diet exhibited an inverse relationship between pendrin expression and changes in Cl$^-$ excretion independently of the administered cation or of acid-base disturbances (59). Other investigators observed a decrease in pendrin in response to high renal Cl$^-$ delivery (60) or to dietary Cl$^-$ restriction (58, 61) in mice. Pendrin expression is also increased by angiotensin II (62) and by aldosterone (63).

Because the total NaCl body content is the critical determinant of the extracellular fluid volume and is under the tight control of the renin-angiotensin-aldosterone system (RAAS), the two latter studies suggested that pendrin is adapted to control vascular volume and thus blood pressure. In line with this hypothesis, pendrin knockout mice were found to be protected against deoxycorticosterone pivalate–induced hypertension (63). Conversely, following NaCl restriction, Cl^- excretion was excessive in pendrin knockout mice, leading to negative balance and hypotension (58). Taken together, the aforementioned studies demonstrate that pendrin is critically involved in maintaining Cl^- balance, responds to prohypertensive agents, and plays a critical role in blood pressure regulation by controlling distal Cl^- transport.

INTERCALATED CELLS MEDIATE THIAZIDE-SENSITIVE NaCl ABSORPTION

Prior to our studies of the role of ICs in extracellular fluid volume regulation, the only Na^+ transporter identified in the CD was ENaC. Because extracellular fluid volume is determined by NaCl and not only by Na^+, we expected ENaC and pendrin to function in parallel to mediate net NaCl absorption. However, we were puzzled by several observations. Using a model of acute thiazide administration to create a state of volume depletion with secondary hyperaldosteronism together with high urinary Cl^- delivery, we observed the expected increase in ENaC abundance. However, in this setting, we detected not an increase but rather a decrease in pendrin abundance (64). Using a mouse model of Liddle syndrome (65), we also did not detect any change in pendrin expression (64), even though ENaC-mediated Na^+ absorption was markedly increased in this model (66). Finally, targeted deletion of α-ENaC in the CD disrupts ENaC activity but does not impair the kidney's ability to maintain normal Na^+ balance (67).

However, previous studies indicated that an additional mechanism for Na^+ absorption might be present in the CD. Thus, bradykinin inhibits net Na^+ absorption by 50% without affecting K^+ secretion or the transepithelial voltage, suggesting that this hormone acts on an electroneutral Na^+ transporter rather than on ENaC (31, 68). Furthermore, pharmacological studies on isolated CCD segments demonstrated that two distinct Na^+ transporters are present in this nephron segment. The first transporter, mediated by ENaC, is highly amiloride sensitive, is electrogenic, and accounts for the transepithelial voltage that drives K^+ secretion. In contrast, the second transporter is electroneutral, is not affected by amiloride, and is thiazide sensitive (69). However, because other researchers did not confirm these observations (70), and because careful localization studies did not confirm the expression of NCC, the only target known to date for thiazide diuretics in CCD cells (71–73), the identity of this electroneutral, thiazide-sensitive Na^+ transporter remained unknown. Because of the availability of many genetically engineered mouse models with specific deletion of transporters, we reexamined the presence of this electroneutral NaCl transport pathway in isolated CCD segments (74).

When we simultaneously measured the transepithelial Na^+ (J_{Na}), K^+ (J_K), and Cl^- (J_{Cl}) fluxes and the transepithelial voltage (V_{te}) in isolated mouse CCDs microperfused in vitro, we detected two components of Na^+ absorption. The first component was abolished either by amiloride or by ENaC genetic deletion and accounted for both the negative V_{te} and K^+ secretion. In contrast, the second component was electroneutral and thiazide sensitive. Importantly, thiazide-sensitive NaCl absorption was present and even stimulated in $NCC^{-/-}$ mice (20). This finding indicates that such NaCl transport is operated by a mechanism that differs from NCC and that it may be one of the compensatory mechanisms that limit the effects of NCC deletion.

In many epithelia, Na^+-dependent Cl^- transport occurs through the functional coupling of a Cl^-/HCO_3^- exchanger and a Na^+/H^+ exchanger. Because genetic ablation of pendrin eliminates Cl^- absorption in the CCD, we hypothesized that thiazide-sensitive NaCl transport results from the coupling of pendrin-mediated Cl^- reabsorption with a H^+/HCO_3^--dependent Na^+ transporter. Indeed, by an approach combining intracellular pH measurements on isolated CCDs and different ion substitutions, we identified a thiazide-sensitive, Na^+-driven Cl^-/HCO_3^-

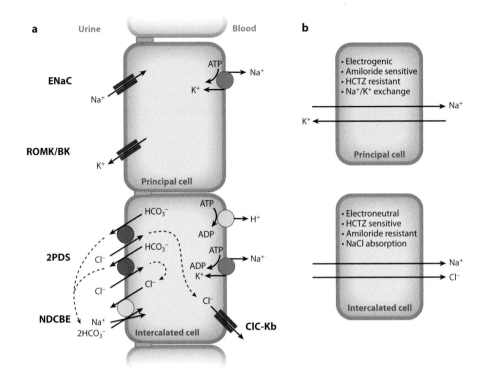

Figure 1

Schematic model of Na$^+$ transport in the collecting duct. (*a*) Cell models with the different transporters of the principal cells and intercalated cells. Two cycles of pendrin for one cycle of NDCBE are predicted to mediate net NaCl uptake with the recycling of one Cl$^-$ and two HCO$_3^-$ ions. Dashed lines indicate ion recycling. (*b*) Cell model summarizing the characteristic features of Na$^+$ transport in principal cells and intercalated cells. Abbreviations: BK, large-conductance, Ca^{2+}-activated potassium channel; ClC-Kb, voltage-gated chloride channel Kb; ENaC, epithelial Na$^+$ channel; NDCBE, Na$^+$-driven Cl$^-$/HCO$_3^-$ exchanger; PDS, pendrin; ROMK, renal outer medullary K$^+$ channel.

exchanger in the apical membrane of ICs. Many HCO$_3^-$ transporters have been reported, but only NDCBE/Slc4a8 and Slc4a10 have been reported to mediate Na$^+$- and Cl$^-$-dependent HCO$_3^-$ transport in mammals (75–77). However, we did not find any evidence for the presence of Slc4a10 in the mouse CCD, whereas we detected both Slc4a8 mRNA and protein in mouse CCD. Furthermore, NDCBE deletion in mice abrogated amiloride-resistant electroneutral NaCl absorption in isolated CCD segments. This confirms that NDCBE operates in tandem with pendrin to mediate NaCl absorption (74). Thus, we identified a specific novel NaCl transport system along the nephron that is expressed by ICs (**Figure 1**). This finding indicates that the role of these cells is not limited to acid-base transport but also encompasses NaCl reabsorption and, accordingly, blood pressure regulation.

Pendrin is critically required to mediate net NaCl uptake by ICs. In vivo experiments in which we administrated thiazide to *NCC*$^{-/-}$ mice demonstrated that this novel system mediates a significant fraction of renal NaCl absorption (see figure 4 in Reference 74). Moreover, pendrin has emerged as an important player in renal NaCl conservation and in blood pressure regulation. Recent data suggest that this is not the only function of pendrin. In fact, pendrin disruption not only affects IC function but also impairs ENaC function and molecular regulation in the adjacent PC (78, 79). This finding indicates strong functional cooperation between the different cell types within the CCD.

PHYSIOLOGY OF ATP-RELEASING MEMBRANE (HEMI)CHANNELS

The aforementioned study of Pech et al. (78) proposed that ICs can modulate PC function through changes in HCO_3^- secretion. However, this is not the only paracrine signaling pathway that links IC and PC activities. One of the emerging new topics in renal transport physiology is the role of purinergic signaling in the regulation of tubular salt and water reabsorption and K^+ secretion. In particular, the field of P2 purinergic receptors that bind extracellular (released) ATP has advanced significantly. According to the current view, a local purinergic system intrinsic to the distal nephron provides a powerful control of ENaC- and aquaporin 2 (AQP2)-mediated Na^+ and water excretion. This purinergic system involves ectonucleotidases (ATP-degrading enzymes) and purinergic receptors. Due to the availability of $P2Y_2$ knockout mice, $P2Y_2$ receptors in the luminal plasma membrane of PCs provide most of our knowledge on purinergic receptors in this context. Mice lacking the $P2Y_2$ receptor have impaired Na^+ excretion, increased blood pressure, and consequently reduced plasma renin, aldosterone, and K^+ (80). The renal physiological role of the $P2Y_2$ receptor and the purinergic regulation of renal salt and water transport were extensively studied and reviewed recently (81–90). Here we focus on a new development in this purinergic system, namely, the mechanisms by which ATP, the ligand, is released from renal epithelial cells into the tubular lumen.

Studies in several organs suggest that connexin (Cx) proteins, the building blocks of gap junctions, can form hemichannels, or connexons, which allow ATP release into the extracellular fluid (91, 92). Inspired by these results, our laboratories first localized Cx30 (93) and then Cx30.3 (94) to the apical, nonjunctional plasma membrane of specific cells in the distal nephron. A detailed review of the localization and function of connexins in the kidney is available (95). The initial studies found that both Cx30 and Cx30.3 immunolabeling in the mouse kidney was particularly strong and limited to the apical membrane of ICs in the CNT and CCD. A high-salt diet caused upregulation of Cx30 expression (93). The distinct, continuous labeling of the IC luminal plasma membrane and upregulation by high salt suggest that Cx30 may function as a hemichannel involved in the paracrine regulation of salt reabsorption in the distal nephron.

Subsequent work by Sipos et al. (96) used an ATP biosensor approach to demonstrate that functional Cx30 hemichannels mediate luminal ATP release in the intact, microperfused CCD and control salt and water transport by purinergic signaling. Cx30 hemichannel opening at the luminal cell membrane of the CCD required mechanical stimulation by either increased tubular fluid flow rate or an interstitium-to-lumen osmotic pressure gradient and resulted in significant amounts of released ATP (up to 50 μM) in the luminal microenvironment. Consistent with preferential Cx30 localization in ICs, ATP biosensor responses were threefold larger in ICs versus PCs. Confirming the physiological significance of Cx30, genetic deletion of Cx30 markedly reduced flow-induced luminal ATP release and impaired salt and water excretion associated with pressure natriuresis, an important mechanism that maintains body fluid and electrolyte balance and blood pressure (96). Thus, $Cx30^{-/-}$ mice express a salt retention phenotype due to the hyperactive Cx30-expressing CD. Benzamil, a specific inhibitor of ENaC, prevents the salt-sensitive hypertension observed in $Cx30^{-/-}$ mice (96).

Additional whole-animal studies using $Cx30^{-/-}$ mice confirmed the importance of Cx30-mediated ATP release in the control of renal salt excretion by testing the effects of changes in dietary salt intake. Studies by Mironova et al. (97) established that urinary ATP levels increase with high Na$^+$ intake in wild-type mice. In contrast, urinary ATP in Cx30$^{-/-}$ mice was lower and less dependent on dietary salt. Loss of inhibitory ATP regulation in Cx30$^{-/-}$ mice also caused high ENaC open probability, particularly with high Na$^+$ intake. The level of ENaC expression in Cx30$^{-/-}$ mice was not altered (96). Together with the observed increase in ENaC open probability rather than in the number of Na$^+$ channels (97), this last finding suggests that ATP functions in a paracrine manner. Importantly, the loss of paracrine ATP feedback regulation of ENaC in Cx30$^{-/-}$ mice disrupts normal responses to changes in Na$^+$ intake. Thus, Cx30$^{-/-}$ mice have lower Na$^+$ excretion in states of positive Na$^+$ balance. Moreover, the loss of the ability of ENaC to respond to changes in Na$^+$ levels causes salt-sensitive hypertension in Cx30$^{-/-}$ mice (97).

The finding that reduced salt excretion in Cx30$^{-/-}$ mice is most apparent in high-salt conditions is consistent with several observations. First, in normal mice, a high-salt diet upregulates Cx30 expression (93), and the greatest feedback inhibition of ENaC by Cx30-mediated ATP release is observed in animals receiving a high-salt diet (97). Moreover, clamping mineralocorticoids high in Cx30$^{-/-}$ mice fed a high-Na$^+$ diet caused a marked decline in renal Na$^+$ excretion. Such responses are in contrast to normal physiological responses in wild-type mice that are capable of undergoing aldosterone escape, which refers to increased salt excretion during high Na$^+$ intake despite inappropriately increased levels of mineralocorticoids (98). These studies not only provide novel insights into aldosterone research but further support the key importance of the local CD purinergic system, including the role of Cx30 in renal control of salt transport.

On the basis of these data, the disruption of Cx30 function seems to be equivalent to the loss of the P2Y$_2$ receptor (80, 84, 85) in terms of ENaC control. This conclusion is not surprising because the P2Y$_2$ receptor is localized downstream of Cx30-mediated ATP release, linking the inhibitory ATP signal to ENaC. Thus, several lines of evidence point to Cx30 as an integral component of a local purinergic signaling system within the distal nephron. It provides insights into a dietary salt–sensitive negative feedback mechanism for the control of renal salt excretion via ATP release from ICs and ATP's paracrine purinergic actions on PCs targeting ENaC.

Purinergic regulation of K$^+$ secretion in the distal nephron is quite complex, with different types of K$^+$ channels involved. Extracellular ATP is a potent inhibitor of the small-conductance K$^+$ (SK) channel located in the apical plasma membrane of PCs, suggesting that ATP in the tubular fluid inhibits net K$^+$ secretion (99). However, the large-conductance, Ca^{2+}-activated K$^+$ (BK) channel, which is localized predominantly in the luminal membrane of ICs, seems also to be regulated by ATP released via connexin hemichannels. Holtzclaw et al. (100) found recently that pharmacological inhibition of connexins or P2 receptors in ICs diminished flow-induced K$^+$ efflux through BK channels as well as ATP secretion. These investigators also showed that K$^+$ efflux and ATP release from ICs are interdependent because downregulation of the BK-β4 subunit using siRNA decreased ATP secretion and both ATP-dependent and flow-induced K$^+$ efflux. In mice with high distal nephron flows induced by a high-K$^+$ diet, the genetic disruption of BK channel function resulted in significantly reduced urinary ATP excretion. The observed Na$^+$ and water retention as well as decreased K$^+$ secretion in BK-β4$^{-/-}$ mice are consistent with reduced purinergic tone (101). Because ICs have not traditionally been considered to play a major role in K$^+$ secretion, further studies are needed to clarify the autocrine/paracrine actions of ATP on ICs and PCs and their possible interactions.

The physiological importance of ATP–BK channel interaction may be related to the need to maintain high distal flow rates; maximal K$^+$ secretion and minimal Na$^+$ reabsorption occur when animals consume a high-K$^+$ diet. High flow may be augmented by the release of ATP, which inhibits Na$^+$ reabsorption (via ENaC) and water reabsorption (via AQP2) yet at the same time stimulates K$^+$ secretion. This newly described BK channel function may help us to understand the so-called aldosterone paradox. This term refers to the apparently opposite effects of aldosterone,

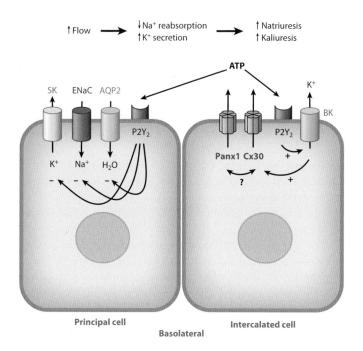

Figure 2

Novel elements and proposed function of the intratubular autocrine/paracrine purinergic system in the collecting duct. Abbreviations: AQP2, aquaporin 2 water channel; BK, large-conductance, Ca^{2+}-activated potassium channel; Cx30, connexin 30 hemichannel; ENaC, epithelial Na^+ channel; $P2Y_2$, ATP receptor subtype $P2Y_2$; Panx1, pannexin 1 channel; SK, small-conductance K^+ channel.

salt reabsorption (e.g., in hypovolemia), and K^+ secretion (e.g., in hyperkalemia). Investigators recently reviewed the molecular mechanisms mediating the aldosterone paradox, that is, the interdependent but differential regulation of renal Na^+ and K^+ transport, including the role of the serine/threonine kinases WNK1, WNK4, SPAK, and OSR1 (102). The local purinergic system in the distal nephron may also be involved via modulation of tubular flow. In volume depletion, aldosterone favors salt retention, with no change in K^+ secretion (low tubular flow, low ATP levels, inactive purinergic inhibition, and high ENaC but low BK channel activity). In contrast, in hyperkalemia aldosterone helps to excrete K^+ without affecting salt reabsorption (high tubular flow, high levels of ATP release, powerful purinergic inhibition of ENaC, high BK-mediated K^+ secretion). The precise mechanism by which ATP affects the activity of ROMK and maxi-K^+ channels as well as purinergically mediated activity needs further investigation.

On the basis of these findings, a more complete picture of the local intratubular purinergic system has emerged (**Figure 2**). Mechanically induced opening of Cx30 hemichannels in ICs (e.g., by tubular flow, interstitial pressure, cell volume changes) allows ATP to enter the tubular fluid and to bind to local P2 receptors in both ICs and PCs in an autocrine fashion and a paracrine fashion, respectively. Activation of these receptors inhibits salt and water reabsorption in PCs via ENaC and AQP2 and facilitates K^+ secretion via BK channels. Sustained activity of this purinergic system appears to participate in the homeostasis of body fluid and electrolytes and blood pressure in a variety of physiological conditions, including pressure natriuresis, high-salt diet, aldosterone escape, the aldosterone paradox, and high-K^+ diet.

However, the idea that functional hemichannels are formed solely by connexins under physiological conditions is controversial (103). Then how does Cx30 mediate ATP release from ICs in the CD? Such release may occur due to the close association of Cx30 with the recently identified plasma membrane ATP channel pannexin 1 (Panx1) (**Figure 2**). Pannexins are a class of proteins that also form hexameric transmembrane channels that are similar to those formed by connexins (104). Despite having similar structures, connexins and pannexins have no sequence homology. In several cell types, including epithelial cells, Panx1 forms functional, mechanosensitive ATP-releasing channels under physiological conditions (104, 105). Interestingly, a preliminary study from our laboratory (106) found strong Panx1 expression in the apical membrane of the CD system, raising a question about its role alongside that of Cx30. Future work will address the exciting possibility of Cx30-Panx1 interactions in the CD.

$Cx30^{-/-}$ mice are deaf due to inner-ear dysfunction (107). Similarities in regulatory mechanisms and common ion transporters in the renal CD and the inner ear are well known (108). Consistent with this, these deaf animals also possess a renal phenotype (i.e., salt retention), as detailed above, on the basis of the dysfunction of Cx30 in the renal CD system. A potential interaction between Cx30 and pendrin may also be part of the local purinergic system, as suggested by their colocalization in non-α ICs (93) and the similar phenotype of pendrin knockout mice (Pendred syndrome, deafness) (12). Recently, a genetic link between Bartter syndrome (which causes renal salt wasting) and sensorineural deafness in humans was established (109). Unlike in the mouse kidney, in other species Cx30 is expressed in the thick ascending limb and distal tubule (93). The findings suggest that Cx30-mediated ATP release is an important physiological regulatory mechanism in many organs and species, including humans. ATP release mediated by connexin hemichannels (and other associated proteins) and the complex regulation of water and electrolyte transport mechanisms by a local purinergic signaling system appear to be a well-preserved mechanosensory and regulatory machinery in various organs and cell types.

THE COLLECTING DUCT RENIN-ANGIOTENSIN SYSTEM AND NOVEL METABOLIC REGULATORS

The RAAS is a key regulator of renal salt and water reabsorption whose molecular and cellular effects in various nephron segments have been extensively studied. Here we briefly emphasize two important developments concerning paracrine regulation of distal nephron function. First, a local, distal nephron's RAAS has been established and characterized (110, 111). This local system is now recognized as a key anatomical site of the overall RAAS in pathological states including diabetes and as the major source of systemic RAAS components, for example, (pro)renin (110, 112). The local generation of angiotensin peptides in the tubular fluid of the CD and their stimulatory effects on salt reabsorption via ENaC have been demonstrated (113, 114). The importance of angiotensin II generated de novo by this local intratubular RAAS in overall renal salt and water excretion as well as its interaction with the systemic RAAS (e.g., aldosterone) need to be further studied.

Another emerging field is the direct metabolic control of distal nephron function by metabolic intermediates in the tubular fluid, for example, succinate and α-ketoglutarate. The special cell metabolism (high anaerobic glycolysis) and tissue environment (hypoxia) under which some cells of the distal nephron operate are highly consistent with the local accumulation of these dicarboxylates (115). The G protein–coupled receptors GPR91 and GPR99, newly identified specific cell membrane receptors for succinate and α-ketoglutarate, respectively, are localized predominantly in the luminal cell membrane of the distal nephron–CD system and have been linked to the activation of RAAS (116–119). GPR91 signaling involves the generation of prostaglandins (e.g., PGE_2), which is highly relevant to the control of renal salt and water transport (117–119). In addition, elements of an olfactory signaling system along the apical surface of the distal nephron have been discovered (120). Although the activating ligand is not known, the idea that cells of the distal nephron and CD can "smell" certain chemical signals in the tubular fluid that may directly regulate electrolyte transport is exciting and will undoubtedly open new areas in kidney research.

ROLE OF TISSUE KALLIKREIN IN POTASSIUM HOMEOSTASIS

Another example of paracrine regulation is the modulation of K^+ transport in the distal nephron by tissue kallikrein (TK, *klk1* gene; number NM_010639), a serine protease involved in kinin generation. In the kidney, TK is synthesized in large amounts by CNT cells and, to a lesser extent, by the DCT and CCD (121, 122). Although TK can be released into the circulation, active TK is found predominantly in the urine (123). TK release leads to kinin formation. TK cleaves locally available kininogen to yield kallidin (lysyl-bradykinin), which is converted through the action of an N-aminopeptidase to bradykinin. Bradykinin in turn activates the bradykinin type 2 (B2) receptor. B2 receptors have been localized to both the luminal side and the basal side of CD cells (124). Kinins have potent vasodilatory properties, and intrarenal infusions of kinins increase renal blood flow and produce diuresis and natriuresis (125). This response is probably due both to kinin-induced redistribution of medullary renal blood flow (medullary washout of solute) and to direct effects on Na^+ and water transport across the CD. In fact, kinins (bradykinin and kallidin) inhibit Na^+ and water transport when applied to the basolateral surface but not to the luminal surface of the CD (31, 68, 126). Although a kinin-dependent effect of TK on renal Na^+ and water transport is well established, the role of the large amount of TK that is secreted into the urinary fluid remains poorly understood.

MECHANISMS FOR THE RELEASE OF RENAL TISSUE KALLIKREIN: EFFECT OF POTASSIUM

A recent study performed in humans by Azizi et al. (127) confirmed earlier observations on the influence of dietary K^+ intake on urinary TK activity. A diet with high K^+ and low Na^+ content increased whereas a diet with low K^+ and high Na^+ content decreased TK activity and excretion (127). Thus, the distal nephron segments responsible for regulation of K^+ excretion (i.e., late distal tubules, CNTs, and CCDs) are exposed to variable amounts of TK. Because maneuvers modulating urinary kallikrein activity alter K^+ balance, it is tempting to postulate that the controlled upstream release of TK from the apical membranes of CNT cells into the tubular fluid by K^+ may lead to autocrine and paracrine action of TK on distal tubular K^+ transport. Early studies indicate that K^+ intake stimulates TK synthesis and excretion directly and/or through secondary mechanisms (125, 128, 129). Aldosterone may also influence the effect of K^+ on the urinary excretion of renal TK because high K^+ intake stimulates the secretion of aldosterone, which has a stimulatory effect on TK synthesis and excretion. However, distinct effects of mineralocorticoids on TK synthesis and excretion have been reported (130, 131). In fact, acute administration of aldosterone does not induce the synthesis of renal TK, whereas chronic treatment with deoxycorticosterone or adrenalectomy increases and decreases, respectively, TK synthesis in the CNTs and TK secretion in the urine.

The concept that K^+ regulates renal TK has been reconsidered. In a recent study, we showed that a high-K^+ diet increases to a similar extent renal TK secretion in both control and aldosterone synthase–deficient mice, indicating that aldosterone is not required for the stimulatory effect of K^+ intake on TK secretion (132). We also showed that a single meal, which represents an acute K^+ load, increases renal K^+ and TK excretion in parallel, whereas aldosterone secretion is not significantly modified. An early increase in renal TK secretion, which occurs after intravenous infusion of K^+, was also reported in rats (133, 134). Administration of ATP-sensitive K^+ channel

(K_{ATP}) blockers also increased renal kallikrein secretion in superfused slices of kidney cortex (134). Thus, K^+ and K_{ATP} channel blockers may increase renal kallikrein secretion through the same mechanism. A decrease in membrane potential caused by high concentrations of extracellular K^+ has been reported for renal tubular cells (135), and K^+ channel blockers such as barium Cl^- produced the same effect (135).

El Moghrabi et al. (132) showed that acute K^+ loading rendered *TK*-deficient mice hyperkalemic, whereas this maneuver did not lead to any significant change in plasma [K^+] in wild-type mice. The authors concluded that TK is regulated by extracellular K^+ independently of aldosterone and in turn protects the organism against an acute K^+ load. A typical meal of a Western-style diet represents a K^+ load equivalent to half of the total amount of K^+ in the extracellular fluid. As a consequence, every meal represents an acute and massive K^+ load with which the body has to cope (136). Renal TK represents a new factor that protects the body against postfeeding hyperkalemia besides insulin or β-adrenergic agonists, both of which stimulate the rapid transfer of K^+ from the extracellular space to the intracellular space (137). Although K^+ balance is under the tight control of the mineralocorticoid hormone aldosterone, which activates and fine-tunes K^+ secretion by the CD, aldosterone, whose effects are mostly transcriptional, is not suitable for such an acute adaptation.

TISSUE KALLIKREIN AND RENAL POTASSIUM HANDLING

K^+ is freely filtered at the glomerulus and is almost totally reabsorbed by the proximal tubule and the loop of Henle. The amount of K^+ excreted by the kidney is determined by events beyond the early distal tubule, where either reabsorption or secretion of K^+ can occur. Generally, a large amount of K^+ has to be eliminated to maintain balance. K^+ secretion occurs mainly through the PCs and involves the following: (*a*) cellular K^+ entry across the basolateral membrane via the Na^+,K^+-ATPase pump and (*b*) K^+ exit across the apical membrane via K^+ channels, allowing K^+ secretion into the lumen. In PCs, Na^+ reabsorption is energized by the basolateral Na^+,K^+-ATPase, which exchanges three Na^+ for two K^+. The resulting electrochemical gradient for Na^+ is dissipated mainly by apical Na^+ entry via ENaC, which depolarizes the apical membrane and facilitates apical K^+ exit via the K^+ channel ROMK. Na^+ transport via ENaC is thus an important determinant for K^+ secretion through ROMK. Accordingly, amiloride administration decreased urinary K^+ excretion. Under conditions in which the kidney must retain K^+, such as a situation of K^+ depletion, the CD reverses the transepithelial net flux of K^+ to yield net K^+ reabsorption (136). K^+ absorption is localized to the ICs in the CCD and is mediated by H^+,K^+-ATPases, which reabsorb K^+ in exchange for H^+ ions (138). Net secretion may be the result of simultaneous bidirectional secretory and reabsorptive K^+ fluxes.

If one does not consider a situation of K^+ depletion in which net K^+ absorption occurs, the CCD is a site of net K^+ secretion. Remarkably, microperfused CCDs isolated from *TK*-deficient mice (139) exhibit net transepithelial K^+ absorption (132). TK may also modulate K^+ transport by favoring ENaC-mediated Na^+ absorption that, in turn, is expected to favor renal K^+ secretion and/or by inhibiting K^+ absorption through H^+,K^+-ATPase.

TISSUE KALLIKREIN AND PROTEOLYTIC ACTIVATION OF THE EPITHELIAL SODIUM CHANNEL

In 1997, the channel-activating protease 1 (CAP1, also referred as prostasin), a membrane-bound serine protease, was identified in *Xenopus* kidney A6 cells on the basis of its ability to activate ENaC when coexpressed in *Xenopus* oocytes (140). From this study emerged the concept that a variety of serine proteases can activate ENaC. Such proteases include (*a*) extracellular serin proteases [e.g., trypsin; plasmin; and elastase, which is a serine protease produced by neutrophils (141–144)], (*b*) intracellular serin proteases [e.g., furin, a serine protease located in the *trans*-Golgi network (145)], and (*c*) membrane-bound serine proteases [e.g., members of the CAP family (146–148)], all of which act through their catalytic activity to increase the open probability of ENaC (141–143, 147).

ENaC is a heteromultimeric protein composed of three related subunits: α, β, and γ. All subunits have an intracytoplasmic N terminus and C terminus and two membrane-spanning domains connected by a large extracellular loop. Proteolytic cleavage of ENaC occurs for the α subunit and the γ subunit (γ-ENaC) in vitro. Hughey et al. (149, 150) demonstrated that furin activates ENaC by cleaving the channel in the ectodomain at two sites in the α subunit and at a single site within γ-ENaC. There is increasing in vitro evidence that ENaC activation by serine proteases is related to proteolytic cleavage of the extracellular domain of γ-ENaC at sites distal to the furin cleavage site (for review, see References 144 and 151–155). Masilamani et al. (156) suggest that endogenous serine proteases may cleave γ-ENaC in vivo. However, the serine protease(s) that mediates this process in vivo has not been identified, including in the kidney.

Endogenous membrane-bound and/or secreted serine proteases such as CAPs are likely candidates for protease-mediated regulation of Na^+ transport in the lung (157, 158). Accordingly, gene disruption of CAP1 specifically in the alveolar epithelium using conditional Cre-loxP-mediated recombination revealed that CAP1 plays a crucial role in the regulation of ENaC-mediated alveolar Na^+ and water transport and in mouse lung fluid balance (158). CAP1 is coexpressed with ENaC in Na^+-transporting epithelia, including the kidney (140), where CAP1 is present in native and cultured CCD cells (148). CAP1 is detectable in mouse exosomes, a preparation that reflects the pool of protein apically expressed in renal tubular cells (159), suggesting that CAP1 is present at the apical surface of renal cells (R. Chambrey & B. Rossier, unpublished data). Activation of ENaC by aldosterone has been consistently associated with a shift in the molecular weight of renal γ-ENaC from 85 to 70 kDa, consistent with physiological cleavage of the extracellular loop of that subunit (156, 160, 161). Aldosterone increases the expression of prostasin, the human ortholog for CAP1, in vitro and in vivo (162). Narikiyo et al. (162) showed that urinary excretion of prostasin is inversely correlated with the urinary $Na^+:K^+$ ratio in patients with primary aldosteronism, suggesting that prostasin may be responsible for aldosterone-dependent proteolytic ENaC activation. Gene disruption of CAP1 specifically in the CD will be necessary for direct assessment of its physiological role in ENaC regulation.

Because CNT cells secrete a large amount of TK into the tubular fluid, the luminal membranes of renal tubular cells starting from the CNT are highly exposed to TK. Thus, TK, through its catalytic activity, may act directly on ENaC in the apical membranes of tubular cells located downstream of the site of TK secretion to modulate ENaC activity.

Picard et al. (161) recently reported that *TK*-deficient mice have decreased ENaC activity in the kidney and the distal colon, two organs that exhibit ENaC-mediated Na^+ transport and produce abundant amounts of TK, whereas in lung alveolae, an organ devoid of TK expression, TK disruption does not impair ENaC activity. In this study, reduced ENaC activity in *TK*-deficient mice correlated with a dramatic decrease in expression of the 70-kDa truncated form of γ-ENaC in both renal cortical membrane and urinary exosome preparations. In contrast, in the lung, TK disruption did not impair ENaC processing, confirming that in lung alveolae a protease other than TK, such as CAP1, regulates ENaC, as recently described by Planès et al. (158).

When applied to the luminal surface of CCDs isolated and microperfused in vitro, TK has a marked stimulatory effect on ENaC-dependent Na^+ entry (161). In vitro experiments using either commercially available TK or mouse TK from desalted urine demonstrated that TK promotes the

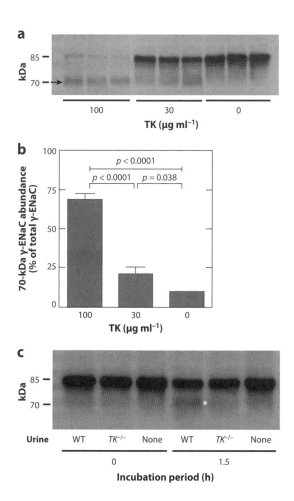

Figure 3

Tissue kallikrein (TK) promotes cleavage of the epithelial Na^+ channel (ENaC) γ subunit (γ-ENaC) in vitro. (*a*) Renal membrane fractions from *TK*-deficient ($TK^{-/-}$) mice were incubated with TK, and appearance of the 70-kDa form of γ-ENaC was studied by immunoblotting. Membrane proteins were mixed with either 0, 30, or 100 μg ml^{-1} TK purified from porcine pancreas and were incubated for 15 min at 37°C. (*b*) Densitometric analyses of data showing the appearance of the 70-kDa form after treatment with TK. (*c*) Renal membrane fractions from $TK^{-/-}$ mice were incubated with desalted and concentrated urine from wild-type (WT) or $TK^{-/-}$ mice for 1.5 h at 37°C. The appearance of the 70-kDa form of γ-ENaC was studied by immunoblotting. Modified from Reference 161.

cleavage of γ-ENaC (**Figure 3**) (161). Thus, urinary TK promotes in vivo and in vitro γ-ENaC cleavage, which is associated with ENaC activation.

Interestingly, we also observed that elevated circulating aldosterone levels did not interfere with ENaC processing and activation in the absence of TK (161). Thus, serine proteases other than TK, presumably CAPs (see above), may promote the cleavage and activation of γ-ENaC in the kidneys of TK-deficient mice with elevated circulating aldosterone. These data suggest that TK is not critical for protease-mediated regulation of Na^+ absorption in the CD.

Although ENaC activity is reduced in the absence of TK, CCDs from *TK*-deficient mice absorb Na^+ without affecting the transepithelial potential difference, indicating that Na^+ absorption occurs through an electroneutral process (74, 132). Early studies showed that bradykinin inhibits Na^+ absorption without affecting net K^+ transport or the transepithelial potential difference in isolated and microperfused CDs through the activation of the basolateral B2 receptor (68). *TK* gene inactivation in the mouse not only disrupts TK production but also impairs local kinin production (139). Thus, the absence of TK upregulates ENaC-independent electroneutral NaCl absorption, presumably through a decrease in the local production of bradykinin, and prevents proteolytic activation of ENaC (161). The combination of these two opposite effects on Na^+ transport explains why *TK*-deficient mice do not have a significantly altered Na^+ balance (139, 161). In conclusion, TK-dependent ENaC activation is not critical for the regulation of renal Na^+ homeostasis but rather plays an important role in maintaining K^+ balance.

TISSUE KALLIKREIN AND THE COLONIC H^+,K^+-ATPASE

As discussed above, CCDs isolated from *TK*-deficient mice exhibit net transepithelial K^+ absorption (132). Because *TK*-deficient mice display decreased ENaC activity, a shutdown of K^+ secretion likely contributes to net K^+ absorption in the CDs of these mice. However, these data do not rule out the possibility that H^+,K^+-ATPase activity is also involved by further reducing urinary K^+ losses.

We refer the reader to the review by Gumz et al. (163) for background on H^+,K^+-ATPases. Briefly, the H^+,K^+-ATPases consist of a catalytic α subunit and a regulatory β subunit. Researchers have identified two renal H^+,K^+-ATPases that contain the catalytic subunit $HK\alpha_1$ or $HK\alpha_2$. The $HK\alpha_1$ H^+,K^+-ATPase, or gastric H^+,K^+-ATPase, has a role in acidifying stomach content. The $HK\alpha_2$ H^+,K^+-ATPase, termed the colonic H^+,K^+-ATPase, is abundant in and cloned from the colon. Using $HK\alpha_2$-deficient mice, Meneton et al. (164) demonstrated that under K^+ deprivation $HK\alpha_2$ plays a critical role in the maintenance of K^+ homeostasis in vivo.

K^+ depletion, which causes renal K^+ absorption, is associated with an increase in kidney $HK\alpha_2$ mRNA, suggesting that the colonic type of the H^+,K^+-ATPase plays an important role in K^+ conservation by the kidney. Accordingly, the colonic type of the H^+,K^+-ATPase mediates increased HCO_3^- absorption in the CD during K^+ depletion (165, 166). H^+,K^+-ATPase activity estimated by the initial recovery rate of intracellular pH after an acute intracellular acid load was higher in CD ICs from *TK*-deficient mice. Acid extrusion in ICs from TK-deficient mice was inhibited either by 30 μM SCH28080, a specific inhibitor of H^+,K^+-ATPases, or by 1 mM ouabain. These pharmacological properties correspond to those described for the colonic type of the H^+,K^+-ATPase (167). Consistently, mRNA for the $HK\alpha_2$ subunit but not that for the $HK\alpha_1$ subunit of the H^+,K^+-ATPase was markedly increased in CNTs and CCDs of *TK*-deficient mice (132). Conversely, in CCDs isolated from *TK*-deficient mice and microperfused in vitro, the addition of TK to the luminal surface but not to the basolateral surface of the tubule caused a 70% inhibition of H^+,K^+-ATPase activity. These results demonstrate that the absence of TK stimulates the expression and activity of the colonic H^+,K^+-ATPase and results in net K^+ absorption by the CCD. Under conditions of K^+ excretion, TK exerts a negative braking effect on K^+ absorption by inhibiting the H^+,K^+-ATPase. The mechanism or signaling pathway by which luminal TK controls H^+,K^+-ATPase remains unknown. As mentioned above, serine proteases act directly on ENaC gating by enhancing the open probability of ENaC. Recent studies show that TK can also directly activate the B2 receptor independently of bradykinin release (168). TK, which colocalizes with the epithelial Ca^{2+} channel TRPV5 in the distal part of the nephron, activates TRPV5 from the luminal side (169). Using primary cultures of CNT/CCD, Gkika et al. (169) showed that TK, like bradykinin, stimulates TRPV5 indirectly via activation of the B2 receptor. This induces the

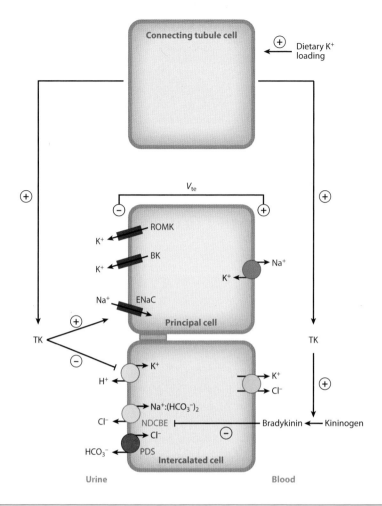

Figure 4

Schematic model depicting how tissue kallikrein (TK) participates in the response to dietary K^+ loading. Dietary K^+ loading stimulates TK production by connecting tubule cells. TK is then released into the urinary fluid and reaches the cortical collecting duct. There, TK may favor K^+ secretion by principal cells through TK's stimulating action on epithelial Na^+ channel (ENaC)-mediated Na^+ absorption, an effect that occurs through proteolytic processing of the γ-ENaC subunit (161). TK also inhibits K^+ absorption by intercalated cells by decreasing colonic H^+,K^+-ATPase expression and activity. Moreover, TK upregulates ENaC-independent electroneutral NaCl absorption, presumably through a decrease in the local production of bradykinin (68). Other abbreviations: BK, large-conductance, Ca^{2+}-activated potassium channel; NDCBE, Na^+-driven Cl^-/HCO_3^- exchanger; PDS, pendrin; ROMK, renal outer medullary K^+ channel. From Reference 132.

phosphorylation of TRPV5 through a protein kinase C–dependent mechanism, with consequent redistribution of TRPV5 channels toward the plasma membrane.

The experimental data summarized in this part indicate that TK regulates K^+ transport by two mechanisms: (*a*) by favoring ENaC-mediated Na^+ transport that, in turn, is expected to favor renal K^+ secretion and (*b*) by inhibiting K^+ absorption through the colonic H^+,K^+-ATPase (**Figure 4**). Thus, TK is a kaliuretic factor that provides rapid and aldosterone-independent protection against hyperkalemia after a dietary K^+ load. Importantly, these data also suggest that K^+ absorption through the colonic H^+,K^+-ATPase is under the negative control of TK in the distal nephron.

CONCLUSIONS

There has been significant progress in our understanding of the molecular mechanisms involved in the renal control of homeostasis. Recent studies in particular have highlighted the emerging importance of paracrine regulation of ion transport and cross talk between nephron segments and different cell types within a single nephron segment. The identification of several novel ion transport mechanisms and local signaling pathways in the distal nephron that we summarize in this review will help us to further understand the pathophysiology of several diseases. Various past studies that focused on the classical hormonal regulation of the distal nephron and its abnormalities repeatedly failed to identify a unifying molecular mechanism. This failure is particularly evident for disorders that affect Na^+ balance, such as hypertension and nephrotic syndrome. Another significant development that will challenge the current concepts in renal homeostasis is the emerging importance of ICs in Na^+ and K^+ balance. This is highlighted by the strong modifier effects of pendrin disruption on blood pressure regulation (61, 63) and by the identification of a novel therapeutic target of thiazide in these cells (74). Another perhaps paradigm-shifting finding is the identification of ICs as an important source of paracrine factors that regulate diverse functions of the CD, particularly those of the adjacent PCs. Therefore, future studies should reevaluate the precise role and involvement of ICs in disorders that were initially thought to originate from PC dysfunction.

SUMMARY POINTS

1. In contrast to the existing paradigm, intercalated cells not only are involved in acid-base transport but also are important for NaCl and K^+ absorption. Accordingly, these cells participate in the regulation of extracellular fluid volume and blood K^+ concentration.

2. Cl^- transport in the collecting duct occurs through intercalated cells rather than through the paracellular pathway and is an important determinant of blood pressure.

3. Intercalated cells are the primary source of ATP that is released into the tubular fluid (urine), and hence intercalated cells play a central role in purinergic signaling in the distal nephron.

4. Disruption of connexin30 impairs ATP release and purinergic signaling within the distal nephron, leading to the loss of negative feedback control of ENaC that results in excessive salt retention and ultimately salt-sensitive hypertension.

5. Tissue kallikrein is a serine protease abundantly released into urine by connecting tubule cells. It stimulates K^+ secretion by principal cells indirectly through proteolytic activation of ENaC and inhibits K^+ absorption by the H^+,K^+-ATPase in intercalated cells.

FUTURE ISSUES

1. In light of the new ion transport and paracrine signaling mechanisms that have been identified in the distal nephron, the role of intercalated cells and principal cells and their interactions in the regulation of NaCl and K^+ transport need to be further investigated.

2. The involvement of intercalated cells in disorders characterized by abnormal regulation of blood pressure (particularly Cl^--dependent hypertension like pseudohypoaldosteronism type II) should be reevaluated.

3. The physiological importance of the local purinergic regulation of collecting duct K^+ secretion (via BK and SK channels) and water reabsorption (via aquaporin 2) needs to be determined in vivo (for example, by using connexin30 knockout mice).

4. On the basis of the functional similarities in molecular mechanisms between the renal collecting duct and the inner ear, future clinical studies may identify the diagnostic potential of newborn hearing tests as a predictor of kidney disease and hypertension.

5. Whether tissue kallikrein modulates H^+,K^+-ATPase activity through proteolysis or through another mechanism remains to be determined.

DISCLOSURE STATEMENT

The authors are not aware of any affiliations, memberships, funding, or financial holdings that might be perceived as affecting the objectivity of this review.

ACKNOWLEDGMENTS

Dominique Eladari and Régine Chambrey are funded by INSERM, by CNRS, by the Transatlantic Network for Hypertension (TNH) from the Fondation Leducq, and by grants ANR PHYSIO 2007-RPV07084 and ANR BLANC 2010-R10164DD to Dr. Eladari from l'Agence Nationale de la Recherche (ANR). Janos Peti-Peterdi is funded by NIH grants DK64324 and DK74754, by American Diabetes Association Research grant 1-11-BS-121, and by American Heart Association Established Investigator award 0640056N.

LITERATURE CITED

1. Kriz W, Kaissling B. 2000. Structural organization of the distal nephron. In *The Kidney: Physiology and Pathophysiology*, ed. DW Seldin, G Giebisch, pp. 587–654. Philadelphia, PA: Lippincott
2. Bachmann S, Bostanjoglo M, Schmitt R, Ellison DH. 1999. Sodium transport-related proteins in the mammalian distal nephron—distribution, ontogeny and functional aspects. *Anat. Embryol.* 200:447–68
3. Loffing J, Kaissling B. 2003. Sodium and calcium transport pathways along the mammalian distal nephron: from rabbit to human. *Am. J. Physiol. Ren. Physiol.* 284:628–43
4. Meneton P, Loffing J, Warnock DG. 2004. Sodium and potassium handling by the aldosterone-sensitive distal nephron: the pivotal role of the distal and connecting tubule. *Am. J. Physiol. Ren. Physiol.* 287:593–601
5. Schwartz GJ, Tsuruoka S, Vijayakumar S, Petrovic S, Mian A, Al-Awqati Q. 2002. Acid incubation reverses the polarity of intercalated cell transporters, an effect mediated by hensin. *J. Clin. Investig.* 109:89–99
6. Schwartz GJ, Barasch J, Al-Awqati Q. 1985. Plasticity of functional epithelial polarity. *Nature* 318:368–71
7. Gao X, Eladari D, Leviel F, Tew BY, Miro-Julia C, et al. 2010. Deletion of hensin/DMBT1 blocks conversion of β- to α-intercalated cells and induces distal renal tubular acidosis. *Proc. Natl. Acad. Sci. USA* 107:21872–77
8. Al-Awqati Q. 2011. Terminal differentiation in epithelia: the role of integrins in hensin polymerization. *Annu. Rev. Physiol.* 73:401–12
9. Brown D, Hirsch S, Gluck S. 1988. An H^+-ATPase in opposite plasma membrane domains in kidney epithelial cell subpopulations. *Nature* 331:622–24
10. Alper SL, Natale J, Gluck S, Lodish HF, Brown D. 1989. Subtypes of intercalated cells in rat kidney collecting duct defined by antibodies against erythroid band 3 and renal vacuolar H^+-ATPase. *Proc. Natl. Acad. Sci. USA* 86:5429–33

11. Kim J, Kim YH, Cha JH, Tisher CC, Madsen KM. 1999. Intercalated cell subtypes in connecting tubule and cortical collecting duct of rat and mouse. *J. Am. Soc. Nephrol.* 10:1–12
12. Royaux IE, Wall SM, Karniski LP, Everett LA, Suzuki K, et al. 2001. Pendrin, encoded by the Pendred syndrome gene, resides in the apical region of renal intercalated cells and mediates bicarbonate secretion. *Proc. Natl. Acad. Sci. USA* 98:4221–26
13. Wall SM, Hassell KA, Royaux IE, Green ED, Chang JY, et al. 2003. Localization of pendrin in mouse kidney. *Am. J. Physiol. Ren. Physiol.* 284:229–41
14. Teng-umnuay P, Verlander JW, Yuan W, Tisher CC, Madsen KM. 1996. Identification of distinct subpopulations of intercalated cells in the mouse collecting duct. *J. Am. Soc. Nephrol.* 7:260–74
15. Gamba G, Saltzberg SN, Lombardi M, Miyanoshita A, Lytton J, et al. 1993. Primary structure and functional expression of a cDNA encoding the thiazide-sensitive, electroneutral sodium-chloride cotransporter. *Proc. Natl. Acad. Sci. USA* 90:2749–53
16. Gamba G. 2005. Molecular physiology and pathophysiology of electroneutral cation-chloride cotransporters. *Physiol. Rev.* 85:423–93
17. Gamba G, Miyanoshita A, Lombardi M, Lytton J, Lee WS, et al. 1994. Molecular cloning, primary structure, and characterization of two members of the mammalian electroneutral sodium-(potassium)-chloride cotransporter family expressed in kidney. *J. Biol. Chem.* 269:17713–22
18. Hebert SC, Gamba G. 1994. Molecular cloning and characterization of the renal diuretic-sensitive electroneutral sodium-(potassium)-chloride cotransporters. *Clin. Investig.* 72:692–94
19. Simon DB, Nelson-Williams C, Bia MJ, Ellison D, Karet FE, et al. 1996. Gitelman's variant of Bartter's syndrome, inherited hypokalaemic alkalosis, is caused by mutations in the thiazide-sensitive Na-Cl cotransporter. *Nat. Genet.* 12:24–30
20. Schultheis PJ, Lorenz JN, Meneton P, Nieman ML, Riddle TM, et al. 1998. Phenotype resembling Gitelman's syndrome in mice lacking the apical Na^+-Cl^- cotransporter of the distal convoluted tubule. *J. Biol. Chem.* 273:29150–55
21. Ji W, Foo JN, O'Roak BJ, Zhao H, Larson MG, et al. 2008. Rare independent mutations in renal salt handling genes contribute to blood pressure variation. *Nat. Genet.* 40:592–99
22. Beyer KH Jr, Baer JE, Russo HF, Noll R. 1958. Electrolyte excretion as influenced by chlorothiazide. *Science* 127:146–47
23. ALLHAT Off. Coord. ALLHAT Collab. Res. Group. 2002. Major outcomes in high-risk hypertensive patients randomized to angiotensin-converting enzyme inhibitor or calcium channel blocker versus diuretic: the Antihypertensive and Lipid-Lowering Treatment to Prevent Heart Attack Trial (ALLHAT). *JAMA* 288:2981–97
24. Appel LJ. 2002. The verdict from ALLHAT–thiazide diuretics are the preferred initial therapy for hypertension. *JAMA* 288:3039–42
25. Chang SS, Grunder S, Hanukoglu A, Rosler A, Mathew PM, et al. 1996. Mutations in subunits of the epithelial sodium channel cause salt wasting with hyperkalaemic acidosis, pseudohypoaldosteronism type 1. *Nat. Genet.* 12:248–53
26. Canessa CM, Schild L, Buell G, Thorens B, Gautschi I, et al. 1994. Amiloride-sensitive epithelial Na^+ channel is made of three homologous subunits. *Nature* 367:463–67
27. Kellenberger S, Schild L. 2002. Epithelial sodium channel/degenerin family of ion channels: a variety of functions for a shared structure. *Physiol. Rev.* 82:735–67
28. Qadri YJ, Song Y, Fuller CM, Benos DJ. 2010. Amiloride docking to acid-sensing ion channel-1. *J. Biol. Chem.* 285:9627–35
29. Bicking JB, Mason JW, Woltersdorf OW Jr, Jones JH, Kwong SF, et al. 1965. Pyrazine diuretics. I. N-Amidino-3-amino-6-halopyrazinecarboxamides. *J. Med. Chem.* 638–42
30. Frindt G, Burg MB. 1972. Effect of vasopressin on sodium transport in renal cortical collecting tubules. *Kidney Int.* 1:224–31
31. Tomita K, Pisano JJ, Knepper MA. 1985. Control of sodium and potassium transport in the cortical collecting duct of the rat. Effects of bradykinin, vasopressin, and deoxycorticosterone. *J. Clin. Investig.* 76:132–36
32. O'Neil RG, Helman SI. 1977. Transport characteristics of renal collecting tubules: influences of DOCA and diet. *Am. J. Physiol. Ren. Physiol.* 233:544–58

33. Sansom SC, Weinman EJ, O'Neil RG. 1984. Microelectrode assessment of chloride-conductive properties of cortical collecting duct. *Am. J. Physiol. Ren. Physiol.* 247:291–302
34. Warden DH, Schuster VL, Stokes JB. 1988. Characteristics of the paracellular pathway of rabbit cortical collecting duct. *Am. J. Physiol. Ren. Physiol.* 255:720–27
35. O'Neil RG, Boulpaep EL. 1982. Ionic conductive properties and electrophysiology of the rabbit cortical collecting tubule. *Am. J. Physiol. Ren. Physiol.* 243:81–95
36. Hou J, Renigunta A, Yang J, Waldegger S. 2010. Claudin-4 forms paracellular chloride channel in the kidney and requires claudin-8 for tight junction localization. *Proc. Natl. Acad. Sci. USA* 107:18010–15
37. Schuster VL, Stokes JB. 1987. Chloride transport by the cortical and outer medullary collecting duct. *Am. J. Physiol. Ren. Physiol.* 253:203–12
38. Tago K, Schuster VL, Stokes JB. 1986. Stimulation of chloride transport by HCO_3^--CO_2 in rabbit cortical collecting tubule. *Am. J. Physiol. Ren. Physiol.* 251:49–56
39. Star RA, Burg MB, Knepper MA. 1985. Bicarbonate secretion and chloride absorption by rabbit cortical collecting ducts. Role of chloride/bicarbonate exchange. *J. Clin. Investig.* 76:1123–30
40. Schuster VL, Bonsib SM, Jennings ML. 1986. Two types of collecting duct mitochondria-rich (intercalated) cells: lectin and band 3 cytochemistry. *Am. J. Physiol. Cell Physiol.* 251:347–55
41. Emmons C, Kurtz I. 1994. Functional characterization of three intercalated cell subtypes in the rabbit outer cortical collecting duct. *J. Clin. Investig.* 93:417–23
42. Matsuzaki K, Stokes JB, Schuster VL. 1989. Stimulation of Cl^- self exchange by intracellular HCO_3^- in rabbit cortical collecting duct. *Am. J. Physiol. Cell Physiol.* 257:94–101
43. Nissant A, Paulais M, Lachheb S, Lourdel S, Teulon J. 2006. Similar chloride channels in the connecting tubule and cortical collecting duct of the mouse kidney. *Am. J. Physiol. Ren. Physiol.* 290:1421–29
44. Pendred V. 1896. Deaf-mutism and goitre. *Lancet* 2:532
45. Everett LA, Glaser B, Beck JC, Idol JR, Buchs A, et al. 1997. Pendred syndrome is caused by mutations in a putative sulphate transporter gene (PDS). *Nat. Genet.* 17:411–22
46. Kraiem Z, Heinrich R, Sadeh O, Shiloni E, Nassir E, et al. 1999. Sulfate transport is not impaired in Pendred syndrome thyrocytes. *J. Clin. Endocrinol. Metab.* 84:2574–76
47. Scott DA, Karniski LP. 2000. Human pendrin expressed in *Xenopus laevis* oocytes mediates chloride/formate exchange. *Am. J. Physiol. Cell Physiol.* 278:207–11
48. Scott DA, Wang R, Kreman TM, Sheffield VC, Karniski LP. 1999. The Pendred syndrome gene encodes a chloride-iodide transport protein. *Nat. Genet.* 21:440–43
49. Soleimani M, Greeley T, Petrovic S, Wang Z, Amlal H, et al. 2001. Pendrin: an apical Cl^-/OH^-/HCO_3^- exchanger in the kidney cortex. *Am. J. Physiol. Ren. Physiol.* 280:356–64
50. Wangemann P, Nakaya K, Wu T, Maganti RJ, Itza EM, et al. 2007. Loss of cochlear HCO_3^- secretion causes deafness via endolymphatic acidification and inhibition of Ca^{2+} reabsorption in a Pendred syndrome mouse model. *Am. J. Physiol. Ren. Physiol.* 292:1345–53
51. Nakaya K, Harbidge DG, Wangemann P, Schultz BD, Green ED, et al. 2007. Lack of pendrin HCO_3^- transport elevates vestibular endolymphatic $[Ca^{2+}]$ by inhibition of acid-sensitive TRPV5 and TRPV6 channels. *Am. J. Physiol. Ren. Physiol.* 292:1314–21
52. Kim HM, Wangemann P. 2011. Epithelial cell stretching and luminal acidification lead to a retarded development of stria vascularis and deafness in mice lacking pendrin. *PLoS ONE* 6:e17949
53. Kim HM, Wangemann P. 2010. Failure of fluid absorption in the endolymphatic sac initiates cochlear enlargement that leads to deafness in mice lacking pendrin expression. *PLoS ONE* 5:e14041
54. Wangemann P, Kim HM, Billings S, Nakaya K, Li X, et al. 2009. Developmental delays consistent with cochlear hypothyroidism contribute to failure to develop hearing in mice lacking Slc26a4/pendrin expression. *Am. J. Physiol. Ren. Physiol.* 297:1435–47
55. Everett LA, Belyantseva IA, Noben-Trauth K, Cantos R, Chen A, et al. 2001. Targeted disruption of mouse *Pds* provides insight about the inner-ear defects encountered in Pendred syndrome. *Hum. Mol. Genet.* 10:153–61
56. Bizhanova A, Kopp P. 2009. Minireview: the sodium-iodide symporter NIS and pendrin in iodide homeostasis of the thyroid. *Endocrinology* 150:1084–90
57. Bizhanova A, Kopp P. 2010. Genetics and phenomics of Pendred syndrome. *Mol. Cell Endocrinol.* 322:83–90

58. Wall SM, Kim YH, Stanley L, Glapion DM, Everett LA, et al. 2004. NaCl restriction upregulates renal Slc26a4 through subcellular redistribution: role in Cl^- conservation. *Hypertension* 44:982–87
59. Quentin F, Chambrey R, Trinh-Trang-Tan MM, Fysekidis M, Cambillau M, et al. 2004. The Cl^-/HCO_3^- exchanger pendrin in the rat kidney is regulated in response to chronic alterations in chloride balance. *Am. J. Physiol. Ren. Physiol.* 287:1179–88
60. Hafner P, Grimaldi R, Capuano P, Capasso G, Wagner CA. 2008. Pendrin in the mouse kidney is primarily regulated by Cl^- excretion but also by systemic metabolic acidosis. *Am. J. Physiol. Cell Physiol.* 295:1658–67
61. Verlander JW, Kim YH, Shin W, Pham TD, Hassell KA, et al. 2006. Dietary Cl^- restriction upregulates pendrin expression within the apical plasma membrane of type B intercalated cells. *Am. J. Physiol. Ren. Physiol.* 291:833–39
62. Pech V, Zheng W, Pham TD, Verlander JW, Wall SM. 2008. Angiotensin II activates H^+-ATPase in type A intercalated cells. *J. Am. Soc. Nephrol.* 19:84–91
63. Verlander JW, Hassell KA, Royaux IE, Glapion DM, Wang ME, et al. 2003. Deoxycorticosterone upregulates PDS (Slc26a4) in mouse kidney: role of pendrin in mineralocorticoid-induced hypertension. *Hypertension* 42:356–62
64. Vallet M, Picard N, Loffing-Cueni D, Fysekidis M, Bloch-Faure M, et al. 2006. Pendrin regulation in mouse kidney primarily is chloride-dependent. *J. Am. Soc. Nephrol.* 17:2153–63
65. Pradervand S, Wang Q, Burnier M, Beermann F, Horisberger JD, et al. 1999. A mouse model for Liddle's syndrome. *J. Am. Soc. Nephrol.* 10:2527–33
66. Pradervand S, Vandewalle A, Bens M, Gautschi I, Loffing J, et al. 2003. Dysfunction of the epithelial sodium channel expressed in the kidney of a mouse model for Liddle syndrome. *J. Am. Soc. Nephrol.* 14:2219–28
67. Rubera I, Loffing J, Palmer LG, Frindt G, Fowler-Jaeger N, et al. 2003. Collecting duct-specific gene inactivation of αENaC in the mouse kidney does not impair sodium and potassium balance. *J. Clin. Investig.* 112:554–65
68. Tomita K, Pisano JJ, Burg MB, Knepper MA. 1986. Effects of vasopressin and bradykinin on anion transport by the rat cortical collecting duct. Evidence for an electroneutral sodium chloride transport pathway. *J. Clin. Investig.* 77:136–41
69. Terada Y, Knepper MA. 1990. Thiazide-sensitive NaCl absorption in rat cortical collecting duct. *Am. J. Physiol. Ren. Physiol.* 259:519–28
70. Rouch AJ, Chen L, Troutman SL, Schafer JA. 1991. Na^+ transport in isolated rat CCD: effects of bradykinin, ANP, clonidine, and hydrochlorothiazide. *Am. J. Physiol. Ren. Physiol.* 260:86–95
71. Plotkin MD, Kaplan MR, Verlander JW, Lee WS, Brown D, et al. 1996. Localization of the thiazide sensitive Na-Cl cotransporter, rTSC1, in the rat kidney. *Kidney Int.* 50:174–83
72. Bachmann S, Velazquez H, Obermuller N, Reilly RF, Moser D, Ellison DH. 1995. Expression of the thiazide-sensitive Na-Cl cotransporter by rabbit distal convoluted tubule cells. *J. Clin. Investig.* 96:2510–14
73. Obermuller N, Bernstein P, Velazquez H, Reilly R, Moser D, et al. 1995. Expression of the thiazide-sensitive Na-Cl cotransporter in rat and human kidney. *Am. J. Physiol. Ren. Physiol.* 269:900–10
74. Leviel F, Hubner CA, Houillier P, Morla L, El Moghrabi S, et al. 2010. The Na^+-dependent chloride-bicarbonate exchanger SLC4A8 mediates an electroneutral Na^+ reabsorption process in the renal cortical collecting ducts of mice. *J. Clin. Investig.* 120:1627–35
75. Damkier HH, Aalkjaer C, Praetorius J. 2010. Na^+-dependent HCO_3^- import by the *slc4a10* gene product involves Cl^- export. *J. Biol. Chem.* 285:26998–7007
76. Grichtchenko II, Choi I, Zhong X, Bray-Ward P, Russell JM, Boron WF. 2001. Cloning, characterization, and chromosomal mapping of a human electroneutral Na^+-driven $Cl-HCO_3$ exchanger. *J. Biol. Chem.* 276:8358–63
77. Parker MD, Musa-Aziz R, Rojas JD, Choi I, Daly CM, Boron WF. 2008. Characterization of human SLC4A10 as an electroneutral Na/HCO_3 cotransporter (NBCn2) with Cl^- self-exchange activity. *J. Biol. Chem.* 283:12777–88
78. Pech V, Pham TD, Hong S, Weinstein AM, Spencer KB, et al. 2010. Pendrin modulates ENaC function by changing luminal HCO_3. *J. Am. Soc. Nephrol.* 21:1928–41

79. Kim YH, Pech V, Spencer KB, Beierwaltes WH, Everett LA, et al. 2007. Reduced ENaC protein abundance contributes to the lower blood pressure observed in pendrin-null mice. *Am. J. Physiol. Ren. Physiol.* 293:1314–24
80. Rieg T, Bundey RA, Chen Y, Deschenes G, Junger W, et al. 2007. Mice lacking $P2Y_2$ receptors have salt-resistant hypertension and facilitated renal Na^+ and water reabsorption. *FASEB J.* 21:3717–26
81. Bailey M, Shirley D. 2009. Effects of extracellular nucleotides on renal tubular solute transport. *Purinergic Signal.* 5:473–80
82. Garvin JL, Herrera M, Ortiz PA. 2011. Regulation of renal NaCl transport by nitric oxide, endothelin, and ATP: clinical implications. *Annu. Rev. Physiol.* 73:359–76
83. Kishore BK, Chou CL, Knepper MA. 1995. Extracellular nucleotide receptor inhibits AVP-stimulated water permeability in inner medullary collecting duct. *Am. J. Physiol. Ren. Physiol.* 269:863–69
84. Pochynyuk O, Bugaj V, Rieg T, Insel PA, Mironova E, et al. 2008. Paracrine regulation of the epithelial Na^+ channel in the mammalian collecting duct by purinergic $P2Y_2$ receptor tone. *J. Biol. Chem.* 283:36599–607
85. Pochynyuk O, Rieg T, Bugaj V, Schroth J, Fridman A, et al. 2010. Dietary Na^+ inhibits the open probability of the epithelial sodium channel in the kidney by enhancing apical $P2Y_2$-receptor tone. *FASEB J.* 24:2056–65
86. Rieg T, Vallon V. 2009. ATP and adenosine in the local regulation of water transport and homeostasis by the kidney. *Am. J. Physiol. Regul. Integr. Comp. Physiol.* 296:419–27
87. Vallon V. 2008. P2 receptors in the regulation of renal transport mechanisms. *Am. J. Physiol. Ren. Physiol.* 294:10–27
88. Wildman SSP, Boone M, Peppiatt-Wildman CM, Contreras-Sanz A, King BF, et al. 2009. Nucleotides downregulate aquaporin 2 via activation of apical P2 receptors. *J. Am. Soc. Nephrol.* 20:1480–90
89. Zhang Y, Kohan DE, Nelson RD, Carlson NG, Kishore BK. 2010. Potential involvement of $P2Y_2$ receptor in diuresis of postobstructive uropathy in rats. *Am. J. Physiol. Ren. Physiol.* 298:634–42
90. Zhang Y, Nelson RD, Carlson NG, Kamerath CD, Kohan DE, Kishore BK. 2009. Potential role of purinergic signaling in lithium-induced nephrogenic diabetes insipidus. *Am. J. Physiol. Ren. Physiol.* 296:1194–201
91. Ebihara L. 2003. New roles for connexons. *Physiology* 18:100–3
92. Evans WH, De Vuyst E, Leybaert L. 2006. The gap junction cellular internet: Connexin hemichannels enter the signalling limelight. *Biochem. J.* 397:1–14
93. McCulloch F, Chambrey R, Eladari D, Peti-Peterdi J. 2005. Localization of connexin 30 in the luminal membrane of cells in the distal nephron. *Am. J. Physiol. Ren. Physiol.* 289:1304–12
94. Hanner F, Schnichels M, Zheng-Fischhöfer Q, Yang LE, Toma I, et al. 2008. Connexin 30.3 is expressed in the kidney but not regulated by dietary salt or high blood pressure. *Cell Commun. Adhes.* 15:219–30
95. Hanner F, Sorensen CM, Holstein-Rathlou N-H, Peti-Peterdi J. 2010. Connexins and the kidney. *Am. J. Physiol. Regul. Integr. Comp. Physiol.* 298:1143–55
96. Sipos A, Vargas SL, Toma I, Hanner F, Willecke K, Peti-Peterdi J. 2009. Connexin 30 deficiency impairs renal tubular ATP release and pressure natriuresis. *J. Am. Soc. Nephrol.* 20:1724–32
97. Mironova E, Peti-Peterdi J, Bugaj V, Stockand JD. 2011. Diminished paracrine regulation of the epithelial Na^+ channel by purinergic signaling in mice lacking connexin 30. *J. Biol. Chem.* 286:1054–60
98. Stockand JD, Mironova E, Bugaj V, Rieg T, Insel PA, et al. 2010. Purinergic inhibition of ENaC produces aldosterone escape. *J. Am. Soc. Nephrol.* 21:1903–11
99. Lu M, MacGregor GG, Wang W, Giebisch G. 2000. Extracellular ATP inhibits the small-conductance K channel on the apical membrane of the cortical collecting duct from mouse kidney. *J. Gen. Physiol.* 116:299–310
100. Holtzclaw JD, Cornelius RB, Hatcher LI, Sansom SC. 2011. Coupled ATP and potassium efflux from intercalated cells. *Am. J. Physiol. Ren. Physiol.* 300:1319–26
101. Holtzclaw JD, Grimm PR, Sansom SC. 2010. Intercalated cell BK-$\alpha/\beta 4$ channels modulate sodium and potassium handling during potassium adaptation. *J. Am. Soc. Nephrol.* 21:634–45
102. Arroyo JP, Ronzaud C, Lagnaz D, Staub O, Gamba G. 2011. Aldosterone paradox: differential regulation of ion transport in distal nephron. *Physiology* 26:115–23

103. Spray DC, Ye Z-C, Ransom BR. 2006. Functional connexin "hemichannels": a critical appraisal. *Glia* 54:758–73
104. Barbe MT, Monyer H, Bruzzone R. 2006. Cell-cell communication beyond connexins: the pannexin channels. *Physiology* 21:103–14
105. Huang Y-J, Maruyama Y, Dvoryanchikov G, Pereira E, Chaudhari N, Roper SD. 2007. The role of pannexin 1 hemichannels in ATP release and cell–cell communication in mouse taste buds. *Proc. Natl. Acad. Sci.* 104:6436–41
106. Hanner F, Peti-Peterdi J. 2010. Pannexin1 is a novel renal ATP release mechanism. *FASEB J.* 24:606.27 (Meet. Abstr. Suppl.)
107. Teubner B, Michel V, Pesch J, Lautermann J, Cohen-Salmon M, et al. 2003. Connexin30 (Gjb6)-deficiency causes severe hearing impairment and lack of endocochlear potential. *Hum. Mol. Genet.* 12:13–21
108. Lang F, Vallon V, Knipper M, Wangemann P. 2007. Functional significance of channels and transporters expressed in the inner ear and kidney. *Am. J. Physiol. Cell Physiol.* 293:1187–208
109. Nozu K, Inagaki T, Fu XJ, Nozu Y, Kaito H, et al. 2008. Molecular analysis of digenic inheritance in Bartter syndrome with sensorineural deafness. *J. Med. Genet.* 45:182–86
110. Prieto-Carrasquero MC, Botros FT, Kobori H, Navar LG. Collecting duct renin: a major player in angiotensin II-dependent hypertension. *J. Am. Soc. Hypertens.* 3:96–104
111. Rohrwasser A, Morgan T, Dillon HF, Zhao L, Callaway CW, et al. 1999. Elements of a paracrine tubular renin-angiotensin system along the entire nephron. *Hypertension* 34:1265–74
112. Kang JJ, Toma I, Sipos A, Meer EJ, Vargas SL, Peti-Peterdi J. 2008. The collecting duct is the major source of prorenin in diabetes. *Hypertension* 51:1597–604
113. Komlosi P, Fuson AL, Fintha A, Peti-Peterdi J, Rosivall L, et al. 2003. Angiotensin I conversion to angiotensin II stimulates cortical collecting duct sodium transport. *Hypertension* 42:195–99
114. Peti-Peterdi J, Warnock DG, Bell PD. 2002. Angiotensin II directly stimulates ENaC activity in the cortical collecting duct via AT1 receptors. *J. Am. Soc. Nephrol.* 13:1131–35
115. Peti-Peterdi J. 2010. High glucose and renin release: the role of succinate and GPR91. *Kidney Int.* 78:1214–17
116. He W, Miao FJP, Lin DCH, Schwandner RT, Wang Z, et al. 2004. Citric acid cycle intermediates as ligands for orphan G-protein-coupled receptors. *Nature* 429:188–93
117. Robben JH, Fenton RA, Vargas SL, Schweer H, Peti-Peterdi J, et al. 2009. Localization of the succinate receptor in the distal nephron and its signaling in polarized MDCK cells. *Kidney Int.* 76:1258–67
118. Toma I, Kang JJ, Sipos A, Vargas S, Bansal E, et al. 2008. Succinate receptor GPR91 provides a direct link between high glucose levels and renin release in murine and rabbit kidney. *J. Clin. Investig.* 118:2526–34
119. Vargas SL, Toma I, Kang JJ, Meer EJ, Peti-Peterdi J. 2009. Activation of the succinate receptor GPR91 in macula densa cells causes renin release. *J. Am. Soc. Nephrol.* 20:1002–11
120. Pluznick JL, Zou D-J, Zhang X, Yan Q, Rodriguez-Gil DJ, et al. 2009. Functional expression of the olfactory signaling system in the kidney. *Proc. Natl. Acad. Sci. USA* 106:2059–64
121. Figueroa CD, MacIver AG, Mackenzie JC, Bhoola KD. 1988. Localisation of immunoreactive kininogen and tissue kallikrein in the human nephron. *Histochemistry* 89:437–42
122. Proud D, Vio CP. 1993. Localization of immunoreactive tissue kallikrein in human trachea. *Am. J. Resp. Cell Mol. Biol.* 8:16–19
123. van Leeuwen BH, Grinblat SM, Johnston CI. 1984. Release of active and inactive kallikrein from the isolated perfused rat kidney. *Clin. Sci.* 66:207–15
124. Figueroa CD, Gonzalez CB, Grigoriev S, Abd Alla SA, Haasemann M, et al. 1995. Probing for the bradykinin B2 receptor in rat kidney by anti-peptide and anti-ligand antibodies. *J. Histochem. Cytochem.* 43:137–48
125. Bhoola KD, Figueroa CD, Worthy K. 1992. Bioregulation of kinins: kallikreins, kininogens, and kininases. *Pharmacol. Rev.* 44:1–80
126. Schuster VL, Kokko JP, Jacobson HR. 1984. Interactions of lysyl-bradykinin and antidiuretic hormone in the rabbit cortical collecting tubule. *J. Clin. Investig.* 73:1659–67

127. Azizi M, Boutouyrie P, Bissery A, Agharazii M, Verbeke F, et al. 2005. Arterial and renal consequences of partial genetic deficiency in tissue kallikrein activity in humans. *J. Clin. Investig.* 115:780–87
128. Horwitz D, Margolius HS, Keiser HR. 1978. Effects of dietary potassium and race on urinary excretion of kallikrein and aldosterone in man. *J. Clin. Endocrinol. Metab.* 47:296–99
129. Vio CP, Figueroa CD. 1987. Evidence for a stimulatory effect of high potassium diet on renal kallikrein. *Kidney Int.* 31:1327–34
130. Marchetti J, Imbert-Teboul M, Alhenc-Gelas F, Allegrini J, Menard J, Morel F. 1984. Kallikrein along the rabbit microdissected nephron: a micromethod for its measurement. Effect of adrenalectomy and DOCA treatment. *Pflüg. Arch.* 401:27–33
131. Miller DH, Lindley JG, Margolius HS. 1985. Tissue kallikrein levels and synthesis rates are not changed by an acute physiological dose of aldosterone. *Proc. Soc. Exp. Biol. Med.* 180:121–25
132. El Moghrabi S, Houillier P, Picard N, Sohet F, Wootla B, et al. 2010. Tissue kallikrein permits early renal adaptation to potassium load. *Proc. Natl. Acad. Sci. USA* 107:13526–31
133. Fujita T, Hayashi I, Kumagai Y, Inamura N, Majima M. 1999. K^+ loading, but not Na^+ loading, and blockade of ATP-sensitive K^+ channels augment renal kallikrein secretion. *Immunopharmacology* 44:169–75
134. Fujita T, Hayashi I, Kumagai Y, Inamura N, Majima M. 1999. Early increases in renal kallikrein secretion on administration of potassium or ATP-sensitive potassium channel blockers in rats. *Br. J. Pharmacol.* 128:1275–83
135. Stanton BA. 1989. Characterization of apical and basolateral membrane conductances of rat inner medullary collecting duct. *Am. J. Physiol. Ren. Physiol.* 256:862–68
136. Wang W. 2004. Regulation of renal K transport by dietary K intake. *Annu. Rev. Physiol.* 66:547–69
137. Youn JH, McDonough AA. 2009. Recent advances in understanding integrative control of potassium homeostasis. *Annu. Rev. Physiol.* 71:381–401
138. Giebisch G, Malnic G, Berliner RW. 1996. Control of renal potassium excretion. In *The Kidney*, ed. BM Brenner, 1:371–407. Philadelphia: W.B. Saunders
139. Meneton P, Bloch-Faure M, Hagege AA, Ruetten H, Huang W, et al. 2001. Cardiovascular abnormalities with normal blood pressure in tissue kallikrein-deficient mice. *Proc. Natl. Acad. Sci. USA* 98:2634–39
140. Vallet V, Chraibi A, Gaeggeler HP, Horisberger JD, Rossier BC. 1997. An epithelial serine protease activates the amiloride-sensitive sodium channel. *Nature* 389:607–10
141. Caldwell RA, Boucher RC, Stutts MJ. 2004. Serine protease activation of near-silent epithelial Na^+ channels. *Am. J. Physiol. Cell Physiol.* 286:190–94
142. Caldwell RA, Boucher RC, Stutts MJ. 2005. Neutrophil elastase activates near-silent epithelial Na^+ channels and increases airway epithelial Na^+ transport. *Am. J. Physiol. Lung Cell Mol. Physiol.* 288:813–19
143. Chraibi A, Vallet V, Firsov D, Hess SK, Horisberger JD. 1998. Protease modulation of the activity of the epithelial sodium channel expressed in *Xenopus* oocytes. *J. Gen. Physiol.* 111:127–38
144. Passero CJ, Mueller GM, Rondon-Berrios H, Tofovic SP, Hughey RP, Kleyman TR. 2008. Plasmin activates epithelial Na^+ channels by cleaving the γ subunit. *J. Biol. Chem.* 283:36586–91
145. Hughey RP, Bruns JB, Kinlough CL, Harkleroad KL, Tong Q, et al. 2004. Epithelial sodium channels are activated by furin-dependent proteolysis. *J. Biol. Chem.* 279:18111–14
146. Guipponi M, Vuagniaux G, Wattenhofer M, Shibuya K, Vazquez M, et al. 2002. The transmembrane serine protease (TMPRSS3) mutated in deafness DFNB8/10 activates the epithelial sodium channel (ENaC) in vitro. *Hum. Mol. Genet.* 11:2829–36
147. Vuagniaux G, Vallet V, Jaeger NF, Hummler E, Rossier BC. 2002. Synergistic activation of ENaC by three membrane-bound channel-activating serine proteases (mCAP1, mCAP2, and mCAP3) and serum- and glucocorticoid-regulated kinase (Sgk1) in *Xenopus* oocytes. *J. Gen. Physiol.* 120:191–201
148. Vuagniaux G, Vallet V, Jaeger NF, Pfister C, Bens M, et al. 2000. Activation of the amiloride-sensitive epithelial sodium channel by the serine protease mCAP1 expressed in a mouse cortical collecting duct cell line. *J. Am. Soc. Nephrol.* 11:828–34
149. Hughey RP, Bruns JB, Kinlough CL, Kleyman TR. 2004. Distinct pools of epithelial sodium channels are expressed at the plasma membrane. *J. Biol. Chem.* 279:48491–94

150. Hughey RP, Mueller GM, Bruns JB, Kinlough CL, Poland PA, et al. 2003. Maturation of the epithelial Na$^+$ channel involves proteolytic processing of the α- and γ-subunits. *J. Biol. Chem.* 278:37073–82
151. Adebamiro A, Cheng Y, Rao US, Danahay H, Bridges RJ. 2007. A segment of gamma ENaC mediates elastase activation of Na$^+$ transport. *J. Gen. Physiol.* 130:611–29
152. Bruns JB, Carattino MD, Sheng S, Maarouf AB, Weisz OA, et al. 2007. Epithelial Na$^+$ channels are fully activated by furin- and prostasin-dependent release of an inhibitory peptide from the gamma-subunit. *J. Biol. Chem.* 282:6153–60
153. Diakov A, Bera K, Mokrushina M, Krueger B, Korbmacher C. 2008. Cleavage in the γ-subunit of the epithelial sodium channel (ENaC) plays an important role in the proteolytic activation of near-silent channels. *J. Physiol.* 586:4587–608
154. Harris M, Firsov D, Vuagniaux G, Stutts MJ, Rossier BC. 2007. A novel neutrophil elastase inhibitor prevents elastase activation and surface cleavage of the epithelial sodium channel expressed in *Xenopus laevis* oocytes. *J. Biol. Chem.* 282:58–64
155. Kleyman TR, Carattino MD, Hughey RP. 2009. ENaC at the cutting edge: regulation of epithelial sodium channels by proteases. *J. Biol. Chem.* 284:20447–51
156. Masilamani S, Kim GH, Mitchell C, Wade JB, Knepper MA. 1999. Aldosterone-mediated regulation of ENaC α, β, and γ subunit proteins in rat kidney. *J. Clin. Investig.* 104:R19–23
157. Donaldson SH, Hirsh A, Li DC, Holloway G, Chao J, et al. 2002. Regulation of the epithelial sodium channel by serine proteases in human airways. *J. Biol. Chem.* 277:8338–45
158. Planès C, Randrianarison NH, Charles RP, Frateschi S, Cluzeaud F, et al. 2010. ENaC-mediated alveolar fluid clearance and lung fluid balance depend on the channel-activating protease 1. *EMBO Mol. Med.* 2:26–37
159. Pisitkun T, Shen RF, Knepper MA. 2004. Identification and proteomic profiling of exosomes in human urine. *Proc. Natl. Acad. Sci. USA* 101:13368–73
160. Ergonul Z, Frindt G, Palmer LG. 2006. Regulation of maturation and processing of ENaC subunits in the rat kidney. *Am. J. Physiol. Ren. Physiol.* 291:683–93
161. Picard N, Eladari D, El Moghrabi S, Planès C, Bourgeois S, et al. 2008. Defective ENaC processing and function in tissue kallikrein-deficient mice. *J. Biol. Chem.* 283:4602–11
162. Narikiyo T, Kitamura K, Adachi M, Miyoshi T, Iwashita K, et al. 2002. Regulation of prostasin by aldosterone in the kidney. *J. Clin. Investig.* 109:401–8
163. Gumz ML, Lynch IJ, Greenlee MM, Cain BD, Wingo CS. 2010. The renal H$^+$-K$^+$-ATPases: physiology, regulation, and structure. *Am. J. Physiol. Ren. Physiol.* 298:12–21
164. Meneton P, Schultheis PJ, Greeb J, Nieman ML, Liu LH, et al. 1998. Increased sensitivity to K$^+$ deprivation in colonic H,K-ATPase-deficient mice. *J. Clin. Investig.* 101:536–42
165. Nakamura S, Amlal H, Galla JH, Soleimani M. 1998. Colonic H$^+$-K$^+$-ATPase is induced and mediates increased HCO$_3^-$ reabsorption in inner medullary collecting duct in potassium depletion. *Kidney Int.* 54:1233–39
166. Wingo CS. 1989. Active proton secretion and potassium absorption in the rabbit outer medullary collecting duct. Functional evidence for proton-potassium-activated adenosine triphosphatase. *J. Clin. Investig.* 84:361–65
167. Dherbecourt O, Cheval L, Bloch-Faure M, Meneton P, Doucet A. 2006. Molecular identification of Sch28080-sensitive K-ATPase activities in the mouse kidney. *Pflüg. Arch.* 451:769–75
168. Hecquet C, Tan F, Marcic BM, Erdos EG. 2000. Human bradykinin B$_2$ receptor is activated by kallikrein and other serine proteases. *Mol. Pharmacol.* 58:828–36
169. Gkika D, Topala CN, Chang Q, Picard N, Thebault S, et al. 2006. Tissue kallikrein stimulates Ca^{2+} reabsorption via PKC-dependent plasma membrane accumulation of TRPV5. *EMBO J.* 25:4707–16

Renal Function in Diabetic Disease Models: The Tubular System in the Pathophysiology of the Diabetic Kidney

Volker Vallon[1,2,3] and Scott C. Thomson[1,3]

[1]Department of Medicine and [2]Department of Pharmacology, University of California San Diego, La Jolla, California 92093; email: vvallon@ucsd.edu

[3]VA San Diego Healthcare System, San Diego, California 92161

Keywords

tubular transport, glomerular hyperfiltration, tubular growth, sodium-glucose cotransport, diabetic nephropathy

Abstract

Diabetes mellitus affects the kidney in stages. At the onset of diabetes mellitus, in a subset of diabetic patients the kidneys grow large, and glomerular filtration rate (GFR) becomes supranormal, which are risk factors for developing diabetic nephropathy later in life. This review outlines a pathophysiological concept that focuses on the tubular system to explain these changes. The concept includes the tubular hypothesis of glomerular filtration, which states that early tubular growth and sodium-glucose cotransport enhance proximal tubule reabsorption and make the GFR supranormal through the physiology of tubuloglomerular feedback. The diabetic milieu triggers early tubular cell proliferation, but the induction of TGF-β and cyclin-dependent kinase inhibitors causes a cell cycle arrest and a switch to tubular hypertrophy and a senescence-like phenotype. Although this growth phenotype explains unusual responses like the salt paradox of the early diabetic kidney, the activated molecular pathways may set the stage for tubulointerstitial injury and diabetic nephropathy.

INTRODUCTION

T1DM: type 1 diabetes mellitus

T2DM: type 2 diabetes mellitus

ESRD: end-stage renal disease

GFR: glomerular filtration rate

SNGFR: single-nephron glomerular filtration rate

MD$_{NaClK}$: concentration and delivery of Na$^+$, Cl$^-$, and K$^+$ at the luminal macula densa

SGLT: Na$^+$-glucose cotransporter

After 10 to 20 years of diabetes mellitus, approximately 20% of patients with either type 1 or type 2 diabetes mellitus (T1DM, T2DM) develop diabetic nephropathy, making diabetes mellitus the leading cause of end-stage renal disease (ESRD). We still do not understand the genetic and environmental factors that determine which patients eventually develop diabetic nephropathy. Thus, there is a need to better understand the pathophysiology and molecular pathways that lead from the onset of hyperglycemia to renal failure. Changes in the vasculature and the glomerulus, including those to mesangial cells, the filtration barrier, and podocytes, play important roles in the pathophysiology of the diabetic kidney (1–3). In addition, the diabetic milieu has primary effects on the tubular system of the kidney, which are the focus of this review.

Diabetes mellitus affects the kidney in stages. At the onset of T1DM or T2DM, a subset of diabetic patients undergo an increase in glomerular filtration rate (GFR) (1, 2). Although there is residual debate on the subject (4), diabetics with early glomerular hyperfiltration appear to be overrepresented among those who develop diabetic nephropathy later in life (5). The early hemodynamic phenotype is imagined to provoke the subsequent demise of a diabetic kidney through glomerular capillary hypertension, although glomerular capillary hypertension is not required for hyperfiltration (6). Investigators have reported many abnormalities that may cause diabetic hyperfiltration through impaired constriction of the afferent arteriole (1). Several years ago, we began encountering situations in diabetic rats in which the concentration and delivery of Na$^+$, Cl$^-$, and K$^+$ at the luminal macula densa (MD$_{NaClK}$) and single-nephron GFR (SNGFR) change in opposite directions. We recognized that isolated defects in vasomotion could cause SNGFR and MD$_{NaClK}$ to change only in the same direction but that reciprocal changes in MD$_{NaClK}$ and SNGFR are the expected result when a primary change in proximal tubule reabsorption affects MD$_{NaClK}$ and subsequently SNGFR by negative feedback through the macula densa through a process known as tubuloglomerular feedback. Hence, we proposed the so-called tubular hypothesis of glomerular filtration as an archetype for the kidney in early diabetes (7, 8). Examples in which this hypothesis applies include diabetic hyperfiltration and the salt paradox, a unique phenomenon of the diabetic kidney that refers to the inverse relationship between changes in dietary NaCl intake and GFR.

Although kidney growth shortly after the onset of diabetes in a subset of diabetic patients has been known for many years and kidney size has been linked to the development of diabetic nephropathy (9–13), little attention has been given to this phenomenon. Tubular growth, however, explains early functional changes in the diabetic kidney, including the primary increase in proximal tubule reabsorption, and is thus relevant for the tubular hypothesis of glomerular filtration. Proximal tubule reabsorption is further enhanced in hyperglycemia due to increased glomerular filtration of glucose, which increases proximal tubule reabsorption of glucose and Na$^+$ through the Na$^+$-glucose cotransporters SGLT2 and SGLT1. The interest in SGLT2 has recently been revived due to the current development of SGLT2 inhibitors as new antidiabetic drugs (14, 15).

The molecular signature of proximal tubule growth in the diabetic kidney is unique and includes elements of cell proliferation, hypertrophy, and cellular senescence. This unique growth pattern explains unusual functional responses observed only in the diabetic kidney, like the salt paradox. Through its effects on GFR and the deleterious consequences of hyperfiltration, the salt paradox may account for the unexpected finding of two recent cohort studies in patients with T1DM and T2DM showing that lower NaCl intake is unexpectedly associated with increased rates of ESRD, cardiovascular death, and all-cause mortality (16, 17). Moreover, the molecular pathways involved in the tubular growth of the diabetic kidney are linked to inflammation and fibrosis and may set the stage for renal damage. Therefore, genetic and/or environmental factors that affect the tubular growth response to the diabetic milieu may determine not only the extent of kidney growth, tubular hyperreabsorption, and glomerular hyperfiltration in early diabetes but also the later progression of renal disease.

This review discusses early changes that occur in the tubular system of the diabetic kidney, illustrates their role in the tubular hypothesis of glomerular filtration, and proposes potential links to the later development of diabetic nephropathy. The interested reader is referred to previous reviews on the tubular hypothesis of glomerular filtration and implications of the salt paradox (7, 8, 18, 19) as well as on the link between early tubular changes and the progression of renal disease in diabetes (19–22); we focus this review on the most recent studies.

GTB: glomerulotubular balance

GLOMERULAR HYPERFILTRATION IN DIABETES MELLITUS AS A PRIMARY TUBULAR EVENT

We begin with a theoretical framework for describing interactions between the glomerulus and tubule. These interactions include a forward effect of SNGFR on the tubule, known as glomerulotubular balance (GTB) or the load dependency of reabsorption (**Figure 1**). The other component

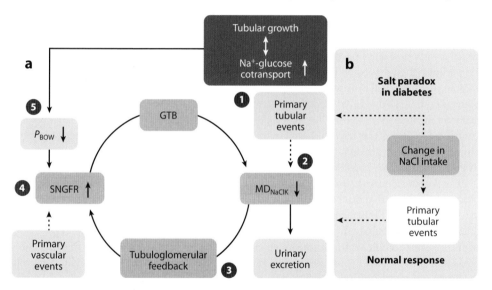

Figure 1

Tubular hypothesis of glomerular filtration in diabetes mellitus. (*a*) Glomerular hyperfiltration in diabetes mellitus as a primary tubular event. The single-nephron glomerular filtration rate (SNGFR) is determined by primary vascular events, tubuloglomerular feedback, and tubuloglomerular feedback resetting. Glomerulotubular balance (GTB) and primary tubular events determine the concentration and delivery of Na^+, Cl^-, and K^+ at the luminal macula densa (MD_{NaClK}). A primary vascular event causes SNGFR and MD_{NaClK} to change in the same direction, whereas a primary tubular event causes SNGFR and MD_{NaClK} to change in opposite directions. Hyperglycemia causes a primary increase in proximal tubule reabsorption (the primary tubular event) through enhanced tubular growth and Na^+-glucose cotransport ❶. This reduces MD_{NaClK} ❷ and, via tubuloglomerular feedback ❸, increases SNGFR ❹. Enhanced growth and tubular reabsorption can also reduce the hydrostatic pressure in Bowman space (P_{BOW}) ❺, which by increasing effective filtration pressure can also increase SNGFR ❹. The resulting increase in SNGFR partly restores the fluid and electrolyte load to the distal nephron. (*b*) The salt paradox. The nondiabetic kidney adjusts NaCl transport to dietary NaCl intake primarily downstream of the macula densa, and thus MD_{NaClK} or SNGFR is not altered. In contrast, diabetes renders reabsorption in the proximal tubule very sensitive to dietary NaCl, with subsequent effects on MD_{NaClK} and SNGFR.

is the tubuloglomerular feedback system, which senses changes in MD_{NaClK} and induces reciprocal changes in SNGFR to stabilize electrolyte delivery to the distal tubule, where fine adjustments of reabsorption and excretion occur. GFR is determined by a balance of forces between primary vascular and primary tubular events. Isolated vascular or tubular events each filtered by the GTB–tubuloglomerular feedback system lead to changes in both MD_{NaClK} and SNGFR. The vascular event causes SNGFR and MD_{NaClK} to change in the same direction, whereas the tubular event causes SNGFR and MD_{NaClK} to change in opposite directions. The tubular component of an outside disturbance dominates the vascular component whenever SNGFR and MD_{NaClK} change in opposite directions (see **Supplemental Figure 1**; follow the **Supplemental Material link** from the Annual Reviews home page at http://www.annualreviews.org). Identifying that a tubular event is the dominant cause of a change in SNGFR does not preclude the existence of a simultaneous vascular event.

> Streptozotocin (STZ): injected to induce a model of type 1 diabetes mellitus

A Primary Increase in Proximal Tubule Reabsorption in Diabetes Mellitus

Lithium clearance is a useful, albeit imperfect, indicator of proximal reabsorption and NaCl delivery to the macula densa. Lithium clearance data published 20 years ago revealed increased proximal reabsorption in patients with early T1DM (23) or T2DM (24). The authors did not consider the macula densa mechanism but posited excessive proximal reabsorption as a cause of systemic volume expansion that led to hemodynamic consequences in diabetes (23). However, extracellular fluid expansion is not required for diabetic hyperfiltration (25). Additional studies showed that fractional proximal reabsorption was elevated and positively correlated with GFR in patients with T1DM (25, 26), and Hannedouche and colleagues (26) speculated that the ensuing decrease in distal Na^+ delivery could deactivate the tubuloglomerular feedback response and contribute to glomerular hyperfiltration in some diabetics. Likewise, investigators recently found a strong correlation between newly discovered T2DM, glomerular hyperfiltration, and decreased lithium clearance in a large trial involving subjects of African descent (27).

The present concern is to distinguish primary changes in tubular reabsorption from secondary changes in reabsorption that begin as primary vascular events and then impact tubular reabsorption via GTB. When GTB operates normally, fractional reabsorption declines as GFR increases. Therefore, when diabetic hyperfiltration is accompanied by increased fractional reabsorption of the proximal tubule or the segments upstream of the macula densa (28–30), this scenario is consistent with primary hyperreabsorption. In addition, we artificially activated tubuloglomerular feedback as a tool to manipulate SNGFR in order to compare tubular reabsorption at similar SNGFRs. Data were thus obtained for expressing proximal reabsorption as a function of SNGFR in individual nephrons. This approach confirmed a major primary increase in proximal reabsorption in rats with early streptozotocin (STZ) diabetes, a model of T1DM (31, 32).

Early Distal NaClK Delivery Is Reduced in Diabetes Mellitus

For this primary increase in proximal reabsorption to be the dominant cause of glomerular hyperfiltration, the diabetic nephron must operate with MD_{NaClK} below normal. In fact, data on the ionic content and NaClK delivery to the early distal nephron in diabetes show values substantially below normal (28–30, 32). Micropuncture in rats with superficial glomeruli allows the collection of tubular fluid close to the macula densa. This approach revealed ambient early distal tubule concentrations of Na^+, Cl^-, and K^+ in nondiabetic rats of 21, 20, and 1.2 mM, respectively; these values (together with their absolute deliveries) were reduced by 20–28% in hyperfiltering STZ-diabetic rats (**Figure 1**) (30).

Resetting of the Tubuloglomerular Feedback Curve in Diabetes Mellitus

SNGFR collected from the proximal tubule also increases in diabetes. Because tubuloglomerular feedback is inoperative during proximal tubule fluid collections, the tubuloglomerular feedback curve must shift upward in diabetes. The curve can be influenced by events outside of the juxtaglomerular apparatus, and the upward shift in the tubuloglomerular feedback curve in diabetes may represent a primary vascular event mediated by any number of the factors affecting the afferent arteriole. Nonetheless, reduced MD_{NaClK} invokes the tubule as the dominant controller of SNGFR, with a primary vascular effect as runner up. Furthermore, the upward shift of the tubuloglomerular feedback curve may be explained by tubuloglomerular feedback resetting from within the juxtaglomerular apparatus. The juxtaglomerular apparatus of each nephron can adjust its own tubuloglomerular feedback response and tends to invoke this capacity to align the steep portion of its tubuloglomerular feedback curve with the ambient tubular flow (33). In accordance and as illustrated in **Figure 2**, the entire tubuloglomerular feedback curve in diabetes resets leftward and upward, and the greatest tubuloglomerular feedback efficiency resides close to the ambient operating point (29).

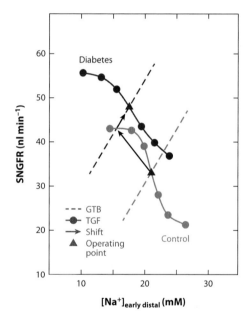

Figure 2

A proposed role for tubuloglomerular feedback (TGF) resetting in the diabetic kidney. TGF is depicted as the inverse relationship between early distal Na$^+$ concentration ([Na$^+$]$_{early\ distal}$) and single-nephron GFR (SNGFR). Data from perturbation analysis of late proximal tubule flow rate were combined with data for fractional proximal reabsorption (29), ambient [Na$^+$]$_{early\ distal}$ (30), and the relationship between late proximal flow rate and early distal conductivity (29) in control and STZ-diabetic rats. Similar curves were derived for early distal Cl$^-$ concentration (not shown). Dashed lines represent glomerular-tubular balance (GTB), which refers to the load dependency of reabsorption between the glomerulus and the macula densa. The operating point (*triangles*) of the nephron, where TGF and GTB intersect, is usually located where TGF is steep. A primary diabetes–induced increase in reabsorption (with a parallel shift in the GTB line) lowers [Na$^+$]$_{early\ distal}$, which increases SNGFR but shifts the operating point to the flatter part of the TGF curve. To operate at high TGF efficiency, the TGF curve in diabetes resets upward such that the operating point is restored to the steeper part of the TGF curve.

Manipulating the Tubuloglomerular Feedback Response Affects Diabetic Hyperfiltration

The tubular hypothesis predicts that diabetic hyperfiltration will be attenuated or blunted in a tubuloglomerular feedback–less mouse. Adenosine mediates the tubuloglomerular feedback response by activation of adenosine A_1 receptors (A_1R), and mice lacking these receptors ($A_1R-/-$ mice) have no acute tubuloglomerular feedback response. Two recent studies reported glomerular hyperfiltration in diabetic $A_1R-/-$ mice (34, 35). As discussed above, the tubular hypothesis invokes feedback from the tubule as the dominant controller of GFR in early diabetes but does not require tubuloglomerular feedback to be the only controller. The theory allows for additional primary defects in afferent arteriolar vasoconstriction and predicts such defects will be unmasked when feedback from the tubule is eliminated. Under these conditions, some degree of hyperfiltration would persist in the absence of A_1R. Moreover, in one of the two studies the nondiabetic but not the alloxan-diabetic $A_1R-/-$ mice were hypotensive compared with wild-type controls during measurements of GFR (34). As a consequence, the higher GFR in normotensive alloxan-diabetic $A_1R-/-$ mice may reflect impaired renal autoregulation, a known trait of the tubuloglomerular feedback–less mouse. The other study used severely hyperglycemic Akita-diabetic $A_1R-/-$ mice, which have blood glucose levels of 600 to 900 mg dl^{-1} (35). At those levels, which far exceed the glucose transport maximum, glucose becomes a net proximal diuretic (36), and the primary increase in proximal reabsorption, which characterizes diabetic hyperfiltration by the tubular hypothesis with modest hyperglycemia (30, 31), disappears. As a consequence, tubuloglomerular feedback activation may limit glomerular hyperfiltration during severe hyperglycemia. Furthermore, the severity of diabetes may affect the tubuloglomerular feedback–independent influence of adenosine on GFR and thereby the net response to A_1R blockade or knockout. In accordance with this discussion and the tubular hypothesis of GFR in diabetes, glomerular hyperfiltration was blunted when A1R$-/-$ mice were exposed to STZ-induced, more moderate hyperglycemia (37). Tubular control of GFR has also been proposed in dogs, in which acute hyperglycemia increased GFR, but only if tubuloglomerular feedback was intact (38).

Determinants of Primary Proximal Tubule Hyperreabsorption in Diabetes Mellitus

A primary increase in proximal reabsorption in diabetes is the natural consequence of tubular growth because a larger tubule reabsorbs more, and of moderate hyperglycemia, which provides more substrate for proximal tubule Na$^+$-glucose cotransport (**Figure 1**). Conversely, inhibiting tubular growth or Na$^+$-glucose cotransport prevents or reduces hyperreabsorption. Moreover, and in accordance with the tubular hypothesis of glomerular filtration, these maneuvers also suppress glomerular hyperfiltration. Proximal tubule glucose transport and tubular growth are important for the pathophysiology of the diabetic kidney, and therefore they are separately and in more detail discussed below, including discussion of their role in primary proximal tubule hyperreabsorption.

GLUCOSE TRANSPORT IN THE DIABETIC PROXIMAL TUBULE

Under normoglycemic conditions, most of the tubular glucose uptake across the apical membrane occurs in the early proximal tubule and is mediated by the high-capacity Na$^+$-glucose cotransporter SGLT2 (SLC5A2) (**Figure 3**). Most of the remaining luminal glucose is taken up in further distal parts of the proximal tubule by low-capacity SGLT1 (SLC5A1). In accordance, SGLT2 and SGLT1 protein expression has been located in the brush border

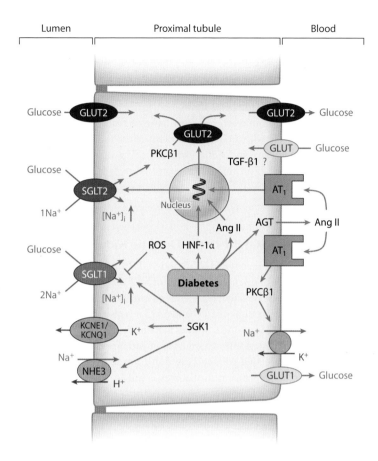

Figure 3

Proximal tubule glucose and Na$^+$ transport in the diabetic kidney. Hyperglycemia increases glomerular filtration of glucose and enhances glucose reabsorption in the proximal tubule via SGLT2 (early segments) and SGLT1 (later segments), thereby enhancing the reabsorption and intracellular concentration of Na$^+$ ([Na$^+$]$_i$). An increase in [Na$^+$]$_i$ and activation of basolateral Na-K-ATPase [via diabetes-induced protein kinase C β1 (PKCβ1)] have been implicated in diabetic tubular growth. Diabetes increases angiotensin II (Ang II) in the cytosol and in the extracellular space; the latter may include enhanced release of angiotensinogen (AGT) (75). Release of AGT and activation of AT$_1$ receptors may also occur at the luminal membrane. Ang II and hepatocyte nuclear factor HNF-1α increase SGLT2 and glucose transporter (GLUT)2 expression. Diabetes also induces the serum- and glucocorticoid-inducible kinase SGK1, which increases SGLT1 activity. Induction of oxidative stress (ROS) can inhibit SGLT. Glucose reabsorption via SGLTs is electrogenic, and luminal K$^+$ channels (e.g., SGK1-activated KCNE1/KCNQ1 in the late proximal tubule) stabilize the membrane potential. Glucose exits via basolateral GLUTs. In diabetes, PKCβ1 can shift part of GLUT2 into the apical membrane, which would equilibrate luminal and basolateral glucose concentrations. The induction of transforming growth factor β1 (TGF-β1) may be particularly sensitive to basolateral glucose uptake. ROS denotes reactive oxygen species. Modified with permission from Reference 19.

membrane of the early and later proximal tubule sections, respectively (39, 40). Expression of the human genes in HEK293 cells confirmed that the Na$^+$:glucose coupling ratio equals a value of 1 for hSGLT2 and 2 for hSGLT1, indicating a greater concentrative power of the latter, whereas hSGLT2 and hSGLT1 transport glucose with similar affinity (5 mM versus 2 mM) (41).

Ang II: angiotensin II

Renal micropuncture experiments in knockout mice demonstrated that SGLT2 is responsible for all glucose reabsorption in the early proximal tubule and overall is the major pathway of renal glucose reabsorption (40). The lack of SGLT2 suppresses renal mRNA and protein expression of SGLT1 by approximately 40%, which may prevent excessive glucose uptake in late proximal tubule segments. Whereas mean fractional renal glucose reabsorption at euglycemia was ~40% in *Sglt2* knockout mice, studies in mice lacking *Sglt1* showed normal renal SGLT2 protein expression and a significant but minor reduction in fractional renal glucose reabsorption from 99.8% to 96.9% (42). Similarly, human subjects with *SGLT1* mutations show intestinal glucose malabsorption with little or no glucosuria, whereas individuals with *SGLT2* gene mutations have persistent and prominent renal glucosuria (43). The glucose reabsorption in *Sglt2* knockout mice is thought to reflect the significant capacity of SGLT1 to increase glucose reabsorption when glucose delivery increases.

Expression of SGLT2 and SGLT1 in the Diabetic Kidney

The capacity of glucose transport through SGLT2 and SGLT1 is determined by expression levels, which can increase in diabetes. For example, induction of STZ diabetes in rats increased mRNA expression for *Sglt1* and *Sglt2* in the renal cortex (44) and increased renal SGLT1 protein expression (45). Enhanced renal mRNA expression of *Sglt1* and *Sglt2* was also found in diabetic obese Zucker rats compared with age-matched leans (46). Studies in primary cultures of human exfoliated proximal tubule epithelial cells harvested from fresh urine of patients with T2DM revealed increased glucose uptake, which was associated with increased mRNA and protein expression of SGLT2 (47). Upregulation of SGLT2 expression in diabetes is linked to the activation of angiotensin II (Ang II) AT_1 receptors (48) and the transcription factor hepatocyte nuclear factor HNF-1α (49), whereas upregulation of SGLT1 is linked to serum- and glucocorticoid-inducible kinase 1 (SGK1) (**Figure 3**). *SGK1* is upregulated in proximal tubules in STZ-diabetic rats and in patients with diabetic nephropathy (50). Studies in knockout mice implicated SGK1 in the stimulation of SGLT1 activity in proximal renal tubules in diabetes (51). SGK1 may also facilitate proximal tubule glucose transport by the stimulation of luminal K^+ channels (e.g., KCNE1/KCNQ1) (52), which counteract the depolarization induced by electrogenic Na^+-glucose cotransport, thereby maintaining the electrical driving force (**Figure 3**) (53, 54). Moreover, SGK1 may upregulate mRNA levels of the Na-H exchanger *Nhe3* in STZ-diabetic rats (55), an important determinant of proximal tubule Na^+ reabsorption. SGK1 effects on intestinal and proximal and distal tubule transport as well as in fibrosis make SGK1 an interesting target in diabetes (50).

Diabetes may not uniformly increase renal expression of SGLTs because some studies found unchanged or even reduced renal SGLT expression and/or activity in diabetic rodent models (19, 56, 57). Studies in primary cultures of renal proximal tubule cells indicated that high-glucose-induced oxidative stress can reduce SGLT expression and activity (**Figure 3**) (58). This may relate to the downregulation of SGLT1 found in mice lacking *Sglt2* (40) and may limit renal glucose uptake and toxicity. Earlier studies reported that hyperglycemia causes the induction and membrane incorporation of a low-affinity Na^+-dependent D-glucose transporter in the proximal tubule of 4-day-old STZ-diabetic rats, an effect that was retained for at least 4 weeks, but only when the animals maintained or increased their body weight (59). This effect was lost in severely ill ketoacidotic and cachectic animals (59). These results indicate the importance of metabolic conditions, which may contribute to the different findings on SGLT expression in diabetes. Even with unchanged expression of SGLT2 and SGLT1, the increase in the tubular glucose load associated with hyperglycemia is expected to increase absolute glucose uptake through SGLT2 and SGLT1 because their capacity is not saturated under normoglycemic conditions.

Sodium-Glucose Cotransport Contributes to Primary Proximal Tubule Hyperreabsorption in Diabetes Mellitus

Modeling the effects of Na^+-linked glucose transport on the active and passive components of proximal reabsorption predicts that modest hyperglycemia enhances Na^+ reabsorption in the proximal tubule (36). Bank & Aynedjian (60) performed microperfusion studies in STZ-diabetic rats and proposed that high glucose in the proximal tubule fluid stimulates Na^+ absorption through Na^+-glucose cotransport. Increasing luminal glucose (from 100 to 500 mg dl^{-1}) induced significantly greater increases in Na^+ versus glucose absorption on a molar basis, which may reflect the Na^+:glucose coupling ratio of 2:1 for SGLT1 (41). Increased SGLT-mediated Na^+ transport was confirmed by micropuncture in moderately hyperglycemic STZ-diabetic rats and by the direct application of the nonselective SGLT inhibitor phlorizin into the free-flowing early proximal tubules of superficial glomeruli. In diabetic rats, phlorizin elicited a greater decline in absolute and fractional reabsorption up to the early distal tubule and abolished hyperreabsorption (30). Moreover, this maneuver increased Na^+, Cl^-, and K^+ concentrations in early distal tubule fluid and reduced diabetic glomerular hyperfiltration, consistent with the tubular hypothesis (30). Studies using the SGLT2 inhibitor dapagliflozin confirmed the acute effects observed with phlorizin on proximal reabsorption and GFR. In addition, studies with continuous SGLT2 blockade for 2 weeks from the onset of STZ diabetes suggested such blockade as a means to normalize NaCl delivery to the macula densa and to thereby attenuate hyperfiltration, consistent with a role of SGLT2 in diabetes-induced hyperreabsorption and hyperfiltration (**Figures 1** and **3**) (61). Finally, the glucose reabsorptive rate in early STZ-diabetic rats increases with kidney weight (62). Tubular growth, the other major cause of proximal hyperreabsorption (see below), may cause proximal hyperreabsorption, in part by enhancing Na^+-glucose cotransport capacity.

TGF-β: transforming growth factor β

Selective inhibitors of SGLT2 are currently in clinical trials to inhibit renal glucose reabsorption and to increase renal excretion, thereby lowering hyperglycemia (14, 15). This approach can reduce plasma glucose without inducing increased insulin secretion, hypoglycemia, or weight gain. This approach would constitute a major advance and appears to have a good safety profile. Under hyperglycemic conditions, such inhibitors are expected to lower proximal tubule Na^+ reabsorption and thereby diabetic glomerular hyperfiltration. Shifting glucose reabsorption from SGLT2 to SGLT1, which has a greater Na^+:glucose coupling ratio, is expected to attenuate the renal Na^+ loss in response to SGLT2 inhibition. Preventing the early proximal tubule from sensing episodes of hyperglycemia through SGLT2 may attenuate the negative effects of glucose on tubular growth and function. However, blocking apical glucose entry via SGLT2 may simply increase basolateral glucose entry when blood glucose is rising and may thus inhibit Na^+ reabsorption and glomerular hyperfiltration, but not the other nefarious effects of high intracellular glucose. Moreover, basolateral glucose uptake may be particularly important for the induction of transforming growth factor β1 (TGF-β1) (**Figure 3**) (21). Studies in mice with a loss of SGLT2 protein function revealed no evidence for tubular injury (63). Induction of STZ diabetes in these mice showed a higher risk for infection and an increased mortality rate, although the influence of the greater total STZ doses required to achieve similar hyperglycemia compared with wild-type mice remained unclear. In patients treated with SGLT2 inhibitors, the increased glucosuria appears to increase the risk of genital infections, but not the risk of ascending urinary tract infections (14). Moreover, patients with familial renal glycosuria due to mutations in *SGLT2* do not show signs of general renal tubular dysfunction or other pathological changes and seem to have normal life expectancies (43). Future studies will help to better define and understand the consequences that occur along the nephron and the urinary system when SGLT2 is inhibited under diabetic conditions, including the role of SGLT-mediated increases in intracellular Na$^+$ concentration as a trigger of proximal tubule growth in diabetes (**Figure 3**) (64).

Luminal Translocation of GLUT2 in the Diabetic Proximal Tubule

Proximal tubule cells do not use glucose for energy production, and glucose that is reabsorbed across the luminal membrane leaves the cell across the basolateral membrane by the low-affinity glucose transporter, GLUT2, in the S1 segment and by high-affinity GLUT1 in the S3 segment. Upregulation of GLUT2 expression occurs in renal proximal tubule cells in diabetic patients (47) and in diabetic rats (65) and has been linked to transcriptional activity of both HNF-1α and HNF-3β (**Figure 3**) (66). The gene for HNF-1α is mutated in maturity-onset diabetes of the young type 3 (MODY3), an autosomal dominant form of non-insulin-dependent diabetes, and mice lacking HNF-1α have a renal glucose reabsorption defect (43). In contrast, GLUT1 is downregulated in cortical tubules in diabetes (65) or is unchanged (57).

Diabetes mellitus also enhances glucose absorption in the small intestine, where high luminal glucose concentrations lead to the rapid insertion of GLUT2 into the brush border membrane to operate in parallel with SGLT1-mediated glucose uptake. This luminal insertion of GLUT2 involves a Ca^{2+}- and protein kinase C (PKC)β2-dependent mechanism (67). Similarly, an increase in the facilitative glucose absorption associated with the translocation of GLUT2 and GLUT5 (but not of GLUT1) to the luminal brush border of the proximal tubule was observed in hyperglycemic STZ-diabetic rats, whereas normalization of blood glucose levels by overnight fasting reversed the translocation of GLUT2, but not that of GLUT5 (57). Similar to the mechanism in enterocytes, the luminal targeting of GLUT2 in the proximal tubule was linked to a Ca^{2+}- and PKCβ-dependent mechanism but involved the PKCβ1 isoform and not PKCβ2 (**Figure 3**) (68). PKCβ1 is expressed in the brush border of the proximal tubule, where its expression and activity increase in STZ-diabetic rats (68–70).

Because GLUT5 is thought to transport fructose in vivo and has a low affinity for glucose, the relevance of its translocation to the luminal brush border in the diabetic kidney is less clear. With a predicted K_m of 20 to 40 mM for glucose, the luminal translocation of GLUT2 implicates a role for GLUT2 in glucose reabsorption in diabetes to the extent that luminal glucose concentrations exceed peritubular concentrations as a consequence of tubular fluid reabsorption and luminal glucose concentrations saturating the SGLTs. Inducing acute hyperglycemia (20 to 35 mM) increased glucose concentrations in late proximal tubule fluid to plasma levels (71), which may reflect equilibration through luminal and basolateral GLUT2. In the small intestine, SGLT1 senses the high glucose concentrations and is required for the insertion of GLUT2 into the brush border membrane (42, 67). If SGLT2 has a similar role in the early proximal tubule (**Figure 3**), then SGLT2 inhibitors may also lower renal glucose reabsorption during hyperglycemia by inhibiting luminal GLUT2 translocation.

EARLY TUBULAR GROWTH IN DIABETES MELLITUS

The kidney in general and the proximal tubules in particular grow large from the onset of diabetes mellitus (19). Proximal tubule growth involves an early period of hyperplasia followed by a shift to hypertrophy (72), as discussed in this section and illustrated in **Figure 4**.

Early Hyperplastic Phase

Numerous growth factors contribute to the early proliferation phase of the diabetic tubular system, including insulin-like growth factor 1 (IGF-1), platelet-derived growth factor (PDGF), vascular

---

GLUT: glucose transporter

PKC: protein kinase C

IGF-1: insulin-like growth factor 1

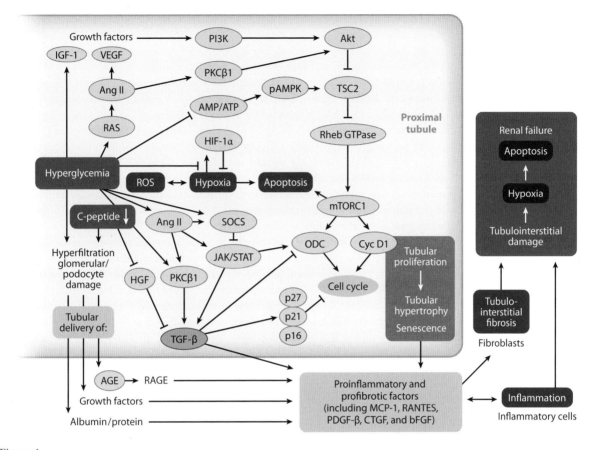

Figure 4
The early growth phenotype of the diabetic proximal tubule and potential links to renal injury and failure. Diabetes mellitus–induced growth of the proximal tubule includes an initial phase of cell proliferation and an early transition to hypertrophy via G_1 cell cycle arrest and the development of a senescence-like phenotype. The molecular pathways involved in this tubular growth phenotype are linked to tubulointerstitial fibrosis and inflammation and may thus set the stage for renal failure later in life. Abbreviations: AGE, advanced glycation end products; Ang II, angiotensin II; bFGF, basic fibroblast growth factor; CTGF, connective tissue growth factor; Cyc D1, cyclin D1; HGF, hepatocyte growth factor; HIF-1α, hypoxia-inducible factor 1α; IGF-1, insulin-like growth factor 1; PKCβ1, protein kinase C β1; JAK/STAT, Janus kinase/signal transducers and activators of transcription; MCP-1, monocyte chemotactic protein-1; mTORC1, mammalian target of rapamycin complex 1; ODC, ornithine decarboxylase; p16, p16^{INK4A}; p21, the CDK inhibitor waf1/cip1; p27, p27^{KIP1}; pAMPK, phosphorylated AMP-activated protein kinase; PI3K, phosphoinositide 3-kinase; PDGF-β, platelet-derived growth factor β; RAGE, receptor for AGE; RAS, renin-angiotensin system; ROS, reactive oxygen species; SOCS, suppressor of cytokine signaling; TGF-β, transforming growth factor β; TSC, tuberous sclerosis complex; VEGF, vascular endothelial growth factor. Modified, with permission, from Reference 1.

endothelial growth factor (VEGF), and epidermal growth factor (EGF) (73, 74). Recent studies proposed that mouse proximal tubule cells express an endogenous renin-angiotensin system (RAS) that is activated by high glucose (see also **Figure 3**) to stimulate VEGF synthesis through activation of the Ang II AT$_1$ receptor and the extracellular signal–regulated kinase (ERK) pathway (75). Glucose and Ang II activate intracellular signaling processes, including the polyol pathway and generation of reactive oxygen species (ROS) (including H_2O_2 and O_2^-), which activate the Janus kinase (JAK)/signal transducers and activators of transcription (STAT) signaling cascades (76). Ang II activates the JAK/STAT pathway via AT$_1$ receptors (77) and increases cell

VEGF: vascular endothelial growth factor

RAS: renin-angiotensin system

ROS: reactive oxygen species

JAK/STAT: Janus kinase/signal transducers and activators of transcription

ODC: ornithine decarboxylase

CDK: cyclin-dependent kinase

proliferation, which is enhanced by high glucose (78). Diabetes and high glucose concentrations also induce suppressors of cytokine signaling (SOCS1, SOCS3) in tubular cells. SOCS1 and SOCS3 are intracellular negative regulators of JAK/STAT activation that inhibit the expression of STAT-dependent genes and high-glucose-induced cell proliferation (79). The JAK/STAT pathway is linked to the induction of immediate early genes like c-*jun* and c-*fos* (76), which can activate ornithine decarboxylase (ODC) (80), the rate-limiting enzyme in polyamine synthesis. In the early diabetic kidney, ODC is required for hyperplasia and most likely also for hypertrophy of the proximal tubule (**Figure 4**) (31, 80, 81). The rapid yet transient renal induction of IGF-1 correlates with the upregulation of renal ODC expression and activity, the induction of intracellular polyamines in the kidney cortex, and the early proliferative phase (80). Deng et al. (82) proposed that the increase in ODC expression in early diabetes occurs mainly in the distal nephron and that polyamines may pass from the distal tubule to the proximal tubule in a paracrine fashion to trigger proximal tubule growth. Further studies are needed to confirm these findings and to determine the mechanisms that induce ODC expression in the distal tubule.

Diabetes mellitus activates PKC, which can produce a myriad of consequences, including a mitogen-induced early proliferation phase (83). In particular, diabetes can enhance proximal tubule activity of the PKCβ1 isoform (68, 70) and PKCβ has been implicated in Akt activation in the renal cortex of diabetic rats (84). In accordance with a role of PKCβ in kidney growth, the early diabetes-induced increase in kidney weight was blunted in mice lacking this PKC isoform (85). ACE inhibition inhibits diabetes-induced activation of renal PKCβ1 (70), consistent with PKCβ1 being downstream of Ang II. Downstream signaling events of IGF-1 and VEGF include activation of the phosphoinositide 3-kinase (PI3K)/Akt pathways and thus merge with PKCβ-activated pathways; both pathways are linked to ODC activation (80). Diabetic renal growth is also associated with reduced phosphorylation of AMP-activated protein kinase (AMPK) (86). Phosphorylated AMPK inhibits the activity of mammalian target of rapamycin complex 1 (mTORC1) by phosphorylating and activating tuberous sclerosis complex (TSC2) (87). As a consequence, mTOR activity is enhanced in the diabetic kidney, and increasing AMPK phosphorylation reverses mTOR activation and inhibits renal growth without affecting hyperglycemia (86). Together these studies propose that the early tubular proliferation phase in diabetes is the consequence of high-glucose-induced oxidative stress and activation of the tubular RAS, enhanced glomerular filtration and tubular expression of growth factors, activation of PKCβ and the JAK/STAT pathway, AMPK inhibition, and activation of both mTORC1 and ODC (**Figure 4**).

Transition from the Hyperplastic Phase to the Hypertrophic Phase

The transition of the diabetic kidney from hyperplastic to hypertrophic growth occurs early (at approximately day 4 in the model of STZ diabetes) (72, 82) and is mediated by TGF-β1 (88). In accordance, primary tubule cells from TGF-β knockout mice respond to high glucose concentrations with an increased rate of proliferation compared with cells from wild-type littermates but show no hypertrophy (89). The JAK/STAT signaling pathway (90) and PKCβ (91) can induce TGF-β expression in the diabetic kidney (**Figure 4**). In addition, ERK and p38 have been implicated in high-glucose-induced TGF-β expression and cellular hypertrophy in renal tubular cells of STZ-diabetic rats (92).

TGF-β can induce a G_1 phase cell cycle arrest by induction of the cyclin-dependent kinase (CDK) inhibitor p27^{KIP1} (p27) (93), which can also be induced in diabetes by PKC (94). Diabetes also increases the renal expression of the CDK inhibitor waf1/cip1 (p21) (95, 96), and loss of p21 increases tubular cell proliferation (96). Consistent with a role of ROS, the antioxidants N-acetylcysteine and taurine attenuated high-glucose-induced activation of the JAK/STAT signaling pathways, p21 and p27 expression, and hypertrophic growth in renal tubular epithelial cells (97). Antioxidants also attenuated the enhanced p21 expression and cellular hypertrophy induced by advanced glycation end products (AGE) and its receptor (RAGE) in human renal proximal tubule cells (98). Thus, signaling pathways that initially induce proliferation subsequently provoke a switch to hypertrophy through the induction of TGF-β and CDK inhibitors in the diabetic tubule (**Figure 4**). Sustaining kidney hypertrophy and size in the long-term diabetic state involves additional mechanisms, including decreased proteolysis (99).

Tubular Growth as a Determinant of Proximal Tubule Hyperreabsorption in Diabetes Mellitus

It is easy to imagine that an increase in proximal tubule length and diameter enhances proximal tubule reabsorptive capacity. Difluoromethylornithine (DFMO), an ODC inhibitor, attenuates kidney growth in early STZ-diabetic rats (81) and was used to test whether tubular growth contributes to the primary increase in proximal reabsorption in early diabetes mellitus. DFMO had no significant effect on kidney weight or GFR in nondiabetic rats (31). In comparison, DFMO attenuated kidney growth and glomerular hyperfiltration in similar proportions in diabetic rats. Moreover, DFMO eliminated the primary increase in proximal reabsorption in STZ-diabetic rats; i.e., at the same level of single-nephron GFR, proximal reabsorption was lower in DFMO-treated than in untreated STZ-diabetic rats (31).

Diabetes stimulates the basolateral Na-K-ATPase activity of the proximal tubule that has been associated with and implicated in renal hypertrophy (64, 100), although the involved mechanisms still remain unclear. PKCβ may be part of this link because the kinase is involved in tubular growth in diabetes (85) and has been implicated in the activation of Na-K-ATPase and Na$^+$ transport in the proximal tubule (**Figure 3**) (101). Because PKCβ inhibition can also lower diabetic hyperfiltration (91), we speculate that this effect involves inhibitory effects on proximal tubule growth and reabsorption (70).

Tubular Senescence in the Early Diabetic Kidney

Senescence is a tumor suppressor mechanism that involves CDK inhibition to halt cells from replicating and passing on a damaged genome (102). Transient induction of p21, p16^{INK4A} (p16), and/or p27 is involved in prototypical senescent arrest or senescence-like growth arrest. Satriano et al. (95) showed an early transient induction of growth-phase components in the kidney followed by their suppression at day 10 after the onset of STZ diabetes. These events were concurrent with the induction of CDK inhibitors p16, p21, and p27 and the expression of senescence-associated β-galactosidase activity in cortical tubules (**Figure 4**) (95). Moreover, they showed that proximal tubule cells in culture transition to senescence in response to oxidative stress. An accelerated senescent phenotype was also found in tubule cells of patients with T2DM and nephropathy (103). Senescent cells are relatively well differentiated but skewed in several aspects, including a striking increase in the secretion of proinflammatory cytokines and the production of growth factors and extracellular matrix (ECM), and are resistant to apoptotic remodeling (102). Whereas the senescent arrest of tubular cells may be triggered by glucotoxic signals to prevent excessive proliferation, we speculate that such arrest contributes to inflammation and fibrosis in the diabetic kidney (**Figure 4**) and alters early proximal tubule function. One example for the latter effect may be the so-called salt paradox of the diabetic kidney, which is discussed in the following section.

THE SALT PARADOX OF THE DIABETIC KIDNEY

In normal subjects, GFR is insensitive to dietary NaCl or to changes in the same direction as the change in NaCl intake (32, 104). In 1995, we reported that a low-NaCl diet reduced renal vascular resistance and increased renal blood flow, GFR, and kidney weight in male STZ-diabetic rats (105). In contrast, female rats with early (1 week) or established (4–5 weeks) STZ diabetes responded to a high-NaCl diet with renal vasoconstriction (106). To describe the inverse relationship between dietary NaCl and GFR that is counterintuitive with regard to NaCl homeostasis, we coined the term salt paradox of the diabetic kidney. The salt paradox was confirmed in STZ-diabetic mice (37); in Long-Evans rats (107); and, most importantly, in diabetic patients, including young patients with uncomplicated T1DM, in whom restriction of dietary Na^+ to 20 mmol day^{-1} decreased renal vascular resistance and increased effective renal plasma flow and GFR (108).

How Can Dietary NaCl Suppress Glomerular Hyperfiltration, and Why Is This Unique to the Diabetic Kidney?

Micropuncture studies established that the salt paradox occurs because diabetes causes proximal tubule reabsorption to become markedly sensitive to changes in dietary NaCl such that eating more NaCl leads to greater suppression of proximal tubule reabsorption and greater MD_{NaClK} and vice versa for a low-NaCl diet, with secondary consequences on GFR via tubuloglomerular feedback (**Figure 1**) (32). In accordance, the salt paradox is absent in the STZ-diabetic, tubuloglomerular feedback–less $A_1R-/-$ mouse (37). In comparison, nondiabetic rats on various NaCl intakes manage NaCl balance primarily in the distal nephron downstream of the macula densa, and thus a tubuloglomerular feedback–mediated inverse effect of dietary NaCl on GFR does not occur (32). Considering the need to maintain effective circulating volume, if dietary NaCl restriction progresses to actual NaCl depletion, the salt paradox will become imperceptible (7).

The Salt Paradox Is Linked to Tubular Growth

Ang II and renal nerves are prominent effectors that link proximal reabsorption to total body NaCl, but neither chronic renal denervation (109) nor chronic Ang II AT_1 receptor blockade (105) prevented the salt paradox in STZ-diabetic rats. Supporting a role of diabetic kidney growth, pharmacological inhibition of ODC, which inhibited tubular growth and hyperfiltration (see above), also prevented the salt paradox (110). As discussed above and as illustrated in **Figure 4**, hypertrophic proximal tubule cells in early diabetes are continuously stimulated by mitogens, at the same time being prevented from entering the cell cycle, and have a senescent phenotype. These tubular cells may have lost the programming of a differentiated proximal tubule cell. For example, normal proximal tubule cells do not respond to moderate changes in dietary NaCl, and nephron segments downstream of the macula dense normally attend to this balance. The diabetic proximal tubule may have lost this characteristic of a differentiated nephron segment and responds strongly to dietary NaCl, forming the basis for the salt paradox. Thus, in addition to tubular hyperreabsorption and glomerular hyperfiltration, the tubular growth phenotype also contributes to this phenomenon in the diabetic kidney.

An Anomalous Role for Dietary NaCl in Diabetes Mellitus Beyond the Salt Paradox?

Many guidelines recommend low NaCl consumption for patients with T1DM or T2DM (111), even though the impact or relationship of dietary NaCl to overall mortality or ESRD has not been firmly established. Two recent prospective studies provided unexpected findings in patients with T1DM and T2DM. Both studies estimated NaCl intake on the basis of 24-h urine collections and followed up on the patients over a median of 10 years. The first study reported that survival tracked monotonically with urinary Na^+ in patients with T2DM such that a daily 100-mmol increase in Na^+ intake predicted 30% lower cardiovascular and all-cause mortality (16). The second study reported that in patients with T1DM the relationship of all-cause mortality to Na^+ intake was biphasic, with a minimum at 100–200 mmol day^{-1}, a steep rise at lower Na^+ intake, and a gradual rise at higher intake. However, the cumulative incidence of ESRD was monotonically associated with Na^+ intake such that reducing Na^+ intake from the ninetieth percentile to the tenth percentile mapped to a tenfold increase in the likelihood of developing ESRD (17). These studies question the notion that a lower NaCl intake is always better for patients with T1DM and T2DM and indicate the need for intervention studies.

The studies also pose the question as to whether there is something unique about diabetes with regard to the response to dietary NaCl. The kidney and cardiovascular system in diabetes may be predisposed to damage by counterregulatory and NaCl-conserving sympathetic nerves and hormones (like Ang II) that are progressively activated by a decline in extracellular volume and thus by a low NaCl intake. Another mechanism that may make the diabetic kidney vulnerable to chronic damage on a low-NaCl diet is the above-described salt paradox, in which a low-NaCl diet can increase GFR and can also augment diabetic kidney growth, which can predispose diabetic patients to renal failure later in life (see above). Further studies are needed to better understand the link between (*a*) dietary NaCl and (*b*) ESRD and mortality in diabetes, including the influence of NaCl intake on tubular growth and the involved molecular pathways.

LINKING THE MOLECULAR SIGNATURE OF TUBULAR GROWTH TO TUBULOINTERSTITIAL INJURY AND PROGRESSION OF DIABETIC KIDNEY DISEASE

The diabetic milieu and the prolonged interaction of albuminuria, AGE, and other factors in the glomerular filtrate with the tubular system trigger renal oxidative stress and cortical interstitial inflammation; the resulting hypoxia and tubulointerstitial fibrosis determine to a great extent the progression of diabetic renal disease (1, 19–22, 112). In addition, the molecular mechanisms involved in the early growth phenotype of the diabetic kidney may set the stage for long-term progression of diabetic kidney disease. This scenario is consistent with the observation that kidney size is linked to diabetic nephropathy (9, 11–13). This principle was recently confirmed in a cross section of older patients with longstanding diabetes mellitus and an estimated GFR of <60 ml min^{-1}. Among these patients, those with larger kidneys were more likely to develop ESRD after 5 years than were those who entered the study with smaller kidneys (10).

TGF-β may have a special role in linking the early tubular growth phenotype of the diabetic kidney to inflammation as well as to fibrotic changes, scarring, and impairment of renal function (21, 113), as illustrated in **Figure 4** and as further discussed below. We have outlined various factors that are upstream of TGF-β, including ROS, Ang II, the JAK/STAK pathway, and PKCβ. New insights into C-peptide have recently been gained. C-peptide is coreleased with insulin, and therefore plasma levels of C-peptide are low in T1DM. Substitution of C-peptide reduces diabetic kidney growth (114), tubular hyperreabsorption, and glomerular hyperfiltration (115) and lowers albuminuria/proteinuria in patients and animal models of T1DM (114, 116). Moreover, in vitro studies in human kidney proximal tubule cells showed that C-peptide can enhance the expression of hepatocyte growth factor, thereby reversing the effects of TGF-β1 (117). Here we summarize recent studies that link elements of the molecular pathways involved in diabetic tubular growth (depicted in **Figure 4**) to tubular injury.

Links to Inflammation

The release of chemokines and macrophage infiltration are important for the initiation of pathological changes in STZ-diabetic rats and human diabetic nephropathy (118–120). Growth factors like IGF-1 and TGF-β are important for diabetic tubular growth, as discussed above, but their interaction with their respective receptors on proximal and distal tubules and on collecting ducts also enhances the levels of cytokines like monocyte chemotactic protein-1 (MCP-1 or CCL2), RANTES (CCL5), and PDGF-β, which activate the proliferation of fibroblasts as well as of macrophages (**Figure 4**) (74, 120, 121). Recent studies implicated TGF-β1 in (a) the high-glucose induction of macrophage inflammatory protein-3α in human proximal tubule cells (122) and in (b) IL-18 overexpression in human tubular epithelial cells in diabetic nephropathy (123). The synergism of high glucose concentrations with cytokines such as PDGF or the proinflammatory macrophage-derived cytokine IL-1β can further stimulate TGF-β1 synthesis in proximal tubule cells (21), indicating local positive feedback loops. ROS stimulate many proinflammatory mediators relevant to chronic kidney disease, including MCP-1 and RANTES (124, 125). Furthermore, the gene expression of osteopontin, which promotes inflammation and cell recruitment, is increased by high glucose in diabetic rat proximal tubules and in rat immortalized renal proximal tubule cells via ROS generation, intrarenal RAS activation, TGF-β1 expression, and PKCβ1 signaling (126), all of which are involved in diabetic tubular growth. In contrast, the small leucine-rich proteoglycan decorin suppressed TGF-β1, connective tissue growth factor (CTGF), and p27 in tubular epithelial cells of STZ-diabetic mice and reduced upregulation of the proinflammatory proteoglycan biglycan, the infiltration of mononuclear cells, and ECM accumulation (127). These effects reflect the critical role of TGF-β1 in the promotion of both inflammation and fibrosis in the diabetic kidney.

Links to Fibrosis

High glucose induces cell growth but also increases the amount of type IV collagen and fibronectin in primary cultures of human renal proximal tubule cells (21, 128) as a consequence of decreased degradation reflecting reduced gelatinolytic activity (21, 129). The regions of active interstitial fibrosis in chronic kidney disease exhibit predominantly a peritubular distribution (130), indicating that proximal tubule cells may release fibrogenic signals to cortical fibroblasts (**Figure 4**). In fact, TGF-β1 can stimulate the release of preformed basic fibroblast growth factor from renal proximal tubule cells (131). In contrast, studies in human renal fibroblasts indicated that they can modulate proximal tubule cell growth and transport via the secretion of IGF-1 and IGF-binding protein-3 (132). Although TGF-β can induce epithelial-mesenchymal transition, the extent to which this process contributes to renal fibrosis in vivo, especially in humans, remains a matter of intense debate and may depend on the experimental and clinical context (for review, see References 1 and 133). CTGF is another prosclerotic cytokine that is induced mainly by TGF-β and IGF-1, particularly in dilated-appearing proximal tubules. CTGF is involved in the regulation of matrix accumulation and determines the outcome of diabetic renal injury in human and animal models (120, 134, 135).

Clinical trials and experimental studies implicated the importance of epigenetic processes in the development of diabetic complications. EGF contributes to diabetic kidney growth, and EGF signaling is altered by the acetylation status of histone proteins. Recent studies revealed that pharmacological inhibition of histone deacetylase (HDAC) reduced early tubular epithelial cell proliferation in diabetic rats and blunted renal growth, which may be mediated in part through downregulation of the EGF receptor (136). Other studies implicated HDAC-2 in the development of ECM accumulation in the diabetic kidney and showed that ROS mediate TGF-β1-induced activation of HDAC-2 (137). These findings indicate that epigenetic processes affect renal growth and fibrosis of the diabetic kidney.

HIF: hypoxia-inducible factor

Links to Hypoxia and Apoptosis

Hypoxia can be due to enhanced tubular oxygen consumption, as shown ex vivo in cortical and medullary tubule cells of STZ-diabetic rats (138). Hypoxia has been implicated as a cause of oxidative stress in the diabetic kidney and in the pathophysiology of diabetic nephropathy (112). Superoxide enhances Na-K-2Cl cotransporter activity in the thick ascending limb (TAL), which can further aggravate renal hypoxia (139). Defense against hypoxia involves hypoxia-inducible factor (HIF). STZ-diabetic rats and Cohen diabetes–sensitive rats, a nonobese normolipidemic genetic model of diet-induced T2DM, transiently upregulated the hypoxia marker pimonidazole and HIF-1α, primarily in the TAL of the renal outer medulla (140). Whether there is any link to the formation of glycogen deposits (Armanni-Ebstein lesions), which are found particularly in the cells of the TAL, is not known (1). Importantly, diabetes or high glucose levels appear to blunt the hypoxia-induced HIF pathway by inducing oxidative stress, as observed in the kidneys of STZ-diabetic rats (141) and in rat proximal tubule cells in vitro (**Figure 4**) (140, 141). In accordance, dietary eicosapentaenoic acid has beneficial effects on STZ-diabetic kidney injury by suppressing ROS generation and mitochondrial apoptosis, partly through augmentation of the HIF-1α response (142). Overexpression of catalase in renal proximal tubule cells not only attenuated apoptosis in STZ diabetes (143) and in *db/db* transgenic mice (144) but also reduced interstitial fibrosis in the latter model (144). Studies in *db/db* mice implicated Nox4-based NADPH oxidase in glucose-induced oxidative stress in proximal tubules and fibrosis (145). Inhibition of JAK2 protected renal endothelial and epithelial cells from oxidative stress (146), and a direct relationship between tubulointerstitial JAK/STAT expression and progression of kidney failure was found in patients with T2DM (147). In accordance, SOCS proteins inhibit the expression of STAT-dependent genes (involved in cell proliferation, inflammation, and fibrosis) and improve renal function in diabetes (79). Besides ROS (148), both PKCβ (149) and the TSC2/mTOR pathway (150) have been implicated in promoting apoptosis of proximal tubule cells in diabetes.

PERSPECTIVES

Inhibition of proximal tubule glucose reabsorption via SGLT2 is a promising new therapeutic approach that can lower blood glucose levels and attenuate glomerular hyperfiltration in diabetes, and ongoing studies are assessing the long-term safety of this approach. Tubular glucose uptake contributes to tubular injury, but further studies are needed to better define the role of GLUT2 translocation and the role of SGLTs and GLUT2 in tubular injury and growth in the diabetic kidney. Further studies are required to elucidate the proposed deleterious effects of dietary NaCl restriction on kidney function and mortality in diabetic patients. The unique early growth phenotype of the proximal tubule is important for early tubular hyperreabsorption, glomerular hyperfiltration, and the salt paradox. However, we speculate that the molecular signature of tubular growth sets the stage for the development of inflammation, fibrosis, tubulointerstitial injury, hypoxia, and apoptosis, which would explain the strong correlation between kidney size and prognosis of kidney outcome in diabetic patients. Genetic and environmental factors may modulate the tubular response to hyperglycemia, thereby contributing to the fact that only some diabetic patients develop large kidneys, tubular hyperreabsorption, glomerular hyperfiltration, and diabetic nephropathy. Finally, we need to better understand why it takes 10 to 20 years for the diabetic milieu to cause renal failure when many of the proposed deleterious molecular pathways can be activated within hours, days, or weeks of hyperglycemia.

SUMMARY POINTS

1. Enhanced Na^+-glucose cotransport and tubular growth cause a primary increase in proximal tubule reabsorption in the diabetic kidney, which through tubuloglomerular feedback induces glomerular hyperfiltration.

2. The kidney in general and the proximal tubules in particular grow large from the onset of diabetes mellitus, and kidney size has been linked to the development of diabetic nephropathy.

3. Hypertrophic proximal tubule cells in early diabetes are continuously stimulated by mitogens, at the same time being prevented from entering the cell cycle, and have a senescent phenotype.

4. The salt paradox, a unique phenomenon of the diabetic kidney that refers to an inverse relationship between changes in dietary NaCl intake and GFR, occurs because diabetes causes proximal tubule reabsorption to become extensively sensitive to changes in dietary NaCl, which is related to the tubular growth phenotype.

5. The molecular signature of tubular growth in the diabetic kidney is linked to tubulointerstitial fibrosis, inflammation, hypoxia, and apoptosis and may set the stage for tubulointerstitial injury and the progression of diabetic kidney disease.

6. Genetic and environmental factors may modulate the tubular response to hyperglycemia, thereby contributing to the fact that only some diabetic patients develop large kidneys, tubular hyperreabsorption, glomerular hyperfiltration, and diabetic nephropathy.

DISCLOSURE STATEMENT

The authors' work was supported by Bristol-Myers Squibb and Astra-Zeneca.

ACKNOWLEDGMENTS

We apologize to all the investigators whose research could not be appropriately cited owing to space constraints. The authors' work was supported by the National Institutes of Health (R01DK56248, R01HL094728, R01DK28602, R01GM66232, P30DK079337), the American Heart Association (GRNT3440038), the Department of Veterans Affairs, Bristol-Myers Squibb, and Astra-Zeneca.

LITERATURE CITED

1. Vallon V, Komers R. 2011. Pathophysiology of the diabetic kidney. *Compr. Physiol.* 1:1175–1232
2. Kanwar YS, Sun L, Xie P, Liu FY, Chen S. 2011. A glimpse of various pathogenetic mechanisms of diabetic nephropathy. *Annu. Rev. Pathol.* 6:395–423

3. Ziyadeh FN, Wolf G. 2008. Pathogenesis of the podocytopathy and proteinuria in diabetic glomerulopathy. *Curr. Diabetes Rev.* 4:39–45
4. Jerums G, Premaratne E, Panagiotopoulos S, Macisaac RJ. 2010. The clinical significance of hyperfiltration in diabetes. *Diabetologia* 53:2093–104
5. Magee GM, Bilous RW, Cardwell CR, Hunter SJ, Kee F, Fogarty DG. 2009. Is hyperfiltration associated with the future risk of developing diabetic nephropathy? A meta-analysis. *Diabetologia* 52:691–97
6. Slomowitz LA, Peterson OW, Thomson SC. 1999. Converting enzyme inhibition and the glomerular hemodynamic response to glycine in diabetic rats. *J. Am. Soc. Nephrol.* 10:1447–54
7. Vallon V, Blantz RC, Thomson S. 2003. Glomerular hyperfiltration and the salt paradox in early type 1 diabetes mellitus: a tubulo-centric view. *J. Am. Soc. Nephrol.* 14:530–37
8. Thomson SC, Vallon V, Blantz RC. 2004. Kidney function in early diabetes: the tubular hypothesis of glomerular filtration. *Am. J. Physiol. Ren. Physiol.* 286:8–15
9. Zerbini G, Bonfanti R, Meschi F, Bognetti E, Paesano PL, et al. 2006. Persistent renal hypertrophy and faster decline of glomerular filtration rate precede the development of microalbuminuria in type 1 diabetes. *Diabetes* 55:2620–25
10. Rigalleau V, Garcia M, Lasseur C, Laurent F, Montaudon M, et al. 2010. Large kidneys predict poor renal outcome in subjects with diabetes and chronic kidney disease. *BMC. Nephrol.* 11:3
11. Lawson ML, Sochett EB, Chait PG, Balfe JW, Daneman D. 1996. Effect of puberty on markers of glomerular hypertrophy and hypertension in IDDM. *Diabetes* 45:51–55
12. Bognetti E, Zoja A, Meschi F, Paesano PL, Chiumello G. 1996. Relationship between kidney volume, microalbuminuria and duration of diabetes mellitus. *Diabetologia* 39:1409
13. Baumgartl HJ, Sigl G, Banholzer P, Haslbeck M, Standl E. 1998. On the prognosis of IDDM patients with large kidneys. *Nephrol. Dial. Transplant.* 13:630–34
14. Vallon V, Sharma K. 2010. Sodium-glucose transport: role in diabetes mellitus and potential clinical implications. *Curr. Opin. Nephrol. Hypertens.* 19:425–31
15. Nair S, Wilding JP. 2010. Sodium glucose cotransporter 2 inhibitors as a new treatment for diabetes mellitus. *J. Clin. Endocrinol. Metab.* 95:34–42
16. Ekinci EI, Clarke S, Thomas MC, Moran JL, Cheong K, et al. 2011. Dietary salt intake and mortality in patients with type 2 diabetes. *Diabetes Care* 34:703–709
17. Thomas MC, Moran J, Forsblom C, Harjutsalo V, Thorn L, et al. 2011. The association between dietary sodium intake, ESRD, and all-cause mortality in patients with type 1 diabetes. *Diabetes Care* 34:861–66
18. Vallon V, Blantz R, Thomson S. 2005. The salt paradox and its possible implications in managing hypertensive diabetic patients. *Curr. Hypertens. Rep.* 7:141–47
19. Vallon V. 2011. The proximal tubule in the pathophysiology of the diabetic kidney. *Am. J. Physiol. Regul. Integr. Comp. Physiol.* 300:1009–22
20. Thomas MC, Burns WC, Cooper ME. 2005. Tubular changes in early diabetic nephropathy. *Adv. Chronic Kidney Dis.* 12:177–86
21. Phillips AO, Steadman R. 2002. Diabetic nephropathy: the central role of renal proximal tubular cells in tubulointerstitial injury. *Histol. Histopathol.* 17:247–52
22. Magri CJ, Fava S. 2009. The role of tubular injury in diabetic nephropathy. *Eur. J. Intern. Med.* 20:551–55
23. Ditzel J, Lervang HH, Brochner-Mortensen J. 1989. Renal sodium metabolism in relation to hypertension in diabetes. *Diabète Métab.* 15:292–95
24. Mbanya JC, Thomas TH, Taylor R, Alberti KG, Wilkinson R. 1989. Increased proximal tubular sodium reabsorption in hypertensive patients with type 2 diabetes. *Diabet. Med.* 6:614–20
25. Vervoort G, Veldman B, Berden JH, Smits P, Wetzels JF. 2005. Glomerular hyperfiltration in type 1 diabetes mellitus results from primary changes in proximal tubular sodium handling without changes in volume expansion. *Eur. J. Clin. Investig.* 35:330–36
26. Hannedouche TP, Delgado AG, Gnionsahe DA, Boitard C, Lacour B, Grunfeld JP. 1990. Renal hemodynamics and segmental tubular reabsorption in early type 1 diabetes. *Kidney Int.* 37:1126–33
27. Pruijm M, Wuerzner G, Maillard M, Bovet P, Renaud C, et al. 2010. Glomerular hyperfiltration and increased proximal sodium reabsorption in subjects with type 2 diabetes or impaired fasting glucose in a population of the African region. *Nephrol. Dial. Transplant.* 25:2225–31

28. Pollock CA, Lawrence JR, Field MJ. 1991. Tubular sodium handling and tubuloglomerular feedback in experimental diabetes mellitus. *Am. J. Physiol. Ren. Physiol.* 260:946–52
29. Vallon V, Blantz RC, Thomson S. 1995. Homeostatic efficiency of tubuloglomerular feedback is reduced in established diabetes mellitus in rats. *Am. J. Physiol. Ren. Physiol.* 269:876–83
30. Vallon V, Richter K, Blantz RC, Thomson S, Osswald H. 1999. Glomerular hyperfiltration in experimental diabetes mellitus: potential role of tubular reabsorption. *J. Am. Soc. Nephrol.* 10:2569–76
31. Thomson SC, Deng A, Bao D, Satriano J, Blantz RC, Vallon V. 2001. Ornithine decarboxylase, kidney size, and the tubular hypothesis of glomerular hyperfiltration in experimental diabetes. *J. Clin. Investig.* 107:217–24
32. Vallon V, Huang DY, Deng A, Richter K, Blantz RC, Thomson S. 2002. Salt-sensitivity of proximal reabsorption alters macula densa salt and explains the paradoxical effect of dietary salt on glomerular filtration rate in diabetes mellitus. *J. Am. Soc. Nephrol.* 13:1865–71
33. Thomson SC, Vallon V, Blantz RC. 1998. Resetting protects efficiency of tubuloglomerular feedback. *Kidney Int. Suppl.* 67:S65–70
34. Sallstrom J, Carlsson PO, Fredholm BB, Larsson E, Persson AE, Palm F. 2007. Diabetes-induced hyperfiltration in adenosine A_1-receptor deficient mice lacking the tubuloglomerular feedback mechanism. *Acta Physiol.* 190:253–59
35. Faulhaber-Walter R, Chen L, Oppermann M, Kim SM, Huang Y, et al. 2008. Lack of A1 adenosine receptors augments diabetic hyperfiltration and glomerular injury. *J. Am. Soc. Nephrol.* 19:722–30
36. Weinstein AM. 1986. Osmotic diuresis in a mathematical model of the rat proximal tubule. *Am. J. Physiol. Ren. Physiol.* 250:874–84
37. Vallon V, Schroth J, Satriano J, Blantz RC, Thomson SC, Rieg T. 2009. Adenosine A_1 receptors determine glomerular hyperfiltration and the salt paradox in early streptozotocin diabetes mellitus. *Nephron. Physiol.* 111:30–38
38. Woods LL, Mizelle HL, Hall JE. 1987. Control of renal hemodynamics in hyperglycemia: possible role of tubuloglomerular feedback. *Am. J. Physiol. Ren. Physiol.* 252:65–73
39. Balen D, Ljubojevic M, Breljak D, Brzica H, Zlender V, et al. 2008. Revised immunolocalization of the Na^+-D-glucose cotransporter SGLT1 in rat organs with an improved antibody. *Am. J. Physiol. Cell Physiol.* 295:475–89
40. Vallon V, Platt KA, Cunard R, Schroth J, Whaley J, et al. 2011. SGLT2 mediates glucose reabsorption in the early proximal tubule. *J. Am. Soc. Nephrol.* 22:104–12
41. Hummel CS, Lu C, Loo DF, Hirayama BA, Voss AA, Wright EM. 2011. Glucose transport by human renal Na^+-D-glucose cotransporters. *Am. J. Physiol. Cell Physiol.* 300:14–21
42. Gorboulev V, Schürmann A, Vallon V, Kipp H, Jaschke A, et al. 2011. Na^+-d-glucose cotransporter SGLT1 is pivotal for intestinal glucose absorption and glucose-dependent incretin secretion. *Diabetes.* doi: 10.2337/db11-1029
43. Santer R, Calado J. 2010. Familial renal glucosuria and SGLT2: from a Mendelian trait to a therapeutic target. *Clin. J. Am. Soc. Nephrol.* 5:133–41
44. Vestri S, Okamoto MM, De Freitas HS, Aparecida Dos Santos R, Nunes MT, et al. 2001. Changes in sodium or glucose filtration rate modulate expression of glucose transporters in renal proximal tubular cells of rat. *J. Membr. Biol.* 182:105–12
45. Vidotti DB, Arnoni CP, Maquigussa E, Boim MA. 2008. Effect of long-term type 1 diabetes on renal sodium and water transporters in rats. *Am. J. Nephrol.* 28:107–14
46. Tabatabai NM, Sharma M, Blumenthal SS, Petering DH. 2009. Enhanced expressions of sodium-glucose cotransporters in the kidneys of diabetic Zucker rats. *Diabetes Res. Clin. Pract.* 83:e27–30
47. Rahmoune H, Thompson PW, Ward JM, Smith CD, Hong G, Brown J. 2005. Glucose transporters in human renal proximal tubular cells isolated from the urine of patients with non-insulin-dependent diabetes. *Diabetes* 54:3427–34
48. Osorio H, Bautista R, Rios A, Franco M, Santamaria J, Escalante B. 2009. Effect of treatment with losartan on salt sensitivity and SGLT2 expression in hypertensive diabetic rats. *Diabetes Res. Clin. Pract.* 86:e46–49

49. Freitas HS, Anhe GF, Melo KF, Okamoto MM, Oliveira-Souza M, et al. 2008. Na$^+$-glucose transporter-2 messenger ribonucleic acid expression in kidney of diabetic rats correlates with glycemic levels: involvement of hepatocyte nuclear factor-1α expression and activity. *Endocrinology* 149:717–24

50. Lang F, Gorlach A, Vallon V. 2009. Targeting SGK1 in diabetes. *Expert Opin. Ther. Targets* 13:1303–11

51. Ackermann TF, Boini KM, Volkl H, Bhandaru M, Bareiss PM, et al. 2009. SGK1-sensitive renal tubular glucose reabsorption in diabetes. *Am. J. Physiol. Ren. Physiol.* 296:859–66

52. Embark HM, Bohmer C, Vallon V, Luft F, Lang F. 2003. Regulation of KCNE1-dependent K$^{(+)}$ current by the serum and glucocorticoid-inducible kinase (SGK) isoforms. *Pflüg. Arch.* 445:601–6

53. Vallon V, Grahammer F, Richter K, Bleich M, Lang F, et al. 2001. Role of KCNE1-dependent K$^+$ fluxes in mouse proximal tubule. *J. Am. Soc. Nephrol.* 12:2003–11

54. Vallon V, Grahammer F, Volkl H, Sandu CD, Richter K, et al. 2005. KCNQ1-dependent transport in renal and gastrointestinal epithelia. *Proc. Natl. Acad. Sci. USA* 102:17864–69

55. Saad S, Stevens VA, Wassef L, Poronnik P, Kelly DJ, et al. 2005. High glucose transactivates the EGF receptor and up-regulates serum glucocorticoid kinase in the proximal tubule. *Kidney Int.* 68:985–97

56. Albertoni Borghese MF, Majowicz MP, Ortiz MC, Passalacqua MR, Sterin Speziale NB, Vidal NA. 2009. Expression and activity of SGLT2 in diabetes induced by streptozotocin: relationship with the lipid environment. *Nephron Physiol.* 112:45–52

57. Marks J, Carvou NJ, Debnam ES, Srai SK, Unwin RJ. 2003. Diabetes increases facilitative glucose uptake and GLUT2 expression at the rat proximal tubule brush border membrane. *J. Physiol.* 553:137–45

58. Han HJ, Lee YJ, Park SH, Lee JH, Taub M. 2005. High glucose-induced oxidative stress inhibits Na$^+$/glucose cotransporter activity in renal proximal tubule cells. *Am. J. Physiol. Ren. Physiol.* 288:988–96

59. Blank ME, Bode F, Huland E, Diedrich DF, Baumann K. 1985. Kinetic studies of D-glucose transport in renal brush-border membrane vesicles of streptozotocin-induced diabetic rats. *Biochim. Biophys. Acta* 844:314–19

60. Bank N, Aynedjian HS. 1990. Progressive increases in luminal glucose stimulate proximal sodium absorption in normal and diabetic rats. *J. Clin. Investig.* 86:309–16

61. Thomson SC, Rieg T, Miracle C, Mansoury H, Whaley J, et al. 2011. Acute and chronic effects of SGLT2 blockade on glomerular and tubular function in the early diabetic rat. *Am. J. Physiol. Regul. Integr. Comp. Physiol.* doi: 10.1152/ajpregu.00357.2011

62. Seyer-Hansen K. 1987. Renal hypertrophy in experimental diabetes: some functional aspects. *J. Diabet. Complicat.* 1:7–10

63. Ly JP, Onay T, Sison K, Sivaskandarajah G, Sabbisetti V, et al. 2011. The Sweet Pee model for Sglt2 mutation. *J. Am. Soc. Nephrol.* 22:113–23

64. Kumar AM, Gupta RK, Spitzer A. 1988. Intracellular sodium in proximal tubules of diabetic rats. Role of glucose. *Kidney Int.* 33:792–97

65. Dominguez JH, Camp K, Maianu L, Feister H, Garvey WT. 1994. Molecular adaptations of GLUT1 and GLUT2 in renal proximal tubules of diabetic rats. *Am. J. Physiol. Ren. Physiol.* 266:283–90

66. Freitas HS, Schaan BD, David-Silva A, Sabino-Silva R, Okamoto MM, et al. 2009. SLC2A2 gene expression in kidney of diabetic rats is regulated by HNF-1α and HNF-3β. *Mol. Cell. Endocrinol.* 305:63–70

67. Kellett GL, Brot-Laroche E, Mace OJ, Leturque A. 2008. Sugar absorption in the intestine: the role of GLUT2. *Annu. Rev. Nutr.* 28:35–54

68. Goestemeyer AK, Marks J, Srai SK, Debnam ES, Unwin RJ. 2007. GLUT2 protein at the rat proximal tubule brush border membrane correlates with protein kinase C (PKC)-β1 and plasma glucose concentration. *Diabetologia* 50:2209–17

69. Pfaff IL, Wagner HJ, Vallon V. 1999. Immunolocalization of protein kinase C isoenzymes α, β1 and βII in rat kidney. *J. Am. Soc. Nephrol.* 10:1861–73

70. Pfaff IL, Vallon V. 2002. Protein kinase Cβ isoenzymes in diabetic kidneys and their relation to nephroprotective actions of the ACE inhibitor lisinopril. *Kidney Blood Press. Res.* 25:329–40

71. Blantz RC, Tucker BJ, Gushwa L, Peterson OW. 1983. Mechanism of diuresis following acute modest hyperglycemia in the rat. *Am. J. Physiol. Ren. Physiol.* 244:185–94

72. Huang HC, Preisig PA. 2000. G1 kinases and transforming growth factor-β signaling are associated with a growth pattern switch in diabetes-induced renal growth. *Kidney Int.* 58:162–72
73. Wolf G, Ziyadeh FN. 1999. Molecular mechanisms of diabetic renal hypertrophy. *Kidney Int.* 56:393–405
74. Chiarelli F, Gaspari S, Marcovecchio ML. 2009. Role of growth factors in diabetic kidney disease. *Horm. Metab. Res.* 41:585–93
75. Feliers D, Kasinath BS. 2010. Mechanism of VEGF expression by high glucose in proximal tubule epithelial cells. *Mol. Cell. Endocrinol.* 314:136–42
76. Marrero MB, Banes-Berceli AK, Stern DM, Eaton DC. 2006. Role of the JAK/STAT signaling pathway in diabetic nephropathy. *Am. J. Physiol. Ren. Physiol.* 290:762–68
77. Hernandez-Vargas P, Lopez-Franco O, Sanjuan G, Ruperez M, Ortiz-Munoz G, et al. 2005. Suppressors of cytokine signaling regulate angiotensin II-activated Janus kinase-signal transducers and activators of transcription pathway in renal cells. *J. Am. Soc. Nephrol.* 16:1673–83
78. Amiri F, Shaw S, Wang X, Tang J, Waller JL, et al. 2002. Angiotensin II activation of the JAK/STAT pathway in mesangial cells is altered by high glucose. *Kidney Int.* 61:1605–16
79. Ortiz-Munoz G, Lopez-Parra V, Lopez-Franco O, Fernandez-Vizarra P, Mallavia B, et al. 2010. Suppressors of cytokine signaling abrogate diabetic nephropathy. *J. Am. Soc. Nephrol.* 21:763–72
80. Satriano J, Vallon V. 2006. Primary kidney growth and its consequences at the onset of diabetes mellitus. *Amino Acids* 31:1–9
81. Pedersen SB, Flyvbjerg A, Richelsen B. 1993. Inhibition of renal ornithine decarboxylase activity prevents kidney hypertrophy in experimental diabetes. *Am. J. Physiol. Cell Physiol.* 264:453–56
82. Deng A, Munger KA, Valdivielso JM, Satriano J, Lortie M, et al. 2003. Increased expression of ornithine decarboxylase in distal tubules of early diabetic rat kidneys: Are polyamines paracrine hypertrophic factors? *Diabetes* 52:1235–39
83. Brownlee M. 2001. Biochemistry and molecular cell biology of diabetic complications. *Nature* 414:813–20
84. Wu D, Peng F, Zhang B, Ingram AJ, Kelly DJ, et al. 2009. PKC-β1 mediates glucose-induced Akt activation and TGF-β1 upregulation in mesangial cells. *J. Am. Soc. Nephrol.* 20:554–66
85. Meier M, Park JK, Overheu D, Kirsch T, Lindschau C, et al. 2007. Deletion of protein kinase C-β isoform in vivo reduces renal hypertrophy but not albuminuria in the streptozotocin-induced diabetic mouse model. *Diabetes* 56:346–54
86. Lee MJ, Feliers D, Mariappan MM, Sataranatarajan K, Mahimainathan L, et al. 2007. A role for AMP-activated protein kinase in diabetes-induced renal hypertrophy. *Am. J. Physiol. Ren. Physiol.* 292:617–27
87. Lieberthal W, Levine JS. 2009. The role of the mammalian target of rapamycin (mTOR) in renal disease. *J. Am. Soc. Nephrol.* 20:2493–502
88. Han DC, Hoffman BB, Hong SW, Guo J, Ziyadeh FN. 2000. Therapy with antisense TGF-β1 oligodeoxynucleotides reduces kidney weight and matrix mRNAs in diabetic mice. *Am. J. Physiol. Ren. Physiol.* 278:628–34
89. Chen S, Hoffman BB, Lee JS, Kasama Y, Jim B, et al. 2004. Cultured tubule cells from TGF-β1 null mice exhibit impaired hypertrophy and fibronectin expression in high glucose. *Kidney Int.* 65:1191–204
90. Wang X, Shaw S, Amiri F, Eaton DC, Marrero MB. 2002. Inhibition of the Jak/STAT signaling pathway prevents the high glucose-induced increase in TGF-β and fibronectin synthesis in mesangial cells. *Diabetes* 51:3505–9
91. Noh H, King GL. 2007. The role of protein kinase C activation in diabetic nephropathy. *Kidney Int. Suppl.* 106:S49–53
92. Fujita H, Omori S, Ishikura K, Hida M, Awazu M. 2004. ERK and p38 mediate high-glucose-induced hypertrophy and TGF-β expression in renal tubular cells. *Am. J. Physiol. Ren. Physiol.* 286:120–26
93. Kamesaki H, Nishizawa K, Michaud GY, Cossman J, Kiyono T. 1998. TGF-$β_1$ induces the cyclin-dependent kinase inhibitor p27[Kip1] mRNA and protein in murine B cells. *J. Immunol.* 160:770–77
94. Wolf G, Schroeder R, Ziyadeh FN, Thaiss F, Zahner G, Stahl RA. 1997. High glucose stimulates expression of p27Kip1 in cultured mouse mesangial cells: relationship to hypertrophy. *Am. J. Physiol. Ren. Physiol.* 273:348–56
95. Satriano J, Mansoury H, Deng A, Sharma K, Vallon V, et al. 2010. Transition of kidney tubule cells to a senescent phenotype in early experimental diabetes. *Am. J. Physiol. Cell Physiol.* 299:374–80

96. Al-Douahji M, Brugarolas J, Brown PA, Stehman-Breen CO, Alpers CE, Shankland SJ. 1999. The cyclin kinase inhibitor p21$^{WAF1/CIP1}$ is required for glomerular hypertrophy in experimental diabetic nephropathy. *Kidney Int.* 56:1691–99

97. Huang JS, Chuang LY, Guh JY, Huang YJ, Hsu MS. 2007. Antioxidants attenuate high glucose-induced hypertrophic growth in renal tubular epithelial cells. *Am. J. Physiol. Ren. Physiol.* 293:1072–82

98. Huang JS, Chuang LY, Guh JY, Huang YJ. 2009. Effects of nitric oxide and antioxidants on advanced glycation end products-induced hypertrophic growth in human renal tubular cells. *Toxicol. Sci.* 111:109–19

99. Franch HA. 2002. Pathways of proteolysis affecting renal cell growth. *Curr. Opin. Nephrol. Hypertens.* 11:445–50

100. Ku DD, Sellers BM, Meezan E. 1986. Development of renal hypertrophy and increased renal Na,K-ATPase in streptozotocin-diabetic rats. *Endocrinology* 119:672–79

101. Efendiev R, Bertorello AM, Pedemonte CH. 1999. PKC-β and PKC-ζ mediate opposing effects on proximal tubule Na$^+$,K$^+$-ATPase activity. *FEBS Lett.* 456:45–48

102. Ren JL, Pan JS, Lu YP, Sun P, Han J. 2009. Inflammatory signaling and cellular senescence. *Cell Signal.* 21:378–83

103. Verzola D, Gandolfo MT, Gaetani G, Ferraris A, Mangerini R, et al. 2008. Accelerated senescence in the kidneys of patients with type 2 diabetic nephropathy. *Am. J. Physiol. Ren. Physiol.* 295:1563–73

104. Thomson SC, Deng A, Wead L, Richter K, Blantz RC, Vallon V. 2006. An unexpected role for angiotensin II in the link between dietary salt and proximal reabsorption. *J. Clin. Investig.* 116:1110–16

105. Vallon V, Wead LM, Blantz RC. 1995. Renal hemodynamics and plasma and kidney angiotensin II in established diabetes mellitus in rats: effect of sodium and salt restriction. *J. Am. Soc. Nephrol.* 5:1761–67

106. Vallon V, Kirschenmann D, Wead LM, Lortie MJ, Satriano J, et al. 1997. Effect of chronic salt loading on kidney function in early and established diabetes mellitus in rats. *J. Lab. Clin. Med.* 130:76–82

107. Lau C, Sudbury I, Thomson M, Howard PL, Magil AB, Cupples WA. 2009. Salt-resistant blood pressure and salt-sensitive renal autoregulation in chronic streptozotocin diabetes. *Am. J. Physiol. Regul. Integr. Comp. Physiol.* 296:1761–70

108. Miller JA. 1997. Renal responses to sodium restriction in patients with early diabetes mellitus. *J. Am. Soc. Nephrol.* 8:749–55

109. Birk C, Richter K, Huang DY, Piesch C, Luippold G, Vallon V. 2003. The salt paradox of the early diabetic kidney is independent of renal innervation. *Kidney Blood Press. Res.* 26:344–50

110. Miracle CM, Rieg T, Mansoury H, Vallon V, Thomson SC. 2008. Ornithine decarboxylase inhibitor eliminates hyperresponsiveness of the early diabetic proximal tubule to dietary salt. *Am. J. Physiol. Ren. Physiol.* 295:995–1002

111. Bantle JP, Wylie-Rosett J, Albright AL, Apovian CM, Clark NG, et al. 2008. Nutrition recommendations and interventions for diabetes: a position statement of the American Diabetes Association. *Diabetes Care* 31(Suppl. 1):61–78

112. Singh DK, Winocour P, Farrington K. 2008. Mechanisms of disease: the hypoxic tubular hypothesis of diabetic nephropathy. *Nat. Clin. Pract. Nephrol.* 4:216–26

113. Ziyadeh FN, Hoffman BB, Han DC, Iglesias–De La Cruz MC, Hong SW, et al. 2000. Long-term prevention of renal insufficiency, excess matrix gene expression, and glomerular mesangial matrix expansion by treatment with monoclonal antitransforming growth factor-β antibody in *db/db* diabetic mice. *Proc. Natl. Acad. Sci. USA* 97:8015–20

114. Samnegard B, Jacobson SH, Jaremko G, Johansson BL, Sjoquist M. 2001. Effects of C-peptide on glomerular and renal size and renal function in diabetic rats. *Kidney Int.* 60:1258–65

115. Nordquist L, Brown R, Fasching A, Persson P, Palm F. 2009. Proinsulin C-peptide reduces diabetes-induced glomerular hyperfiltration via efferent arteriole dilation and inhibition of tubular sodium reabsorption. *Am. J. Physiol. Ren. Physiol.* 297:1265–72

116. Johansson BL, Borg K, Fernqvist-Forbes E, Kernell A, Odergren T, Wahren J. 2000. Beneficial effects of C-peptide on incipient nephropathy and neuropathy in patients with Type 1 diabetes mellitus. *Diabet. Med.* 17:181–89

117. Hills CE, Willars GB, Brunskill NJ. 2010. Proinsulin C-peptide antagonizes the profibrotic effects of TGF-β1 via up-regulation of retinoic acid and HGF-related signaling pathways. *Mol. Endocrinol.* 24:822–31
118. Ruster C, Wolf G. 2008. The role of chemokines and chemokine receptors in diabetic nephropathy. *Front. Biosci.* 13:944–55
119. Navarro-Gonzalez JF, Mora-Fernandez C, de Fuentes MM, Garcia-Perez J. 2011. Inflammatory molecules and pathways in the pathogenesis of diabetic nephropathy. *Nat. Rev. Nephrol.* 7:327–40
120. Hirschberg R, Wang S. 2005. Proteinuria and growth factors in the development of tubulointerstitial injury and scarring in kidney disease. *Curr. Opin. Nephrol. Hypertens.* 14:43–52
121. Chow FY, Nikolic-Paterson DJ, Ozols E, Atkins RC, Rollin BJ, Tesch GH. 2006. Monocyte chemoattractant protein-1 promotes the development of diabetic renal injury in streptozotocin-treated mice. *Kidney Int.* 69:73–80
122. Qi W, Chen X, Zhang Y, Holian J, Mreich E, et al. 2007. High glucose induces macrophage inflammatory protein-3α in renal proximal tubule cells via a transforming growth factor-β1 dependent mechanism. *Nephrol. Dial. Transplant.* 22:3147–53
123. Miyauchi K, Takiyama Y, Honjyo J, Tateno M, Haneda M. 2009. Upregulated IL-18 expression in type 2 diabetic subjects with nephropathy: TGF-β1 enhanced IL-18 expression in human renal proximal tubular epithelial cells. *Diabetes Res. Clin. Pract.* 83:190–99
124. Okamura DM, Himmelfarb J. 2009. Tipping the redox balance of oxidative stress in fibrogenic pathways in chronic kidney disease. *Pediatr. Nephrol.* 24:2309–19
125. Shah SV, Baliga R, Rajapurkar M, Fonseca VA. 2007. Oxidants in chronic kidney disease. *J. Am. Soc. Nephrol.* 18:16–28
126. Hsieh TJ, Chen R, Zhang SL, Liu F, Brezniceanu ML, et al. 2006. Upregulation of osteopontin gene expression in diabetic rat proximal tubular cells revealed by microarray profiling. *Kidney Int.* 69:1005–15
127. Merline R, Lazaroski S, Babelova A, Tsalastra-Greul W, Pfeilschifter J, et al. 2009. Decorin deficiency in diabetic mice: aggravation of nephropathy due to overexpression of profibrotic factors, enhanced apoptosis and mononuclear cell infiltration. *J. Physiol. Pharmacol.* 60(Suppl. 4):5–13
128. Jones SC, Saunders HJ, Pollock CA. 1999. High glucose increases growth and collagen synthesis in cultured human tubulointerstitial cells. *Diabet. Med.* 16:932–38
129. Phillips AO, Morrisey K, Steadman R, Williams JD. 1999. Decreased degradation of collagen and fibronectin following exposure of proximal cells to glucose. *Exp. Nephrol.* 7:449–62
130. Alpers CE, Hudkins KL, Floege J, Johnson RJ. 1994. Human renal cortical interstitial cells with some features of smooth muscle cells participate in tubulointerstitial and crescentic glomerular injury. *J. Am. Soc. Nephrol.* 5:201–9
131. Jones SG, Morrisey K, Williams JD, Phillips AO. 1999. TGF-β1 stimulates the release of pre-formed bFGF from renal proximal tubular cells. *Kidney Int.* 56:83–91
132. Johnson DW, Saunders HJ, Brew BK, Ganesan A, Baxter RC, et al. 1997. Human renal fibroblasts modulate proximal tubule cell growth and transport via the IGF-I axis. *Kidney Int.* 52:1486–96
133. Liu Y. 2010. New insights into epithelial-mesenchymal transition in kidney fibrosis. *J. Am. Soc. Nephrol.* 21:212–22
134. Wang S, Denichilo M, Brubaker C, Hirschberg R. 2001. Connective tissue growth factor in tubulointerstitial injury of diabetic nephropathy. *Kidney Int.* 60:96–105
135. Guha M, Xu ZG, Tung D, Lanting L, Natarajan R. 2007. Specific down-regulation of connective tissue growth factor attenuates progression of nephropathy in mouse models of type 1 and type 2 diabetes. *FASEB J.* 21:3355–68
136. Gilbert RE, Huang Q, Thai K, Advani SL, Lee K, et al. 2011. Histone deacetylase inhibition attenuates diabetes-associated kidney growth: potential role for epigenetic modification of the epidermal growth factor receptor. *Kidney Int.* 79:1312–21
137. Noh H, Oh EY, Seo JY, Yu MR, Kim YO, et al. 2009. Histone deacetylase-2 is a key regulator of diabetes- and transforming growth factor-β1-induced renal injury. *Am. J. Physiol. Ren. Physiol.* 297:729–39
138. Palm F, Cederberg J, Hansell P, Liss P, Carlsson PO. 2003. Reactive oxygen species cause diabetes-induced decrease in renal oxygen tension. *Diabetologia* 46:1153–60

139. Juncos R, Garvin JL. 2005. Superoxide enhances Na-K-2Cl cotransporter activity in the thick ascending limb. *Am. J. Physiol. Ren. Physiol.* 288:982–87
140. Rosenberger C, Khamaisi M, Abassi Z, Shilo V, Weksler-Zangen S, et al. 2008. Adaptation to hypoxia in the diabetic rat kidney. *Kidney Int.* 73:34–42
141. Katavetin P, Miyata T, Inagi R, Tanaka T, Sassa R, et al. 2006. High glucose blunts vascular endothelial growth factor response to hypoxia via the oxidative stress-regulated hypoxia-inducible factor/hypoxia-responsible element pathway. *J. Am. Soc. Nephrol.* 17:1405–13
142. Taneda S, Honda K, Tomidokoro K, Uto K, Nitta K, Oda H. 2010. Eicosapentaenoic acid restores diabetic tubular injury through regulating oxidative stress and mitochondrial apoptosis. *Am. J. Physiol. Ren. Physiol.* 299:1451–61
143. Brezniceanu ML, Liu F, Wei CC, Tran S, Sachetelli S, et al. 2007. Catalase overexpression attenuates angiotensinogen expression and apoptosis in diabetic mice. *Kidney Int.* 71:912–23
144. Brezniceanu ML, Liu F, Wei CC, Chenier I, Godin N, et al. 2008. Attenuation of interstitial fibrosis and tubular apoptosis in *db/db* transgenic mice overexpressing catalase in renal proximal tubular cells. *Diabetes* 57:451–59
145. Sedeek M, Callera G, Montezano A, Gutsol A, Heitz F, et al. 2010. Critical role of Nox4-based NADPH oxidase in glucose-induced oxidative stress in the kidney: implications in type 2 diabetic nephropathy. *Am. J. Physiol. Ren. Physiol.* 299:1348–58
146. Neria F, Castilla MA, Sanchez RF, Gonzalez Pacheco FR, Deudero JJ, et al. 2009. Inhibition of JAK2 protects renal endothelial and epithelial cells from oxidative stress and cyclosporin A toxicity. *Kidney Int.* 75:227–34
147. Berthier CC, Zhang H, Schin M, Henger A, Nelson RG, et al. 2009. Enhanced expression of Janus kinase-signal transducer and activator of transcription pathway members in human diabetic nephropathy. *Diabetes* 58:469–77
148. Wagener FA, Dekker D, Berden JH, Scharstuhl A, van der Vlag J. 2009. The role of reactive oxygen species in apoptosis of the diabetic kidney. *Apoptosis* 14:1451–58
149. Sun L, Xiao L, Nie J, Liu FY, Ling GH, et al. 2010. p66Shc mediates high-glucose and angiotensin II-induced oxidative stress renal tubular injury via mitochondrial-dependent apoptotic pathway. *Am. J. Physiol. Ren. Physiol.* 299:1014–25
150. Velagapudi C, Bhandari BS, Abboud-Werner S, Simone S, Abboud HE, Habib SL. 2011. The tuberin/mTOR pathway promotes apoptosis of tubular epithelial cells in diabetes. *J. Am. Soc. Nephrol.* 22:262–73

Autophagy in Pulmonary Diseases

Stefan W. Ryter, Kiichi Nakahira, Jeffrey A. Haspel, and Augustine M.K. Choi*

Pulmonary and Critical Care Medicine, Brigham and Women's Hospital, Harvard Medical School, Boston, Massachusetts 02115; email: sryter@partners.org, knakahira@rics.bwh.harvard.edu, jhaspel@partners.org, amchoi@rics.bwh.harvard.edu

Keywords

apoptosis, autophagosome, cell proliferation, chronic obstructive pulmonary disease, cigarette smoke, hypoxia, inflammation, pulmonary hypertension

Abstract

(Macro)autophagy provides a membrane-dependent mechanism for the sequestration, transport, and lysosomal turnover of subcellular components, including proteins and organelles. In this capacity, autophagy maintains basal cellular homeostasis and healthy organelle populations such as mitochondria. During starvation, autophagy prolongs cell survival by recycling metabolic precursors from intracellular macromolecules. Furthermore, autophagy represents an inducible response to chemical and physical cellular stress. Increasing evidence suggests that autophagy, and its regulatory proteins, may critically influence vital cellular processes such as programmed cell death, cell proliferation, inflammation, and innate immune functions and thereby may play a critical role in the pathogenesis of human disease. The function of autophagy in disease pathogenesis remains unclear and may involve either impaired or accelerated autophagic activity or imbalances in the activation of autophagic proteins. This review examines the roles of autophagy in the pathogenesis of pulmonary diseases, with emphasis on pulmonary vascular disease and acute and chronic lung diseases.

INTRODUCTION

Macroautophagy (autophagy) refers to an evolutionarily conserved and genetically programmed pathway for the lysosome-dependent autodigestion of subcellular components (1). During autophagy, newly formed double-membraned vacuoles, termed autophagosomes, encapsulate cytosolic material such as dysfunctional or damaged organelles or proteins. The autophagosomes subsequently fuse to lysosomes, delivering their sequestered cargo to this compartment for enzymatic digestion (**Figure 1**).

In recent years, autophagy has emerged as a vital cellular process involved in cellular homeostasis and survival mechanisms (2–4). This process serves an essential function in basal homeostasis in normal respiring cells and furthermore confers survival during nutrient deficiency by recycling metabolic precursors from cytosolic substrates (5). Increasing evidence suggests that autophagy can impact either the pathogenesis or the progression of human diseases (2, 3, 6), including cancer (7), neurodegenerative diseases (8), cardiovascular diseases (9, 10), and inflammatory diseases (11–13). The relationships between autophagy and disease pathogenesis remain incompletely characterized and may involve alterations in the activation of the regulatory proteins underlying autophagy, as well as positive or negative changes in autophagic activity or flux (2, 14). To date, only a few studies have sought to understand the roles of autophagy in the contexts of lung biology and human pulmonary disease (15–21).

The lung functions primarily in gas exchange, including the intake and transfer of molecular oxygen (O_2) to the pulmonary vasculature, as well as the elimination of waste gases (22). In

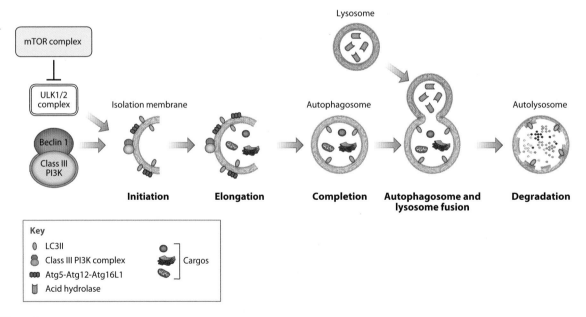

Figure 1

Schemata of the autophagic pathway. The autophagic pathway responds to environmental cues, including nutrient levels and stress signals, through a complex network of regulatory factors. These include three major macromolecular complexes: the mTOR complex 1, the ULK1/2 complex (i.e., the mTOR substrate complex), and the Beclin 1–class III PI3K complex. The autophagic pathway proceeds through several phases, including initiation (formation of an isolation membrane), elongation, autophagosome formation and completion, and autophagosome/lysosome fusion. In the final stage, lysosomal acid hydrolases degrade autophagosomal contents, which are released for metabolic recycling. Abbreviations: LC3II, microtubule-associated protein-1 light chain 3-II (lipidated form); mTOR, mammalian target of rapamycin; PI3K, phosphatidylinositol-3-kinase; ULK, uncoordinated-51-like protein kinase.

performing these functions, the lung can encounter various insults of environmental origin, including the inspiration of foreign matter and pathogens (e.g., viruses, bacteria, particles, smoke), mechanical stresses (e.g., mechanical ventilation), and extreme changes in ambient O_2 tension and/or reactive oxygen species (ROS) production. Acute or chronic exposure to such stimuli, in combination with predisposing genetic factors, can lead to the development of pathological states (22). The lung employs endogenous inducible defense mechanisms to protect itself against such environmental derangements. These include constitutive and inducible stress protein and antioxidant defenses, innate immune responses, and pro- and antiapoptotic mechanisms (23, 24). We predict that in addition to these mechanisms, autophagy, whether acting alone or as an intimate coactivator or coregulator of these processes, serves a critical role in lung homeostasis (25). The proteins that regulate autophagy may have pleiotropic roles in cellular signaling, which in turn may also impact cellular processes independently of autophagy (26). This review considers the role of autophagy as a component of the lung's inducible defense system against environmental insult. We also consider the underlying mechanisms by which autophagy and the proteins that regulate this process may act as either protective or contributory factors in the pathogenesis of human pulmonary diseases.

ROS: reactive oxygen species

Atg: autophagy-related gene

mTOR: mammalian target of rapamycin

Mammalian target of rapamycin complex 1 (mTORC1): a macromolecular complex containing mTOR, Raptor, Deptor, PRAS40, and GβL

SEQUENCE AND MOLECULAR REGULATION OF THE AUTOPHAGIC PATHWAY

The macroautophagic pathway proceeds through a series of membrane restructuring and translocation events (1). An initiation phase involves the activation and assembly of signaling components that trigger the process in response to environmental cues, including nutrient and energy levels and the accumulation of damaged substrates (2–5). In the nucleation phase, a preautophagosomal structure develops from subcellular membranes and further evolves into the phagophore or isolation membrane. The origins of this structure remain incompletely understood, but the structure may arise from the endoplasmic reticulum and possibly from other membrane compartments (27). In a subsequent elongation phase, the isolation membrane expands to surround and engulf a cytoplasmic cargo of material targeted for degradation to form complete autophagic vacuoles or autophagosomes with a double-membraned structure (1, 2). Ultimately, the fusion of autophagosomes with lysosomes results in the formation of autolysosomes with degradative capacity. Mature autophagosomes can also fuse with endocytic vacuoles (endosomes) to form amphisomes, which also progress to the lysosomes. During the degradative phase, a series of lysosomal degradative enzymes (e.g., cathepsins and other acid hydrolases) digest the encapsulated contents of autolysosomes. The digested contents are then released to the cytosol for reutilization in biosynthetic pathways (1–5).

A number of autophagy-related genes (Atgs) have been identified from studies of lower organisms, including yeast. Homologs of many of these Atg have been identified and cloned in mammals. Atg gene products interact with cellular signaling pathways in a complex regulatory network that governs the initiation and execution of the autophagic program (2, 3, 28).

Several macromolecular signaling complexes regulate the activation of autophagy. Of these, the mammalian target of rapamycin (mTOR) negatively regulates autophagy during nutrient-rich conditions (29). mTOR resides in a multiprotein complex, mTOR complex 1 (mTORC1) (30, 31), which includes the regulatory associated protein of mTOR (Raptor), mammalian G protein β-subunit-like protein (GβL), DEP domain–containing mTOR-interacting protein (Deptor), and proline-rich Akt/PKB substrate of 40 kDa (PRAS40) (29). Through the mTOR pathway, starvation or stimulation with the antibiotic rapamycin acts as a potent inducer of autophagy by inhibiting mTOR serine/threonine kinase activity (29). mTORC1 represents the target of regulatory cascades responsive to nutrient availability, such as amino acid bioavailability

PI3K:
phosphatidylinositol-3-kinase

mTOR substrate complex:
a macromolecular complex consisting of ULK1, Atg13, Atg101, and FIP200

Class III PI3K complex:
a macromolecular complex consisting of class III PI3K and Beclin 1 (Atg6); Atg14; and alternatively other regulatory proteins, including UVRAG and Rubicon

Beclin 1:
Bcl-2-interacting protein 1

PI3P:
phosphatidylinositol-3-phosphate

LC3: microtubule-associated protein-1 light chain 3

and energy charge (29). These regulatory pathways include class I phosphatidylinositol-3-kinase (PI3K)/Akt, which activates mTOR in response to insulin signaling, thereby negatively regulating autophagy (30). The adenosine 5′-monophosphate-activated protein kinase (AMPK), which is regulated by adenosine 5′-monophosphate levels, negatively regulates mTORC1 and thus positively regulates autophagy in response to energy exhaustion (31).

Among the known downstream regulatory targets of mTORC1, the mammalian uncoordinated-51-like protein kinase ULK1 (Atg1 in yeast) is a major regulator of autophagic initiation (32). ULK1 forms a macromolecular complex, the mTOR substrate complex, which includes Atg13, Atg101, and the 200-kDa focal adhesion kinase family–interacting protein (FIP200) (33–35). Under nutrient-replete conditions, the interaction between mTORC1 and ULK1 suppresses ULK1 kinase activity, thereby inhibiting the activation of autophagy (36). Recent studies indicate that AMPK, which negatively regulates mTORC1, also directly interacts with and regulates ULK1 in a nutrient-sensitive fashion, thus providing additional regulatory complexity (37–39). Starvation or rapamycin treatment inhibits mTOR kinase activity, resulting in global dephosphorylation of ULK1 (40), activation of ULK1 kinase activity, and the ULK1-dependent phosphorylation of FIP200 and Atg13 (34, 40–42). The activation of ULK1 kinase is a critical step in the initiation of starvation-induced autophagy in mammalian cells (41, 42). However, several mTORC1/ULK1-independent pathways to autophagy regulation have also been described (2).

The nucleation step of autophagosome formation requires an additional regulatory complex involving Vps34L (Vps34 in yeast), which is a class III PI3K complex, in association with p150 (Vps15 in yeast), Atg14L, and the Bcl-2-interacting protein Beclin 1 (Atg6 in yeast). Beclin 1 is a major autophagic regulator and tumor suppressor protein (43, 44). The activation of ULK1 leads to the recruitment of the class III PI3K complex to autophagosomes. The UV radiation resistance–associated tumor suppressor gene protein (UVRAG) (45) forms an alternate Beclin 1–Vps34L complex, which also promotes autophagy (43, 46, 47). The newly discovered factor Rubicon negatively regulates autophagy by binding to the class III PI3K complex (48, 49). Several additional factors, including Ambra-1, Bif-1, and the antiapoptotic proteins Bcl-2 and Bcl-X_L, interface with the class III PI3K complex and influence its activity (50–52).

The stimulation of class III PI3K activity results in the increased production of phosphatidylinositol-3-phosphate (PI3P), which regulates the formation of autophagosomes through the recruitment of protein factors. Several of these newly discovered PI3P-sensitive factors localize to preautophagosomal structures; such factors include WD-repeat proteins interacting with phosphoinositides (WIPI-1/2; Atg18 in yeast) and double-FYVE-containing protein 1 (DFCP1) (2, 53, 54).

Downstream of mTORC1 and class III PI3K complexes, the elongation phase of autophagosome formation requires two ubiquitin-like conjugation systems (2, 3). In the first system, the ubiquitin-like protein Atg12 is conjugated to Atg5 by Atg7 (E1 ubiquitin–activating enzyme–like) and Atg10 (E2 ubiquitin-conjugase-like) enzymes. The resulting Atg5-Atg12 complex forms a complex with Atg16L, which participates in elongation of the autophagic membrane. These factors subsequently dissociate from the autophagosome during maturation (55).

The second conjugation system requires the ubiquitin-like protein microtubule-associated protein-1 light chain 3 (LC3; Atg8 in yeast) (56). The homologs of Atg8 include several isozymes of LC3 (A, B, C) and related proteins (e.g., GABARAP). LC3 and its homologs are modified with the cellular lipid phosphatidylethanolamine (PE) (57–59). The protease Atg4B cleaves the precursor form of LC3 to generate the LC3-I form, which has an exposed lipid conjugation site at the C-terminal glycine residue. Conjugation of PE with LC3-I occurs from the sequential action of Atg7 (E1-like) and Atg3 (E2-like) activities. In mammals, the conversion of LC3-I (the free form) to LC3-II (the PE-conjugated form) is a key regulatory step in autophagosome formation

(2, 3). The cytosolic redistribution of LC3-II, as evidenced by a shift in the staining pattern of LC3 from diffuse to punctate staining, is an indicator of autophagosome formation (57, 58). The recruitment of LC3-II to the autophagosome is mediated by the Atg5-Atg12-Atg16L complex, which also facilitates LC3 conjugation (60).

In the final stages of autophagy, the autophagosome fuses with the lysosome. LC3B remains integrated in the autophagosomal membrane until the fusion step, following which the LC3B-II is delipidated at the membrane surface by Atg4B or is degraded within the autolysosome by proteolytic activity. Autophagosome maturation and fusion are assisted by several proteins, including small GTPases (e.g., Rab7), class C Vps proteins, UVRAG, and lysosome-associated membrane proteins (e.g., LAMP2) (57, 61–63).

In conclusion, the definition of the molecular mechanisms underlying the regulation of autophagy is evolving, with currently more than 33 Atg and related proteins identified to date (57). The continued elucidation of the molecular machinery of autophagy may be useful in the targeted design of therapies.

METHODS AND CONTROVERSIES FOR MEASURING AUTOPHAGIC ACTIVITY

Given the rapid pace at which new information about autophagy is being published, one might assume that a broad consensus exists on how to analyze this process. However, despite consensus statements (64–66), there remains significant controversy about which are the most valid approaches to analyze autophagy in vitro and in vivo. Part of the difficulty arises from the fact that autophagy is a process with more than one functional output: It simultaneously (*a*) serves as a source of amino acids for metabolism, (*b*) mediates protein and organelle quality control, (*c*) determines cell fate, and (*d*) constitutively produces autophagosomes that may serve as signaling platforms in their own right (2). Depending on the experimental model, it is natural for investigators to emphasize the measurement of those aspects of the autophagic pathway most relevant to their studies. **Supplemental Table 1** (follow the **Supplemental Material link** from the Annual Reviews home page at **http://www.annualreviews.org**) lists the most common approaches for examining the macroautophagy pathway. In general, the methods can be segregated into two approaches: static measurements of the autophagic machinery and dynamic measurements of the throughput of the autophagic pathway.

Static measures of autophagy include quantifying the number of autophagosomes within cells by electron microscopy (67), measuring steady-state levels of pathway constituents such as LC3B-II by Western blot analysis (65), and localizing macroautophagic proteins to punctate cytosolic structures (64). The appearance of punctate staining in GFP-LC3-expressing cells and tissues is an indicator of autophagosome formation (66).

Until recently, such static measurements were considered a sufficient indication of the rate of autophagic activity: The greater the number of autophagosomes or amount of LC3B-II, the higher is the rate of proteolytic activity (or flux) generated by the system. However, there has been increasing recognition that autophagosome or LC3B protein abundance represents a balance between the relative rates of synthesis and degradation of these species, which does not necessarily coincide with relative autophagic activity. For example, it is possible to have situations in which the steady-state abundance of autophagosomes is high but autophagic flux is low (due to a downstream blockade in autophagosome-lysosome fusion) or even unchanged (reflecting an earlier imbalance in the rate of synthesis and turnover that has since normalized) (66, 68, 69). Therefore, dynamic measurements of autophagic flux are required to properly interpret the significance of changes in the static levels of autophagic proteins or autophagosomes (66).

The classical method of directly measuring macroautophagic flux is the bulk protein degradation assay (64), which measures the liberation of radioisotope-tagged amino acids into the tissue culture medium in the presence or absence of 3-methyladenine. However, the most commonly used flux assay is the LC3B turnover assay, which relies on the fact that LC3B-II is targeted to the interior and the exterior of the autophagosomal membrane. LC3B-II on the luminal autophagosomal surface is normally degraded. By administration of an inhibitor of lysosomal proteolysis such as chloroquine (a chemical inhibitor of autophagic processing) (70), bafilomycin (66, 68), or leupeptin (71–73), luminal LC3B-II can be rescued. Autophagic flux can then be expressed as the difference in LC3B-II signal on a Western blot obtained in the presence versus the absence of lysosomal protein inhibitors. Recently, multiple groups developed automated fluorescence- and luminescence-based strategies that measure the turnover of specific macroautophagy substrates, including LC3B (74–78), p62/SQSTM1 (76, 79), next to BRCA1 gene 1 protein (NBR1) (76), betaine-homocysteine S-methyltransferase (80), and polyglutamine-protein aggregates (81). At present there are no reports of autophagic flux assays that have been applied to human subjects or tissues.

In summary, there now exist multiple valid approaches to analyzing autophagy both in tissue culture systems and in vivo. The current controversy surrounding the measurement of autophagy is not due primarily to methodological shortcomings in the available assays but rather reflects the fact that autophagy simultaneously produces multiple outputs and is therefore hard to quantify with a single variable. A common approach is to combine static measurements of autophagosomes and autophagic proteins with flux measurements to obtain a broad perspective of how particular experimental conditions affect the status of autophagy pathway constituents as well as their cooperative behavior as an integrated process.

MODES AND FUNCTIONS OF AUTOPHAGY

Macroautophagy, the most frequently described category of autophagy, generally involves the autophagosomal delivery and lysosomal degradation of cytosolic constituents (4). Although macroautophagy has traditionally been viewed as a random event involving the nonspecific encapsulation of cytosol, recent advances have uncovered the increasing selectivity and specificity of this process (2). The term selective autophagy refers to the specific molecular targeting of subcellular components for digestion (82; see sidebar entitled "Selective Autophagy"). Selective autophagy may include

SELECTIVE AUTOPHAGY

Recent advances in autophagy research reveal molecular mechanisms underlying the specificity of autophagic processing (82). The term selective autophagy refers to the specific targeting of cellular subcomponents for autophagic degradation. p62/sequestosome-1 and NBR1 have emerged as selective autophagy substrates that are degraded by the autophagosome (147, 148). These proteins have been found in association with polyubiquitinated proteins and assist in their delivery to autophagosomes (147, 148). p62 also facilitates the autophagosomal delivery of ubiquitinated protein and histone deacetylase-6 complexes (aggresomes) (82). NBR1, p62, and related selective autophagy substrates are marked in their ability to interact with LC3 proteins through a specific sequence motif, the LC3/Atg8-interacting region (82). Mitophagy, the selective degradation of depolarized mitochondria by autophagy (83), is orchestrated by several proteins, including PINK1 and Parkin. PINK1 facilitates the mitochondrial recruitment of Parkin, which in turn catalyzes the polyubiquitination of damaged mitochondria (149). Xenophagy refers to the mobilization of these systems for the autophagosomal delivery of bacteria.

various classes of organellophagy, in which organelles are selected for turnover (e.g., mitophagy refers to selective digestion of mitochondria) (83), and aggrephagy, which involves the degradation of ubiquitinated protein aggregates or aggresomes (82). Furthermore, xenophagy refers to the autophagic digestion of foreign matter such as invading pathogens or bacteria (84). In addition to the macroautophagic processes discussed primarily in this review, there exist additional subtypes of autophagy, including chaperone-mediated autophagy and microautophagy (2). In chaperone-mediated autophagy, proteins marked with a specific amino acid recognition sequence (KFERQ) are delivered directly to lysosomes by chaperone-assisted (i.e., via Hsc70, Hsp90) and lysosomal receptor–dependent (i.e., via LAMP2a) processes (85). Microautophagy refers to the direct endocytosis of cytosolic material by lysosomes through a membrane invagination process independent of autophagosomal delivery (86).

Aggresome: a macromolecular assimilation of denatured polyubiquitinated proteins encased in vimentin

PH: pulmonary hypertension

HIF: hypoxia-inducible factor

Autophagy serves a basal role in cellular homeostasis by providing a housekeeping function (2). In this capacity, autophagy maintains a healthy pool of organelles by removing damaged or dysfunctional specimens (e.g., depolarized mitochondria) (83). Autophagy also provides a major mechanism for intracellular protein turnover, especially of long-lived proteins; this mechanism is second in importance only to the proteasome for protein turnover (2). Under stress conditions, in which the probability of the accumulation of damaged organelles or ubiquitinated protein aggregates increases, autophagy acts as an inducible cellular defense mechanism by facilitating the removal of such damaged cellular components (26, 82). In the specific case of starvation or nutrient depletion, autophagy is an alternative provider of basic metabolic building blocks, largely through the bulk recycling of protein, lipids, and other organic molecules via lysosome-dependent degradation of macromolecules (5). Finally, autophagy can assist in host defense and innate immune responses by degrading invading pathogens and/or by dampening inflammatory cascades (11, 87).

FUNCTIONAL ROLE OF AUTOPHAGY IN HYPOXIC VASCULAR RESPONSES

Hypoxia, or reduced pO_2, can represent a critical component of cardiopulmonary diseases such as ischemia-reperfusion injury, pulmonary hypertension (PH), and atherosclerosis (88). Cellular responses to hypoxia are regulated through the activation of hypoxia-inducible factors (HIFs), which are heterodimers of the Per-Arnt-Sim/basic helix-loop-helix family of transcription factors. The HIF-1 heterodimer consists of a constitutively expressed subunit (HIF-1β) and an O_2-sensitive subunit (HIF-1α). HIF-1α is continuously degraded under normoxia by the proteasome, which is facilitated by the action of HIF prolyl hydroxylase and the von Hippel–Lindau E3 ubiquitin ligase (88).

Recent studies have identified autophagy as a process that can be regulated by hypoxia in cells and tissues and that may be important for mitochondrial turnover during hypoxia-induced cell injury (15, 89–91). The activation of autophagy by hypoxia in vitro requires Beclin 1 and an inducible component involving increased intracellular ROS production and HIF-1α stabilization (89). Furthermore, the downstream target of HIF-1α, the Bcl-2 family member BNIP3 (Bcl-2/adenovirus E1B 19 kDa–interacting protein-3) participates in the regulation of hypoxia-inducible autophagy (89). In tumor cell lines, activation of autophagy by hypoxia also requires BNIP3 (90, 91). Autophagy may promote cell survival during hypoxia through selective mitophagy (89). However, genetic studies have not reached a consensus on whether BNIP3 operates a cell survival or a cell death pathway during hypoxia.

Using primary cells of pulmonary vascular origin, we observed the induction of autophagic proteins as the consequence of hypoxia exposure. Both the accumulation of LC3B-II and the

siRNA: small interfering ribonucleic acid

expression of Beclin 1 were enhanced in human pulmonary vascular endothelial cells and in smooth muscle cells exposed to hypoxia in vitro. Similar results were observed in mouse lung capillary endothelial cells. Furthermore, hypoxia induces the formation of GFP-LC3 puncta, a marker of autophagosome formation in GFP-LC3-transfected endothelial cells. Ultrastructural analyses revealed that hypoxia caused the increased formation of autophagosomes after 24 h of exposure in pulmonary vascular cells (15). Experiments conducted using bafilomycin A1, an inhibitor of autophagosomal maturation, suggested that, at least in vitro, the increases in LC3-II and autophagosome formation were associated with increased autophagic activity (15).

Several proteins involved in autophagic regulation (e.g., UVRAG, Beclin 1) have been characterized as tumor suppressor proteins, implying negative regulatory effects on cell growth (69, 92). In our studies, the small interfering RNA (siRNA)-dependent knockdown of the autophagic protein LC3B accelerated hypoxia-induced proliferation of human pulmonary vascular endothelial and smooth muscle cells. The endothelial cells subjected to LC3B knockdown displayed increases in the basal tone of intracellular ROS and, concurrently, increases in the stabilization of the hypoxic regulator HIF-1α. Conversely, overexpression of LC3B in vascular cells suppressed mitogen- and hypoxia-dependent proliferative responses in cultured vascular cells. Interestingly, knockdown of LC3B inhibited lung vascular smooth muscle cell apoptosis induced by the proapoptotic agent staurosporine. These results are consistent with a hypothesis that LC3B is required for protection against harmful pathogenic responses to hypoxia and that LC3B does so by impairing the proliferation and promoting the apoptosis of smooth muscle cells, two processes implicated in pathological vascular remodeling (15). We also examined the hypoxia response in mouse lung endothelial cells derived from mice heterozygous for deletion of the $Becn1$ locus ($Becn1^{+/-}$ mice) (93). $Becn1^{+/-}$ endothelial cells displayed enhanced proliferation in vitro in response to hypoxia relative to wild-type cells (93). In conclusion, by this line of evidence, autophagic proteins Beclin 1 and LC3B may have a beneficial regulatory impact on pathological hyperproliferative processes in the setting of lung vascular disease.

Hypoxic vascular responses can promote angiogenesis, a complex process involving cell proliferation, migration, and maturation of vessels. Hypoxia-driven angiogenesis supports tumor growth and survival. Rapidly growing tumors outpace the growth of their supporting vasculature, resulting in tissue hypoxia, which in turn stimulates tumor neovascularization (94). The ability of hypoxia to stimulate angiogenesis involves the increased production of proangiogenic factors and mediators through HIF-dependent pathways. However, the potential interplay between hypoxia-induced autophagy and angiogenic processes remains unclear.

The major autophagic regulator protein Beclin is a tumor suppressor protein that inhibits cell growth and tumorigenesis (44). $Becn1^{+/-}$ mice display increased tumor formation in vivo (44). However, the specific mechanisms by which Beclin 1 acts as a tumor suppressor are incompletely understood. We observed a proangiogenic phenotype of $Becn1^{+/-}$ mice using a primary mouse B16F10 melanoma tumor model. Tumor-implanted $Becn1^{+/-}$ mice displayed an aggressive tumor growth phenotype after three weeks of chronic hypoxia exposure relative to corresponding wild-type mice. Significant increases in vascularization were observed in tumor tissue recovered from $Becn1^{+/-}$ mice relative to $Becn1^{+/+}$ mice after chronic hypoxia. In addition to increased angiogenesis in implanted tumors, $Becn1^{+/-}$ mice also exhibited increased metastasis to the lung. The proangiogenic phenotypes of $Becn1^{+/-}$ mice were associated with enhanced levels of systemic erythropoietin production, but not of vascular endothelial growth factor, relative to wild-type mice (93).

A proangiogenic phenotype was also characterized in pulmonary endothelial cells derived from $Becn1^{+/-}$ mice. The $Becn1^{+/-}$ cells, when subjected to hypoxia, displayed an altered balance of HIF isoform expression relative to wild-type cells, namely increased HIF-2α expression relative to HIF-1α expression (93). In addition to an enhanced proliferative response, $Becn1^{+/-}$ endothelial cells displayed enhanced migration and tube-forming activity in vitro in response to hypoxia relative to wild-type cells. Moreover, siRNA-dependent genetic interference of HIF-2α, but not that of HIF-1α, dramatically reduced hypoxia-inducible proliferation, migration, and tube formation in these $Becn1^{+/-}$ endothelial cells (93). We conclude that the preferential expression of HIF-2α is partially responsible for the proangiogenic phenotype of $Becn1^{+/-}$ endothelial cells (93). A better understanding of how autophagic proteins can regulate the angiogenic process may lead to novel therapeutic strategies in the treatment of cancer or vascular disorders.

AUTOPHAGY AND APOPTOSIS: A VOLATILE RELATIONSHIP

The relationship between autophagy and cell death pathways may have relevance to pulmonary diseases, in particular those that have a component of cell death implicated in their progression. We discuss this relationship in this section.

Autophagy is a cell survival mechanism during nutrient deficiency states (95). In this capacity, autophagy functions as a mechanism to delay the onset of cell death by necrosis by providing alternate sources of precursor molecules for biosynthetic reactions (5, 96). Successful escape from necrosis may be permissive of apoptosis (type I programmed cell death), thus allowing partially protected cells to die in a less catastrophic and more regulated fashion (97). However, complete cytoprotection conferred by autophagy would imply survival as the final outcome, rather than progression to any form of death. The current conundrum arises from the fact that elevations in autophagic markers and/or in autophagosome accumulation are frequently observed in dying cells (97–99). Autophagy has paradoxically been implicated as the basis for the development of resistance to chemotherapeutic agents but also as an effector of drug-induced toxicity (99). Alternatively, knockdown of critical autophagic regulators such as Atg5 confers cell survival or promotes cell death under chemical stress (100, 101). Taken together, such observations have begged the question as to whether autophagy associated with dying cells plays an effector role in the death process or represents an insufficient counterregulatory or protective process (96). The term autophagic cell death has been used to describe autophagy occurring in the context of cell death and has led to the classification of autophagy as an alternative form of programmed cell death (102). The validity of these classifications has been debated extensively and remains controversial (96, 97, 102). Accumulations of autophagosomes can represent increased autophagic activity yet in some cases may reflect impaired autophagic activity by blockage of downstream events such as lysosomal activity and autophagosome-lysosome fusion (98). Thus, the accumulation of autophagosomes in association with cell death may sometimes reflect deficiency of autophagic processes. Because autophagy represents primarily a degradative process, autophagy in a deregulated or hyperactive state, as may be elicited by chemical or toxin exposure, may exhaust cellular resources and lead to cell death (99).

Apoptosis involves the targeted and sequential activation of caspases that leads to DNA fragmentation and cell death without loss of membrane integrity (97). Recent literature suggests an intimate relationship between autophagy and apoptosis (96, 103). Both anti- and proapoptotic roles of autophagy have been implicated, depending on the experimental model (104). Under conditions, usually nonphysiological, in which apoptosis is chemically impaired (such as exposure to chemical inhibitors of caspase activity) or genetically impaired (as in $Bax^{-/-}$ or $Bak^{-/-}$ cells), autophagy provides an alternative pathway to cell death (101, 105–108). For example, cells treated with z-VAD-*fmk*, a general inhibitor of caspases, die essentially by a nonapoptotic pathway characterized by dramatic accumulations of autophagic vacuoles (105). An emerging viewpoint is that such examples may represent necrotic cell death occurring independently of increased vacuolization (96). There is also evidence that autophagy is also protective in the context of chemically induced, nonapoptotic cell death (109).

CSE: aqueous cigarette smoke extract

The signaling mechanisms that regulate apoptosis and autophagy share some overlapping or interactive elements. For example, several Atg proteins participate in the regulation of apoptosis, and conversely, apoptotic proteins influence the regulation of autophagy (110). The autophagic protein Beclin 1 (Atg6) can directly interact with antiapoptotic Bcl-2 family members, including Bcl-2 and Bcl-X_L, through interactions with its BH3 domain (74, 75). Bcl-2 family proteins exert antiautophagic effects by acting as negative regulators of the Beclin 1–class III PI3K complex (111). Recent studies have uncovered additional factors that regulate these interactions (111). The antiautophagic function of Bcl-2 takes place predominantly in the endoplasmic reticulum and is facilitated by the protein NAF1 (CDGSH iron sulfur domain 2). NAF1 binds to Bcl-2 through its 2Fe-2S cluster and stabilizes the Beclin 1–Bcl-2 interaction (111).

The autophagic protein Atg5 also interacts with apoptosis signaling pathways. Specifically, Atg5 forms a complex with the apoptogenic Fas-associated death domain (FADD) protein (112). Furthermore, a proteolytic fragment of Atg5 exerts proapoptotic effects through the inhibition of Bcl-X_L (113). Atg5 knockdown had variable roles in cell survival, depending on the experimental conditions. For example, siRNA-dependent Atg5 knockdown increased fibroblast apoptosis when induced with death receptor pathway agonists but inhibited apoptosis in response to xenobiotics (104). Several studies have also proposed that activation of apoptosis can downregulate autophagy through the targeted proteolytic cleavage of Beclin 1 by activated caspase-8 (114–116).

The c-Jun-NH_2-terminal kinase (JNK) pathway, commonly associated with prodeath signaling, can exert several regulatory functions on autophagy (117). JNK positively regulates autophagy by inhibiting the interaction of Beclin 1 with Bcl-2 family proteins (118). Additional apoptosis-related pathway proteins implicated in apoptotic signaling include p53 and DRAM, the death-associated-related protein kinase (DAPK) family proteins, and the IP_3 receptor (IP_3R) (96, 119).

These examples make evident that a complex panoply of factors, some with dual function, regulate both apoptosis and autophagy. Experimental evidence has shown that autophagy and apoptosis may be regulated separately, concurrently, or in a reciprocal fashion, depending on the experimental context. Further research will delineate the molecular mechanisms by which cells decide among these various outcomes.

ROLE OF AUTOPHAGY IN CIGARETTE SMOKE–INDUCED APOPTOSIS

Recently, we studied a model system of cigarette smoke exposure in which autophagy and apoptosis appear to be concurrently regulated toward a functional outcome of cell death. Cigarette smoke is a complex mixture that contains ~4,700 chemical constituents, including carbon monoxide, heavy metals, aldehydes, aromatic hydrocarbons, free radicals, and other oxidizing compounds (120).

The application of aqueous cigarette smoke extract (CSE) to cultured cells is a commonly used model of cigarette smoke exposure for in vitro experimentation. CSE treatment of various types of epithelial cells increased the accumulation of autophagic markers concurrently with apoptotic markers (16, 121). CSE exposure in epithelial cells triggered dose-dependent and time-dependent increases in total LC3B expression as well as the accumulation of the active form of LC3B, LC3B-II, in primary human bronchial epithelial cells (16). Exposure of various types of epithelial cells to CSE caused dose-dependent increases in autophagosome formation (16), as evidenced by electron microscopy or by the manifestation of GFP-LC3 punctate structures in GFP-LC3-transfected cells (16, 121). The increases of LC3B-II by CSE in primary human bronchial epithelial cells were further enhanced by bafilomycin A1, an inhibitor of autophagosome-lysosome fusion, and by lysosomal protease inhibitors (16, 121). These studies suggest that the increases in autophagic markers and autophagosome formation caused by CSE treatment occurred in the context of active rather than impaired autophagic flux in epithelial cells.

These indicators of autophagic pathway activation in bronchial epithelial cells in response to CSE exposure were observed under conditions that culminated in increased apoptotic cell death. The apoptosis observed under these conditions was characteristic of an extrinsic pathway involving the formation of the Fas-dependent death-inducing signaling complex (DISC) and the activation of caspase-8 and -3 (16, 17, 121). The initiating events in apoptosis signaling such as DISC activation preceded the activation of LC3B, whereas the activation of executioner caspases occurred later than the incidence of LC3B conversion. Thus, autophagy and apoptosis were concurrently activated in CSE-treated epithelial cells, although the relative temporal relationship depended on the specific markers being analyzed.

DISC: death-inducing signaling complex

Under conditions of CSE exposure, siRNA-dependent genetic interference of autophagic protein Beclin 1 or LC3B in epithelial cells resulted in the inhibition of apoptotic cell death and the suppression of DISC formation (16). Furthermore, LC3B formed reversible binding interactions with the death effector molecule Fas/CD95, the principal component of DISC (17). The formation of this complex required the caveolae scaffolding protein caveolin-1 because genetic deletion of caveolin-1 abolished Fas-LC3B complex formation. Both LC3B and Fas also assimilated in reversible complexes with caveolin-1. These binding interactions between the three proteins were detectable in epithelial cells under basal conditions. However, exposure of epithelial cells to CSE caused the time-dependent dissociation of these complexes. Mutation analysis revealed that the interaction between LC3B and caveolin-1 required the caveolin-1-binding motif of LC3B and the scaffolding domain of caveolin-1, whereas the binding of Fas to caveolin-1 occurred independently of the scaffolding domain. The mechanism by which cellular stress caused the dissociation of complexes between LC3B, caveolin-1, and Fas remains unclear (17).

Further genetic experiments revealed additional functional relationships between these interactions. Genetic interference of LC3B resulted in protection against CSE-induced apoptosis, which was associated with an increase in Fas-caveolin-1 complex formation and a decrease in Fas-caspase-8 complex (DISC) formation. Genetic interference of caveolin-1 increased CSE-induced apoptosis in epithelial cells. Conversely, forced expression of LC3B in epithelial cells promoted CSE-induced apoptosis and DISC formation. The proapoptotic effect of LC3B expression was diminished by mutation of the caveolin-1-binding motif of LC3B (17).

These observations suggested that LC3B plays an antiapoptotic role in healthy cells through binding interactions with Fas and caveolin-1. The sequestration of Fas by caveolin-1 and LC3B inhibits the activation of the extrinsic apoptosis pathway. However, LC3B plays an apparently proapoptotic role during cigarette smoke exposure. Our studies conclude that LC3B may facilitate the mobilization of Fas for extrinsic pathway activation during stress conditions. The mechanism remains incompletely clear but may involve changes in LC3B or caveolin-1 conformation and/or posttranslational modification (17).

In conclusion, the stimulation of the autophagic pathway promotes apoptotic cell death during cigarette smoke exposure. In this model, which challenges cells with a complex mixture of chemicals and particles, autophagy may represent a frustrated mechanism that fails to prevent against cell death. However, molecular studies reveal that autophagic proteins can cross-regulate the programmed cell death pathways that are activated during cigarette smoke exposure, leading to the promotion of epithelial cell death. A limitation of the study was the reliance on genetic knockdown of autophagic proteins such as LC3B to arrive at this conclusion. Furthermore, these effects do not necessarily depend on the functional outcome of autophagy in its degradative capacity but may involve direct protein-protein interactions affecting signaling pathways independently of autophagic function. In either case, therapeutic targeting of autophagic proteins could be harnessed for cytoprotection.

AUTOPHAGY IN INFLAMMATION AND INNATE IMMUNITY

Autophagy can exert a critical influence on immune and inflammatory responses during the pathogenesis of various diseases (11). Recent human studies link autophagy to inflammatory diseases such as Crohn's disease (13, 122) and systemic lupus erythematosus (123), which are characterized by chronic inflammation. Of note, the role of autophagy in Crohn's disease, a chronic inflammatory bowel disease, has been rigorously studied in an animal model of colitis because mutations of autophagy regulator genes such as *ATG16L1*, *NOD2* (nucleotide oligomerization domain 2), and *IGRM* have been found in Crohn's patients (122). Although the precise pathophysiological mechanisms of Crohn's disease still remain unclear, several key factors, including excess inflammatory responses, abnormal Paneth cell granule secretion, and the failure of intracellular bacterial clearance (as observed in autophagy gene–deficient mice), are involved in the development of intestinal inflammation (124–126). In a manner similar to that of the intestine, the respiratory airways also represent a site of direct contact with pathogens and environmental stressors. Therefore, it is important for these tissues to maintain proper cellular homeostasis from such insults. Autophagy can exert an important antipathogenic function, as discovered in infectious disease models using *Mycobacterium tuberculosis* and group A *Streptococcus* (127, 128). The critical role of autophagy in host defense against various microbes, including bacteria, viruses, and parasites, has since been more broadly demonstrated (11).

The involvement of autophagy in inflammatory responses has also been studied in vitro and in vivo. Recent observations have revealed a link between autophagic protein deficiency and proinflammatory cytokine secretion in macrophages. The early clues to the involvement of autophagy in inflammation were provided by the enhancement of proinflammatory cytokine release in autophagy gene–deleted macrophages in response to pathogen-associated molecular patterns (PAMPs) such as endotoxin (124). Interestingly, *atg16l1* or *atg7* deficiency enhanced the secretion of interleukin (IL)-1β and IL-18 but not that of tumor necrosis factor and interferon (IFN)-β in response to endotoxin (124). In contrast, the activation of nuclear factor-κB, a critical transcriptional factor for proinflammatory cytokine gene expression, as well as the activation of mitogen-activated protein kinase (MAPK) pathways (i.e., ERK1/2) in response to lipopolysaccharide were similar in *atg16l1*-deficient cells and in wild-type cells. Inflammasomes, a newly discovered inflammatory signaling platform, tightly regulate the secretion of IL-1β and IL-18 (129). The inflammasome is a multiprotein complex that mediates the cleavage and activation of caspase-1, leading to the maturation and release of IL-1β and IL-18 (129). Cytosolic receptors of the NLR (NOD-like receptor) family are critical components of the inflammasome and interact with the apoptosis-associated adaptor molecule protein ASC, which recruits the precursor form of caspase-1 and initiates its cleavage. Increased activation of IL-1β and IL-18 has been observed in macrophages and monocytes genetically deficient in Beclin 1 and LC3B; such increased activation occurs through the increased activation of the NLRP3 (NALP3) inflammasome pathway (130, 131). In addition to endotoxin, IFN-β production induced by synthetic double-stranded DNA is regulated by Atg9a, but not by Atg7, suggesting unique immune functions for specific autophagic proteins (132). Autophagy is also involved in cytokine production triggered by viral infection. Atg5 is required to activate type 1 IFN production in plasmacytoid dendritic cells stimulated with single-stranded RNA viruses (133). In contrast, deletion of *Atg5* in macrophages enhances retinoic acid–inducible gene-I–like receptor (RLR)-mediated type 1 IFN production in response to single-stranded RNA viruses (134). Importantly, the mechanism by which autophagy deficiency enhances the

---

Inflammasome: a macromolecular complex responsible for the activation of caspase-1 and downstream proinflammatory responses

NLRP3 inflammasome and the RLR signaling pathway involves the deregulation of mitochondrial homeostasis, including the enhanced production of mitochondrial ROS and increased mitochondrial membrane permeability transition (130, 131, 134).

Furthermore, autophagy can exert an influence on adaptive immunity, including antigen presentation, and on the maintenance of lymphocyte function and homeostasis. Autophagy restricts pathogen growth by autolysosome-mediated degradation, which initiates after autophagosome-lysosome fusion (11). The immune system surveys microbes by detecting their antigens by two systems: MHC class I and II loading compartments. The pathogen-derived peptides generated by lysosomal degradation are displayed on MHC class I and II molecules. *Atg12* gene knockdown by siRNA decreases MHC class II antigen presentation to $CD4^+$ T cells in response to Epstein-Barr virus infection (135). Genetic deletion of *Atg5* also suppresses processing and presentation of phagocytosed antigen on MHC class II molecules and increases susceptibility in response to herpes simplex virus (HSV)-2 in vivo (136). Furthermore, autophagy induced during HSV-1 infection enhances antigen presentation on MHC class I molecules (137). Another crucial role of autophagy is the generation of a self-tolerant T cell repertoire. Constitutively highly expressed autophagy in thymic epithelial cells delivers endogenous proteins to MHC class II molecules and contributes to CD4 T cell selection. After grafting thymi from $Atg5^{-/-}$ or $Atg5^{+/+}$ embryos under the kidney capsule of athymic nude mice, systemic lymphoid infiltration including the lung is observed in the recipients of $Atg5^{-/-}$ thymi (138).

Given the important role of the autophagic machinery in regulating immune responses, autophagy may be critically involved in immune or inflammatory responses in lung diseases. Pneumonia and tuberculosis caused by airway infection remain life-threatening lung diseases despite the development of antibiotics and antituberculosis drugs. Loss of *atg9* results in the increased growth of *Legionella pneumophila*, implying an important role of autophagy in *Legionella* pneumonia (139, 140). In addition, extensive studies of autophagy and *M. tuberculosis* suggest a potential therapeutic strategy for tuberculosis by manipulating the autophagy machinery (128, 141, 142). Alveolar macrophages provide the first line of defense against invading microbes or environmental insults in the airways. By sensing pathogens or PAMPs, alveolar macrophages immediately activate the innate immune system and induce proinflammatory responses for host defense. However, an uncontrolled excessive and persistent inflammatory response may cause lung tissue injury, which allows the invasion of the microbes, leading to systemic severe infection such as sepsis.

Loss of autophagic proteins exaggerates sepsis-induced inflammatory responses in the cecal-ligation and puncture model of polymicrobial sepsis, as well as in endotoxemia, and compromises survival during septic shock in mice (130). Importantly, marked autophagic vacuolization is observed in livers from patients who die from sepsis (143). However, it is unclear that this observation represents increased autophagic activity in sepsis patients or downstream inhibition of autophagy that leads to the inappropriate accumulation of autophagosomes. IL-18, one of the inflammasome-dependent cytokines, is significantly elevated in plasma from septic patients (130). These observations suggest the potential link between autophagy and the inflammasome in the pathogenesis of sepsis. Given the potential impact of autophagic processes on the regulation of inflammation, further research is required to determine the significance of these associations in the context of chronic lung diseases.

AUTOPHAGY IN LUNG DISEASES

Autophagy in Emphysema and Chronic Obstructive Pulmonary Disease

Emphysema is defined by the destruction of lung parenchyma in the context of an aberrant inflammatory response and increased matrix proteolysis resulting in enlarged airspace and impaired

COPD: chronic obstructive pulmonary disease

respiratory function (144). Mice chronically exposed to cigarette smoke develop airspace enlargement after 3–6 months of exposure and are frequently used as an animal model of emphysema. We recently reported increased markers of autophagy in the lungs of C57Bl/6 mice exposed to environmental cigarette smoke for 12 weeks. Lungs isolated from cigarette smoke–exposed animals displayed increased autophagosomal numbers in lung tissue sections and increased LC3B-II accumulation in lung tissue homogenates (16). Using this chronic model of ambient cigarette smoke exposure in mice, we observed that LC3B status affects the development of lung cell death in situ and airspace enlargement. Mice genetically deficient in LC3B ($LC3B^{-/-}$) resisted changes to airspace induced by cigarette smoke, as detected by mean-linear intercept and alveolar diameter measurements, relative to wild-type mice. We could not exclude the possibility that LC3B is essential for the development of lung structure. We also could not discount that these phenotypic effects could be related to nonautophagic functions of LC3B, which impact fibronectin translation and microtubule networks (145).

Interestingly, mice genetically deleted for early growth response-1 (Egr-1), a molecule that regulates LC3B transcription, also displayed basal airspace enlargement but resisted cigarette smoke–induced changes in airspace. In contrast, caveolin-1 knockout ($Cav-1^{-/-}$) mice were sensitized to changes in airspace enlargement and lung apoptosis after chronic cigarette smoke exposure and displayed no basal airspace enlargement. Collectively, these data suggest that the autophagic protein LC3B and its related regulatory molecule Egr-1 promote airspace enlargement, whereas caveolin-1 inhibits airspace enlargement in the chronic cigarette smoke exposure model in mice (16, 17). However, we cannot exclude the possibility that other factors, including imbalances in circulating mediators and/or inflammatory cell influx, in addition to the demonstrated effects on lung cell apoptosis, contribute to airspace enlargement phenotypes.

In conclusion, these studies identify LC3B as an important regulator of autophagic signaling and apoptosis within the specific context of the polychemical stress presented by cigarette smoke/CSE exposure (16, 17). These studies also support the conclusions of others that autophagic proteins can participate in a dynamic equilibrium between cell survival and death.

In humans, long-term exposure to cigarette smoke (usually decades), coupled with the potential involvement of genetic predisposing factors, can lead to the development of chronic obstructive pulmonary disease (COPD). By definition, COPD is a complex lung disease process involving adverse tissue responses to inhaled particles; such adverse reactions include tissue destruction (emphysema), bronchitis, and fibrogenesis (146). Recent studies from this laboratory demonstrate increases in autophagic proteins in lung tissue derived from COPD patients (16). The expression level of the LC3B-II protein, an indicator of autophagosome formation, as well as the expression of autophagic proteins Atg4, Atg5-Atg12, and Atg7 were increased in COPD lung tissue relative to control lung tissue (16). Ultrastructural analysis of lung tissue from these COPD patients revealed increased abundance of autophagosomes in these specimens relative to normal lung tissues (16). The autophagic markers were elevated at all stages of disease severity (GOLD 0–4) relative to control lungs of nonsmokers without airway disease. In contrast, caspase-3 activation was detectable only at very late stages of the disease progression. We speculate that autophagy represents a general response to smoke exposure of which the morphological and biochemical indices are detectable earlier in the course of disease progression relative to the indicators of cell death in situ, although these processes appear simultaneously with direct smoke exposure in vitro (16). Limitations of this clinical study are that, by nature, it involved frozen tissue sections analyzed retrospectively for expression of autophagic markers and that no mechanistic studies in human disease samples were possible.

Similar to the observations made in COPD patients, autophagic markers are increased in lung tissues from patients with genetic emphysema caused by $\alpha 1$ antitrypsin deficiency. These observations suggest that intrinsic factors such as increased matrix proteolysis or altered signaling pathways may contribute to the activation of autophagy in COPD in addition to the direct cellular responses to cigarette smoke. In contrast, increased markers of autophagy were not observed in specimens from other lung diseases such as idiopathic pulmonary fibrosis, cystic fibrosis (CF), sarcoidosis, and systemic sclerosis. These observations suggest that autophagic proteins and/or autophagosome formation may serve as tissue biomarkers of certain pulmonary diseases (16).

Interestingly, a recent study analyzed autophagic markers in alveolar macrophages isolated from the lungs of active smokers (18). In this study, activation of autophagic markers, including LC3-II and autophagosome formation, was observed in smokers' macrophages. The macrophages also displayed elevated autophagosome formation by ultrastructural analysis. On the basis of chemical inhibitor studies using chloroquine, the authors conclude that autophagic activity is impaired in alveolar macrophages from smokers' lungs (18). Taken together, these studies suggest that autophagic processes are important in COPD development and may impact several cell types, including airway and alveolar epithelium and alveolar macrophages. Further studies will be needed to clarify whether autophagic flux is enhanced or impaired in lung parenchyma in response to smoke exposure, as well as to determine the relationships between autophagic flux, cell death, and emphysema development.

These studies also implicate autophagy as a general toxicological response to cigarette smoke exposure in vitro (16, 17, 121) and imply that cigarette smoking in humans may impair autophagic processing in alveolar macrophages (18). On the basis of these findings, it is tempting to speculate that the selective targeting of autophagic proteins using pharmacological or genetic strategies may serve as the basis for developing novel therapies for COPD. The current data in model systems suggest that autophagic proteins can cross-regulate inflammatory processes, although this has not been rigorously studied in the context of cigarette smoke–induced inflammation or COPD. Deregulated immune and inflammatory responses have been implicated in the pathogenesis of COPD, and autophagy may impact these processes as well (144).

Autophagy in Pulmonary Hypertension

To determine the involvement of autophagy in pulmonary vascular disease, we examined this process in an experimental mouse model of chronic hypoxia–induced PH. Exposure to chronic hypoxia in mice resulted in increased expression of lung autophagic markers, including LC3B in small pulmonary vessels and LC3B-II in lung tissue. Furthermore, hypoxic lung tissues displayed a dramatic increase in autophagic vacuoles relative to lung tissue from normoxic animals. Our recent studies indicated that after chronic hypoxia $LC3B^{-/-}$ mice display increased indices of PH, including increased right ventricular systolic pressure and Fulton's Index, relative to wild-type mice (15). These results identified an endogenous role for the autophagic protein LC3B in the regulation of protective processes during the development of PH. Genetic deletion of LC3B aggravated the hypertensive phenotype as the result of hypoxia exposure.

In vitro studies revealed that Egr-1 regulates $LC3B$ gene expression. Consequently, we evaluated the phenotype of $Egr-1^{-/-}$ mice in the chronic hypoxia model. Similar to the results found with the $LC3B^{-/-}$ mice, the $Egr-1^{-/-}$ mice displayed increased indices of PH after chronic hypoxia, including right ventricular systolic pressure and vascular wall thickness. In contrast, $Becn1^{+/-}$ mice displayed no apparent phenotype with respect to indices of PH after chronic hypoxia. Furthermore, systemic administration of chloroquine in vivo did not replicate the effects of LC3B deletion on the development of functional indices of PH in hypoxia (15). These results suggest that the phenotypic effects observed following LC3B deletion may be more relevant to the signaling effects of this protein than to the autophagic process per se.

Finally, we examined the relevance of autophagy in human clinical vascular disease. We analyzed lung tissue sections from normal and PH patients for the expression of autophagic markers. In human lung tissue isolated from patients with various forms of PH, the total expression of LC3B as well as the accumulation of its activated (lipidated) form, LC3B-II, were increased relative to normal lung tissue. The expression of LC3B was markedly increased in the endothelial cell layer, as well as in the adventitial and medial regions of large and small pulmonary resistance vessels from the PH lung relative to normal vascular tissue. These experiments suggested that autophagy may have a role in vascular disease in humans. These observations were based on static measurements, however, and further experiments are needed to deduce the direction of autophagic flux associated with this disease (15).

Autophagy in Cystic Fibrosis

CF, a life-threatening lung disease, is a genetic disorder caused by a mutation in the gene encoding the cystic fibrosis transmembrane conductance regulator (CFTR). CF is characterized by abnormal mucous accumulation in the airways. Such accumulation can cause pulmonary damage associated with secondary infections. Recent studies show that the CFTR mutation in human airway epithelial cells from CF patients was associated with defective autophagy, as evidenced by reduced numbers of autophagosomes and LC3 puncta during starvation and the accumulation of p62 (20). Consequently, cells with the CFTR mutation displayed accumulation of polyubiquitinated proteins and the decreased processing of aggresomes. Similar findings were reported in nasal biopsies from CF patients (20). In epithelial cells, mutation and/or loss of function of CFTR caused increased ROS production and, consequently, increased tissue transglutaminase 2 levels, an important factor of the inflammatory response in CF. Activation of these pathways caused a loss of function in Beclin 1 and in the Beclin 1–class III PI3K complex and, as a result, loss of autophagic function. In cells or mice expressing mutant CFTR, the restoration of autophagy by Beclin 1 overexpression improved disease phenotypes in vitro and in vivo (20, 21). Reconstitution of Beclin 1 levels also restored membrane trafficking of mutant CFTR and reduced its accumulation in aggresomes. Improvement of airway phenotypes was also seen with application of the antioxidant *N*-acetyl-L-cysteine to CFTR mutant mice. Taken together, these studies suggest that autophagy deficiency is linked to the CFTR defect and to the lung inflammation associated with CF (20, 21). Thus, therapeutics targeting autophagy, including antioxidants, may find applications in CF.

Figure 2

Role of autophagy in pulmonary diseases. Autophagy has been implicated in various pulmonary diseases, including pulmonary arterial hypertension, chronic obstructive pulmonary disease, sepsis, cystic fibrosis, and infectious diseases (e.g., pneumonia). (*a*) In pulmonary arterial hypertension, autophagy plays a protective role by limiting hypoxia-dependent vascular cell proliferation. (*b*) In chronic obstructive pulmonary disease and cigarette smoke–induced lung injury, autophagy exerts a deleterious role by promoting epithelial cell apoptosis. (*c*) In cystic fibrosis, impaired autophagy leads to decreased aggresome clearance. (*d,e*) Autophagy may also influence inflammation and host defense, thus possibly playing roles in (*d*) pneumonia and (*e*) sepsis, respectively. Other abbreviations: CFTR, cystic fibrosis transmembrane conductance regulator; DAMP, damage-associated molecular patterns; HIF, hypoxia-inducible factor; IFN, interferon; IL, interleukin; PAMP, pathogen-associated molecular pattern; RLR, retinoic acid–inducible gene-I-like receptor; ROS, reactive oxygen species.

a Inhibition of cell proliferation   **b** Promotion of cell death

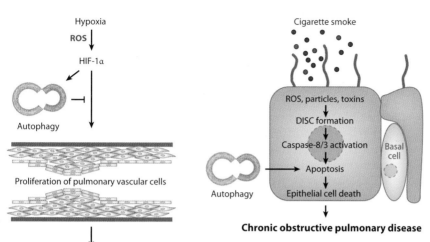

c Removal of protein aggregates   **d** Pathogen degradation

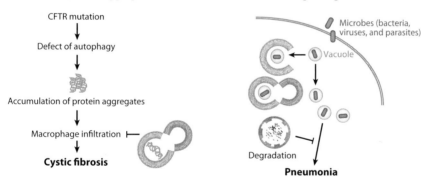

e Regulation of inflammatory response

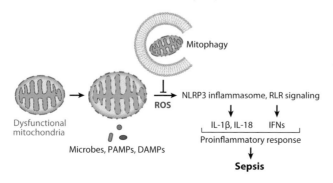

CONCLUSIONS

Autophagy and the signaling pathways that regulate this process may exert pleiotropic roles in normal homeostatic as well as pathological cellular functions. Our recent studies and those of others have uncovered functional roles for autophagy in the development of several pulmonary diseases (**Figure 2**). In particular, we and others have found evidence for deregulated autophagic responses in smoking and COPD. Additional studies have linked autophagic proteins with the aggravation of apoptotic responses to cigarette smoke exposure, which may contribute to tissue loss during emphysema development.

Genetic studies involving deletion of select Atg genes have shown a critical involvement of autophagic proteins in the regulation of pulmonary vascular cell growth, apoptosis, and angiogenesis; such involvement may impact proliferative disorders such as PH. With the emerging roles of autophagy in immunity and inflammation, it will be of intense interest to also investigate the role of autophagy in the inflammatory responses associated with lung diseases. Ongoing investigations in several laboratories will determine the role that autophagic processes play in other lung diseases such as CF, idiopathic pulmonary fibrosis, and lung cancer.

SUMMARY POINTS

1. Autophagy is a fundamental homeostatic process that is involved in the intracellular turnover of organelles and proteins and that prolongs survival in nutrient deficiency states.

2. Autophagy has been implicated in the pathogenesis of human diseases, including cancer and neurodegenerative and inflammatory diseases. The relative importance of this process in pulmonary diseases has recently been recognized.

3. The specific role of autophagy in disease may involve altered activation of autophagic proteins or altered autophagic activity (flux).

4. The autophagic protein LC3B contributes to protective processes during hypoxic vascular remodeling by limiting cell proliferation.

5. Autophagic markers increase in patients with chronic obstructive pulmonary disease. In contrast, the alveolar macrophages of smokers exhibit impaired autophagic flux.

6. Activated autophagy may promote cell death in cigarette smoke–induced epithelial cell injury. LC3B may regulate extrinsic apoptosis through interactions with death-regulatory proteins and caveolin-1.

7. Macrophages deficient in autophagic proteins are sensitized to the secretion of inflammasome-associated cytokines through a mechanism involving mitochondrial dysfunction and activation of the NLRP3 inflammasome.

8. Cross-regulation of inflammatory pathways by autophagy may play an important role in diseases involving chronic inflammation.

DISCLOSURE STATEMENT

The authors are not aware of any affiliations, memberships, funding, or financial holdings that might be perceived as affecting the objectivity of this review.

ACKNOWLEDGMENTS

This work was supported by NIH grants R01-HL60234, R01-HL55330, and R01-HL079904 and by a FAMRI Clinical Innovator Award to A.M.K. Choi. S.W. Ryter received salary support from the Lovelace Respiratory Research Institute.

LITERATURE CITED

1. Kelekar A. 2006. Autophagy. *Ann. N. Y. Acad. Sci.* 1066:259–71
2. Ravikumar B, Sarkar S, Davies JE, Futter M, Garcia-Arencibia M, et al. 2010. Regulation of mammalian autophagy in physiology and pathophysiology. *Physiol. Rev.* 90:1383–435
3. Eskelinen EL, Saftig P. 2009. Autophagy: a lysosomal degradation pathway with a central role in health and disease. *Biochim. Biophys. Acta* 1793:664–73
4. Klionsky DJ. 2007. Autophagy: from phenomenology to molecular understanding in less than a decade. *Nat. Rev. Mol. Cell Biol.* 8:931–37
5. Rabinowitz JD, White E. 2010. Autophagy and metabolism. *Science* 330:1344–48
6. Mizushima N, Levine B, Cuervo AM, Klionsky DJ. 2008. Autophagy fights disease through cellular self-digestion. *Nature* 451:1069–75
7. Rosenfeldt MT, Ryan KM. 2011. The multiple roles of autophagy in cancer. *Carcinogenesis* 32:955–63
8. Lee JA. 2009. Autophagy in neurodegeneration: two sides of the same coin. *BMB Rep.* 42:324–30
9. Martinet W, De Meyer GR. 2009. Autophagy in atherosclerosis: a cell survival and death phenomenon with therapeutic potential. *Circ. Res.* 104:304–17
10. Gottlieb RA, Mentzer RM. 2010. Autophagy during cardiac stress: joys and frustrations of autophagy. *Annu. Rev. Physiol.* 72:45–59
11. Levine B, Mizushima N, Virgin HW. 2011. Autophagy in immunity and inflammation. *Nature* 469:323–35
12. Brest P, Corcelle EA, Cesaro A, Chargui A, Belaïd A, et al. 2010. Autophagy and Crohn's disease: at the crossroads of infection, inflammation, immunity, and cancer. *Curr. Mol. Med.* 10:486–502
13. Rioux JD, Xavier RJ, Taylor KD, Silverberg MS, Goyette P, et al. 2007. Genome-wide association study identifies new susceptibility loci for Crohn disease and implicates autophagy in disease pathogenesis. *Nat. Genet.* 39:596–604
14. Liang C. 2010. Negative regulation of autophagy. *Cell Death Differ.* 17:1807–15
15. Lee SJ, Smith A, Guo L, Alastalo T-P, Li M, et al. 2011. Autophagic protein LC3B confers resistance against hypoxia-induced pulmonary hypertension. *Am. J. Resp. Crit. Care Med.* 183:649–58
16. Chen ZH, Kim HP, Sciurba FC, Lee SJ, Feghali-Bostwick C, et al. 2008. Egr-1 regulates autophagy in cigarette smoke-induced chronic obstructive pulmonary disease. *PLoS ONE* 3(10):e3316
17. Chen ZH, Lam HC, Jin Y, Kim HP, Cao J, et al. 2010. Autophagy protein LC3B activates extrinsic apoptosis during cigarette-smoke induced emphysema. *Proc. Natl. Acad. Sci. USA* 107:18880–85
18. Monick MM, Powers LS, Walters K, Lovan N, Zhang M, et al. 2010. Identification of an autophagy defect in smokers' alveolar macrophages. *J. Immunol.* 185:5425–35
19. Ryter SW, Lee SJ, Choi AM. 2010. Autophagy in cigarette smoke-induced chronic obstructive pulmonary disease. *Expert Rev. Respir. Med.* 4:573–84
20. Luciani A, Villella VR, Esposito S, Brunetti-Pierri N, Medina D, et al. 2010. Defective CFTR induces aggresome formation and lung inflammation in cystic fibrosis through ROS-mediated autophagy inhibition. *Nat. Cell Biol.* 12:863–75
21. Luciani A, Villella VR, Esposito S, Brunetti-Pierri N, Medina DL, et al. 2011. Cystic fibrosis: a disorder with defective autophagy. *Autophagy* 7:104–6
22. Menzel DB, Amdur MO. 1986. Toxic response of the respiratory system. In *Casarett and Doull's Toxicology: The Basic Science of Poisons*, ed. K Klaassen, MO Amdur, J Doull, pp. 330–58. New York, NY: MacMillan. 3rd ed.

---

2. A comprehensive review on the molecular regulation and function of autophagy and its role in disease.

15. Demonstrates that LC3B regulates hypoxic vascular cell proliferation.

16. Shows for the first time increased morphological and biochemical indices of autophagy in the human COPD lung.

17. Shows that LC3B regulates epithelial cell apoptosis through interactions with Fas and caveolin-1.

18. Autophagic flux was impaired in the alveolar macrophages of smokers.

20. Identifies a role for autophagic dysfunction and impaired aggresome clearance in cystic fibrosis.

23. Ryter SW, Choi AM. 2009. Heme oxygenase-1/carbon monoxide: from metabolism to molecular therapy. *Am. J. Respir. Cell Mol. Biol.* 41:251–60
24. Yao H, Rahman I. 2011. Current concepts on oxidative/carbonyl stress, inflammation and epigenetics in pathogenesis of chronic obstructive pulmonary disease. *Toxicol. Appl. Pharmacol.* 254:72–85
25. Ryter SW, Choi AM. 2010. Autophagy in the lung. *Proc. Am. Thorac. Soc.* 7:13–21
26. Kroemer G, Mariño G, Levine B. 2010. Autophagy and the integrated stress response. *Mol. Cell.* 40:280–93
27. Tooze SA, Yoshimori T. 2010. The origin of the autophagosomal membrane. *Nat. Cell Biol.* 12:831–35
28. Yang Z, Klionsky DJ. 2010. Mammalian autophagy: core molecular machinery and signaling regulation. *Curr. Opin. Cell Biol.* 22:124–31
29. Jung CH, Ro SH, Cao J, Otto NM, Kim DH. 2010. mTOR regulation of autophagy. *FEBS Lett.* 584:1287–95
30. Vander Haar E, Lee SI, Bandhakavi S, Griffin TJ, Kim DH. 2007. Insulin signalling to mTOR mediated by the Akt/PKB substrate PRAS40. *Nat. Cell Biol.* 9:316–23
31. Gwinn DM, Shackelford DB, Egan DF, Mihaylova MM, Mery A, et al. 2008. AMPK phosphorylation of raptor mediates a metabolic checkpoint. *Mol. Cell* 30:214–26
32. Chan EY, Kir S, Tooze SA. 2007. siRNA screening of the kinome identifies ULK1 as a multidomain modulator of autophagy. *J. Biol. Chem.* 282:25464–74
33. Ganley IG, Lam DH, Wang J, Ding X, Chen S, Jiang X. 2009. ULK1·ATG13·FIP200 complex mediates mTOR signaling and is essential for autophagy. *J. Biol. Chem.* 284:12297–305
34. Jung CH, Jun CB, Ro SH, Kim YM, Otto NM, et al. 2009. ULK-Atg13–FIP200 complexes mediate mTOR signaling to the autophagy machinery. *Mol. Biol. Cell* 20:1992–2003
35. Hosokawa N, Sasaki T, Iemura S, Natsume T, Hara T, et al. 2009. Atg101, a novel mammalian autophagy protein interacting with Atg13. *Autophagy* 5:973–79
36. Hosokawa N, Hara T, Kaizuka T, Kishi C, Takamura A, et al. 2009. Nutrient-dependent mTORC1 association with the ULK1–Atg13–FIP200 complex required for autophagy. *Mol. Biol. Cell* 20:1981–91
37. Lee JW, Park S, Takahashi Y, Wang HG. 2010. The association of AMPK with ULK1 regulates autophagy. *PLoS ONE* 5:e15394
38. Shang L, Chen S, Du F, Li S, Zhao L, Wang X. 2011. Nutrient starvation elicits an acute autophagic response mediated by Ulk1 dephosphorylation and its subsequent dissociation from AMPK. *Proc. Natl. Acad. Sci. USA* 108:4788–93
39. Kim J, Kundu M, Viollet B, Guan KL. 2011. AMPK and mTOR regulate autophagy through direct phosphorylation of Ulk1. *Nat. Cell Biol.* 13:132–41
40. Chan EY, Longatti A, McKnight NC, Tooze SA. 2009. Kinase-inactivated ULK proteins inhibit autophagy via their conserved C-terminal domains using an Atg13-independent mechanism. *Mol. Cell. Biol.* 29:157–71
41. Mizushima N. 2010. The role of the Atg1/ULK1 complex in autophagy regulation. *Curr. Opin. Cell Biol.* 22:132–39
42. Chang YY, Neufeld T. 2009. An Atg1/Atg13 complex with multiple roles in TOR-mediated autophagy regulation. *Mol. Biol. Cell* 20:2004–14
43. He C, Levine B. 2010. The Beclin 1 interactome. *Curr. Opin. Cell Biol.* 22:140–49
44. Liang XH, Jackson S, Seaman M, Brown K, Kempkes B, et al. 1999. Induction of autophagy and inhibition of tumorigenesis by beclin 1. *Nature* 402:672–76
45. Liang C, Feng P, Ku B, Dotan I, Canaani D, et al. 2006. Autophagic and tumour suppressor activity of a novel Beclin1-binding protein UVRAG. *Nat. Cell Biol.* 8:688–99
46. Itakura E, Kishi C, Inoue K, Mizushima N. 2008. Beclin 1 forms two distinct phosphatidylinositol 3-kinase complexes with mammalian Atg14 and UVRAG. *Mol. Biol. Cell* 19:5360–72
47. Itakura E, Mizushima N. 2009. Atg14 and UVRAG: mutually exclusive subunits of mammalian Beclin 1-PI3K complexes. *Autophagy* 5:534–36
48. Zhong Y, Wang QJ, Li X, Yan Y, Backer JM, et al. 2009. Distinct regulation of autophagic activity by Atg14L and Rubicon associated with Beclin 1–phosphatidylinositol-3-kinase complex. *Nat. Cell Biol.* 11:468–76

49. Matsunaga K, Saitoh T, Tabata K, Omori H, Satoh T, et al. 2009. Two Beclin 1-binding proteins, Atg14L and Rubicon, reciprocally regulate autophagy at different stages. *Nat. Cell Biol.* 11:385–96
50. Maiuri MC, Le Toumelin G, Criollo A, Ran JC, Gautier F, et al. 2007. Functional and physical interaction between Bcl-X_L and a BH3-like domain in Beclin-1. *EMBO J.* 26:2527–39
51. Pattingre S, Tassa A, Qu X, Garuti R, Liang XH, et al. 2005. Bcl-2 antiapoptotic proteins inhibit Beclin 1-dependent autophagy. *Cell* 122:927–39
52. Fimia GM, Stoykova A, Romagnoli A, Giunta L, Di Bartolomeo S, et al. 2007. Ambra1 regulates autophagy and development of the nervous system. *Nature* 447:1121–25
53. Proikas-Cezanne T, Waddell S, Gaugel A, Frickey T, Lupas A, Nordheim A. 2004. WIPI-1α (WIPI49), a member of the novel 7-bladed WIPI protein family, is aberrantly expressed in human cancer and is linked to starvation-induced autophagy. *Oncogene* 23:9314–25
54. Polson HE, de Lartigue J, Rigden DJ, Reedijk M, Urbé S, et al. 2010. Mammalian Atg18 (WIPI2) localizes to omegasome-anchored phagophores and positively regulates LC3 lipidation. *Autophagy* 6(4):506–22
55. Ohsumi Y. 2001. Molecular dissection of autophagy: two ubiquitin-like systems. *Nat. Rev. Mol. Cell Biol.* 2:211–16
56. Tanida I, Ueno T, Kominami E. 2004. LC3 conjugation system in mammalian autophagy. *Int. J. Biochem. Cell Biol.* 36:2503–18
57. Tanida I. 2011 Autophagy basics. *Microbiol. Immunol.* 55:1–11
58. Kabeya Y, Mizushima N, Ueno T, Yamamoto A, Kirisako T, et al. 2000. LC3, a mammalian homologue of yeast Apg8p, is localized in autophagosome membranes after processing. *EMBO J.* 19:5720–28
59. Kabeya Y, Mizushima N, Yamamoto A, Oshitani-Okamoto S, Ohsumi Y, et al. 2004. LC3, GABARAP and GATE16 localize to autophagosomal membrane depending on form-II formation. *J. Cell Sci.* 117:2805–12
60. Hanada T, Noda NN, Satomi Y, Ichimura Y, Fujioka Y, et al. 2007. The Atg12-Atg5 conjugate has a novel E3-like activity for protein lipidation in autophagy. *J. Biol. Chem.* 282:37298–302
61. Gutierrez MG, Munafo DB, Beron W, Colombo MI. 2004. Rab7 is required for the normal progression of the autophagic pathway in mammalian cells. *J. Cell Sci.* 117:2687–97
62. Jager S, Bucci C, Tanida I, Ueno T, Kominami E, et al. 2004. Role for Rab7 in maturation of late autophagic vacuoles. *J. Cell Sci.* 117:4837–48
63. Liang C, Lee JS, Inn KS, Gack MU, Li Q, et al. 2008. Beclin1-binding UVRAG targets the class C Vps complex to coordinate autophagosome maturation and endocytic trafficking. *Nat. Cell Biol.* 10:776–87
64. Klionsky DJ, Abeliovich H, Agostinis P, Agrawal DK, Aliev G, et al. 2008. Guidelines for the use and interpretation of assays for monitoring autophagy in higher eukaryotes. *Autophagy* 4:151–75
65. Mizushima N, Yoshimori T. 2007. How to interpret LC3 immunoblotting. *Autophagy* 3:542–45
66. **Mizushima N, Yoshimori T, Levine B. 2010. Methods in mammalian autophagy research. *Cell* 140:313–26**
67. Swanlund JM, Kregel KC, Oberley TD. 2010. Investigating autophagy: quantitative morphometric analysis using electron microscopy. *Autophagy* 6:270–27
68. Rubinsztein DC, Cuervo AM, Ravikumar B, Sarkar S, Korolchuk V, et al. 2009. In search of an "autophagomometer." *Autophagy* 5:585–89
69. Tanida I, Minematsu-Ikeguchi N, Ueno T, Kominami E. 2005. Lysosomal turnover, but not a cellular level, of endogenous LC3 is a marker for autophagy. *Autophagy* 1:84–91
70. Iwai-Kanai E, Yuan H, Huang C, Sayen MR, Perry-Garza CN, et al. 2008. A method to measure cardiac autophagic flux in vivo. *Autophagy* 4:322–29
71. Berg TO, Fengsrud M, Stromhaug PE, Berg T, Seglen PO. 1998. Isolation and characterization of rat liver amphisomes. Evidence for fusion of autophagosomes with both early and late endosomes. *J. Biol. Chem.* 273:21883–92
72. Furuno K, Ishikawa T, Kato K. 1982. Appearance of autolysosomes in rat liver after leupeptin treatment. *J. Biochem.* 91:1485–94
73. Haspel J, Rahamthulla SS, Ifedigbo E, Nakahira K, Dolinay T, et al. 2011. Characterization of macroautophagic flux in vivo using a leupeptin-based assay. *Autophagy* 7:629–42

66. Critically surveys the methodologies used to measure autophagy in mammalian cells and their caveats.

74. Shvets E, Elazar Z. 2009. Flow cytometric analysis of autophagy in living mammalian cells. *Methods Enzymol.* 452:131–41
75. Farkas T, Hoyer-Hansen M, Jaattela M. 2009. Identification of novel autophagy regulators by a luciferase-based assay for the kinetics of autophagic flux. *Autophagy* 5:1018–25
76. Larsen KB, Lamark T, Øvervatn A, Harneshaug I, Johansen T, et al. 2010. A reporter cell system to monitor autophagy based on p62/SQSTM1. *Autophagy* 6:784–93
77. Kimura S, Noda T, Yoshimori T. 2007. Dissection of the autophagosome maturation process by a novel reporter protein, tandem fluorescent-tagged LC3. *Autophagy* 3:452–60
78. Eng KE, Panas MD, Karlsson Hedestam GB, McInerney GM. 2010. A novel quantitative flow cytometry-based assay for autophagy. *Autophagy* 6:634–41
79. Bjorkoy G, Lamark T, Pankiv S, Overvatn A, Brech A, et al. 2009. Monitoring autophagic degradation of p62/SQSTM1. *Methods Enzymol.* 452:181–97
80. Dennis PB, Mercer CA. 2009. The GST-BHMT assay and related assays for autophagy. *Methods Enzymol.* 452:97–118
81. Ju JS, Miller SE, Jackson E, Cadwell K, Piwnica-Worms D, et al. 2009. Quantitation of selective autophagic protein aggregate degradation in vitro and in vivo using luciferase reporters. *Autophagy* 5:511–19
82. Johansen T, Lamark T. 2011. Selective autophagy mediated by autophagic adapter proteins. *Autophagy* 7(3):279–96
83. Kim I, Rodriguez-Enriquez S, Lemasters JJ. 2007. Selective degradation of mitochondria by mitophagy. *Arch. Biochem. Biophys.* 462:245–53
84. Levine B. 2005. Eating oneself and uninvited guests: autophagy-related pathways in cellular defense. *Cell* 120:159–62
85. Koga H, Cuervo AM. 2011. Chaperone-mediated autophagy dysfunction in the pathogenesis of neurodegeneration. *Neurobiol. Disease* 43:29–37
86. Shpilka T, Elazar Z. 2011. Shedding light on mammalian microautophagy. *Dev. Cell.* 20:1–2
87. Virgin HW, Levine B. 2009. Autophagy genes in immunity. *Nat. Immunol.* 10:461–70
88. Semenza GL. 2005. Involvement of hypoxia-inducible factor 1 in pulmonary pathophysiology. *Chest* 128:S592–94
89. Zhang H, Bosch-Marce M, Shimoda LA, Tan YS, Baek JH, et al. 2008. Mitochondrial autophagy is an HIF-1-dependent adaptive metabolic response to hypoxia. *J. Biol. Chem.* 283:10892–903
90. Bellot G, Garcia-Medina R, Gounon P, Chiche J, Roux D, et al. 2009. Hypoxia-induced autophagy is mediated through hypoxia-inducible factor induction of BNIP3 and BNIP3L via their BH3 domains. *Mol. Cell. Biol.* 29:2570–81
91. Azad MB, Chen Y, Henson ES, Cizeau J, McMillan-Ward E, et al. 2008. Hypoxia induces autophagic cell death in apoptosis-competent cells through a mechanism involving BNIP3. *Autophagy* 4:195–204
92. Qu X, Yu J, Bhagat G, Furuya N, Hibshoosh H, et al. 2003. Promotion of tumorigenesis by heterozygous disruption of the beclin 1 autophagy gene. *J. Clin. Investig.* 112:1809–20
93. Lee SJ, Kim HP, Jin Y, Choi AM, Ryter S. 2011. Beclin 1 deficiency is associated with increased hypoxia-induced angiogenesis. *Autophagy* 7:829–39
94. Semenza GL. 2010. Vascular responses to hypoxia and ischemia. *Arterioscler. Thromb. Vasc. Biol.* 30:648–52
95. Kanamori H, Takemura G, Maruyama R, Goto K, Tsujimoto A, et al. 2009. Functional significance and morphological characterization of starvation-induced autophagy in the adult heart. *Am. J. Pathol.* 174:1705–14
96. Scarlatti F, Granata R, Meijer AJ, Codogno P. 2009. Does autophagy have a license to kill mammalian cells? *Cell Death Differ.* 16:12–20
97. Kroemer G, Galluzzi L, Vandenabeele P, Abrams J, Alnemri ES, et al. 2009. Classification of cell death: recommendations of the nomenclature committee on cell death. *Cell Death Differ.* 16:3–11
98. Boya P, Gonzalez-Polo RA, Casares N, Perfettini JL, Dessen P, et al. 2005. Inhibition of macroautophagy triggers apoptosis. *Mol. Cell. Biol.* 25:1025–40

99. Dalby KN, Tekedereli I, Lopez-Berestein G, Ozpolat B. 2010. Targeting the prodeath and prosurvival functions of autophagy as novel therapeutic strategies in cancer. *Autophagy* 6:322–29
100. Ding WX, Ni HM, Gao W, Hou YF, Melan MA, et al. 2007. Differential effects of endoplasmic reticulum stress-induced autophagy on cell survival. *J. Biol. Chem.* 282:4702–10
101. Ullman E, Fan Y, Stawowczyk M, Chen HM, Yue Z, Zong WX. 2008. Autophagy promotes necrosis in apoptosis-deficient cells in response to ER stress. *Cell Death Differ.* 15:422–25
102. Kroemer G, Levine B. 2008. Autophagic cell death: the story of a misnomer. *Nat. Rev. Mol. Cell Biol.* 9:1004–10
103. Galluzzi L, Vicencio JM, Kepp O, Tasdemir E, Maiuri MC, Kroemer G. 2008. To die or not to die: That is the autophagic question. *Curr. Mol. Med.* 8:78–91
104. Wang Y, Singh R, Massey AC, Kane SS, Kaushik S, et al. 2008. Loss of macroautophagy promotes or prevents fibroblast apoptosis depending on the death stimulus. *J. Biol. Chem.* 283:4766–77
105. Xu Y, Kim SO, Li Y, Han J. 2006. Autophagy contributes to caspase-independent macrophage cell death. *J. Biol. Chem.* 281:19179–87
106. Yu L, Wan F, Dutta S, Liu Z, Freundt E, et al. 2006. Autophagic programmed cell death by selective catalase degradation. *Proc. Natl. Acad. Sci. USA* 103:4952–57
107. Madden DT, Egger L, Bredesen DE. 2007. A calpain-like protease inhibits autophagic cell death. *Autophagy* 3:519–22
108. Shimizu S, Kanaseki T, Mizushima N, Mizuta T, Arakawa-Kobayashi S, et al. 2004. Role of Bcl-2 family proteins in a non-apoptotic programmed cell death dependent on autophagy genes. *Nat. Cell Biol.* 6:1221–28
109. Wu YT, Tan HL, Huang Q, Kim YS, Pan N, et al. 2008. Autophagy plays a protective role during zVAD-induced necrotic cell death. *Autophagy* 4:457–66
110. Maiuri MC, Zalckvar E, Kimchi A, Kroemer G. 2007. Self-eating and self-killing: crosstalk between autophagy and apoptosis. *Nat. Rev. Mol. Cell Biol.* 8:741–52
111. Chang NC, Nguyen M, Germain M, Shore GC. 2010. Antagonism of Beclin 1-dependent autophagy by BCL-2 at the endoplasmic reticulum requires NAF-1. *EMBO J.* 29:606–18
112. Pyo JO, Jang MH, Kwon YK, Lee HJ, Jun JI, et al. 2005. Essential roles of Atg5 and FADD in autophagic cell death: dissection of autophagic cell death into vacuole formation and cell death. *J. Biol. Chem.* 280:20722–29
113. Yousefi S, Perozzo R, Schmid I, Ziemiecki A, Schaffner T, et al. 2006. Calpain-mediated cleavage of Atg5 switches autophagy to apoptosis. *Nat. Cell Biol.* 8:1124–32
114. Wirawan E, Vande Walle L, Kersse K, Cornelis S, Claerhout S, et al. 2010. Caspase-mediated cleavage of Beclin-1 inactivates Beclin-1-induced autophagy and enhances apoptosis by promoting the release of proapoptotic factors from mitochondria. *Cell Death Disease* 1:e18
115. Li H, Wang P, Sun Q, Ding WX, Yin XM, et al. 2011. Following cytochrome c release, autophagy is inhibited during chemotherapy-induced apoptosis by caspase-8-mediated cleavage of Beclin-1. *Cancer Res.* 71:3625–34
116. Yu L, Alva A, Su H, Dutt P, Freundt E, et al. 2004. Regulation of an ATG7-beclin 1 program of autophagic cell death by caspase-8. *Science* 304:1500–2
117. Shimizu S, Konishi A, Nishida Y, Mizuta T, Nishina H, et al. 2010. Involvement of JNK in the regulation of autophagic cell death. *Oncogene* 29:2070–82
118. Wei Y, Sinha S, Levine B. 2008. Dual role of JNK1-mediated phosphorylation of Bcl-2 in autophagy and apoptosis regulation. *Autophagy* 4:949–51
119. Crighton D, Wilkinson S, O'Prey J, Syed N, Smith P, et al. 2006. DRAM, a p53-induced modulator of autophagy, is critical for apoptosis. *Cell* 126:121–34
120. Church DF, Pryor WA. 1985. Free-radical chemistry of cigarette smoke and its toxicological implications. *Environ. Health Perspect.* 64:111–26
121. Kim HP, Wang X, Chen Z-H, Lee SJ, Huang M-H, et al. 2008. Autophagic proteins regulate cigarette smoke induced apoptosis: protective role of heme oxygenase-1. *Autophagy* 4:887–95
122. Barrett JC, Hansoul S, Nicolae DL, Cho JH, Duerr RH, et al. 2008. Genome-wide association defines more than 30 distinct susceptibility loci for Crohn's disease. *Nat. Genet.* 40:955–62

123. Gateva V, Sandling JK, Hom G, Taylor KE, Chung SA, et al. 2009. A large-scale replication study identifies *TNIP1*, *PRDM1*, *JAZF1*, *UHRF1BP1* and *IL10* as risk loci for systemic lupus erythematosus. *Nat. Genet.* 41:1228–33

124. **Saitoh T, Fujita N, Jang MH, Uematsu S, Yang BG. 2008. Loss of the autophagy protein Atg16L1 enhances endotoxin-induced IL-1β production. *Nature* 456:264–68**

    > 124. The first report to link autophagy and inflammation.

125. Cadwell K, Liu JY, Brown SL, Miyoshi H, Loh J. 2008. A key role for autophagy and the autophagy gene *Atg16l1* in mouse and human intestinal Paneth cells. *Nature* 456:259–63

126. Travassos LH, Carneiro LA, Ramjeet M, Hussey S, Kim YG, et al. 2010. Nod1 and Nod2 direct autophagy by recruiting ATG16L1 to the plasma membrane at the site of bacterial entry. *Nat. Immunol.* 11:55–62

127. Nakagawa I, Amano A, Mizushima N, Yamamoto A, Yamaguchi H, et al. 2004. Autophagy defends cells against invading group A *Streptococcus*. *Science* 306:1037–40

128. Gutierrez MG, Master SS, Singh SB, Taylor GA, Colombo MI, Deretic V. 2004. Autophagy is a defense mechanism inhibiting BCG and *Mycobacterium tuberculosis* survival in infected macrophages. *Cell* 119:753–66

129. Schroder K, Tschopp J. 2010. The inflammasomes. *Cell* 140:821–32

130. **Nakahira K, Haspel JA, Rathinam VA, Lee SJ, Dolinay T, et al. 2011. Autophagy proteins regulate innate immune responses by inhibiting the release of mitochondrial DNA mediated by the NALP3 inflammasome. *Nat. Immunol.* 12:222–30**

    > 130. Autophagy-deficient macrophages are sensitized to activation of NALP3 inflammasome–associated cytokine production.

131. Zhou R, Yazdi AS, Menu P, Tschopp J. 2011. A role for mitochondria in NLRP3 inflammasome activation. *Nature* 469:221–25

132. Saitoh T, Fujita N, Hayashi T, Takahara K, Satoh T, et al. 2009. Atg9a controls dsDNA-driven dynamic translocation of STING and the innate immune response. *Proc. Natl. Acad. Sci. USA* 106:20842–46

133. Lee HK, Lund JM, Ramanathan B, Mizushima N, Iwasaki A. 2007. Autophagy-dependent viral recognition by plasmacytoid dendritic cells. *Science* 315:1398–401

134. Tal MC, Sasai M, Lee HK, Yordy B, Shadel GS, Iwasaki A. 2009. Absence of autophagy results in reactive oxygen species-dependent amplification of RLR signaling. *Proc. Natl. Acad. Sci. USA* 106:2770–75

135. Paludan C, Schmid D, Landthaler M, Vockerodt M, Kube D, et al. 2005. Endogenous MHC class II processing of a viral nuclear antigen after autophagy. *Science* 307:593–96

136. Lee HK, Mattei LM, Steinberg BE, Alberts P, Lee YH, et al. 2010. *Immunity* 32:227–39

137. English L, Chemali M, Duron J, Rondeau C, Laplante A, et al. 2009. *Nat. Immunol.* 10:480–87

138. Nedjic J, Aichinger M, Emmerich J, Mizushima N, Klein L. 2008. Autophagy in thymic epithelium shapes the T-cell repertoire and is essential for tolerance. *Nature* 455:396–400

139. Tung SM, Unal C, Ley A, Peña C, Tunggal B, et al. 2010. Loss of *Dictyostelium* ATG9 results in a pleiotropic phenotype affecting growth, development, phagocytosis and clearance and replication of *Legionella pneumophila*. *Cell. Microbiol.* 12:765–80

140. Dubuisson JF, Swanson MS. 2006. Mouse infection by *Legionella*, a model to analyze autophagy. *Autophagy* 2:179–82

141. Behr M, Schurr E, Gros P. 2010. TB: screening for responses to a vile visitor. *Cell* 140:615–18

142. Deretic V, Delgado M, Vergne I, Master S, De Haro S, et al. 2009. Autophagy in immunity against mycobacterium tuberculosis: a model system to dissect immunological roles of autophagy. *Curr. Top. Microbiol. Immunol.* 335:169–88

143. Watanabe E, Muenzer JT, Hawkins WG, Davis CG, Dixon DJ. 2009. Sepsis induces extensive autophagic vacuolization in hepatocytes: a clinical and laboratory-based study. *Lab. Investig.* 89:549–61

144. Celli BR. 1995. Pathophysiology of chronic obstructive pulmonary disease. *Chest Surg. Clin. N. Am.* 5:623–34

145. Zhou B, Boudreau N, Coulber C, Hammarback J, Rabinovitch M. 1997. Microtubule-associated protein 1 light chain 3 is a fibronectin mRNA-binding protein linked to mRNA translation in lamb vascular smooth muscle cells. *J. Clin. Investig.* 100:3070–82

146. Rabe KF, Hurd S, Anzueto A, Barnes PJ, Buist SA, et al. 2007. Global Initiative for Chronic Obstructive Lung Disease. Global strategy for the diagnosis, management, and prevention of chronic obstructive pulmonary disease: GOLD executive summary. *Am. J. Respir. Crit. Care Med.* 176:532–55

147. Pankiv S, Clausen TH, Lamark T, Brech A, Bruun JA, et al. 2007. p62/SQSTM1 binds directly to Atg8/LC3 to facilitate degradation of ubiquitinated protein aggregates by autophagy. *J. Biol. Chem.* 282:24131–45
148. Kirkin V, Lamark T, Sou YS, Bjørkøy G, Nunn JL, et al. 2009. A role for NBR1 in autophagosomal degradation of ubiquitinated substrates. *Mol. Cell* 33:505–16
149. **Matsuda N, Sato S, Shiba K, Okatsu K, Saisho K, et al. 2010. PINK1 stabilized by mitochondrial depolarization recruits Parkin to damaged mitochondria and activates latent Parkin for mitophagy.** *J. Cell Biol.* **189:211–21**

149. Describes the fine molecular regulation of mitochondrial selective autophagy (mitophagy) by PINK1 and Parkin.

RELATED RESOURCES

1. Kim HP, Chen ZH, Choi AM, Ryter SW. 2009. Analyzing autophagy in clinical tissues of lung and vascular diseases. *Methods Enzymol.* 453:197–216
2. Klionsky DJ, Codogno P, Cuervo AM, Deretic V, Elazar Z, et al. 2010. A comprehensive glossary of autophagy-related molecules and processes. *Autophagy* 6(4):438–48
3. Tanida I. 2011. Autophagosome formation and molecular mechanism of autophagy. *Antioxid. Redox Signal.* 14(11):2201–14

Stop the Flow: A Paradigm for Cell Signaling Mediated by Reactive Oxygen Species in the Pulmonary Endothelium

Elizabeth A. Browning, Shampa Chatterjee, and Aron B. Fisher

Institute for Environmental Medicine, University of Pennsylvania School of Medicine, Philadelphia, Pennsylvania 19104; email: abf@mail.med.upenn.edu

Keywords

chemotransduction, mechanotransduction, K_{ATP} channel, NADPH oxidase, endothelial cell membrane potential, nitric oxide

Abstract

The lung endothelium is exposed to mechanical stimuli through shear stress arising from blood flow and responds to altered shear by activation of NADPH (NOX2) to generate reactive oxygen species (ROS). This review describes the pathway for NOX2 activation and the downstream ROS-mediated signaling events on the basis of studies of isolated lungs and flow-adapted endothelial cells in vitro that are subjected to acute flow cessation (ischemia). Altered mechanical stress is detected by a cell-associated complex involving caveolae and other membrane proteins that results in endothelial cell membrane depolarization and then the activation of specific kinases that lead to the assembly of NOX2 components. ROS generated by this enzyme amplify the mechanosignal within the endothelial cell to regulate activation and/or synthesis of proteins that participate in cell growth, proliferation, differentiation, apoptosis, and vascular remodeling. These responses indicate an important role for NOX2-derived ROS associated with mechanotransduction in promoting vascular homeostasis.

INTRODUCTION

ROS: reactive oxygen species

Reactive oxygen species (ROS) are well established as a cause of tissue injury via oxidative damage to lipids, proteins, DNA, and other biomolecules. The toxic effects of ROS constitute a major mechanism for inflammatory lung injury of diverse etiology and following exposure to various exogenous agents such as hyperoxia, paraquat, and bleomycin. ROS are generated normally in cells by the activity of several enzymes or through auto-oxidation of endogenous compounds. The endogenous generation of ROS was first recognized as a component in the bactericidal activity of phagocytic cells (1), and only several decades later did the scientific community realize that these molecules could also serve physiological signaling functions. This belated understanding of the physiological role of ROS contrasts with that of another free radical, ·NO, whose signaling function was recognized long before the appreciation of its potential toxicity. The lag in recognizing ROS as signaling molecules is related to their evanescence in tissues and the cumbersome and indirect methods available for their detection. Furthermore, the biological effects of ROS have generally been determined by the response to exogenous application of H_2O_2 or compounds that generate the superoxide anion ($O_2^{·-}$), almost always in concentrations that greatly exceed potential physiological levels. Finally, availability of and cell membrane permeability to detoxifying agents greatly influence the biological response to oxidants. Increasing evidence over the past two decades has indicated that ROS exert subtle effects when produced in a controlled and regulated manner in a specific subcellular location or compartment (2). In such a setting, ROS participate as second messengers to regulate cellular signaling pathways that control cellular and tissue homeostasis through the modulation of cell growth and proliferation, apoptosis, differentiation, and a myriad of other cellular processes.

This review focuses on the physiological role of ROS in the pulmonary vasculature and emphasizes our model of altered lung (endothelial) shear stress in which the endogenous generation of ROS and their physiological effects can be readily measured. Altered (acutely decreased) shear stress represents a physiological model for pulmonary ischemia.

REACTIVE OXYGEN SPECIES (ROS)

We now define ROS from a chemical standpoint and consider their sources in the intact cell. These sources include both endogenous production within the cell as well as exogenously administered drugs and poisons.

Definition of ROS

The term ROS describes a broad group of molecules that includes both free radical (e.g., $O_2^{·-}$, ·OH) and nonradical (e.g., H_2O_2) species derived from the reduction of O_2. These species are of varying reactivity with tissue components. For example, the hydroxyl radical (·OH) is extremely reactive and oxidizes almost any biomolecule with which it comes in contact. $O_2^{·-}$ and H_2O_2 are less reactive and interact with chemically defined targets. Several members of the ROS family have properties that enable them to function in signaling; i.e., they are generated at low levels in a precise location and show both specificity and reversibility in their reactivity with biomolecules. H_2O_2, which is present normally at $\sim 10^{-8}$ M in the cytosol, can reversibly oxidize protein cysteines and is regarded as the major ROS-related signaling molecule. Less well understood is the role of $O_2^{·-}$ in cell signaling; although this molecule is relatively short lived, it may participate in a limited range of signaling functions.

Cellular Sources of ROS

ROS generation in vivo can occur via various enzymatic pathways and via nonenzymatic processes. The former generate either $O_2^{·-}$ or H_2O_2 by the transfer of one or two electrons, respectively, whereas the nonenzymatic pathway generates $O_2^{\cdot-}$ by auto-oxidation of a reduced compound. An important difference between these two general mechanisms is that enzymatic pathways, but not auto-oxidation, represent a potentially controlled reaction that can participate in cellular homeostasis.

Major enzymatic pathways for generation of ROS are as follows:

1. NADPH oxidase (NOX): This family of proteins (NOX2 is the prototype) utilizes NADPH to reduce molecular oxygen to $O_2^{\cdot-}$ (3). This is the only family of enzymes with the known primary function of generating ROS.
2. Xanthine oxidase: This enzyme catalyzes the oxidation of hypoxanthine to xanthine and then to uric acid. These reactions use molecular oxygen as an electron acceptor, resulting in the generation of $O_2^{\cdot-}$ or H_2O_2. Activity of the similar xanthine dehydrogenase enzyme does not generate ROS.
3. Cyclo-oxygenase/lipoxygenase (COX/LOX): These enzymes produce $O_2^{\cdot-}$ during the metabolism of arachidonate in the presence of NAD(P)H.
4. Nitric oxide synthase (NOS): With deficiency of the important cofactor tetrahydrobiopterin or in a low-pH medium, this enzyme can generate $O_2^{\cdot-}$ instead of $\cdot NO$.
5. Various amine oxidases: Monoamine oxidase, for example, generates H_2O_2 during the oxidation of dopamine.

NOX2: NADPH oxidase, type 2

NOS: nitric oxide synthase

ETC: electron transport chain

Common nonenzymatic sources of ROS are as follows:

1. Mitochondrial electron transport chain (ETC): Under normal physiological conditions, approximately 1–3% of electrons carried by the mitochondrial ETC leak out of the pathway and pass directly to oxygen, generating $O_2^{\cdot-}$. This leakage represents an auto-oxidation reaction. There are several sites of possible auto-oxidation during the sequential transfer of electrons to ETC components in the process of oxidative phosphorylation. A major site is ubisemiquinone, the component at the interface between one- and two-electron transfer in the ETC.
2. Other cytochromes: A similar auto-oxidation reaction can occur with reduced members of the cytochrome P-450 family of enzymes.
3. Free iron (Fe^{2+}): The presence of Fe^{2+} can catalyze the generation of $\cdot OH$ from H_2O_2 and $O_2^{\cdot-}$, thereby generating a more potent electrophile.
4. Quinones and other auto-oxidizable chemicals: A variety of endogenously produced compounds, especially those with a quinone moiety, can auto-oxidize on exposure to O_2 with the generation of $O_2^{\cdot-}$. An example is the metal-catalyzed oxidation of dopamine to a semiquinone; this compound can further auto-oxidize, thereby generating $O_2^{\cdot-}$. Other endogenous auto-oxidizable chemicals include epinephrine, ascorbic acid (vitamin C), menadione (vitamin K), and certain thiols.
5. Exogenous agents: A variety of exogenously administered agents, including drugs and xenobiotics, demonstrate chemistry similar to that described for quinones. For example, the herbicide paraquat is reduced by NADPH–cytochrome P-450 reductase and auto-oxidizes to generate $O_2^{\cdot-}$ as the basis for its toxicity. The cancer therapeutic agents doxorubicin and bleomycin can also generate ROS.

ROS-MEDIATED SIGNALING IN THE PULMONARY CIRCULATION

ROS generated in endothelial cells exert their physiological effects by interaction with transcription factors, protein tyrosine phosphatases, protein tyrosine kinases, mitogen-activated protein kinases, and other components of various signaling cascades. Many of these modified proteins transmit their signal through phosphorylation and dephosphorylation of specific amino acid residues

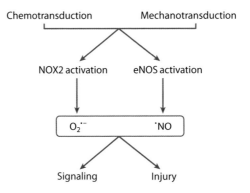

Figure 1

Reactive oxygen species (ROS) generation upon chemotransduction or altered mechanotransduction. Chemical stimuli or stop of shear activate NADPH oxidase, type 2 (NOX2) and endothelial nitric oxide synthase (eNOS). The resultant generation of oxygen ($O_2^{·-}$)- and nitrogen ($·NO$)-centered radicals results in signaling or injury, depending on the level of production and antioxidant defenses of the cell.

(serine, threonine, tyrosine, and histidine). ROS modify these signaling-related proteins through the oxidation of specific residues that result in reversible activation or inactivation of enzymatic activity. Excess ROS produced by these pathways can result in cell injury, and organisms have developed a variety of enzymatic and nonenzymatic defense mechanisms to protect against this possibility. The vast majority of studies of cell signaling have evaluated the response to various chemical agents (chemotransduction), many of which initiate the signaling process through cell membrane–associated receptors. Recent publications have reported that cells respond similarly to mechanical stresses (mechanotransduction), which may initiate the signaling process through forces acting on the cell membrane (4). The similarity between the chemo- and mechanotransduction processes is shown in **Figure 1**.

NADPH Oxidases

The predominant and most-studied source of ROS within the pulmonary vasculature is the NOX pathway. The NOX family has seven members, of which NOX2 and -4 are most highly expressed in the pulmonary endothelium. NOX2 was the first member of the NOX family to be described and is the best understood (5, 6). Active NOX2 is composed of two integral membrane proteins (gp91^{phox} and p22^{phox}) and four cytosolic proteins (Rac, p40^{phox}, p47^{phox}, and p67^{phox}); the latter proteins translocate to the membrane upon activation and associate with the membrane subunits (**Figure 2**). Current usage identifies the gp91^{phox} flavoprotein as NOX2. Phosphorylation of Rac (Rac1 in pulmonary endothelium) is the key step in initiating translocation of the cytosolic components. The activation step regulates $O_2^{·-}$ production by the enzyme complex. In contrast, NOX4 appears to be constitutively active and does not require translocation of cytosolic subunits for ROS generation, although this point remains controversial (7).

Chemotransduction

Cell signaling can be initiated by soluble factors that act through either receptor-mediated or receptor-independent mechanisms to promote ROS generation. This process, termed chemotransduction, is described in this section. Similarly, cells can respond to an altered physical environment to initiate a signaling cascade; this latter process is termed mechanotransduction and is considered subsequently.

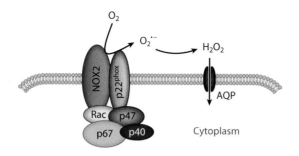

Figure 2

The assembled NADPH oxidase, type 2 (NOX2) complex. gp91^{phox} (NOX2) and p22^{phox} are integral plasma membrane components and together constitute cytochrome b_{558}. The remaining components are normally cytosolic proteins that translocate to the plasma membrane following their phosphorylation. $O_2^{\cdot-}$ that is generated by NOX2 is converted to H_2O_2, which can diffuse across the plasma membrane, possibly through aquaporin (AQP) channels, to initiate cell signaling.

Activation of NOX2. Cell signaling mediated by ROS can occur via the activation of chemoreceptors on the endothelial plasma membrane by various agonists. Representative agonists include angiotensin, thrombin, transforming growth factor-β, and platelet-derived growth factor. Each of these agonists initiates ROS generation through the activation of NOX2 activity. The interaction of angiotensin II (Ang II) with its receptor activates phospholipase C, which in turn hydrolyzes phosphatidylinositol-4,5-biphosphate to inositol 1,4,5-triphosphate (InsP$_3$) and diacylglycerol (DAG) (8). DAG and Ca^{2+} released by InsP$_3$ then activate protein kinase C (PKC) (8, 9), with subsequent phosphorylation of the cytosolic and possibly membrane components of NOX2 (10). NOX2 activation can also occur through nonreceptor mechanisms, such as stimulation by phorbol esters, that directly activate PKC (11, 12). Thus, PKC activation through either receptor-mediated or non-receptor-mediated mechanisms represents a key step in NOX2 activation, at least in response to some agonists.

Role of peroxiredoxin 6. Our recent studies using pulmonary microvascular endothelial cells have shown that peroxiredoxin 6 (Prdx6) serves as a link between PKC activity and NOX2 activation upon either Ang II or PMA (phorbol-12-myristate-13 acetate; also known as 12-O-tetradecanoylphorbol-13-acetate) treatment (13). The peroxiredoxins are a ubiquitously distributed family of peroxidases that function in cellular redox reactions (14, 15). Most of the peroxiredoxins utilize thioredoxin as the redox cofactor, although Prdx6 is GSH dependent (16). Prdx6 uniquely possesses phospholipase A$_2$ (PLA$_2$) in addition to its peroxidase activity (15, 16). Translocation of cytosolic NOX2 components and enzyme activation did not occur in endothelial cells that were Prdx6 null; transfection of Prdx6-null cells with constructs expressing the PLA$_2$ activity of Prdx6 rescued the ROS response to agonist treatment (13). NOX2 activation was also suppressed by the treatment of lungs and cells with an agent [MJ33 (1-hecadecyl-3-(trifluoroethyl)-sn-glycero-2-phosphomethanol)] that inhibits the PLA$_2$ but not the peroxidase activity of Prdx6 (13). Thus, the PLA$_2$ activity of Prdx6 is essential for NOX2 activation in pulmonary endothelial cells.

Prdx6 is a cytosolic enzyme that requires phosphorylation for binding to substrate lipids, such as those of the cell membrane, and its subsequent PLA$_2$ activity (17). Prdx6 can be phosphorylated by mitogen-activated protein kinases (MAP kinases) in vitro (18), and MAP kinase inhibitors block Prdx6 phosphorylation in Ang II–treated endothelial cells (13). Erk and p38 MAP kinase are able to phosphorylate Prdx6 in vitro. Thus, we propose that agonist-mediated PKC activation in turn

Ang II: angiotensin II

InsP$_3$: inositol 1,4,5-triphosphate

PKC: protein kinase C

Prdx6: peroxiredoxin 6

PLA$_2$: phospholipase A$_2$

MAP kinases: mitogen-activated protein kinases

activates MAP kinases, resulting in Prdx6 phosphorylation and Prdx6 binding (translocation) to the plasma membrane (13). This paradigm constitutes the upstream events leading to translocation of the cytosolic components of NOX2 and activation of ROS generation. A Prdx6-dependent mechanism similar to that described for endothelium is also required for NOX activation in pulmonary alveolar macrophages (13).

Although the role of PLA_2 activity in NOX2 activation is clear, the product of the PLA_2 reaction that is responsible for promoting NOX assembly and activation is less so. PLA_2 activity liberates both a free fatty acid and a lysophospholipid, and there is evidence that either metabolite may be directly or indirectly responsible for activation of the NOX2 pathway (19–21). Prdx6 does not show a preference for arachidonate-containing phospholipids such that this eicosanoid precursor would not be preferentially liberated by Prdx6 activity (22).

Mechanotransduction

Mechanotransduction is the process by which altered physical forces are sensed by the cell and converted into biochemical signals. The concept is analogous to chemotransduction. The endothelium is normally exposed to dynamic fluid shear stress as a result of viscous drag of blood flow and to cyclic strain resulting from hydrostatic pressure. Shear stress is defined as the fluid frictional force per unit area and acts at the luminal surface (endothelial lining) of blood vessels; it can be calculated as $4\mu Q/\pi r^3$, where μ is the dynamic viscosity, Q is the volumetric flow rate, and r is the radius of the conducting vessel. Thus, the magnitude of a change of shear, in a perfused vessel of given radius, is directly proportional to the change in perfusate flow rate (ΔQ). The wall strain associated with altered hydrostatic (perfusion) pressure is the ratio of the change in matrix unit length or area to the original length ($\Delta l/l_0$) or original area ($\Delta A/A_0$). Endothelial cells, located at the interface between blood and tissue, can sense changes in hemodynamic forces and respond appropriately to maintain homeostasis (23–26). These responses to pulsatile fluid shear stress and to cyclic circumferential strain contribute to the phenotype of endothelial cells.

Mechanotransduction in the lung. In the lung, the endothelium of major vessels is exposed to both shear stress as well as cyclic circumferential strain due to pulsations associated with the cardiac cycle. Both of these forces generated by the cardiac cycle can result in cell-mediated signaling responses. Although the effect of cyclic strain on lung cells has been studied relative to acute lung injury (27–29), less is known about strain-related signaling in the pulmonary endothelium. Expansion of the alveolus during ventilation is thought to be more of a change in conformation (unfolding) rather than one in surface area and thus does not significantly contribute to strain of the matrix and vascular structures. However, the relationship between respiration and circumferential strain at the capillary level is complex, depending on alteration of cardiac output with ventilation and the zones of the upright lung; these considerations are beyond the scope of this review.

The effects of shear stress on the pulmonary endothelium have received greater attention. The endothelium of the major pulmonary vessels is exposed to cardiac-induced cyclic changes in shear, whereas blood flow in the capillaries approaches a more constant laminar flow. Under conditions of altered flow (altered shear stress), increased ROS production mediates signal transduction, leading to downstream physiological responses in pulmonary as well as systemic endothelial cells (30, 31). This review concentrates on the role of ROS in pulmonary vascular signaling, with a focus on the generation of ROS associated with a model of acutely altered shear stress. The review emphasizes changes in laminar flow rate and does not discuss the responses to oscillatory or turbulent flow that are more relevant to pathophysiology associated with vascular abnormalities in the systemic vasculature (32).

Shear stresses in the circulation. The shear stresses in human pulmonary arteries have been determined by magnetic resonance imaging. Time-averaged axial near-wall shear stress was estimated at approximately 5–7 dyn cm^{-2} in the main pulmonary arteries (33), although calculations based on physiological parameters of flow and vessel dimensions suggest a lesser value of $\sim$2 dyn cm^{-2} (34). The measured near-wall shear in the main pulmonary artery is similar to that in the aorta ($\sim$8 dyn cm^{-2}) (35). Thus, although the driving pressure in the aorta is significantly greater than in the pulmonary artery (consistent with the greater vascular resistance in the systemic versus pulmonary circulation), the flow velocity and shear in the major vessels appear to be not significantly different. The blood flow velocity in the pulmonary capillaries is significantly lower than in the large vessels, but because shear stress varies with the cube of the radius, the shear stress in the capillaries is estimated to be significantly greater than in the major vessels (34).

Response of pulmonary endothelium to altered shear stress. Altered endothelial mechanotransduction associated with altered shear stress or cyclic strain results in ROS-dependent signaling (31, 36–38). The model used in our studies of altered mechanotransduction in the lung is the abrupt reduction of shear stress, i.e., an acute decrease (lung ischemia) that can occur physiologically due to vascular obstruction or during lung transplantation (30). In the isolated rat lung, a decrease in perfusion of approximately 80% represents the threshold for eliciting a response (39). Unlike the systemic circulation, loss of blood flow (ischemia) in the lung is not equated with tissue hypoxia, provided that ventilation is unimpaired (40–42). This physiological property has enabled us to study the endothelial response to decreased perfusion (altered shear stress) without the confounding variables of diminished oxygen supply and the metabolic responses to hypoxia.

In an ex vivo intact lung model (rat and mouse), loss of flow induces pulmonary endothelial membrane depolarization followed by ROS generation due to activation of NOX2 (43, 44). These studies have been carried out primarily by imaging of the lung by epifluorescence, confocal, or multiphoton techniques following cell labeling with various fluorophores (45). This response of the perfused lung to altered shear has been reproduced with an in vitro model by subjecting endothelial cells to a flow adaptation regimen (physiological shear stresses of 5–10 dyn cm^{-2} for 24–72 h) followed by abrupt flow cessation (46–48). Exposure of endothelial cells in culture to shear leads over time to a range of biochemical and structural alterations that constitute a flow-adapted state (26, 49). Several systems that we have used for flow adaptation of cells in vitro along with some examples of data that were generated with the flow cessation model are described in Chatterjee & Fisher (45) and are shown schematically in **Figure 3**. Because endothelial cells in vivo are presumably flow adapted, we believe that flow-adapting cells in culture results in a more physiological in vitro model for the study of the effects of altered shear stress compared with the commonly used model of acutely increased shear with non-flow-adapted cells.

ROS AS A SIGNAL TRANSDUCER WITH ALTERED SHEAR

Studies of chemotransduction involving Ang II and other signaling agonists have established the role of the NOX enzymes in signaling (11, 50–52). Studies using our mechanotransduction model of acute interruption of perfusion (acute loss of shear stress) to the isolated lung or to flow-adapted endothelial cells in culture reveal the rapid (within-seconds) appearance of ROS (43, 44). This reaction is inhibited by the presence of diphenyleneiodonium, a nonspecific flavoprotein inhibitor, and is abolished by knock out of gp91^{phox} (NOX2) (44, 46, 53). Studies described below document the important signaling role in the pulmonary endothelium of NOX2-derived ROS subsequent to loss of shear. ROS generation in these pathways represents a mechanism to transduce the signal generated in response to altered mechanical forces into a physiological response. ROS function as transducing agents by specifically and reversibly reacting with proteins to alter their structure, activity, and function.

Specificity of ROS in the Signaling Paradigm

NOX2 activation generates O_2^- as its only product: $NADPH + 2O_2 \rightarrow NADP^+ + 2O_2^- + H^+$. On the basis of its location as a plasma membrane–spanning protein, gp91^{phox} (NOX2) transfers its electron across the cell membrane and generates O_2^- at the extracellular face of the endothelial cell (54). The reaction generates a proton in the cytoplasm that requires neutralization to avoid cellular acidification (54); discussion of the potential mechanisms for proton removal is beyond the scope of this review. Superoxide is a short-lived species that dismutes to H_2O_2, either rapidly in a reaction catalyzed by superoxide dismutase (SOD) (rate constant $>10^9$ $M^{-1}s^{-1}$) or spontaneously at a somewhat slower ($>10^5$ $M^{-1}s^{-1}$) but still significant rate: $O_2^- + 2H^+ \rightarrow H_2O_2$. H_2O_2 can cross biological membranes, thereby enabling its interaction with intracellular signaling pathways. However, recent evidence suggests that H_2O_2 does not directly cross the lipid bilayer but rather passes through membranes via aquaporins, similarly to H_2O (55). Because H_2O_2 is highly diffusible, it has a steep gradient from its sites of production. Furthermore, H_2O_2 is relatively specific in its chemical reactions, and its concentration is well regulated by various specific enzymes (catalase, glutathione peroxidase, peroxiredoxins). These characteristics make H_2O_2 an ideal signaling molecule (2, 54). H_2O_2 functions in signaling through a reversible reaction with low-pK_a cysteine residues on proteins to initially form a disulfide bond (–SS–) or sulfenic acid (–SOH) (56). Further oxidation of –SOH yields sulfinic (–SO$_2$H) and sulfonic (–SO$_3$H) acids, which are considered to be irreversible reactions (2, 57). These oxidation products in some proteins (e.g., most peroxiredoxins, but not Prdx6) can be reduced by an energy (ATP)-dependent enzyme, sulfiredoxin (58, 59).

Although it is generally believed that H_2O_2 represents the primary ROS-related signal transducer, recent evidence suggests that O_2^- may activate discrete signaling pathways independently of H_2O_2 (54, 60). Because it is short lived, O_2^- must be produced in very close proximity to its target to be effective as a signaling molecule (61). Thus, if extracellular O_2^- functions as a signal transducer, it most likely acts on receptor proteins on the extracellular face of the endothelial

---

Figure 3

Apparatus for flow-adapting endothelial cells in vitro to shear stress. Cells are seeded, are allowed to attach for 24 h, and are then subjected to flow (generally at shear stresses of 5–10 dyn cm^{-2}) for 24–72 h. (*a*) Procedure for flow-adapting cells for fluorescence imaging. The upper-panel chamber (Warner Instruments, Hamden, CT) shows the parallel plate. The middle panel shows a schematic of the perfusion circuit. Cells are seeded on cover slips inserted into the chamber. Flow-adapted wild-type, caveolin-1-null, and K$_{IR}$6.2 (K$_{ATP}$ channel)-null cells were evaluated under continuous-flow (control) or stopped-flow (ischemia) conditions in the presence of the membrane potential–sensitive fluorophore bisoxonol; images were acquired by confocal microscopy (*lower panel*). Depolarization indicated by increased fluorescence was observed in wild-type but not in caveolin-1-null and K$_{ATP}$ channel–null cells (91). (*b*) Procedure for cellular and biochemical analysis of flow-adapted cells. Cells are seeded on fibronectin-coated polycarbonate capillaries encased in a housing for perfusion via luminal or abluminal ports (*upper panel*). For the initial cell attachment and the subsequent ischemia phases of the experiment, medium is perfused through the abluminal ports, which removes the shear stress but allows adequate oxygenation and provision of nutrients (*black arrows*). For flow adaptation, medium is perfused through the luminal ports (*red arrows*). The lower panel shows images of cells that have been trypsinized from the capillaries and labeled with PKH for analysis by fluorescence-activated cell sorting; cell generations are indicated by the differently colored peaks. Continuous laminar flow results in an antiproliferative state. Increased proliferation is shown by wild-type cells that were flow adapted and then subjected to flow cessation; increased proliferation is not seen in cells that do not generate ROS, either NOX2-null cells or cells treated with a K$_{ATP}$ channel agonist (cromakalim) to prevent NOX2 activation (91).

cell plasma membrane. Recent studies (see below) indicate that $O_2^{·-}$ may penetrate the plasma membrane via Cl^- channels, although the travel distances are likely quite short (62).

Mitochondria-derived ROS and signal amplification. Mitochondria generate $O_2^{·-}$ essentially by leakage of electrons from the ETC, as described above. ROS production by this mechanism is nonenzymatic and is unlikely to constitute a regulated signaling process, although this is an area of active investigation, and the disagreements may be largely semantic (63, 64). The amount of $O_2^{·-}$ generated by mitochondria varies significantly with the mitochondrial metabolic state, and

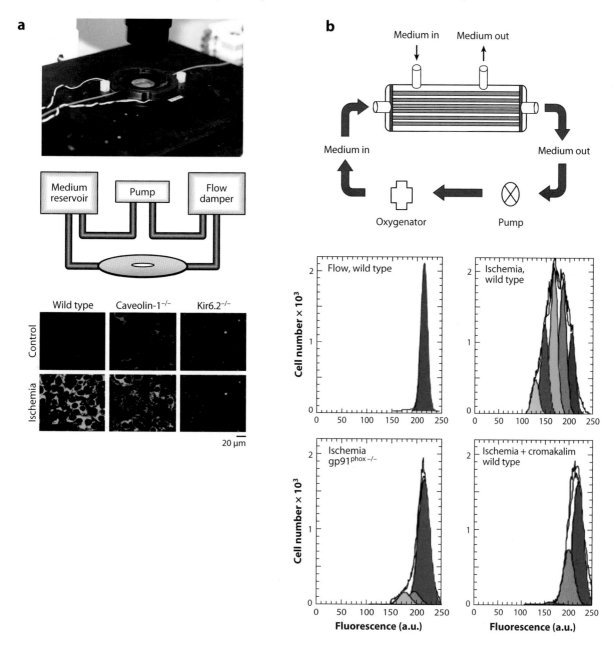

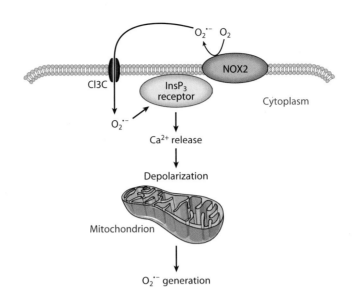

Figure 4

Possible mechanism for $O_2^{\cdot-}$-induced $O_2^{\cdot-}$ release by mitochondria. $O_2^{\cdot-}$ generated extracellularly by NOX2 can penetrate the pulmonary endothelial cell membrane through Cl^- 3 channels (Cl3C) and can interact with the inositol 1,4,5-triphosphate (InsP$_3$) receptor, resulting in Ca^{2+} release, depolarization of the mitochondrial inner membrane, and mitochondrial $O_2^{\cdot-}$ generation by auto-oxidation of components of the electron transport chain (62).

physiological changes in mitochondrial metabolism may result in increased $O_2^{\cdot-}$ generation (64). With respect to cell signaling, mitochondria can also be a secondary source of $O_2^{\cdot-}$ as a response to a primary ROS signal from the plasma membrane or from other intracellular sources (65). This process has been termed $O_2^{\cdot-}$-induced $O_2^{\cdot-}$ generation. Our studies in endothelial cells have shown that the transmembrane flux of extracellularly generated $O_2^{\cdot-}$ (by the addition of Ang II, xanthine/xanthine oxidase, or KO_2 to the medium) leads to a transient increase in cytosolic Ca^{2+} and intracellular ROS generation (60, 62). The proposed mechanism is the activation of plasma membrane InsP$_3$ receptors resulting in Ca^{2+} release from intracellular stores, depolarization of the mitochondrial membrane, and activation of mitochondrial $O_2^{\cdot-}$ production (**Figure 4**). Another possible mechanism for secondary mitochondrial ROS production is H_2O_2-mediated oxidation of Fe^{3+}–cytochrome *c*, with sequential transfer of the electron to generate $NAD^{\cdot}$ and then $O_2^{\cdot-}$ (66). This amplification process represents another potential mechanism (in addition to H_2O_2 diffusion) for transmission of the initial extracellular ROS signal across the plasma membrane to serve an intracellular signaling function.

UPSTREAM EVENTS IN THE ACTIVATION OF ROS-MEDIATED SIGNALING

Sensors for Altered Shear

Although cells sense and respond to altered shear, the mechanism(s) by which endothelial cells sense mechanical forces is less clear. Several models have been proposed. As one possibility, proteins that are integral to the cell membrane may link the extracellular matrix to the cytoskeleton; altered external forces may lead to cytoskeletal distortion, which generates a biochemical response.

Alternatively, flow sensors situated on the cell surface may respond to mechanical forces (e.g., shear stress) at the cell membrane. These views of flow sensing are not mutually exclusive, and both mechanisms or an additional mechanism may cooperatively regulate the response to flow. The cell membrane proteins that have been considered responsible for mechanosensing by the endothelium include G protein–coupled receptors, ion channels, the glycocalyx, platelet–endothelial cell adhesion molecule-1 (PECAM-1), vascular endothelial cadherin (VE-cadherin), receptor tyrosine kinases, and tight junctions. Although several such proteins have been demonstrated to transduce strain, most require forces of supraphysiological magnitude that are unlikely to be relevant to normal signaling.

PECAM: platelet–endothelial cell adhesion molecule

VE-cadherin: vascular endothelial cadherin

VEGFR2: vascular endothelial growth factor receptor 2

Akt: also termed protein kinase B

PI3K: phosphatidyl inositol-3 kinase

Adhesion receptor complex. Initial studies demonstrated that PECAM is involved in cell sensing of altered shear stress and requires phosphorylation for this effect (67). Subsequently, significant evidence indicates that shear sensing by endothelium involves a mechanosensory complex composed of vascular endothelial growth factor receptor 2 (VEGFR2) (Flk-1), VE-cadherin, and PECAM-1 (68–70). VEGFR2 is a transmembrane receptor tyrosine kinase. VE-cadherin and PECAM-1 are transmembrane cell adhesion molecules that form homophilic dimers at cell-cell junctions but also complex with VEGFR2. The role of VE-cadherin does not require interaction with juxtaposed cadherins, and VE-cadherin is thus considered to be primarily an adaptor molecule in this complex (68). PECAM-1 can mediate Src-dependent transactivation of VEGFR2, which in turn can activate phosphatidylinositol-3 kinase (PI3K) and Akt, thereby initiating downstream signaling in response to shear stress (69, 70). The components of this complex are conserved in pulmonary endothelial cells, although its role in shear sensing has not yet been characterized in the pulmonary circulation.

Glycocalyx. The endothelial glycocalyx is an extracellular polymeric surface layer of variable thickness containing glycoproteins, proteoglycans, and their associated glycosaminoglycan side chains. Heparan sulfates are the predominant glycosaminoglycans found on the endothelial surface and are important in ˙NO generation following mechanical stress (71), although no studies specifically deal with ROS generation and pulmonary circulation.

Caveolae. Caveolae, the flask-like invaginations within the plasma membrane of the endothelium, may also play an important role in sensing and/or coordinating the response to altered shear. These structures (or similar lipid-rich lipid rafts) are signaling hubs localizing several redox-sensitive and ROS-generating molecules in endothelial cells, including the membrane-associated subunits of the NOX2 complex (gp91^{phox}, p22^{phox}) (72). Ex vivo–perfused lungs and flow-adapted endothelial cells from caveolin-1-null mice that have caveolar disruption show a marked attenuation of the rate of ROS generation in response to loss of shear compared with the wild type (73). Similar attenuation of the response to loss of shear was observed following the treatment of wild-type cells with an agent (filipin) that disrupts caveolae (73); here, loss of caveolin may disrupt the NOX2 assembly mechanism (74). However, the ROS response (via NOX2) following thrombin treatment of caveolin-1-null cells was similar to that of wild type, indicating that the ROS-generating pathway was intact (73). Thus, caveolae appear to be involved in the sensing of altered shear rather than merely serving as a platform for enzymatic activity of the ROS-generating complex. VEGFR2 and possibly PECAM-1 also localize to caveolae so that the caveolar role in mechanosensing may be through PECAM-1-dependent, ligand-independent activation of VEGFR2, which then activates downstream signaling pathways (75). Thus, caveolae may function along with the adhesion receptor complex (VEGFR, VE-cadherin, PECAM) to constitute a mechanosensory receptor complex.

Role of Ion Channels

K_{IR} channels: inwardly rectifying K^+ channels

Studies using isolated cells suggest that an alteration of ion channel activity represents an early response to altered shear stress (26, 76). To determine possible changes in channel activity with altered shear under physiological conditions, we used intravital microscopy to monitor endothelial membrane potential in the isolated lung that was perfused with membrane potential–sensitive fluorescent dyes (45). Depolarization of the pulmonary microvascular endothelial cells in the intact lung was observed almost immediately following the cessation of flow (39, 43, 44). Perfusion of intact lungs with increasing K^+ concentrations as calibrating signals indicated that endothelial cell membrane depolarization with flow cessation was equivalent to that observed by perfusion with ∼12 mM KCl; this represents a depolarization of ∼17 mV, assuming a resting membrane potential of −70 mV (77, 78). A similar membrane depolarization with stop of flow was observed using fluorescent dyes as well as whole-cell patch-clamp analysis of pulmonary endothelial cells that had been flow adapted in vitro (44, 73, 79). Thus, altered (decreased) shear results in membrane depolarization, suggesting a decreased open probability of channels that set the cell membrane potential.

The membrane potential of endothelial cells, as for most nonexcitable cells, is controlled predominantly by inwardly rectifying K^+ (K_{IR}) channels. On the basis of studies of isolated aortic endothelial cells, a K_{IR} current was identified with increased shear, although the molecular identity of the responsible channel was not determined (80). Subsequently, an outwardly rectifying Cl^- channel and various nonselective cation channels were also proposed to play a role in the response to altered shear (76, 81–83). Of these, a K_{IR} channel has the most support for a role in the early response to altered shear, although other channels may be responsible for subsequent effects.

K_{ATP} channels. The K_{IR} family consists of seven subgroups (84, 85). A K_{IR} of the $K_{IR}2.1$ family (cloned from bovine aortic endothelial cells) resulted in a shear-activated K^+ current when expressed in *Xenopus* oocytes or mammalian HEK293 cells and was proposed as a mechanosensitive K_{IR} channel (86). However, this result has not been confirmed under physiological conditions. Our studies have identified K_{ATP} channels as having a major role in the endothelial cell response to acutely decreased shear (87). A K_{ATP} channel is a heteromeric complex of a K_{IR} protein of the 6 family, i.e., $K_{IR}6.1$ or $K_{IR}6.2$ with a sulfonylurea receptor (88). $K_{IR}6.2$ is the pore-forming component of K_{ATP} channels in endothelial cells of the pulmonary macro- and microvasculature (44, 87). With flow-adapted cells subjected to whole-cell patch-clamp studies, K_{ATP} channels deactivated upon the cessation of flow, providing a basis for the observed cell membrane depolarization (79). In the absence of K_{ATP} channels ($K_{IR}6.2$-null cells), the K_{IR} currents were very low and were unaffected upon the removal of shear (79). Pretreatment of lungs with a K_{ATP} channel antagonist (glyburide) led to endothelial cell membrane depolarization, even with continued flow (36). Membrane depolarization with stop of flow was prevented in isolated lungs and flow-adapted cells by pretreatment with cromakalim, a classic agonist for K_{ATP} channels (43, 87, 89); this agonist is expected to maintain the K_{ATP} channels in an open configuration, even with the loss of shear. These results provide strong evidence that K_{ATP} channel closure (decreased open probability) is responsible for membrane depolarization in the stop-of-flow model.

Physiological role for K_{ATP} channels. An important question is whether the membrane depolarization observed with altered shear results in a downstream signaling response or is an epiphenomenon. Depolarization of the endothelial cell membrane by the addition of high K^+ led to ROS generation during continued flow in both intact isolated lungs and endothelial cells in vitro, an effect that was abolished in NOX2-null cells (77, 78, 90). Treatment of lungs and cells with cromakalim to prevent depolarization associated with stop of flow also prevented ROS generation (36, 43, 91). These results indicate that a change in membrane potential triggers NOX2 activation and support the role of membrane depolarization in initiating the signaling cascade.

Although K_{ATP} channels can initiate the signaling cascade, what is their role in shear sensing? K_{ATP} channels are induced during in vitro flow adaption of pulmonary macro- and microvascular endothelial cells, an observation indicating that the K_{ATP} channels are not the primary sensor for altered shear but rather respond to an upstream signal (87). To support this observation, membrane depolarization with stop of flow was not observed in lungs or cells from caveolin-1-null mice (73). Thus, membrane depolarization due to K_{ATP} channel inactivation associated with loss of shear would represent a response to altered mechanical stress that is sensed by the mechanosensory complex.

Signal Transmission from the Membrane Potential to Activation of NOX2

The signaling pathway described thus far involves sensing of altered flow by a mechanosensory complex coupled to caveolae, resulting in endothelial cell membrane depolarization and then NOX2 assembly with ROS generation. But what are the signals that are activated by the change in endothelial cell membrane potential? On the basis of the above-described studies dealing with chemoreception, we evaluated PI3K and Akt as the signaling links. Our studies in this area are preliminary and have been published only in abstract form (92). Cessation of flow in our in vitro model of flow-adapted endothelial cells resulted in PI3K and Akt activation. PI3K and Akt activation was not seen in cells that were treated with cromakalim and that therefore did not depolarize with flow cessation. Similar observations were made in cells with knock out of the K_{ATP} channel. NOX2 activation with altered shear stress was greatly diminished by the presence of the PI3K inhibitor wortmannin. Lungs that were null for Akt-1, but not for Akt-2, failed to show ROS generation with stop of flow. Thus, Akt-1 is required for NOX2 activation in the mechanotransduction pathway. PI3K and Akt activation was also observed in cells that were null for NOX2, and thus such activation does not represent an effect of ROS. These results indicate that cell membrane depolarization with stop of flow precedes PI3K and Akt activation, which in turn is upstream of NOX2 generation. The mechanism for PI3K activation subsequent to membrane depolarization is not apparent, but an electrical potential–sensitive site in either this enzyme or a relevant upstream G protein is a possibility.

Activation of PI3K and Akt represents an initial (or early) phosphorylation event in the mechanotransduction signaling cascade. The effects of MAP kinase inhibitors indicate that these proteins also participate in the NOX2 activation process. Loss of shear resulted in Erk activation in flow-adapted bovine pulmonary artery endothelial cells (93). MAP kinases are responsible for phosphorylation of Prdx6 (18); the PLA_2 activity of this protein is required for NOX2 activation with stop of flow (94), as described above for the response to Ang II (13). The major elements that lead from altered shear to NOX2 activation and ROS production are shown in **Figure 5**.

CELLULAR AND PHYSIOLOGICAL RESPONSES TO THE ROS-DELIVERED SIGNAL

ROS can influence a broad spectrum of physiological processes through the modification of enzymes, transcription factors, and other metabolic intermediates. With respect to the endothelium, ROS modulates adhesion/migration, growth/proliferation, and survival/apoptosis. Some of these effects, e.g., modulation of apoptosis through the activation of caspase 3, may require supraphysiological concentrations of H_2O_2 and thus may represent a toxic response rather than a cell-signaling

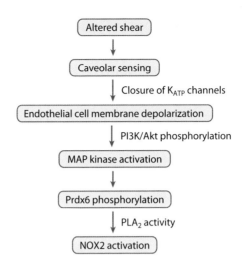

Figure 5
Key elements in endothelial cell sensing and signal transduction following an abrupt decrease in flow in the pulmonary circulation. The signaling cascade leads to NOX2 activation and ROS generation. Abbreviations: MAP kinase, mitogen-activated protein kinase; PI3K, phosphatidylinositol-3 kinase; PLA_2, phospholipase A_2; Prdx6, peroxiredoxin 6.

process (72). Studies with models of altered shear stress demonstrate that physiological levels of ROS can modulate several cell functions and may play an important role in vascular remodeling and angiogenesis.

Regulation of Gene Transcription

The modification of transcription factors by interaction with ROS can affect their translocation and binding to elements/promoters in the genome, representing an important redox-sensitive function (**Figure 5**). Nrf2 is a good example of a transcription factor that is ROS sensitive. Nrf2 binds to the ARE (antioxidant response element, also termed the electrophilic response element) in DNA, driving the transcription of several antioxidant enzymes, including Prdx6 (95). The role of Nrf2 in the response to altered shear stress has not yet been evaluated. Activator protein-1 (AP-1) and nuclear factor κB (NFκB), two other redox-sensitive transcription factors, are upregulated by ROS that are generated subsequent to acute flow cessation (46). These studies, carried out with bovine pulmonary artery endothelial cells, showed an increased expression in the p50 and p65 subunits of NFκB and in the c-jun and c-fos subunits of AP-1 (**Figure 6**). Additionally, the p50 subunit of NFκB requires cysteine oxidation to bind DNA as the basis for its redox sensitivity (96). Both NFκB and AP-1 have the potential to drive a proinflammatory phenotype in the endothelium through the upregulation of adhesion molecules such as vascular cell adhesion molecule-1, intercellular adhesion molecule-1, and E-selectin (97, 98).

Cell Proliferation and Angiogenesis

ROS induce endothelial cell proliferation in various isolated cell models. An increase in the ROS-dependent proliferative profile was confirmed by fluorescence-activated cell sorting analysis with cells obtained using our model of altered shear stress in flow-adapted pulmonary microvascular endothelium (73, 91, 99). This proliferative phenotype was blocked by the administration of

AP-1: activator protein-1

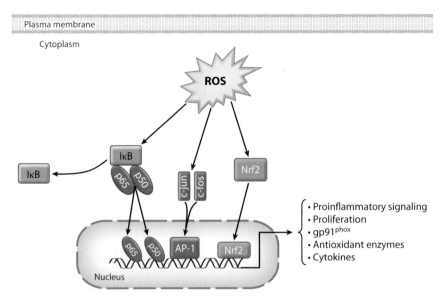

Figure 6
Downstream signaling by NOX2-generated ROS. ROS can activate several transcription factors that result in a diverse array of physiological effects. AP-1, activator protein-1; IκB, inhibitor of NFκB activation.

antioxidants or by knock out of NOX2, the source of ROS in this model (91, 99). Likewise, preventing cell membrane depolarization (K_{ATP} channel–null or cromakalim pretreatment) with altered shear in flow-adapted cells prevented the proliferative phenotype (91, 99). A critical question is whether the cell proliferation associated with altered mechanotransduction is random or represents an attempt at angiogenesis. Our studies have only recently begun to address this issue. Preliminary experiments with cells grown in Matrigel™ in vitro or placed subcutaneously in mice indicate new vessel formation compatible with angiogenesis (100). Ligation of the left pulmonary artery in mice resulted in proliferation of the systemic vasculature, arising mainly from intercostal vessels (101). This angiogenesis appears to be ROS dependent, although how the signal is transmitted from the pulmonary to the systemic circulation is not clear. From a teleological standpoint, angiogenesis would be a logical response to the acute loss of blood flow.

MECHANOSIGNALING BEYOND ROS

Intracellular Ca^{2+} Flux

Intracellular Ca^{2+} concentrations as detected by fluorescent imaging rapidly increase upon the cessation of flow in flow-adapted endothelial cells in vitro and in lung endothelium in situ (43, 89, 102). This rise in Ca^{2+} was significantly inhibited by pretreatment with thapsigargin, an agent that depletes Ca^{2+} from intracellular stores. Thus, an increase in intracellular Ca^{2+} with stop of flow can occur by Ca^{2+} entry from the extracellular space, presumably via endothelial cell membrane Ca^{2+} channels. Using blockers or inhibitors, we identified this channel to be a T-type Ca^{2+} channel; Ca^{2+} influx in flow-adapted cells with loss of shear was blocked by mibefradil (which is semispecific as a T-type Ca^{2+} channel blocker), but not by nifedipine (an L-type Ca^{2+} channel blocker) (103). Furthermore, the estimated change in membrane potential with stop of flow

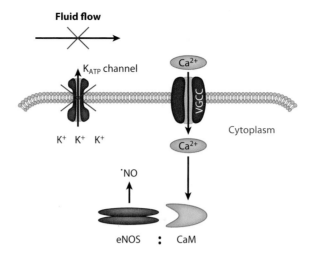

Figure 7

Mechanism for increased intracellular Ca^{2+} following an abrupt cessation of flow. The acute decrease in shear results in K_{ATP} channel closure, leading to endothelial cell membrane depolarization. Such depolarization in turn results in the opening of T-type voltage-gated Ca^{2+} channels (VGCC) in the plasma membrane, permitting Ca^{2+} entry (103). The increased intracellular Ca^{2+} can associate with calmodulin (CaM) to activate endothelial nitric oxide synthase (eNOS), resulting in ·NO generation (89, 93).

($\sim$20 mV) is compatible with the known voltage window for T-type channels. Ca^{2+} influx was blocked by a K_{ATP} channel agonist (cromakalim), as expected, because this treatment prevents cell membrane depolarization with flow cessation. Mibefradil had no effect on cell membrane depolarization with flow cessation, indicating that K_{ATP} channel closure preceded (and was responsible for) Ca^{2+} channel activation (103). Thus, activation of T-type Ca^{2+} channels as a result of endothelial cell membrane depolarization can allow Ca^{2+} influx with mechanotransduction (**Figure 7**).

In the absence of thapsigargin treatment, the abrupt stop of flow can result in the release of Ca^{2+} from its intracellular stores. Although there may be more than one mechanism for this effect, we demonstrated that one possibility is associated with extracellular $O_2^{·-}$ generated by NOX2. Extracellular $O_2^{·-}$ can penetrate the endothelial cell through Cl^- channels and interact with membrane-associated proteins such as the IP_3 receptor, an effect that results in intracellular Ca^{2+} release (60, 62). This effect has been described above as a stimulus for $O_2^{·-}$-mediated $O_2^{·-}$ release by mitochondria, a Ca^{2+}-dependent mechanism (**Figure 3**).

Nitric Oxide Synthase and ·NO Generation

In addition to ROS, flow-adapted endothelial cells and lung endothelium in situ produce ·NO with cessation of flow. In isolated rat lungs, ·NO production by the pulmonary endothelium was detected during the first minute after stop of flow and was preceded by a rise in intracellular Ca^{2+} (43, 89). Perfusion with a Ca^{2+}-free medium inhibited ·NO production, implicating the activation of the Ca^{2+}-dependent enzyme endothelial nitric oxide synthase (eNOS). Inhibitors of PI3K, Akt, and MAP kinases block ·NO generation with stop flow (89, 93), indicating that the upstream activating pathway is similar for eNOS and NOX2 (**Figure 4**). Thus, both $O_2^{·-}$ and ·NO are generated intracellularly following flow cessation due to increased intracellular Ca^{2+}, ·NO from eNOS activation (93), and $O_2^{·-}$ from mitochondrial depolarization (62). ·NO and $O_2^{·-}$ can react to form the highly reactive and potent nitrating agent $ONOO^-$ (the peroxynitrite anion), which effectively reduces the ˙NO concentration. Extensive generation of nitrotyrosine moieties in lung proteins has been demonstrated as a result of lung ischemia (104). Whether such generation represents part of the signaling paradigm or is just an unfortunate (toxic) by-product is not known at this time. The generation of a vasodilator (NO) as part of the mechanosensory response can be considered an appropriate physiological response to the loss of blood flow.

SUMMARY AND CONCLUSIONS

ROS have now been recognized as important physiological mediators (in addition to their microbicidal activity) for the transmission of cellular signals. This review discusses results for ROS-mediated cell signaling by the pulmonary endothelium using the models of an intact lung and flow-adapted pulmonary microvascular endothelial cells in culture subjected to the physiological stimulus of acute reduction in perfusate flow. These models may correspond to the expected response to acute ischemia associated with a pulmonary artery embolism, lung transplantation, or other etiology for altered pulmonary blood flow. A key component of this response is endothelial NOX (NOX2). Although discovered in phagocytic cells, NOX2 is now well characterized in the endothelium. ROS generated by NOX2 participate in cell signaling, but overactivity of this enzyme can contribute to the oxidative stress that is evident in various vascular pathologies. Thus, information on NOX2 expression, activation, and assembly in response to various agonists and stimuli is critical for understanding the regulation of this enzyme in lung physiology and pathophysiology.

Our work on mechano- and chemotransduction of pulmonary endothelial cells sheds light on the seemingly diverse mechanisms that drive assembly of the NOX complex. Mechanotransduction activates NOX2 via depolarization of the endothelial cell plasma membrane, which in turn triggers the activation of kinases (PI3K and Akt) that drive NOX2 assembly. NOX2 agonists that activate PKC via either receptor-dependent or receptor-independent mechanisms work through pathways similar to mechanotransduction pathways, although the inciting signal is membrane perturbation of a different sort. An important new finding is that the PLA_2 activity of Prdx6 serves as a link between the activation of intracellular kinases and the assembly of the oxidase in both chemotransduction as well as mechanotransduction pathways. The complex roles played by ROS in redox signaling and oxidative stress and the highly specific regulation of NOX2 may serve as the basis for the development of novel therapies that target redox pathways in a cell-specific manner at appropriate time points in the disease process.

DISCLOSURE STATEMENT

The authors are not aware of any affiliations, memberships, funding, or financial holdings that might be perceived as affecting the objectivity of this review.

LITERATURE CITED

1. Babior BM, Kipnes RS, Curnutte JT. 1973. Biological defense mechanisms. The production by leukocytes of superoxide, a potential bactericidal agent. *J. Clin. Investig.* 52:741–44
2. Forman HJ, Torres M, Fukuto J. 2002. Redox signaling. *Mol. Cell. Biochem.* 234/235:49–62
3. Lambeth JD. 2004. NOX enzymes and the biology of reactive oxygen. *Nat. Rev. Immunol.* 4:181–89
4. Davies PF, Barbee KA, Volin MV, Robotewskyj A, Chen J, et al. 1997. Spatial relationships in early signaling events of flow-mediated endothelial mechanotransduction. *Annu. Rev. Physiol.* 59:527–49
5. Babior BM. 1999. NADPH oxidase: an update. *Blood* 93:1464–76

6. Lassegue B, Griendling KK. 2010. NADPH oxidases: functions and pathologies in the vasculature. *Arterioscler. Thromb. Vasc. Biol.* 30:653–61
7. Ellmark SH, Dusting GJ, Fui MN, Guzzo-Pernell N, Drummond GR. 2005. The contribution of Nox4 to NADPH oxidase activity in mouse vascular smooth muscle. *Cardiovasc. Res.* 65:495–504
8. Seshiah PN, Weber DS, Rocic P, Valppu L, Taniyama Y, Griendling KK. 2002. Angiotensin II stimulation of NAD(P)H oxidase activity: upstream mediators. *Circ. Res.* 91:406–13
9. Touyz RM. 2005. Intracellular mechanisms involved in vascular remodelling of resistance arteries in hypertension: role of angiotensin II. *Exp. Physiol.* 90:449–55
10. Raad H, Paclet MH, Boussetta T, Kroviarski Y, Morel F, et al. 2009. Regulation of the phagocyte NADPH oxidase activity: Phosphorylation of gp91^{phox}/NOX2 by protein kinase C enhances its diaphorase activity and binding to Rac2, p67^{phox}, and p47^{phox}. *FASEB J.* 23:1011–22
11. Frey RS, Ushio-Fukai M, Malik AB. 2009. NADPH oxidase-dependent signaling in endothelial cells: role in physiology and pathophysiology. *Antioxid. Redox Signal.* 11:791–810
12. Pendyala S, Usatyuk PV, Gorshkova IA, Garcia JG, Natarajan V. 2009. Regulation of NADPH oxidase in vascular endothelium: the role of phospholipases, protein kinases, and cytoskeletal proteins. *Antioxid. Redox Signal.* 11:841–60
13. Chatterjee S, Feinstein SI, Dodia C, Sorokina E, Lien YC, et al. 2011. Peroxiredoxin 6 phosphorylation and subsequent phospholipase A2 activity are required for agonist-mediated activation of NADPH oxidase in mouse pulmonary microvascular endothelium and alveolar macrophages. *J. Biol. Chem.* 286:11696–701
14. Rhee SG, Chae HZ, Kim K. 2005. Peroxiredoxins: a historical overview and speculative preview of novel mechanisms and emerging concepts in cell signaling. *Free Radic. Biol. Med.* 38:1543–52
15. Fisher AB. 2011. Peroxiredoxin 6: a bifunctional enzyme with glutathione peroxidase and phospholipase A$_2$ activities. *Antioxid. Redox Signal.* 15(3):831–44
16. Chen JW, Dodia C, Feinstein SI, Jain MK, Fisher AB. 2000. 1-Cys peroxiredoxin, a bifunctional enzyme with glutathione peroxidase and phospholipase A2 activities. *J. Biol. Chem.* 275:28421–27
17. Manevich Y, Shuvaeva T, Dodia C, Kazi A, Feinstein SI, Fisher AB. 2009. Binding of peroxiredoxin 6 to substrate determines differential phospholipid hydroperoxide peroxidase and phospholipase A$_2$ activities. *Arch. Biochem. Biophys.* 485:139–49
18. Wu Y, Feinstein SI, Manevich Y, Chowdhury I, Pak JH, et al. 2009. Mitogen activated protein kinase-mediated phosphorylation of peroxiredoxin 6 regulates its phospholipid A2 activity. *Biochem. J.* 419:669–79
19. Bostan M, Galatiuc C, Hirt M, Constantin MC, Brasoveanu LI, Iordachescu D. 2003. Phospholipase A2 modulates respiratory burst developed by neutrophils in patients with rheumatoid arthritis. *J. Cell Mol. Med.* 7:57–66
20. Silliman CC, Elzi DJ, Ambruso DR, Musters RJ, Hamiel C, et al. 2003. Lysophosphatidylcholines prime the NADPH oxidase and stimulate multiple neutrophil functions through changes in cytosolic calcium. *J. Leukoc. Biol.* 73:511–24
21. Dana R, Leto TL, Malech HL, Levy R. 1998. Essential requirement of cytosolic phospholipase A2 for activation of the phagocyte NADPH oxidase. *J. Biol. Chem.* 273:441–45
22. Akiba S, Dodia C, Chen X, Fisher AB. 1998. Characterization of acidic Ca^{2+}-independent phospholipase A2 of bovine lung. *Comp. Biochem. Physiol. B* 120:393–404
23. Barbee KA, Mundel T, Lal R, Davies PF. 1995. Subcellular distribution of shear stress at the surface of flow-aligned and nonaligned endothelial monolayers. *Am. J. Physiol. Heart Circ. Physiol.* 268:1765–72
24. Frangos J, McIntire L, Eskin S. 1988. Shear stress induced stimulation of mammalian cell metabolism. *Biotechnol. Bioeng.* 32:1053–60
25. Langille BL, Adamson SL. 1981. Relationship between blood flow direction and endothelial cell orientation at arterial branch sites in rabbits and mice. *Circ. Res.* 48:481–88
26. Davies PF. 1995. Flow-mediated endothelial mechanotransduction. *Physiol. Rev.* 75:519–60
27. Yerrapureddy A, Tobias J, Margulies SS. 2010. Cyclic stretch magnitude and duration affect rat alveolar epithelial gene expression. *Cell. Physiol. Biochem.* 25:113–22

28. Papaiahgari S, Yerrapureddy A, Hassoun PM, Garcia JG, Birukov KG, Reddy SP. 2007. EGFR-activated signaling and actin remodeling regulate cyclic stretch-induced NRF2-ARE activation. *Am. J. Respir. Cell Mol. Biol.* 36:304–12

29. Birukova AA, Chatchavalvanich S, Rios A, Kawkitinarong K, Garcia JG, Birukov KG. 2006. Differential regulation of pulmonary endothelial monolayer integrity by varying degrees of cyclic stretch. *Am. J. Pathol.* 168:1749–61

30. Chatterjee S, Chapman KE, Fisher AB. 2008. Lung ischemia: a model for endothelial mechanotransduction. *Cell Biochem. Biophys.* 52:125–38

31. Chiu JJ, Wung BS, Shyy JY, Hsieh HJ, Wang DL. 1997. Reactive oxygen species are involved in shear stress-induced intercellular adhesion molecule-1 expression in endothelial cells. *Arterioscler. Thromb. Vasc. Biol.* 17:3570–77

32. Jo H, Song H, Mowbray A. 2006. Role of NADPH oxidases in disturbed flow- and BMP4-induced inflammation and atherosclerosis. *Antioxid. Redox Signal.* 8:1609–19

33. Morgan VL, Graham TP Jr, Roselli RJ, Lorenz CH. 1998. Alterations in pulmonary artery flow patterns and shear stress determined with three-dimensional phase-contrast magnetic resonance imaging in Fontan patients. *J. Thorac. Cardiovasc. Surg.* 116:294–304

34. Ochoa CD, Wu S, Stevens T. 2010. New developments in lung endothelial heterogeneity: Von Willebrand factor, P-selectin, and the Weibel-Palade body. *Semin. Thromb. Hemost.* 36:301–8

35. Gelfand BD, Epstein FH, Blackman BR. 2006. Spatial and spectral heterogeneity of time-varying shear stress profiles in the carotid bifurcation by phase-contrast MRI. *J. Magn. Reson. Imaging* 24:1386–92

36. Al-Mehdi AB, Shuman H, Fisher AB. 1997. Intracellular generation of reactive oxygen species during nonhypoxic lung ischemia. *Am. J. Physiol. Lung Cell. Mol. Physiol.* 272:294–300

37. Matsuzaki I, Chatterjee S, Debolt K, Manevich Y, Zhang Q, Fisher AB. 2005. Membrane depolarization and NADPH oxidase activation in aortic endothelium during ischemia reflect altered mechanotransduction. *Am. J. Physiol. Heart Circ. Physiol.* 288:336–43

38. Ali MH, Pearlstein DP, Mathieu CE, Schumacker PT. 2004. Mitochondrial requirement for endothelial responses to cyclic strain: implications for mechanotransduction. *Am. J. Physiol. Lung Cell. Mol. Physiol.* 287:486–96

39. Al-Mehdi AB, Zhao G, Fisher AB. 1998. ATP-independent membrane depolarization with ischemia in the oxygen-ventilated isolated rat lung. *Am. J. Respir. Cell Mol. Biol.* 18:653–61

40. Fisher AB, Dodia C, Tan ZT, Ayene I, Eckenhoff RG. 1991. Oxygen-dependent lipid peroxidation during lung ischemia. *J. Clin. Investig.* 88:674–79

41. Ayene IS, Dodia C, Fisher AB. 1992. Role of oxygen in oxidation of lipid and protein during ischemia/reperfusion in isolated perfused rat lung. *Arch. Biochem. Biophys.* 296:183–89

42. Al-Mehdi A, Shuman H, Fisher AB. 1994. Fluorescence microtopography of oxidative stress in lung ischemia-reperfusion. *Lab. Investig.* 70:579–87

43. Song C, Al-Mehdi AB, Fisher AB. 2001. An immediate endothelial cell signaling response to lung ischemia. *Am. J. Physiol. Lung Cell. Mol. Physiol.* 281:993–1000. Erratum. 2002. *Am. J. Physiol. Lung Cell. Mol. Physiol.* 282:167

44. Zhang Q, Matsuzaki I, Chatterjee S, Fisher AB. 2005. Activation of endothelial NADPH oxidase during normoxic lung ischemia is K_{ATP} channel dependent. *Am. J. Physiol. Lung Cell. Mol. Physiol.* 289:954–61

45. Chatterjee S, Fisher AB. 2010. Detection of ROS with altered shear stress in lung endothelium. In *Methods in Redox Signaling*, ed. DK Das, p. 213. New Rochelle, NY: Mary Ann Liebert

46. Wei Z, Costa K, Al-Mehdi AB, Dodia C, Muzykantov V, Fisher AB. 1999. Simulated ischemia in flow-adapted endothelial cells leads to generation of reactive oxygen species and cell signaling. *Circ. Res.* 85:682–89

47. Manevich Y, Al-Mehdi A, Muzykantov V, Fisher AB. 2001. Oxidative burst and NO generation as initial response to ischemia in flow-adapted endothelial cells. *Am. J. Physiol. Heart Circ. Physiol.* 280:2126–35

48. Zhang Q, Chatterjee S, Wei Z, Liu WD, Fisher AB. 2008. Rac and PI3 kinase mediate endothelial cell-reactive oxygen species generation during normoxic lung ischemia. *Antioxid. Redox Signal.* 10:679–89

49. Boyd NL, Park H, Yi H, Boo YC, Sorescu GP, et al. 2003. Chronic shear induces caveolae formation and alters ERK and Akt responses in endothelial cells. *Am. J. Physiol. Heart Circ. Physiol.* 285:1113–22

50. Nakashima H, Suzuki H, Ohtsu H, Chao JY, Utsunomiya H, et al. 2006. Angiotensin II regulates vascular and endothelial dysfunction: recent topics of angiotensin II type-1 receptor signaling in the vasculature. *Curr. Vasc. Pharmacol.* 4:67–78
51. Mehta PK, Griendling KK. 2007. Angiotensin II cell signaling: physiological and pathological effects in the cardiovascular system. *Am. J. Physiol. Cell Physiol.* 292:82–97
52. Brown DI, Griendling KK. 2009. Nox proteins in signal transduction. *Free Radic. Biol. Med.* 47:1239–53
53. Zhao G, Al-Mehdi AB, Fisher AB. 1997. Anoxia-reoxygenation versus ischemia in isolated rat lungs. *Am. J. Physiol. Lung Cell. Mol. Physiol.* 273:1112–17
54. Fisher AB. 2009. Redox signaling across cell membranes. *Antioxid. Redox Signal.* 11:1349–56
55. Bienert GP, Moller AL, Kristiansen KA, Schulz A, Moller IM, et al. 2007. Specific aquaporins facilitate the diffusion of hydrogen peroxide across membranes. *J. Biol. Chem.* 282:1183–92
56. Winterbourn CC, Metodiewa D. 1999. Reactivity of biologically important thiol compounds with superoxide and hydrogen peroxide. *Free Radic. Biol. Med.* 27:322–28
57. Barford D. 2004. The role of cysteine residues as redox-sensitive regulatory switches. *Curr. Opin. Struct. Biol.* 14:679–86
58. Rhee SG, Yang KS, Kang SW, Woo HA, Chang TS. 2005. Controlled elimination of intracellular H_2O_2: regulation of peroxiredoxin, catalase, and glutathione peroxidase via post-translational modification. *Antioxid. Redox Signal.* 7:619–26
59. Biteau B, Labarre J, Toledano MB. 2003. ATP-dependent reduction of cysteine-sulphinic acid by *S. cerevisiae* sulphiredoxin. *Nature* 425:980–84
60. Madesh M, Hawkins BJ, Milovanova T, Bhanumathy CD, Joseph SK, et al. 2005. Selective role for superoxide in $InsP_3$ receptor-mediated mitochondrial dysfunction and endothelial apoptosis. *J. Cell Biol.* 170:1079–90
61. Mikkelsen RB, Wardman P. 2003. Biological chemistry of reactive oxygen and nitrogen and radiation-induced signal transduction mechanisms. *Oncogene* 22:5734–54
62. Hawkins BJ, Madesh M, Kirkpatrick CJ, Fisher AB. 2007. Superoxide flux in endothelial cells via the chloride channel-3 mediates intracellular signaling. *Mol. Biol. Cell* 18:2002–12
63. Matsuzaki S, Szweda PA, Szweda LI, Humphries KM. 2009. Regulated production of free radicals by the mitochondrial electron transport chain: cardiac ischemic preconditioning. *Adv. Drug Deliv. Rev.* 61:1324–31
64. Widlansky ME, Gutterman DD. 2011. Regulation of endothelial function by mitochondrial reactive oxygen species. *Antioxid. Redox Signal.* 15:1517–30
65. Zorov DB, Filburn CR, Klotz LO, Zweier JL, Sollott SJ. 2000. Reactive oxygen species (ROS)-induced ROS release: a new phenomenon accompanying induction of the mitochondrial permeability transition in cardiac myocytes. *J. Exp. Med.* 192:1001–14
66. Velayutham M, Hermann C, Zweier JL. 2011. Removal of H_2O_2 and generation of superoxide radical: role of cytochrome *c* and NADH. *Free Radic. Biol. Med.* 51:160–70
67. Osawa M, Masuda M, Kusano K, Fujiwara K. 2002. Evidence for a role of platelet endothelial cell adhesion molecule-1 in endothelial cell mechanosignal transduction: Is it a mechanoresponsive molecule? *J. Cell Biol.* 158:773–85
68. Tzima E, Irani-Tehrani M, Kiosses WB, Dejana E, Schultz DA, et al. 2005. A mechanosensory complex that mediates the endothelial cell response to fluid shear stress. *Nature* 437:426–31
69. Shay-Salit A, Shushy M, Wolfovitz E, Yahav H, Breviario F, et al. 2002. VEGF receptor 2 and the adherens junction as a mechanical transducer in vascular endothelial cells. *Proc. Natl. Acad. Sci. USA* 99:9462–67
70. Jin ZG, Ueba H, Tanimoto T, Lungu AO, Frame MD, Berk BC. 2003. Ligand-independent activation of vascular endothelial growth factor receptor 2 by fluid shear stress regulates activation of endothelial nitric oxide synthase. *Circ. Res.* 93:354–63
71. Florian JA, Kosky JR, Ainslie K, Pang Z, Dull RO, Tarbell JM. 2003. Heparan sulfate proteoglycan is a mechanosensor on endothelial cells. *Circ. Res.* 93:e136–42
72. Yang B, Rizzo V. 2007. TNF-α potentiates protein-tyrosine nitration through activation of NADPH oxidase and eNOS localized in membrane rafts and caveolae of bovine aortic endothelial cells. *Am. J. Physiol. Heart Circ. Physiol.* 292:954–62

73. Milovanova T, Chatterjee S, Hawkins BJ, Hong N, Sorokina EM, et al. 2008. Caveolae are an essential component of the pathway for endothelial cell signaling associated with abrupt reduction of shear stress. *Biochim. Biophys. Acta* 1783:1866–75
74. Durr E, Yu J, Krasinska KM, Carver LA, Yates JR, et al. 2004. Direct proteomic mapping of the lung microvascular endothelial cell surface in vivo and in cell culture. *Nat. Biotechnol.* 22:985–92
75. Oshikawa J, Urao N, Kim HW, Kaplan N, Razvi M, et al. 2010. Extracellular SOD-derived H_2O_2 promotes VEGF signaling in caveolae/lipid rafts and post-ischemic angiogenesis in mice. *PLoS ONE* 5:e10189
76. Fisher AB, Chien S, Barakat AI, Nerem RM. 2001. Endothelial cellular response to altered shear stress. *Am. J. Physiol. Lung Cell. Mol. Physiol.* 281:529–33
77. Al-Mehdi AB, Zhao G, Dodia C, Tozawa K, Costa K, et al. 1998. Endothelial NADPH oxidase as the source of oxidants in lungs exposed to ischemia or high K^+. *Circ. Res.* 83:730–37
78. Al-Mehdi AB, Shuman H, Fisher AB. 1997. Oxidant generation with K^+-induced depolarization in the isolated perfused lung. *Free Radic. Biol. Med.* 23:47–56
79. Chatterjee S, Levitan I, Wei Z, Fisher AB. 2006. K_{ATP} channels are an important component of the shear-sensing mechanism in the pulmonary microvasculature. *Microcirculation* 13:633–44
80. Olesen SP, Clapham DE, Davies PF. 1988. Haemodynamic shear stress activates a K^+ current in vascular endothelial cells. *Nature* 331:168–70
81. Barakat AI, Leaver EV, Pappone PA, Davies PF. 1999. A flow-activated chloride-selective membrane current in vascular endothelial cells. *Circ. Res.* 85:820–28
82. Ohno M, Cooke JP, Dzau VJ, Gibbons GH. 1995. Fluid shear stress induces endothelial transforming growth factor β-1 transcription and production. Modulation by potassium channel blockade. *J. Clin. Investig.* 95:1363–69
83. Suvatne J, Barakat AI, O'Donnell ME. 2001. Flow-induced expression of endothelial Na-K-Cl cotransport: dependence on K^+ and Cl^- channels. *Am. J. Physiol. Cell Physiol.* 280:216–27
84. Doupnik CA, Davidson N, Lester HA. 1995. The inward rectifier potassium channel family. *Curr. Opin. Neurobiol.* 5:268–77
85. Nichols CG, Lopatin AN. 1997. Inward rectifier potassium channels. *Annu. Rev. Physiol.* 59:171–91
86. Hoger JH, Ilyin VI, Forsyth S, Hoger A. 2002. Shear stress regulates the endothelial Kir2.1 ion channel. *Proc. Natl. Acad. Sci. USA* 99:7780–85
87. Chatterjee S, Al-Mehdi AB, Levitan I, Stevens T, Fisher AB. 2003. Shear stress increases expression of a K_{ATP} channel in rat and bovine pulmonary vascular endothelial cells. *Am. J. Physiol. Cell Physiol.* 285:959–67
88. Yokoshiki H, Sunagawa M, Seki T, Sperelakis N. 1998. ATP-sensitive K^+ channels in pancreatic, cardiac, and vascular smooth muscle cells. *Am. J. Physiol. Cell Physiol.* 274:25–37
89. Al-Mehdi AB, Song C, Tozawa K, Fisher AB. 2000. Ca^{2+}- and phosphatidylinositol 3-kinase-dependent nitric oxide generation in lung endothelial cells in situ with ischemia. *J. Biol. Chem.* 275:39807–10
90. Al-Mehdi AB, Ischiropoulos H, Fisher AB. 1996. Endothelial cell oxidant generation during K^+-induced membrane depolarization. *J. Cell. Physiol.* 166:274–80
91. Milovanova T, Chatterjee S, Manevich Y, Kotelnikova I, Debolt K, et al. 2006. Lung endothelial cell proliferation with decreased shear stress is mediated by reactive oxygen species. *Am. J. Physiol. Cell Physiol.* 290:66–76
92. Chatterjee S, Browning E, Hong NK, DeBolt K, Sorokina E, et al. 2010. PI3 kinase/Akt activation trigger ROS production in a model of pulmonary ischemia. *FASEB J.* 24:796.6
93. Wei Z, Al-Mehdi AB, Fisher AB. 2001. Signaling pathway for nitric oxide generation with simulated ischemia in flow-adapted endothelial cells. *Am. J. Physiol. Heart Circ. Physiol.* 281:2226–32
94. Chatterjee S, Feinstein S, Hong NK, Debolt K, Fisher AB. 2007. Paradoxical response of ROS production in peroxiredoxin 6 null mice to ischemia. *FASEB J.* 21:894.5
95. Chowdhury I, Mo Y, Gao L, Kazi A, Fisher AB, Feinstein SI. 2009. Oxidant stress stimulates expression of the human peroxiredoxin 6 gene by a transcriptional mechanism involving an antioxidant response element. *Free Radic. Biol. Med.* 46:146–53
96. Sen CK, Packer L. 1996. Antioxidant and redox regulation of gene transcription. *FASEB J.* 10:709–20

97. Closa D, Folch-Puy E. 2004. Oxygen free radicals and the systemic inflammatory response. *IUBMB Life* 56:185–91
98. Wiggins JE, Patel SR, Shedden KA, Goyal M, Wharram BL, et al. 2010. NFκB promotes inflammation, coagulation, and fibrosis in the aging glomerulus. *J. Am. Soc. Nephrol.* 21:587–97
99. Milovanova T, Manevich Y, Haddad A, Chatterjee S, Moore JS, Fisher AB. 2004. Endothelial cell proliferation associated with abrupt reduction in shear stress is dependent on reactive oxygen species. *Antioxid. Redox Signal.* 6:245–58
100. Chatterjee S, Hong NK, Yu K, Fisher AB. 2010. Endothelial mechanotransduction with loss of shear is a signal for angiogenesis. *FASEB J.* 24:602.3
101. Nijmeh J, Moldobaeva A, Wagner EM. 2010. Role of ROS in ischemia-induced lung angiogenesis. *Am. J. Physiol. Lung Cell. Mol. Physiol.* 299:535–41
102. Tozawa K, Al-Mehdi AB, Muzykantov V, Fisher AB. 1999. In situ imaging of intracellular calcium with ischemia in lung subpleural microvascular endothelial cells. *Antioxid. Redox Signal.* 1:145–54
103. Wei Z, Manevich Y, Al-Mehdi AB, Chatterjee S, Fisher AB. 2004. Ca^{2+} flux through voltage-gated channels with flow cessation in pulmonary microvascular endothelial cells. *Microcirculation* 11:517–26
104. Ischiropoulos H, Al-Mehdi AB, Fisher AB. 1995. Reactive species in ischemic rat lung injury: contribution of peroxynitrite. *Am. J. Physiol. Lung Cell. Mol. Physiol.* 269:158–64

The Molecular Control of Meiotic Chromosomal Behavior: Events in Early Meiotic Prophase in *Drosophila* Oocytes

Cathleen M. Lake[1] and R. Scott Hawley[1,2]

[1]Stowers Institute for Medical Research, Kansas City, Missouri 64110; email: rsh@stowers.org
[2]Department of Molecular and Integrative Physiology, University of Kansas Medical Center, Kansas City, Kansas 66160

Keywords

pachytene, synaptonemal complex, meiotic recombination, centromere, meiotic checkpoints

Abstract

We review the critical events in early meiotic prophase in *Drosophila melanogaster* oocytes. We focus on four aspects of this process: the formation of the synaptonemal complex (SC) and its role in maintaining homologous chromosome pairings, the critical roles of the meiosis-specific process of centromere clustering in the formation of a full-length SC, the mechanisms by which preprogrammed double-strand breaks initiate meiotic recombination, and the checkpoints that govern the progression and coordination of these processes. Central to this discussion are the roles that somatic pairing events play in establishing the necessary conditions for proper SC formation, the roles of centromere pairing in synapsis initiation, and the mechanisms by which oocytes detect failures in SC formation and/or recombination. Finally, we correlate what is known in *Drosophila* oocytes with our understanding of these processes in other systems.

INTRODUCTION: OUR UNDERSTANDING OF MEIOSIS IN *DROSOPHILA* FEMALES

SC: synaptonemal complex

RN: recombination nodule

DSB: double-strand break

The analysis of meiosis really began with Bridges's (1) proof of the chromosome theory of heredity by the analysis of failed meiotic segregation in *Drosophila melanogaster* females. During the next 50 years *D. melanogaster* females continued to be a prime system for the genetic analysis of meiosis, mostly through the study of the meiotic behavior of chromosome aberrations (2). These studies provided us with key insights into the critical roles of chromosomal sites such as the centromere (3) and the telomere (4, 5). This pure-genetics approach, which began with the construction of the first meiotic map based on frequencies of recombination by Sturtevant (6) and the demonstration of genetic interference, also revealed the number and distribution of recombination events throughout the genome (7). Further work demonstrated both the critical role of recombination in ensuring segregation (8–10) and the existence of a backup system that ensures the proper segregation of those chromosomes that fail to recombine (11–13).

The middle of the twentieth century saw the beginning of the mutational analysis of meiosis in *Drosophila*, starting with the isolation of the first meiotic mutant [*c(3)G*] (14) and culminating with the first large screens for meiotic mutants by Sandler et al. (15) and Baker & Carpenter (16). In the past 40 years additional screens for meiotic mutants have been completed (17, 18); most, if not all, of the genes defined by these screens have been characterized at the molecular level; and the corresponding protein products have been identified (19). Many of these genes turned out to have homologs in other systems, and the identification of mutants in these genes has contributed greatly to understanding processes such as arm and centromere cohesion, the DNA repair checkpoint, and the role of microtubule motors in spindle assembly and chromosome segregation (19).

However, despite this progress in genetic analysis, the study of meiosis in *Drosophila* females was long hampered by the lack of available cytological tools. Although excellent descriptions of the stages of oogenesis were available (20–22), it was difficult to visualize meiotic events such as pairing, synapsis [the formation of a tripartite proteinaceous structure known as the synaptonemal complex (SC)], chiasma formation, spindle assembly, metaphase arrest, and homolog segregation at anaphase I [but see Nokkala & Puro (23)].

This roadblock was largely ameliorated by the work of Carpenter in the 1970s and by the efforts of Dernburg and Theurkauf in the 1990s. Carpenter's (24, 25) thorough analysis of the SC in *Drosophila* oocytes by serial section electron microscopy (EM) not only characterized the formation and structure of the SC during zygotene and pachytene in *Drosophila* females but also identified structures known as recombination nodules (RNs) that mark the physical sites of recombination (also known as crossing over and exchange). Carpenter's comparison of SC structure and RN number between wild-type oocytes and oocytes carrying a number of well-characterized recombination-defective mutants provided a cytological view of recombination with which to understand the effects of those mutations on the number and distribution of exchange events. The demonstration by Page & Hawley (26) that the product of the *c(3)G* gene mentioned above, C(3)G, is a major component of the SC and the isolation of antibodies against native and fluorescently tagged versions of this protein allowed SC structure to be studied in both wild-type and mutant oocytes by standard microscopy of fixed tissue. Equally critical, the creation of *Drosophila*-specific antibodies that recognize the chromatin changes associated with the programmed double-strand breaks (DSBs) that initiate meiotic recombination (27, 28) also allowed the analysis of the timing, number, and distribution of potential recombination events by immunofluorescence in whole-mount tissue.

In 1996 Dernburg et al. (29) introduced fluorescence in situ hybridization (FISH) as a tool for assessing meiotic chromosome pairing in *Drosophila* oocytes. In doing so, they demonstrated that prophase in *Drosophila* oocytes is unusual in that upon SC disassembly at the end of pachytene the euchromatin desynapses (as it does in other meiotic systems). However, the pericentromeric heterochromatin regions of the chromosomes remain paired with their homologs, clustered with other heterochromatic regions in a structure referred to as the chromocenter. Along with genetic studies by Hawley et al. (13) and Karpen et al. (30), Dernburg et al. (29) further showed that the perdurance of these heterochromatic pairings underlies the oocyte's ability to segregate those chromosomes that fail to recombine with their homologs. Although, with the exception of the euchromatin histone gene cluster, Dernburg et al. (29) focused primarily on the analysis of heterochromatic pairing, Sherizen et al. (31) and others (32–35) have extended this approach to characterize the pairing of numerous euchromatic sites.

Last, in 1992 Theurkauf & Hawley (36) utilized confocal microscopy to visualize the events of spindle assembly, prometaphase I, metaphase I arrest, and anaphase I chromosome segregation. Although Gilliland et al. (37) subsequently modified the staging of these events relative to chromosome positioning and spindle morphology, the studies by Theurkauf & Hawley (36) and their subsequent extension to include the analysis of postprophase events in living oocytes by Matthies et al. (38) and others (39–41) allowed the process of chromosome segregation in *Drosophila* oocytes to evolve from blackboard flowcharts of segregational events to events that could be seen and described in living oocytes.

These technological advances have vastly increased both our understanding of the meiotic process in *Drosophila* oocytes and the utility of this system for future studies. In this review we focus primarily on those events that underlie the early stages of meiotic prophase, namely pairing, synapsis, and the initiation of recombination, as well as on the checkpoints that exist to monitor the success of those events. Although we attempt to faithfully convey the state of our art in terms of what is known about these processes, we are equally concerned with identifying areas of ignorance and promising new approaches that may help us fill in those gaps.

BASIC BIOLOGY OF OOCYTE DEVELOPMENT: THE BIOLOGY OF THE GERMARIUM

Drosophila is an excellent model organism to study early events of meiosis because the female ovary is arranged in accordance with developmental age. Each ovary consists of 14–16 ovarioles, and each ovariole contains a chain of developing egg chambers that become more mature as they reach the posterior end of the ovary. This review focuses on events that occur at the tip of the ovariole in a region known as the germarium, where meiosis begins (**Figure 1**).

The germarium can be divided into four regions; the germline stem cells reside in the anterior-most part of region 1. Each germline stem cell divides asymmetrically to produce a cystoblast that undergoes four rounds of synchronized mitotic cell divisions with incomplete cytokinesis to produce a 16-cell interconnected cyst. All 16 cells of the cyst share the same cytoplasm, and several cells within the cyst enter meiosis. In region 2A, up to four cells within the cyst initiate assembly of the SC. However, only two pro-oocytes within the cyst enter into pachytene and form a full-length SC. As pachytene progresses, only one of the two cells is selected to become the oocyte, and the other pro-oocyte disassembles the SC and becomes a nurse cell. The oocyte is positioned at the posterior end of the developing egg chamber within the germarium.

As shown in **Figure 1**, in this review early pachytene is denoted as the stage at which a full-length SC is detected and DSBs can be visualized, and mid-pachytene is the stage at which DSB repair occurs. Using γ-H2AV [phosphorylated H2A variant (H2AV)] foci as a marker for DSBs, along with the shape and position of the developing cyst within the germarium, allows for visualization of the process of DSB repair over time. Therefore, region 2B is denoted as early/mid-pachytene and

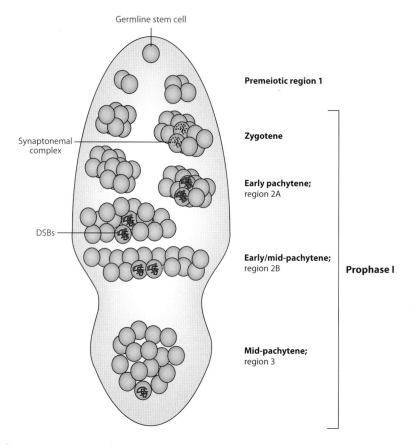

Figure 1
Schematic depiction of the *Drosophila* germarium. At the anterior tip of the germarium, a germline stem cell divides to produce a cystoblast (premeiotic region 1). The cystoblast undergoes four rounds of mitotic divisions to produce a 16-cell germline cyst. Up to four cells within the cyst can initiate the formation of the synaptonemal complex (SC) (*red*) in zygotene, but only two pro-oocytes continue into meiosis and form a full-length SC in early pachytene (region 2A). At this stage programmed meiotic DSBs are induced and can be detected with an antibody to a phosphorylated form of a histone H2A variant, H2AV, at serine 137 (γ-H2AV) (*blue circles* denote DSBs detected with a γ-H2AV antibody). DSBs are also formed in nurse cells (not shown). By early/mid-pachytene (region 2B), fewer DSBs (γ-H2AV foci) can be detected within the pro-oocytes. By mid-pachytene (region 3), one cell has been selected to become the oocyte, and few to no γ-H2AV foci remain. The remainder of the ovariole (stages 2–14) (not shown) is termed the vitellarium. Pachytene persists until stage 6.

region 3 as mid-pachytene. The remainder of the ovariole (stage 2–14) is termed the vitellarium. The later events of pachytene (stage 2–6) are not discussed in this review.

PAIRING: THE TRANSITION FROM SOMATIC PAIRING TO MEIOTIC SYNAPSIS

Although the structure of the SC and the mechanisms that underlie its nucleation and extension are becoming progressively well understood (see below), very little is understood about the mechanisms of meiotic chromosome pairing and alignment in *Drosophila* oocytes. There are four primary reasons for this current lack of understanding. First, during very early meiotic prophase the chromosomes are small and refractory to the types of cytological analysis that have allowed pairing to be studied in other organisms, most notably plants (42–44).

Second, early inferences regarding the pairing process were based on the effects of heterozygosity for chromosome rearrangements on meiotic recombination. The basis for this analysis came from studies in organisms, such as corn, in which rearrangements disrupt pairing for some distance around their breakpoint (45), and those regions of disrupted pairing correspond to regions of exchange suppression. However, we now know that, although such aberrations inhibit recombination in *Drosophila*, they apparently do not impair the completion of either pairing or synapsis as assayed at pachytene (31, 32). (However, these aberrations may inhibit the progression of pairing and/or synapsis during zygotene.) This issue is of particular interest because of the suggestion of Hawley (46) that pairing (or synapsis) may be initiated at specific chromosomal sites along the length of each chromosome, as has been demonstrated in *Caenorhabditis elegans* (47). Until pairing and/or synapsis can be visualized in living oocytes bearing such aberrations, directly testing the validity of this hypothesis will be difficult.

Third, as discussed in the next section, *Drosophila* has noncanonical telomeres (48) that do not abut the nuclear envelope during early prophase, and thus the types of telomere clustering–based models of pairing initiation that are now well accepted in other organisms are not applicable to *Drosophila*. Moreover, *Drosophila* oocytes are also noncanonical in the sense that DSB formation is not required for either pairing or synapsis (49). Thus, applying the knowledge gleaned regarding pairing mechanisms in other organisms to *Drosophila* oocytes has been problematic.

However, the fourth reason, the ubiquitous presence of somatic pairing in *Drosophila*, is perhaps the toughest problem to address. The biology of somatic pairing was reviewed thoroughly by Williams et al. (50) and is only briefly described here. As Williams et al. (50) note, "homologous chromosomes are essentially paired in all somatic cells throughout development." Both cytological studies of condensed chromosomes in late prophase and FISH analysis of cells in various points in the cell cycle suggest that pairing is continuous throughout both G1 and G2, except perhaps during S phase (51–55). The mechanisms that underlie somatic pairing are unclear. Both searches for the usual suspects (33) and genome-wide screens (56) have failed to identify critical genes whose products mediate somatic pairing, although a recent RNAi-based screen in cultured cells implicated topoisomerase II in promoting somatic pairing (50).

The issue of somatic pairing is critical to this discussion because somatic pairing is observed in the oogonial cells that generate the 16-cell cysts. FISH studies of the cells in 16-cell cysts show clear evidence of strong somatic pairing in all cells of the cyst prior to the onset of SC formation at both the centromeres (57) and multiple euchromatic sites (31). Although it is unlikely that existing somatic pairing is dissolved and a new meiosis-specific pairing process initiated in early meiosis, somatic pairings may be modified prior to the onset of synapsis. For example, evidence suggests that some aberrations, most notably insertional duplications that pair in somatic polytene chromosomes or that undergo somatic association as defined by their effects of gene expression, do not facilitate recombination during meiosis (58). Similarly, although all three homologs pair efficiently in somatic cells in triploids, Sturtevant's (6) analysis of recombination in triploids is most consistent with pair-wise recombination events.

Although only live analysis of the events occurring between the end of premeiotic S phase and the completion of synapsis can answer these questions directly, we think it most prudent at this time to presume that a distinct process of meiotic pairing does not exist in *Drosophila* but rather that synapsis is simply superimposed onto chromosomes that were initially paired by a mechanism that mediates somatic pairing (59). Such a model explains both the absence of pairing-defective mutants and the absence of a requirement for DSBs to promote pairing and synapsis.

TELOMERE ASSOCIATION WITH THE NUCLEAR ENVELOPE IN *DROSOPHILA* MEIOCYTES

LE: lateral element
CR: central region
CE: central element
TF: transverse filament
AE: axial element

In most organisms studied, an early event in meiotic prophase includes the anchoring and clustering of telomeres on a small region of the inner nuclear envelope of the meiotic cell. This clustering is a hallmark of the bouquet stage (60–63). Bouquet formation occurs at the leptotene-to-zygotene transition. However, the telomeres remain tightly associated with the inner nuclear envelope throughout pachytene (64). Due to the timing of telomere bouquet formation, which initiates just prior to chromosome pairing and synapsis, bouquet formation may play a major role in synapsis initiation and perhaps in recombination (61, 65–69). Most recently, work from the Alani lab (70) showed that in budding yeast the rapid movement of the telomere-led chromosomes is critical for homolog pairing and recombination. Additional insight into a role for the bouquet has been gained from studies in fission yeast that showed that the bouquet plays a role in controlling the behavior of the microtubule-organizing center (or spindle pole body) and the meiotic spindle (71).

Drosophila and *C. elegans* do not have a bouquet stage. Therefore, if one of the roles of the bouquet is to aid in pairing and synapsis, flies and worms likely employ another mechanism to ensure that these processes occur. In *C. elegans*, pairing is initiated at pairing centers that are enriched in heterochromatic repeats at one end of each chromosome (47, 72, 73). Pairing depends on the interaction of these pairing centers with the microtubule cytoplasmic network; this interaction occurs through protein bridges that span the nuclear envelope (74–76). In *Drosophila* pairing centers have not yet been proven to exist. Early EM work by Carpenter (25) failed to detect telomeres clustered at the nuclear envelope; however, some telomeres were distributed nonrandomly at or near the nuclear envelope. Work by Takeo et al. (57) has shown by immunofluorescence analysis that throughout pachytene telomeres are commonly found close to, but are never clustered at, the nuclear membrane (an additional image is shown in **Figure 2**). However, whether telomeres are physically anchored there is unknown. Thus, it seems very unlikely that telomere clustering, or indeed any telomere-related mechanism, plays a role in initiating synapsis in *Drosophila*. Such a view is consistent with considerable evidence that ring chromosomes recombine with and segregate well from normal sequence homologs.

THE BASIC STRUCTURE OF THE SYNAPTONEMAL COMPLEX

First identified by Fawcett (77) and Moses (78) in 1956, the SC is a tripartite chromosome structure that forms between homologous chromosomes during pachytene of prophase I. Most sexually reproducing organisms form an SC that consists of two lateral elements (LEs) and a central region (CR) that contains the central element (CE) and the transverse filament (TF). The TF runs perpendicular to the LEs and connects them to the CE (79). After premeiotic DNA replication, axial element (AE) components are loaded onto chromosome cores (the scaffold established by the shortening of paired sister chromatids along their longitudinal axes) (80–82) and function in part to connect pairs of sister chromatids. Eventually the AEs form LEs when a full-length SC is formed (see **Figure 3a** for a schematic of the *Drosophila* SC).

COMPONENTS OF THE AXIAL AND LATERAL ELEMENTS OF THE SYNAPTONEMAL COMPLEX

In *Drosophila*, AE/LE components assemble along the pairs of sister chromatids using and colocalizing with cohesion proteins. Meiotic cohesion is established through the cohesin complex, which in most organisms contains two Smc subunits (Smc1/Smc1β and Smc3) and two non-Smc subunits

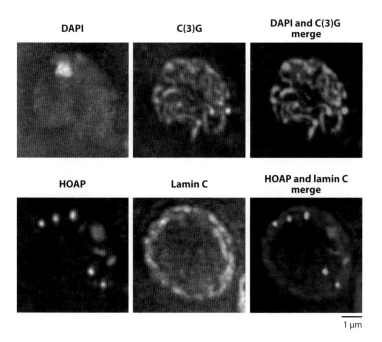

Figure 2

Telomeres appear to associate with the nuclear envelope at pachytene. An oocyte nuclei from region 3 (mid-pachytene) stained with antibodies to C(3)G (91) to mark the synaptonemal complex (SC), lamin C (Iowa Hybridoma Bank) (194) to mark the nuclear envelope, and HOAP (HP1/ORC-associated protein) (195) to mark the telomeres. The majority of the telomeres are located at or near the nuclear envelope. Microscopy was conducted using a DeltaVision microscopy system (Applied Precision, Issaquah, WA) equipped with an Olympus 1,370 inverted microscope and a high-resolution charge-coupled-device camera. Image was deconvolved using softWoRx version 3.5.1 software (Applied Precision). Image is shown as a maximum-intensity projection.

(the α-kleisin Rec8 and Stag3/Rec11) (83). In flies, known components of the meiotic cohesin complex are Smc1 and Smc3. Both Smc1 and Smc3 localize to chromosome arms and centromeres of meiocytes (82). Non-Smc subunits have not yet been identified. Initially, it was proposed that C(2)M might be the α-kleisin of the cohesin complex (84), in part because it physically interacts with Smc3 (85). However, several lines of evidence now suggest that C(2)M is unlikely to function as part of the cohesin complex (82, 86). First, C(2)M accumulates on chromosomes after premeiotic S phase and disappears prior to metaphase I (85). Second, C(2)M expression is restricted to cells that build the SC (87), and third, mutations affecting C(2)M function do not result in high levels of meiosis II segregation defects (87). Finally, C(2)M does not appear to play a role outside of female meiosis (87). Indeed, in flies a noncanonical mechanism may be used to establish cohesion by employing two separate proteins to fill the role of the Rec8 protein in closing the cohesin ring (82).

In addition to the cohesin complex, other non-Smc proteins are critical to the process of cohesion and the formation of the LEs of the SC. One of the best studied of these proteins is the orientation disruptor (Ord) protein, which localizes to chromosome arms and centromeres and is required for meiotic chromosome cohesion, normal chromosome disjunction at meiosis I and II, and normal levels of homologous recombination (88–90). One proposed function of Ord (and of the LEs of the SC) is suppression of sister chromatid exchange, as the absence of Ord leads to an elevated level of repair of DSBs off the sister chromatid, resulting in an increased level of sister

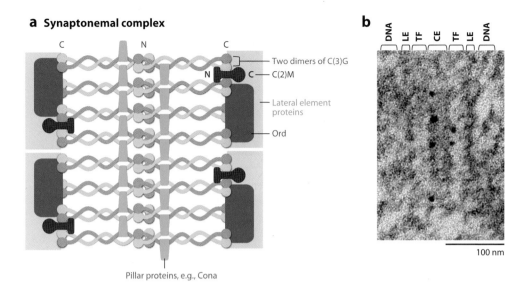

Figure 3

Schematic of the synaptonemal complex (SC) in *Drosophila* oocytes. (*a*) The SC is composed of two lateral elements (LEs) and a central region containing both the transverse filament (TF) and the central element (CE). After premeiotic DNA replication and loading of the cohesion proteins Smc1/3 of the cohesin complex, Nipped-B (not shown), and orientation disruptor (Ord), the chromosome core is formed by shortening of the axes. This process requires the axial element (AE)/LE component C(2)M. The AE will become the LEs of the SC. Immunoelectron microscopy studies show that the N terminus of C(2)M is located at the inner side of the LEs (91). Although the C terminus is depicted as embedded inside the LE, this orientation has not been confirmed. C(3)G proteins are arranged as dimers that are connected in a head-to-head orientation with the N terminus at the middle of the CE. The C terminus of C(3)G is located at or adjacent to the LEs. CE proteins or pillar proteins help to make up the CE of the SC. Corona (Cona) is predicted to be a CE/pillar protein. Panel *a* adapted from Reference 91. (*b*) Electron microscopy micrograph of a section of the SC from a wild-type oocyte expressing a *cona-venus* transgene labeled with anti–green fluorescent protein antibody (Invitrogen). Cona localizes along the outer edge of the CE of the 100-nm SC, indicating that Cona is a bona fide CE protein.

chromatid exchange (89). As assayed by EM analysis, the SC was rarely detected in the oocytes of *ord* mutant females, and the observed SC structures were abnormal and disassembled prematurely (89). As a result, although a normal-looking SC is visible by immunofluorescence in fixed whole-mount tissue in early pachytene (region 2A), the SC is unstable, and chromosome cores begin to fragment at this stage (82). As discussed below, although Ord is required for the accumulation of cohesin onto centromeres, Ord is not essential for the initial loading of the cohesin subunits Smc1/3 to chromosome arms, as this loading process appears normal in the absence of Ord (82).

C(2)M, shown to be associated with the inner edge of the LEs by immuno-EM (91), is thought to control an early step in the formation of chromosome cores (82) and may be involved in the early decision that directs the repair of meiotically induced DSBs to an SC-dependent pathway (83, 87). In the absence of *c(2)M*, chromosome cores fail to form, although this failure does not result in the inability of the cohesion proteins Smc1/3 and Ord to associate with chromosome arms or centromeres (82). Likely as a result of failed core formation, *c(2)M* mutant oocytes fail to form the CE of the SC and in early pachytene assemble only short patches of SC that never extend into a full-length SC (87). Interestingly, in both the absence of chromosome cores and synapsis, *c(2)M* mutants show residual crossing over (25% of wild type) (87), indicating that the absence of C(2)M licenses the activity of a normally inactive SC-independent exchange pathway.

Finally, Nipped-B, a second non-Smc protein, is a cohesin loading factor (92) linked to the regulation of gene transcription by stabilizing the DNA/cohesin association in somatic cells (93).

Immunofluorescence studies reveal that Nipped-B localizes along meiotic chromosome arms in a fashion similar to that of Smc1/3. The colocalization of Nipped-B with Smc1/3 in *Drosophila* oocytes was seen only along chromosome arms, as Nipped-B did not localize to centromeres (94). Nipped-B plays a role in chromosome core maintenance, as females carrying only one copy of *nipped-B* show early disassembly of the chromosome cores and of the SC in the majority of oocytes in region 3 of the germarium (94). The observed defect in premature chromosome core fragmentation occurs later than what has been reported in the absence of the Ord protein. In *ord* mutants, chromosome core fragmentation occurs in late region 2A and before defects in SC structure can be seen by immunofluorescence (89). Unlike in *ord* mutant females (88), the later chromosome core disassembly defect in Nipped-B heterozygotes does not result in meiotic chromosome missegregation (94). Taken together, these proteins function in forming the AE/LE, with which the central region of the SC assembles.

COMPONENTS OF THE CENTRAL REGION OF THE SYNAPTONEMAL COMPLEX

The TFs run perpendicular to the LEs of the SC and thus connect the LEs like ties on a railroad tract. Halfway between the LEs is the CE, which appears as a linear, electron-dense structure that runs parallel to the LEs (83). Although little sequence conservation exists between TF proteins, they have been identified in budding yeast (Zip1) (95), *Drosophila* [C(3)G] (26), *C. elegans* (SYP1 and SYP2) (96, 97, 98), and mammals (SYCP1) (99). Analysis of TF proteins from yeast, *Drosophila*, and mammals indicates that these proteins are structurally similar; each has a globular N-terminal domain, a coiled-coil central domain, and a globular C-terminal domain (83). EM analysis shows these domains are organized in the SC in a similar fashion. The N-terminal region of the TF protein is located in the CR of the SC, and the C-terminal region is located adjacent to the LEs (see **Figure 3a**) (91, 100–103). In addition, deletion mutagenesis of C(3)G indicates that the N-terminal globular domain, as well as a small region of the coiled-coil domain adjacent to the N terminus, is required to form the antiparallel tetramers that make up the SC. The tetramers are composed of parallel homodimers that associate via their coiled-coil region in a head-to-head fashion at their N termini. The C-terminal globular domains are dispensable for tetramer formation but are required to connect the tetramers to the LEs (104). Similar to studies in yeast for Zip1 (105), in the absence of the C-terminal globular domain, C(3)G organizes into a polycomplex structure within the nucleus of the meiotic cell (91, 104).

The CE of the SC is thought to be a highly ordered structure that contains pillar-like proteins that link multiple layers of CR together (see **Figure 3a**) (91, 106). Proteins that colocalize to the CE have been identified in mice [SYCE1 (107), SYCE2 (107), TEX12 (108), and SYCE3 (109)]. Biochemical studies (107–109), along with data acquired from the analysis of individual knockout mice for all four CE proteins (109–112), are beginning to shed light on how this highly ordered structure assembles. SYCP1 (the TF protein), which is required for the localization of all four CE proteins, provides the molecular framework through which the CE forms (109). Although no direct interaction between SYCP1 and SYCE3 has been observed, SYCE3 is required to recruit the remaining CE proteins to the CR (109). Both SYCE3 and SYCE1 are required for the initiation of synapsis in mice (109, 112). Loading of the SYCE2/TEX12 complex allows for the propagation of synapsis by stabilizing the N termini of SYCP1 molecules and thereby anchoring the CE and TF together (108). Deletion of any of the CE proteins leads to infertility and disrupts the progression of recombination (109–112).

In *Drosophila* only one such putative CE protein has been identified. Corona (Cona) colocalizes with C(3)G in a mutually dependent fashion (34). Immunogold labeling of a functional tagged version of Cona by EM analysis indicates that Cona is a CE component that localizes along the outside of the electron-dense linear CE of the SC (C.M. Lake, R.J. Nielsen & R.S. Hawley, unpublished observation; see also **Figure 3b**). In the absence of *cona*, synapsis does not occur, nor do polycomplexes form when the C terminus of C(3)G has been removed (34). These findings indicate that the CE protein Cona is required for the self-association of C(3)G molecules, which is necessary to form a full-length SC. We have yet to determine whether Cona directly associates with C(3)G and/or if Cona acts alone or in concert with other unknown proteins to link the CE to the TF in *Drosophila* oocytes.

POSSIBLE FUNCTIONS OF THE SYNAPTONEMAL COMPLEX

Perhaps the best-understood function of the SC lies in mediating the maturation of DSBs into crossovers. In most organisms SC formation depends on the occurrence of DSBs. However, in *Drosophila* oocytes DSBs are not required for SC formation (49), presumably because of the perdurance of somatic pairing. Additionally, the initiation of synapsis is not absolutely required for DSB induction (28, 34, 89). Rather, the processes of DSB induction and SC formation appear to act independently. In *Drosophila* females the SC proteins that compose the TFs [C(3)G] and a component of the CE (Cona) are absolutely required for the processing of DSBs into meiotic crossover events (26, 34, 113). In addition, one study provides suggestive evidence that C(3)G is required for the formation of gene conversion events (114). The observation that TF and CE proteins are critical to create crossover events is consistent with similar findings in multiple other organisms (110, 111, 115).

The role of the LE in promoting crossovers is less clear. Studies of two proteins that reside within the AE and the LE [Ord and C(2)M] suggest a more complex picture in terms of their relationship to crossover production. The number and timing of the appearance of DSBs appear to be unaffected in *ord* mutants (89). However, crossing over is strongly reduced; in a female bearing a null mutation of *ord*, recombination is reduced to 16% of normal (88; but see also References 116 and 117). As noted above, elegant studies of ring chromosome loss performed by Webber et al. (89) suggest that some of the DSBs that might otherwise become crossovers are instead resolved as sister chromatid exchange events. Similarly, a null mutation in *c(2)M* reduces recombination to approximately 25% of normal but fails to abolish it (87). These observations suggest that, although the LEs may play a role in promoting crossover production, perhaps by stabilizing the CR of the SC, LEs are not required for crossover production. However, the LEs may promote homolog bias. In addition to the observation that in *Drosophila* the Ord protein is required to ensure that repair occurs through an interaction with a homolog, recent studies in yeast implicate a role for both a cohesin protein and components of the AE in mediating homolog bias (118).

Extensive studies in *C. elegans* suggest that the CR components of the SC may also play a role in the maintenance of meiotic pairing (96, 97). A similar picture of the *Drosophila* SC emerges in terms of the ability of the CR components to maintain pairing following mid-zygotene. In both *c(3)G* (31, 34) and *cona* (34) mutant oocytes, pairing (as assayed by FISH to follow the pairing of discrete euchromatic sites) appears greatly reduced or absent. However, mutations either in the LE protein C(2)M or in Ord appear to have minimal effects on pairing, consistent with the view that the primary functions of the LEs are to inhibit sister exchange and to promote homolog bias. In *c(2)M* mutant oocytes, at least for one euchromatic locus analyzed, the homologs were paired throughout all regions of the germarium, even when patches of C(3)G were far from the FISH signal (E. Joyce & K. McKim, personal communication). Similarly, in *ord* mutants, homolog pairing appears normal at least until early pachytene, consistent with the observation that C(3)G localization as assayed by immunofluorescence appears normal during zygotene and early pachytene but becomes fragmented into short stretches by early/mid-pachytene (region 2B) (89). Although EM studies suggest that the SC structure is highly aberrant at both zygotene and pachytene in *ord* oocytes, even such malformed SC structures are likely sufficient to recruit enough C(3)G both to be detected by immunofluorescence and to maintain synapsis [as defined by the total length of C(3)G staining] and homolog pairing (as directly assayed by FISH analysis) (89).

At least one line of evidence suggests that the roles of some CE proteins in mediating crossover formation may be functionally distinct from their roles in mediating pairing. Page & Hawley (26) described an internal deletion of C(3)G that maintained high levels of recombination despite exhibiting high levels of failed synapsis. This finding suggests that the role of the N terminus of C(3)G in promoting exchange (via its participation in CE formation) may be distinct from the role of the coiled-coil domain in linking the LEs of homologous chromosomes. This suggestion is also supported by the work of Storlazzi et al. (119) in yeast, in which the SC component Zip1 plays a role in meiotic recombination independently of SC polymerization along the chromosomes.

In addition to a role in maintaining homolog pairing, at least some components of the SC play more specific roles in mediating pairing at the centromeres. Specifically, in yeast, the TF protein Zip1 is critical in mediating both nonhomologous centromere pairing during early prophase (120) and the resolution of those pairings to homologous couplings following DSB formation and the initiation of full synapsis (121). The conversion from nonhomologous to homologous centromere interactions following the appearance of DSBs requires the phosphorylation of a specific site (serine 75) on Zip1 by the ATR-like Mec1 kinase. This phosphorylation appears to disrupt early nonhomologous pairings and thus allow homologous centromeres to engage in homologous synapsis. The persistence of Zip1 at the centromeres following this realignment helps to facilitate meiotic segregation (122, 123). As Gladstone et al. (122) show, both exchange and nonexchange chromosomes undergo Zip1-dependent centromeric pairing, and centromeric localization of Zip1 persists after SC breakdown, even to the point of microtubule attachment. Loss of this Zip1-dependent centromere pairing induces high levels of nondisjunction, even for homolog pairs that undergo exchange (122). These data suggest that Zip1-dependent pairings help to facilitate proper centromere biorientation at the first meiotic division. Curiously, Zip1-dependent pairs that persist (or reoccur after DSB formation) also play critical roles in mediating the segregation of nonexchange chromosome pairs that may be homologous or nonhomologous (113, 124).

The ability of Zip1 to facilitate such nonexchange segregations may imply a role for the SC in mediating nonexchange, or distributive, segregation in *Drosophila*. A useful application of this hypothesis may explain the process of secondary nonexchange, whereby in *XXY* females the two *X* chromosomes segregate away from the *Y* chromosome, but only if they fail to undergo homologous exchange (1, 125). Xiang & Hawley (126) observed that in early prophase the pericentromeric regions of all three chromosomes are tightly associated. However, after pachytene, the two *X* chromosomes exclude the *Y* if they have undergone exchange. In the absence of such exchange, the three chromosomes remain associated via their pericentromeric regions until the early stages of spindle formation. We can imagine that in flies, in which the centromeres are larger and more complex than in yeast (127), such early centromeric associations, along with homologous heterochromatic pairings, facilitate the initial *XXY* associations. But a destabilization of the SC in the centromeric or pericentromeric regions, similar to those observed in yeast by Falk et al. (121), may allow those *X* chromosomes that have recombined, and thus initiated synapsis along their arms (see below), to limit such centromeric or pericentromeric interactions to their homolog, thus excluding the *Y*. We further propose, again by analogy to the Falk et al. (121) study in yeast, that in the absence of such exchanges, the associations between the *X* and *Y* chromosomes are free both to reform and to persist until spindle formation. Such processes may underlie other forms of achiasmate chromosome segregation (12, 13) such that in the absence of exchange heterochromatic homology is sufficient to support homologous centromeric couplings.

Finally, at least in yeast and *C. elegans*, the SC may play a role in mediating crossover interference (105, 128–130). The term crossover interference refers to the process that controls crossover distribution by preventing the formation of a second crossover event in the vicinity of an already established crossover event. That the SC extends the full length of each bivalent makes it an ideal candidate for mediating such a process. Nonetheless, for yeast considerable evidence suggests that the full extension of the SC is not required for the establishment of interference (131, 132). Studies in mice (133) and *Arabidopsis* (134) also suggest that full synapsis may not be required for normal levels of interference. In *Drosophila*, a weak allele of *c(3)G* studied by Page & Hawley (26), which allows only poor synapsis and reduces, but does not eliminate, crossing over, still exhibits near-wild-type interference. The most parsimonious explanation of these results may be that some SC regions or components are required for interference, but full extension of the SC is likely not required for this process.

ROLE OF SYNAPTONEMAL COMPLEX PROTEINS IN CENTROMERE AGGREGATION AND IN INITIATING SYNAPSIS

Early work by Carpenter (25) and Dernburg et al. (29), and more recent work by the Bickel lab (82, 89), showed that the centromeres of pro-oocytes are clustered into one or two masses rather than paired just with their homolog. As assayed by the number of foci recognized by an antibody to CID, the *Drosophila* CENP-A homolog, pro-oocytes and oocytes usually exhibit only one or two large CID foci rather than the four discrete CID foci that should be exhibited in a diploid cell containing four pairs of homologous chromosomes. As described below, the SC is required for centromere clustering from early zygotene to very late prophase. Centromere clustering is not observed in the nonmeiotic nurse cells that share the 16-cell cyst with the pro-oocytes.

Centromere clustering is specific to nuclei that form the SC (i.e., this process does not occur in premeiotic nuclei or in nurse cell nuclei) and does not require the initiation of homologous recombination (57). Moreover, recent work by Takeo et al. (57) and Tanneti et al. (134a) supports the idea that in *Drosophila* synapsis initiates at the centromeres, as it does in yeast (see **Figure 4**). Both groups analyzed the very early stages of meiotic prophase with regard to SC formation (see **Figure 1**, zygotene stage). In early zygotene, the earliest formation of SC components (both AE/LE and CE proteins) occurs at the centromeres. As noted above, both the presence of the SC at the cluster and the clustering itself persist long past the end of pachytene and the dissolution of the heterochromatic and euchromatic SC (57).

Tanneti et al. (134a) further show that following this initial assembly of the SC at the centromere cluster, there are additional synapsis initiation sites within euchromatin. At least initially, euchromatic synapsis initiation appears to be restricted to relatively few sites within euchromatin and, like centromere clustering, depends on Ord. The extension of these initial euchromatic initiation events to other regions within euchromatin requires functional C(2)M. In *c(2)M* mutant oocytes, SC initiation is blocked at the mid-zygotene stage, and the short patches of SC that are initially formed cannot extend into a full-length SC. These researchers also find that the short patches of the SC that are able to assemble in euchromatin in the absence of *c(2)M* are often associated with the chromatin mark of DSBs. Whether specific sites within the genome couple synapsis initiation within euchromatin to the sites of recombination initiation is unknown.

The mechanism by which the SC is assembled during centromere clustering may be mechanistically different from that responsible for SC initiation along the arms. This possibility is supported by the observation that Ord is specifically required to load centromeric Smc1/3 but is not required

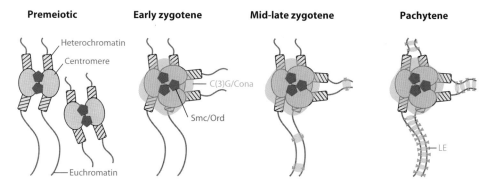

Figure 4

Mechanism of homologous chromosome pairing and synapsis. A model for euchromatic pairing and centromeric pairing and clustering during prophase I. Only two sets of homologous chromosomes are shown, and cohesion established by Smc1/3, Ord, and Nipped-B between sister chromatid arms is not represented. In wild-type premeiotic cells, homologous chromosomes are aligned along the entire region, presumably by somatic pairing, and Smc proteins and Ord have accumulated on centromeres (82). During early zygotene, the synaptonemal complex (SC) proteins C(3)G and Cona are first loaded at centromeres, and centromere pairs of different homologs cluster together. As zygotene progresses, the SC begins to form first at a few euchromatic sites and then at additional sites until a full-length SC is formed between homologous chromosomes at pachytene. LE denotes lateral element.

to load Smc1/3 along the euchromatic arms (82). However, the core components of the SC [the LE component C(2)M, the CE protein Cona, and the TF protein C(3)G] are not required to load Smc1/3 or Ord at centromeres (34, 82).

An increase in the number of CID foci in a representative image of *ord* mutant oocytes suggests that the localization of these SC proteins to the centromeres is required for meiotic centromere clustering (82). Takeo et al. (57) and Tanneti et al. (134a) extend this observation with quantitative analysis, thus demonstrating that, although the majority of wild-type oocytes cluster their centromeres into one or two masses throughout pachytene, *ord* mutant oocytes do not cluster their centromeres. In addition to failing to cluster their centromeres, *ord* mutants also exhibit a defect in maintaining centromere pairing between homologs, showing an average number of CID foci that is greater than six. These reports also demonstrate that other known components of the CR of the SC, C(3)G and Cona, are both required for centromere cluster formation and maintenance and may also be required to some degree for homologous centromere pairing. In oocytes homozygous for null mutations in genes encoding either C(3)G or Cona, the average number of CID foci is approximately four, as would be expected if homologous centromeres were paired but unclustered; however, some fraction (18–33%) of mutant oocytes show more than four foci (57).

These observations are consistent with the view that two mechanistically distinct processes may initiate synapsis in *Drosophila* oocytes. The first mechanism requires only the nucleation of the SC at the centromeres and the processive extension of that synapsis along the arms. The second mechanism involves SC initiation at multiple euchromatic sites by a mechanism that remains unclear.

INITIATION OF HOMOLOGOUS RECOMBINATION

Homologous recombination begins during pachytene with the induction of programmed meiotic DSBs. It is thought that in most, if not all, organisms the enzyme required to catalyze the DSB is Spo11, a type II topoisomerase–like protein (135). DSB formation is considered to be a tightly controlled process in terms of both timing and number. However, the mechanism(s) that controls DSB formation is not completely understood. In *Saccharomyces cerevisiae*, in addition to Spo11, at least nine other proteins (Rec102, Rec104, Ski8, Rec114, Mei4, Mer2, Mre11, Rad50, and Xrs2) are essential for DSB formation (135). Although most of these proteins do not share sequence homology outside of *S. cerevisiae*, additional DSB accessory proteins have also been identified in *Schizosaccharomyces pombe*. These include Rec12 (the Spo11 ortholog), Rec6, Rec7, Rec10, Rec14, Rec15, Rec24, and Mde2 (136). Until recently, in *Drosophila* the only other protein known to be required for DSB formation, other than the Spo11 ortholog Mei-W68 (137), was Mei-P22, and the mechanism of action for Mei-P22 is unknown (138). However, despite the studies in *S. pombe* and *Drosophila*, most of what we know regarding control of the timing and localization of DSB formation comes from studies in budding yeast, in which DSB hot spots exist. In yeast, two kinases, CDK-S and Cdc7/Dbf4, control the timing of DSB formation (139, 140). Both of these kinases act through Mer2 to recruit other proteins and complexes to future DSB sites (139–141).

In *Drosophila* detailed studies using immunological techniques have been undertaken to address when DSB formation occurs and when the DSBs are subsequently repaired (see **Figure 1**) (28). Using an antibody that recognizes γ-H2AV at DSB sites, Mehrotra & McKim (28) showed that γ-H2AV foci first appear in early pachytene after SC formation. By counting the number of γ-H2AV foci, these researchers showed a gradual increase in the number of DSBs as cysts mature in region 2A. This increase is followed by a decline in number as the cyst progresses into early/mid-pachytene (region 2B). The γ-H2AV foci virtually disappear in the oocyte by the time the cyst reaches mid-pachytene (region 3). Because the gradual decrease in the number of γ-H2AV foci coincides with progression through pachytene, this decrease likely represents the process of DSB repair. Whether the removal of γ-H2AV indicates that repair is complete is unknown.

The events leading up to DSB formation are not well understood in *Drosophila*. Mei-P22 is a chromatin-associated protein required for DSB formation. Immunofluorescence analysis of an epitope-tagged Mei-P22 showed that during early pachytene after SC formation, Mei-P22 localizes to discrete foci prior to the time DSBs can be detected (28, 138). A percentage of these Mei-P22 foci show partial overlap with the γ-H2AV marker for DSBs (28). In the absence of *mei-W68*, Mei-P22 is able to load onto the chromosomes, which indicates that such Mei-P22 localization is independent of DSB formation (138).

We recently reported a third gene required for the induction of meiotically induced DSBs (35). *trade embargo* (*trem*), a 5-C2H2 zinc finger protein expressed in early zygotene, localizes to chromatin in a thread-like fashion and is required for normal DSB levels. Analysis of a point mutation in *trem* (*trem^{F9}*), which disrupts a conserved residue in the linker domain between the first and second zinc fingers, failed to localize Mei-P22 to discrete foci during pachytene. Because Mei-P22 can associate with chromatin in the absence of DSBs, these data suggest that Trem acts upstream of both Mei-P22 and Mei-W68 in the initiation of DSBs (see **Figure 5***a*). The mechanism by which Trem localizes Mei-P22 to discrete foci on the chromatin is unknown. Trem may be physically required to load or stabilize Mei-P22 at future DSB sites, or Trem may alter the chromatin structure at or around these sites as a prerequisite for Mei-P22 loading.

Data from other model organisms help to model possible ways in which Trem may function. In both fission and budding yeast, several of the accessory proteins required for DSB formation form complexes that may or may not depend on Spo11 function. For example, in fission yeast the recently identified ortholog of mouse Mei4 (142), Rec24, forms foci on linear elements in a Spo11-independent manner, and Rec7 stabilizes this localization (136). Investigators speculated that only when this stable complex is formed is the Spo11 ortholog loaded onto chromosomes (136).

The model that Trem may act to alter chromatin structure in a manner that allows for loading of other DSB accessory proteins comes from studies of another multiple C2H2 zinc finger

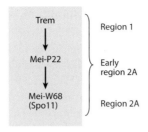

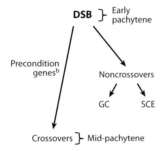

Figure 5

Double-strand break (DSB) formation and repair during prophase in *Drosophila*. (*a*) Temporal function of proteins required for DSB formation. Trem is expressed in region 1 of the germarium and appears to localize to chromatin in a thread-like pattern. Epitope-tagged Mei-P22 localizes to discrete foci on meiotic chromosomes in early pachytene cysts. Mei-P22 foci in wild-type oocytes partially overlap with γ-H2AV foci (modification at DSB sites) in pachytene. DSBs are created by the conserved Spo11 homolog Mei-W68. (*b*) Processing of meiotically induced DSBs. DSBs that are induced in pachytene are processed into crossovers or noncrossovers [by gene conversion (GC) or by sister chromatid exchange (SCE)]. A set of genes known as precondition genes are required to direct the DSBs toward the path of crossing over. Mutations in genes in this class process the DSBs primarily into noncrossovers. The exchange-class genes are involved in the resolution of the recombination intermediate. Immunofluorescence analysis shows that the phosphorylation of H2AV is removed at mid-pachytene, indicating that repair is either complete or in the process of being completed. [a]DSB repair genes include the spindle class of genes (*spn-A*, *spn-B*, *spn-C*, and *spn-D*), *okra*, *hus1*, and *brca2*. [b]Precondition genes include *mei-217*, *mei-218*, *rec*, and *mcm5*. [c]Exchange genes include *mei-9*, *ercc1*, *mus-312*, and *hdm*.

protein, Prdm9. Prdm9 localizes to DSB hot spots in mammalian systems and modifies chromatin through its histone methyltransferase activity, which promotes the binding of proteins required for recombination initiation, as well as targeting recombination to specific sites in the genome (143–149).

Although we do not know the exact mechanism by which Trem acts in initiating DSBs during meiosis, a comparison of a weak allele of *trem* to weak alleles of *mei-P22* or *mei-W68* that reduce but do not eliminate DSBs indicates that Trem may also have a direct role in crossover formation. Weak alleles of *mei-P22* and *mei-W68* appear to process residual DSBs into crossovers more efficiently than does an allele of *trem* that reduces the DSBs to a similar level (28, 35).

MATURATION OF DOUBLE-STRAND BREAKS

Although the initial step in the pathway leading to homologous recombination is the induction of meiotic DSBs, not all DSBs are repaired into crossovers. This leaves a subset of DSBs repaired as noncrossovers, either through gene conversions or through perhaps the repair of a small percentage of DSBs off the sister chromatid (see below and **Figure 5b**). In *Drosophila*, studies of genes required for DSB repair indicate that up to 24 DSBs per meiosis are induced (28, 150, 151), yet on average only one crossover per chromosome arm occurs. Studies have shown that mutants with greatly reduced levels of meiotic recombination display an elongated prometaphase followed by failed homolog segregation (152). This segregational failure is due to a failure to establish the necessary chiasmata (the physical manifestation of an exchange event) needed to ensure the co-orientation of homologous centromeres. Therefore, normal levels of homologous recombination are required for proper chromosome segregation during meiosis (153).

Genes required for repair of meiotic DSBs, and thus for crossover formation, can be divided into three classes on the basis of the ability of mutants that affect gene function to repair DSBs. The first class of genes appears to be required to repair all meiotically induced DSBs. Mutations in this class of genes result in sterility (or significantly reduced fertility) and lead to the accumulation of DSBs in mid-pachytene. This class of genes includes *spn-A* (a Rad-51 ortholog) (154), *spn-B* (a Rad51 paralog) (155), *mus-301/spn-C* (human HEL308-like protein) (156), *spn-D* (a Rad51 paralog) (155, 157), *okra* (a Rad54 ortholog) (155), *brca2* (151), and *hus1* (158). *spn-A* encodes an enzyme that mediates strand invasion and strand exchange during DSB repair (159), whereas *okra* encodes a DNA helicase required for DSB repair (155). Both *spn-A* and *okra* function in somatic as well as meiotic repair processes. *spn-B* and *spn-D* also encode components of DSB repair machinery, but unlike their mammalian counterparts (XRCC3 and Rad51C, respectively), these gene products function primarily in meiotic homologous recombination repair (157). *brca2* and *hus1* encode gene products that have dual roles in that they not only are required for DNA damage checkpoint activity (see below) but also have a direct role in DSB repair (151, 158).

Genes in the second class, known as the precondition class, direct meiotic DSBs toward the process of crossing over. Mutations in this class of genes not only alter the frequency of crossover events but also disrupt the mechanism that controls the distribution of the remaining crossovers (7, 153). In addition, these genes may play a role in crossover interference. Members include *mei-217* (160), *mei-218* (161), *rec/mcm8* (162, 163), and *mcm5* (164). Mcm5 and Rec/Mcm8 are bona fide MCM proteins (162–164), whereas Mei-217 and Mei-218 have similarity to the N- and C-terminal portions of the MCM domain, respectively (J. Sekelsky, K. Kohl & K. McKim, personal communications). The mechanism by which each of these proteins controls DSB repair is not understood. Mei-218 likely does not have a direct role, as it localizes to the cytoplasm (165). Rec/Mcm8 may facilitate the DNA synthesis that is associated with the formation of recombination intermediates because gene conversion tracts associated with defective Rec/Mcm8 are approximately half the length of wild-type gene conversion tracts (162). Similar results of shorter-than-normal gene conversion tracts were reported for *mei-218* mutants as well (166, 167). However, other reports show that most of the conversion tracts associated with *mei-218* are indistinguishable from those of wild type (168). Whether Mcm5 plays a role in processive DNA synthesis during recombination remains to be investigated.

The third class of genes is required to resolve recombination intermediates into crossovers and is known as the exchange class of genes. Mutations in this class of genes do not alter the overall distribution of crossovers, only the frequency at which they occur (169–171). This group includes *mei-9* (170), *ercc1* (172), *mus-312* (173), and *hdm* (171). Yeast two-hybrid studies suggest that all four of these members form a complex (171, 173). A model has been proposed, on the basis of the homology of *mei-9* to human xeroderma pigmentosum factor F, that this complex is required for the exonuclease activity needed to resolve recombination intermediates during the homologous recombination process (170, 171, 174). In addition, studies by the Sekelsky lab (175) identified BTBD12 as the vertebrate ortholog of *mus-312*. BTBD12 is required during meiotic prophase I for the proper repair of DSB events leading to crossovers (176), and therefore the role of this subunit in the resolution of recombination intermediates may be similar in a variety of organisms (175).

MEIOTIC CHECKPOINTS DURING FEMALE MEOSIS SURVEY BOTH DOUBLE-STRAND BREAK–INDEPENDENT AND DOUBLE-STRAND BREAK–DEPENDENT EVENTS

To ensure that errors that have occurred during normal meioses are resolved prior to causing deleterious effects, checkpoints delay or stop meiotic cell cycle progression until the error can be resolved. Failure to take time to pause and fix the error could result in apoptosis, cell cycle arrest, and/or aneuploidy of the resulting meiotic products. At least two surveillance mechanisms have been established in *Drosophila* to monitor the repair of meiotically induced DSBs: the canonical DNA damage checkpoint (also known as the ATR/Mei-41-dependent meiotic checkpoint) (177) and the more recently discovered pachytene checkpoint (178). Since many genes required for resolving DSBs into crossovers are also involved in the pachytene checkpoint, these two distinct pathways (by virtue of their dependence on DSBs) can be confusing. Therefore, here we discuss what is known about each of these checkpoints.

THE DNA DAMAGE CHECKPOINT

In *Drosophila* failure to repair meiotically induced DSBs activates the canonical DNA damage checkpoint, resulting in meiotic arrest (155, 177). Mutations in most, if not all, genes required for the proper repair of DSBs [*spn-A* (154), *spn-B* (155), *spn-D* (155, 157), *okra* (155), *hus1* (158), and *brca2* (151)] result in the activation of this checkpoint. Gene products that are required to sense and signal the response of checkpoint activation include the 9-1-1 complex of Rad9, Rad1, and Hus1, as well as Brca2, which in addition to its role in repair is also required to transduce the signal required for the checkpoint (151). As a result of checkpoint activation, at least two conserved kinases, Mei-41 and Chk2, are activated. This activation leads both to defects in the organization of meiotic chromosomes and to oocyte polarization defects (failure to establish the dorsal-ventral fate of the eggshell) (177, 179). Each of these defects is a result of the effect of the transducer, Chk2, on specific effector proteins. Dorsal-ventral patterning defects result from insufficient translation of Gurken, which is required to establish the dorsal fate of the embryo (179), and defects in chromosomal organization result from a suppression of Nhk-1 activity, which is required to release the meiotic chromosomes from the nuclear envelope (180). The result in both cases is meiotic arrest.

One major difference in the DNA damage checkpoint compared with the pachytene checkpoint described below is the requirement of DSB induction. The DNA damage checkpoint can be suppressed by simultaneously abolishing DSB formation, which indicates that this checkpoint depends on DSB formation (138, 177). In addition, activation of the DNA damage checkpoint through the disruption of genes required for DNA repair can be suppressed when the downstream sensor or transducer of the checkpoint is inactive.

THE PACHYTENE CHECKPOINT

The pachytene checkpoint is used to monitor the events of meiotic recombination and to delay or stop meiotic prophase until meiotic recombination intermediates can be resolved. In budding yeast and worms, checkpoint activation results from a failure to complete synapsis and/or from a failure to complete recombination. Activation of this checkpoint leads to meiotic arrest in yeast and to apoptosis in worms and mammals. The McKim lab (178) recently identified the pachytene checkpoint in *Drosophila*, and in contrast to the checkpoint in other organisms, this checkpoint appears to have both conserved and unique aspects.

In yeast, incomplete synapsis or failure to complete recombination activates checkpoint proteins (including Rad17, Rad24, and Mec1), which presumably generate an inhibitory signal indicating that meiotic prophase should pause until the meiotic recombination process is complete (181, 182). The ability to stop meiotic prophase has been linked to the limited production of the cyclin Clb1 and to the inactivation of Cdc28 in yeast (181). It is thought that upon completion of meiotic synapsis and/or recombination, the inhibitory signal is removed, which leads to cell cycle progression.

In addition to the proteins listed above, Pch2 (a meiosis-specific AAA+ ATPase) is required for checkpoint activation in yeast and worms, but its mode of action is unclear (183, 184). In yeast, the Pch2 requirement for the pachytene checkpoint depends specifically on a subpopulation of the protein within the nucleolus. However, a subpopulation also localizes to the chromatin (183). In both yeast and mammals, Pch2 has been linked to roles in homologous chromosome synapsis, DSB repair, and the regulation of both the normal number and the distribution of crossovers (185–192). However, in mice no function has been identified for Pch2 in the pachytene checkpoint (185, 186) as has been reported in yeast and worms.

In flies, pachytene arrest is not triggered solely by incomplete synapsis, as it is in yeast and worms, but rather by defects in the pathway leading to crossover formation. The data suggest that in flies the monitored event is synapsis-independent changes in chromosome structure (193). In flies, as in worms (184) and probably yeast (189), the checkpoint depends on Pch2 but is independent of DSB formation (178). Upon activation of the checkpoint, there are delays (and not an arrest) in prophase events, e.g., delays both in oocyte selection and in the chromatin remodeling response to DSBs (178).

There appear to be two separate activators of the pachytene checkpoint in flies. Mutations in genes required for DSB repair (*spn-A*, *spn-B*, *spn-D*, and *okra*) and for exchange (*mei-9*, *mus-312*, and *hdm*) cause checkpoint delays. This activation requires, along with Pch2, genes whose products play a role in establishing crossover distribution (the precondition class, including *mei-218* and *rec*) (178). A second independent activator of the checkpoint does not depend on the precondition genes but is Pch2 dependent and monitors defects in chromosome axis formation. This defect may be due either to mutations in genes required for normal axis formation [*ord* and *c(2)M*] or to chromosomal aberrations created in balancer heterozygotes (193). In either case, as in yeast, Pch2 activity in checkpoint activation depends on Sir2 (193), a histone deacetylase.

As discussed above, in yeast Pch2 localizes both to the chromosomes and to the nucleolus. Localization to the nucleolus depends on Sir2 (183). Sir2 is required for Pch2 to exert its activity in yeast and in flies (193). In addition, timely expression of Pch2 is important for function in that prolonged expression of Pch2 by Sir2 causes checkpoint activation (193). However, one major difference between budding yeast and flies is the localization of Pch2. Studies have shown that in flies Pch2 is excluded from the nucleus and localizes to the cytoplasmic side of the nuclear envelope (193). The absence of Pch2 on chromosomes in *Drosophila* may explain the failure to detect a direct role for Pch2 in homologous chromosome synapsis or in crossover formation. Surprisingly, although Pch2 does not appear to localize to chromosomes, it is indirectly required to promote an optimal number of crossovers in flies (193). How Pch2 exerts its effects within the nucleus during the process of crossing over remains to be determined.

SUMMARY POINTS

1. Synapsis in *Drosophila* is required for the meiosis-specific process of centromere clustering and appears to be initiated at clustered centromeres. Centromeric aggregation and pairing in *Drosophila* may thus mirror a role similar to that of the formation of telomeric bouquets in other organisms by providing the initial sites of synapsis initiation.

2. The complete SC has multiple roles in mediating the meiotic process, including a critical role of the CE in the formation of crossover events; a role of the LEs in promoting homolog bias; and, by analogy with work done by others in yeast (121–124), a role in allowing crossover events to inhibit heterologous centromeric associations.

3. At least two proteins (Trem and Mei-P22) are required for the *Drosophila* Spo11 homolog (Mei-W68) to initiate DSBs. Much remains to be learned about the mechanisms by which DSB distribution is controlled.

4. At least two different checkpoints monitor the creation of mature crossover events; one checkpoint is DSB dependent, and the other checkpoint is DSB independent.

DISCLOSURE STATEMENT

The authors are not aware of any affiliations, memberships, funding, or financial holdings that might be perceived as affecting the objectivity of this review.

ACKNOWLEDGMENTS

The authors would like to thank Sharon Bickel, Kim McKim, and members of the Hawley lab for critical reading of the manuscript. We would also like to thank Angela Seat for her graphic design assistance, Fengli Guo in the Stowers Histology and EM facility for immuno-EM analysis, and Satomi Takeo for providing us with an additional image from her recent work. R.S.H. is supported by the Stowers Institute for Medical Research and by an American Cancer Society Research Professor Award (RP-05-086-06-DDC).

LITERATURE CITED

1. Bridges CB. 1916. Non-disjunction as proof of the chromosome theory of heredity. *Genetics* 1:1–52, 1:107–63
2. Ashburner M, Golic K, Hawley RS. 2005. *Drosophila—A Laboratory Handbook*. Cold Spring Harbor, NY: Cold Spring Harb. Lab. Press. 2nd ed.
3. Anderson EG. 1925. Crossing over in a case of attached X chromosomes in *Drosophila melanogaster*. *Genetics* 10:403–17
4. Herskowitz IH, Muller HJ. 1954. Evidence against a straight end-to-end alignment of chromosomes in *Drosophila* spermatozoa. *Genetics* 39:836–50
5. Parker DR, McCrone J. 1958. A genetic analysis of some rearrangements induced in oocytes of *Drosophila*. *Genetics* 43:172–86
6. Sturtevant AH. 1936. Preferential segregation in triplo-IV females of *Drosophila melanogaster*. *Genetics* 21:444–66
7. Lindsley DL, Sandler L. 1977. The genetic analysis of meiosis in female *Drosophila melanogaster*. *Philos. Trans. R. Soc. Lond. Ser. B* 277:295–312

8. Koehler KE, Boulton CL, Collins HE, French RL, Herman KC, et al. 1996. Spontaneous X chromosome MI and MII nondisjunction events in *Drosophila melanogaster* oocytes have different recombinational histories. *Nat. Genet.* 14:406–14
9. Merriam JR. 1967. The initiation of nonhomologous chromosome pairing before exchange in female *Drosophila melanogaster*. *Genetics* 57:409–25
10. Merriam JR, Frost JN. 1964. Exchange and nondisjunction of the *X* chromosomes in female *Drosophila melanogaster*. *Genetics* 49:109–22
11. Grell RF. 1962. A new hypothesis on the nature and sequence of meiotic events in the female of *Drosophila melanogaster*. *Proc. Natl. Acad. Sci. USA* 48:165–72
12. Grell RF. 1976. Distributive pairing. In *The Genetics and Biology of* Drosophila, ed. M Ashburner, E Novitski, 1a:436–86. London: Academic
13. Hawley RS, Irick H, Zitron AE, Haddox DA, Lohe A, et al. 1993. There are two mechanisms of achiasmate segregation in *Drosophila* females, one of which requires heterochromatic homology. *Dev. Genet.* 13:440–67
14. Gowen MS, Gowen JW. 1922. Complete linkage in *Drosophila melanogaster*. *Am. Nat.* 61:286–88
15. Sandler L, Lindsley DL, Nicoletti B, Trippa G. 1968. Mutants affecting meiosis in natural populations of *Drosophila melanogaster*. *Genetics* 60:525–58
16. Baker BS, Carpenter AT. 1972. Genetic analysis of sex chromosomal meiotic mutants in *Drosophilia melanogaster*. *Genetics* 71:255–86
17. Sekelsky JJ, McKim KS, Messina L, French RL, Hurley WD, et al. 1999. Identification of novel *Drosophila* meiotic genes recovered in a P-element screen. *Genetics* 152:529–42
18. Page SL, Nielsen R, Teeter K, Lake CM, Ong S, et al. 2007. A germline clone screen for meiotic mutants in *Drosophila melanogaster*. *Fly* 1:172–81
19. McKim KS, Jang JK, Manheim EA. 2002. Meiotic recombination and chromosome segregation in *Drosophila* females. *Annu. Rev. Genet.* 36:205–32
20. King RC. 1970. *Ovarian Development in* Drosophila melanogaster. London: Academic. 227 pp.
21. Mahowald AP, Kambysellis MP. 1980. Oogenesis. In *Genetics and Biology of* Drosophila, ed. M Ashburner, TR Wright, 2-D:141–224. New York: Academic
22. Spradling AC. 1993. Developmental genetics of oogenesis. In *The Development of* Drosophila melanogaster, ed. M Bate, AA Martinez, 1:1–70. Cold Spring Harbor, NY: Cold Spring Harb. Lab. Press
23. Nokkala S, Puro J. 1976. Cytological evidence for a chromocenter in *Drosophila melanogaster* oocytes. *Hereditas* 83:265–68
24. Carpenter AT. 1979. Synaptonemal complex and recombination nodules in wild-type *Drosophila melanogaster* females. *Genetics* 92:511–41
25. Carpenter AT. 1975. Electron microscopy of meiosis in *Drosophila melanogaster* females. *Chromosoma* 51:157–82
26. Page SL, Hawley RS. 2001. *c(3)G* encodes a *Drosophila* synaptonemal complex protein. *Genes Dev.* 15:3130–43
27. Madigan JP, Chotkowski HL, Glaser RL. 2002. DNA double-strand break-induced phosphorylation of *Drosophila* histone variant H2Av helps prevent radiation-induced apoptosis. *Nucleic Acids Res.* 30:3698–705
28. Mehrotra S, McKim KS. 2006. Temporal analysis of meiotic DNA double-strand break formation and repair in *Drosophila* females. *PLoS Genet.* 2:e200
29. Dernburg AF, Sedat JW, Hawley RS. 1996. Direct evidence of a role for heterochromatin in meiotic chromosome segregation. *Cell* 86:135–46
30. Karpen GH, Le MH, Le H. 1996. Centric heterochromatin and the efficiency of achiasmate disjunction in *Drosophila* female meiosis. *Science* 273:118–22
31. Sherizen D, Jang JK, Bhagat R, Kato N, McKim KS. 2005. Meiotic recombination in *Drosophila* females depends on chromosome continuity between genetically defined boundaries. *Genetics* 169:767–81
32. Gong WJ, McKim KS, Hawley RS. 2005. All paired up with no place to go: pairing, synapsis, and DSB formation in a balancer heterozygote. *PLoS Genet.* 1:e67

33. Blumenstiel JP, Fu R, Theurkauf WE, Hawley RS. 2008. Components of the RNAi machinery that mediate long distance chromosomal associations are dispensable for meiotic and early somatic homolog pairing in *Drosophila melanogaster*. *Genetics* 180(3):1355–65
34. Page SL, Khetani RS, Lake CM, Nielsen RJ, Jeffress JK, et al. 2008. Corona is required for higher-order assembly of transverse filaments into full-length synaptonemal complex in *Drosophila* oocytes. *PLoS Genet.* 4:e1000194
35. Lake CM, Nielsen RJ, Hawley RS. 2011. The *Drosophila* zinc finger protein Trade Embargo is required for double strand break formation in meiosis. *PLoS Genet.* 7:e1002005
36. Theurkauf WE, Hawley RS. 1992. Meiotic spindle assembly in *Drosophila* females: behavior of nonexchange chromosomes and the effects of mutations in the nod kinesin-like protein. *J. Cell Biol.* 116:1167–80
37. Gilliland WD, Hughes SF, Vietti DR, Hawley RS. 2009. Congression of achiasmate chromosomes to the metaphase plate in *Drosophila melanogaster* oocytes. *Dev. Biol.* 325:122–28
38. Matthies HJ, McDonald HB, Goldstein LS, Theurkauf WE. 1996. Anastral meiotic spindle morphogenesis: role of the non-claret disjunctional kinesin-like protein. *J. Cell Biol.* 134:455–64
39. Xiang Y, Takeo S, Florens L, Hughes SE, Huo LJ, et al. 2007. The inhibition of polo kinase by matrimony maintains G2 arrest in the meiotic cell cycle. *PLoS Biol.* 5:e323
40. Hughes SE, Gilliland WD, Cotitta JL, Takeo S, Collins KA, Hawley RS. 2009. Heterochromatic threads connect oscillating chromosomes during prometaphase I in *Drosophila* oocytes. *PLoS Genet.* 5:e1000348
41. Colombié N, Cullen CF, Brittle AL, Jang JK, Earnshaw WC, et al. 2008. Dual roles of Incenp crucial to the assembly of the acentrosomal metaphase spindle in female meiosis. *Development* 135:3239–46
42. McClintock B. 1933. The association of non-homologous parts of chromosomes in the mid-prophase of meiosis in *Zea mays*. *Z. Zellforsch. Mikroskop. Anat.* 19:191–237
43. Albini SM, Jones GH. 1987. Synaptonemal complex spreading in *Allium cepa* and *A. fistulosum*. I. The initiation and sequence of pairing. *Chromosoma* 95:324–38
44. Bozza CG, Pawlowski WP. 2008. The cytogenetics of homologous chromosome pairing in meiosis in plants. *Cytogenet. Genome Res.* 120:313–19
45. Burnham CR, Stout JT, Weinheimer WH, Kowles RV, Phillips RL. 1972. Chromosome pairing in maize. *Genetics* 71:111–26
46. Hawley RS. 1980. Chromosomal sites necessary for normal levels of meiotic recombination in *Drosophila melanogaster*. I. Evidence for and mapping of the sites. *Genetics* 94:625–46
47. Phillips CM, Wong C, Bhalla N, Carlton PM, Weiser P, et al. 2005. HIM-8 binds to the X chromosome pairing center and mediates chromosome-specific meiotic synapsis. *Cell* 123:1051–63
48. Pardue ML, DeBaryshe PG. 1999. *Drosophila* telomeres: two transposable elements with important roles in chromosomes. *Genetica* 107:189–96
49. McKim KS, Green-Marroquin BL, Sekelsky JJ, Chin G, Steinberg C, et al. 1998. Meiotic synapsis in the absence of recombination. *Science* 279:876–78
50. Williams BR, Bateman JR, Novikov ND, Wu C-T. 2007. Disruption of topoisomerase II perturbs pairing in *Drosophila* cell culture. *Genetics* 177:31–46
51. Csink AK, Henikoff S. 1998. Large-scale chromosomal movements during interphase progression in *Drosophila*. *J. Cell Biol.* 143:13–22
52. Fung JC, Marshall WF, Dernburg A, Agard DA, Sedat JW. 1998. Homologous chromosome pairing in *Drosophila melanogaster* proceeds through multiple independent initiations. *J. Cell Biol.* 141:5–20
53. Duncan IW. 2002. Transvection effects in *Drosophila*. *Annu. Rev. Genet.* 36:521–56
54. McKee BD. 2004. Homologous pairing and chromosome dynamics in meiosis and mitosis. *Biochim. Biophys. Acta* 1677:165–80
55. Fritsch C, Ploeger G, Arndt-Jovin D. 2006. *Drosophila* under the lens: imaging from chromosomes to whole embryos. *Chromosome Res.* 14:451–64
56. Bateman JR, Wu C-t. 2008. A genomewide survey argues that every zygotic gene product is dispensable for the initiation of somatic homolog pairing in *Drosophila*. *Genetics* 180:1329–42
57. Takeo S, Lake CM, Morais-de-Sá E, Sunkel CE, Hawley RS. 2011. Synaptonemal complex-dependent centromeric clustering and the initiation of synapsis in *Drosophila* oocytes. *Curr. Biol.* 21:1845–51

58. Grell EH. 1964. Influence of the location of a chromosomal duplication on crossing over in *Drosophila melanogaster*. *Genetics* 50:251–52 (Abstr.)
59. Roeder GS. 1997. Meiotic chromosomes: It takes two to tango. *Genes Dev.* 11:2600–21
60. Zickler D, Kleckner N. 1998. The leptotene-zygotene transition of meiosis. *Annu. Rev. Genet.* 32:619–97
61. Bass HW. 2003. Telomere dynamics unique to meiotic prophase: formation and significance of the bouquet. *Cell. Mol. Life Sci.* 60:2319–24
62. Harper L, Golubovskaya I, Cande WZ. 2004. A bouquet of chromosomes. *J. Cell Sci.* 117:4025–32
63. Scherthan H. 2007. Telomeres and meiosis in health and disease. *Cell. Mol. Life Sci.* 64:117–24
64. Siderakis M, Tarsounas M. 2007. Telomere regulation and function during meiosis. *Chromosome Res.* 15:667–79
65. Trelles-Sticken E, Dresser ME, Scherthan H. 2000. Meiotic telomere protein Ndj1p is required for meiosis-specific telomere distribution, bouquet formation and efficient homologue pairing. *J. Cell Biol.* 151:95–106
66. Koszul R, Kim KP, Prentiss M, Kleckner N, Kameoka S. 2008. Meiotic chromosomes move by linkage to dynamic actin cables with transduction of force through the nuclear envelope. *Cell* 133:1188–201
67. Wanat JJ, Kim KP, Koszul R, Zanders S, Weiner B, et al. 2008. Csm4, in collaboration with Ndj1, mediates telomere-led chromosome dynamics and recombination during yeast meiosis. *PLoS Genet.* 4:e1000188
68. Kosaka H, Shinohara M, Shinohara A. 2008. Csm4-dependent telomere movement on nuclear envelope promotes meiotic recombination. *PLoS Genet.* 4:e1000196
69. Rockmill B, Roeder GS. 1998. Telomere-mediated chromosome pairing during meiosis in budding yeast. *Genes Dev.* 12:2574–86
70. Sonntag Brown M, Zanders S, Alani E. 2011. Sustained and rapid chromosome movements are critical for chromosome pairing and meiotic progression in budding yeast. *Genetics* 188:21–32
71. Tomita K, Cooper JP. 2007. The telomere bouquet controls the meiotic spindle. *Cell* 130:113–26
72. Phillips CM, Dernburg AF. 2006. A family of zinc-finger proteins is required for chromosome-specific pairing and synapsis during meiosis in *C. elegans*. *Dev. Cell* 11:817–29
73. Phillips CM, Meng X, Zhang L, Chretien JH, Urnov FD, Dernburg AF. 2009. Identification of chromosome sequence motifs that mediate meiotic pairing and synapsis in *C. elegans*. *Nat. Cell Biol.* 11:934–42
74. Jaspersen SL, Hawley RS. 2009. Pushing and pulling: Microtubules mediate meiotic pairing and synapsis. *Cell* 139:861–63
75. Penkner AM, Fridkin A, Gloggnitzer J, Baudrimont A, Machacek T, et al. 2009. Meiotic chromosome homology search involves modifications of the nuclear envelope protein Matefin/SUN-1. *Cell* 139:920–33
76. Sato A, Isaac B, Phillips CM, Rillo R, Carlton PM, et al. 2009. Cytoskeletal forces span the nuclear envelope to coordinate meiotic chromosome pairing and synapsis. *Cell* 139:907–19
77. Fawcett DW. 1956. The fine structure of chromosomes in the meiotic prophase of vertebrate spermatocytes. *J. Biophys. Biochem. Cytol.* 2:403–6
78. Moses MJ. 1956. Chromosomal structures in crayfish spermatocytes. *J. Biophys. Biochem. Cytol.* 2:215–18
79. Zickler D, Kleckner N. 1999. Meiotic chromosomes: integrating structure and function. *Annu. Rev. Genet.* 33:603–754
80. Revenkova E, Jessberger R. 2006. Shaping meiotic prophase chromosomes: cohesins and synaptonemal complex proteins. *Chromosoma* 115:235–40
81. Stack SM, Anderson LK. 2001. A model for chromosome structure during the mitotic and meiotic cell cycles. *Chromosome Res.* 9:175–98
82. Khetani RS, Bickel SE. 2007. Regulation of meiotic cohesion and chromosome core morphogenesis during pachytene in *Drosophila* oocytes. *J. Cell Sci.* 120:3123–37
83. Page SL, Hawley RS. 2004. The genetics and molecular biology of the synaptonemal complex. *Annu. Rev. Cell Dev. Biol.* 20:525–58
84. Schleiffer A, Kaitna S, Maurer-Stroh S, Glotzer M, Nasmyth K, Eisenhaber F. 2003. Kleisins: a superfamily of bacterial and eukaryotic SMC protein partners. *Mol. Cell* 11:571–75
85. Heidmann D, Horn S, Heidmann S, Schleiffer A, Nasmyth K, Lehner CF. 2004. The *Drosophila* meiotic kleisin C(2)M functions before the meiotic divisions. *Chromosoma* 113:177–87

86. Yan R, Thomas SE, Tsai J-H, Yamada Y, McKee BD. 2010. SOLO: a meiotic protein required for centromere cohesion, coorientation, and SMC1 localization in *Drosophila melanogaster*. *J. Cell Biol.* 188:335–49
87. Manheim EA, McKim KS. 2003. The synaptonemal complex component C(2)M regulates meiotic crossing over in *Drosophila*. *Curr. Biol.* 13:276–85
88. Bickel SE, Wyman DW, Orr-Weaver TL. 1997. Mutational analysis of the *Drosophila* sister-chromatid cohesion protein ORD and its role in the maintenance of centromeric cohesion. *Genetics* 146:1319–31
89. Webber HA, Howard L, Bickel SE. 2004. The cohesion protein ORD is required for homologue bias during meiotic recombination. *J. Cell Biol.* 164:819–29
90. Bickel SE, Wyman DW, Miyazaki WY, Moore DP, Orr-Weaver TL. 1996. Identification of ORD, a *Drosophila* protein essential for sister chromatid cohesion. *EMBO J.* 15:1451–59
91. Anderson LK, Royer SM, Page SL, McKim KS, Lai A, et al. 2005. Juxtaposition of C(2)M and the transverse filament protein C(3)G within the central region of *Drosophila* synaptonemal complex. *Proc. Natl. Acad. Sci. USA* 102:4482–87
92. Dorsett D. 2009. Cohesin, gene expression and development: lessons from *Drosophila*. *Chromosome Res.* 17:185–200
93. Gause M, Misulovin Z, Bilyeu A, Dorsett D. 2010. Dosage-sensitive regulation of cohesin chromosome binding and dynamics by Nipped-B, Pds5, and Wapl. *Mol. Cell. Biol.* 30:4940–51
94. Gause M, Webber H, Misulovin Z, Haller G, Rollins R, et al. 2008. Functional links between *Drosophila* Nipped-B and cohesin in somatic and meiotic cells. *Chromosoma* 117:51–66
95. Sym M, Engebrecht JA, Roeder GS. 1993. ZIP1 is a synaptonemal complex protein required for meiotic chromosome synapsis. *Cell* 72:365–78
96. MacQueen AJ, Colaiácovo MP, McDonald K, Villeneuve AM. 2002. Synapsis-dependent and -independent mechanisms stabilize homolog pairing during meiotic prophase in *C. elegans*. *Genes Dev.* 16:2428–42
97. Colaiácovo MP, MacQueen AJ, Martinez-Perez E, McDonald K, Adamo A, et al. 2003. Synaptonemal complex assembly in *C. elegans* is dispensable for loading strand-exchange proteins but critical for proper completion of recombination. *Dev. Cell* 5:463–74
98. Hawley RS. 2011. Solving a meiotic LEGO puzzle: transverse filaments and the assembly of the synaptonemal complex in *Caenorhabditis elegans*. *Genetics* 189:405–9
99. Meuwissen RL, Offenberg HH, Dietrich AJ, van Iersel M, Heyting C. 1992. A coiled-coil related protein specific for synapsed regions of meiotic prophase chromosomes. *EMBO J.* 11:5091–100
100. Dobson MJ, Pearlman RE, Karaiskakis A, Spyropoulos B, Moens PB. 1994. Synaptonemal complex proteins: occurrence, epitope mapping and chromosome disjunction. *J. Cell Sci.* 107:2749–60
101. Liu J-G, Yuan L, Brundell E, Björkroth B, Danehold B, Höög C. 1996. Localization of the N-terminus of SCP1 to the central element of the synaptonemal complex and evidence for direct interactions between the N-termini of SCP1 molecules organized head-to-head. *Exp. Cell Res.* 226:11–19
102. Schmekel K, Meuwissen RLJ, Dietrich AJJ, Vink ACG, van Marle J, et al. 1996. Organization of SCP1 protein molecules within synaptonemal complexes of the rat. *Exp. Cell Res.* 226:20–30
103. Dong H, Roeder GS. 2000. Organization of the yeast Zip1 protein within the central region of the synaptonemal complex. *J. Cell Biol.* 148:417–26
104. Jeffress JK, Page SL, Royer SK, Belden ED, Blumenstiel JP, et al. 2007. The formation of the central element of the synaptonemal complex may occur by multiple mechanisms: the roles of the N- and C-terminal domains of the *Drosophila* C(3)G protein in mediating synapsis and recombination. *Genetics* 177:2445–56
105. Tung KS, Roeder GS. 1998. Meiotic chromosome morphology and behavior in zip1 mutants of *Saccharomyces cerevisiae*. *Genetics* 149:817–32
106. Schmekel K, Danehold B. 1995. The central region of the synaptonemal complex revealed in three dimensions. *Trends Cell Biol.* 5:239–42
107. Costa Y, Speed R, Ollinger R, Alsheimer M, Semple CA, et al. 2005. Two novel proteins recruited by synaptonemal complex protein 1 (SYCP1) are at the centre of meiosis. *J. Cell Sci.* 118:2755–62

108. Hamer G, Gell K, Kouznetsova A, Novak I, Benavente R, Hoog C. 2006. Characterization of a novel meiosis-specific protein within the central element of the synaptonemal complex. *J. Cell Sci.* 119:4025–32
109. Schramm S, Fraune J, Naumann R, Hernandez-Hernandez A, Höög C, et al. 2011. A novel mouse synaptonemal complex protein is essential for loading of central element proteins, recombination, and fertility. *PLoS Genet.* 7:e1002088
110. Bolcun-Filas E, Costa Y, Speed R, Taggart M, Benavente R, et al. 2007. SYCE2 is required for synaptonemal complex assembly, double strand break repair, and homologous recombination. *J. Cell Biol.* 176:741–47
111. Hamer G, Wang H, Bolcun-Filas E, Cooke HJ, Benavente R, Höög C. 2008. Progression of meiotic recombination requires structural maturation of the central element of the synaptonemal complex. *J. Cell Sci.* 121:2445–51
112. Bolcun-Filas E, Speed R, Taggart M, Grey C, de Massy B, et al. 2009. Mutation of the mouse *Syce1* gene disrupts synapsis and suggests a link between synaptonemal complex structural components and DNA repair. *PLoS Genet.* 5:e1000393
113. Hall JC. 1972. Chromosome segregation influenced by two alleles of the meiotic mutant *c(3)G* in *Drosophila melanogaster*. *Genetics* 71:367–400
114. Carlson PS. 1972. The effects of inversions and the *c(3)G* mutation on intragenic recombination in *Drosophila*. *Genet. Res.* 19:129–32
115. de Boer E, Heyting C. 2006. The diverse roles of transverse filaments of synaptonemal complexes in meiosis. *Chromosoma* 115:220–34
116. Miyazaki WY, Orr-Weaver TL. 1992. Sister-chromatid misbehavior in *Drosophila ord* mutants. *Genetics* 132:1047–61
117. Mason JM. 1976. Orientation disruptor (*ord*): a recombination-defective and disjunction-defective meiotic mutant in *Drosophila melanogaster*. *Genetics* 84:545–72
118. Kim KP, Weiner BM, Zhang L, Jordan A, Dekker J, Kleckner N. 2010. Sister cohesion and structural axis components mediate homolog bias of meiotic recombination. *Cell* 143:924–37
119. Storlazzi A, Xu L, Schwacha A, Kleckner N. 1996. Synaptonemal complex (SC) component Zip1 plays a role in meiotic recombination independent of SC polymerization along the chromosomes. *Proc. Natl. Acad. Sci. USA* 93:9043–48
120. Tsubouchi T, Roeder GS. 2005. A synaptonemal complex protein promotes homology-independent centromere coupling. *Science* 308:870–73
121. Falk JE, Chan AC-h, Hoffmann E, Hochwagen A. 2010. A Mec1- and PP4-dependent checkpoint couples centromere pairing to meiotic recombination. *Dev. Cell* 19:599–611
122. Gladstone MN, Obeso D, Chuong H, Dawson DS. 2009. The synaptonemal complex protein Zip1 promotes bi-orientation of centromeres at meiosis I. *PLoS Genet.* 5:e1000771
123. Newnham L, Jordan P, Rockmill B, Roeder GS, Hoffmann E. 2010. The synaptonemal complex protein, Zip1, promotes the segregation of nonexchange chromosomes at meiosis I. *Proc. Natl. Acad. Sci. USA* 107:781–85
124. Bardhan A, Chuong H, Dawson DS. 2010. Meiotic cohesin promotes pairing of nonhomologous centromeres in early meiotic prophase. *Mol. Biol. Cell* 21:1799–809
125. Cooper KW. 1948. A new theory of secondary nondisjunction in female *Drosophila melanogaster*. *Proc. Natl. Acad. Sci. USA* 34:179–87
126. Xiang Y, Hawley RS. 2006. The mechanism of secondary nondisjunction in *Drosophila melanogaster* females. *Genetics* 174:67–78
127. Karpen GH, Allshire RC. 1997. The case for epigenetic effects on centromere identity and function. *Trends Genet.* 13:489–96
128. Sym M, Roeder GS. 1994. Crossover interference is abolished in the absence of a synaptonemal complex protein. *Cell* 79:283–92
129. Tsubouchi T, Zhao H, Roeder GS. 2006. The meiosis-specific Zip4 protein regulates crossover distribution by promoting synaptonemal complex formation together with Zip2. *Dev. Cell* 10:809–19

130. Hayashi M, Mlynarczyk-Evans S, Villeneuve AM. 2010. The synaptonemal complex shapes the crossover landscape through cooperative assembly, crossover promotion and crossover inhibition during *Caenorhabditis elegans* meiosis. *Genetics* 186:45–58

131. Shinohara M, Oh SD, Hunter N, Shinohara A. 2008. Crossover assurance and crossover interference are distinctly regulated by the ZMM proteins during yeast meiosis. *Nat. Genet.* 40:299–309

132. Fung JC, Rockmill B, Odell M, Roeder GS. 2004. Imposition of crossover interference through the nonrandom distribution of synapsis initiation complexes. *Cell* 116:795–802

133. de Boer E, Dietrich AJJ, Höög C, Stam P, Heyting C. 2007. Meiotic interference among MLH1 foci requires neither an intact axial element structure nor full synapsis. *J. Cell Sci.* 120:731–36

134. Osman K, Sanchez-Moran E, Higgins J, Jones G, Franklin F. 2006. Chromosome synapsis in *Arabidopsis*: Analysis of the transverse filament protein ZYP1 reveals novel functions for the synaptonemal complex. *Chromosoma* 115:212–19

134a. Tanneti NS, Landy K, Joyce EF, McKim KS. 2011. A pathway for synapsis initiation during zygotene in *Drosophila* oocytes. *Curr. Biol.* 21:1852–57

135. Keeney S. 2001. Mechanism and control of meiotic recombination initiation. *Curr. Top. Dev. Biol.* 52:1–53

136. Bonfils S, Rozalen AE, Smith GR, Moreno S, Martin-Castellanos C. 2011. Functional interactions of Rec24, the fission yeast ortholog of mouse Mei4, with the meiotic recombination-initiation complex. *J. Cell Sci.* 124:1328–38

137. McKim KS, Hayashi-Hagihara A. 1998. *mei-W68* in *Drosophila melanogaster* encodes a Spo11 homolog: evidence that the mechanism for initiating meiotic recombination is conserved. *Genes Dev.* 12:2932–42

138. Liu H, Jang JK, Kato N, McKim KS. 2002. *mei-P22* encodes a chromosome-associated protein required for the initiation of meiotic recombination in *Drosophila melanogaster*. *Genetics* 162:245–58

139. Sasanuma H, Hirota K, Fukuda T, Kakusho N, Kugou K, et al. 2008. Cdc7-dependent phosphorylation of Mer2 facilitates initiation of yeast meiotic recombination. *Genes Dev.* 22:398–410

140. Wan L, Niu H, Futcher B, Zhang C, Shokat KM, et al. 2008. Cdc28–Clb5 (CDK-S) and Cdc7–Dbf4 (DDK) collaborate to initiate meiotic recombination in yeast. *Genes Dev.* 22:386–97

141. Henderson KA, Kee K, Maleki S, Santini PA, Keeney S. 2006. Cyclin-dependent kinase directly regulates initiation of meiotic recombination. *Cell* 125:1321–32

142. Kumar R, Bourbon H-M, de Massy B. 2010. Functional conservation of Mei4 for meiotic DNA double-strand break formation from yeasts to mice. *Genes Dev.* 24:1266–80

143. Baudat F, Buard J, Grey C, Fledel-Alon A, Ober C, et al. 2010. PRDM9 is a major determinant of meiotic recombination hotspots in humans and mice. *Science* 327:836–40

144. Myers S, Bowden R, Tumian A, Bontrop RE, Freeman C, et al. 2010. Drive against hotspot motifs in primates implicates the *PRDM9* gene in meiotic recombination. *Science* 327:876–79

145. Parvanov ED, Petkov PM, Paigen K. 2010. *Prdm9* controls activation of mammalian recombination hotspots. *Science* 327:835

146. Myers S, Freeman C, Auton A, Donnelly P, McVean G. 2008. A common sequence motif associated with recombination hot spots and genome instability in humans. *Nat. Genet.* 40:1124–29

147. Buard J, Barthes P, Grey C, de Massy B. 2009. Distinct histone modifications define initiation and repair of meiotic recombination in the mouse. *EMBO J.* 28:2616–24

148. Berg IL, Neumann R, Lam K-WG, Sarbajna S, Odenthal-Hesse L, et al. 2010. PRDM9 variation strongly influences recombination hot-spot activity and meiotic instability in humans. *Nat. Genet.* 42:859–63

149. Borde V, Robine N, Lin W, Bonfils S, Geli V, Nicolas A. 2009. Histone H3 lysine 4 trimethylation marks meiotic recombination initiation sites. *EMBO J.* 28:99–111

150. Jang JK, Sherizen DE, Bhagat R, Manheim EA, McKim KS. 2003. Relationship of DNA double-strand breaks to synapsis in *Drosophila*. *J. Cell Sci.* 116:3069–77

151. Klovstad M, Abdu U, Schüpbach T. 2008. *Drosophila brca2* is required for mitotic and meiotic DNA repair and efficient activation of the meiotic recombination checkpoint. *PLoS Genet.* 4:e31

152. McKim KS, Jang JK, Theurkauf WE, Hawley RS. 1993. Mechanical basis of meiotic metaphase arrest. *Nature* 362:364–66

153. Baker BS, Hall JC. 1976. Meiotic mutants: genetic control of meiotic recombination and chromosome segregation. In *The Genetics and Biology of* Drosophila, ed. M Ashburner, E Novitski, 1a:352–434. London: Academic
154. Staeva-Vieira E, Yoo S, Lehmann R. 2003. An essential role of DmRad51/SpnA in DNA repair and meiotic checkpoint control. *EMBO J.* 22:5863–74
155. Ghabrial A, Ray RP, Schupbach T. 1998. *okra* and *spindle-B* encode components of the *RAD52* DNA repair pathway and affect meiosis and patterning in *Drosophila* oogenesis. *Genes Dev.* 12:2711–23
156. McCaffrey R, St Johnston D, González-Reyes A. 2006. *Drosophila mus301/spindle-C* encodes a helicase with an essential role in double-strand DNA break repair and meiotic progression. *Genetics* 174:1273–85
157. Abdu U, Gonzalez-Reyes A, Ghabrial A, Schupbach T. 2003. The *Drosophila spn-D* gene encodes a RAD51C-like protein that is required exclusively during meiosis. *Genetics* 165:197–204
158. Peretz G, Arie LG, Bakhrat A, Abdu U. 2009. The *Drosophila hus1* gene is required for homologous recombination repair during meiosis. *Mech. Dev.* 126:677–86
159. Yoo S, McKee BD. 2005. Functional analysis of the *Drosophila Rad51* gene (*spn-A*) in repair of DNA damage and meiotic chromosome segregation. *DNA Repair* 4:231–42
160. Liu H, Jang JK, Graham J, Nycz K, McKim KS. 2000. Two genes required for meiotic recombination in *Drosophila* are expressed from a dicistronic message. *Genetics* 154:1735–46
161. McKim KS, Dahmus JB, Hawley RS. 1996. Cloning of the *Drosophila melanogaster* meiotic recombination gene *mei-218*: a genetic and molecular analysis of interval 15E. *Genetics* 144:215–28
162. Blanton HL, Radford SJ, McMahan S, Kearney HM, Ibrahim JG, Sekelsky J. 2005. REC, *Drosophila* MCM8, drives formation of meiotic crossovers. *PLoS Genet.* 1:e40
163. Matsubayashi H, Yamamoto M-T. 2003. REC, a new member of the MCM-related protein family, is required for meiotic recombination in *Drosophila*. *Genes Genet. Syst.* 78:363–71
164. Lake CM, Teeter K, Page SL, Nielsen R, Hawley RS. 2007. A genetic analysis of the *Drosophila mcm5* gene defines a domain specifically required for meiotic recombination. *Genetics* 176:2151–63
165. Manheim EA, Jang JK, Dominic D, McKim KS. 2002. Cytoplasmic localization and evolutionary conservation of MEI-218, a protein required for meiotic crossing-over in *Drosophila*. *Mol. Biol. Cell* 13:84–95
166. Carpenter AT. 1982. Mismatch repair, gene conversion, and crossing-over in two recombination-defective mutants of *Drosophila melanogaster*. *Proc. Natl. Acad. Sci. USA* 79:5961–65
167. Carpenter AT. 1989. Are there morphologically abnormal early recombination nodules in the *Drosophila melanogaster* meiotic mutant *mei-218*? *Genome* 31:74–80
168. Curtis D, Bender W. 1991. Gene conversion in *Drosophila* and the effects of meiotic mutants *mei-9* and *mei-218*. *Genetics* 127:739–46
169. Carpenter AT, Sandler L. 1974. On recombination-defective meiotic mutants in *Drosophila melanogaster*. *Genetics* 76:453–75
170. Sekelsky JJ, McKim KS, Chin GM, Hawley RS. 1995. The *Drosophila* meiotic recombination gene *mei-9* encodes a homologue of the yeast excision repair protein Rad1. *Genetics* 141:619–27
171. Joyce EF, Tanneti SN, McKim KS. 2009. *Drosophila* hold'em is required for a subset of meiotic crossovers and interacts with the DNA repair endonuclease complex subunits MEI-9 and ERCC1. *Genetics* 181:335–40
172. Sekelsky JJ, Hollis KJ, Eimerl AI, Burtis KC, Hawley RS. 2000. Nucleotide excision repair endonuclease genes in *Drosophila melanogaster*. *Mutat. Res.* 459:219–28
173. Yildiz O, Majumder S, Kramer B, Sekelsky JJ. 2002. *Drosophila* MUS312 interacts with the nucleotide excision repair endonuclease MEI-9 to generate meiotic crossovers. *Mol. Cell* 10:1503–9
174. Yildiz O, Kearney H, Kramer BC, Sekelsky JJ. 2004. Mutational analysis of the *Drosophila* DNA repair and recombination gene *mei-9*. *Genetics* 167:263–73
175. Andersen SL, Bergstralh DT, Kohl KP, LaRocque JR, Moore CB, Sekelsky J. 2009. *Drosophila* MUS312 and the vertebrate ortholog BTBD12 interact with DNA structure-specific endonucleases in DNA repair and recombination. *Mol. Cell* 35:128–35
176. Holloway JK, Mohan S, Balmus G, Sun X, Modzelewski A, et al. 2011. Mammalian BTBD12 (SLX4) protects against genomic instability during mammalian spermatogenesis. *PLoS Genet.* 7:e1002094

177. Ghabrial A, Schupbach T. 1999. Activation of a meiotic checkpoint regulates translation of Gurken during *Drosophila* oogenesis. *Nat. Cell Biol.* 1:354–57
178. Joyce EF, McKim KS. 2009. *Drosophila* PCH2 is required for a pachytene checkpoint that monitors double-strand-break-independent events leading to meiotic crossover formation. *Genetics* 181:39–51
179. Abdu U, Brodsky M, Schüpbach T. 2002. Activation of a meiotic checkpoint during *Drosophila* oogenesis regulates the translation of Gurken through Chk2/Mnk. *Curr. Biol.* 12:1645–51
180. Lancaster OM, Breuer M, Cullen CF, Ito T, Ohkura H. 2010. The meiotic recombination checkpoint suppresses NHK-1 kinase to prevent reorganisation of the oocyte nucleus in *Drosophila*. *PLoS Genet.* 6:e1001179
181. Roeder GS, Bailis JM. 2000. The pachytene checkpoint. *Trends Genet.* 16:395–403
182. Hochwagen A, Amon A. 2006. Checking your breaks: surveillance mechanisms of meiotic recombination. *Curr. Biol.* 16:R217–28
183. San-Segundo PA, Roeder GS. 1999. Pch2 links chromatin silencing to meiotic checkpoint control. *Cell* 97:313–24
184. Bhalla N, Dernburg AF. 2005. A conserved checkpoint monitors meiotic chromosome synapsis in *Caenorhabditis elegans*. *Science* 310:1683–86
185. Li X, Schimenti JC. 2007. Mouse pachytene checkpoint 2 (*Trip13*) is required for completing meiotic recombination but not synapsis. *PLoS Genet.* 3:e130
186. Roig I, Dowdle JA, Toth A, de Rooij DG, Jasin M, Keeney S. 2010. Mouse TRIP13/PCH2 is required for recombination and normal higher-order chromosome structure during meiosis. *PLoS Genet.* 6:e1001062
187. Börner GV, Barot A, Kleckner N. 2008. Yeast Pch2 promotes domainal axis organization, timely recombination progression, and arrest of defective recombinosomes during meiosis. *Proc. Natl. Acad. Sci. USA* 105:3327–32
188. Hochwagen A, Tham W-H, Brar GA, Amon A. 2005. The FK506 binding protein Fpr3 counteracts protein phosphatase 1 to maintain meiotic recombination checkpoint activity. *Cell* 122:861–73
189. Wu H-Y, Burgess SM. 2006. Two distinct surveillance mechanisms monitor meiotic chromosome metabolism in budding yeast. *Curr. Biol.* 16:2473–79
190. Joshi N, Barot A, Jamison C, Börner GV. 2009. Pch2 links chromosome axis remodeling at future crossover sites and crossover distribution during yeast meiosis. *PLoS Genet.* 5:e1000557
191. Zanders S, Alani E. 2009. The *pch2Δ* mutation in baker's yeast alters meiotic crossover levels and confers a defect in crossover interference. *PLoS Genet.* 5:e1000571
192. Wojtasz L, Daniel K, Roig I, Bolcun-Filas E, Xu H, et al. 2009. Mouse HORMAD1 and HORMAD2, two conserved meiotic chromosomal proteins, are depleted from synapsed chromosome axes with the help of TRIP13 AAA-ATPase. *PLoS Genet.* 5:e1000702
193. Joyce EF, McKim KS. 2010. Chromosome axis defects induce a checkpoint-mediated delay and interchromosomal effect on crossing over during *Drosophila* meiosis. *PLoS Genet.* 6:e1001059
194. Riemer D, Stuurman N, Berrios M, Hunter C, Fisher P, Weber K. 1995. Expression of *Drosophila* lamin C is developmentally regulated: analogies with vertebrate A-type lamins. *J. Cell Sci.* 108:3189–98
195. Klattenhoff C, Xi H, Li C, Lee S, Xu J, et al. 2009. The *Drosophila* HP1 homolog Rhino is required for transposon silencing and piRNA production by dual-strand clusters. *Cell* 138:1137–49

The Control of Male Fertility by Spermatozoan Ion Channels

Polina V. Lishko,[1] Yuriy Kirichok,[2] Dejian Ren,[3] Betsy Navarro,[4,5] Jean-Ju Chung,[4,5] and David E. Clapham[4,5,*]

[1]Department of Molecular and Cell Biology, University of California, Berkeley, California 94720

[2]Department of Physiology, University of California, San Francisco, California 94158

[3]Department of Biology, University of Pennsylvania, Philadelphia, Pennsylvania 19104

[4]Howard Hughes Medical Institute, Department of Cardiology, Manton Center for Orphan Disease, Children's Hospital, Boston, Massachusetts 02115

[5]Department of Neurobiology, Harvard Medical School, Boston, Massachusetts 02115; email: dclapham@enders.tch.harvard.edu

Keywords

sperm ion channels, intracellular pH, capacitation, patch clamp, hyperactivation, chemotaxis, male fertility, CatSper, Hv1, KSper, acrosome reaction

Abstract

Ion channels control the sperm ability to fertilize the egg by regulating sperm maturation in the female reproductive tract and by triggering key sperm physiological responses required for successful fertilization such as hyperactivated motility, chemotaxis, and the acrosome reaction. CatSper, a pH-regulated, calcium-selective ion channel, and KSper (Slo3) are core regulators of sperm tail calcium entry and sperm hyperactivated motility. Many other channels had been proposed as regulating sperm activity without direct measurements. With the development of the sperm patch-clamp technique, CatSper and KSper have been confirmed as the primary spermatozoan ion channels. In addition, the voltage-gated proton channel Hv1 has been identified in human sperm tail, and the P2X2 ion channel has been identified in the midpiece of mouse sperm. Mutations and deletions in sperm-specific ion channels affect male fertility in both mice and humans without affecting other physiological functions. The uniqueness of sperm ion channels makes them ideal pharmaceutical targets for contraception. In this review we discuss how ion channels regulate sperm physiology.

INTRODUCTION

In sexual reproduction two haploid gametes (spermatozoon and egg) fuse and restore the original number of chromosomes, resulting in the zygote and the development of a new organism. In many aquatic organisms, mature sperm cells sense the egg and swim toward it in an almost infinite unregulated environment. In contrast, spermatozoa of terrestrial animals are delivered directly into the confined, strictly regulated environment of the female reproductive tract, where they must undergo final maturation before fertilization can occur. Thus, the female reproductive tract has the capacity to select and orient sperm, making it an active recipient of male gametes.

Ion channels control sperm membrane potential, cytoplasmic Ca^{2+}, and intracellular pH (pH_i), which in turn regulate motility, the acrosome reaction, and other diverse physiological processes essential for successful fertilization (1–3). The dramatic improvement in our understanding of the sperm ion channels was triggered by the discovery of CatSper (*cat*ionic channel of *sper*m), a novel and complex ion channel that mediates Ca^{2+} entry in sperm flagellum and is required for sperm hyperactivation and male fertility (4). The interest in functional characterization of the CatSper channel and the mechanisms of its regulation led to the first successful application of the whole-cell patch-clamp technique to mice and then to human spermatozoa (5, 6). The scope of this review is to discuss the role of plasma membrane ion channels in normal sperm physiology, to provide an update of recent important developments in sperm ion channel research, and to discuss sperm channelopathies that cause male infertility.

pH_i: intracellular pH

Acrosome: a cap-like vesicle covering the anterior portion of the head of a spermatozoon. The acrosome contains hydrolytic enzymes for penetrating through the protective vestments of the oocyte

Acrosome reaction: exocytosis of the acrosomal vesicle upon the spermatozoon's contact with egg's protective vestments. Hydrolytic enzymes released by the acrosome digest the zona pellucida and help spermatozoa to reach the egg's surface

Cationic channel of sperm (CatSper): a pH-regulated, calcium-selective ion channel required for sperm hyperactivation

Hyperactivation: a whip-like, high-amplitude asymmetrical beat of the sperm flagellum that helps spermatozoa overcome the egg's protective vestments. Hyperactivation is different from the low-amplitude symmetrical beat observed in normal motility

SPERM MORPHOLOGY

Spermatozoa are terminally differentiated motile cells with a clear cell polarity determined by the two main structural elements: the head, which contains tightly packed DNA, and the motile flagellum, which delivers the genetic material of the sperm head into the egg (**Figure 1a**). Structurally similar flagella are present in all spermatozoa across the animal and plant kingdoms (**Figure 1b**). The sperm head consists of the nucleus, the tiny residual nuclear envelope vestiges, and the acrosome (a Golgi-derived vesicle that helps spermatozoa to penetrate egg's protective vestments). The mammalian flagellum has a central axoneme surrounded by specialized structural components and is composed of three parts: the midpiece, which contains mitochondria wrapped in a spiral pattern around the axoneme; the principal piece, which is primarily responsible for motility; and the endpiece, which contains few structural elements (**Figure 1a,c**). The midpiece and the principal piece are separated by a ringed septin structure termed the annulus, which prevents diffusion of plasma membrane proteins between these two flagellar domains (**Figure 1d**) (7).

The sperm plasma membrane is tightly attached to the underlying cellular structures along the whole sperm body to provide stiffness. Many membrane proteins of the sperm principal piece appear to be anchored to the underlying fibrous sheath to ensure their strict compartmentalization. The fibrous sheath functions as a scaffold for proteins in signaling pathways that regulate sperm maturation, motility, capacitation, hyperactivation, and/or the acrosome reaction. Interestingly, the fibrous sheath also anchors enzymes of the sperm-specific glycolytic pathway that provide ATP for motility (8). The cytoplasmic droplet, the remnant of the precursor cell's cytoplasm, is the only region of the plasma membrane loosely attached to the intracellular structures (**Figure 1d**). The cytoplasmic droplet is likely to serve as a reservoir for adaptation to osmotic changes occurring at ejaculation (9) but in many species is shed from spermatozoa after ejaculation.

The sperm axoneme is composed of microtubules: Nine outer doublet microtubules surround a central pair of singlet microtubules (a 9+2 arrangement) and the associated proteins such as the

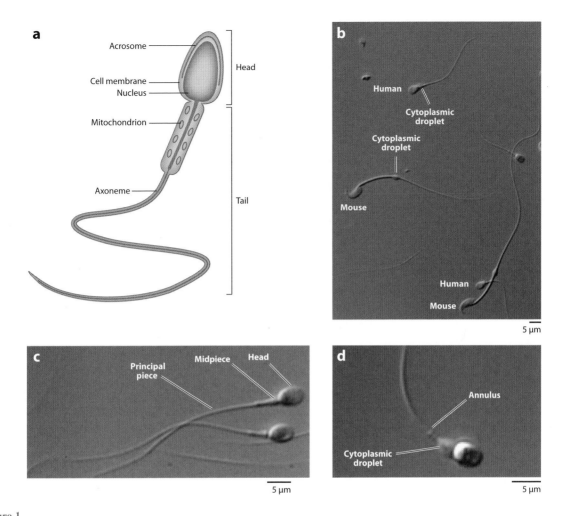

Figure 1

Mammalian spermatozoa. (*a*) Schematic representation of mammalian sperm. (*b*) Comparison between human and mouse spermatozoa. (*c*) Human spermatozoa with head, midpiece, and principal piece as indicated. (*d*) Cytoplasmic droplet and annulus are labeled.

molecular motor dynein. Axonemal bending is produced by sliding between pairs of outer doublet microtubules (10), and this sliding is powered by ATP hydrolysis by dynein's heavy chains. This active sliding of microtubules is a linear phenomenon, but the bending and propagation of the wave of motion down the flagellum are not well understood. One theory for the generation of flagellar motion is the geometric clutch hypothesis, whereby dynein engagement alternates between sides of the axoneme as the flagellum bends (11). Sperm axoneme bending is sensitive to intracellular alkalinization (12) and intracellular Ca^{2+} ($[Ca^{2+}]_i$) (13) so that increasing pH_i above 7 stimulates dynein activity and promotes flagellar beating, whereas increasing intracellular Ca^{2+} enhances asymmetrical flagellum bending (14, 15). As spermatozoa travel through the environment of changing pH, osmolarity and sense extracellular cues such as progesterone and chemoattractants, sperm ion channels, and transporters regulate ion concentrations within the sperm's cytoplasm to control motility and to trigger physiological responses such as hyperactivation of motility (hyperactivation) and the acrosome reaction.

Fibrous sheath: a unique cytoskeletal structure of two longitudinal columns, connected by closely arrayed semicircular ribs. Fibrous sheath surrounds the outer dense fibers and the axoneme. The fibrous sheath influences the degree of flexibility, the plane of flagellar motion, and the shape of the flagellar beat

ACTIVATION OF MOTILITY, CAPACITATION, AND HYPERACTIVATION

Capacitation: spermatozoa's acquisition of fertilizing capacity upon exposure to the fluids of the female reproductive tract for several hours. Capacitation results in sperm hyperactivation and the ability to undergo the acrosome reaction

Epididymis: a part of the male reproductive tract in which spermatozoa continue maturation and are stored

cAMP: cyclic adenosine monophosphate

pH$_o$: extracellular pH

Cumulus oophorus: the mass of cells, derived from granulosa cells of the Graafian follicle, that surrounds the oocyte upon its release during ovulation. To fertilize the oocyte, spermatozoon must first penetrate the cumulus oophorus

Zona pellucida (ZP): the dense extracellular matrix surrounding the developing oocyte. The ZP prevents fertilization by multiple spermatozoa (polyspermy), prevents premature implantation, and protects the embryo during its first week of development

Mammalian spermatozoa from all portions of the epididymis have an acidic pH$_i$ (~6.8) and are essentially quiescent (16). When spermatozoa are mixed with seminal plasma (pH > 7.0) upon ejaculation, the sperm cytoplasm is alkalinized (17), and sperm become motile. Despite being motile, freshly ejaculated mammalian spermatozoa are unable or poorly able to fertilize the oocyte. To become competent to fertilize the egg, they must undergo capacitation: a phenomenon reported in 1951 by Austin (18) and Chang (19).

Capacitation results in the removal of noncovalently attached glycoproteins acquired in the epididymis, in the removal of adherent seminal plasma proteins, and in the depletion of the membrane cholesterol and other sterols (20, 21). Moreover, during capacitation, intracellular Ca^{2+}, pH, and cyclic adenosine monophosphate (cAMP) increase, and sperm membrane proteins are phosphorylated on tyrosines (22–24). The first steps of capacitation may begin anywhere extracellular pH (pH$_o$) is elevated, such as at the cervical mucus, but the ampulla of Fallopian tubes is critical for the completion of the process. Motility is hyperactivated during capacitation. Hyperactivation is defined by an increase in the angle of the flagellar bend, which results in more asymmetrical (whip-like) movements and more powerful swimming force (14, 15, 25). The second major change is that sperm acquire the ability to undergo the acrosome reaction (18, 19, 26). Capacitation increases the fluidity of the plasma membrane and sensitizes sperm to fertilization cues.

When capacitated spermatazoa encounter the cumulus oophorus (27) or bind the glycoproteins of the egg's zona pellucida (ZP), there are additional steep increases in sperm pH$_i$ and Ca^{2+}, resulting in the acrosome reaction (28–32). The sperm plasma membrane contains specific ion channels and transporters that initiate changes in these ions in the sperm cytoplasm (4–6, 33–39). During the acrosome reaction, hydrolytic enzymes are expelled from the sperm acrosome to facilitate penetration through the egg's protective vestments (29).

As in all cells, sperm Na$^+$/K$^+$-ATPases establish the high K$^+$ and low Na$^+$ concentration of the sperm cytoplasm (38). As in serum, the extracellular fluids surrounding sperm cells contain 1–2 mM [Ca^{2+}]. Cells maintain a remarkable 20,000-fold gradient from outside the cell to inside the cell, with resting cytosolic Ca^{2+} concentrations ranging from ~50–100 nM [Ca^{2+}]. In somatic cells, these gradients are maintained by the export of cytoplasmic Ca^{2+} across the plasma membrane and by the import of Ca^{2+} into the endoplasmic reticulum and mitochondria. In sperm, these gradients are maintained primarily by a plasma membrane Ca^{2+}-ATPase pump (PMCA4) that extrudes Ca^{2+} (40–42). Male mice deficient in PMCA4 have impaired sperm motility and are infertile (40, 41). Unlike other cells, spermatozoa do not contain significant amounts of endoplasmic reticulum. How much Ca^{2+} is stored in sperm mitochondria remains unexplored.

The proton gradient across the sperm plasma membrane is the inverse of the Ca^{2+} gradient. Serum [H$^+$] (40 nM, pH ~ 7.4) is fourfold lower than the intracellular [H$^+$] of ejaculated spermatozoa (160 nM, pH ~ 6.8) (17, 43, 44). In epididymis, the extracellular fluid (pH 5.5 to 6.8; [H$^+$] from 3160 to 160 nM) is even more acidic (16). Epididymal sperm pH$_i$ drops below 6.0 due to the activity of different exchangers, including Na$^+$/H$^+$ and bicarbonate exchangers (45–48). Thus, there is always a concentration gradient in protons between cytoplasm and extracellular fluid. Low epididymal pH (and thus low pH$_i$) appears to be a major factor in rendering spermatozoa quiescent before ejaculation by inhibiting axonemal dynein activity (16, 17, 49). Also, the high viscosity of the cauda epididymal fluid (50, 51) and proteins such as semenogelin (52) inhibits sperm motility. Upon ejaculation, sperm cells are mixed with seminal plasma of much higher pH (~7.4; [H$^+$] = 40 nM), and as sperm pH$_i$ rises to ~6.5 ([H$^+$] = 316 nM), sperm become motile for the first time (1, 44, 47). Lactobacilli and other vaginal flora acidify the vagina (pH ~ 4; [H$^+$] = 100 μM); seminal plasma transiently increases female vaginal pH from 4.3 to 7.2 after intercourse (53), alkalinizing the environment and thus enabling spermatozoa to begin swimming. During subsequent transit through the female reproductive tract, pH$_i$ increases further but still lags behind pH$_o$. Interestingly, at the peak of fertility in the middle of the menstrual cycle, cervical mucus becomes less viscous, and its pH can reach 9.0, making it less of a barrier to sperm (54). The pH of follicular fluid varies between 7 and 8, depending on the species and the phase of the menstrual cycle (55).

sAC: soluble adenylyl cyclase

PKA: protein kinase A

Upon ejaculation, sperm intracellular cAMP is elevated due to HCO_3^- activation of sperm soluble adenylyl cyclase (sAC) in a pH-independent manner (56). HCO_3^- concentration is higher in the seminal plasma/female reproductive tract than in the epididymal fluid (57), and HCO_3^- transporters deliver HCO_3^- into sperm cells. Moreover, the female oviduct is enriched in CO_2, which is converted into HCO_3^- by sperm extracellular glycosyl phosphatidylinositol–anchored carbonic anhydrase IV (58). Intracellular cAMP induces phosphorylation of axonemal dynein by protein kinase A (PKA) (59–61) to increase flagellar beating and sperm motility (62). sAC is also activated by Ca^{2+} (63, 64), and extracellular Ca^{2+} is required for the sAC-dependent increase in the frequency of flagellar beat triggered by HCO_3^- (65). Not surprisingly, male sAC and PKA knockout mice have impaired sperm motility and are infertile (66–68). PKA and sAC do not seem to be required for initiation of sperm motility but rather increase the frequency of sperm tail beating and improve progressive motility (61, 68, 69).

As mentioned above, the high proton and low Ca^{2+} concentrations in the sperm cytoplasm suppress sperm motility. Activation of spermatazoa requires alkalinization of sperm cytoplasm by the extrusion of protons and the elevation of intracellular Ca^{2+} concentration $[Ca^{2+}]_i$. However, because transporters pump ions much more slowly than do channels, changes in intracellular ion concentrations are relatively slow compared with the rapid changes elicited by ion channels. As such, ion channels are primarily responsible for rapid signaling events (70). A more rapid change in sperm motility is achieved by fast diffusion of K^+, protons, and Ca^{2+} down their K^+, H^+, and Ca^{2+} concentration gradients across the sperm plasma membrane through selective ion channels. The opening of such channels is controlled by specific cues in the female reproductive tract that regulate the activity of the sperm cells (**Figure 2**). This regulation is both spatial and temporal in accordance with female anatomy and the phase of the menstrual cycle. Known cues for human spermatozoa are H^+ concentration, progesterone, and anandamide released by the cumulus oophorus; glycoproteins of the ZP; and proteins of the oviductal fluid such as serum albumin (31, 32, 71–75). However, there are probably more yet-to-be-discovered factors in the female reproductive tract that directly or indirectly control activity of sperm ion channels that synchronize arrival of the egg and the sperm at the fertilization site.

SPERM ION CHANNELS

Calcium Channels

Ca^{2+} is critical to the initiation of cellular motion of all kinds (76). In spermatozoa, however, normal swimming behavior does not require the elevation of Ca^{2+}, a fact that at first surprised physiologists who focus on muscle and nerve. Sperm can swim over a range of $[Ca^{2+}]_i$, even in the absence of a plasma membrane, because they are essentially ciliary dynein ATPase motors. Just as in muscle and nerve, changing Ca^{2+} triggers changes in the behavior of motor proteins. Nonetheless, the elevation of intracellular Ca^{2+} is essential for changes in flagellar function that are

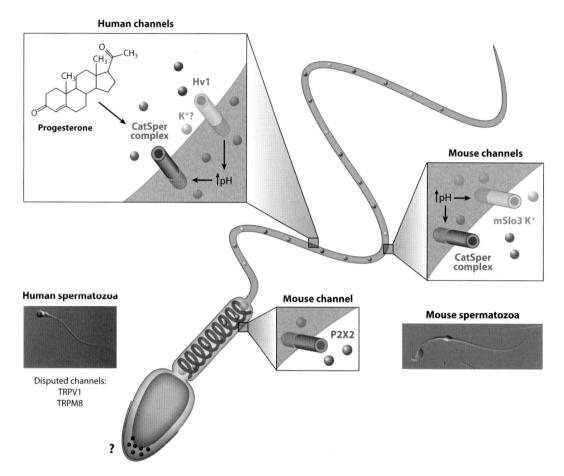

Figure 2

Sperm ion channel localization and function. Flagellar beating is regulated by at least three ion channels: alkaline-sensitive CatSper (Ca^{2+} entry), pH-regulated Slo3 (K^+ exit), and Hv1 (H^+ exit; human sperm only). The upper half of the figure depicts ion channels and their regulation as detected in human sperm (the regulation of the CatSper complex by progesterone and Hv1), whereas the lower half of the figure shows ion channels found in mouse spermatozoa (CatSper complex, Slo3, and P2X2).

manifested by capacitation, chemotaxis, and hyperactivated motility. Moreover, Ca^{2+} is required for initiation of the acrosome reaction.

Increases in $[Ca^{2+}]_i$ regulate the sperm's flagellar waveform and promote its asymmetrical bending (77). Before 2001, the channel responsible for sperm Ca^{2+} elevation was believed to be a voltage-gated Ca^{2+} channel (Ca_v; VGCC) and was perceived as the principal Ca^{2+} conductance of sperm (78–80). This notion was supported by electrophysiological identification of Ca_v channels in testicular spermatocytes (immature spermatogenic cells) using the patch-clamp technique (81–83) and by observation of a putative voltage-gated Ca^{2+} influx into mature sperm cells in response to the application of a high-K^+/high-pH extracellular medium (84). Yet, male mice deficient in $Ca_v2.2$, $Ca_v2.3$, and $Ca_v3.1$ were fertile, indicating that these VGCC channels were not essential for sperm physiology or functioned redundantly (85–87). Knockouts of $Ca_v1.2$ and $Ca_v2.1$ were lethal either at embryonic stages or soon after birth, thus precluding assessment of Ca^{2+} elevation in sperm (88, 89).

Chemotaxis: directional migration of the spermatozoa toward higher concentrations of chemoattractant released by the egg or cumulus oophorus

VGCC: voltage-gated calcium channel

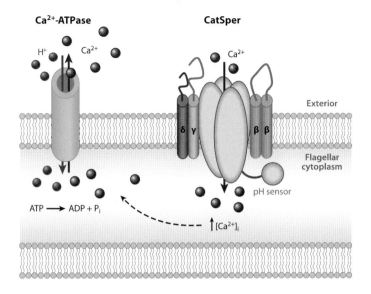

Figure 3

Regulation of flagellar Ca^{2+}. Ca^{2+} enters the sperm flagellum via the alkaline-activated CatSper channel and is extruded from the flagellum by a plasma membrane Ca^{2+}-ATPase (42). Ca^{2+}-ATPase pumps hydrolyze ATP to export a cytoplasmic Ca^{2+} ion and to import extracellular protons. The resulting acidification of flagellar cytoplasm must be prevented by proton extrusion via channels or transporters.

In 2001, the first member of a completely new family of Ca^{2+}-selective ion channel subunits was discovered. Termed CatSper1, it was found to be only in sperm cells and to be required for male fertility (4). Since then, seven CatSper subunits composing the heteromeric CatSper channel have been identified, and at least five of them—CatSper1–4 and CatSperδ—have been shown to be indispensible for proper channel formation and function (**Figure 3**) (**Table 1**) (4, 90–96). CatSper's pore is formed by four α subunits, the products of four distinct genes: *Catsper1*, *Catsper2*, *Catsper3*, and *Catsper4*. The channel contains three auxiliary subunits—CatSperβ, CatSperγ, and CatSperδ—of unknown stoichiometry (32, 93, 95, 96). All CatSper subunits are sperm-specific proteins and are located in the principal piece of the sperm flagellum. CatSperβ is predicted to have two transmembrane helices connected by a large extracellular loop. CatSperγ and CatSperδ have single predicted transmembrane helices and large extracellular domains. The function of the auxiliary subunits after assembly of the CatSper channel complex is not known.

Although interaction of CatSperβ and CatSperγ with the CatSper complex was clearly demonstrated biochemically (93, 95), whether they are required for the functional CatSper channel assembly is not clear. In contrast, CatSperδ not only interacts with the CatSper complex but is essential for functional CatSper (96). Like mice lacking any of the four CatSper α subunits (94), *CatSperδ*$^{-/-}$ mice have no measurable CatSper current ($I_{CatSper}$) and have the identical phenotype of male infertility due to loss of hyperactivated motility (96). Interestingly, CatSperβ, CatSperγ, CatSperδ, CatSper2, CatSper3, and CatSper4 are all undetectable on CatSper1 knockout sperm plasma membranes (92, 93, 95, 96), suggesting that all CatSper subunits are required for proper channel assembly; the absence of a single subunit may lead to degradation of remaining CatSper proteins. Humans with mutations or deletions in *CatSper1* and *CatSper2* are infertile (97–100). We suspect that loss-of-function mutations in any of the seven known CatSper subunits result in male infertility.

$I_{CatSper}$: CatSper current

Table 1 Ion channels of mammalian spermatozoa

Channel name	Gene name and chromosomal location	Ion selectivity	Subunit(s)/composition	Localization in sperm	Role in sperm physiology, including sperm specificity	Endogenous regulators	Knockout phenotype
CatSper	*CatSper1* (11q13.1) *CatSper2* (15q15.3), (15q15.3-pseudogene) *CatSper3* (5q31.1) *CatSper4* (1p36.11) *CatSperβ* (14q32.11) *CatSperγ* (19p13.2) *CatSperδ* (19p13.3); locations shown are for human CatSper	Ca^{2+}	Seven subunits total: heteromeric assembly of CatSper1–4 (pore subunits), CatSperβ, -γ, -δ (auxiliary subunits); stoichiometry is unknown	Principal piece	Ca^{2+} influx into flagella, hyperactivated motility, sperm specific	Progesterone (human sperm only), pH$_i$, egg coat proteins, albumin	Male sterility; sperm from CatSper$^{-/-}$ mice are unable to hyperactivate
KSper	Mouse: *Slo3* (8A3); human: *Slo3* (8p11.23)	K^+	Probable tetramer	Principal piece	Hyperpolarizes sperm membrane, regulates membrane potential; sperm specific	pH$_i$	Male sterility—increased bent hairpin morphology of spermatozoa
Hv1	Human: *Hvcn1* (12q24.11)	H^+	Probable homomeric dimer	Principal piece	Extrudes protons from flagellum, alkalinizes cytcplasm; primarily in phagocytic cells (e.g., leukocytes), alveolar type II cells, and spermatozoa (human)	pH$_i$, membrane voltage, removal of zinc, anandamide	I_{Hv1} is not present in murine epididymal sperm; Hvcn1$^{-/-}$ mice are fertile
I_{ATP}	Human: *P2RX2* (12q24.33); mouse: *P2rx2* (5)	Na^+, K^+, Ca^{2+}	Probable homotrimer	Midpiece	Widespread		P2rx2$^{-/-}$ mice are fertile

Direct electrophysiological characterization of CatSper1 was achieved with the whole-cell patch-clamp technique applied to mouse spermatozoa (6). Comparison of ion currents recorded from wild-type and *CatSper1*-deficient spermatozoa confirmed that CatSper1 is required for a highly selective Ca^{2+} current. Recording from fragments of mouse spermatozoa established that $I_{CatSper}$ originated from the principal piece of the sperm flagellum, corresponding to antibody localization of the CatSper1 protein. $I_{CatSper}$ is weakly voltage dependent (the slope factor of the voltage activation curve $k = 30$) in comparison to strongly voltage-activated channels ($k = 4$) (6). Interestingly, the S4 transmembrane helix of CatSper1 contains six positively charged lysine/arginine residues aligned in the same manner as in strongly voltage-sensitive channels. However, CatSper2 has only four such residues, and only two are preserved in CatSper3 and -4. Because the pore of this heteromeric channel is formed by all four CatSpers, the voltage sensitivity of the complete channel is weak (34, 94).

> **G-V curve:** plot of membrane conductance (G) versus voltage (V) that is used to show the percentage of activated ion channels in relation to membrane potential

The mouse CatSper channel is gated by changes in pH_i: The current is increased approximately sevenfold when pH_i is increased from 6.0 to 7.0 (6), corresponding to a (leftward) shift in the G-V curve of -70 mV. The abundance of histidines in the mouse CatSper1 N-terminal domain (51 His in the 250-residue N terminus) is one possible mechanism for this pH sensitivity (4, 34). Intracellular alkalinization by extracellular application of NH_4Cl not only causes $[Ca^{2+}]_i$ elevation by activating the CatSper channel but also triggers sperm hyperactivation (101).

Another hallmark of capacitation is reduction of sperm membrane cholesterol. Albumin, the main protein of the tubular fluid and an important component of in vitro capacitation media, also causes CatSper-dependent Ca^{2+} influx into mouse spermatozoa (74), perhaps affecting CatSper gating by modification of the lipid composition of the sperm plasma membrane. Finally, Ca^{2+} influx into mouse spermatozoa induced by the glycoproteins of the egg's ZP requires the CatSper channel (31), a property formerly assigned to the putative sperm Ca_v channels (78, 79). In this regard, the Ca_v current present in spermatocytes is not detected in mature spermatozoa—all channel-mediated Ca^{2+} entry in mature spermatozoa is via CatSper.

The subunits of CatSper channel are present in all mammalian genomes and some invertebrate species, such as the sea urchin and the freshwater mold *Allomyces macrogynus* (102, 103), but not in genomes of birds, amphibians, insects, and worms. The rapid disappearance of CatSper from some intermediate species over millions of years reflects the strong evolutionary pressure on gamete genes (102).

The CatSper channel is present in human sperm (5, 104) and, like mouse CatSper, is weakly voltage dependent but potently activated by intracellular alkalinization. The voltage dependency of human CatSper is slightly steeper ($k = 20$ compared with $k = 30$ in mice) than in mouse CatSper. Importantly, the $V_{1/2}$ (the voltage at which half of the channels are activated) of human CatSper is $+85$ mV versus $+11$ mV of mouse CatSper at the same pH_i ($pH_i = 7.5$) (6, 104), leading to the question of how human CatSper might be activated at such high membrane potentials.

Progesterone, a major steroid hormone released by the ovaries and the cumulus cells surrounding the egg, induces robust Ca^{2+} influx into human sperm cells (105, 106), triggers sperm hyperactivation, and initiates the acrosome reaction. These rapid effects are not via the nuclear progesterone receptor (107, 108). Progesterone exerts its effect on *Xenopus laevis* oocyte maturation and affects neural function without binding nuclear DNA or regulating gene expression. For example, *X. laevis* oocytes that are arrested in the G2 phase undergo maturation after the addition of extracellular progesterone. This phenomenon can occur even in enucleated cells. Also, progesterone can modulate γ-aminobutyric acid–mediated, glycine-mediated, and 5-hydroxytryptamine-mediated currents in neurons. However, the elusive progesterone receptor associated with human spermatozoa is probably the best-known example of a nongenomic progesterone receptor (107, 108).

The mystery of progesterone's short-term responses on sperm was recently solved. Progesterone activates human CatSper at low concentrations [$EC_{50} \approx 7.7$ nM (104)] by shifting the voltage dependency of the human CatSper channel into the physiological range (104). The action of progesterone is rapid (latency <36 ms) and does not depend on intracellular ATP, GDP, cyclic nucleotides, Ca^{2+}, or other soluble intracellular messengers (104, 109). The simplest explanation of these results is that the progesterone-binding site may be located on one of the CatSper subunits or on a currently unidentified protein associated with the CatSper complex. The binding site associated with the CatSper channel for this hormone has not been identified but appears to be accessible from the extracellular space (104).

Prostaglandins are abundant in the seminal plasma (110) and are secreted by the oviduct and cumulus cells surrounding the oocyte (111). Nanomolar concentrations of select prostaglandins, including PGE_1, evoke intracellular Ca^{2+} transients similar in amplitude and waveform to those induced by progesterone (112–114). The relative potency of the human CatSper activators is as follows: progesterone > $PGF_1 \geq PGE_1 > PGA_1 > PGE_2 > PGD_2$ (104). Prostaglandin effects are additive to those of progesterone and thus may be mediated through a different receptor (104, 109). High levels of Zn^{2+} in seminal plasma (115) are likely to block the CatSper channel and to prevent its activation in the seminal plasma, but once spermatozoa are in the female-dominant environment, Zn^{2+} should be diluted or chelated (116–118).

In conclusion, the CatSper complex is encoded by at least seven genes, making it the most biochemically complex of all ion channels. This complexity may be required for its assembly, trafficking, and localization to the flagella and for its sensitivity to pH_i, progesterone, prostaglandins, and perhaps other proteins. Because orthologs of CatSper subunits present in different species have low identity (50% or less) (32, 93, 102), regulation of the CatSper channel may differ significantly between species. The CatSper channel of murine epididymal sperm cells, for example, is not sensitive to the activators of human CatSper such as progesterone and prostaglandins (104). This difference in CatSper channel regulation and even the absence of *CatSper* genes in some species highlight the common trend in evolutionary pressure on gamete genes, which applies also to critical genes in sex determination pathways such as *SRY* and *DAX1*.

KSper (Slo3): The Principal K+ Channel of Spermatozoa

Incapacitated murine sperm hyperpolarize to approximately −60 mV during capacitation (119), an effect attributed to an increase in K^+ permeability. In a series of experiments in which voltage and intracellular and extracellular solutions were controlled, Navarro et al. (120) determined that pH_i sets the sperm membrane potential primarily by modifying the K^+ conductance. Under direct voltage clamp of mouse epididymal spermatozoa, resting membrane potential hyperpolarized to −45 mV within a few seconds after alkalinization. This hyperpolarization was due to a weakly outwardly rectifying K^+ current (I_{KSper}). I_{KSper} exhibited minimal time and voltage dependence, was relatively K^+ selective, and originated from the principal piece of the sperm flagellum. Intracellular alkalinization strongly potentiated I_{KSper} independent of extracellular $[K^+]$. I_{KSper} was not affected by 2 mM membrane-permeant cAMP and cGMP analogs, by increasing extracellular $[Ca^{2+}]$, or by changes in bath osmolarity. Barium, quinine, clofilium, EIPA [a Na^+/H^+ exchanger (NHE) antagonist], and mibefradil reversibly inhibited I_{KSper}. Thus, I_{KSper} is the only detectable hyperpolarizing current in spermatozoa and largely sets its resting membrane potential. These authors suggested that I_{KSper} is encoded by Slo3 on the basis of its tissue localization and other properties (**Table 1**). Like I_{KSper}, *Xenopus* oocyte–expressed Slo3 is also weakly voltage sensitive

---

Slo3: a K^+-permeant ion channel

($\sim$16 mV/e-fold), has relaxed K$^+$ selectivity, and is insensitive to [Ca^{2+}]$_i$ and to external tetraethyl ammonium chloride (36, 121–123). This prediction was confirmed by recordings from mice lacking the *Slo3* gene (124, 125). However, heterologously expressed mSlo3 had different pH sensitivity, which suggests that Slo3 in spermatozoa may be regulated by other subunits or mechanisms that are absent from heterologous expression systems.

In a carefully done study, genetic deletion of *Slo3* abolished all pH-dependent K$^+$ current at physiological membrane potentials in mouse corpus epididymal sperm (125). *Slo3*$^{-/-}$ mice are infertile and do not exhibit capacitation-dependent membrane hyperpolarization, and Slo3-deficient sperm morphological abnormalities are accentuated by hypotonic challenge. Solutions of lower osmolality (230–310 mOsm kg^{-1}) resulted in an increase in bent and hairpin shapes, whereas spermatozoa kept in a hyperosmolar solution were protected against these changes. Incapacitated *Slo3*$^{-/-}$ sperm also have modest defects in motility, which may be related to a requirement for osmolar adaptation during spermatogenesis and sperm maturation (125). Only 10% of *Slo3*$^{-/-}$ sperm were able to fertilize oocytes during in vitro fertilization experiments. In summary, mSlo3 accounts for KSper, the dominant, if not the only, K$^+$-selective channel in mouse epididymal spermatozoa.

The protein responsible for K$^+$ current in human spermatozoa has not been identified but is likely to be the human homolog of Slo3, KCNU1. However, in contrast to the situation in murine sperm cells, human KSper seems to be independent of intracellular alkalinization (P. Lishko & Y. Kirichok, unpublished observation). Murine Slo3 and human Slo3 proteins are 65% identical; mouse Slo3 is more enriched in histidines in the cytoplasmic C terminus. The difference between human and murine sperm K$^+$ current represents another discrepancy in physiology between human and mouse spermatozoa.

sNHE: sperm Na$^+$/H$^+$ exchanger

Hv1: the voltage-gated, proton-selective ion channel found usually in phagocytes but also present in human sperm

The Voltage-Gated Proton Channel Hv1: A Fast Regulator of Intracellular pH in Human Sperm

Intracellular alkalinization is essential for the initiation of motility, capacitation, hyperactivation, and the acrosome reaction. On the basis of experiments with pH$_i$-sensitive fluorescent probes that detected changes in pH$_i$, the NHE (45, 126, 127) and a Na$^+$-dependent Cl$^-$/HCO$_3^-$ exchanger (46, 47) were proposed to participate in sperm alkalinization. Upon ejaculation, mammalian spermatozoa are exposed to 100–150 mM [Na$^+$] in seminal plasma, a much higher Na$^+$ concentration than the 30 mM [Na$^+$] found in the cauda epididymis. In the female reproductive tract, Na$^+$ levels are similar to those in sera (140–150 mM) (128, 129). Thus, in the exchange of Na$^+$ for H$^+$, spermatozoan pH$_i$ should increase. Sperm-specific molecules homologous to known Na$^+$/H$^+$ exchangers (sNHE) (130) are found in the principal piece of sperm flagellum. *sNHE* knock-out mouse spermatozoa have impaired motility, and these males were completely infertile (130). Unfortunately, it has been difficult to demonstrate that sNHE actually functions as an NHE, as no significant difference in pH was found between wild-type and sNHE$^{-/-}$ spermatozoa (130). Complicating matters are the findings that sAC expression levels are significantly reduced in sNHE-deficient spermatozoa and that the sperm motility defect could be rescued by the addition of membrane-permeable cAMP analogs (130, 131).

The proton-selective, voltage-gated ion channel Hv1 (HVCN1) was cloned in 2006. This unusual channel is composed of a voltage sensor domain homologous to the voltage sensor of voltage-gated cation channels (132, 133). In contrast to the conventional ion channel, Hv1 lacks a classical pore region. The permeation pathway seems to be formed by an internal water wire completed by a movement of the charged S4 helix (134). Hv1 molecules dimerize, but each Hv1 subunit can function independently as a voltage-gated proton channel (135–137). The primary

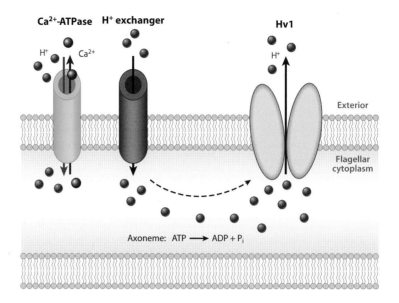

Figure 4

Regulation of flagellar pH. Protons may accumulate in the sperm flagellum via proton exchange, ATP hydrolysis by axonemal dynein, and active glycolysis. Rapid proton extrusion from the human sperm flagella may be carried out by Hv1 proteins, which form a proton-selective, voltage-gated ion channel restricted to the sperm's principal piece.

function of Hv1 in phagocytes is to allow intracellular protons to flow down their electrochemical gradient as electrons are extruded from cells via NADPH oxidase (NOX); block of Hv1 inhibits the innate immunity function of NOX (138, 139). Hv1 is characterized by strong voltage dependence, activation by high intracellular $[H^+]$, unidirectional proton extrusion (Hv1 is physiologically unidirectional), and inhibition by low micromolar concentrations of zinc and potentiation by fatty acids (140).

A voltage-gated proton channel was recorded in human spermatozoa (**Table 1**) (5). Although it has the electrophysiological and pharmacological properties of Hv1 (5, 141), its function in sperm may not be simply to support NOX. Hv1 is abundantly expressed in human sperm cells within the principal piece of the sperm flagellum, making it ideally positioned to activate pH-dependent proteins of the axoneme and thus to control sperm motility (5, 25, 141). The normal to alkaline pH of the upper female reproductive tract (∼7.4) may rapidly alkalinize the acidic intracellular compartments of sperm as they leave the acidic environment of the cauda epididymis and vagina. However, these changes need not be rapid and may easily be accomplished by exchangers, leaving one to wonder about the need for fast H^+ adaptation.

Human sperm flagella are long (40 μm), thin (<2 μm), and filled with axonemal structures. Because diffusion is inversely proportional to the area through which a substance diffuses, molecules take many seconds to travel within the extremely narrow flagellum from the midpiece to the endpiece. Thus, ATP, generated in the mitochondria of the midpiece, is slow to reach the end of the flagellum. Therefore, flagellar movement, especially at the distal parts of the sperm tail, is powered mainly by glycolysis (142–144), which results in cytoplasm acidification. Moreover, axonemal dynein hydrolyzes ATP to produce ADP, P_i, and H^+, all of which also contribute to intracellular acidification. The prompt removal of protons is thus vital to dynein function

(**Figure 4**). Sperm Hv1 conducts protons much more rapidly and efficiently than do exchangers or transporters and conducts them unidirectionally to the extracellular space.

Another possible role assigned to Hv1 is the regulation of intracellular Ca^{2+} homeostasis. Ca^{2+} delivered through CatSper is pumped out by a flagellar Ca^{2+}-ATPase that exports a cytoplasmic Ca^{2+} ion and imports extracellular protons. Its functioning results in decreasing flagellar pH_i, potentially inhibiting the CatSper channel. To prevent this scenario and to return the system to the status quo, Hv1 may balance pH_i by proton extrusion (**Figure 3**).

Hv1 is activated by the combination of the pH gradient and membrane depolarization (5, 132, 133, 140). However, because there is always a H^+ gradient out of the spermatozoa, the sperm's membrane potential is an important unknown and changes during sperm travel through the female reproductive tract. Membrane potential is set by Na^+/K^+-ATPases, which distribute ions over long durations, but is rapidly changed by the opening of ion channels.

Sperm Hv1 can be activated by the removal of extracellular zinc (5, 141). Zinc in humans is highest in seminal plasma (total 2.2 ± 1.1 mM compared with 14 ± 3 μM in serum) (115). Seminal zinc should inhibit Hv1, but as sperm travel through the female reproductive tract, any bound zinc is released through dilution, absorption by the uterine epithelium, and chelation by albumin and other molecules (116–118). Upon arrival at the Fallopian tube, spermatozoa should be essentially free from zinc inhibition. In addition, low micromolar concentrations of the endogenous cannabinoid anandamide strongly potentiate sperm Hv1 (5). The effect of anandamide is not mediated by CB1 or CB2 cannabinoid receptors and is likely due to a direct interaction of anandamide with Hv1 (5). Bulk concentrations of anandamide in the fluids of the male and female reproductive tracts are in the nanomolar range (145). However, because cumulus cells also synthesize and release anandamide, spermatozoa may experience much higher anandamide concentrations during the sperm's penetration of the cumulus oophorus (75). Finally, Hv1 is activated during in vitro capacitation (5), a time when tyrosine phosphorylation is very active. The mechanism of this potentiation remains unknown, but one hypothesis is that Hv1 is phosphorylated, especially because phosphorylation is the primary mechanism of Hv1 regulation in other tissues (146, 147). Moreover, intracellular alkalinization is considered to be a key factor during capacitation (15), and the coincidence of capacitation and the enhancement of Hv1 activity suggest a strong connection between these two events.

To date, patch-clamp experiments with mouse epididymal spermatozoa have not detected proton currents (5), and Hv1-deficient mice do not exhibit fertility defects (138, 139). Unfortunately, the NHE is electroneutral, and its activity cannot be recorded with patch-clamp techniques. Thus, the identification of all components of H^+ exchange in spermatozoa in mammals is an area for future detailed exploration.

In conclusion, sperm Hv1 may play an important role in the regulation of human sperm pH_i. By doing so, it could potentially influence almost every aspect of sperm behavior in the female reproductive tract, including initiation of motility, capacitation, hyperactivation, and the acrosome reaction. However, the physiological function of sperm Hv1 remains to be established. To date, the only correlation between human infertility and Hv1 is low levels of sperm HVCN1 mRNA in some infertility patients (148). Studies of genetic infertility in humans may thus help us understand the exact role of Hv1 in male fertility.

The ATP-Gated P2X2 Channel of Mammalian Sperm

To date, transmitter-mediated currents have not been reported in mouse spermatozoa. After screening a number of neurotransmitters and other biological molecules for their ability to induce ion channel currents in the whole spermatazoon, Navarro et al. (37) found a cation-nonselective,

CNG channel: cyclic nucleotide–gated channel

HCN channel: hyperpolarization-activated and cyclic nucleotide–gated channel

Ca^{2+}-permeable current originating from the midpiece of mouse epididymal spermatozoa that is activated by external ATP (I_{ATP}) (**Table 1**). Various plasma membrane purinergic receptors for ATP (purinergic receptors) were found in sperm by immunocytochemical studies (149), and ATP was reported to mediate an increase in intracellular Ca^{2+} (74). Navarro et al. (37) show that the behavior of this slowly desensitizing and strongly inwardly rectifying ATP-gated current has biophysical and pharmacological properties that mimic those of the heterologously expressed P2X2 oligomeric cation channel. Moreover, I_{ATP} is absent in spermatozoa of mice lacking the *P2rx2* gene. Despite the loss of I_{ATP}, *P2rx2*-deficient mice are fertile and have normal sperm morphology, sperm count, motility, and percent of sperm undergoing the acrosome reaction. However, the fertility of $P2rx2^{-/-}$ males declines with frequent mating over days, suggesting that the P2X2 receptor may confer a selection advantage under these conditions, perhaps through energizing mitochondria in the midpiece. ATP reportedly triggers the acrosome reaction in ejaculated bovine and human spermatozoa, reportedly via an uncharacterized sperm ATP-gated Na^+ channel (150).

Other Spermatozoan Ion Channels

In addition to the four sperm ion channels reviewed above, less evidence exists for other functional ion channels. Before 2001, the VGCCs were perceived as the principal Ca^{2+} conductance of sperm (78–80), but patch-clamp recording from mature sperm did not reveal any functional VGCCs and established that the CatSper channel is the principal sperm Ca^{2+} channel (6). In addition, several stimuli (e.g., increase in pH_i, depolarization, bovine serum albumin, and egg coat proteins) that trigger sperm Ca^{2+} influx previously assigned to VGCCs were later found to do so via the CatSper channel (31, 32). Finally, as discussed above, male mice deficient in $Ca_v2.2$, $Ca_v2.3$, and $Ca_v3.1$, three proteins detected only by antibodies in sperm, are fertile (85–87), indicating that these channels are not essential for sperm physiology or function redundantly (85–87). Our opinion is that VGCCs do not function in mature spermatozoa and do not have a significant role in mature sperm.

Cyclic nucleotide–gated (CNG) channels have also been proposed as mediating sperm Ca^{2+} influx. Both cAMP and cGMP elicit increases in $[Ca^{2+}]_i$ in sperm, as demonstrated in assays in which cell-permeable cAMP or cGMP is applied or when caged cGMP is uncaged (151). Thus, similar to photoreceptors and olfactory neurons, CNG channels may be responsible for the cyclic nucleotide–induced Ca^{2+} influx in sperm (151). A more recent variation of this model proposes that cyclic nucleotides activate the hyperpolarization-activated and cyclic nucleotide–gated (HCN) channels, resulting in the depolarization and subsequent opening of Ca^{2+} channels (79, 152). Although CNG and HCN channels may be present in sea urchin spermatozoa, mice and humans deficient in the CNG and HCN channels are fertile and have not been shown to exhibit defects in sperm function despite deficiencies in vision and cardiac function. Furthermore, no CNG or HCN currents in mouse or human sperm have been detected to date. Finally, because CatSper is responsible for the cAMP/cGMP-induced Ca^{2+} influx into sperm (32), cyclic nucleotides may activate CatSper indirectly, possibly via a PKA-dependent mechanism.

Several of the 28 members of the transient receptor potential (TRP) ion channels were recently proposed to function in mature spermatozoa. These include TRPM8, TRPV1, TRPC2, and others (153–155). For example, the TRPC2 protein was detected in sperm, and an anti-TRPC2 antibody reduced the sustained Ca^{2+} response elicited by egg coat proteins in mouse sperm and the ZP-induced acrosome reaction (154). However, mice deficient in *TRPC2* (as well as in the genes encoding TRPC1–7, TRPV1–4, TRPA1, TRPM1–4, and TRPM8) have no obvious defects in sperm physiology or male fertility. Indeed, in humans, *TRPC2* is a pseudogene. Therefore, the contribution of TRPC2, TRPM8, and TRPV1 in sperm physiology, if any, remains to be clarified.

SPERM CHEMOTAXIS

In the search for the egg, spermatozoa of many species are aided by chemotactic factors (73, 152, 156). Chemotaxis was first discovered in invertebrate marine animals such as the sea urchin, starfish, and sea squirt (157–159). Most marine animals produce and release sperm cells and eggs into seawater. To reach the egg in time, sperm cells must navigate a gradient of the chemoattractant(s) released by the egg and swim toward it.

The first putative chemoattractant of sea urchin, *Strongylocentrotus purpuratus* spermatozoa, a small peptide speract, was isolated from egg coat in 1981 by Hansbrough & Garbers (159). Later, picomolar concentrations of speract were found to activate a K^+ channel of *S. purpuratus* sperm (160). Interestingly, speract binding to the sea urchin spermatozoa also resulted in an increase in pH_i (161). Resact, a peptide from the egg coat of another species, *Arbacia punctulata*, was discovered in 1985 and was clearly shown to attract spermatozoa (162). The current model for sea urchin chemotaxis is built around the actions of resact on spermatozoa from *A. punctulata*, which suggested that resact activates flagellar guanylyl cyclase (GC) and triggers a signal transduction pathway leading to Ca^{2+} influx into the sperm flagellum. In short, the sea urchin's sperm flagellum contains a high density of membrane GC that produces cGMP from GTP in response to resact binding. cGMP is proposed to open K^+-selective cyclic nucleotide–gated (KCNG) channels; such opening briefly hyperpolarizes the sperm membrane (163). As a result, HCN channels open and allow Na^+ entry into the sperm flagellum. The resulting depolarization opens VGCCs, which conduct Ca^{2+} into flagellum and change the beating pattern. This sequence of events will require direct confirmation by recording under voltage clamp. Because all the CatSper α genes are present in the sea urchin (102, 103), we suspect that the story of sea urchin chemotaxis is not yet complete.

In contrast to the well-studied chemotaxis of sea urchin spermatozoa, much less is known about the chemotaxis of mammalian sperm. Mammalian spermatozoa are not likely to be engaged in the competitive-race model. Out of millions of spermatozoa delivered into the female reproductive tract, only one of every million succeeds in entering the Fallopian tubes, and <100 are able to reach the ampulla at any given time (73). Spermatozoa may be directed to the oocyte by specific cues or by chemicals released by the cumulus oophorus. Indeed, in 1991 researchers showed that human spermatozoa tend to accumulate in the follicular fluid (164) and that there is a positive correlation between sperm accumulation in the fluid and fertilization rate. Sperm chemotaxis was proposed in frogs, mice, and rabbits (for review see Reference 73). The discovery that human and rabbit spermatozoa are sensitive to picomolar concentrations of female hormone progesterone (165) established progesterone as a potential chemoattractant for human spermatozoa. Progesterone secreted from the cumulus oophorus peaks at midcycle and is present in oviductal and follicular fluid. As mentioned above, picomolar concentrations of progesterone activate CatSper and may regulate directional movement of the spermatozoa (104, 109). Mouse epididymal sperm CatSper is insensitive to progesterone (104), but this hypothesis should also be tested in ejaculated mouse spermatozoa.

Interestingly, *Ciona intestinalis* (sea squirt) eggs release sperm-attracting and -activating factor (SAAF), a molecule structurally similar to progesterone, and SAAF is a potent chemoattractant for *Ciona* spermatozoa (166). The molecular target for SAAF is not known, but because the *Ciona* genome contains *CatSper* and *Hvcn1* genes, the mechanism may be similar to that in human spermatozoa.

CONCLUSIONS

Ion channels of the sperm plasma membrane control the sperm membrane potential, establish intracellular Ca^{2+} and proton concentrations, direct cell movement, and, most importantly, are required for male fertility. With the ability to patch-clamp sperm, more light will be shed on the molecular identities and physiological regulation of sperm ion channels, resulting in new tools to control the behavior of spermatozoa and to increase or decrease male fertility. Given the enormous evolutionary pressure on genes optimizing gamete performance, there are likely many modifications or fine-tuning of the basic framework discussed above.

> **SUMMARY POINTS**
>
> 1. Sperm are free-swimming gametes that must adapt to changes in local environments on their journey to the egg. Ion channels of sperm enable sperm to respond and adjust to constantly changing environments and are required for male fertility.
> 2. Spermatozoa are compartmentalized cells, and plasma membrane ion channels are found primarily in the flagella. CatSper, KSper, and, in humans, Hv1 are localized in the principal piece of sperm flagellum, where they regulate sperm motility.
> 3. Direct recordings of spermatozoan ion currents under voltage clamp are essential for the proper identification of putative ion channel proteins found in sperm by other techniques.

DISCLOSURE STATEMENT

The authors are not aware of any affiliations, memberships, funding, or financial holdings that might be perceived as affecting the objectivity of this review.

ACKNOWLEDGMENTS

This work was supported by grant R01HD068914 (NICHD) to Y.K., by grant 5R01HD47578 to D.R., and by grant U01HD045857 and a grant from the Gates Foundation to D.E.C. The content is solely the responsibility of the authors and does not necessarily represent the official views of NICHD or the NIH.

LITERATURE CITED

1. Babcock DF, Rufo GA Jr, Lardy HA. 1983. Potassium-dependent increases in cytosolic pH stimulate metabolism and motility of mammalian sperm. *Proc. Natl. Acad. Sci. USA* 80:1327–31
2. Yanagimachi R, Usui N. 1974. Calcium dependence of the acrosome reaction and activation of guinea pig spermatozoa. *Exp. Cell Res.* 89:161–74
3. Dan JC. 1954. Studies on the acrosome. III. Effect of Ca^{2+} deficiency. *Biol. Bull.* 107:335–49
4. Ren D, Navarro B, Perez G, Jackson AC, Hsu S, et al. 2001. A sperm ion channel required for sperm motility and male fertility. *Nature* 413:603–9
5. Lishko PV, Botchkina IL, Fedorenko A, Kirichok Y. 2010. Acid extrusion from human spermatozoa is mediated by flagellar voltage-gated proton channel. *Cell* 140:327–37
6. Kirichok Y, Navarro B, Clapham DE. 2006. Whole-cell patch-clamp measurements of spermatozoa reveal an alkaline-activated Ca^{2+} channel. *Nature* 439:737–40
7. Kwitny S, Klaus AV, Hunnicutt GR. 2010. The annulus of the mouse sperm tail is required to establish a membrane diffusion barrier that is engaged during the late steps of spermiogenesis. *Biol. Reprod.* 82:669–78
8. Eddy EM, Toshimori K, O'Brien DA. 2003. Fibrous sheath of mammalian spermatozoa. *Microsc. Res. Tech.* 61:103–15
9. Cooper TG. 2011. The epididymis, cytoplasmic droplets and male fertility. *Asian J. Androl.* 13:130–38

10. Summers KE, Gibbons IR. 1971. Adenosine triphosphate-induced sliding of tubules in trypsin-treated flagella of sea-urchin sperm. *Proc. Natl. Acad. Sci. USA* 68:3092–96
11. Lindemann CB. 2011. Experimental evidence for the geometric clutch hypothesis. *Curr. Top. Dev. Biol.* 95:1–31
12. Brokaw CJ, Kamiya R. 1987. Bending patterns of *Chlamydomonas* flagella. IV. Mutants with defects in inner and outer dynein arms indicate differences in dynein arm function. *Cell Motil. Cytoskelet.* 8:68–75
13. Brokaw CJ. 1987. Regulation of sperm flagellar motility by calcium and cAMP-dependent phosphorylation. *J. Cell Biochem.* 35:175–84
14. White DR, Aitken RJ. 1989. Relationship between calcium, cyclic AMP, ATP, and intracellular pH and the capacity of hamster spermatozoa to express hyperactivated motility. *Gamete Res.* 22:163–77
15. Suarez SS. 2008. Control of hyperactivation in sperm. *Hum. Reprod. Update* 14:647–57
16. Acott TS, Carr DW. 1984. Inhibition of bovine spermatozoa by caudal epididymal fluid. II. Interaction of pH and a quiescence factor. *Biol. Reprod.* 30:926–35
17. Hamamah S, Gatti JL. 1998. Role of the ionic environment and internal pH on sperm activity. *Hum. Reprod.* 13(Suppl. 4):20–30
18. Austin CR. 1951. Observations on the penetration of the sperm in the mammalian egg. *Aust. J. Sci. Res. B* 4:581–96
19. Chang MC. 1951. Fertilizing capacity of spermatozoa deposited into the fallopian tubes. *Nature* 168:697–98
20. Eliasson R. 1966. Cholesterol in human semen. *Biochem. J.* 98:242–43
21. De Jonge C. 2005. Biological basis for human capacitation. *Hum. Reprod. Update* 11:205–14
22. Carr DW, Acott TS. 1989. Intracellular pH regulates bovine sperm motility and protein phosphorylation. *Biol. Reprod.* 41:907–20
23. Visconti PE, Moore GD, Bailey JL, Leclerc P, Connors SA, et al. 1995. Capacitation of mouse spermatozoa. II. Protein tyrosine phosphorylation and capacitation are regulated by a cAMP-dependent pathway. *Development* 121:1139–50
24. Visconti PE, Westbrook VA, Chertihin O, Demarco I, Sleight S, Diekman AB. 2002. Novel signaling pathways involved in sperm acquisition of fertilizing capacity. *J. Reprod. Immunol.* 53:133–50
25. Kirichok Y, Lishko PV. 2011. Rediscovering sperm ion channels with the patch-clamp technique. *Mol. Hum. Reprod.* 17:478–99
26. Mahi CA, Yanagimachi R. 1973. The effects of temperature, osmolality and hydrogen ion concentration on the activation and acrosome reaction of golden hamster spermatozoa. *J. Reprod. Fertil.* 35:55–66
27. Jin M, Fujiwara E, Kakiuchi Y, Okabe M, Satouh Y, et al. 2011. Most fertilizing mouse spermatozoa begin their acrosome reaction before contact with the zona pellucida during in vitro fertilization. *Proc. Natl. Acad. Sci. USA* 108:4892–96
28. Florman HM, Tombes RM, First NL, Babcock DF. 1989. An adhesion-associated agonist from the zona pellucida activates G protein-promoted elevations of internal Ca^{2+} and pH that mediate mammalian sperm acrosomal exocytosis. *Dev. Biol.* 135:133–46
29. Roldan ER, Murase T, Shi QX. 1994. Exocytosis in spermatozoa in response to progesterone and zona pellucida. *Science* 266:1578–81
30. Arnoult C, Zeng Y, Florman HM. 1996. ZP3-dependent activation of sperm cation channels regulates acrosomal secretion during mammalian fertilization. *J. Cell Biol.* 134:637–45
31. Xia J, Ren D. 2009. Egg coat proteins activate calcium entry into mouse sperm via CATSPER channels. *Biol. Reprod.* 80:1092–98
32. Ren D, Xia J. 2010. Calcium signaling through CatSper channels in mammalian fertilization. *Physiology* 25:165–75
33. Lee HC, Garbers DL. 1986. Modulation of the voltage-sensitive Na^+/H^+ exchange in sea urchin spermatozoa through membrane potential changes induced by the egg peptide speract. *J. Biol. Chem.* 261:16026–32
34. Navarro B, Kirichok Y, Chung JJ, Clapham DE. 2008. Ion channels that control fertility in mammalian spermatozoa. *Int. J. Dev. Biol.* 52:607–13
35. Sanchez D, Labarca P, Darszon A. 2001. Sea urchin sperm cation-selective channels directly modulated by cAMP. *FEBS Lett.* 503:111–15

36. Schreiber M, Wei A, Yuan A, Gaut J, Saito M, Salkoff L. 1998. Slo3, a novel pH-sensitive K^+ channel from mammalian spermatocytes. *J. Biol. Chem.* 273:3509–16
37. Navarro B, Miki K, Clapham DE. 2011. ATP-activated P2·2 current in mouse spermatozoa. *Proc. Natl. Acad. Sci. USA* 108:14342–47
38. Jimenez T, McDermott JP, Sanchez G, Blanco G. 2011. Na,K-ATPase α4 isoform is essential for sperm fertility. *Proc. Natl. Acad. Sci. USA* 108:644–49
39. Linares-Hernandez L, Guzman-Grenfell AM, Hicks-Gomez JJ, Gonzalez-Martinez MT. 1998. Voltage-dependent calcium influx in human sperm assessed by simultaneous optical detection of intracellular calcium and membrane potential. *Biochim. Biophys. Acta* 1372:1–12
40. Okunade GW, Miller ML, Pyne GJ, Sutliff RL, O'Connor KT, et al. 2004. Targeted ablation of plasma membrane Ca^{2+}-ATPase (PMCA) 1 and 4 indicates a major housekeeping function for PMCA1 and a critical role in hyperactivated sperm motility and male fertility for PMCA4. *J. Biol. Chem.* 279:33742–50
41. Schuh K, Cartwright EJ, Jankevics E, Bundschu K, Liebermann J, et al. 2004. Plasma membrane Ca^{2+} ATPase 4 is required for sperm motility and male fertility. *J. Biol. Chem.* 279:28220–26
42. Wennemuth G, Babcock DF, Hille B. 2003. Calcium clearance mechanisms of mouse sperm. *J. Gen. Physiol.* 122:115–28
43. Babcock DF, Pfeiffer DR. 1987. Independent elevation of cytosolic $[Ca^{2+}]$ and pH of mammalian sperm by voltage-dependent and pH-sensitive mechanisms. *J. Biol. Chem.* 262:15041–47
44. Hamamah S, Magnoux E, Royere D, Barthelemy C, Dacheux JL, Gatti JL. 1996. Internal pH of human spermatozoa. effect of ions, human follicular fluid and progesterone. *Mol. Hum. Reprod.* 2:219–24
45. Garcia MA, Meizel S. 1999. Regulation of intracellular pH in capacitated human spermatozoa by a Na^+/H^+ exchanger. *Mol. Reprod. Dev.* 52:189–95
46. Tajima Y, Okamura N. 1990. The enhancing effects of anion channel blockers on sperm activation by bicarbonate. *Biochim. Biophys. Acta* 1034:326–32
47. Zeng Y, Oberdorf JA, Florman HM. 1996. pH regulation in mouse sperm: identification of Na^+-, Cl^--, and HCO_3^--dependent and arylaminobenzoate-dependent regulatory mechanisms and characterization of their roles in sperm capacitation. *Dev. Biol.* 173:510–20
48. Jiang D, Zhao L, Clapham DE. 2009. Genome-wide RNAi screen identifies Letm1 as a mitochondrial Ca^{2+}/H^+ antiporter. *Science* 326:144–47
49. Carr DW, Acott TS. 1984. Inhibition of bovine spermatozoa by caudal epididymal fluid. I. Studies of a sperm motility quiescence factor. *Biol. Reprod.* 30:913–25
50. Usselman MC, Cone RA. 1983. Rat sperm are mechanically immobilized in the caudal epididymis by "immobilin," a high molecular weight glycoprotein. *Biol. Reprod.* 29:1241–53
51. Carr DW, Usselman MC, Acott TS. 1985. Effects of pH, lactate, and viscoelastic drag on sperm motility: a species comparison. *Biol. Reprod.* 33:588–95
52. Mitra A, Richardson RT, O'Rand MG. 2010. Analysis of recombinant human semenogelin as an inhibitor of human sperm motility. *Biol. Reprod.* 82:489–96
53. Fox CA, Meldrum SJ, Watson BW. 1973. Continuous measurement by radio-telemetry of vaginal pH during human coitus. *J. Reprod. Fertil.* 33:69–75
54. Eggert-Kruse W, Kohler A, Rohr G, Runnebaum B. 1993. The pH as an important determinant of sperm-mucus interaction. *Fertil. Steril.* 59:617–28
55. Maas DH, Storey BT, Mastroianni L Jr. 1977. Hydrogen ion and carbon dioxide content of the oviductal fluid of the rhesus monkey (*Macaca mulatta*). *Fertil. Steril.* 28:981–85
56. Chen Y, Cann MJ, Litvin TN, Iourgenko V, Sinclair ML, et al. 2000. Soluble adenylyl cyclase as an evolutionarily conserved bicarbonate sensor. *Science* 289:625–28
57. Okamura N, Tajima Y, Soejima A, Masuda H, Sugita Y. 1985. Sodium bicarbonate in seminal plasma stimulates the motility of mammalian spermatozoa through direct activation of adenylate cyclase. *J. Biol. Chem.* 260:9699–705
58. Wandernoth PM, Raubuch M, Mannowetz N, Becker HM, Deitmer JW, et al. 2010. Role of carbonic anhydrase IV in the bicarbonate-mediated activation of murine and human sperm. *PLoS ONE* 5:e15061
59. Goltz JS, Gardner TK, Kanous KS, Lindemann CB. 1988. The interaction of pH and cyclic adenosine 3′,5′-monophosphate on activation of motility in Triton X-100 extracted bull sperm. *Biol. Reprod.* 39:1129–36

60. Harrison RA. 2004. Rapid PKA-catalysed phosphorylation of boar sperm proteins induced by the capacitating agent bicarbonate. *Mol. Reprod. Dev.* 67:337–52
61. Nolan MA, Babcock DF, Wennemuth G, Brown W, Burton KA, McKnight GS. 2004. Sperm-specific protein kinase A catalytic subunit Cα2 orchestrates cAMP signaling for male fertility. *Proc. Natl. Acad. Sci. USA* 101:13483–88
62. Salathe M. 2007. Regulation of mammalian ciliary beating. *Annu. Rev. Physiol.* 69:401–22
63. Jaiswal BS, Conti M. 2003. Calcium regulation of the soluble adenylyl cyclase expressed in mammalian spermatozoa. *Proc. Natl. Acad. Sci. USA* 100:10676–81
64. Litvin TN, Kamenetsky M, Zarifyan A, Buck J, Levin LR. 2003. Kinetic properties of "soluble" adenylyl cyclase. Synergism between calcium and bicarbonate. *J. Biol. Chem.* 278:15922–26
65. Carlson AE, Hille B, Babcock DF. 2007. External Ca^{2+} acts upstream of adenylyl cyclase SACY in the bicarbonate signaled activation of sperm motility. *Dev. Biol.* 312:183–92
66. Esposito G, Jaiswal BS, Xie F, Krajnc-Franken MA, Robben TJ, et al. 2004. Mice deficient for soluble adenylyl cyclase are infertile because of a severe sperm-motility defect. *Proc. Natl. Acad. Sci. USA* 101:2993–98
67. Hess KC, Jones BH, Marquez B, Chen Y, Ord TS, et al. 2005. The "soluble" adenylyl cyclase in sperm mediates multiple signaling events required for fertilization. *Dev. Cell* 9:249–59
68. Xie F, Garcia MA, Carlson AE, Schuh SM, Babcock DF, et al. 2006. Soluble adenylyl cyclase (sAC) is indispensable for sperm function and fertilization. *Dev. Biol.* 296:353–62
69. Wennemuth G, Carlson AE, Harper AJ, Babcock DF. 2003. Bicarbonate actions on flagellar and Ca^{2+}-channel responses: initial events in sperm activation. *Development* 130:1317–26
70. Hille B. 1992. Elementary properties of pores. In *Ionic Channels of Excitable Membranes*, ed. B Hille, pp. 291–314. 2nd ed. Sunderland, MA: Sinauer Assoc.
71. Fraser LR. 2010. The "switching on" of mammalian spermatozoa: molecular events involved in promotion and regulation of capacitation. *Mol. Reprod. Dev.* 77:197–208
72. Publicover S, Harper CV, Barratt C. 2007. $[Ca^{2+}]_i$ signalling in sperm—making the most of what you've got. *Nat. Cell Biol.* 9:235–42
73. Eisenbach M, Giojalas LC. 2006. Sperm guidance in mammals—an unpaved road to the egg. *Nat. Rev. Mol. Cell Biol.* 7:276–85
74. Xia J, Ren D. 2009. The BSA-induced Ca^{2+} influx during sperm capacitation is CATSPER channel-dependent. *Reprod. Biol. Endocrinol.* 7:119
75. El-Talatini MR, Taylor AH, Elson JC, Brown L, Davidson AC, Konje JC. 2009. Localisation and function of the endocannabinoid system in the human ovary. *PLoS ONE* 4:e4579
76. Clapham DE. 2007. Calcium signaling. *Cell* 131:1047–58
77. Brokaw CJ. 1979. Calcium-induced asymmetrical beating of triton-demembranated sea urchin sperm flagella. *J. Cell Biol.* 82:401–11
78. Florman HM, Arnoult C, Kazam IG, Li C, O'Toole CM. 1998. A perspective on the control of mammalian fertilization by egg-activated ion channels in sperm: a tale of two channels. *Biol. Reprod.* 59:12–16
79. Darszon A, Labarca P, Nishigaki T, Espinosa F. 1999. Ion channels in sperm physiology. *Physiol. Rev.* 79:481–510
80. Publicover SJ, Barratt CL. 1999. Voltage-operated Ca^{2+} channels and the acrosome reaction: Which channels are present and what do they do? *Hum. Reprod.* 14:873–79
81. Hagiwara S, Kawa K. 1984. Calcium and potassium currents in spermatogenic cells dissociated from rat seminiferous tubules. *J. Physiol.* 356:135–49
82. Arnoult C, Cardullo RA, Lemos JR, Florman HM. 1996. Activation of mouse sperm T-type Ca^{2+} channels by adhesion to the egg zona pellucida. *Proc. Natl. Acad. Sci. USA* 93:13004–9
83. Santi CM, Darszon A, Hernandez-Cruz A. 1996. A dihydropyridine-sensitive T-type Ca^{2+} current is the main Ca^{2+} current carrier in mouse primary spermatocytes. *Am. J. Physiol. Cell Physiol.* 271:1583–93
84. Wennemuth G, Westenbroek RE, Xu T, Hille B, Babcock DF. 2000. $Ca_V 2.2$ and $Ca_V 2.3$ (N- and R-type) Ca^{2+} channels in depolarization-evoked entry of Ca^{2+} into mouse sperm. *J. Biol. Chem.* 275:21210–17
85. Beuckmann CT, Sinton CM, Miyamoto N, Ino M, Yanagisawa M. 2003. N-type calcium channel α_{1B} subunit ($Ca_V 2.2$) knock-out mice display hyperactivity and vigilance state differences. *J. Neurosci.* 23:6793–97

86. Saegusa H, Kurihara T, Zong S, Minowa O, Kazuno A, et al. 2000. Altered pain responses in mice lacking α_{1E} subunit of the voltage-dependent Ca^{2+} channel. *Proc. Natl. Acad. Sci. USA* 97:6132–37

87. Kim D, Song I, Keum S, Lee T, Jeong MJ, et al. 2001. Lack of the burst firing of thalamocortical relay neurons and resistance to absence seizures in mice lacking α_{1G} T-type Ca^{2+} channels. *Neuron* 31:35–45

88. Jun K, Piedras-Renteria ES, Smith SM, Wheeler DB, Lee SB, et al. 1999. Ablation of P/Q-type Ca^{2+} channel currents, altered synaptic transmission, and progressive ataxia in mice lacking the α_{1A}-subunit. *Proc. Natl. Acad. Sci. USA* 96:15245–50

89. Seisenberger C, Specht V, Welling A, Platzer J, Pfeifer A, et al. 2000. Functional embryonic cardiomyocytes after disruption of the L-type α_{1C} ($Ca_v1.2$) calcium channel gene in the mouse. *J. Biol. Chem.* 275:39193–99

90. Lobley A, Pierron V, Reynolds L, Allen L, Michalovich D. 2003. Identification of human and mouse CatSper3 and CatSper4 genes: characterisation of a common interaction domain and evidence for expression in testis. *Reprod. Biol. Endocrinol.* 1:53

91. Quill TA, Sugden SA, Rossi KL, Doolittle LK, Hammer RE, Garbers DL. 2003. Hyperactivated sperm motility driven by CatSper2 is required for fertilization. *Proc. Natl. Acad. Sci. USA* 100:14869–74

92. Carlson AE, Quill TA, Westenbroek RE, Schuh SM, Hille B, Babcock DF. 2005. Identical phenotypes of CatSper1 and CatSper2 null sperm. *J. Biol. Chem.* 280:32238–44

93. Liu J, Xia J, Cho KH, Clapham DE, Ren D. 2007. CatSperβ, a novel transmembrane protein in the CatSper channel complex. *J. Biol. Chem.* 282:18945–52

94. Qi H, Moran MM, Navarro B, Chong JA, Krapivinsky G, et al. 2007. All four CatSper ion channel proteins are required for male fertility and sperm cell hyperactivated motility. *Proc. Natl. Acad. Sci. USA* 104:1219–23

95. Wang H, Liu J, Cho KH, Ren D. 2009. A novel, single, transmembrane protein CATSPERG is associated with CATSPER1 channel protein. *Biol. Reprod.* 81:539–44

96. Chung JJ, Navarro B, Krapivinsky G, Krapivinsky L, Clapham DE. 2011. A novel gene required for male fertility and functional CATSPER channel formation in spermatozoa. *Nat. Commun.* 2:153

97. Hildebrand MS, Avenarius MR, Fellous M, Zhang Y, Meyer NC, et al. 2010. Genetic male infertility and mutation of CATSPER ion channels. *Eur. J. Hum. Genet.* 18:1178–84

98. Avenarius MR, Hildebrand MS, Zhang Y, Meyer NC, Smith LL, et al. 2009. Human male infertility caused by mutations in the CATSPER1 channel protein. *Am. J. Hum. Genet.* 84:505–10

99. Zhang Y, Malekpour M, Al-Madani N, Kahrizi K, Zanganeh M, et al. 2007. Sensorineural deafness and male infertility: a contiguous gene deletion syndrome. *J. Med. Genet.* 44:233–40

100. Avidan N, Tamary H, Dgany O, Cattan D, Pariente A, et al. 2003. CATSPER2, a human autosomal nonsyndromic male infertility gene. *Eur. J. Hum. Genet.* 11:497–502

101. Marquez B, Suarez SS. 2007. Bovine sperm hyperactivation is promoted by alkaline-stimulated Ca^{2+} influx. *Biol. Reprod.* 76:660–65

102. Cai X, Clapham DE. 2008. Evolutionary genomics reveals lineage-specific gene loss and rapid evolution of a sperm-specific ion channel complex: CatSpers and CatSperβ. *PLoS ONE* 3:e3569

103. Cai X, Clapham DE. 2011. Ancestral Ca^{2+} signaling machinery in early animal and fungal evolution. *Mol. Biol. Evol.* In press

104. Lishko PV, Botchkina IL, Kirichok Y. 2011. Progesterone activates the principal Ca^{2+} channel of human sperm. *Nature* 471:387–91

105. Blackmore PF, Beebe SJ, Danforth DR, Alexander N. 1990. Progesterone and 17α-hydroxyprogesterone. Novel stimulators of calcium influx in human sperm. *J. Biol. Chem.* 265:1376–80

106. Thomas P, Meizel S. 1989. Phosphatidylinositol 4,5-bisphosphate hydrolysis in human sperm stimulated with follicular fluid or progesterone is dependent upon Ca^{2+} influx. *Biochem. J.* 264:539–46

107. Losel R, Wehling M. 2003. Nongenomic actions of steroid hormones. *Nat. Rev. Mol. Cell Biol.* 4:46–56

108. Luconi M, Francavilla F, Porazzi I, Macerola B, Forti G, Baldi E. 2004. Human spermatozoa as a model for studying membrane receptors mediating rapid nongenomic effects of progesterone and estrogens. *Steroids* 69:553–59

109. Strünker T, Goodwin N, Brenker C, Kashikar N, Weyand I, et al. 2011. The CatSper channel mediates progesterone-induced Ca^{2+} influx in human sperm. *Nature* 471:382–86

110. Mann T, Lutwak-Mann C. 1981. Biochemistry of seminal plasma and male accessory fluids: application to andrological problems. In *Male Reproductive Function and Semen: Themes and Trends in Physiology, Biochemistry and Investigative Andrology*, ed. T Mann, C Lutwak-Mann, pp. 269–336. Berlin: Springer-Verlag

111. Espey LL, Richards JS. 2006. Ovulation. In *Knobil and Neill's The Physiology of Reproduction*, ed. DJ Neill, 1:425–75. St. Louis: Elsevier

112. Aitken RJ, Irvine S, Kelly RW. 1986. Significance of intracellular calcium and cyclic adenosine 3′,5′-monophosphate in the mechanisms by which prostaglandins influence human sperm function. *J. Reprod. Fertil.* 77:451–62

113. Schaefer M, Hofmann T, Schultz G, Gudermann T. 1998. A new prostaglandin E receptor mediates calcium influx and acrosome reaction in human spermatozoa. *Proc. Natl. Acad. Sci. USA* 95:3008–13

114. Shimizu Y, Yorimitsu A, Maruyama Y, Kubota T, Aso T, Bronson RA. 1998. Prostaglandins induce calcium influx in human spermatozoa. *Mol. Hum. Reprod.* 4:555–61

115. Saaranen M, Suistomaa U, Kantola M, Saarikoski S, Vanha-Perttula T. 1987. Lead, magnesium, selenium and zinc in human seminal fluid: comparison with semen parameters and fertility. *Hum. Reprod.* 2:475–79

116. Ehrenwald E, Foote RH, Parks JE. 1990. Bovine oviductal fluid components and their potential role in sperm cholesterol efflux. *Mol. Reprod. Dev.* 25:195–204

117. Gunn SA, Gould TC. 1958. Role of zinc in fertility and fecundity in the rat. *Am. J. Physiol.* 193:505–8

118. Lu J, Stewart AJ, Sadler PJ, Pinheiro TJ, Blindauer CA. 2008. Albumin as a zinc carrier: properties of its high-affinity zinc-binding site. *Biochem. Soc. Trans.* 36:1317–21

119. Muñoz-Garay C, de la Vega-Beltrán JL, Delgado R, Labarca P, Felix R, Darszon A. 2001. Inwardly rectifying K$^+$ channels in spermatogenic cells: functional expression and implication in sperm capacitation. *Dev. Biol.* 234:261–74

120. Navarro B, Kirichok Y, Clapham DE. 2007. KSper, a pH-sensitive K$^+$ current that controls sperm membrane potential. *Proc. Natl. Acad. Sci. USA* 104:7688–92

121. Xia XM, Zhang X, Lingle CJ. 2004. Ligand-dependent activation of Slo family channels is defined by interchangeable cytosolic domains. *J. Neurosci.* 24:5585–91

122. Tang QY, Zhang Z, Xia XM, Lingle CJ. 2010. Block of mouse Slo1 and Slo3 K$^+$ channels by CTX, IbTX, TEA, 4-AP and quinidine. *Channels* 4:22–41

123. Zhang X, Zeng X, Lingle CJ. 2006. Slo3 K$^+$ channels: voltage and pH dependence of macroscopic currents. *J. Gen. Physiol.* 128:317–36

124. Santi CM, Martínez-López P, de la Vega-Beltrán JL, Butler A, Alisio A, et al. 2010. The SLO3 sperm-specific potassium channel plays a vital role in male fertility. *FEBS Lett.* 584:1041–46

125. Zeng XH, Yang C, Kim ST, Lingle CJ, Xia XM. 2011. Deletion of the Slo3 gene abolishes alkalization-activated K$^+$ current in mouse spermatozoa. *Proc. Natl. Acad. Sci. USA* 108:5879–84

126. Woo AL, James PF, Lingrel JB. 2002. Roles of the Na,K-ATPase α4 isoform and the Na$^+$/H$^+$ exchanger in sperm motility. *Mol. Reprod. Dev.* 62:348–56

127. Bibring T, Baxandall J, Harter CC. 1984. Sodium-dependent pH regulation in active sea urchin sperm. *Dev. Biol.* 101:425–35

128. Borland RM, Biggers JD, Lechene CP, Taymor ML. 1980. Elemental composition of fluid in the human Fallopian tube. *J. Reprod. Fertil.* 58:479–82

129. Mann T. 1964. *The Biochemistry of Semen and of the Male Reproductive Tract*. London/New York: Methuen/Wiley

130. Wang D, King SM, Quill TA, Doolittle LK, Garbers DL. 2003. A new sperm-specific Na$^+$/H$^+$ exchanger required for sperm motility and fertility. *Nat. Cell Biol.* 5:1117–22

131. Wang D, Hu J, Bobulescu IA, Quill TA, McLeroy P, et al. 2007. A sperm-specific Na$^+$/H$^+$ exchanger (sNHE) is critical for expression and in vivo bicarbonate regulation of the soluble adenylyl cyclase (sAC). *Proc. Natl. Acad. Sci. USA* 104:9325–30

132. Ramsey IS, Moran MM, Chong JA, Clapham DE. 2006. A voltage-gated proton-selective channel lacking the pore domain. *Nature* 440:1213–16

133. Sasaki M, Takagi M, Okamura Y. 2006. A voltage sensor-domain protein is a voltage-gated proton channel. *Science* 312:589–92

134. Ramsey IS, Mokrab Y, Carvacho I, Sands ZA, Sansom MS, Clapham DE. 2010. An aqueous H^+ permeation pathway in the voltage-gated proton channel Hv1. *Nat. Struct. Mol. Biol.* 17:869–75
135. Koch HP, Kurokawa T, Okochi Y, Sasaki M, Okamura Y, Larsson HP. 2008. Multimeric nature of voltage-gated proton channels. *Proc. Natl. Acad. Sci. USA* 105:9111–16
136. Lee SY, Letts JA, Mackinnon R. 2008. Dimeric subunit stoichiometry of the human voltage-dependent proton channel Hv1. *Proc. Natl. Acad. Sci. USA* 105:7692–95
137. Tombola F, Ulbrich MH, Isacoff EY. 2008. The voltage-gated proton channel Hv1 has two pores, each controlled by one voltage sensor. *Neuron* 58:546–56
138. Okochi Y, Sasaki M, Iwasaki H, Okamura Y. 2009. Voltage-gated proton channel is expressed on phagosomes. *Biochem. Biophys. Res. Commun.* 382:274–79
139. Ramsey IS, Ruchti E, Kaczmarek JS, Clapham DE. 2009. Hv1 proton channels are required for high-level NADPH oxidase-dependent superoxide production during the phagocyte respiratory burst. *Proc. Natl. Acad. Sci. USA* 106:7642–47
140. DeCoursey TE. 2008. Voltage-gated proton channels. *Cell Mol. Life Sci.* 65:2554–73
141. Lishko PV, Kirichok Y. 2010. The role of Hv1 and CatSper channels in sperm activation. *J. Physiol.* 588:4667–72
142. Miki K, Qu W, Goulding EH, Willis WD, Bunch DO, et al. 2004. Glyceraldehyde 3-phosphate dehydrogenase-S, a sperm-specific glycolytic enzyme, is required for sperm motility and male fertility. *Proc. Natl. Acad. Sci. USA* 101:16501–6
143. Mukai C, Okuno M. 2004. Glycolysis plays a major role for adenosine triphosphate supplementation in mouse sperm flagellar movement. *Biol. Reprod.* 71:540–47
144. Williams AC, Ford WC. 2001. The role of glucose in supporting motility and capacitation in human spermatozoa. *J. Androl.* 22:680–95
145. Schuel H, Burkman LJ. 2005. A tale of two cells: Endocannabinoid-signaling regulates functions of neurons and sperm. *Biol. Reprod.* 73:1078–86
146. Decoursey TE. 2003. Voltage-gated proton channels and other proton transfer pathways. *Physiol. Rev.* 83:475–579
147. Musset B, Capasso M, Cherny VV, Morgan D, Bhamrah M, et al. 2010. Identification of Thr29 as a critical phosphorylation site that activates the human proton channel Hvcn1 in leukocytes. *J. Biol. Chem.* 285:5117–21
148. Platts AE, Dix DJ, Chemes HE, Thompson KE, Goodrich R, et al. 2007. Success and failure in human spermatogenesis as revealed by teratozoospermic RNAs. *Hum. Mol. Genet.* 16:763–73
149. Banks FC, Calvert RC, Burnstock G. 2010. Changing P2X receptor localization on maturing sperm in the epididymides of mice, hamsters, rats, and humans: a preliminary study. *Fertil. Steril.* 93:1415–20
150. Foresta C, Rossato M, Chiozzi P, Di Virgilio F. 1996. Mechanism of human sperm activation by extracellular ATP. *Am. J. Physiol. Cell Physiol.* 270:1709–14
151. Wiesner B, Weiner J, Middendorff R, Hagen V, Kaupp UB, Weyand I. 1998. Cyclic nucleotide-gated channels on the flagellum control Ca^{2+} entry into sperm. *J. Cell Biol.* 142:473–84
152. Kaupp UB, Kashikar ND, Weyand I. 2008. Mechanisms of sperm chemotaxis. *Annu. Rev. Physiol.* 70:93–117
153. Francavilla F, Battista N, Barbonetti A, Vassallo MR, Rapino C, et al. 2009. Characterization of the endocannabinoid system in human spermatozoa and involvement of transient receptor potential vanilloid 1 receptor in their fertilizing ability. *Endocrinology* 150:4692–700
154. Jungnickel MK, Marrero H, Birnbaumer L, Lemos JR, Florman HM. 2001. Trp2 regulates entry of Ca^{2+} into mouse sperm triggered by egg ZP3. *Nat. Cell Biol.* 3:499–502
155. Martínez-López P, Treviño CL, de la Vega-Beltrán JL, De Blas G, Monroy E, et al. 2011. TRPM8 in mouse sperm detects temperature changes and may influence the acrosome reaction. *J. Cell Physiol.* 226:1620–31
156. Cook SP, Brokaw CJ, Muller CH, Babcock DF. 1994. Sperm chemotaxis: Egg peptides control cytosolic calcium to regulate flagellar responses. *Dev. Biol.* 165:10–19
157. Miller RL. 1975. Chemotaxis of the spermatozoa of *Ciona intestinalis*. *Nature* 254:244–45
158. Matsumoto M, Solzin J, Helbig A, Hagen V, Ueno S, et al. 2003. A sperm-activating peptide controls a cGMP-signaling pathway in starfish sperm. *Dev. Biol.* 260:314–24

159. Hansbrough JR, Garbers DL. 1981. Speract. Purification and characterization of a peptide associated with eggs that activates spermatozoa. *J. Biol. Chem.* 256:1447–52
160. Babcock DF, Bosma MM, Battaglia DE, Darszon A. 1992. Early persistent activation of sperm K$^+$ channels by the egg peptide speract. *Proc. Natl. Acad. Sci. USA* 89:6001–5
161. Cook SP, Babcock DF. 1993. Activation of Ca^{2+} permeability by cAMP is coordinated through the pH$_i$ increase induced by speract. *J. Biol. Chem.* 268:22408–13
162. Ward GE, Brokaw CJ, Garbers DL, Vacquier VD. 1985. Chemotaxis of *Arbacia punctulata* spermatozoa to resact, a peptide from the egg jelly layer. *J. Cell Biol.* 101:2324–29
163. Strunker T, Weyand I, Bonigk W, Van Q, Loogen A, et al. 2006. A K$^+$-selective cGMP-gated ion channel controls chemosensation of sperm. *Nat. Cell Biol.* 8:1149–54
164. Ralt D, Goldenberg M, Fetterolf P, Thompson D, Dor J, et al. 1991. Sperm attraction to a follicular factor(s) correlates with human egg fertilizability. *Proc. Natl. Acad. Sci. USA* 88:2840–44
165. Teves ME, Barbano F, Guidobaldi HA, Sanchez R, Miska W, Giojalas LC. 2006. Progesterone at the picomolar range is a chemoattractant for mammalian spermatozoa. *Fertil. Steril.* 86:745–49
166. Shiba K, Baba SA, Inoue T, Yoshida M. 2008. Ca^{2+} bursts occur around a local minimal concentration of attractant and trigger sperm chemotactic response. *Proc. Natl. Acad. Sci. USA* 105:19312–17

Sperm-Egg Interaction

Janice P. Evans

Department of Biochemistry and Molecular Biology, Bloomberg School of Public Health, Johns Hopkins University, Baltimore, Maryland 21205; email: jpevans@jhsph.edu

Keywords

fertilization, cell-cell fusion, cell adhesion, tetraspanin, CD9, IZUMO

Abstract

A crucial step of fertilization is the sperm-egg interaction that allows the two gametes to fuse and create the zygote. In the mouse, CD9 on the egg and IZUMO1 on the sperm stand out as critical players, as $Cd9^{-/-}$ and $Izumo1^{-/-}$ mice are healthy but infertile or severely subfertile due to defective sperm-egg interaction. Moreover, work on several nonmammalian organisms has identified some of the most intriguing candidates implicated in sperm-egg interaction. Understanding of gamete membrane interactions is advancing through characterization of in vivo and in vitro fertilization phenotypes, including insights from less robust phenotypes that highlight potential supporting (albeit not absolutely essential) players. An emerging theme is that there are varied roles for gamete molecules that participate in sperm-egg interactions. Such roles include not only functioning as fusogens, or as adhesion molecules for the opposite gamete, but also functioning through interactions in *cis* with other proteins to regulate membrane order and functionality.

OVERVIEW, DEFINITIONS, AND SETTING THE STAGE

Sperm-egg fusion: the bona fide membrane fusion event; results in cytoplasmic continuity between the gametes (**Figure 1d**)

Sperm-egg binding (or sperm-egg adhesion): cell-cell interaction that precedes membrane fusion and that is defined experimentally as maintained after a specific series of washes (distinguishing adhesion from mere sperm attachment)

Sperm-egg interaction (or sperm-egg membrane interaction): the interaction of the sperm membrane with the egg membrane, including sperm-egg adhesion and membrane fusion

Fertilization, the union of the gametes (sperm and egg in heterogametic species), is a critical component of sexual reproduction. This process brings together two terminally differentiated cells that carry the genomes of two different individuals and creates a totipotent cell, the zygote, and a new, genetically distinct individual. Fertilization occurs through a series of coordinated steps, culminating in the merger of the two cells. This merger is achieved through gamete interactions, specifically cell adhesion and then membrane fusion of the gamete plasma membranes.

The term sperm-egg fusion has been used to refer collectively to the entire process of gamete membrane interactions that culminate in sperm incorporation into the egg. Wherever possible, this review strives for specificity in describing cellular events, using the terms sperm-egg binding and sperm-egg fusion as distinct steps and the term sperm-egg membrane interaction more generically where appropriate. Indeed, the two proteins most strongly implicated in mouse sperm-egg interaction, IZUMO1 on the sperm and CD9 on the egg, are often described as having roles in so-called sperm-egg fusion, although these proteins lack features associated with a fusogenic molecule, and these proteins may have other functions (addressed below). Distinct events lead to the fusion of two membranes. For cell-cell fusion (including sperm-egg fusion), cell adhesion is a precursor step, progressing from interaction of the membranes to the close apposition of the membranes. This membrane interaction ultimately leads to the actual fusion event, with mixing of the lipid bilayers, formation of fusion pores, and establishment of cytoplasmic continuity (**Figure 1d**). Analogies exist in other systems: Intracellular vesicles dock or tether prior to fusion, myoblasts undergo a process of adhesion and alignment that leads to fusion, and membrane-coated virus particles bind to host cells and then progress to fusion (1–4). Sperm binding to the egg is a step-wise process as well, starting with loose attachment and progressing to firm adhesion that can lead to membrane fusion. In in vitro fertilization (IVF) assays, sperm readily attach to the zona pellucida (ZP) or to the egg membrane of ZP-free eggs (**Figure 1c**). Washes of inseminated eggs are required to remove loosely attached sperm, allowing identification of the subpopulation of sperm that are truly bound (5, 6) (see also sidebar entitled "Technical Aspects of Assessing Sperm-Egg Membrane Interactions").

TECHNICAL ASPECTS OF ASSESSING SPERM-EGG MEMBRANE INTERACTIONS

Just as it is important to appreciate the distinct steps of sperm-egg interaction (see text and **Figure 1d**), it is similarly valuable to understand the assays used to assess these interactions. Sperm-egg fusion is relatively easy to assess through observation of the characteristic morphological changes in the sperm head that occur when sperm DNA is exposed to the egg cytoplasm or through use of eggs that are loaded with a dye such as Hoechst to observe dye transfer to the sperm DNA (7, 8). Sperm-egg adhesion needs to be distinguished from the mere attachment of sperm. As is done in other cell adhesion assays, which define an adherent cell as one that stays bound after a defined series of washes (9), washes are required to remove loosely attached sperm to distinguish the sperm that are truly bound (5, 6). A key variable in IVF assays is sperm concentration (or sperm-egg ratio) used in the insemination, which can impact the extent of binding and fusion. Although a dramatic reduction in gamete interaction observed with eggs challenged with high numbers of sperm is an exciting result, there are also cases in which a difference can be detected with experimental conditions that less heavily favor fertilization [i.e., lower sperm-egg ratios (10, 11)]. Such assay modifications detect more subtle abnormalities in gamete interaction, but as noted in the main text, both robust and subtle phenotypes likely have implications for human reproductive health.

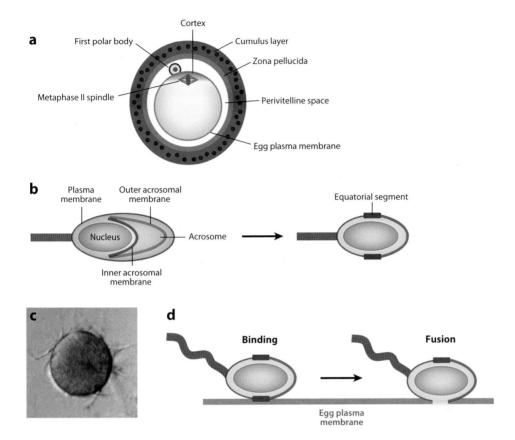

Figure 1

Diagrams illustrating the basics of sperm-egg interaction. (*a*) Schematic diagram of an ovulated mammalian egg. The egg is arrested in metaphase II (with the product of cytokinesis of meiosis I, the first polar body, nearby), and the spindle is sequestered near the cortex (the specialized region of the cortex involved in sequestering the spindle is in *orange*). The egg is surrounded by two extracellular coats: the cumulus layer, which develops coincident with ovulation, and the zona pellucida, which is synthesized by the oocyte during oogenesis. The sperm penetrates through these two layers to gain access to the perivitelline space and to the egg plasma membrane. Sperm-egg interactions take place on the microvillar domain, the region of the egg membrane away from the metaphase II spindle. (*b*) Schematic diagrams of the sperm head before and after acrosome exocytosis. The acrosome is a large secretory vesicle that is bordered by the outer acrosomal membrane and the inner acrosomal membrane. During acrosome exocytosis, the outer acrosomal membrane fuses with the overlying plasma membrane. Such fusion leads to the release of the acrosome contents and ultimately to the exposure of the inner acrosomal membrane as well as of the equatorial segment, both regions of the sperm surface that participate in interactions with the egg plasma membrane. (*c*) Micrograph from time-lapse imaging of in vitro fertilization of zona pellucida–free mouse eggs. The micrograph illustrates the extent to which sperm attach to the egg plasma membrane, as this egg was photographed prior to washes to remove the loosely attached sperm. (*d*) Schematic diagrams illustrating the distinct steps of sperm binding and fusion with the egg plasma membrane. Sperm binding (or adhesion), mediated by adhesion molecules on the sperm and egg, brings the membranes in close apposition. Then, by a mechanism that is not fully understood, membrane fusion occurs; the fusion event is distinguished by the establishment of cytoplasmic continuity between the two cells.

Fusogen/fusogenic molecule: a molecule that mediates the actual membrane fusion process specifically (as opposed to the plasma membrane interaction that precedes and leads to membrane fusion)

IVF: in vitro fertilization

ZP: zona pellucida

Perivitelline space: the space between the egg plasma membrane and the zona pellucida (**Figure 1***a*)

ADAM: a disintegrin and a metalloprotease

The term sperm-egg membrane interaction also distinguishes this interaction of the sperm with the egg proper from sperm interactions with the egg's extracellular vestments, the cumulus layer (cumulus cells embedded in a hyaluronic acid extracellular matrix) and the egg coat (the ZP in mammals and the vitelline envelope in amphibian species and numerous marine invertebrates) (**Figure 1***a*). Gamete plasma membrane interactions occur following sperm penetration of the cumulus layer and the ZP once the sperm gains access to the perivitelline space between the egg plasma membrane and the ZP (**Figure 1***a*). For an egg fertilized with these extracellular coats still on the egg, a sperm-egg interaction deficiency would manifest itself as sperm accumulating in the perivitelline space and sometimes even haplessly circling the egg membrane due to failure to form productive interactions between the gamete membranes. Sperm-cumulus and sperm-ZP interactions are key steps of the fertilization process, and the cumulus may also have additional functions in fertilization, but these are distinct events from sperm-egg membrane interactions. [The reader is referred to other overviews on these steps of fertilization (12–14).]

Mammalian sperm-egg membrane interactions occur in a spatially restricted manner (**Figure 1***b*). The sperm head has a large secretory vesicle, the acrosome, overlying the nucleus. Acrosome exocytosis has two important results: (*a*) release of the contents of the acrosome, which facilitates sperm gaining access to the perivitelline space, and (*b*) exposure and/or modification of the regions of the sperm head that participate in gamete binding and membrane fusion, the inner acrosomal membrane and the equatorial segment (12, 15–17). (See also figures 3 and 4 in Reference 12 for illustrations of sperm morphology before and after acrosome exocytosis.) Rodent eggs also have morphological and functional polarity. Sperm bind and fuse with the portion of the egg plasma membrane termed the microvillar domain, which is located away from the meiotic spindle (**Figure 1***a*). This asymmetric localization of microvilli has not been observed on human eggs (18), but eggs from nonrodent species may have other, less obvious asymmetries that may affect gamete membrane interactions. Finally, a number of beautiful microscopic studies provide insights into these membrane interaction events (please see References 12, 19, and 20 for citations to these previous studies).

APPROACHES TO AND CONSIDERATIONS FOR IDENTIFYING CANDIDATES

The original unbiased approach used to elucidate the molecular foundations of fertilization was the development of batteries of antigamete monoclonal antibodies (usually against sperm) that were tested in IVF assays for function-blocking activity and/or were used to examine antigen localization (e.g., References 21–23). This approach identified notable sperm proteins, including IZUMO1 (24) and ADAM1 and ADAM2 (where ADAM denotes "a disintegrin and a metalloprotease"; discussed more below) (25). A related and more recent approach has been to characterize the sperm proteome, particularly subpopulations of proteins, such as glycoslyated proteins, proteins that partition into a Triton X-114 detergent phase, or proteins in certain subcellular fractions (26–28). Such proteomic approaches have identified sperm equatorial segment protein 1 (SPESP1) (27, 29), transmembrane protein 190 (TMEM190) (28, 30), and three sperm acrosome-associated (SPACA) proteins [SPACA1, also known as sperm acrosomal membrane-associated 32 (SAMP32); SPACA3, also known as sperm lysosomal-like protein 1 (SLLP1); and SPACA4, also known as sperm acrosomal membrane-associated 14 (SAMP14)] (**Table 1**) (26, 31–33).

Forward genetics is the mainstay unbiased approach in systems such as *Drosophila, Caenorhabditis elegans, Arabidopsis, Chlamydomonas,* and more recently mouse, with programs such as the Jackson Laboratory's ReproGenomics program (34). The endpoint(s) assessed in such a screen

Table 1 Candidate sperm proteins for participation in sperm-egg membrane interactions

Protein	Summary of data	Still to be characterized or determined
Proteins associated with structures, fractions, functions of interest		
Equatorin	Novel protein localized in the equatorial segment (149). Monoclonal antibody MN-9 reduces sperm-egg fusion (23).	Knockout is still pending.
SPACA1 (SAMP32)	Inner acrosomal membrane protein. Antibodies reduce binding and fusion of human sperm to ZP-free hamster eggs (26).	Knockout is still pending.
SPACA3 (SLLP1)	Acrosomal matrix protein. Antibodies and recombinant protein reduce sperm-egg binding and fusion (32, 33); a putative egg partner was recently reported (162).	Knockout is still pending.
SPACA4 (SAMP14)	Inner acrosomal membrane protein. Antibodies reduce binding and fusion of human sperm to ZP-free hamster eggs (31).	Knockout is still pending.
TMEM190	Identified in a fraction of surface and vesicle proteins (28); colocalizes with IZUMO1 but does not coimmunoprecipitate with IZUMO1; knockout mice are fertile (30).	Knockout has been tested only in basic mating trials; additional analyses may reveal phenotypes.
CRISP1	Epididymal protein. Knockouts sire litters of normal sizes in normal time frames in conventional mating trials; null sperm have moderate deficiencies in sperm-egg interaction in in vitro assays (146).	CRISP1 and CRISP4 may have redundant functions (20), and thus a $Crisp1^{-/-}/Crisp4^{-/-}$ mouse (150) may have a more robust phenotype.
Known adhesion molecules in other cell types		
E-cadherin	Identified on human sperm. Antibodies reduce the fusion of human sperm to ZP-free hamster eggs (151).	Effects of male germ cell deletion of E-cadherin
N-cadherin	Identified on human sperm. Antibodies reduce the fusion of human sperm to ZP-free hamster eggs (152).	Effects of male germ cell deletion of N-cadherin
$\alpha_v \beta_3$ or $\alpha_6 \beta_1$ integrin	Characterized in mouse sperm; antibody-based inhibition (153)	Whether knockout/conditional knockout mice have male reproductive phenotypes
Enzyme activities		
Zinc metalloprotease activity	Inhibitors and zinc chelators reduce sperm-egg fusion (154).	The molecular identity of the protease(s) involved
Protein disulfide isomerases (PDIs)	Identified in a fraction of surface and vesicle proteins (28); inhibitors reduce sperm-egg fusion (155, 156).	The molecular identity of the PDI(s) involved

is largely dependent on the model system. The unicellular alga *Chlamydomonas* is easily observed during fertilization and thus is commonly used to identify mutations linked with defective gamete interactions (35, 36). In *C. elegans*, assessing brood size and certain gamete functions is also straightforward. Ovulation, oocyte production and morphology, and sperm motility and ability to reach the spermatheca of the hermaphrodite reproductive tract (where fertilization occurs in this species) can be observed in the worm's transparent body, making it possible to determine (*a*) whether normal gametes are produced and (*b*) whether the sperm and oocyte make contact but few or none of the oocytes appear to be fertilized or begin development (37). In contrast, mammals present broader challenges. Determining whether a male or female is infertile is clear-cut (i.e., a knockout or mutant produces no pups), but more subtle reproductive phenotypes require more effort to identify and characterize; this issue is discussed in more detail below.

Unbiased approaches have been complemented by candidate approaches; the reasons behind hypothesized candidates are varied [e.g., integrins on eggs became of interest with the discovery of an integrin ligand-like domain in a mammalian sperm protein (38)]. A more recent variation has been a combination of unbiased approaches and candidate approaches in mining databases for gamete-enriched genes with motifs of interest, such as transmembrane domains (e.g., Reference 39), with a subsequent focus on candidates of interest in this subpopulation. Such approaches have been especially useful with oocytes. Oocyte proteomics (40) is more challenging than sperm proteomics from the practical standpoint of collecting sufficient biological material, although this is certain to change as mass spectrometric methods advance. There is also serendipity; the tetraspanin CD9 is the best example (41–43), as this protein was never hypothesized to have a specific role in sperm-egg interaction.

Whether an unbiased approach or a candidate approach identifies a molecule of potential interest, a key test is assessing the effects of genetic deletion or mutation on reproductive function. As noted above, this has added complexity for mammalian models. A knockout mouse with a failure to produce any offspring demonstrates an essential, indispensable role in reproductive function for the molecule in question. On the flip side, if a mouse is fertile and does produce/sire a pup, then the molecule in question is not completely essential and/or has complementary molecules that work in redundant pathways. However, in these cases, there is value in considering not just production of any pups but also related endpoints such as time to pregnancy, litter size, and numbers of litters. Although these are more subtle phenotypes, they can still provide significant insights. Another type of subtle phenotype that has long been appreciated in classic genetics is synthetic lethality, in which a mutation in a single gene has little to no effect on viability but combining that first mutation with a mutation in another gene(s) results in a lethal phenotype. An analogous phenomenon of synthetic infertility also likely exists in some cases in reproductive function. Specifically, a deficiency in a single gene has a modest effect on reproduction, whereas the combination of that genetic deficiency with another genetic deficiency results in complete infertility. Mouse *Cd9* and *Cd81* are such an example. *Cd9*-null female mice are severely subfertile, *Cd81*-null females have a moderate loss of reproductive function, and $Cd9^{-/-}/Cd81^{-/-}$ females are completely infertile (44). Another example of synthetic infertility involves the *egg-1* and *egg-2* genes in *C. elegans*, with combined loss of function (by mutation or RNAi-mediated knockdown) of both *egg-1* and *egg-2*, rendering hermaphrodite animals infertile (37, 39). This foundation provides a rationale for the consideration of subtle (and still fertile) reproductive phenotypes as well as for the possibility of multigene disruptions and synthetic infertility in future analyses of reproductive function.

Infertility is a practical endpoint for large mutagenesis project (versus screening for a more subtle phenotype), and complete infertility is the endpoint of most interest for identification of possible contraceptive targets. For the purposes of dissecting reproductive function, the binary assessment of infertile versus fertile (i.e., no offspring versus any offspring) could result in missing or underappreciating certain molecular contributors to reproductive success, as some subtle phenotypes are likely to prove important. For example, male mice deficient in the polycystin-1 *Pkdrej* are fertile in conventional mating trials, but modified assays of in vivo sperm function as well as in vitro analyses reveal a role for PKDREJ in sperm competition, an evolutionarily conserved and significant component of reproductive function (45). Additionally, a genetic deficiency that does not have a dramatic in vivo effect on murine reproduction could have an appreciable effect on fertility in humans, especially considering human female reproductive physiology. Mice have much more frequent estrous cycles and ovulate more eggs per cycle than do human females, and thus a genetic deficiency that causes only a modest reproductive defect in mice (e.g., reduced litter size, longer time to achieve pregnancy) might manifest in humans as a failure to conceive after 12 months of unprotected intercourse (the clinical definition of infertility). Therefore, despite the experimental challenges in identification and analysis, there is significant value in considering subtle phenotypes, both those of subfertility in vivo and those that are unveiled in in vitro assays, in particular to make breakthroughs relevant to human reproduction.

THE CAST OF CHARACTERS

IZUMO1 on Sperm

The sperm protein IZUMO1 was first characterized through screening antisperm monoclonal antibodies for disruption of fertilization. IZUMO1 was identified by liquid chromatography tandem mass spectrometry as the antigen of the monoclonal antibody OBF13, which inhibits sperm-egg binding with ZP-free eggs (24, 46). IZUMO1 is a novel member of the immunoglobulin superfamily (IgSF) of proteins. Mouse IZUMO1 is an ∼56-kDa protein that appears to be testis specific. IZUMO contains one immunoglobulin-like domain with an N-glycosylation site (**Figure 2**) (47). There also appears to be a human ortholog (HGNC:28,539); antibodies to the putative human IZUMO1 cross-react with an ∼37-kDa protein in human sperm lysates and inhibit fusion of human sperm to ZP-free hamster eggs (47). The possible association between IZUMO1 abnormality and human infertility has been only preliminarily investigated (48, 49) and is certain to be a future area of interest.

The most significant data on IZUMO1 come from the knockout mouse. These mice appear healthy, and females have normal fertility. In contrast, $Izumo1^{-/-}$ males are infertile, despite the fact these males have normal mating behavior and ejaculation, as well as normal sperm motility and migration into the oviduct (47). IVF assays with $Izumo1$-deficient sperm show that the sperm penetrate the ZP and then accumulate in the perivitelline space, suggestive of defective interaction with the egg plasma membrane. Sperm from $Izumo1^{-/-}$ males attach to the membranes of ZP-free eggs, and eggs fertilized by intracytoplasmic sperm injection with $Izumo^{-/-}$ sperm develop into normal embryos that implant and develop to term. These results provide evidence that IZUMO1 is essential for sperm-egg fusion, although the precise function of IZUMO1—as a fusogen, as a regulator of a fusogen, and/or as an adhesion molecule—is still to be determined. IZUMO1 has no stereotypical features of a fusogen, but the immunoglobulin-like domain likely interacts with other proteins, either on other cells (*trans* interactions for function as an adhesion molecule) or on the sperm (*cis* interactions functioning through associated proteins in the membrane) (50). Structure-function analyses of IZUMO1 will be exciting but also demanding, as the most persuasive test is attempting transgenic rescue in the $Izumo1^{-/-}$ background (51). Thus far, researchers have demonstrated that mice expressing an N-glycosylation mutant form of IZUMO1 (N204Q) on the $Izumo1$-null background were not completely infertile, but these males sired fewer than 50% of the number of pups than did $Izumo1$-null mice that expressed wild-type IZUMO1 from a transgene. Sperm from these N204Q-$Izumo1$ males showed greatly reduced levels of sperm-egg fusion in IVF assays, as well as reduced amounts of the mutant IZUMO1 protein after epididymal transit (51). Thus, this N204-glycosylation site appears to be important for expression, organization, stability, or function of IZUMO1.

Proteins Working with or Affecting IZUMO1

Attention has turned to what sperm proteins might function with IZUMO1. The working hypothesis is that IZUMO1 is not necessarily a fusogen but may be associated with a fusogen. As of this writing, angiotensin-converting enzyme 3 (ACE3) has been identified as an IZUMO1-associated protein (**Figure 2**), although *Ace3* deletion does not have major effects on male reproductive function (52). *Ace3*$^{-/-}$ males sire pups in conventional matings, and *Ace3*-null sperm fertilize cumulus-intact, ZP-intact, and ZP-free eggs in IVF assays; the most noteworthy abnormality observed is that *Ace3*-deficient sperm have a modest mislocalization of IZUMO1 (52).

IZUMO1 localization is also affected by the deletion of the testis-specific serine kinase encoded by *Tssk6*, and *Tssk6*$^{-/-}$ mice have a more dramatic reproductive phenotype than do *Ace3*$^{-/-}$ mice. In *Tssk6*-null sperm, IZUMO1 fails to undergo its characteristic change in localization, leading to exposure on the sperm after acrosome exocytosis (47, 53). Moreover, *Tssk6*-null sperm are not able to fertilize eggs in vitro, and *Tssk6*-null males do not sire pups, although this male infertility is the result of multiple deficiencies (reduced sperm count, sperm motility, and morphological abnormalities) (53, 54).

The characterization of IZUMO1 has led to another unexpected discovery: the identification of a novel ∼150-amino-acid domain dubbed the Izumo domain. This domain has been found in four other proteins, some of which form heteromultimeric complexes on sperm (**Figure 2**) (55). At this time, it is unknown whether these proteins play any role in gamete interactions directly

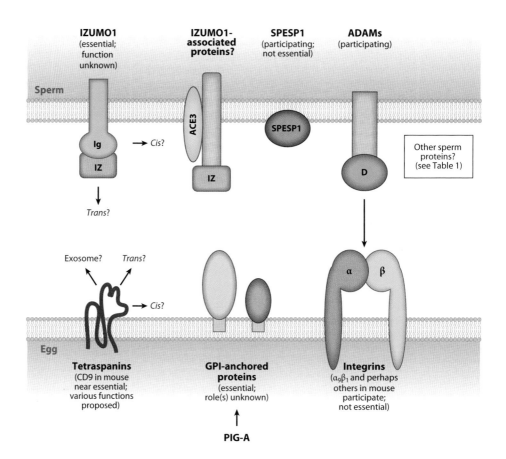

(i.e., through interacting with an egg protein) or indirectly (e.g., organizing key sperm proteins such as IZUMO1). However, the discovery of this motif may someday shed light on potential modes of IZUMO1 function and the possible actions of these proteins with IZUMO1, as well as uncover findings applicable to cell types beyond the gametes.

Vitelline envelope receptor for lysin (VERL): a protein in the vitelline envelope of abalone eggs

Additional Sperm Molecules of Note

The ADAM family has been of interest, building on the identification of sperm ADAM2 (fertilin β) in studies with a fertilization-blocking antibody and characterization as one of the founding members of the ADAM family (25, 38). Sperm ADAMs are binding partners for several members of the integrin family (table 4 in Reference 56); a number of these integrins are expressed in eggs and can participate in sperm-egg interactions (see below). This finding puts specific ADAM-integrin pairs, especially ADAM2 and the integrin $\alpha_9\beta_1$ (**Figure 2**) (10, 57–59), in a category with abalone sperm lysin and egg VERL (vitelline envelope receptor for lysin) as cognate binding partners on the two gametes. Although these data are notable, we are still far from a complete picture of mammalian fertilization, particularly considering the robustness of the *Izumo1*-null phenotype. Multiple ADAMs are expressed on male germ cells, and several *Adam*-null mice have been generated (60–64). In several of these *Adam* knockouts, the sperm show reduced migration

Figure 2

Sperm and egg molecules implicated in gamete membrane interactions. This diagram illustrates the molecules proposed to participate in sperm-egg membrane interactions. IZUMO1 on the sperm is an immunoglobulin superfamily member [with an immunoglobulin-like domain (Ig)] that is essential for sperm-oocyte fusion (47). As noted in the text, the function of IZUMO1 is not entirely clear; IZUMO1 may function by interacting with a molecule on the egg (in *trans*) and/or may act through IZUMO1-associated proteins (in *cis*). IZUMO1-associated proteins are beginning to be identified; ACE3 (angiotensin-converting enzyme 3) is one such protein, and the *Ace3* knockout shows a slight abnormality in the localization of IZUMO1 (52). IZUMO1 has an ∼150-amino-acid domain termed the Izumo domain (IZ). This domain has been found in other proteins, and these proteins form multimeric complexes (55), although the functions of these proteins in gamete fusion are not known. The tetraspanin CD9 is the major player identified thus far on the mouse egg (41–43) and is likely to function in conjunction with another tetraspanin, CD81, as $Cd9^{-/-}/Cd81^{-/-}$ female mice are completely infertile (44). Multiple functions for CD9 have been proposed (see text). CD9 may work by interacting with a sperm protein in *trans* (although data supporting this mode of action are minimal), by regulating other egg membrane proteins in *cis*, and/or through exosome-mediated release. Glycosyl phosphatidylinositol (GPI)-anchored proteins on the egg are also implicated on the basis of the phenotype of the oocyte-specific conditional knockout of phosphatidylinositol glycan anchor biosynthesis, class A (PIG-A), and the function(s) of egg GPI-anchored proteins is unknown. A few other players are noted here. Sperm equatorial segment protein 1 (SPESP1) is one, as $Spesp1^{-/-}$ males produce sperm with reduced ability to undergo sperm-egg fusion (74). SPESP1 does not have an obvious hydrophobic transmembrane domain but may bind to membranes (27) and thus is depicted here as a peripheral membrane protein. Several sperm ADAMs (where ADAM denotes "a disintegrin and a metalloprotease") have been implicated in sperm-egg interaction. Although no single ADAM is essential, there appears to be a correlation between the presence/levels of certain ADAM proteins and the ability of sperm to interact with the egg membrane (see Reference 20 for more information). Thus, ADAMs may function in redundant roles, consistent with the observation that ADAMs have similar adhesion-mediating motifs for interacting with integrins via their integrin ligand-like disintegrin domain (D) (57, 157–159). Integrins are heterodimeric membrane proteins composed of an α subunit and a β subunit; 18 α subunits and eight β subunits form at least 24 different combinations. In vitro studies of eggs with reduced amounts of α_9 revealed reduced sperm-egg binding and fusion, in agreement with the finding that several ADAMs can interact with $\alpha_9\beta_1$ (10, 57). α_9's likely β-subunit partner is β_1, although α_9 can also dimerize with at least one other β subunit (10, 59, 109, 160). Data on other sperm proteins not pictured here are summarized in **Table 1**.

into the oviduct through the uterotubal junction, reduced binding to the ZP and/or reduced binding and fusion to the egg plasma membrane, and abnormalities in the sperm surface proteome with loss of multiple ADAMs (summaries in References 20 and 65). However, deficiencies in sperm-egg interaction in sperm from *Adam* knockouts are not as dramatic as in *Izumo1*-null sperm. The *Adam2*$^{-/-}$ knockout has the most serious defects in gamete membrane interactions and has the lowest overall levels of several ADAM proteins on the sperm surface (60, 62, 66), but other *Adam* knockouts have little to no apparent effect on male reproduction (e.g., References 64 and 67).

Loss of ADAM2 or ADAM3 correlates with sperm functional deficiencies. Additionally, ADAM3 is reduced or lost in unrelated, non-*Adam* knockouts, and these knockouts have various abnormalities in sperm function (68–70). Thus, ADAM3 may have a role in regulating aspects of mouse sperm function or, at the very least, may be a marker of sperm function. The loss of ADAM3 is associated with sperm having abnormal functionality [although, in a cautionary comment, the human *ADAM3A* (HGNC:209) and *ADAM3B* (HGNC:210) genes are pseudogenes that do not code for a functional protein (71, 72)]. Insights may be gleaned from comparing the surface proteome of wild-type sperm and *Adam3*-null sperm. One report did not detect any differences (28), but more subtle changes may present on *Adam3*-null sperm, such as in the localization or functionality of key proteins. ADAM3 has been proposed to be a sperm protein that mediates sperm-ZP interaction on the basis of findings that *Adam3*-null sperm bind poorly to the ZP (61, 62) and that incubation of solubilized ZP proteins with sperm lysates pulls down ADAM3 (73). Although these data may indicate that ADAM3 binds a ZP component(s) directly, it is also possible that *Adam3*-null sperm lack critical proteins for ZP interaction [as *Adam3*-null sperm have an altered surface proteome, with reduced amounts of several ADAMs (62, 65, 66), and thus may lack other proteins as well]. In this model, ADAM3 would be associated with this molecule(s) on wild-type sperm and thus would pull down in a complex with ZP proteins.

A number of sperm proteins associated with sperm structures or biochemical fractions of interest have been investigated for their roles in gamete membrane interaction. Of these, SPESP1 has been studied to the greatest extent, with characterization of a knockout mouse (**Figure 2**). *Spesp1*$^{-/-}$ males produce slightly smaller litters than do wild-type controls. Sperm of *Spesp1*$^{-/-}$ males have reduced ability to undergo sperm-egg fusion and display delayed migration through the female reproductive tract compared with sperm of wild-type controls (74). The deletion of *Spesp1* also affects biochemical and localization characteristics of IZUMO1 and of equatorin, another sperm protein possibly involved in sperm-egg membrane interaction (**Table 1**). In addition, *Spesp1* deletion affects overall membrane morphology; scanning electron microscopy reveals loss of the equatorial segment membrane in these sperm. **Table 1** summarizes studies of other sperm proteins implicated in sperm-egg interaction, including equatorin, SPACA1, SPACA3, SPACA4, TMEM190, and CRISP1, and various adhesion molecules and enzyme activities.

Tetraspanins on Eggs

The phenotype of the *Cd9* knockout mouse was a surprise as well as a breakthrough for the gamete interaction field. CD9 is a member of the tetraspanin family (so named because members have four transmembrane domains) (75, 76). *Cd9*$^{-/-}$ females, although not completely infertile, are severely subfertile. Mating trials reveal that *Cd9*$^{-/-}$ females produce few litters, few pups, and, in some cases, no litters (41–44). *Cd9*$^{-/-}$ females that are able to become pregnant have delayed onset of pregnancy (42, 44). In IVF, very few *Cd9*-null eggs are fertilized (41–44). The *Cd9* knockout mouse phenotype is remarkable in several ways. CD9 is widely expressed in the body and yet has been identified to have an essential function only in the egg; the *Cd9* knockout mouse as a whole remains viable and healthy, except for compromised female fertility. CD9 is one of more than

30 mammalian tetraspanins, and multiple tetraspanins are expressed in mouse eggs, yet *Cd9* deletion still leads to significant reduction of female fertility, indicating that the remaining tetraspanins on the egg do not compensate for the loss of CD9. These two aspects of *Cd9*/CD9 biology combine for a striking and likely uncommon knockout phenotype: A widely expressed protein has a highly specific effect in a genetic knockout with major effects manifested only in the egg, and a member of a protein family with other members expressed in the same cell type has a specific, nonredundant function.

CD81 is a related tetraspanin that is 45% identical to CD9. The $Cd81^{-/-}$ mouse also shows defects in female fertility and sperm-egg interaction with in vivo–fertilized and in vitro–fertilized eggs, although several endpoints were not as severely affected as in *Cd9*-null females/eggs (44). $Cd9^{-/-}/Cd81^{-/-}$ female mice are completely infertile, and thus this combination of gene disruptions represents a case of synthetic infertility (see above). Expression of mouse CD81 or human CD9 in *Cd9*-null eggs, achieved through microinjection of mRNA, partially rescues the $Cd9^{-/-}$ fusion-defective phenotype (77). Little is known of tetraspanin involvement in human fertilization, although there are data from antibody inhibition studies. Two different anti-CD9 antibodies have no effect on the fusion of human sperm with human ZP-free eggs (78), whereas anti-CD9 antibodies have inhibitory effects on sperm binding and fusion with mouse or pig ZP-free eggs (11, 79, 80). Human sperm-egg fusion is partially inhibited by treating eggs with an antibody to a different tetraspanin, CD151 (78). No reproduction defects have been reported in $Cd151^{-/-}$ mice or in humans with mutated forms of *CD151*, although reproductive function parameters have not been extensively assessed (81–84). These data preliminarily raise the possibility that sperm-egg interaction in different mammalian species may rely on different members of the tetraspanin family.

Although the importance of CD9 in mouse sperm-egg interaction is clearly established (41–44), the exact function(s) of CD9 in sperm-egg interaction is not known. Several tetraspanins, including CD9 and CD81, have apparently indirect (and still poorly defined) roles in membrane fusion processes (85). One tetraspanin, peripherin-2, has membrane fusion–promoting activity as a purified protein in cell-free membrane fusion assays, with this activity ascribed to the amphiphilic fusion peptide–like domain on its C terminus (86), but other tetraspanins lack this domain. The large extracellular loop in multiple tetraspanins is important for interactions with other membrane proteins (**Figure 3a**). Amino acids 173–175 (with the sequence SFQ) in this large extracellular loop of mouse CD9 are crucial for CD9 function in sperm-egg interaction (87).

Although a direct role of CD9 in membrane fusion is uncertain, there are nevertheless intriguing possibilities for the role(s) tetraspanins could play in gamete interactions. Through interactions with other proteins in *cis*, tetraspanins create tetraspanin-enriched microdomains, which are a distinct type of membrane microdomain (**Figure 3b**) (88, 89). Thus, *Cd9* depletion (and perhaps *Cd81* depletion to a lesser extent) could have broad effects on mouse egg membrane functionality (**Figure 3c**). Different lines of experimental evidence suggest various potential functions for egg CD9. The morphology of egg microvilli is disrupted in *Cd9*-null eggs (90), in agreement with a general role of CD9 in egg membrane function and order (**Figure 3**). Investigators have also proposed that CD9 is released from the egg surface in exosome-like structures [40–100-nm vesicles released from the cell surface (91)] and that these CD9-containing exosomes associate with sperm to facilitate gamete interaction (92, 93), although data from a very similar study do not support this hypothesis (94). Nevertheless, the possible role of egg-based exosomes in fertilization is intriguing, and multiple members of the tetraspanin family are reported to be components of exosomes [91; ExoCarta database (**http://www.exocarta.org**)]. Whether exosome release from eggs occurs and whether exosome-based release of CD9 from the egg surface is a component of how CD9 functions (and/or is a mode of how CD9 affects egg membrane order) merit further investigation. Finally, recent innovative analysis of sperm-egg interaction reveals that *Cd9*-deficient eggs have a reduced ability to strongly adhere to sperm (95). This finding is in agreement with (*a*) observations of sperm accumulating in the perivitelline space of *Cd9*-null eggs and only transiently attaching to the egg surface (44) and with (*b*) data from studies with function-blocking anti-CD9 antibodies that suggest a role of egg CD9 in adhesion strengthening (96). These data

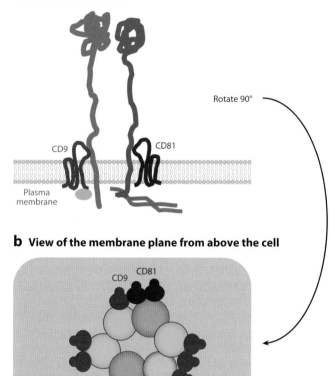

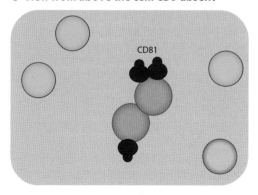

suggest that egg CD9 participates in some way in sperm-egg binding rather than in fusion. These possible modes of CD9 action (in microvillar structure, in exosomes, and in adhesion strengthening) are not necessarily mutually exclusive.

CD9 and other tetraspanins are likely involved in the regulation of membrane order by associating with and/or facilitating the activity of other membrane proteins. CD9-associated proteins in other cell types include several integrins and other adhesion molecules, IgSF members, ectoenzymes, and various intracellular signaling molecules (97–101), but there is only limited knowledge about CD9-associated proteins in eggs. IgSF8 (also known as EWI-2, KAI/CD82-associated protein, and prostaglandin-regulatory-like protein; MGI:2,154,090) coimmunoprecipitates with CD9 in egg lysates (11) and is absent from the surfaces of *Cd9*-null eggs (102). IgSF8 was not detected in anti-CD81 egg immunoprecipitates (11), so IgSF8 may associate specifically with CD9, or CD81-associated IgSF8 may be present in amounts below the threshold of detection. Finally, an anti-IgSF8 antibody has a subtle inhibitory effect on sperm-egg binding (11). A second egg CD9-associated protein is PTGFRN [prostaglandin F2 receptor negative regulator; also known as EWI-F or CD9 partner 1 (100, 103); MGI:1,277,114) (90)]. Preliminary studies did not identify an effect of an anti-PTGFRN antibody on mouse fertilization (44), although impaired gamete interactions could be detectable under certain IVF conditions with perturbation of PTGFRN, as was observed for IgSF8 (11). *Igsf8*-null and *Ptgfrn*-null mice have not been analyzed.

Glycosyl Phosphatidylinositol–Anchored Proteins on Eggs

Research that identified glycosyl phosphatidylinositol (GPI)-anchored proteins as potential fertilization proteins began as a candidate approach (104), testing whether treating mouse eggs with phosphatidylinositol-specific phospholipase C (PI-PLC), which cleaves GPI-anchored proteins, would impair fertilization. PI-PLC-treated ZP-free eggs show dramatically reduced levels of reduced sperm-egg binding and fusion, and this effect was not observed when sperm were treated with PI-PLC (104). Subsequent studies used genetic means to create oocytes lacking GPI-anchored proteins through an oocyte-specific knockout of the *Piga* gene (phosphatidylinositol glycan anchor biosynthesis, class A); PIG-A is a subunit of an N-acetylglucosaminyl transferase that participates in the first steps of GPI synthesis (105). These females produced no pups in mating trials in which they were housed with a male for three weeks (106). Moreover, very few two-cell embryos were produced from in vivo fertilization, and sperm were observed in the perivitelline space. Fertilization of *Piga*-deficient eggs can occur but is dramatically reduced; sperm fusion with ZP-free *Piga*-deficient eggs in vitro was ∼90% reduced, and binding was reduced by ∼40% (106).

Figure 3

Possible function of tetraspanins in membrane order. (*a*) CD9 (*red*) and CD81 (*purple*) in a side view through the egg membrane. CD9 and CD81 are shown associated in *cis* with other membrane proteins (*green, dark blue*) and with unspecified cytoplasmic proteins (*light blue, orange*). This view in panel *a* is roughly to scale; tetraspanin proteins are small and likely extend up from the plane of the bilayer only ∼5 nm (75, 161), and other membrane proteins can extend to a height of 100+ nm. CD9 and CD81 are drawn with their characteristic four transmembrane domains and small and large extracellular loops, with interactions with other membrane proteins mediated by the large extracellular loop. (*b*) The image has been rotated 90° so that the view is from above the cell, looking down on the plane of the membrane. The "snowman" appearance here of CD9 (*red*) and CD81 (*purple*) is reflective of the large and small extracellular loops. This perspective illustrates how tetraspanins and their associated proteins (*green, blue,* and *orange*) form a tetraspanin-enriched microdomain in the membrane. (*c*) The hypothesized effects of CD9 depletion, with disruption of the organization and composition of the tetraspanin-enriched microdomain.

The major remaining questions are what GPI-anchored proteins on the egg surface are important and whether gamete interaction deficiency is due to the introduction of membrane disorder through the depletion of this class of proteins. The question of what proteins are lost has been only minimally investigated. Two-dimensional gel analysis of proteins that were released from PI-PLC-treated eggs (104) has so far shown that one known GPI-anchored protein, CD55, is decreased on *Piga*-deficient eggs (106). A membrane lacking all GPI-anchored proteins may be generally nonfunctional, particularly with the *Piga* conditional knockout producing long-term, chronic depletion of GPI-anchored proteins for nearly the entire lifetime of the oocytes (i.e., while the *Zp3* promoter drives Cre recombinase expression and leads to *Piga* excision) (106). The long-term loss of GPI-anchored proteins may alter the composition, organization, and/or microdomain structure of the egg membrane so that sperm interactions are not favored, particularly considering that GPI-anchored proteins are enriched in specific membrane microdomains (detergent-resistant membranes) (89).

Integrins on Eggs

The identification of an integrin ligand-like domain in ADAM2, an antigen of a function-blocking antisperm antibody (38), increased interest in the potential role of egg integrins in sperm-egg interaction. Knockout mouse studies have since downgraded the significance of integrins in fertilization, although the nuances here merit brief comment. Mouse eggs express 3 of 8 integrin β subunits and 8 of 18 integrin α subunits (*Itgb1*, *Itgb3*, *Itgb5*, *Itga1*, *Itga2*, *Itga3*, *Itga5*, *Itga6*, *Itga8*, *Itga9*, and *Itgav*) (59 and references therein) and thus can express at least 10 different α-β integrin combinations based on known heterodimer pairs (59). Several of these, especially ITGA9-ITGB1 ($\alpha_9\beta_1$), interact with several ADAMs (56). No fertility defect is reported in the viable knockouts of egg-expressed integrins (*Itga1*, *Itga2*, *Itgb3*, *Itgb5*). The other 7 in this list are embryonic or neonatal lethal, but 4 of these integrin subunits have been investigated in eggs by oocyte-specific conditional knockout [*Itgb1* (107)], by kidney capsule transplants of ovaries from knockout neonates [*Itga3*, *Itga6* (108)], or by RNAi-mediated knockdown [*Itga9* (10)]. Of these, the *Itga9*-deficient eggs have the most obvious deficiency. Eggs deficient in *Itga3*, *Itga6*, or *Itgb1* can be fertilized in vitro, and female mice with *Itgb1*-deficient eggs have no deficits in fertility (107, 108). However, *Itgb1*-deficient eggs show a slight delay in sperm-egg binding in time-lapse video analysis (109), so subtle defects may also be detectable with *Itga3*- and *Itga6*-deficient eggs in modified assays (10, 95). Eggs with 50% knockdown of ITGA9 on the egg surface show a reduction in sperm-egg binding and fusion and in the percentage of eggs fertilized compared with controls (10). ITGA9 knockdown did not result in a complete loss of fertilizability, which may be because of only partial reduction of ITGA9 on the egg surface and because other molecules on the egg surface likely function in gamete membrane interaction. Nevertheless, egg integrins, particularly the ITGA9 α subunit, have the potential to contribute to sperm-egg interactions with wild-type eggs, even though data from knockouts demonstrate that eggs deficient in various integrins are still able to be fertilized.

Insights on Candidates from Nonmammalian Organisms

Although the mouse is the experimental model system that is most widely used and that has the closest relevance to humans, there is tremendous value in considering what can be gleaned outside the class Mammalia. Fertilization biology has an incredibly rich history; gametes were among the first cells ever studied under the microscope (witness van Leeuwenhoek's animalcules). Marine invertebrates, such as echinoderms, mollusks, and ascidians, have been used classically because of the relative ease of collecting gametes or early embryos. More recently, genetic model systems have become major contributors to the field, owing to conservation of the general cellular processes occurring during fertilization. In terms of sperm-egg interaction, the outlier may be *Drosophila*, as *Drosophila* gametes use a mechanism of sperm-egg merger that involves sperm entering the egg with their membranes intact; the sperm membrane is then broken down in the egg cytoplasm. Nevertheless, discoveries in this system can still elucidate the underpinnings of gamete membrane functions. Sperm plasma membrane breakdown requires a sperm protein termed Sneaky, which, interestingly, has some features of other proteins noted here [such features include multiple transmembrane domains, as do the tetraspanins, and a dendritic cell–specific transmembrane protein (DC-STAMP) domain, as does SPE-42, discussed below] (110); dissecting Sneaky functions could provide key insights for fertilization as well as for membrane fusion.

Forward genetics studies in *C. elegans* are producing an interesting list of molecules. Seven *C. elegans* genes (*spe-9*, *spe-13*, *spe-36*, *spe-38*, *spe-41/trp-3*, *spe-42*, *fer-14*), when mutated, produce animals with sperm appearing normal in most respects (e.g., morphology and motility), but the sperm fail in fertilization, in spite of making contact with eggs in the spermatheca (37). The product of *spe-41/trp-3* is a TRPC channel (111), which underscores how this phenotype may not necessarily translate to the gene product having a direct function in sperm-egg interaction. A TRPC protein may function as a gamete interaction molecule (adhesive or fusogenic), but another possible function is that channel activity somehow primes the sperm for sperm-egg interaction. [Additionally, worms with null alleles of *spe-41* are severely subfertile, rather than completely infertile (111).] Other worm candidates that have been cloned thus far are intriguing. *Spe-42* encodes a protein with a DC-STAMP motif (112, 113), as was found in *Drosophila* Sneaky (110). The DC-STAMP protein, also known as TM7SF4, is essential for osteoclast cell fusion (114), positioning SPE-42 as an exciting candidate to mediate worm sperm-egg fusion. Other worm sperm proteins have interest as well: SPE-9 bears ten EGF-like repeats (115), FER-14 is a novel membrane protein (unpublished but noted in Reference 37), and SPE-38 is a protein with four transmembrane domains (although not a bona fide tetraspanin) (116).

On the oocyte side, *C. elegans egg* mutants produce oocytes that cannot be fertilized and/or cannot be activated (37). Two of the proteins encoded by *egg* genes, EGG-1 and EGG-2, are transmembrane proteins with extracellular domains containing eight low-density lipoprotein (LDL) receptor repeats (39), and thus are candidates to mediate sperm-oocyte interaction or perhaps oocyte membrane functionality. Hermaphrodites lacking functional *egg-1* or *egg-2* are subfertile. Lack of both *egg-1* and *egg-2* renders hermaphrodites infertile, and thus lack of *egg-1* and *egg-2* represents a case of synthetic infertility. The other three, EGG-3, EGG-4, and EGG-5, are protein tyrosine phosphatase–like proteins, also known as pseudophosphatases; they do not overtly participate in gamete interaction but, in the oocyte, appear to mediate fertilization-triggered responses that are part of the oocyte-to-embryo transition (117–120).

Some of the most impressive findings have come from two plants (*Arabidopsis thaliana* and *Lilium longiflorum*) and from the green algae *Chlamydomonas reinhardtii*, with the characterization of the protein HAP2/GSC1 (121). [HAP originates from hapless in *Arabidopsis* (122), and GSC denotes generative cell-specific protein, originally characterized in the lily (123).] Impressively, these results extend to other organisms as well. In *Arabidopsis*, *hap2/gsc1* mutants have abnormal pollen tube growth such that sperm show reduced targeting to the ovule; the *hap2* sperm that do reach the ovules fail to initiate fertilization (122–124). The HAP2 homolog in *Chlamydomonas* also functions in fertilization, having been identified in a forward genetics screen (125). There are HAP2/GCS1 orthologs in numerous species (reviewed in Reference 121), and HAP2 in the protozoan *Plasmodium berghei* (the *Plasmodium* species that causes malaria in rodents) functions in fertilization as well (125). Added evidence for the importance of this protein comes from

HAP2/GSC1: a protein implicated in gamete interaction in a variety of organisms (*Arabidopsis*, *Chlamydomonas*, lily, and a *Plasmodium* species); the name HAP is for hapless in *Arabidopsis*, and GSC denotes generative cell-specific protein, originally characterized in the lily

the finding that *Chlamydomonas* HAP2 [as well as a second fertilization-involved protein, FUS1 (35)] is degraded following gamete fusion, constituting a block to polygametic fertilization (36). HAP2/GSC1 is a membrane-anchored protein but otherwise has no domains of known function; recent structure-function analysis has produced some mixed results, although the extracellular N terminus appears critical for function (126, 127). There are still elusive pieces of information (the reader is referred to box 3 in Reference 121), including the apparent lack of HAP2/GCS1 orthologs in the deuterostome lineage, fungi, and *C. elegans* [although HAP2/GCS1 orthologs are present in other animal species (see figure 1 in Reference 121)]. This exact protein may not be the only answer to how gamete fusion occurs for all organisms, but the elucidation of mechanism will be informative, and that mechanism may span phyla.

Last but not least, we would be wise to keep in mind a number of other molecules such as bindin, lysin, and sp18 (which is also referred as the 18-kDa sperm protein) (128–130). Such molecules have rich histories in powerful experimental systems that have been used for decades and that continue to provide intriguing insights. This is particularly evident as structural data augment sequence data. One example is an intriguing case involving ZP domain proteins, which were originally characterized in mammalian ZP proteins (131). There are structural similarities between vertebrate ZP domains, the abalone VERL protein (a component of the vitelline envelope), and even *Saccharomyces cerevisiae* α-agglutinin/Sag1p (a protein involved in yeast mating) (132). Such similarities suggest conservation of the mechanisms underlying gamete interactions. Although these data on ZP proteins, VERL, and α-agglutinin are more directly relevant to sperm–egg coat interaction, there may also be connections to sperm–egg membrane interaction. The sperm protein that binds to VERL is lysin; lysin dissolves a pore in the vitelline envelope, allowing the sperm to penetrate this egg coat and to gain access to the egg membrane. Another abalone sperm protein, sp18, is a paralog of lysin, with sequence similarity indicative of duplication from a common ancestral gene but divergence of functionality. Sp18 does not appear to act on the vitelline envelope as lysin does. Sp18 may mediate sperm-egg fusion on the basis of in vitro studies with purified sp18 and liposomes (133). Multiple vitelline envelope ZP domain proteins (VEZPs) have been identified in the abalone vitelline envelope, and interactions between certain VEZPs and lysin and/or sp18 are detected in affinity chromatography assays (134), raising speculation that an interaction between sp18 and a VEZP-like protein on the egg membrane may function in abalone gamete membrane interaction.

MOVING FORWARD

One theme emerging from this body of work is that membrane order and organization appear important for gamete membrane functionality. CD9 on the egg is the most obvious case, but data also point to sperm proteins IZUMO1, ADAMs, and SPESP1 participating in membrane organization and function. It has long been appreciated that macroscale membrane order (membrane topography) on the egg contributes to sperm-egg interactions. Electron microscopy studies show that sperm interact intimately with the microvilli on the egg's surface (135, 136), and *Cd9* deletion alters egg membrane topography (90). Analogously, deletion of *Spesp1* alters the sperm surface, apparently rendering the sperm somewhat less capable of undergoing fertilization (74).

Moreover, more microscale membrane order—such as organization of protein complexes on the cell surface—is also critical for gamete membrane functionality. Part of the function of CD9 and CD81 in mouse eggs is likely to be through the well-known role of these proteins in tetraspanin-enriched microdomains. Tetraspanin-enriched microdomains are impressive in size, with areas of up to 200–400 nm^2 (88), and may affect membrane topography as well as the proposed exosome-associated function of CD9 (92, 93). In sperm, deletion of certain ADAMs [as well as of chaperones in their biosynthetic pathway (70, 137, 138)] produces pleiotropic defects in sperm function. Additionally, IZUMO1's function in the sperm may be linked to the proteins with which IZUMO1 is associated, and research is ongoing to identify IZUMO1-associated proteins and to assess whether phenotypes associated with loss of that gene product mimic those of *Izumo1* deletion.

These effects on gamete membrane function may occur as a result of disorder in the membrane and/or alteration of the gamete surface proteome through defects during protein biosynthesis that cause inappropriate complex formation or through protein instability and posttranslational loss. For proteins that form heteromultimers or are parts of complexes (as is the case for many of these gamete proteins), the loss of one member can affect the functionality or surface expression of its partner(s). Furthermore, deletion of secretory pathway proteins, chaperones, and proteins involved in posttranslational modifications leads to defects in fertility and gamete function (68, 70, 106, 137–139). Characterization of the gamete surface proteome as well as of the organization of those proteins on the gamete surface will be illuminating. Both gamete surface proteome and surface protein organization could be affected in knockouts with defects in sperm-egg membrane interaction. Proteomic methods, as well as the assessment of proteins in situ through advanced microscopic techniques, such as super-resolution microscopy, will be valuable.

As discussed above, there is significant value in considering mouse knockout phenotypes other than complete infertility. One key issue is that molecular loss of function that produces a minimal reproductive phenotype in a mouse may have a more significant effect on fecundity in humans. Moreover, certain gene products may not be essential for reproduction but may be advantageous for reproductive success; such genes would be maintained in the genome as a result of conveying this reproductive advantage to individuals that have them. Another simple approach is to pay detailed attention to the assessments used to conclude that an animal is fertile or infertile. Knockout studies show that different results can be obtained in colony maintenance versus mating trials of various designs (45, 140), so statements about fertility and infertility merit a careful look at the methods used. Fertility and fecundity depend heavily on the circumstances in which animals are offered the opportunity to reproduce, and it is worth emphasizing how much these circumstances can vary, even in a laboratory setting. Clearly there are differences between matings that are used for colony maintenance and matings that are part of a specifically designed mating trial—and these living conditions in turn differ from murine reproduction in a natural setting. Another leap in logic is then required to draw conclusions for how mating trials for experimental mice apply and provide insight relevant to human reproductive health.

Additionally, the field is making advances with more careful phenotypic analyses, as knockouts that were once deemed to have no reproductive phenotype are being found to have a reproductive phenotype upon closer investigation. For example, the knockout of the Src family kinase encoded by *Fyn* was originally characterized in 1992 to have hippocampal and olfactory bulb abnormalities (141). However, nearly 20 years later, *Fyn*-null mice were found to have a female subfertile phenotype (142). Such open-mindedness is valuable for the study of all reproductive functions [e.g., deficiencies uncovered by specific challenges of in vivo gamete function (45, 143)]. There are numerous examples of knockout mice that are not completely infertile and that perhaps even produce litters with size and timing comparable to those of controls but that have gametes with defects revealed in more specific assays (e.g., References 44, 144–146).

Genetic approaches have unquestionably revolutionized reproductive biology, but there are nevertheless gaps remaining. First, no mammalian protein involved in gamete membrane interactions has been identified by an unbiased genetic approach (although proteins involved in other reproductive functions, like meiosis and spermatogenesis, have). Second, genetic approaches have paid off in identifying molecules on the male/sperm side, but there are fewer big wins on the female/egg side. [This parallels what has been observed in the ReproGenomics study, with the identification of substantially more male infertile lines than female infertile lines (**http://reproductivegenomics.jax.org**).] Third, thus far there are relatively few knockouts—*Izumo1* and *Cd9*—with dramatic and highly specific losses of function in gamete membrane interaction. These differences in discovery are probably linked to multiple factors: (*a*) Many of the genes involved in sperm development and function are specific to or are highly enriched in male germ cells; (*b*) proteins involved in gamete membrane interactions, especially on the egg, may have functions in other cell types (and thus a knockout/mutant animal dies before reaching reproductive age or has a pathological problem that indirectly precludes reproductive function); and/or (*c*) other proteins can compensate for the loss of a single protein. As discussed above, the *Cd9* knockout may be one in a million; it is remarkable in that CD9 is expressed throughout the body and that there are other tetraspanins expressed by eggs, but the *Cd9* knockout affects the egg membrane's ability to support sperm interactions. Females in the ReproGenomics studies that became pregnant (presumably with any number of pups, even one or a few) were not used for further analysis (147). Subfertility might be a more likely outcome from mutation or knockout of a single gene, and complete female infertility would be a rarer occurrence; multiple gene disruptions are required to produce full infertility (i.e., synthetic infertility). On the flip side, the *Cd9* and *Piga* knockouts may mimic a multigene knockout through the disruption of the function of multiple proteins on the egg surface, which may be why their null phenotypes are so robust.

Elucidating the function of a protein of interest could require varied methods of experimental manipulation, particularly considering that assembly of a gamete membrane of proper composition and organization seems to be important for sperm-egg interactions. Different phenotypes could be observed with chronic depletion of a gene product over the lifetime of the organism (e.g., conventional gene knockout) versus more acute, short-term disruption (e.g., a temporally controlled conditional knockout through tetracycline-controlled transcriptional regulation, RNAi, or small-molecule inhibitors/activators). The chronic depletion of proteins broadly involved in generating a functional membrane (e.g., tetraspanins, GPI-anchored proteins, chaperones, and even subunits of multimeric complexes) could have severe effects, whereas acute manipulation could be a more fine-tuned alteration of protein function. The acute intervention could also allow less time for redundant, backup pathways to fall into place and could overcome the challenges of embryonic/neonatal lethality.

With all the progress that has been made, a significant, lingering, unanswered question is what the mammalian gamete fusogen is—or, even more generally, precisely how mammalian gamete fusion occurs. This is a question not just of whether the fusogen is on the sperm and/or on the egg but of why, after decades of fertilization research, a mammalian gamete fusogen has not been identified. Such a molecule might be difficult to identify by genetic approaches because of the challenges noted above. Such a molecule might also be difficult to identify by biochemical or antibody-based methods if it were present in low amounts and/or were only transiently exposed in a functional conformation. Fusogenic proteins that mediate other types of membrane fusion events have been characterized, but the strongest candidates thus far for gamete fusogens are sp18 in abalone and HAP2/GSC1 in *Arabidopsis*, *Chlamydomonas*, and other species, as well as possibly SPE-42 in worms. More information on all of these is needed. Orthologs of sp18 and HAP2/GSC1 have not been identified in mammals purely on the basis of sequence homology. More sophisticated computational approaches, such as protein threading and sequence-structure homology recognition [as was used to uncover similarities between ZP domains and motifs in abalone VERL (132)], could provide advances here.

The field of gamete biology is at an interesting junction; certain components of fertilization, such as the functions of the cumulus and the ZP, are being actively revisited and debated (e.g., Reference 148). But although some fertilization processes are being reevaluated, sperm interaction with the egg membrane and entry into the egg cytoplasm are unquestionably essential parts of fertilization. Although we still do not know the full mechanistic basis for sperm-egg interaction, work to date provides a strong foundation. Further steps in this exciting research area should exploit various state-of-the-art technologies to yield the next breakthroughs to understand the merger of sperm and egg.

SUMMARY POINTS

1. Gamete proteins that participate in sperm-egg interactions may do so in multiple ways. These proteins not only may act as a binding partner for a molecule on the opposite gamete or as a fusogen but also may interact in *cis* with other membrane proteins and regulate those proteins' function and/or organization or overall membrane order.

2. The field is moving forward through characterization of mouse models—from knockout models that appear to be completely infertile (*Izumo1*) and severely subfertile (*Cd9*) due to a specific defect in gamete interaction, to other mouse knockouts with a range of fertility phenotypes that vary in robustness and that are detected both in vivo and in vitro.

3. Given the small number of single knockouts that prove to be completely infertile due to a defect specifically in sperm-egg interaction, there is value in considering synthetic infertility. This concept of synthetic infertility means that certain genetic deficiencies can individually result in an organism that is still capable of reproduction (perhaps to a reduced extent) but, when combined with another gene deficiency, can result in a fully infertile phenotype (e.g., $Cd9^{-/-}$ and $Cd81^{-/-}$ in the mouse and *egg-1* and *egg-2* in *C. elegans*).

4. Work in nonmammalian systems has identified some of the most intriguing candidates involved in gamete fusion.

DISCLOSURE STATEMENT

The author is not aware of any affiliations, memberships, funding, or financial holdings that might be perceived as affecting the objectivity of this review.

ACKNOWLEDGMENTS

The author thanks the NIH/NICHD for support of research in her lab and colleagues in the field for discussions helpful for the preparation of this review.

LITERATURE CITED

1. Taylor MV. 2003. Muscle differentiation: signalling cell fusion. *Curr. Biol.* 13:R964–66
2. Chen EH, Olson EN. 2005. Unveiling the mechanisms of cell-cell fusion. *Science* 308:369–73
3. Sztul E, Lupashin V. 2006. Role of tethering factors in secretory membrane traffic. *Am. J. Physiol. Cell Physiol.* 290:11–26
4. Zhou X, Platt JL. 2011. Molecular and cellular mechanisms of mammalian cell fusion. *Adv. Exp. Med. Biol.* 713:33–64
5. Hartmann JF, Gwatkin RBL, Hutchinson CF. 1972. Early contact interactions between mammalian gametes in vitro: evidence that the vitellus influences adherence between sperm and zona pellucida. *Proc. Natl. Acad. Sci. USA* 69:2767–69

6. Redkar AA, Olds-Clarke PJ. 1999. An improved mouse sperm-oocyte plasmalemma binding assay: studies on characteristics of sperm binding in medium with or without glucose. *J. Androl.* 20:500–8
7. Wortzman GB, Gardner AJ, Evans JP. 2006. Analysis of mammalian sperm-egg membrane interactions during in vitro fertilization. *Methods Mol. Biol.* 341:89–101
8. Conover JC, Gwatkin RBL. 1988. Pre-loading of mouse oocytes with DNA-specific fluorochrome (Hoechst 33342) permits rapid detection of sperm oocyte-fusion. *J. Reprod. Fertil.* 82:681–90
9. Humphries MJ. 2001. Cell adhesion assays. *Mol. Biotechnol.* 18:57–61
10. Vjugina U, Zhu X, Oh E, Bracero NJ, Evans JP. 2009. Reduction of mouse egg surface integrin $\alpha 9$ subunit (ITGA9) reduces the egg's ability to support sperm-egg binding and fusion. *Biol. Reprod.* 80:833–41
11. Glazar AI, Evans JP. 2009. IgSF8 (EWI-2) and CD9 in fertilization: evidence of distinct functions for CD9 and a CD9-associated protein in mammalian sperm-egg interaction. *Reprod. Fertil. Dev.* 21:293–303
12. Florman HM, Ducibella T. 2006. Fertilization in mammals. In *Knobil and Neill's Physiology of Reproduction*, ed. JD Neill, pp. 55–112. San Diego, CA: Elsevier
13. Visconti PE, Florman HM. 2010. Mechanisms of sperm-egg interactions: between sugars and broken bonds. *Sci. Signal.* 3:pe35
14. Kim E, Yamashita M, Kimura M, Honda A, Kashiwabara S, Baba T. 2008. Sperm penetration through cumulus mass and zona pellucida. *Int. J. Dev. Biol.* 52:677–82
15. Huang TT Jr, Yanagimachi R. 1985. Inner acrosomal membrane of mammalian spermatozoa: its properties and possible functions in fertilization. *Am. J. Anat.* 174:249–68
16. Yanagimachi R. 1988. Sperm-egg fusion. *Curr. Top. Membr. Transp.* 32:3–43
17. Yanagimachi R. 1994. Mammalian fertilization. In *The Physiology of Reproduction*, ed. E Knobil, JD Neill, pp. 189–317. New York: Raven
18. Santella L, Alikani M, Talansky BE, Cohen J, Dale B. 1992. Is the human oocyte plasma membrane polarized? *Hum. Reprod.* 7:999–1003
19. Evans JP. 1999. Sperm disintegrins, egg integrins, and other cell adhesion molecules of mammalian gamete plasma membrane interactions. *Front. Biosci.* 4:D114–31
20. Vjugina U, Evans JP. 2008. New insights into the molecular basis of mammalian sperm-egg membrane interactions. *Front. Biosci.* 13:462–76
21. Primakoff P, Myles DG. 1983. A map of the guinea pig sperm surface constructed with monoclonal antibodies. *Dev. Biol.* 98:417–28
22. Allen CA, Green DP. 1995. Monoclonal antibodies which recognize equatorial segment epitops de novo following the A23187-induced acrosome reaction of guinea pig sperm. *J. Cell Sci.* 108:767–77
23. Toshimori K, Saxena DK, Tanii I, Yoshinaga K. 1998. An MN9 antigenic molecule, equatorin, is required for successful sperm-oocyte fusion in mice. *Biol. Reprod.* 59:22–29
24. Okabe M, Yagasaki M, Oda H, Matzno S, Kohama Y, Mimura T. 1988. Effect of a monoclonal anti-mouse sperm antibody (OBF13) on the interaction of mouse sperm with zona-free mouse and hamster eggs. *J. Reprod. Immunol.* 13:211–19
25. Primakoff P, Hyatt H, Tredick-Kline J. 1987. Identification and purification of a sperm surface protein with a potential role in sperm-egg membrane fusion. *J. Cell Biol.* 104:141–49
26. Hao Z, Wolkowicz MJ, Shetty J, Klotz K, Bolling L, et al. 2002. SAMP32, a testis-specific, isoantigenic sperm acrosomal membrane-associated protein. *Biol. Reprod.* 66:735–44
27. Wolkowicz MJ, Shetty J, Westbrook A, Klotz K, Jayes F, et al. 2003. Equatorial segment protein defines a discrete acrosomal subcompartment persisting throughout acrosomal biogenesis. *Biol. Reprod.* 69:735–45
28. Stein KK, Go JC, Lane WS, Primakoff P, Myles DG. 2006. Proteomic analysis of sperm regions that mediate sperm-egg interactions. *Proteomics* 6:3533–43
29. Wolkowicz MJ, Digilio L, Klotz K, Shetty J, Flickinger CJ, Herr JC. 2008. Equatorial Segment Protein (ESP) is a human alloantigen involved in sperm-egg binding and fusion. *J. Androl.* 29:272–82
30. Nishimura H, Gupta S, Myles DG, Primakoff P. 2011. Characterization of mouse sperm TMEM190, a small transmembrane protein with the trefoil domain: evidence for co-localization with IZUMO1 and complex formation with other sperm proteins. *Reproduction* 141:437–51
31. Shetty J, Wolkowicz MJ, Digilio LC, Klotz KL, Jayes FL, et al. 2003. SAMP14, a novel, acrosomal membrane-associated, glycosylphosphatidylinositol-anchored member of the Ly-6/urokinase-type plasminogen activator receptor superfamily with a role in sperm-egg interaction. *J. Biol. Chem.* 278:30506–15

32. Mandal A, Klotz KL, Shetty J, Jayes FL, Wolkowicz MJ, et al. 2003. SLLP1, a unique, intra-acrosomal, non-bacteriolytic, C lysozyme-like protein of human spermatozoa. *Biol. Reprod.* 68:1525–37
33. Herrero MB, Mandal A, Digilio LC, Coonrod SA, Maier B, Herr JC. 2005. Mouse SLLP1, a sperm lysozyme-like protein involved in sperm-egg binding and fertilization. *Dev. Biol.* 284:126–42
34. Handel MA, Lessard C, Reinholdt L, Schimenti J, Eppig JJ. 2006. Mutagenesis as an unbiased approach to identify novel contraceptive targets. *Mol. Cell. Endocrinol.* 250:201–5
35. Misamore MJ, Gupta S, Snell WJ. 2003. The *Chlamydomonas* Fus1 protein is present on the mating type plus fusion organelle and required for a critical membrane adhesion event during fusion with minus gametes. *Mol. Biol. Cell* 14:2530–42
36. Liu Y, Misamore MJ, Snell WJ. 2010. Membrane fusion triggers a rapid degradation of two gamete-specific fusion-essential proteins in a membrane block to polygamy in *Chlamydomonas*. *Development* 137:1473–81
37. Singson A, Hang JS, Parry JM. 2008. Genes required for the common miracle of fertilization in *Caenorhabditis elegans*. *Int. J. Dev. Biol.* 52:647–56
38. Blobel CP, Wolfsberg TG, Turck CW, Myles DG, Primakoff P, White JM. 1992. A potential fusion peptide and an integrin ligand domain in a protein active in sperm-egg fusion. *Nature* 356:248–52
39. Kanandale P, Stewart-Michaelis A, Gordon S, Rubin J, Klancer R, et al. 2005. The egg surface LDL receptor repeat-containing proteins EGG-1 and EGG-2 are required for fertilization in *Caenorhabditis elegans*. *Curr. Biol.* 15:2222–29
40. Coonrod SA, Wright PW, Herr JC. 2002. Oolemmal proteomics. *J. Reprod. Immunol.* 53:55–65
41. Kaji K, Oda S, Shikano T, Ohnuki T, Uematsu Y, et al. 2000. The gamete fusion process is defective in eggs of CD9-deficient mice. *Nat. Genet.* 24:279–82
42. Le Naour F, Rubinstein E, Jasmin C, Prenant M, Boucheix C. 2000. Severely reduced female fertility in CD9-deficient mice. *Science* 287:319–21
43. Miyado K, Yamada G, Yamada S, Hasuwa H, Nakamura Y, et al. 2000. Requirement of CD9 on the egg plasma membrane for fertilization. *Science* 287:321–24
44. Rubinstein E, Ziyyat A, Prenant M, Wrobel E, Wolf JP, et al. 2006. Reduced fertility of female mice lacking CD81. *Dev. Biol.* 290:351–58
45. Sutton KA, Jungnickel MK, Florman HM. 2008. A polycystin-1 controls postcopulatory reproductive selection in mice. *Proc. Natl. Acad. Sci. USA* 105:8661–66
46. Okabe M, Adachi T, Takada K, Oda H, Yagasaki M, et al. 1987. Capacitation-related changes in antigen distribution on mouse sperm heads and its relation to fertilization rate in vitro. *J. Reprod. Immunol.* 11:91–100
47. Inoue N, Ikawa M, Isotani A, Okabe M. 2005. The immunoglobulin superfamily protein Izumo is required for sperm to fuse with eggs. *Nature* 434:234–38
48. Hayasaka S, Terada Y, Inoue N, Okabe M, Yaegashi N, Okamura K. 2007. Positive expression of the immunoglobulin superfamily protein IZUMO on human sperm of severely infertile male patients. *Fertil. Steril.* 88:214–16
49. Granados-Gonzalez V, Aknin-Selfer I, Touraine RL, Chouteau J, Wolf JP, Levy R. 2008. Preliminary study on the role of the human *IZUMO* gene in oocyte-spermatozoa fusion failure. *Fertil. Steril.* 90:1246–48
50. Brummendorf T, Lemmon V. 2001. Immunoglobulin superfamily receptors: *cis*-interactions, intracellular adapters and alternative splicing regulate adhesion. *Curr. Opin. Cell Biol.* 13:611–18
51. Inoue N, Ikawa M, Okabe M. 2008. Putative sperm fusion protein IZUMO and the role of N-glycosylation. *Biochem. Biophys. Res. Commun.* 377:910–14
52. Inoue N, Kasahara T, Ikawa M, Okabe M. 2010. Identification and disruption of sperm-specific angiotensin converting enzyme-3 (ACE3) in mouse. *PLoS ONE* 5:e10301
53. Sosnik J, Miranda PV, Spiridonov NA, Yoon S-Y, Fissore RA, et al. 2009. Tssk6 is required for Izumo relocalization and gamete fusion in the mouse. *J. Cell Sci.* 122:2741–49
54. Spiridonov NA, Wong L, Zerfas PM, Starost MF, Pack SD, et al. 2005. Identification and characterization of SSTK, a serine/threonine protein kinase essential for male fertility. *Mol. Cell. Biol.* 25:4250–61
55. Ellerman DA, Pei J, Gupta S, Snell WJ, Myles DG, Primakoff P. 2009. Izumo is part of a multiprotein family whose members form large complexes on mammalian sperm. *Mol. Reprod. Dev.* 76:1188–99

56. Edwards DR, Handsely MM, Pennington CJ. 2008. The ADAM metalloproteinases. *Mol. Aspects Med.* 29:258–89
57. Eto K, Huet C, Tarui T, Kupriyanov S, Liu HZ, et al. 2002. Functional classification of ADAMs based on a conserved motif for binding to integrin $\alpha_9\beta_1$: implications for sperm-egg binding and other cell interactions. *J. Biol. Chem.* 277:17804–10
58. Tomczuk M, Takahashi Y, Huang J, Murase S, Mistretta M, et al. 2003. Role of multiple β_1 integrins in cell adhesion to the disintegrin domains of ADAMs 2 and 3. *Exp. Cell Res.* 290:68–81
59. Desiderio UV, Zhu X, Evans JP. 2010. ADAM2 interactions with mouse eggs and cell lines expressing $\alpha 4/\alpha 9$ (ITGA4/ITGA9) integrins: implications for integrin-based adhesion and fertilization. *PLoS ONE* 5:e13744
60. Cho C, Bunch DO, Faure J-E, Goulding EH, Eddy EM, et al. 1998. Fertilization defects in sperm from mice lacking fertilin β. *Science* 281:1857–59
61. Shamsadin R, Adham IM, Nayernia K, Heinlein UAO, Oberwinkler H, Engel W. 1999. Male mice deficient for germ-cell cyritestin are infertile. *Biol. Reprod.* 61:1445–51
62. Nishimura H, Cho C, Branciforte DR, Myles DG, Primakoff P. 2001. Analysis of loss of adhesive function in sperm lacking cyritestin or fertilin β. *Dev. Biol.* 233:204–13
63. Nishimura H, Kim E, Nakanishi T, Baba T. 2004. Possible function of the ADAM1a/ADAM2 fertilin complex in the appearance of ADAM3 on the sperm surface. *J. Biol. Chem.* 279:34957–62
64. Kim E, Yamashita M, Nakanishi T, Park KE, Kimura M, et al. 2006. Mouse sperm lacking ADAM1b/ADAM2 fertilin can fuse with the egg plasma membrane and effect fertilization. *J. Biol. Chem.* 281:5634–39
65. Kim T, Oh J, Woo JM, Choi E, Im SH, et al. 2006. Expression and relationship of male reproductive ADAMs in mouse. *Biol. Reprod.* 74:744–50
66. Han C, Choi E, Park I, Lee B, Jin S, et al. 2009. Comprehensive analysis of reproductive ADAMs: relationship of ADAM4 and ADAM6 with an ADAM complex required for fertilization in mice. *Biol. Reprod.* 80:1001–8
67. Horiuchi K, Weskamp G, Lum L, Hammes HP, Cai H, et al. 2003. Potential role for ADAM15 in pathological neovascularization in mice. *Mol. Cell. Biol.* 23:5614–24
68. Marcello MR, Jia W, Leary JA, Moore KL, Evans JP. 2011. Lack of tyrosylprotein sulfotransferase-2 activity results in altered sperm-egg interactions and loss of ADAM3 and ADAM6 in epididymal sperm. *J. Biol. Chem.* 286:13060–70
69. Yamaguchi R, Yamagata K, Ikawa M, Moss SB, Okabe M. 2006. Aberrant distribution of ADAM3 in sperm from both Angiotensin-converting enzyme (Ace)- and Calmegin (Clgn)-deficient mice. *Biol. Reprod.* 75:760–66
70. Ikawa M, Tokuhiro K, Yamaguchi R, Benham AM, Tamura T, et al. 2011. Calsperin is a testis specific chaperone required for sperm fertility. *J. Biol. Chem.* 286:5639–46
71. Grzmil P, Kim Y, Shamsadin R, Neesen J, Adham IM, et al. 2001. Human cyritestin genes (*CYRN1* and *CYRN2*) are non-functional. *Biochem. J.* 357:551–56
72. Frayne J, Hall L. 1998. The gene for the human tMDC I sperm surface protein is non-functional: implications for its proposed role in mammalian sperm-egg recognition. *Biochem. J.* 334:171–76
73. Kim E, Baba D, Kimura M, Yamashita M, Kashiwabara S, Baba T. 2005. Identification of a hyaluronidase, Hyal5, involved in penetration of mouse sperm through cumulus mass. *Proc. Natl. Acad. Sci. USA* 102:18028–33
74. Fujihara Y, Murakami M, Inoue N, Satouh Y, Kaseda K, et al. 2010. Sperm equatorial segment protein 1, SPESP1, is required for fully fertile sperm in the mouse. *J. Cell Sci.* 123:1531–36
75. Hemler ME. 2005. Tetraspanin functions and associated microdomains. *Nat. Rev. Mol. Cell Biol.* 6:801–11
76. Rubinstein E. 2011. The complexity of tetraspanins. *Biochem. Soc. Trans.* 39:501–5
77. Kaji K, Oda S, Miyazaki S, Kudo A. 2002. Infertility of CD9-deficient mouse eggs is reversed by mouse CD9, human CD9, or mouse CD81; polyadenylated mRNA injection developed for molecular analysis of sperm-egg fusion. *Dev. Biol.* 247:327–34
78. Ziyyat A, Rubinstein E, Monier-Gavelle F, Barraud V, Kulski O, et al. 2006. CD9 controls the formation of clusters that contain tetraspanins and the integrin $\alpha_6\beta_1$, which are involved in human and mouse gamete fusion. *J. Cell Sci.* 119:416–24

79. Chen MS, Tung KSK, Coonrod SA, Takahashi Y, Bigler D, et al. 1999. Role of the integrin associated protein CD9 in binding between sperm ADAM 2 and the egg integrin $\alpha_6\beta_1$: implications for murine fertilization. *Proc. Natl. Acad. Sci. USA* 96:11830–35
80. Li YH, Hou Y, Ma W, Yuan JX, Zhang D, et al. 2004. Localization of CD9 in pig oocytes and its effects on sperm-egg interaction. *Reproduction* 127:151–57
81. Wright MD, Geary SM, Fitter S, Moseley GW, Lau LM, et al. 2004. Characterization of mice lacking the tetraspanin superfamily member CD151. *Mol. Cell. Biol.* 24:5978–88
82. Takeda Y, Kazarov AR, Butterfield CE, Hopkins BD, Benjamin LE, et al. 2007. Deletion of tetraspanin Cd151 results in decreased pathologic angiogenesis in vivo and in vitro. *Blood* 109:1524–32
83. Sachs N, Krfet M, van den Bergh Weerman MA, Beymon AJ, Peters TA, et al. 2006. Kidney failure in mice lacking the tetraspanin CD151. *J. Cell Biol.* 175:33–39
84. Karamatic Crew V, Burton N, Kagan A, Green CA, Levene C, et al. 2004. CD151, the first member of the tetraspanin (TM4) superfamily detected on erythrocytes, is essential for the correct assembly of human basement membranes in kidney and skin. *Blood* 104:2217–23
85. Fanaei M, Monk PN, Partridge LJ. 2011. The role of tetraspanins in fusion. *Biochem. Soc. Trans.* 39:524–28
86. Damek-Poprawa M, Krouse J, Gretzula C, Boesze-Battaglia K. 2005. A novel tetraspanin fusion protein, peripherin-2, requires a region upstream of the fusion domain for activity. *J. Biol. Chem.* 280:9217–24
87. Zhu G-Z, Miller BJ, Boucheix C, Rubinstein E, Liu CC, et al. 2002. Residues SFQ (173–175) in the large extracellular loop of CD9 are required for gamete fusion. *Development* 129:1995–2002
88. Yáñez-Mó M, Barreiro O, Gordon-Alonso M, Sala-Valdés M, Sánchez-Madrid F. 2009. Tetraspanin-enriched microdomains: a functional unit in cell plasma membranes. *Trends Cell Biol.* 19:434–46
89. Levental I, Grzybek M, Simons K. 2010. Greasing their way: Lipid modifications determine protein association with membrane rafts. *Biochemistry* 49:6305–16
90. Runge KE, Evans JE, He ZY, Gupta S, McDonald KL, et al. 2007. Oocyte CD9 is enriched on the microvillar membrane and required for normal microvillar shape and distribution. *Dev. Biol.* 304:317–25
91. Simons M, Raposo G. 2009. Exosomes—vesicular carriers for intercellular communication. *Curr. Opin. Cell Biol.* 21:575–81
92. Barraud-Lange V, Naud-Barriant N, Bomsel M, Wolf JP, Ziyyat A. 2007. Transfer of oocyte membrane fragments to fertilizing spermatozoa. *FASEB J.* 21:3446–49
93. Miyado K, Yoshida K, Yamagata K, Sakakibara K, Okabe M, et al. 2008. The fusing ability of sperm is bestowed by CD9-containing vesicles released from eggs in mice. *Proc. Natl. Acad. Sci. USA* 105:12921–26
94. Gupta S, Primakoff P, Myles DG. 2009. Can the presence of wild-type oocytes during insemination rescue the fusion defect of CD9 null oocytes? *Mol. Reprod. Dev.* 76:602
95. Jégou A, Ziyyat A, Barraud-Lange V, Perez E, Wolf JP, et al. 2011. CD9 tetraspanin generates fusion competent sites on the egg membrane for mammalian fertilization. *Proc. Natl. Acad. Sci. USA* 108:10946–51
96. Zhu X, Evans JP. 2002. Analysis of the roles of RGD-binding integrins, α_4/α_9 integrins, α_6 integrins, and CD9 in the interaction of the fertilin β (ADAM2) disintegrin domain with the mouse egg membrane. *Biol. Reprod.* 66:1193–202
97. Little KD, Hemler ME, Stipp CS. 2004. Dynamic regulation of a GPCR–tetraspanin–G protein complex on intact cells: central role of CD81 in facilitating GPR56-G$\alpha_{q/11}$ association. *Mol. Biol. Cell* 15:2375–87
98. Kovalenko OV, Yang XH, Hemler ME. 2007. A novel cysteine cross-linking method reveals a direct association between claudin-1 and tetraspanin CD9. *Mol. Cell Proteomics* 6:1855–67
99. Le Naour F, Andre M, Greco C, Billard M, Sordat B, et al. 2006. Profiling of the tetraspanin web of human colon cancer cells. *Mol. Cell Proteomics* 5:845–57
100. Stipp CS, Kolesnikova TV, Hemler ME. 2001. EWI-2 is a major CD9 and CD81 partner and member of a novel Ig protein subfamily. *J. Biol. Chem.* 276:40545–54
101. Stipp CS, Orlicky D, Hemler ME. 2001. FPRP, a major, highly stoichiometric, highly specific CD81- and CD9-associated protein. *J. Biol. Chem.* 276:4853–62
102. He ZY, Gupta S, Myles D, Primakoff P. 2009. Loss of surface EWI-2 on CD9 null oocytes. *Mol. Reprod. Dev.* 76:629–36

103. Charrin S, Le Naour F, Oualid M, Billard M, Faure G, et al. 2001. The major CD9 and CD81 molecular partner: identification and characterization of the complexes. *J. Biol. Chem.* 276:14329–37
104. Coonrod SA, Naaby-Hansen S, Shetty J, Shibahara H, Chen M, et al. 1999. Treatment of mouse oocytes with PI-PLC releases 70-kDa (pI 5) and 35- to 45-kDa (pI 5.5) protein clusters from the egg surface and inhibits sperm-oolemma binding and fusion. *Dev. Biol.* 207:334–49
105. Tiede A, Nischan C, Schubert J, Schmidt RE. 2000. Characterisation of the enzymatic complex for the first step in glycosylphosphatidylinositol biosynthesis. *Int. J. Biochem. Cell Biol.* 32:339–50
106. Alfieri JA, Martin AD, Takeda J, Kondoh G, Myles DG, Primakoff P. 2003. Infertility in female mice with an oocyte-specific knockout of GPI-anchored proteins. *J. Cell Sci.* 116:2149–55
107. He Z-Y, Brakebusch C, Fassler R, Kreidberg JA, Primakoff P, Myles DG. 2003. None of the integrins known to be present on the mouse egg or to be ADAM receptors are essential for sperm–egg binding and fusion. *Dev. Biol.* 254:226–37
108. Miller BJ, Georges-Labouesse E, Primakoff P, Myles DG. 2000. Normal fertilization occurs with eggs lacking the integrin $\alpha_6\beta_1$ and is CD9-dependent. *J. Cell Biol.* 149:1289–95
109. Baessler K, Lee Y, Sampson N. 2009. β1 integrin is an adhesion protein for sperm binding to eggs. *ACS Chem. Biol.* 4:357–66
110. Wilson KL, Fitch KR, Bafus BT, Wakimoto BT. 2006. Sperm plasma membrane breakdown during *Drosophila* fertilization requires Sneaky, an acrosomal membrane protein. *Development* 133:4871–79
111. Xu X-ZS, Sternberg PW. 2003. A *C. elegans* sperm TRP protein required for sperm-egg interactions during fertilization. *Cell* 114:285–97
112. Kroft TL, Gleason EJ, L'Hernault SW. 2005. The *spe-42* gene is required for sperm-egg interactions during *C. elegans* fertilization and encodes a sperm-specific transmembrane protein. *Dev. Biol.* 286:169–81
113. Wilson LD, Sackett JM, Mieczkowski BD, Richie AL, Thoemke K, et al. 2011. Fertilization in *C. elegans* requires an intact C-terminal RING finger in sperm protein SPE-42. *BMC Dev. Biol.* 11:10
114. Yagi M, Miyamoto T, Sawatani Y, Iwamoto K, Hosogane N, et al. 2005. DC-STAMP is essential for cell-cell fusion in osteoclasts and foreign body giant cells. *J. Exp. Med.* 202:345–51
115. Singson A, Mercer KB, L'Hernault SW. 1998. The *C. elegans spe-9* gene encodes a sperm transmembrane protein that contains EGF-like repeats and is required for fertilization. *Cell* 93:71–79
116. Chatterjee I, Richmond A, Putiri E, Shakes DC, Singson A. 2005. The *Caenorhabditis elegans spe-38* gene encodes a novel four-pass integral membrane protein required for sperm function at fertilization. *Development* 132:2795–808
117. Maruyama R, Velarde NV, Klancer R, Gordon S, Kadandale P, et al. 2007. EGG-3 regulates cell-surface and cortex rearrangements during egg activation in *Caenorhabditis elegans*. *Curr. Biol.* 17:1555–60
118. Stitzel ML, Cheng KC, Seydoux G. 2007. Regulation of MBK-2/Dyrk kinase by dynamic cortical anchoring during the oocyte-to-zygote transition. *Curr. Biol.* 17:1545–54
119. Parry JM, Velarde NV, Lefkovith AJ, Zegarek Mh, Hang JS, et al. 2009. EGG-4 and EGG-5 link events of the oocyte-to-embryo transition with meiotic progression in *C. elegans*. *Curr. Biol.* 19:1752–57
120. Cheng KC-C, Klancer R, Singson A, Seydoux G. 2009. Regulation of MBK-2/DYRK by CDK-1 and the pseudophosphatases EGG-4 and EGG-5 during the oocyte-to-embryo transition. *Cell* 139:560–72
121. Wong JL, Johnson MA. 2010. Is HAP2-GSC1 an ancestral gamete fusogen? *Trends Cell Biol.* 20:134–41
122. Johnson MA, von Besser K, Zhou Q, Smith E, Aux G, et al. 2004. *Arabidopsis hapless* mutations define essential gametophytic functions. *Genetics* 168:971–82
123. Mori T, Kuriowa H, Higashiyama T, Kuriowa T. 2006. Generative Cell Specific 1 is essential for angiosperm fertilization. *Nat. Cell Biol.* 8:64–71
124. von Besser K, Frank AC, Johnson MA, Preuss D. 2006. *Arabidopsis HAP2 (GCS1)* is a sperm-specific gene required for pollen tube guidance and fertilization. *Development* 133:4761–69
125. Liu Y, Tewari R, Ning J, Blagborough AM, Garbom S, et al. 2008. The conserved plant sterility gene *HAP2* functions after attachment of fusogenic membranes in *Chlamydomonas* and *Plasmodium* gametes. *Genes Dev.* 22:1051–68
126. Wong JL, Leydon AR, Johnson MA. 2010. HAP2(GCS1)-dependent gamete fusion requires a positively charged carboxy-terminal domain. *PLoS Genet.* 6:e1000882
127. Mori T, Hirai M, Kuriowa T, Miyagishima S. 2010. The functional domain of GCS1-based gamete fusion resides in the amino terminus in plant and parasite species. *PLoS ONE* 5:e15957

128. Ulrich AS, Otter M, Glabe CG, Hoekstra D. 1998. Membrane fusion is induced by a distinct peptide sequence of the sea urchin fertilization protein bindin. *J. Biol. Chem.* 273:16748–55
129. Kresge N, Vacquier VD, Stout CD. 2001. The crystal structure of a fusagenic sperm protein reveals extreme surface properties. *Biochemistry* 40:5407–13
130. Zigler KS. 2008. The evolution of sea urchin bindin. *Int. J. Dev. Biol.* 52:791–96
131. Monne M, Han L, Jovine L. 2006. Tracking down the ZP domain: from the mammalian zona pellucida to the molluscan vitelline envelope. *Semin. Reprod. Med.* 24:204–16
132. Swanson WJ, Aagaard JE, Vacquier VD, Monne M, Al Hosseini HS, Jovine L. 2011. The molecular basis of sex: linking yeast to human. *Mol. Biol. Evol.* 28:1963–66
133. Swanson WJ, Vacquier VD. 1995. Liposome fusion induced by a M_r 18000 protein localized to the acrosomal region of acrosome-reacted abalone spermatozoa. *Biochemistry* 34:14202–8
134. Aagaard JE, Vacquier VD, MacCoss MJ, Swanson WJ. 2010. ZP domain proteins in the abalone egg coat include a paralog of VERL under positive selection that binds lysin and 18-kDa sperm proteins. *Mol. Biol. Evol.* 27:193–203
135. Shalgi R, Phillips DM. 1980. Mechanics of in vitro fertilization in the hamster. *Biol. Reprod.* 23:433–344
136. Phillips DM, Shalgi R. 1980. Surface architecture of the mouse and hamster zona pellucida and oocyte. *J. Ultrastruct. Res.* 72:1–12
137. Ikawa M, Wada I, Kominami K, Watanabe D, Toshimuri K, et al. 1997. The putative chaperone calmegin is required for sperm fertility. *Nature* 387:607–11
138. Ikawa M, Nakanishi T, Yamada S, Wada I, Kominami K, et al. 2001. Calmegin is required for fertilin α/β heterodimerization and sperm fertility. *Dev. Biol.* 240:254–61
139. Ueda Y, Yamaguchi R, Ikawa M, Okabe M, Morii E, et al. 2007. PGAP1 knock-out mice show otocephaly and male infertility. *J. Biol. Chem.* 282:30373–80
140. Marcello MR, Evans JP. 2010. Multivariate analysis of male reproductive function in $Inpp5b^{-/-}$ mice reveals heterogeneity in defects in fertility, sperm-egg membrane interaction, and proteolytic cleavage of sperm ADAMs. *Mol. Hum. Reprod.* 16:492–505
141. Grant SG, O'Dell TJ, Karl KA, Stein PL, Soriano P, Kandel E. 1992. Impaired long-term potentiation, spatial learning, and hippocampal development in *fyn* mutant mice. *Science* 258:1903–10
142. Luo J, McGinnis LK, Kinsey WH. 2010. Role of Fyn kinase in oocyte developmental potential. *Reprod. Fertil. Dev.* 22:966–76
143. Navarro B, Miki K, Clapham DE. 2011. ATP-activated P2X2 current in mouse spermatozoa. *Proc. Natl. Acad. Sci. USA* 108:14342–47
144. Baba D, Kashiwabara S, Honda A, Yamagata K, Wu Q, et al. 2002. Mouse sperm lacking cell surface hyaluronidase PH-20 can pass through the layer of cumulus cells and fertilize the egg. *J. Biol. Chem.* 277:30310–14
145. Ensslin MA, Shur BD. 2003. Identification of mouse sperm SED1, a bimotif EGF repeat and discoidin-domain protein involved in sperm-egg binding. *Cell* 114:405–17
146. Da Ros VG, Maldera JA, Willis WD, Cohen DJ, Goulding EH, et al. 2008. Impaired sperm fertilizing ability in mice lacking Cysteine-RIch Secretory Protein 1 (CRISP1). *Dev. Biol.* 320:12–18
147. Lessard C, Pendola JK, Hartford SA, Schimenti JC, Handel MA, Eppig JJ. 2004. New mouse genetic models for contraceptive development. *Cytogenet. Cell Genet.* 105:222–27
148. Jin M, Fujiwara E, Kakiuchi Y, Okabe M, Satouh Y, et al. 2011. Most fertilizing mouse spermatozoa begin their acrosome reaction before contact with the zona pellucida during in vitro fertilization. *Proc. Natl. Acad. Sci. USA* 108:4892–96
149. Yamatoya K, Yoshida K, Ito C, Maekawa M, Yanagida M, et al. 2009. Equatorin: identification and characterization of the epitope of the MN9 antibody in mouse. *Biol. Reprod.* 81:889–97
150. Gibbs GM, Orta G, Reddy T, Koppers AJ, Martínez-López P, et al. 2011. Cysteine-rich secretory protein 4 is an inhibitor of transient receptor potential M8 with a role in establishing sperm function. *Proc. Natl. Acad. Sci. USA* 108:7034–39
151. Marín-Briggiler CI, Velga MF, Matos ML, Echeverria MF, Furlong LI, Vazquez-Levin MH. 2008. Expression of epithelial cadherin in the human male reproductive tract and gametes and evidence of its participation in fertilization. *Mol. Hum. Reprod.* 14:561–71

152. Marín-Briggiler CI, Lapyckj L, González-Echeverría MF, Rawe VY, Alvarez Sedó C, Vazquez-Levin MH. 2010. Neural cadherin is expressed in human gametes and participates in sperm-oocyte interaction events. *Int. J. Androl.* 33:e228–39
153. Boissonnas CC, Montjean D, Lesaffre C, Auer J, Vaiman D, et al. 2010. Role of sperm $\alpha v \beta 3$ integrin in mouse fertilization. *Dev. Dyn.* 239:773–83
154. Correa LM, Cho C, Myles DG, Primakoff P. 2000. A role for a TIMP-3-sensitive Zn^{2+}-dependent metalloprotease in mammalian gamete membrane fusion. *Dev. Biol.* 225:124–34
155. Mammoto A, Masumoto N, Tahara M, Yoneda M, Nishizaki T, et al. 1997. Involvement of a sperm protein sensitive to sulfhydry-depleting reagents in mouse sperm-egg fusion. *J. Exp. Zool.* 278:178–88
156. Ellerman DA, Myles DG, Primakoff P. 2006. A role for sperm surface protein disulfide isomerase activity in gamete fusion: evidence for the participation of ERp57. *Dev. Cell* 10:831–37
157. Bigler D, Takahashi Y, Chen MS, Almeida EAC, Osburne L, White JM. 2000. Sequence-specific interaction between the disintegrin domain of mouse ADAM2 (fertilin β) and murine eggs: role of the α_6 integrin subunit. *J. Biol. Chem.* 275:11576–84
158. Zhu X, Bansal NP, Evans JP. 2000. Identification of key functional amino acids of the mouse fertilin β (ADAM2) disintegrin loop for cell-cell adhesion during fertilization. *J. Biol. Chem.* 275:7677–83
159. Takahashi Y, Bigler D, Ito Y, White JM. 2001. Sequence-specific interaction between the disintegrin domain of mouse ADAM3 and murine eggs: role of the β_1 integrin-associated proteins CD9, CD81, and CD98. *Mol. Biol. Cell* 12:809–20
160. Evans JP. 2009. Egg integrins: back in the game of mammalian fertilization. *ACS Chem. Biol.* 4:321–23
161. Seigneuret M. 2006. Complete predicted three-dimensional structure of the facilitator transmembrane protein and hepatitis C virus receptor CD81: conserved and variable structural domains in the tetraspanin superfamily. *Biophys. J.* 90:212–27
162. Mandal A, Sachdev M, Digilio L, Panneerdoss S, Suryavathi V, et al. 2011. *SAS1B is an egg specific high affinity oolemmal binding partner for sperm specific acrosomal SLLP1 during fertilization.* Presented at Meet. Soc. Study Reprod., Portland, OR

RELATED RESOURCE

1. Ikawa M, Inoue N, Benham AM, Okabe M. 2010. Fertilization: a sperm's journey to and interaction with the oocyte. *J. Clin. Investig.* 120:984–94

Genetics of Mammalian Reproduction: Modeling the End of the Germline

Martin M. Matzuk[1] and Kathleen H. Burns[2]

[1]Departments of Pathology and Immunology, Molecular and Cellular Biology, and Molecular and Human Genetics, Baylor College of Medicine, Houston, Texas 77030; email: mmatzuk@bcm.edu

[2]Departments of Pathology and Oncology, McKusick-Nathans Institute of Genetic Medicine, and High Throughput Biology Center, Johns Hopkins University School of Medicine, Baltimore, Maryland 21205; email: kburns@jhmi.edu

Keywords

transgenic mouse, gametogenesis, preimplantation embryonic development

Abstract

Our understanding of reproduction and early embryonic development has directly enabled our manipulation of the mouse genome for tests of gene function in vivo. In this review, we reflect on the 30 years of work that followed this singular accomplishment. We profile murine models that have given us memorable insights into fundamental processes of male and female gametogenesis and the earliest phases of embryonic life reliant on oocyte-transmitted maternal gene products. We highlight intercellular endocrine and paracrine communications essential to gamete development as well as mechanisms essential for passing the genome, with integrity and appropriate epigenetic marks, on to the next generation. Finally, we reflect on the future of reproductive biology: how advances in clinical genetics will provide special opportunities to understand which reproductive processes are affected by genetic lesions and which may allow mutations to originate.

INTRODUCTION

Genome-wide association study (GWAS): a case and control comparison of single-nucleotide-polymorphism allele frequencies across the genome

Transgenic: describes an animal model wherein exogenous recombinant DNA products have been introduced to the genome

There have been amazing advances over the past three decades in the development of genetic technologies to probe mammalian genomes. Technical advances have enabled impressive progress in our ability to sequence DNA, assemble reference genomes, interrogate changes in gene expression, and perform genome-wide association studies (GWAS). In parallel, the development of transgenic methods has permitted functional analysis of gene products in mammals, allowing us to interrogate the in vivo significance of coding and noncoding pieces of DNA. This genetic revolution has impacted every area of biology and has allowed us to appreciate reproduction as more than the fusion of an oocyte and a spermatozoon and the delivery of a baby. We are increasingly informed about the genes and genetic pathways involved in sex determination and the formation and function of the hypothalamic-pituitary-ovary/testis axes and the female and male reproductive tracts.

In this review, we provide perspectives on where we have been, where we are now, and where we may be in the future in understanding the genetics of mammalian reproduction. Most of the review focuses on mice, the major model used for in vivo manipulation, and on our species, the most relevant mammal whose naturally occurring genetic changes are best understood and ascertained. Over the past decade, we have written differently directed and, in some cases, more comprehensive or detailed reviews (1–5). At times, we refer the reader to aspects of these other reviews and primary literature that describe relevant topics in reproductive genetics in greater detail.

EARLY GENETICS OF THE Y CHROMOSOME AND HORMONES

The first genes shown to play roles in reproduction were identified on the basis of their abilities to influence the sexual differentiation process or to alter hormone production or hormone sensitivity. As shown in the timeline in **Figure 1**, although the X and Y chromosomes were hypothesized in 1905 to influence sex determination in insects, a 1959 analysis of the chromosome status in Turner syndrome females (XO) and Klinefelter males (XXY) was the first to show that the Y chromosome contains the mammalian male-determining gene (6, 7). Another three decades passed before the sex-determining region Y (*SRY*) gene was identified in humans and mice (8–10). These discoveries were critical for defining the sex-determining factors that are upstream of SRY and are involved in formation of the bipotential gonad as well as factors downstream of SRY. Both mouse and human genetics have contributed to the identification of testis-determining genes [e.g., SRY-box 9 (*SOX9*); fibroblast growth factor 9 (*FGF9*); and nuclear receptor subfamily 0, group B, member 1 (*NR0B1*), also known as dosage-sensitive sex reversal, adrenal hypoplasia critical region, on chromosome X, gene 1 (*Dax1*)] and ovary-determining genes [e.g., R-spondin homolog 1 (*RSPO1*), *WNT4*, and β-*catenin*].

Follow-up studies indicated a model for sex determination in which formation of a testis in males and formation of an ovary in females were both active processes rather than a default pathway. Although mutations in *SRY* were shown to result in XY male-to-female sex reversal, a major breakthrough came when loss-of-function mutations in *RSPO1* resulted in complete sex reversal (11). These and subsequent studies helped to define RSPO1, WNT4, β-catenin, and FOXL2 as being relevant to ovary determination and SRY, SOX9, and Müllerian-inhibiting substance [also known as anti-Müllerian hormone (AMH)] as functioning in testis determination. Key to the establishment of these pathways is the ability of SOX9 to antagonize β-catenin and ovarian development in males and the ability of β-catenin and FOXL2 to antagonize SOX9 and testis formation in females (4, 5, 12). The genetic basis of disorders in sexual development is reviewed elsewhere (13, 14).

1905: X and Y chromosomes in insects are proposed to influence sex (Edmund B. Wilson and Nettie Wilson).

1959: Chromosomal analysis of Turner syndrome and Klinefelter syndrome pinpoints the Y chromosome as determining male sex (Ford et al. 1959, Jacobs & Strong 1959).

1970: A gene on the mouse X chromosome is found to cause insensitivity to androgens (testicular feminization) (Lyon & Hawkes 1970).

1976: Tiepolo & Zuffardi (1976) reported Y chromosome deletions in a subset of infertile men.
1977: Gonadotropin-releasing hormone (GnRH) deficiency is identified in mice (Cattanach et al. 1977).

1980 : The first maternal-effect gene is discovered in *Drosophila*; maternal *dorsal* is found to be critical for establishing polarity in the embryo (Nusslein-Volhard et al. 1980).
1981: Egg-injected genes are found to be transmitted through the germline (Brinster et al. 1981, Costantini & Lacy 1981, Gordon & Ruddle 1981).
1982, 1983: Transgenic mice are created to overexpress rat and human growth hormone (Palmiter 1982, 1983).
1984: GnRH precursor cDNA is cloned (Seeburg & Adelman 1984).
1986: The *hpg* mouse is shown to have a mutation in the GnRH gene (Mason, Hayflick et al. 1986), with genetic correction of the defect in transgenic mice (Mason, Pitts et al. 1986).
1988: The Human Genome Project begins.

1990: Human and mouse *SRY* genes are discovered (Berta et al. 1990, Koopman et al. 1990, Sinclair et al. 1990).
1990: Overexpression of anti-Müllerian hormone (AMH) confirms its key role in regression of the Müllerian ducts in males (Behringer et al. 1990).
1991: XX female mice with a mouse *Sry* transgene develop as males (Koopman et al. 1991).
1992: Knockout of the inhibin α (*Inha*) gene causes gonadal cancers and reproductive defects (Matzuk et al. 1992). Knockout of the leukemia inhibitory factor (*Lif*) gene leads to female sterility due to implantation failure (Stewart et al. 1992).

1996: Knockout of growth differentiation factor 9 (*Gdf9*) identifies *Gdf9* as the first gene involved in oocyte–somatic cell communication (Dong et al. 1996).

2000: Mater (*Nalp5*) is identified as the first maternal-effect gene in mice, demonstrating a block at the two-cell to four-cell stage (Tong et al. 2000).
2000: Mutations in the *BMP15* gene are identified for fertility defects in the Inverdale and Hanna breeds of sheep (Galloway et al. 2000).
2002: Zygote arrest 1 (*Zar1*) is identified as the first oocyte gene required for the oocyte-embryo transition (Wu et al. 2003).
2003: Mutations in the human and mouse kisspeptin receptor (*GPR54*) cause infertility secondary to hypogonadotropic hypogonadism.

2005: Massively parallel DNA sequencing (genome sequencing in microfabricated high-density picoliter reactors) is described (Margulies et al. 2005).
2007: The International Knockout Mouse Consortium begins.
2007: The Nobel Prize in Physiology or Medicine is awarded to Mario R. Capecchi, Martin J. Evans, and Oliver Smithies "for their discoveries of principles for introducing specific gene modifications in mouse by the use of embryonic stem cells."

2010: The Nobel Prize in Physiology or Medicine is awarded to Robert G. Edwards "for the development of *in vitro* fertilization."

Figure 1
Reproductive genetics timeline.

Genes related to the synthesis or signaling of peptide hormones were among the first identified and among the first in which mutations were discovered to alter function. The importance of these signaling pathways in the hypothalamus and the pituitary to ovarian and testicular function and steroid production postnatally and the ability to precisely measure serum follicle-stimulating hormone (FSH), luteinizing hormone (LH), and steroids were key factors in advancing this field of

Knockout mouse: a transgenic mouse in which the introduced DNA is targeted to a specific locus by homologous recombination and replaces the functional gene with a loss-of-function allele

Cre-loxP recombination: DNA recombination between 34-base-pair loxP sites is induced by expression of the Cre recombinase; when loxP sites are present in tandem, recombination causes excision of the intervening sequence

N-ethyl-N-nitrosourea (ENU): an alkylating DNA mutagen

research. The description of the cloning of the human gonadotropin-releasing hormone (GnRH) precursor DNA was published in 1984 (15), whereas mutations in the Kallmann syndrome 1 (*KAL1*) gene (16, 17), LHβ (18), FSHβ (19), luteinizing hormone/choriogonadotropin receptor (LHCGR) (20), and the FSH receptor (21) were discovered within the following decade. These findings led to the identification of additional pathways linked in men and women: the synthesis and regulation of GnRH in the hypothalamus, the effects of GnRH on the pituitary, and the responsiveness of the ovary and testis to FSH and LH (4, 5, 22).

Like human patients, mice developed spontaneous mutations in endocrine pathway genes that were among the first genes to be recognized as fundamental for sexual development. In 1970, Lyon & Hawkes (23) discovered a mouse strain exhibiting insensitivity to androgens, a model termed testicular feminization (*tfm*). After the discovery of the human and rat androgen receptor genes in 1988 (24, 25), a frameshift mutation in the androgen receptor was shown to cause the androgen insensitivity in the *tfm* mouse (26). Absence of GnRH was identified in the hypogonadal (*hpg*) mouse described in 1977 (27) and was later shown to be secondary to a large deletion in the mouse *Gnrh* gene (28).

THE MOUSE AS A FUNCTIONAL TOOL FOR IN VIVO DISCOVERIES

Major papers in mouse genetics were reported in 1981–1983, heralding the transgenic age (**Figure 1**). These papers described the first germline transmission of foreign DNA (29–31) and the functional consequences of transgenic overexpression of a rat or human protein (in this case, the dramatic effects of increased expression of rat and human growth hormone that graced the covers of *Nature* and *Science*) (32, 33). These genetic technologies took advantage of important aspects of reproduction that had been developed over the preceding decades.

Reproductive biologists and geneticists did not ignore these transgenic advances. In an article accompanying the *GnRH* mutational analysis described above, Mason and colleagues (34) showed that the GnRH mutation could be corrected with transgenic technology, the first example of such gene therapy in vivo. Our group later showed that a human FSHβ transgene (with its own promoter, exons, and introns) could precisely correct the infertility defect in mice lacking (mouse) FSHβ (35).

The 1980s saw the initial uses of overexpression models to study in vivo function, and the first knockout models were created by the end of the decade. Several groups had begun to work with mouse embryonic stem (ES) cells, and retroviral infection of these ES cells had even been used to create a mouse with a mutation in the hypoxanthine-guanine phosphoribosyltransferase (*Hprt*) gene in an attempt to create a mouse with Lesch-Nyhan syndrome (36). Taking advantage of homologous recombination strategies, researchers were able to create knockouts in vitro in ES cells and to transmit these mutant ES cells through the germline, events that were first described in 1989 (37–40). Increased efficiency of homologous recombination and the establishment of ES cells that could better contribute to the germline made knockout technologies feasible as a tool for studying reproductive processes. To date, there have been more than 500 engineered mouse models with null mutations (see supplemental table 1 that accompanies Reference 4 for a comprehensive list of mutations). The first mouse reproductive genetic models that were created using homologous recombination in ES cells were leukemia-inhibitory factor (*Lif*) knockout mice, which display implantation defects secondary to absence of uterine LIF (41), and inhibin α (*Inha*)-null mice, which display ovarian and testicular cancers and reproductive defects (42). Some of these models (e.g., *Rec8*, *Rxfp2*) have also been created through the use of *N*-ethyl-*N*-nitrosourea (ENU) mutagenesis and transgenic insertional mutagenesis. Additional conditional knockout mouse models have been generated using Cre-loxP recombination. **Table 1** summarizes the reproductive genetics models

Table 1 Matzuk & Lamb (4) created mouse mutations with reproductive defects. Only single-mutant defects are described

Mutant gene	Sex affected	Reproductive phenotype	Fertility status	Reference(s)	Mouse chromosome
TGFβ superfamily signaling					
Activin/inhibin βB subunit (*Inhbb*)	F	Delivery and nursing defects	Subfertile	52	1
Activin/inhibin βA subunit (*Inhba*)	Both	cKO: increased corpora lutea; KI: delayed onset of fertility (M) and ovarian (follicle) failure (F)	Subfertile (cKO) Subfertile (M, KI) Infertile (F, KI)	53, 130	13
Anti-Müllerian hormone receptor (*Amhr2*)	M	Uterus development causes obstruction and secondary infertility (M)	Secondary infertility	85	15
Bone morphogenetic protein 15 (*Bmp15*)	F	Defects in cumulus–oocyte complex formation and ovulation	Subfertile	131	X
Activin receptor type 2A (*Acvr2*)	Both	Antral-follicle block (F); small testes and delayed fertility (M)	Infertile (F) Subfertile (M)	54	2
Follistatin (*Fst*)	F	Null, M-specific coelomic vessel in ovary and loss of oocytes; cKO: premature ovarian failure	Perinatal lethal (null) Subfertile (cKO)	46	13
Growth differentiation factor 9 (*Gdf9*)	F	Folliculogenesis arrest at the one-layer follicle stage	Infertile	61	11
Inhibin α (*Inha*)	Both	Granulosa/Sertoli tumors, gonadotropin hormone dependent	Infertile (F) Secondary infertility (M)	56	1
Bone morphogenetic protein receptor 1A (*Bmpr1a*)	F	cKO (Amhr2-Cre): premature ovarian failure	Subfertile	132	14
Sma/MAD homolog 1 (*Smad1*)	F	cKO (Amhr2-Cre): normal as single cKO	Fertile	133	8
Sma/MAD homolog 2 (*Smad2*)	F	cKO (Amhr2-Cre): normal as single cKO	Fertile	79	18
Sma/MAD homolog 3 (*Smad3*)	F	cKO (Amhr2-Cre): normal as single cKO	Fertile	79	9
Sma/MAD homolog 4 (*Smad4*)	F	cKO (Amhr2-Cre): ovarian defects	Subfertile	134	18
Sma/MAD homolog 5 (*Smad5*)	F	Null, developing embryos lose primordial germ cells; cKO (Amhr2-Cre): normal as single cKO	Lethal (null) Fertile (cKO)	133, 135	13
Germ cell mutants					
Cyclin A1 (*Ccna1*)	M	Block in spermatogenesis before the first meiotic division	Infertile	156	3
Germ cell specific with ankyrin repeat, SAM and basic leucine zipper domain (*Gasz*)	M	Block in spermatogenesis before the first meiotic division	Infertile	136	6

(*Continued*)

Table 1 (*Continued*)

Mutant gene	Sex affected	Reproductive phenotype	Fertility status	Reference(s)	Mouse chromosome
G protein–coupled receptor 149 (*Gpr149*)	F	Enhanced fertility due to increased ovulation	Superfertile	137	3
Kelch-like 10 (*Klhl10*)	M (Hetero)	Dysmorphic spermatozoa and impaired motility	Infertile	157	11
Lysine (K)-specific demethylase 4D (*Kdmd4*)	M	Trimethylated H3K9 accumulation in round spermatids	Fertile	138	9
Mixed-lineage leukemia 2 (*Mll2*)	F	cKO (Gdf9-Cre): anovulation and oocyte death	Infertile	139	15
Newborn ovary homeobox (*Nobox*)	F	Primordial- to primary-follicle block and oocyte loss	Infertile	140	6
Nucleoplasmin 2 (*Npm2*)	F	Maternal-effect gene; partial block at one-cell to two-cell embryo stage	Subfertile	96	14
Oligoadenylate synthetase–like 1D (*Oas1d*)	F	Defects in folliculogenesis and ovulatory efficiency	Subfertile	158	5
Platelet-activating factor acetylhydrolase 1b, subunit 2 (*Pafah1b2*)	M	Pachytene spermatocyte meiosis I block	Infertile	159, 160	9
RNA-binding motif protein 44 (*Rbm44*)	M	Increased sperm, enhanced fertility	Fertile	141	1
Spermatogenesis- and oogenesis-specific basic helix-loop-helix 1 (*Sohlh1*)	Both	Spermatogenesis block at meiotic entry; primordial- to primary-follicle block and oocyte loss	Infertile	142, 143	2
Tektin 3 (*Tekt3*)	M	Progressive sperm motility defects	Fertile	144	11
Tektin 4 (*Tekt4*)	M	Sperm motility defects and ultrastructural defects in the flagellum	Subfertile	161	17
Testis-expressed gene 14 (*TEX14*)	M	Absence of germ cell intercellular bridges; meiotic defect and block	Infertile	145	11
Y box protein 2 (*Ybx2*; *Msy2*)	Both	Elongating spermatid arrest (M); follicular atresia, oocyte loss, anovulation (F)	Infertile	162	11
Zona pellucida–binding protein 1 (*Zpbp1*)	M	Abnormal acrosome and globozoospermia	Infertile	146	11
Zona pellucida–binding protein 2 (*Zpbp2*)	M	Abnormal acrosome formation/sperm head defects	Subfertile	146	11
Zygote arrest 1 (*Zar1*)	F	Maternal-effect gene; block at one-cell to two-cell embryo stage	Infertile	112	5

(*Continued*)

Table 1 (Continued)

Mutant gene	Sex affected	Reproductive phenotype	Fertility status	Reference(s)	Mouse chromosome
Non–germ cell/pituitary mutants					
Dicer 1 (*Dicer1*)	F	cKO (Amhr2-Cre): oviductal diverticuli block embryo transit	Infertile	147	12
Follicle-stimulating hormone β (*Fshb*)	Both	Preantral block in folliculogenesis (F); decreased testis size (M)	Infertile (F)	55	2
Gamma-glutamyl transferase 1 (*Ggt1*)	Both	Males and females are hypogonadal and infertile; phenotype is corrected by feeding mice N-acetylcysteine	Infertile	163	10
Luteinizing hormone β subunit (*Lhb*)	Both	Failure of spermatogenesis and folliculogenesis in null animals	Infertile	164	7
Nuclear receptor subfamily 0, group B, member 1 (*Nr0b1*; *Dax1*)	M	Progressive degeneration of the germinal epithelium	Infertile	165	X
Oxytocin (*Oxt*)	Both	Maternal nursing defects (F); social memory defect and increased aggression (M)	Fertile	148, 149	19
Oxytocin receptor (*Oxtr*)	Both	Maternal nursing defects (F); social discrimination and increased aggression (M)	Fertile	150	6
Pentraxin 3 (*Ptx3*)	F	Defects in cumulus–oocyte complex integrity and ovulation	Subfertile	151	3
Retinoblastoma 1 (*Rb1*)	Both	cKO (Amh-Cre): progressive Sertoli cell dysfunction and germ cell loss; cKO (Amhr2-Cre): progressive follicular atresia and apoptosis	Progressive infertility (M) Subfertility (F)	152, 153	14
Superoxide dismutase 1 (*Sod1*)	F	Folliculogenesis defect, failure to maintain pregnancy	Subfertile	166, 167	16

^aAbbreviations: cKO, conditional knockout; F, female; Hetero, heterozygote; KI, knockin; M, male.

that Matzuk, Burns, and colleagues have created. Some of the models are described in greater detail below.

RELEVANCE OF TRANSFORMING GROWTH FACTOR β FAMILY SIGNALING PATHWAYS TO REPRODUCTION

The transforming growth factor β (TGFβ) superfamily of dimeric ligands is the largest family of secreted proteins in mammals, containing ~40 proteins of related homology (3). Most of these TGFβ family proteins signal through a canonical pathway that involves type 1 and type 2 serine threonine kinase receptors (**Figure 2**). Ligand binding to type 1 permits recruitment of the type 2 receptor (and vice versa, depending on the ligand) to form a complex that results in type 2 phosphorylation of the type 1 receptor and subsequent type 1 phosphorylation of

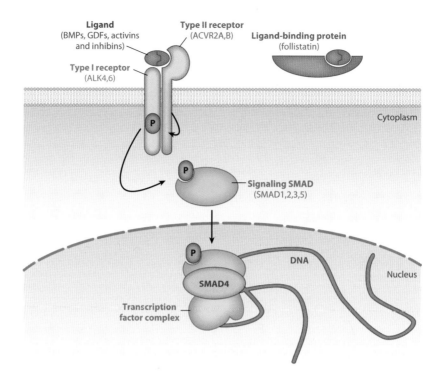

Figure 2

Transforming growth factor β (TGFβ) signaling. The schematic shows a canonical TGFβ signaling pathway. Ligand binding results in heterodimerization of cell surface receptors and their activation through phosphorylation. Activation of downstream SMAD transcription factors occurs through their phosphorylation and association with SMAD4.

downstream SMAD proteins. Activin and TGFβ ligands bind receptors that phosphorylate SMAD2 and SMAD3, whereas bone morphogenetic proteins (BMPs) bind receptors that phosphorylate SMAD1 and SMAD5. These phosphorylated, receptor-regulated SMADs then bind the common SMAD, SMAD4. Mutations in essentially every major component of these signaling pathways have been created in mice and, along with additional studies in humans and other species, have shown that these signaling pathways are essential to nearly every developmental and physiological process (3). Thus, these pathways are also critical for reproductive development and function, and relevant studies are described in additional detail in the next few sections.

Bone Morphogenetic Protein Pathways in Primordial Germ Cell Formation

Beginning with the earliest stages of the mammalian reproductive cycle—the formation of primordial germ cells (PGCs)—BMP2, BMP4, and BMP8 signal between embryonic day (E)5.5 and E6.0 through SMAD1, SMAD5, and SMAD4 in the proximal epiblast to allow these cells to become PGC precursors (5). By E6.25, the PGC precursors are defined by the expression of *PRDM1* (also known as *BLIMP1*) and *PRDM14*, the two major genes required for PGC specification (43, 44). Thus, BMP signaling pathways are required for formation of the first committed cells of the early embryo and the first reproductive cells.

Activins and Follistatin: Opposing Forces in Later Stages of Sex Determination

As mentioned above, follistatin, an activin-binding protein and antagonist, is a downstream player in the ovary determination cascade. Although *Fst* knockout mice die at birth due to craniofacial and muscle defects (45), investigators were able to examine the ovaries of *Fst*-null females. Many characteristics of the ovaries were still present; however, a coelomic vessel was also present, a finding typical of testes (46). This phenotype is likely secondary to unopposed activin B in the absence of its antagonist. However, further proof of this finding has not been demonstrated.

Paracrine Control of Granulosa Cell Function by Inhibin/Activins

Activins, dimers of βA and/or βB subunits, share common β subunits with inhibin A (α:βA) and inhibin B (α:βB) and appear to signal through complexes of type 2 (i.e., ACVR2A and ACVR2B) and type 1 (i.e., ALK4) receptors (47–49). Alternatively, the inhibin heterodimers bind to a complex of activin type 2 receptors and betaglycan to competitively antagonize activin signaling (50). Knockout mouse models (including some conditional knockouts) have been produced for these various subunits and receptors, but some of their roles in reproductive physiology are still somewhat unclear. Mice lacking inhibin/activin βA (*Inhba*) die at birth secondary to craniofacial defects (52), whereas mice lacking inhibin/activin βB (*Inhbb*) are viable and fertile, although these females show parturition defects (53). Mice with a conditional *Inhba* allele have been used to demonstrate that the two activin subunits have some redundant functions in ovarian folliculogenesis; complete activin β subunit deficiency leads to sterility and increased corpora lutea, or ruptured follicles (53).

Mice lacking *Acvr2b* die during embryogenesis or in the neonatal period, precluding analysis of the roles of *Acvr2b* in reproduction. Although ~25% of *Acvr2a*-null mice also die at birth due to craniofacial defects, the majority of these mice survive to the adult period, allowing extensive analysis of the roles of *Acvr2A* in several reproductive and disease processes (54). Both *Acvr2a*-null males and females show reduced pituitary and serum FSH levels, consistent with the roles for which activins were discovered. The reduced FSH levels result in smaller testis size and a block in ovarian folliculogenesis, findings similar to those in FSHβ-null mice (55). However, these similarities suggest that ACVR2A is not involved in gonadal activin signaling, whereas ACVR2B is the dominant receptor, or alternatively that ACVR2A and ACVR2B play redundant roles. Although ALK4 (ACVR1B) is postulated as the activin type 1 receptor on the basis of in vitro studies (49), embryonic lethality of *Acvr1b* knockout mice precludes defining the in vivo roles of *Acvr1b* in activin signaling. Conditional knockouts of *Acvr2b* and *Acvr1b* should be able to define their roles in ovarian and testicular activin signaling pathways.

In contrast to the roles of activins in stimulating pituitary FSH, the inhibins (true to their name) inhibit FSH synthesis and secretion. These endocrine effects (from the gonads) and autocrine (intrapituitary) effects were confirmed in mice lacking the *Inha* subunit; these mice show elevated FSH levels (42). In addition, these mice developed ovarian granulosa cell tumors and testicular Sertoli cell tumors, demonstrating gonadal autocrine and paracrine roles of inhibins in antagonizing activin signaling. Furthermore, follow-up studies revealed pathophysiological roles of tumor-produced activins in signaling specifically through ACVR2A in the liver and glandular stomach (56, 57). Because betaglycan appears to be a major receptor for the inhibins (50), conditional knockout of the betaglycan gene in the gonads or pituitary would be anticipated to downregulate inhibin actions and to potentiate activin signaling, possibly leading to gonadal cancers.

Although the TGFβ family member AMH induces regression of the Müllerian ducts during male fetal sex differentiation (58), AMH and its type II receptor, AMHR2, are also expressed postnatally in granulosa cells of primary and growing preantral follicles. This paracrine signaling pathway directly or indirectly suppresses primordial-follicle recruitment; absence of AMH leads to an increase in growing follicles in the juvenile period, a subsequent reduction of primordial follicles by 4 months, and few primordial and growing follicles by 13 months (59). The expression of AMH in early follicles also has clinical importance. Serum levels of AMH decline with age in women, and thus serum AMH has become a useful biomarker of ovarian reserve and the ability of women to get pregnant with their own oocytes (60).

Dominant negative: describes a gain-of-function mutation that leads to a gene product that counters activity of the functional allele in *trans*

Growth Differentiation Factors in Oocyte–Somatic Cell Communication

For several decades, communication between the somatic granulosa cells and oocyte has been known. Early studies, however, suggested that the oocyte may be just an innocent bystander in the process, responding to signals from the granulosa cells (e.g., granulosa cell KIT ligand signaling through the oocyte KIT receptor). However, in 1996, we published knockout work showing that a unique oocyte-secreted TGFβ family member, growth differentiation factor 9 (GDF9), was required for progression of folliculogenesis beyond the primary-follicle stage (61). These studies established that the communication of the ovary was a dialogue. Follow-up studies by our group and others showed that another oocyte-secreted and structurally related family member, BMP15 (GDF9B), was also involved in this dialogue. We cloned the full-length *BMP15* genes in human and mice and demonstrated that they were X linked (62). This chromosomal localization was a big clue to the sheep reproductive geneticists, who had found that a specific X-linked mutation in the Inverdale (FecX^I) and Hanna (FecX^H) breeds of sheep gave rise to either infertile or superfertile females (63, 64). These findings lead to the discovery that the FecX^I mutation mapped to a syntenic position within the *BMP15* gene and that the FecX^I sheep contained a point mutation in the mature sequence of *BMP15* that resulted in a V31D change, whereas the FecX^H was a truncation mutation in the mature sequence (Q23Ter) (65). In the heterozygous state, these mutations acted to increase the rate of ovulation, whereas in the homozygous state they resulted in sterility. Additional mutations in the genes encoding BMP15 and GDF9 and one of the BMP15 receptors, ALK6, influence fertility in additional breeds of sheep (summarized in Reference 66). These genetic studies indicate that GDF9 is also involved in ovarian folliculogenesis in sheep, as in mice, and that ALK6 is a receptor linked to ovarian function (see below).

Parallel studies have also been performed in humans to attempt to identify mutations in these genes associated with fertility phenotypes. Although a number of studies have identified *GDF9* and *BMP15* polymorphisms that are linked to fertility phenotypes, few studies have performed functional follow-up to characterize the significance of these polymorphisms. In two studies, *GDF9* polymorphisms were discovered in mothers of dizygotic twins (67, 68). In one study, heterozygous mutations in the *BMP15* gene were discovered in sisters who exhibited ovarian failure (69). Such mutations in the proregion-encoding part of the *BMP15* precursor causes an apparent dominant negative effect, leading to streak gonads.

Although the above discussion focuses on GDF9 and BMP15 in signaling between the oocyte and granulosa cells at the primary to secondary (preantral)-follicle stages, these oocyte-secreted proteins also play major roles in the preovulatory follicle. Prior to ovulation, the follicle contains a large antral cavity (**Figure 3**). The wall of the cavity contains mural granulosa cells, whereas the oocyte is surrounded by cumulus granulosa cells that are connected to the wall and the mural granulosa cells by a cellular stalk. Both direct effects of the GDF9/BMP15 (paracrine) and indirect effects of the LH (endocrine) signaling pathway through the mural granulosa cells influence the major ovarian follicular process of cumulus expansion. This process involves laying down an extracellular matrix that is rich in hyaluronic acid, which the cumulus cells synthesize by hyaluranon

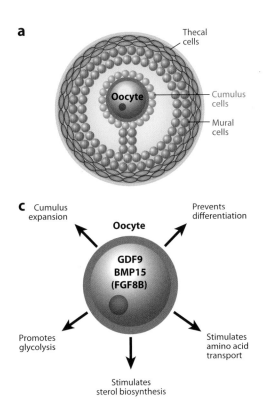
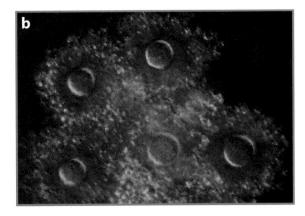

Figure 3

The preovulatory ovarian follicle and regulation of cumulus cell functions. (*a*) Major cell types in the preovulatory ovarian follicle. (*b*) Cumulus cells in the hyaluronin-rich matrix around mouse oocytes isolated after ovulation. (*c*) Cumulus granulosa cell functions that are regulated by oocyte growth factors.

synthase 2, additional cumulus cell–derived proteins (e.g., pentraxin 3 and tumor necrosis factor–induced factor 6), and heavy-chain subunits of interalpha trypsin inhibitor, which are derived from the serum after the LH surge. LH stimulates the (LH receptor–positive) mural granulosa cells to produce epidermal growth factor–like peptides that act on the LH receptor–negative cumulus cells to produce the cumulus cell products in concert with GDF9/BMP15 (70). This cumulus expansion is necessary for proper release of the cumulus cell–oocyte complex from the follicle upon rupture, for capture by the oviductal fimbriae, and for successful fertilization in the oviduct. Not only is this process important for cumulus expansion, but cumulus cell–oocyte signaling is also key for regulating a number of additional events within the complex such as glycolysis and cholesterol synthesis (71, 72). Recently, Eppig and colleagues (73) showed that mural granulosa cell–derived natriuretic peptide type C signals to its cumulus cell receptor, natriuretic peptide receptor 2; this is an important determinant of cGMP levels in the cumulus and maintains oocytes in meiotic arrest. It is truly remarkable that these endocrine, paracrine, and autocrine signaling pathways orchestrate the events of follicular development to produce a healthy, fertilizable oocyte and how mouse genetics have contributed to our understanding of these events.

ALK6 (BMPR1B) has been implicated as one of the type 1 receptors downstream of BMP15 in sheep, mouse, and rat granulosa cells. The Booroola sheep mutation *FecB* is a Q249R mutation in the kinase domain of ALK6, thereby increasing the fertility of these females (74). *Alk6*-null mice are infertile and have cumulus expansion defects (75). Last, ALK6 may be the type 1 receptor for BMP15, whereas BMPR2 is the type 2 receptor for BMP15 in vitro (76).

Germ Cell–Somatic Cell Cross Talk in the Testis

Whereas mutations in BMP signaling pathways during embryogenesis result in defects in PGCs in both sexes, the BMP and activin/inhibin pathways are also functional in the postnatal testes. Playing roles analogous to those of GDF9/BMP15 in the oocyte, BMP7, BMP8a, and BMP8b are expressed in male germ cells and appear to act redundantly to regulate somatic cell (Sertoli cell) functions, roles that are critical for spermatogenesis and male fertility (3, 77). The signaling pathways for these ligands have yet to be elucidated.

Activins and inhibins play paracrine roles in the testis, and the absence of *Inha* in the testes results universally in Sertoli cell tumors that are reminiscent of granulosa cell tumors in females because nearly all these male (Sertoli cell) tumors produce estradiol (42). However, unlike in females that require both SMAD2 and SMAD3 (78, 79), the testicular tumors derive from unopposed activin signaling specifically through SMAD3 (78). These results are somewhat surprising and suggest either that SMAD3 levels are much greater than SMAD2 levels in Sertoli cells or that SMAD3 targets a unique set of transcriptional targets.

Additional Roles of Transforming Growth Factor β Signaling in the Male and Female Reproductive Tracts

In addition to the roles of BMP7, BMP8a, and BMP8b in testicular signaling pathways, these ligands, along with BMP4, play roles in the integrity of the epididymis and functionally affect sperm motility (80–82). Whereas mutations in *BMP7*, *BMP8a*, and *BMP8b* affect the caput and cauda regions of the epididymis, heterozygous mutations in *BMP4* result in degeneration of the epididymal epithelium in the corpus region. The receptor and SMAD signaling pathways for these BMP ligands in the different regions of the epididymis are unknown. Likewise, mutations in the human orthologs of these ligands have yet to be reported but would be expected in populations of men and women with primary infertility and premature ovarian failure. Persistent Müllerian duct syndrome (PMDS; MIM ID#261550) is a defect of sexual differentiation in which the Müllerian duct fails to regress, resulting in a uterus in otherwise normal males. Mutations in AMH (type 1) and AMHR2 (type 2) have been identified in the majority of PMDS families (83, 84). Behringer and colleagues (58, 85–87) used null mice and conditional knockout of genes in the Müllerian duct mesenchyme to genetically confirm these findings with AMH and AMHR2; to identify ALK2 and ALK3 as the AMH type 1 receptors; and to confirm redundancy of SMAD1, SMAD5, and SMAD8 downstream of these receptors. Growth differentiation factor 7 (GDF7) is another TGFβ family member that is required in males only. GDF7 is expressed in the mesenchyme and is essential for the differentiation of the seminal vesicle epithelium (88). Because GDF7 is a member of the BMP subfamily, GDF7 likely signals through BMP receptors and SMADs.

In addition to LIF (41), a large number of proteins as well as estradiol and progesterone function in the preparation of the uterus for pregnancy, including attachment of an embryo, implantation, and maintenance of pregnancy (reviewed in Reference 89). BMP2 not only is required for PGC precursor formation and early embryonic development but also has been identified as a key signaling protein in uterine biology (90). Conditional knockout of *Bmp2* using PR-Cre led to sterility due to failure of the uterine endometrial stromal cells to differentiate into decidual cells, and WNT4 is a major downstream target of BMP2 in both mice and humans. Our group is investigating the key BMP receptors and SMAD proteins in the BMP2 signaling cascade (**Table 1**).

MATERNAL EFFECT IN EARLY DEVELOPMENT

Development of the embryo from fertilization through the early preimplantation stages is heavily dependent on maternal-effect gene products, mRNA, and proteins readied during oogenesis. Discovery of maternal-effect genes first took place in 1980 in *Drosophila* with the recognition that maternal *dorsal* is a critical determinant of polarity in the embryo (91).

Maternal-effect gene: a maternal gene that when dysfunctional affects the phenotype of offspring, regardless of their genotype

In mammals as well, the fertilized oocyte is replete with maternal factors, and some specific examples are fundamental to preimplantation viability. This is true even beyond initiation of zygote transcription [termed zygote genome activation (ZGA)]; inheritance of a functional paternal gene copy does not rescue a maternal-effect phenotype. The maternal-effect window is a time interval of extraordinary cytological and nuclear transitions (92). Experiments in mice have allowed the most detailed descriptions of these changes, and targeted mutations in the model have identified several critical participating factors (**Figure 4** and **Table 2**). The first mammalian maternal-effect gene identified was termed *Mater* [Maternal antigen that embryos require; officially of the NLR family, pyrin domain–containing 5 (*NLRP5*)]; it is a component of a protein complex that forms subcortical cytoplasmic lattices (93) in oocytes and early embryos. Engineering a null mutation in *Nalp5* led to female-specific infertility in homozygous mice that is characterized by normal ovarian histology and oocyte ovulation but loss of embryos at the two-cell stage (94).

In the sections to follow, we discuss observations and models with particular relevance to the continuity of germline DNA and its higher-level organization as chromatin. However, required factors acting in this interval cannot be comprehensively elucidated by traditional knockout models. Essential roles for many exist in development or in oogenesis, precluding phenotyping at the maternal-embryo interface. Complementary approaches, such as RNA and protein inhibition through oocyte injections, are helpful as we dissect the biological pathways of preimplantation embryos.

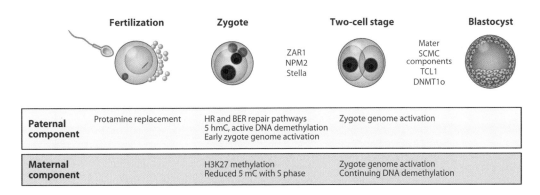

Figure 4

Early embryonic development from fertilization through the blastocyst stages. Fertilization is closely followed by the replacement of protamine proteins for maternal nucleosomes in the male pronucleus. Although morphologically similar, the two pronuclei have several differences with respect to activity of DNA repair pathways, DNA and histone methylation status, and transcription. These are less pronounced in later peri-implantation stage embryos, as the maternal DNA contribution reaches its methylation nadir, and there are sustained high levels of RNA expression from both zygote alleles at active, unimprinted loci. Several maternal-effect gene products provided by the oocyte are key in these transitions, including zygote arrest 1 (ZAR1), nucleoplasmin 2 (NPM2), Stella, Mater and other components of the subcortical maternal complex (SCMC), T cell leukemia/lymphoma 1 (TCL1), and an oocyte-specific isoform of DNA methyltransferase 1 (DNMT1o). In early embryos, there is active demethylation of the paternal genome, whereas maternal-genome methylation is lost throughout the early cleavage stages with successive rounds of replication. Other abbreviations: 5 hmC, 5-hydroxymethylcytosine modification; 5 mC, 5-methylcytosine modification; BER, base excision repair; HR, homologous recombination.

Table 2 Maternal-effect genes identified in mouse models

Gene product	Common abbreviation and gene symbol	Phenotype seen in embryos of null females	Reference(s)
Brg1 (SWI/SNF-related, matrix-associated, actin-dependent regulator of chromatin, subfamily a, member 4)	*Smarca4*	Conditional deletion of the gene in oocytes leads to an embryonic genome activation defect in two-cell embryos	168
DNA methyltransferase 3a	*Dnmt3a*	Conditional mutation in the female germline causes pups to die in utero with lack of methylation and allele-specific expression at maternally imprinted loci	169
DNA methyltransferase 3L	*Dnmt3L*	Pups die in midgestation; as above, their phenotype is associated with a loss of maternal DNA methylation at imprinted loci	170
Filia	*Filia, Ecat1*	Subcortical maternal complex (SCMC) member (with MATER and FLOPED, both below); *Filia*-null females are subfertile, with high embryonic aneuploidy rates and loss of preimplantation-stage embryos	113
FLOPED; officially oocyte-expressed protein homolog (dog)	*Ooep*	*Ooep*-null females are infertile, with embryonic arrest at the two- or four-cell stages; SCMC disappears in these embryos without the maternal protein	154
Lysine demethylase 1B; amine oxidase (flavin-containing) domain 1	*Kdm1b, Aof1*	KDM1B-deficient oocytes showed increased H3K4 methylation; embryos showed loss of imprinting and died before midgestation	117
Maternal antigen that embryos require; officially NLR family, pyrin domain–containing 5	*Mater; Nlrp5*	The SCMC disappears in embryos without the maternal protein; embryos are lost in early cleavage stages	94
Nucleoplasmin 2	*Npm2*	This nuclear protein is critical for chromatin condensation surrounding nucleolar structures and for survival to the two-cell stage	96
p73 (a TAp73 isoform)	*Tap73*	Null females for this isoform have defects in oogenesis but then an absence of the maternal protein in early cleavage embryos leads to loss postfertilization and abnormal cell divisions and numbers of nuclei in blastomeres	155
Stella; officially developmental pluripotency–associated 3	*Dppa3*	Embryo loss at approximately the four-cell stage occurs; Stella is a nuclear protein important for some asymmetries appreciable in maternal and paternal pronuclei (see text)	103, 104
T cell leukemia/lymphoma	*Tcl1*	The TCL1 protein shuttles between nucleus and cytoplasm in early mitoses; loss of the maternal protein compromises early embryonic mitoses	171
Zygote arrest 1	*Zar1*	Pronuclei form, and DNA replication initiates, but there is no syngamy; maternal and paternal genomes remain separate in arrested zygotes	112
ZFP57 (a KRAB zinc finger protein)	*Zfp57*	Roles are described for both embryonic and maternal ZFP57 in establishment and maintenance of DNA methylation patterns	118

Reorganization and Damage Repair in the Paternal Pronucleus

At the time of entry into an oocyte, the paternal haploid genome is packaged with protamine proteins and the male germline–specific pericentric histones H2AL1 and H2AL2; these are rapidly removed from their associated DNA in the fertilized zygote (95). Closely coupled to this chromatin expansion is increasing intensity of staining for nucleosomal structures by immunofluorescence and association of the H3.3–H4 dimer chaperone Hira (95). Over the ensuing hours, the DNA is prepared for the first round of replication, which occurs in the physically separate male and female pronuclei of one-cell zygotes, and is followed by syngamy, the process of pronuclear breakdown and alignment of maternal and paternal genetic complements as chromosomes are condensed during the first mitosis. Histone variant H3.1 becomes associated with paternal DNA in conjunction with the first S phase prior to syngamy with the female pronucleus (95). The ooplasm provides these histones. The maternal genome decondenses from meiosis II without an analogous repackaging.

We described nucleoplasmin 2 (*Npm2*) as a maternal-effect gene using a targeted null allele in mice (96). Immunodepletion and add-back studies in *Xenopus* oocytes suggest that the nucleoplasmin ortholog in the frog is needed for paternal DNA decondensation, including loss of the sperm-specific basic proteins in that species (X and Y) and gain of nucleosomes as marked by acquisition of histones H2A and H2B (97, 98). In contrast, the mammalian protein appears dispensable for these processes, although we did recognize marked subfertility in *Npm2*-null females. Although ovulation of oocytes competent to be fertilized, the unpackaging of sperm DNA, and progress of zygotes through the first replication were all intact, thereafter many embryos were lost in the transition to the two-cell stage. The phenotype was accompanied by striking changes morphologically in nuclear organization, including the dispersal of nucleolar proteins (i.e., loss of the singular germinal vesicle) in oocytes and embryos and perinucleolar heterochromatin as marked by deacetylated histone H3.

DNA damage is incurred during these physiological transitions, during spermiogenesis in the male, or subsequently during paternal chromatin reorganization in the zygote. This is thought to prompt recruitment of maternally encoded markers of damage and repair proteins; such recruitment occurs selectively in the paternal pronucleus in one-cell zygotes. Accumulation of γH2A.X foci and poly(ADP-ribose) polymerase 1 (PARP1) follows this asymmetric pattern in early one-cell embryos (99, 100). γH2A.X and PARP1 are best known as markers of double-stranded and single-stranded DNA breaks, respectively, although neither is exclusive, and there appears to be colocalization of the two signals. Nonhomologous end joining (NHEJ), homologous recombination, and base excision repair (BER) are active in one-cell zygotes. NHEJ is immediately available with sperm unpackaging to resolve damage; homologous recombination becomes available at S phase. Homologous recombination appears to preferentially occur in the paternal pattern, as marked by the presence of RAD51 protein. Experiments using severe combined immunodeficient (*scid*) mice and *Rad54*, *Rad54b* double-mutant mice demonstrate the dependency on maternal genotype of these respective zygotic pathways (99). Activation of BER pathways in the paternal pronucleus, marked by selectively bound X-ray repair complementing in Chinese hamster cells 1 (XRCC1) protein there, attends an active DNA demethylation occurring before replication and congress of the parental genomes (101).

Pronuclear Asymmetries in One-Cell Zygotes

Besides the distribution of DNA repair pathway proteins, there are other interesting asymmetries in comparing male and female pronuclei in one-cell zygotes. These include differential levels of DNA methylation [5-methylcytosine (5 mC)], histone modifications, and transcription.

Imprinted gene: a gene locus with parent-of-origin-specific methylation patterns and expression

As alluded to in the above section, DNA methylation in the male pronucleus is actively removed. This is in contrast to the maternal genome, which is passively demethylated during replication in cleavage-stage embryos. The transition in the paternal pronucleus is accompanied by 5-hydroxymethylcytosine (5 hmC) modification (102), which accumulates with a reduction of 5 mC. STELLA [also known as DPPA3 (developmental pluripotency–associated 3)], a key protein involved in the pronuclear asymmetries, actively counters some pathways in the female pronucleus. STELLA is encoded by a maternal-effect gene required for zygote development beyond the early cleavage stages; most deficient embryos are lost at approximately the four-cell stage (103, 104). Absence of STELLA leads to zygotes with aberrant BER pathway progression and active DNA demethylation in both pronuclei (101, 105) rather than in solely the male pronucleus. The reason for the differential effect when STELLA is seen in both pronuclei, however, is unclear.

A subset of histone modifications are differentially distributed between male and female zygote pronuclei. The latter contains the bulk of methylated H3K9, both dimethylated (H3K9me2) and trimethylated (H3K9me3), as well as the majority of methylated H3K27 and tightly bound heterochromatin protein 1β (95, 106). In contrast, the male pronucleus has a higher proportion of acetylated histone H4 (107). Histone modifications associated with heterochromatin, such as H3K9me3 and H3K9me2, and proteins involved in placing this mark, such as ERG-associated protein with SET domain (ESET/Setdb1) and KRAB-associated protein 1 (KAP1/Trim28), appear important for silencing of endogenous retroviruses and L1 LINE retrotransposons in ES and embryonic carcinoma cell culture models (108–110). These are separable from DNA methylation effects at this stage. The roles of these marks in earlier development and how the male genome in particular is protected during active DNA demethylation in the absence of these heterochromatin modifications remain to be well understood.

Finally, there are notable differences in zygote genome activation on the basis of parent of origin. Early transcription from the zygote genome, known as phase I or minor ZGA, begins in the male pronucleus (111). Such transcription precedes more robust subsequent phases of gene expression and the translation of the first zygote-encoded mRNAs seen in the two-cell stage. At this point, paternal alleles have the theoretical potential to rescue otherwise embryonic lethal deficiencies in oocyte proteins; however, the functions of many maternal-effect genes cannot be fulfilled by expression of these gene products from the paternal genome once irrecoverable damage to early developmental pathways has occurred.

Syngamy and Mitotic Competence

Asymmetries appreciable at the microscopic level in male and female pronuclei are lessened with the first replication (S phase) and are ultimately lost with the first mitosis. However, at the molecular level DNA methylation differences persist between parental alleles globally until peri-implantation and at imprinted loci essentially permanently in cells contributing to somatic tissues. Few maternal-effect genes are known to be critical for syngamy; one is a cytosolic protein known as zygote arrest 1, which is dispensable for oogenesis, fertilization, and pronuclear formation but is required for entry into the first mitosis (112).

Most maternal-effect genes determining early embryonic developmental potential result in lethality somewhat later in development, so fragmentation and embryo loss occur between the two-cell stage and compaction of eight-cell morula. Several proteins encoded by such maternal-effect genes, including *Nlrp5* (MATER) and *Ooep* (FLOPED), are members of a protein complex termed the subcortical maternal complex (SCMC) (**Table 2**). The complex appears vital, given its loss in the absence of MATER or FLOPED and the uniform demise of deficient embryos in these early stages. Although the SCMC may have a myriad of functions, knockout of one component has highlighted a role in chromosomal segregation in early embryos. This gene product, Filia, was recently discovered in mice; *Filia*-null females are markedly subfertile and infertile and have a loss of early embryos typified by abnormalities in chromosome segregation. Early-cleavage-stage embryos undergoing mitosis in the absence of maternal Filia demonstrate an absence of the centrosome-associated spindle assembly regulator aurora kinase A (AURKA), delay in progression of mitosis, and the development of chromosomal aneuploidies and micronuclei (113). Interestingly, the phenotype recapitulates some aspects of the *Aurka*-null mouse. Absence of *Aurka* is manifested by an early embryonic lethal phenotype rather than by a maternal effect, although maternal stores presumably meet a likely need for *Aurka* function in the earliest embryo. In the absence of a competent embryonic *Aurka* locus, null mice succumb to similar defects in chromosome alignment and segregation around the blastocyst stage (114). Thus, maternally encoded gene products are key to embryonic survival and to faithful genome copy distribution to blastomeres.

Phenocopy: a phenotype that recapitulates features of another mutant phenotype; suggests related roles for the two gene products

Methylation at Imprinted Loci

Imprinted loci—those with parent-specific patterns of expression—are substrates for parent-of-origin-specific methylation during gametogenesis. Despite alterations in global chromatin organization with fertilization and early development, patterns of methylation at most imprinted loci are specifically conserved during early development and are rewritten only in the germline. An oocyte and early embryonic isoform of a replication-associated DNA methyltransferase 1 (DNMT1o) (115) and STELLA (105) are responsible for protecting methylation at these loci during genome-wide demethylation in early development.

Embryonic life is also when defects in writing these imprints appear, so pathways not functioning during oogenesis are reflected as a maternal-effect phenotype. This is the case with the de novo DNA methyltransferases DNMT3a and DNMT3L. Conditional depletion of *Dnmt3a* in the female germline (using TNAP-Cre) or homozygosity for a targeted null allele of *Dnmt3L* in a mother causes embryonic loss before midgestation. This is characterized by a loss of maternal DNA methylation at imprinted loci. Structural studies have demonstrated that a tetrameric complex composed of two Dnmt3a DNA methyltransferases and two Dnmt3L proteins interacts with unmethylated H3K4 through Dnmt3L; thus, the complex positions DNMT3a to create maternal methylation patterns (116). Importantly, activity of DNMT3a seems predicated on H3K4 being unmethylated. In females lacking the lysine demethylase KDM1B, oocytes show increases in H3K4 methylation and fail to establish maternal imprints; there is a phenocopy of the DNMT3a/DNMT3L maternal-effect phenotype (117). Similar defects have been described as a maternal effect of ZFP57, a KRAB zinc finger protein thought to be involved in corepressor complex recruitment to chromatin (118).

Relatively fewer imprinted genes are paternally methylated. Interestingly, though, DNMT3a and DNMT3L appear required for spermatogenesis, so no paternal-effect phenotypes are realized with these models. KDM1B expression has not been described in the male germline.

THE GENETICS OF HUMAN INFERTILITY

A wide range of genes have been identified to play roles in male and female fertility; most of these data come from genetic manipulation of the mouse (reviewed in References 2 and 4). From these mouse genetics studies, we have learned much about the factors involved in the hypothalamic-pituitary-gonadal axis and the regulation of other aspects of the male and female reproductive tracts. Two major surprises have come from this work. First, whereas we have been able to easily translate many of the data involving hormonal and sexual differentiation pathway genes between mice and humans, few orthologs of genes specific and essential to gametogenesis in the mouse have been uncovered as mutated in patients. This dearth of gene mutations may be because there are too many candidates; Schultz et al. (119) suggested that ~4% of the genome encodes genes that are expressed postmeiotically in male germ cells, and thus mutations in patients with infertility may be spread among >1,000 genes. Alternatively, deleterious mutations may arise somatically or specifically in the germline, meaning that the cells harboring these are not sampled for most clinical studies.

A second major surprise has been that there are more identified mouse genes that are essential for testicular phenotypes than those involved in ovarian phenotypes. Although there may be ascertainment biases, there is likely either greater genetic complexity or susceptibility inherent in male gametogenesis. The latter has been suggested by comparisons of meiotic gene phenotypes (120).

These genetic findings in mice have implications not only for infertility research but also for contraception. Drugs that could phenocopy infertility in women and men and thus cause a contraceptive effect and no other side effect (especially if the gene were oocyte specific or spermatocyte specific) would be highly desirable. However, to date, no germ cell–specific drugs have been developed.

Most causes of clinical infertility in humans are idiopathic, although genetic etiologies are appreciated in some cases. In 1976, Tiepolo & Zuffardi (121) drew attention to structural abnormalities on the Y chromosome as responsible for defective spermatogenesis. Now, several azoospermia factor (AZF) intervals are recognized on Yq. The most commonly lost in microdeletions characterized today is AZFc, followed by AZFb and combined deletions of AZFbc (122). The major gene loci involved at AZFc are copies of the deleted in azoospermia (*DAZ*) gene. The murine ortholog is autosomal and is critical to germ cell development in both sexes (123), and studies in human embryonic stem cells suggest a requirement for differentiation to PGCs (124). In women, genetic causes of infertility or premature ovarian failure are more heterogeneous, although examples of heterozygous missense mutations in *BMP15* have been uncovered.

Two pilot GWAS published in the summer of 2009 attempted to define common genetic variants responsible for infertility phenotypes in humans. The first examined approximately 50 men with oligospermia and 40 with nonobstructive azoospermia, comparing single-nucleotide polymorphism (SNP) genotypes genome wide to controls. These studies tentatively identified SNPs in or near phosphodiesterase 3A (*PDE3A*), sal-like 4 (*SALL4*), ADP-ribosylation factor–like 6 (*ARL6*), neurexophilin 2 precursor (*NXPH2*), and EF-hand calcium-binding domain 4B isoform b (*EFCAB4B*) as tagging haplotypes containing causative variants (associative p values of 10^{-7}) (125). Similarly, a study of 90 women with premature ovarian failure suggested a genetic etiology in the vicinity of an intronic SNP in ADAM metallopeptidase with thrombospondin type 1 motif, 19 (*ADAMTS19*) (126). If follow-up studies can verify these findings and identify causative lesions affecting these or neighboring genes, this approach may lead to an understanding of common genetic variants participating in the complexities of human reproduction.

TECHNOLOGICAL ADVANCES IN HUMAN GENETICS AND OPPORTUNITIES FOR UNDERSTANDING REPRODUCTIVE BIOLOGY

Meanwhile, massively parallel genome sequencing is changing the landscape of what is possible in human clinical genetics. Whole-exome sequencing is expected to be applied in impressive numbers of clinical cases in the next few years, and this approach should lead to a more comprehensive picture of the role of rare and de novo genetic lesions in a broad spectrum of human pathologies. In many cases, the genomes of a patient's parents are also being sequenced to more easily distinguish causative from rare incidental variants. When highly penetrant, monogeneic lesions are considered, some lesions found in a child will reflect inheritance of recessively acting alleles from each parent, leading to homozygosity or compound heterozygosity. Other lesions will reflect de novo mutations, and these lesions should be especially valuable to those of us with interests in how DNA is passed with fidelity from generation to generation.

With effective data sharing, one very fundamental question that can be addressed is the mutation rate in humans. Empirically determining global rates of base pair substitution of different kinds should be valuable and may indicate genes or motifs that are especially susceptible or protected. A second major type of question we might approach to leverage these new tools is a better determination of the originating cell types and timing of mutations. Are mutations occurring preferentially in maternal or paternal alleles, reflecting perhaps differential opportunities for mutation by specific mechanisms in germline and early embryonic development? Germline genetic mosaicism in males should be possible to directly assess. How the incurrence of mutations may relate to epigenetic marks or locations of binding proteins in parental germ cell development and in early embryos may also become possible to better understand.

DISCLOSURE STATEMENT

The authors are not aware of any affiliations, memberships, funding, or financial holdings that might be perceived as affecting the objectivity of this review.

ACKNOWLEDGMENTS

Dr. Matzuk is indebted to the Eunice Kennedy Shriver National Institute of Child Health and Human Development (HD07495, HD32067, HD33438, HD36289, HD57880, and HD60496), the National Cancer Institute (CA60651), the Ovarian Cancer Research Fund, and the Mary Kay Ash Charitable Foundation for their two decades of support of our studies in reproductive genetics and ovarian cancer. Research in the Burns laboratory is supported by a K08 award from the National Cancer Institute and a Career Award for Medical Scientists from the Burroughs Wellcome Foundation. We are beholden to our outstanding mentors and close colleagues who have given us personal and critical guidance over the years, and we respectfully acknowledge Drs. Irving Boime, Allan Bradley, Franco DeMayo, John Eppig, Milton Finegold, Raj Kumar, Dolores Lamb, Stephanie Pangas, and Richard Schultz. Last, we thank our parents, who instilled in us the quest for scientific discovery.

LITERATURE CITED

1. Matzuk MM, Burns K, Viveiros MM, Eppig J. 2002. Intercellular communication in the mammalian ovary: Oocytes carry the conversation. *Science* 296:2178–80
2. Matzuk MM, Lamb DJ. 2002. Genetic dissection of mammalian fertility pathways. *Nat. Med.* 8(Suppl. 1):41–49
3. Chang H, Brown CW, Matzuk MM. 2002. Genetic analysis of the mammalian TGF-β superfamily. *Endocr. Rev.* 23:787–823
4. Matzuk MM, Lamb DJ. 2008. The biology of infertility: research advances and clinical challenges. *Nat. Med.* 14:1197–213
5. Edson MA, Nagaraja AK, Matzuk MM. 2009. The mammalian ovary from genesis to revelation. *Endocr. Rev.* 30:624–712
6. Ford CE, Jones KW, Polani PE, de Almeida JC, Briggs JH. 1959. A sex-chromosome anomaly in a case of gonadal dysgenesis (Turner's syndrome). *Lancet* 1:711–13

7. Jacobs PA, Strong JA. 1959. A case of human intersexuality having a possible XXY sex-determining mechanism. *Nature* 183:302–3
8. Sinclair AH, Berta P, Palmer MS, Hawkins JR, Griffiths BL, et al. 1990. A gene from the human sex-determining region encodes a protein with homology to a conserved DNA-binding motif. *Nature* 346:240–44
9. Berta P, Hawkins JR, Sinclair AH, Taylor A, Griffiths BL, et al. 1990. Genetic evidence equating *SRY* and the testis-determining factor. *Nature* 348:448–50
10. Koopman P, Munsterberg A, Capel B, Vivian N, Lovell-Badge R. 1990. Expression of a candidate sex-determining gene during mouse testis differentiation. *Nature* 348:450–52
11. Parma P, Radi O, Vidal V, Chaboissier MC, Dellambra E, et al. 2006. R-spondin1 is essential in sex determination, skin differentiation and malignancy. *Nat. Genet.* 38:1304–9
12. Uhlenhaut NH, Jakob S, Anlag K, Eisenberger T, Sekido R, et al. 2009. Somatic sex reprogramming of adult ovaries to testes by FOXL2 ablation. *Cell* 139:1130–42
13. MacLaughlin DT, Donahoe PK. 2004. Sex determination and differentiation. *N. Engl. J. Med.* 350:367–78
14. Veitia RA. 2010. FOXL2 versus SOX9: a lifelong "battle of the sexes." *BioEssays* 32:375–80
15. Seeburg PH, Adelman JP. 1984. Characterization of cDNA for precursor of human luteinizing hormone releasing hormone. *Nature* 311:666–68
16. Franco B, Guioli S, Pragliola A, Incerti B, Bardoni B, et al. 1991. A gene deleted in Kallmann's syndrome shares homology with neural cell adhesion and axonal path-finding molecules. *Nature* 353:529–36
17. Legouis R, Hardelin JP, Levilliers J, Claverie JM, Compain S, et al. 1991. The candidate gene for the X-linked Kallmann syndrome encodes a protein related to adhesion molecules. *Cell* 67:423–35
18. Weiss J, Axelrod L, Whitcomb RW, Harris PE, Crowley WF, Jameson JL. 1992. Hypogonadism caused by a single amino acid substitution in the β subunit of luteinizing hormone. *N. Engl. J. Med.* 326:179–83
19. Matthews CH, Borgato S, Beck-Peccoz P, Adams M, Tone Y, et al. 1993. Primary amenorrhoea and infertility due to a mutation in the β-subunit of follicle-stimulating hormone. *Nat. Genet.* 5:83–86
20. Shenker A, Laue L, Kosugi S, Merendino JJ Jr, Minegishi T, Cutler GB Jr. 1993. A constitutively activating mutation of the luteinizing hormone receptor in familial male precocious puberty. *Nature* 365:652–54
21. Aittomaki K, Lucena JLD, Pakarinen P, Sistonen P, Tapanainen J, et al. 1995. Mutation in the follicle stimulating hormone receptor gene causes hereditary hypergonadotropic ovarian failure. *Cell* 82:959–68
22. Bianco SD, Kaiser UB. 2009. The genetic and molecular basis of idiopathic hypogonadotropic hypogonadism. *Nat. Rev. Endocrinol.* 5:569–76
23. Lyon MF, Hawkes SG. 1970. X-linked gene for testicular feminization in the mouse. *Nature* 227:1217–19
24. Chang CS, Kokontis J, Liao ST. 1988. Molecular cloning of human and rat complementary DNA encoding androgen receptors. *Science* 240:324–26
25. Lubahn DB, Joseph DR, Sar M, Tan J, Higgs HN, et al. 1988. The human androgen receptor: complementary deoxyribonucleic acid cloning, sequence analysis and gene expression in prostate. *Mol. Endocrinol.* 2:1265–75
26. Charest NJ, Zhou ZX, Lubahn DB, Olsen KL, Wilson EM, French FS. 1991. A frameshift mutation destabilizes androgen receptor messenger RNA in the *Tfm* mouse. *Mol. Endocrinol.* 5:573–81
27. Cattanach BM, Iddon CA, Charlton HM, Chiappa SA, Fink G. 1977. Gonadotrophin-releasing hormone deficiency in a mutant mouse with hypogonadism. *Nature* 269:338–40
28. Mason AJ, Hayflick JS, Zoeller RT, Young WS III, Phillips HS, et al. 1986. A deletion truncating the gonadotropin-releasing hormone gene is responsible for hypogonadism in the *hpg* mouse. *Science* 234:1366–71
29. Gordon JW, Ruddle FH. 1981. Integration and stable germ line transmission of genes injected into mouse pronuclei. *Science* 214:1244–46
30. Costantini F, Lacy E. 1981. Introduction of a rabbit β-globin gene into the mouse germ line. *Nature* 294:92–94
31. Brinster RL, Chen HY, Trumbauer M, Senear AW, Warren R, Palmiter RD. 1981. Somatic expression of herpes thymidine kinase in mice following injection of a fusion gene into eggs. *Cell* 27:223–31

32. Palmiter RD, Brinster RL, Hammer RE, Trumbauer ME, Rosenfeld MG, et al. 1982. Dramatic growth of mice that develop from eggs microinjected with metallothionein-growth hormone fusion genes. *Nature* 300:611–15
33. Palmiter RD, Norstedt G, Gelinas RE, Hammer RE, Brinster RL. 1983. Metallothionein-human GH fusion genes stimulate growth of mice. *Science* 222:809–14
34. Mason AJ, Pitts SL, Nikolics K, Szonyi E, Wilcox JN, et al. 1986. The hypogonadal mouse: reproductive functions restored by gene therapy. *Science* 234:1372–78
35. Kumar TR, Low MJ, Matzuk MM. 1998. Genetic rescue of follicle-stimulating hormone β-deficient mice. *Endocrinology* 139:289–95
36. Kuehn MR, Bradley A, Robertson EJ, Evans MJ. 1987. A potential animal model for Lesch-Nyhan syndrome through introduction of HPRT mutations into mice. *Nature* 326:295–98
37. Thompson S, Clarke AR, Pow AM, Hooper ML, Melton DW. 1989. Germ line transmission and expression of a corrected HPRT gene produced by gene targeting in embryonic stem cells. *Cell* 56:313–21
38. Schwartzberg PL, Goff SP, Robertson EJ. 1989. Germ-line transmission of a c-abl mutation produced by targeted gene disruption in ES cells. *Science* 246:799–803
39. Zijlstra M, Li E, Sajjadi F, Subramani S, Jaenisch R. 1989. Germ-line transmission of a disrupted β2-microglobulin gene produced by homologous recombination in embryonic stem cells. *Nature* 342:435–38
40. Koller BH, Hagemann LJ, Doetschman T, Hagaman JR, Huang S, et al. 1989. Germ-line transmission of a planned alteration made in a hypoxanthine phosphoribosyltransferase gene by homologous recombination in embryonic stem cells. *Proc. Natl. Acad. Sci. USA* 86:8927–31
41. Stewart CL, Kaspar P, Brunet LJ, Bhatt H, Gadi I, et al. 1992. Blastocyst implantation depends on maternal expression of leukaemia inhibitory factor. *Nature* 359:76–79
42. Matzuk MM, Finegold MJ, Su JG, Hsueh AJ, Bradley A. 1992. α-Inhibin is a tumour-suppressor gene with gonadal specificity in mice. *Nature* 360:313–19
43. Ohinata Y, Payer B, O'Carroll D, Ancelin K, Ono Y, et al. 2005. Blimp1 is a critical determinant of the germ cell lineage in mice. *Nature* 436:207–13
44. Yamaji M, Seki Y, Kurimoto K, Yabuta Y, Yuasa M, et al. 2008. Critical function of Prdm14 for the establishment of the germ cell lineage in mice. *Nat. Genet.* 40:1016–22
45. Matzuk MM, Lu H, Vogel H, Sellheyer K, Roop DR, Bradley A. 1995. Multiple defects and perinatally death in mice deficient in follistatin. *Nature* 372:360–63
46. Yao HHC, Matzuk MM, Jorgez CJ, Menke DB, Page DC, et al. 2004. Follistatin operates downstream of Wnt4 in mammalian ovary organogenesis. *Dev. Dyn.* 230:210–15
47. Mathews LS, Vale WW. 1991. Expression cloning of an activin receptor, a predicted transmembrane serine kinase. *Cell* 65:973–82
48. Attisano L, Wrana JL, Cheifetz S, Massague J. 1992. Novel activin receptors: Distinct genes and alternative mRNA splicing generate a repertoire of serine/threonine kinase receptors. *Cell* 68:97–108
49. Willis SA, Zimmerman CM, Li LI, Mathews LS. 1996. Formation and activation by phosphorylation of activin receptor complexes. *Mol. Endocrinol.* 10:367–79
50. Lewis KA, Gray PC, Blount AL, MacConell LA, Wiater E, et al. 2000. Betaglycan binds inhibin and can mediate functional antagonism of activin signalling. *Nature* 404:411–14
51. Matzuk MM, Kumar TR, Vassalli A, Bickenbach JR, Roop DR, et al. 1995. Functional analysis of activins in mammalian development. *Nature* 374:354–56
52. Vassalli A, Matzuk MM, Gardner HA, Lee KF, Jaenisch R. 1994. Activin/inhibin βB subunit gene disruption leads to defects in eyelid development and female reproduction. *Genes Dev.* 8:414–27
53. Pangas SA, Jorgez CJ, Tran M, Agno J, Li X, et al. 2007. Intraovarian activins are required for female fertility. *Mol. Endocrinol.* 21:2458–71
54. Matzuk MM, Kumar TR, Bradley A. 1995. Different phenotypes for mice deficient in either activins or activin receptor type II. *Nature* 374:356–60
55. Kumar TR, Wang Y, Lu N, Matzuk MM. 1997. Follicle stimulating hormone is required for ovarian follicle maturation but not male fertility. *Nat. Genet.* 15:201–4

56. Matzuk MM, Finegold MJ, Mather JP, Krummen L, Lu H, Bradley A. 1994. Development of cancer cachexia-like syndrome and adrenal tumors in inhibin-deficient mice. *Proc. Natl. Acad. Sci. USA* 91:8817–21
57. Coerver KA, Woodruff TK, Finegold MJ, Mather J, Bradley A, Matzuk MM. 1996. Activin signaling through activin receptor type II causes the cachexia-like symptoms in inhibin-deficient mice. *Mol. Endocrinol.* 10:534–43
58. Behringer RR, Finegold MJ, Cate RL. 1994. Müllerian-inhibiting substance function during mammalian sexual development. *Cell* 79:415–25
59. Durlinger AL, Kramer P, Karels B, de Jong FH, Uilenbroek JT, et al. 1999. Control of primordial follicle recruitment by anti-Müllerian hormone in the mouse ovary. *Endocrinology* 140:5789–96
60. Visser JA, de Jong FH, Laven JS, Themmen AP. 2006. Anti-Müllerian hormone: a new marker for ovarian function. *Reproduction* 131:1–9
61. Dong J, Albertini DF, Nishimori K, Kumar TR, Lu N, Matzuk MM. 1996. Growth differentiation factor-9 is required during early ovarian folliculogenesis. *Nature* 383:531–35
62. Dube JL, Wang P, Elvin J, Lyons KM, Celeste AJ, Matzuk MM. 1998. The bone morphogenetic protein 15 gene is X-linked and expressed in oocytes. *Mol. Endocrinol.* 12:1809–17
63. Davis GH, McEwan JC, Fennessy PF, Dodds KG, Farquhar PA. 1991. Evidence for the presence of a major gene influencing ovulation rate on the X chromosome of sheep. *Biol. Reprod.* 44:620–24
64. Davis GH, Bruce GD, Reid PJ. 1994. Breeding implications of the streak ovary condition in homozygous FecX^I/FecX^I Inverdale sheep. *Proc. World Congr. Genet. Appl. Livest. Prod., 5th*, Guelph, Can., Aug., 19:249–52
65. Galloway SM, McNatty KP, Cambridge LM, Laitinen MP, Juengel JL, et al. 2000. Mutations in an oocyte-derived growth factor gene (*BMP15*) cause increased ovulation rate and infertility in a dosage-sensitive manner. *Nat. Genet.* 25:279–83
66. Otsuka F, McTavish KJ, Shimasaki S. 2011. Integral role of GDF-9 and BMP-15 in ovarian function. *Mol. Reprod. Dev.* 78:9–21
67. Montgomery GW, Zhao ZZ, Marsh AJ, Mayne R, Treloar SA, et al. 2004. A deletion mutation in GDF9 in sisters with spontaneous DZ twins. *Twin Res.* 7:548–55
68. Palmer JS, Zhao ZZ, Hoekstra C, Hayward NK, Webb PM, et al. 2006. Novel variants in growth differentiation factor 9 in mothers of dizygotic twins. *J. Clin. Endocrinol. Metab.* 91:4713–16
69. Di Pasquale E, Beck-Peccoz P, Persani L. 2004. Hypergonadotropic ovarian failure associated with an inherited mutation of human bone morphogenetic protein-15 (*BMP15*) gene. *Am. J. Hum. Genet.* 75:106–11
70. Park JY, Su YQ, Ariga M, Law E, Jin SL, Conti M. 2004. EGF-like growth factors as mediators of LH action in the ovulatory follicle. *Science* 303:682–84
71. Sugiura K, Su YQ, Diaz FJ, Pangas SA, Sharma S, et al. 2007. Oocyte-derived BMP15 and FGFs cooperate to promote glycolysis in cumulus cells. *Development* 134:2593–603
72. Su YQ, Sugiura K, Wigglesworth K, O'Brien MJ, Affourtit JP, et al. 2008. Oocyte regulation of metabolic cooperativity between mouse cumulus cells and oocytes: BMP15 and GDF9 control cholesterol biosynthesis in cumulus cells. *Development* 135:111–21
73. Zhang M, Su YQ, Sugiura K, Xia G, Eppig JJ. Granulosa cell ligand NPPC and its receptor NPR2 maintain meiotic arrest in mouse oocytes. *Science* 330:366–69
74. Wilson T, Wu XY, Juengel JL, Ross IK, Lumsden JM, et al. 2001. Highly prolific Booroola sheep have a mutation in the intracellular kinase domain of bone morphogenetic protein IB receptor (ALK-6) that is expressed in both oocytes and granulosa cells. *Biol. Reprod.* 64:1225–35
75. Yi SE, LaPolt PS, Yoon BS, Chen JY, Lu JK, Lyons KM. 2001. The type I BMP receptor BmprIB is essential for female reproductive function. *Proc. Natl. Acad. Sci. USA* 98:7994–99
76. Moore RK, Otsuka F, Shimasaki S. 2003. Molecular basis of bone morphogenetic protein-15 signaling in granulosa cells. *J. Biol. Chem.* 278:304–10
77. Itman C, Mendis S, Barakat B, Loveland KL. 2006. All in the family: TGF-β family action in testis development. *Reproduction* 132:233–46
78. Li Q, Graff JM, O'Connor AE, Loveland KL, Matzuk MM. 2007. SMAD3 regulates gonadal tumorigenesis. *Mol. Endocrinol.* 21:2472–86

79. Li Q, Pangas SA, Jorgez CJ, Graff JM, Weinstein M, Matzuk MM. 2008. Redundant roles of SMAD2 and SMAD3 in ovarian granulosa cells in vivo. *Mol. Cell. Biol.* 28:7001–11
80. Zhao G-Q, Liaw L, Hogan BLM. 1998. Bone morphogenetic protein 8A plays a role in the maintenance of spermatogenesis and the integrity of the epididymis. *Development* 125:1103–12
81. Zhao GQ, Chen YX, Liu XM, Xu Z, Qi X. 2001. Mutation in *Bmp7* exacerbates the phenotype of *Bmp8a* mutants in spermatogenesis and epididymis. *Dev. Biol.* 240:212–22
82. Hu J, Chen YX, Wang D, Qi X, Li TG, et al. 2004. Developmental expression and function of Bmp4 in spermatogenesis and in maintaining epididymal integrity. *Dev. Biol.* 276:158–71
83. Imbeaud S, Carre-Eusebe D, Rey R, Belville C, Josso N, Picard JY. 1994. Molecular genetics of the persistent Müllerian duct syndrome: a study of 19 families. *Hum. Mol. Genet.* 3:125–31
84. Imbeaud S, Belville C, Messika-Zeitoun L, Rey R, di Clemente N, et al. 1996. A 27 base-pair deletion of the anti-Müllerian type II receptor gene is the most common cause of the persistent Müllerian duct syndrome. *Hum. Mol. Genet.* 5:1269–77
85. Mishina Y, Rey R, Finegold MJ, Matzuk MM, Josso N, et al. 1996. Genetic analysis of the Müllerian-inhibiting substance signal transduction pathway in mammalian sexual differentiation. *Genes Dev.* 10:2577–87
86. Jamin SP, Arango NA, Mishina Y, Hanks MC, Behringer RR. 2002. Requirement of Bmpr1a for Müllerian duct regression during male sexual development. *Nat. Genet.* 32:408–10
87. Orvis GD, Jamin SP, Kwan KM, Mishina Y, Kaartinen VM, et al. 2008. Functional redundancy of TGF-β family type I receptors and receptor-Smads in mediating anti-Müllerian hormone-induced Müllerian duct regression in the mouse. *Biol. Reprod.* 78:994–1001
88. Settle S, Marker P, Gurley K, Sinha A, Thacker A, et al. 2001. The BMP family member *Gdf7* is required for seminal vesicle growth, branching morphogenesis, and cytodifferentiation. *Dev. Biol.* 234:138–50
89. Wang H, Dey SK. 2006. Roadmap to embryo implantation: clues from mouse models. *Nat. Rev. Genet.* 7:185–99
90. Lee KY, Jeong JW, Wang J, Ma L, Martin JF, et al. 2007. Bmp2 is critical for the murine uterine decidual response. *Mol. Cell. Biol.* 27:5468–78
91. Nusslein-Volhard C, Lohs-Schardin M, Sander K, Cremer C. 1980. A dorso-ventral shift of embryonic primordia in a new maternal-effect mutant of *Drosophila*. *Nature* 283:474–76
92. Burns KH, Matzuk MM. 2006. Pre-implantation embryogenesis. In *The Physiology of Reproduction*, ed. J Neill, pp. 261–310. San Diego, CA: Academic
93. Li L, Baibakov B, Dean J. 2008. A subcortical maternal complex essential for preimplantation mouse embryogenesis. *Dev. Cell* 15:416–25
94. Tong ZB, Gold L, Pfeifer KE, Dorward H, Lee E, et al. 2000. *Mater*, a maternal effect gene required for early embryonic development in mice. *Nat. Genet.* 26:267–68
95. van der Heijden GW, Dieker JW, Derijck AA, Muller S, Berden JH, et al. 2005. Asymmetry in histone H3 variants and lysine methylation between paternal and maternal chromatin of the early mouse zygote. *Mech. Dev.* 122:1008–22
96. Burns KH, Viveiros MM, Ren Y, Wang P, DeMayo FJ, et al. 2003. Roles of NPM2 in chromatin and nucleolar organization in oocytes and embryos. *Science* 300:633–36
97. Philpott A, Leno GH. 1992. Nucleoplasmin remodels sperm chromatin in *Xenopus* egg extracts. *Cell* 69:759–67
98. Philpott A, Leno GH, Laskey RA. 1991. Sperm decondensation in *Xenopus* egg cytoplasm is mediated by nucleoplasmin. *Cell* 65:569–78
99. Derijck A, van der Heijden G, Giele M, Philippens M, de Boer P. 2008. DNA double-strand break repair in parental chromatin of mouse zygotes, the first cell cycle as an origin of de novo mutation. *Hum. Mol. Genet.* 17:1922–37
100. Wossidlo M, Arand J, Sebastiano V, Lepikhov K, Boiani M, et al. 2010. Dynamic link of DNA demethylation, DNA strand breaks and repair in mouse zygotes. *EMBO J.* 29:1877–88
101. Hajkova P, Jeffries SJ, Lee C, Miller N, Jackson SP, Surani MA. 2010. Genome-wide reprogramming in the mouse germ line entails the base excision repair pathway. *Science* 329:78–82

102. Wossidlo M, Nakamura T, Lepikhov K, Marques CJ, Zakhartchenko V, et al. 2011. 5-Hydroxymethylcytosine in the mammalian zygote is linked with epigenetic reprogramming. *Nat. Commun.* 2:241
103. Payer B, Saitou M, Barton SC, Thresher R, Dixon JP, et al. 2003. *stella* is a maternal effect gene required for normal early development in mice. *Curr. Biol.* 13:2110–17
104. Bortvin A, Goodheart M, Liao M, Page DC. 2004. *Dppa3/Pgc7/stella* is a maternal factor and is not required for germ cell specification in mice. *BMC Dev. Biol.* 4:2
105. Nakamura T, Arai Y, Umehara H, Masuhara M, Kimura T, et al. 2007. PGC7/Stella protects against DNA demethylation in early embryogenesis. *Nat. Cell Biol.* 9:64–71
106. Liu H, Kim JM, Aoki F. 2004. Regulation of histone H3 lysine 9 methylation in oocytes and early pre-implantation embryos. *Development* 131:2269–80
107. Adenot PG, Mercier Y, Renard JP, Thompson EM. 1997. Differential H4 acetylation of paternal and maternal chromatin precedes DNA replication and differential transcriptional activity in pronuclei of 1-cell mouse embryos. *Development* 124:4615–25
108. Matsui T, Leung D, Miyashita H, Maksakova IA, Miyachi H, et al. 2010. Proviral silencing in embryonic stem cells requires the histone methyltransferase ESET. *Nature* 464:927–31
109. Rowe HM, Jakobsson J, Mesnard D, Rougemont J, Reynard S, et al. 2010. KAP1 controls endogenous retroviruses in embryonic stem cells. *Nature* 463:237–40
110. Garcia-Perez JL, Morell M, Scheys JO, Kulpa DA, Morell S, et al. 2010. Epigenetic silencing of engineered L1 retrotransposition events in human embryonic carcinoma cells. *Nature* 466:769–73
111. Nothias JY, Miranda M, DePamphilis ML. 1996. Uncoupling of transcription and translation during zygotic gene activation in the mouse. *EMBO J.* 15:5715–25
112. Wu X, Viveiros MM, Eppig JJ, Bai Y, Fitzpatrick SL, Matzuk MM. 2003. Zygote arrest 1 (*Zar1*) is a novel maternal-effect gene critical for the oocyte-to-embryo transition. *Nat. Genet.* 33:187–91
113. Zheng P, Dean J. 2009. Role of *Filia*, a maternal effect gene, in maintaining euploidy during cleavage-stage mouse embryogenesis. *Proc. Natl. Acad. Sci. USA* 106:7473–78
114. Sasai K, Parant JM, Brandt ME, Carter J, Adams HP, et al. 2008. Targeted disruption of Aurora A causes abnormal mitotic spindle assembly, chromosome misalignment and embryonic lethality. *Oncogene* 27:4122–27
115. Howell CY, Bestor TH, Ding F, Latham KE, Mertineit C, et al. 2001. Genomic imprinting disrupted by a maternal effect mutation in the *Dnmt1* gene. *Cell* 104:829–38
116. Jia D, Jurkowska RZ, Zhang X, Jeltsch A, Cheng X. 2007. Structure of Dnmt3a bound to Dnmt3L suggests a model for de novo DNA methylation. *Nature* 449:248–51
117. Ciccone DN, Su H, Hevi S, Gay F, Lei H, et al. 2009. KDM1B is a histone H3K4 demethylase required to establish maternal genomic imprints. *Nature* 461:415–18
118. Li X, Ito M, Zhou F, Youngson N, Zuo X, et al. 2008. A maternal-zygotic effect gene, *Zfp57*, maintains both maternal and paternal imprints. *Dev. Cell* 15:547–57
119. Schultz N, Hamra FK, Garbers DL. 2003. A multitude of genes expressed solely in meiotic or postmeiotic spermatogenic cells offers a myriad of contraceptive targets. *Proc. Natl. Acad. Sci. USA* 100:12201–6
120. Hunt PA, Hassold TJ. 2002. Sex matters in meiosis. *Science* 296:2181–83
121. Tiepolo L, Zuffardi O. 1976. Localization of factors controlling spermatogenesis in the nonfluorescent portion of the human Y chromosome long arm. *Hum. Genet.* 34:119–24
122. Foresta C, Moro E, Ferlin A. 2001. Y chromosome microdeletions and alterations of spermatogenesis. *Endocr. Rev.* 22:226–39
123. Ruggiu M, Speed R, Taggart M, McKay SJ, Kilanowski F, et al. 1997. The mouse *Dazla* gene encodes a cytoplasmic protein essential for gametogenesis. *Nature* 389:73–77
124. Kee K, Angeles VT, Flores M, Nguyen HN, Reijo Pera RA. 2009. Human *DAZL*, *DAZ* and *BOULE* genes modulate primordial germ-cell and haploid gamete formation. *Nature* 462:222–25
125. Aston KI, Carrell DT. 2009. Genome-wide study of single-nucleotide polymorphisms associated with azoospermia and severe oligozoospermia. *J. Androl.* 30:711–25
126. Knauff EA, Franke L, van Es MA, van den Berg LH, van der Schouw YT, et al. 2009. Genome-wide association study in premature ovarian failure patients suggests *ADAMTS19* as a possible candidate gene. *Hum. Reprod.* 24:2372–78

127. Behringer RR, Cate RL, Froelick GJ, Palmiter RD, Brinster RL. 1990. Abnormal sexual development in transgenic mice chronically expressing Müllerian-inhibiting substance. *Nature* 345:167–70
128. Koopman P, Gubbay J, Vivian N, Goodfellow P, Lovell-Badge R. 1991. Male development of chromosomally female mice transgenic for *Sry*. *Nature* 351:117–21
129. Margulies M, Egholm M, Altman WE, Attiya S, Bader JS, et al. 2005. Genome sequencing in microfabricated high-density picolitre reactors. *Nature* 437:376–80
130. Brown CW, Houston-Hawkins DE, Woodruff TK, Matzuk MM. 2000. Insertion of *Inhbb* into the *Inhba* locus rescues the *Inhba*-null phenotype and reveals new activin functions. *Nat. Genet.* 25:453–57
131. Yan C, Wang P, DeMayo J, DeMayo F, Elvin J, et al. 2001. Synergistic roles of bone morphogenetic protein 15 and growth differentiation factor 9 in ovarian function. *Mol. Endocrinol.* 15:854–66
132. Edson MA, Nalam RL, Clementi C, Franco HL, Demayo FJ, et al. 2010. Granulosa cell-expressed BMPR1A and BMPR1B have unique functions in regulating fertility but act redundantly to suppress ovarian tumor development. *Mol. Endocrinol.* 24:1251–66
133. Pangas SA, Li X, Umans L, Zwijsen A, Huylebroeck D, et al. 2008. Conditional deletion of *Smad1* and *Smad5* in somatic cells of male and female gonads leads to metastatic tumor development in mice. *Mol. Cell. Biol.* 28:248–57
134. Pangas SA, Li X, Robertson EJ, Matzuk MM. 2006. Premature luteinization and cumulus cell defects in ovarian-specific *Smad4* knockout mice. *Mol. Endocrinol.* 20:1406–22
135. Chang H, Matzuk MM. 2001. *Smad5* is required for mouse primordial germ cell development. *Mech. Dev.* 104:61–67
136. Ma L, Buchold GM, Greenbaum MP, Roy A, Burns KH, et al. 2009. GASZ is essential for male meiosis and suppression of retrotransposon expression in the male germline. *PLoS Genet.* 5:e1000635
137. Edson MA, Lin YN, Matzuk MM. 2010. Deletion of the novel oocyte-enriched gene, *Gpr149*, leads to increased fertility in mice. *Endocrinology* 151:358–68
138. Iwamori N, Zhao M, Meistrich ML, Matzuk MM. 2011. The testis-enriched histone demethylase, KDM4D, regulates methylation of histone H3 lysine 9 during spermatogenesis in the mouse but is dispensable for fertility. *Biol. Reprod.* 84:1225–34
139. Andreu-Vieyra CV, Chen R, Agno JE, Glaser S, Anastassiadis K, et al. 2010. MLL2 is required in oocytes for bulk histone 3 lysine 4 trimethylation and transcriptional silencing. *PLoS Biol.* 8:e1000453
140. Rajkovic A, Pangas SA, Ballow D, Suzumori N, Matzuk MM. 2004. *NOBOX* deficiency disrupts early folliculogenesis and oocyte-specific gene expression. *Science* 305:1157–59
141. Iwamori T, Lin YN, Ma L, Iwamori N, Matzuk MM. 2011. Identification and characterization of RBM44 as a novel intercellular bridge protein. *PLoS ONE* 6:e17066
142. Ballow D, Meistrich ML, Matzuk M, Rajkovic A. 2006. *Sohlh1* is essential for spermatogonial differentiation. *Dev. Biol.* 294:161–67
143. Pangas SA, Choi Y, Ballow DJ, Zhao Y, Westphal H, et al. 2006. Oogenesis requires germ cell-specific transcriptional regulators *Sohlh1* and *Lhx8*. *Proc. Natl. Acad. Sci. USA* 103:8090–95
144. Roy A, Lin YN, Agno JE, DeMayo FJ, Matzuk MM. 2009. Tektin 3 is required for progressive sperm motility in mice. *Mol. Reprod. Dev.* 76:453–59
145. Greenbaum MP, Yan W, Wu MH, Lin YN, Agno JE, et al. 2006. TEX14 is essential for intercellular bridges and fertility in male mice. *Proc. Natl. Acad. Sci. USA* 103:4982–87
146. Lin YN, Roy A, Yan W, Burns KH, Matzuk MM. 2007. Loss of zona pellucida binding proteins in the acrosomal matrix disrupts acrosome biogenesis and sperm morphogenesis. *Mol. Cell. Biol.* 27:6794–805
147. Nagaraja AK, Andreu-Vieyra C, Franco HL, Ma L, Chen R, et al. 2008. Deletion of *Dicer* in somatic cells of the female reproductive tract causes sterility. *Mol. Endocrinol.* 22:2336–52
148. Nishimori K, Young LJ, Guo Q, Wang Z, Insel TR, Matzuk MM. 1996. Oxytocin is required for nursing but is not essential for parturition or reproductive behavior. *Proc. Natl. Acad. Sci. USA* 93:11699–704
149. Ferguson JN, Young LJ, Hearn EF, Matzuk MM, Insel TR, Winslow JT. 2000. Social amnesia in mice lacking the oxytocin gene. *Nat. Genet.* 25:284–88
150. Takayanagi Y, Yoshida M, Bielsky IF, Ross HE, Kawamata M, et al. 2005. Pervasive social deficits, but normal parturition, in oxytocin receptor-deficient mice. *Proc. Natl. Acad. Sci. USA* 102:16096–101
151. Varani S, Elvin JA, Yan C, DeMayo J, DeMayo FJ, et al. 2002. Knockout of pentraxin 3, a downstream target of growth differentiation factor-9, causes female subfertility. *Mol. Endocrinol.* 16:1154–67

152. Nalam RL, Andreu-Vieyra C, Braun RE, Akiyama H, Matzuk MM. 2009. Retinoblastoma protein plays multiple essential roles in the terminal differentiation of Sertoli cells. *Mol. Endocrinol.* 23:1900–13
153. Andreu-Vieyra C, Chen R, Matzuk MM. 2008. Conditional deletion of the retinoblastoma (*Rb*) gene in ovarian granulosa cells leads to premature ovarian failure. *Mol. Endocrinol.* 22:2141–61
154. Tashiro F, Kanai-Azuma M, Miyazaki S, Kato M, Tanaka T, et al. 2010. Maternal-effect gene *Ces5/Ooep/Moep19/Floped* is essential for oocyte cytoplasmic lattice formation and embryonic development at the maternal-zygotic stage transition. *Genes Cells* 15:813–28
155. Tomasini R, Tsuchihara K, Wilhelm M, Fujitani M, Rufini A, et al. 2008. TAp73 knockout shows genomic instability with infertility and tumor suppressor functions. *Genes Dev.* 22:2677–91
156. Liu D, Matzuk MM, Sung WK, Guo Q, Wang P, Wolgemuth DJ. 1998. Cyclin A1 is required for meiosis in the male mouse. *Nat. Genet.* 20(4):377–88
157. Yan W, Ma L, Burns KH, Matzuk MM. 2004. Haploinsufficiency of kelch-like protein homolog 10 causes infertility in male mice. *Proc. Natl. Acad. Sci. USA* 101(20):7793–98
158. Yan W, Ma L, Pangas SA, Burns KH, Bai Y, et al. 2005. Mice deficient in oocyte-specific oligoadenylate synthase-like protein OAS1D display reduced fertility. *Mol. Cell. Biol.* 25(11):4615–24
159. Yan W, Assadi AH, Wynshaw-Boris A, Eichele G, Matzuk MM, Clark GD. 2003. Previously uncharacterized roles of platelet-activating factor acetylhydrolase 1b complex in mouse spermatogenesis. *Proc. Natl. Acad. Sci. USA* 100(12):7189–94
160. Koizumi H, Yamaguchi N, Hattori M, Ishikawa TO, Aoki J, et al. 2003. Targeted disruption of intracellular type I platelet activating factor-acetylhydrolase catalytic subunits causes severe impairment in spermatogenesis. *J. Biol. Chem.* 278(14):12489–94
161. Roy A, Lin YN, Agno JE, DeMayo FJ, Matzuk MM. 2007. Absence of tektin 4 causes asthenozoospermia and subfertility in male mice. *FASEB J.* 21(4):1013–25
162. Yang J, Medvedev S, Yu J, Tang LC, Agno JE, et al. 2005. Absence of the DNA-/RNA-binding protein MSY2 results in male and female infertility. *Proc. Natl. Acad. Sci. USA* 102(16):5755–60
163. Kumar TR, Wiseman AL, Kala G, Kala SV, Matzuk MM, Lieberman MW. 2000. Reproductive defects in γ-glutamyl transpeptidase-deficient mice. *Endocrinology* 141(11):4270–77
164. Ma X, Dong Y, Matzuk MM, Kumar TR. 2004. Targeted disruption of luteinizing hormone β-subunit leads to hypogonadism, defects in gonadal steroidogenesis and infertility. *Proc. Natl. Acad. Sci. USA* 101(49):17294–99
165. Yu RN, Ito M, Saunders TL, Camper SA, Jameson J. 1998. Role of Ahch in gonadal development and gametogenesis. *Nat. Genet.* 20(4):353–57
166. Matzuk MM, Dionne L, Guo Q, Kumar TR, Lebovitz RM. 1998. Ovarian function in superoxide dismutase 1 and 2 knockout mice. *Endocrinology* 139:4008–11
167. Ho Y-S, Gargano M, Cao J, Bronson RT, Heimler I, Hutz RJ. 1998. Reduced fertility in female mice lacking copper-zinc superoxide dismutase. *J. Biol. Chem.* 273(13):7765–69
168. Bultman SJ, Gebuhr TC, Pan H, Svoboda P, Schultz RM, Magnuson T. 2006. Maternal BRG1 regulates zygotic genome activation in the mouse. *Genes Dev.* 20(13):1744–54
169. Kaneda M, Hirasawa R, Chiba H, Okano M, Li E, Sasaki H. 2010. Genetic evidence for Dnmt3a-dependent imprinting during oocyte growth obtained by conditional knockout with Zp3-Cre and complete exclusion of Dnmt3b by chimera formation. *Genes Cells* 15(3):169–79
170. Hata K, Okano M, Lei H, Li E. 2002. Dnmt3L cooperates with the Dnmt3 family of de novo DNA methyltransferases to establish maternal imprints in mice. *Development* 129(8):1983–93
171. Narducci MG, Fiorenza MT, Kang SM, Bevilacqua A, Di Giacomo M, et al. 2002. TCL1 participates in early embryonic development and is overexpressed in human seminomas. *Proc. Natl. Acad. Sci. USA* 99(18):11712–17

Cumulative Indexes

Contributing Authors, Volumes 70–74

A

Adelman JP, 74:245–70
Ahangari F, 73:479–501
Aitken JD, 74:177–98
Al-Awqati Q, 73:401–12
Albrecht U, 72:517–49
Alcorn JF, 72:495–516
Aldrich RW, 71:19–36
Alexander MR, 74:13–40
Allada R, 72:605–24
Arterburn JB, 70:165–90
Asante A, 73:163–82

B

Baines CP, 72:61–80
Bao H-F, 71:403–23
Barnes PJ, 71:451–64
Barnhart BC, 70:51–71
Baum MJ, 71:141–60
Bers DM, 70:23–49
Betters JL, 73:239–59
Bhatt DH, 71:261–82
Binder HJ, 72:297–313
Bloom SR, 70:239–55
Boles J, 73:515–25
Borst JGG 74:199–224
Bossé Y, 72:437–62
Boswell L, 73:115–34
Boutet de Monvel J, 73:311–34
Bovill EG, 73:527–45
Bradfield CA, 72:625–45

Bradshaw WE, 72:147–66
Brett TJ, 71:425–49
Breyer MD, 70:357–77
Brody JS, 73:437–56
Browning EA, 74:403–24
Bruneau BG, 74:41–68
Burns KH, 74:503–28

C

Cantley LG, 72:357–76
Carvalho FA, 74:177–98
Casals-Casas C, 73:135–62
Castillo PE, 71:283–306
Caviedes-Vidal E, 73:69–93
Chambrey R, 74:325–50
Chandra V, 72:247–72
Chang C-P, 74:41–68
Chao MV, 72:1–13
Chapman HA, 73:413–35
Chatterjee S, 74:403–24
Chaudhri OB, 70:239–55
Chen X, 73:213–37
Cheung E, 72:191–218
Choi AMK, 74:377–402
Chung BY, 72:605–24
Chung J-J, 74:453–76
Clapham DE, 74:453–76
Clevers H, 71:241–60
Clifton DK, 70:213–38
Colgan SP, 74:153–76
Cota D, 70:513–35

Cowley MA, 70:537–56
Crowe CR, 72:495–516

D

Da Silva CA, 73:479–501
Davis CW, 70:487–512
Debnam ES, 70:379–403
Dela Cruz CS, 73:479–501
Desvergne B, 73:135–62
Dibner C, 72:517–49
Dickey BF, 70:487–512;
  72:413–35
Distelhorst CW, 70:73–91
Dixon JA, 73:47–68
Duran C, 72:95–121

E

Eaton DC, 71:403–23
Edwards RH, 74:225–44
Eladari D, 74:325–50
Elias JA, 73:479–501
Eltzschig HK, 74:153–76
Esmon CT, 73:503–14
Esmon NL, 73:503–14
Evans JP, 74:477–502
Evans SE, 72:413–35

F

Farooque SP, 71:465–87
Feder ME, 72:123–25, 167–90

Fenton RA, 70:301–27
Fisher AB, 74:403–24
Fodor AA, 71:19–36
Forte JG, 72:273–96
Frøkiaer J, 70:301–27

G

Gan W-B, 71:261–82
Garland T Jr, 72:167–90
Garvin JL, 73:359–76
Geibel JP, 71:205–17
Gendler SJ, 70:431–57
Gewirtz AT, 74:177–98
Giebisch GH, 73:1–28
Glass CK, 72:219–46
Gottlieb RA, 72:45–60
Greka A, 74:299–324
Guo J-K, 72:357–76

H

Haddad GG, 73:95–113
Halassa MM, 72:335–55
Hamid Q, 71:489–507
Hammerschmidt M, 73:183–211
Hand SC, 73:115–34
Hao C-M, 70:357–77
Harrison JF, 73:95–113
Hartzell HC, 72:95–121
Harvey BJ, 73:335–57
Haspel JA, 74:377–402
Hathaway HJ, 70:165–90
Hattrup CL, 70:431–57
Hawley RS 74:425–52
Haydon PG, 72:335–55
He C-H, 73:479–501
Hebert SC, 71:205–17
Hediger MA, 70:257–71
Heifets BD, 71:283–306
Helms MN, 71:403–23
Herrera M, 73:359–76
Hinton RB, 73:29–46
Hnasko TS, 74:225–44
Hock MB, 71:177–203
Hoffman JF, 70:1–22
Hofmann GE, 72:127–45
Hogenesch JB, 72:625–45
Holtzman MJ, 71:425–49
Holzapfel CM, 72:147–66
Houck LD, 71:161–76
Huang P, 72:247–72

I

Irvin CG, 72:437–62

J

Jain L, 71:403–23
Jia L, 73:239–59

K

Kahle KT, 70:329–55
Kang M-J, 73:479–501
Kaplinskiy V, 72:19–44
Karasov WH, 73:69–93
Karsenty G, 74:87–106
Kashikar ND, 70:93–117
Kato A, 73:261–81
Kaupp UB, 70:93–117
Kay SA, 72:551–77
Kelliher KR, 71:141–60
Kemp PJ, 74:271–98
Kirichok Y, 74:453–76
Kitsis RN, 72:19–44
Kolls JK, 72:495–516
Komarova Y, 72:463–93
Koval M, 71:403–23
Kralli A, 71:177–203
Kraus WL, 72:191–218
Kurtz A, 73:377–99

L

Lake CM, 74:425–52
Landowski CP, 70:257–71
Lauder GV, 70:143–63
Lee CG, 73:479–501
Lee TH, 71:465–87
Lee W-K, 70:119–42
Leinders-Zufall T, 71:115–40
Leinwand LA, 71:1–18
Leipziger J, 72:377–93
Lifton RP, 70:329–55
Lingrel JB, 72:395–412
Lishko PV, 74:453–76
Liu L, 70:51–71
Löhr H, 73:183–211
Loukoianov A, 70:405–29
Lucki NC, 74:131–52
Luczak ED, 71:1–18

M

Ma B, 73:479–501
Mackman N, 73:515–25
Madden PGA, 70:143–63
Malik AB, 72:463–93
Manly DA, 73:515–25
Mansbach CM, 72:315–33
Marden JH, 72:167–90
Martínez del Rio C, 73:69–93
Mäser P, 71:59–82
Matzuk MM, 74:503–28
Maylie J, 74:245–70
McDonough AA, 71:381–401
McGuckin MA, 70:459–86
McIntosh BE, 72:625–45
McNally EM, 71:37–57
Mentzer RM Jr, 72:45–60
Menze MA, 73:115–34
Michell AR, 70:379–403
Mifflin RC, 73:213–37
Mindell JA, 74:69–86
Moore D, 73:115–34
Moore DD, 74:1–12
Mundel P, 74:299–324
Munger SD, 71:115–40
Münzberg H, 70:537–56
Murphy KG, 70:239–55
Myers MG, 70:537–56

N

Nakahira K, 74:377–402
Navarro B, 74:453–76
Nielsen S, 70:301–27
North RA, 71:333–59

O

Olefsky JM, 72:219–46
Oprea TI, 70:165–90
Ortiz PA, 73:359–76
Oury F, 74:87–106
Owens GK, 74:13–40

P

Panettieri RA Jr, 71:509–35
Paré PD, 72:437–62
Patel AC, 71:425–49
Patterson C, 72:81–94
Petersen OH, 70:273–99
Peti-Peterdi J, 74:325–50
Petit C, 73:311–34
Pinchuk IV, 73:213–37
Popa SM, 70:213–38
Powell DW, 73:213–37

Praetorius HA, 72:377–93
Praetorius J, 70:301–27
Prossnitz ER, 70:165–90

R

Raleigh DR, 73:283–309
Rastinejad F, 72:247–72
Ren D, 74:453–76
Riccardi D, 74:271–98
Richardson GP, 73:311–34
Riesenfeld EP, 72:437–62
Rinaudo P, 74:107–30
Ring AM, 70:329–55
Robbins J, 72:15–17
Rojek A, 70:301–27
Romero MF, 73:261–81
Rong Y, 70:73–91
Ronnebaum SM, 72:81–94
Rossier BC, 71:361–79
Rousseau K, 70:459–86
Ryter SW, 74:377–402

S

Saada JI, 73:213–37
Sah P, 74:245–70
Salem V, 70:239–55
Sandoval D, 70:513–35
Schibler U, 72:517–49
Schroeder JI, 71:59–82
Schwartz DA, 73:457–78
Seeley RJ, 70:513–35
Seibold MA, 73:457–78
Sewer MB, 74:131–52
Shen L, 73:283–309
Siddiqi SA, 72:315–33
Simon MC, 70:51–71
Sklar LA, 70:165–90
Smith HO, 70:165–90
Socha JJ, 70:119–42

Soria van Hoeve J, 74:199–224
Spinale FG, 73:47–68
Steiling K, 73:437–56
Steiner RA, 70:213–38
Stengel A, 71:219–39
Stutts MJ, 71:361–79
Surprenant A, 71:333–59
Suzuki Y, 70:257–71

T

Taché Y, 71:219–39
Taguchi A, 70:191–212
Takahashi JS, 72:551–77
Takyar S, 73:479–501
Taylor RN, 73:163–82
Tepikin AV, 70:273–99
Terrell D, 72:15–17
Thai P, 70:405–29
Thomas W, 73:335–57
Thompson CH, 72:95–121
Thomson SC, 74:351–76
Thornton DJ, 70:459–86
Tliba O, 71:509–35
Todgham AE, 72:127–45
Toner M, 73:115–34
Touhara K, 71:307–32
Tulic M, 71:489–507
Turner JR, 73:283–309
Tuvim MJ, 72:413–35

U

Ueda HR, 72:579–603
Ukai H, 72:579–603
Unwin RJ, 70:379–403

V

Vallon V, 74:351–76
van der Flier LG, 71:241–60

van der Vliet A, 73:527–45
Vijay-Kumar M, 74:177–98
Vosshall LB, 71:307–32

W

Wachi S, 70:405–29
Wallace GQ, 71:37–57
Wang E, 74:107–30
Ward JM, 71:59–82
Weber CR, 73:283–309
Welsh DK, 72:551–77
Westneat MW, 70:119–42
Weyand I, 70:93–117
Whelan RS, 72:19–44
White MF, 70:191–212
Wu R, 70:405–29

X

Xiao Q, 72:95–121
Xu Y, 72:413–35

Y

Youn JH, 71:381–401
Young RM, 70:51–71
Yu D, 73:283–309
Yu L, 73:239–59
Yutzey KE, 73:29–46

Z

Zera AJ, 72:167–90
Zhang S, 71:261–82
Zhou J, 71:83–113
Zhu L, 72:273–96
Zufall F, 71:115–40

Chapter Titles, Volumes 70–74

Perspectives

My Passion and Passages with Red Blood Cells	JF Hoffman	70:1–22
A Conversation with Rita Levi-Montalcini	MV Chao	72:1–13
A Long Affair with Renal Tubules	GH Giebisch	73:1–28
A Conversation with Elwood Jensen	DD Moore	74:1–12

Cardiovascular Physiology

Calcium Cycling and Signaling in Cardiac Myocytes	DM Bers	70:23–49
Hypoxia-Induced Signaling in the Cardiovascular System	MC Simon, L Liu, BC Barnhart, RM Young	70:51–71
Sex-Based Cardiac Physiology	ED Luczak, LA Leinwand	71:1–18
Protein Conformation–Based Disease: Getting to the Heart of the Matter	D Terrell, J Robbins	72:15–17
Cell Death in the Pathogenesis of Heart Disease: Mechanisms and Significance	RS Whelan, V Kaplinskiy, RN Kitsis	72:19–44
Autophagy During Cardiac Stress: Joys and Frustrations of Autophagy	RA Gottlieb, RM Mentzer, Jr.	72:45–60
The Cardiac Mitochondrion: Nexus of Stress	CP Baines	72:61–80
The FoxO Family in Cardiac Function and Dysfunction	SM Ronnebaum, C Patterson	72:81–94
Heart Valve Structure and Function in Development and Disease	RB Hinton, KE Yutzey	73:29–46
Myocardial Remodeling: Cellular and Extracellular Events and Targets	JA Dixon, FG Spinale	73:47–68

Epigenetic Control of Smooth Muscle Cell Differentiation and Phenotypic Switching in Vascular Development and Disease	MR Alexander, GK Owens	74:13–40
Epigenetics and Cardiovascular Development	C-P Chang, BG Bruneau	74:41–68

Cell Physiology

Bcl-2 Protein Family Members: Versatile Regulators of Calcium Signaling in Cell Survival and Apoptosis	Y Rong, CW Distelhorst	70:73–91
Mechanisms of Sperm Chemotaxis	UB Kaupp, ND Kashikar, I Weyand	70:93–117
Convergent Evolution of Alternative Splices at Domain Boundaries of the BK Channel	AA Fodor, RW Aldrich	71:19–36
Mechanisms of Muscle Degeneration, Regeneration, and Repair in the Muscular Dystrophies	GQ Wallace, EM McNally	71:37–57
Plant Ion Channels: Gene Families, Physiology, and Functional Genomics Analyses	JM Ward, P Mäser, JI Schroeder	71:59–82
Polycystins and Primary Cilia: Primers for Cell Cycle Progression	J Zhou	71:83–113
Subsystem Organization of the Mammalian Sense of Smell	SD Munger, T Leinders-Zufall, F Zufall	71:115–40
Chloride Channels: Often Enigmatic, Rarely Predictable	C Duran, CH Thompson, Q Xiao, HC Hartzell	72:95–121
Lysosomal Acidification Mechanisms	JA Mindell	74:69–86

Ecological, Evolutionary, and Comparative Physiology

Advances in Biological Structure, Function, and Physiology Using Synchrotron X-Ray Imaging	MW Westneat, JJ Socha, W-K Lee	70:119–42
Advances in Comparative Physiology from High-Speed Imaging of Animal and Fluid Motion	GV Lauder, PGA Madden	70:143–63
Complementary Roles of the Main and Accessory Olfactory Systems in Mammalian Mate Recognition	MJ Baum, KR Kelliher	71:141–60
Pheromone Communication in Amphibians and Reptiles	LD Houck	71:161–76
Physiology and Global Climate Change	ME Feder	72:123–25

Living in the Now: Physiological Mechanisms to Tolerate a Rapidly Changing Environment	GE Hofmann, AE Todgham	72:127–45
Light, Time, and the Physiology of Biotic Response to Rapid Climate Change in Animals	WE Bradshaw, CM Holzapfel	72:147–66
Locomotion in Response to Shifting Climate Zones: Not So Fast	ME Feder, T Garland, Jr, JH Marden, AJ Zera	72:167–90
Ecological Physiology of Diet and Digestive Systems	WH Karasov, C Martínez del Rio, E Caviedes-Vidal	73:69–93
Effects of Oxygen on Growth and Size: Synthesis of Molecular, Organismal, and Evolutionary Studies with *Drosophila melanogaster*	JF Harrison, GG Haddad	73:95–113
LEA Proteins During Water Stress: Not Just for Plants Anymore	SC Hand, MA Menze, M Toner, L Boswell, D Moore	73:115–34

Endocrinology

Estrogen Signaling through the Transmembrane G Protein–Coupled Receptor GPR30	ER Prossnitz, JB Arterburn, HO Smith, TI Oprea, LA Sklar, HJ Hathaway	70:165–90
Insulin-Like Signaling, Nutrient Homeostasis, and Life Span	A Taguchi, MF White	70:191–212
The Role of Kisspeptins and GPR54 in the Neuroendocrine Regulation of Reproduction	SM Popa, DK Clifton, RA Steiner	70:213–38
Transcriptional Control of Mitochondrial Biogenesis and Function	MB Hock, A Kralli	71:177–203
Genomic Analyses of Hormone Signaling and Gene Regulation	E Cheung, WL Kraus	72:191–218
Macrophages, Inflammation, and Insulin Resistance	JM Olefsky, CK Glass	72:219–46
Structural Overview of the Nuclear Receptor Superfamily: Insights into Physiology and Therapeutics	P Huang, V Chandra, F Rastinejad	72:247–72

Endocrine Disruptors: From Endocrine to Metabolic Disruption	C Casals-Casas, B Desvergne	73:135–62
Endometriosis: The Role of Neuroangiogenesis	A Asante, RN Taylor	73:163–82
Zebrafish in Endocrine Systems: Recent Advances and Implications for Human Disease	H Löhr, M Hammerschmidt	73:183–211
Biology Without Walls: The Novel Endocrinology of Bone	G Karsenty, F Oury	74:87–106
Fetal Programming and Metabolic Syndrome	P Rinaudo, E Wang	74:107–30
Nuclear Sphingolipid Metabolism	NC Lucki, MB Sewer	74:131–52

Gastrointestinal Physiology

Gastrointestinal Satiety Signals	OB Chaudhri, V Salem, KG Murphy, SR Bloom	70:239–55
Mechanisms and Regulation of Epithelial Ca^{2+} Absorption in Health and Disease	Y Suzuki, CP Landowski, MA Hediger	70:257–71
Polarized Calcium Signaling in Exocrine Gland Cells	OH Petersen, AV Tepikin	70:273–99
The Functions and Roles of the Extracellular Ca^{2+}–Sensing Receptor along the Gastrointestinal Tract	JP Geibel, SC Hebert	71:205–17
Neuroendocrine Control of the Gut During Stress: Corticotropin-Releasing Factor Signaling Pathways in the Spotlight	A Stengel, Y Taché	71:219–39
Stem Cells, Self-Renewal, and Differentiation in the Intestinal Epithelium	LG van der Flier, H Clevers	71:241–60
Apical Recycling of the Gastric Parietal Cell H,K-ATPase	JG Forte, L Zhu	72:273–96
Role of Colonic Short-Chain Fatty Acid Transport in Diarrhea	HJ Binder	72:297–313
The Biogenesis of Chylomicrons	CM Mansbach, SA Siddiqi	72:315–33
Mesenchymal Cells of the Intestinal Lamina Propria	DW Powell, IV Pinchuk, JI Saada, X Chen, RC Mifflin	73:213–37
Niemann-Pick C1-Like 1 (NPC1L1) Protein in Intestinal and Hepatic Cholesterol Transport	L Jia, JL Betters, L Yu	73:239–59
Regulation of Electroneutral NaCl Absorption by the Small Intestine	A Kato, MF Romero	73:261–81

Tight Junction Pore and Leak Pathways: A Dynamic Duo	L Shen, CR Weber, DR Raleigh, D Yu, JR Turner	73:283–309
Adenosine and Hypoxia-Inducible Factor Signaling in Intestinal Injury and Recovery	SP Colgan, HK Eltzschig	74:153–76
Toll-Like Receptor–Gut Microbiota Interactions: Perturb at Your Own Risk!	FA Carvalho, JD Aitken, M Vijay-Kumar, AT Gewirtz	74:177–98

Neurophysiology

Dendritic Spine Dynamics	DH Bhatt, S Zhang, W-B Gan	71:261–82
Endocannabinoid Signaling and Long-Term Synaptic Plasticity	BD Heifets, PE Castillo	71:283–306
Sensing Odorants and Pheromones with Chemosensory Receptors	K Touhara, LB Vosshall	71:307–32
Signaling at Purinergic P2X Receptors	A Surprenant, RA North	71:333–59
Integrated Brain Circuits: Astrocytic Networks Modulate Neuronal Activity and Behavior	MM Halassa, PG Haydon	72:335–55
How the Genetics of Deafness Illuminates Auditory Physiology	GP Richardson, J Boutet de Monvel, C Petit	73:311–34
The Calyx of Held Synapse: From Model Synapse to Auditory Relay	JGG Borst, J Soria van Hoeve	74:199–224
Neurotransmitter Corelease: Mechanism and Physiological Role	TS Hnasko, RH Edwards	74:225–44
Small-Conductance Ca^{2+}-Activated K^+ Channels: Form and Function	JP Adelman, J Maylie, P Sah	74:245–70

Renal and Electrolyte Physiology

A Current View of the Mammalian Aquaglyceroporins	A Rojek, J Praetorius, J Frøkiaer, S Nielsen, RA Fenton	70:301–27
Molecular Physiology of the WNK Kinases	KT Kahle, AM Ring, RP Lifton	70:329–55
Physiological Regulation of Prostaglandins in the Kidney	C-M Hao, MD Breyer	70:357–77
Regulation of Renal Function by the Gastrointestinal Tract: Potential Role of Gut-Derived Peptides and Hormones	AR Michell, ES Debnam, RJ Unwin	70:379–403

Activation of the Epithelial Sodium Channel (ENaC) by Serine Proteases	BC Rossier, MJ Stutts	71:361–79
Recent Advances in Understanding Integrative Control of Potassium Homeostasis	JH Youn, AA McDonough	71:381–401
Cellular Maintenance and Repair of the Kidney	J-K Guo, LG Cantley	72:357–76
Intrarenal Purinergic Signaling in the Control of Renal Tubular Transport	HA Praetorius, J Leipziger	72:377–93
The Physiological Significance of the Cardiotonic Steroid/Ouabain-Binding Site of the Na,K-ATPase	JB Lingrel	72:395–412
Mechanisms Underlying Rapid Aldosterone Effects in the Kidney	W Thomas, BJ Harvey	73:335–57
Regulation of Renal NaCl Transport by Nitric Oxide, Endothelin, and ATP: Clinical Implications	JL Garvin, M Herrera, PA Ortiz	73:359–76
Renin Release: Sites, Mechanisms, and Control	A Kurtz	73:377–99
Terminal Differentiation in Epithelia: The Role of Integrins in Hensin Polymerization	Q Al-Awqati	73:401–12
The Calcium-Sensing Receptor Beyond Extracellular Calcium Homeostasis: Conception, Development, Adult Physiology, and Disease	D Riccardi, PJ Kemp	74:271–98
Cell Biology and Pathology of Podocytes	P Mundel, A Greka	74:299–324
A New Look at Electrolyte Transport in the Distal Tubule	D Eladari, R Chambrey, J Peti-Peterdi	74:325–50
Renal Function in Diabetic Disease Models: The Tubular System in the Pathophysiology of the Diabetic Kidney	V Vallon, SC Thomson	74:351–76

Respiratory Physiology

Regulation of Airway Mucin Gene Expression	P Thai, A Loukoianov, S Wachi, R Wu	70:405–29
Structure and Function of the Cell Surface (Tethered) Mucins	CL Hattrup, SJ Gendler	70:431–57
Structure and Function of the Polymeric Mucins in Airways Mucus	DJ Thornton, K Rousseau, MA McGuckin	70:459–86
Regulated Airway Goblet Cell Mucin Secretion	CW Davis, BF Dickey	70:487–512
The Contribution of Epithelial Sodium Channels to Alveolar Function in Health and Disease	DC Eaton, MN Helms, M Koval, HF Bao, L Jain	71:403–23
The Role of CLCA Proteins in Inflammatory Airway Disease	AC Patel, TJ Brett, MJ Holtzman	71:425–49

Role of HDAC2 in the Pathophysiology of COPD	PJ Barnes	71:451–64
Inducible Innate Resistance of Lung Epithelium to Infection	SE Evans, Y Xu, MJ Tuvim, BF Dickey	72:413–35
It's Not all Smooth Muscle: Non-Smooth-Muscle Elements in Control of Resistance to Airflow	Y Bossé, EP Riesenfeld, PD Paré, CG Irvin	72:437–62
Regulation of Endothelial Permeability via Paracellular and Transcellular Transport Pathways	Y Komarova, AB Malik	72:463–93
$T_H 17$ Cells in Asthma and COPD	JF Alcorn, CR Crowe, JK Kolls	72:495–516
Epithelial-Mesenchymal Interactions in Pulmonary Fibrosis	HA Chapman	73:413–35
Interaction of Cigarette Exposure and Airway Epithelial Cell Gene Expression	JS Brody, K Steiling	73:437–56
The Lung: The Natural Boundary Between Nature and Nurture	MA Seibold, DA Schwartz	73:457–78
Role of Chitin and Chitinase/Chitinase-Like Proteins in Inflammation, Tissue Remodeling, and Injury	CG Lee, CA Da Silva, CS Dela Cruz, F Ahangari, B Ma, M-J Kang, C-H He, S Takyar, JA Elias	73:479–501
Autophagy in Pulmonary Diseases	SW Ryter, K Nakahira, JA Haspel, AMK Choi	74:377–402
Stop the Flow: A Paradigm for Cell Signaling Mediated by Reactive Oxygen Species in the Pulmonary Endothelium	EA Browning, S Chatterjee, AB Fisher	74:403–24

Special Topics

Asthma

Aspirin-Sensitive Respiratory Disease	SP Farooque, TH Lee	71:465–87
Immunobiology of Asthma	Q Hamid, M Tulic	71:489–507
Noncontractile Functions of Airway Smooth Muscle Cells in Asthma	O Tliba, RA Panettieri, Jr.	71:509–35

Cellular and Molecukar Mechanisms of Circadian Clocks in Animals

The Mammalian Circadian Timing System: Organization and Coordination of Central and Peripheral Clocks	C Dibner, U Schibler, U Albrecht	72:517–49

Suprachiasmatic Nucleus: Cell Autonomy and Network Properties	DK Welsh, JS Takahashi, SA Kay	72:551–77
Systems Biology of Mammalian Circadian Clocks	H Ukai, HR Ueda	72:579–603
Circadian Organization of Behavior and Physiology in *Drosophila*	R Allada, BY Chung	72:605–24
Mammalian Per-Arnt-Sim Proteins in Environmental Adaptation	BE McIntosh, JB Hogenesch, CA Bradfield	72:625–45

Germ Cells in Reproduction

The Molecular Control of Meiotic Chromosomal Behavior: Events in Early Meiotic Prophase in *Drosophila* Oocytes	CM Lake, RS Hawley	74:425–52
The Control of Male Fertility by Spermatozoan Ion Channels	PV Lishko, Y Kirichok, D Ren, B Navarro, J-J Chung, DE Clapham	74:453–76
Sperm-Egg Interaction	JP Evans	74:477–502
Genetics of Mammalian Reproduction: Modeling the End of the Germline	MM Matzuk, KH Burns	74:503–28

Obesity

The Integrative Role of CNS Fuel-Sensing Mechanisms in Energy Balance and Glucose Regulation	D Sandoval, D Cota, RJ Seeley	70:513–35
Mechanisms of Leptin Action and Leptin Resistance	MG Myers, MA Cowley, H Münzberg	70:537–56

Thrombosis

The Link Between Vascular Features and Thrombosis	CT Esmon, NL Esmon	73:503–14
Role of Tissue Factor in Venous Thrombosis	DA Manly, J Boles, N Mackman	73:515–25
Venous Valvular Stasis–Associated Hypoxia and Thrombosis: What Is the Link?	EG Bovill, A van der Vliet	73:527–45